Lecture Notes in Mathematics 1603

Editors:
A. Dold, Heidelberg
F. Takens, Groningen

Springer
Berlin
Heidelberg
New York
Barcelona
Budapest
Hong Kong
London
Milan
Paris
Tokyo

Vasile Ene

Real Functions – Current Topics

Springer

Author

Vasile Ene
23 August, 8717
Judeţul Constanţa, Romania

Library of Congress Cataloging-in-Publication Data

Ene, Vasile, 1957-
Real functions : contemporary aspects / Vasile Ene.
p. cm. -- (Lecture notes in mathematics ; 1603)
Includes bibliographical references and index.
ISBN 0-387-60008-6
1. Functions of real variables. I. Title. II. Series: Lecture notes in mathematics (Springer-Verlag) ; 1603.
QA3.L28 no. 1603
[QA331.5]
515'.83--dc20 95-164
CIP

Mathematics Subject Classification (1991):
Primary: 26A24
Secondary: 26A21, 26A27, 26A30, 26A39, 26A45, 26A48, 26A51

ISBN 3-540-60008-6 Springer-Verlag Berlin Heidelberg New York

Printed in Great Britain

SPIN: 10130302 46/3142-543210 - Printed on acid-free paper

Preface

The Lebesgue integral, introduced at the beginning of our century marked a turning point in the development of the mathematical analysis in general and for the evolution of the notion of integral in particular. Most 20th century books devoted to the theory of the integral are dominated by ideas having their source in Lebesgue's theory. One of these ideas is the property of absolute convergence of the Lebesgue integral (f is integrable if and only if $|f|$ is integrable), very convenient in most fields of functional analysis.

However it is well-known that the theory of trigonometric series and other lines of thinking in mathematical analysis lead to the introduction of some generalizations of the Lebesgue integral that are no longer absolutely convergent. The first monograph giving a clear account of these nonabsolute integrals (for instance, the Perron and Denjoy integrals) was due to Stanislaw Saks [S3]. But this remained an isolated case; most subsequent books devoted to the theory of the integral ignored nonabsolute integrals, despite the fact that the research literature devoted to them became richer and richer. The aim of the present monograph is just to fill this gap.

We don't ignore the existence of several books concerned in the last decade, with the Kurzweil-Henstock integral, one of the most important nonabsolute integrals; but as is well known, this integral is equivalent to the Perron integral, while our aim is to go further and to cover the next steps in the literature devoted to nonabsolute integrals. In this respect, an outstanding result is the Foran integral (1975), which is a classical generalization of the Denjoy integral in the wide sense. In 1986, Iseki gave another generalization of the Denjoy integral, and called it the sparse integral. However neither of the two generalizations is contained in the other. Our purpose is to give a descriptive definition of an integral which contains both, Foran's integral and Iseki's integral. In order to do this, and also to classify the so many existing integrals, it seems natural to perform a very deep study on the large number of classes of real functions that have been introduced in this context (many of them by the author himself).

Since many examples and counterexamples (66) are necessary to illustrate the properties of various classes of functions, we considered that it would be useful to gather them in a separate chapter, at the end of the book.

To show the uniqueness of the integrations we need several monotonicity theorems, which are studied in Chapter IV. This part of the book is also interesting in its own right, because it presents strong generalizations of many monotonicity theorems, studied in the books of Saks (Theory of the Integral [S3]), Bruckner (Differentiation of Real Functions [Br2]) and Thomson (Real Functions [T5]) (often Lusin's condition (N) is replaced by Foran's condition (M)).

Thomson introduced, in 1982, the notion of local systems, trying to unify the theory of the derivation and integration processes, and to indicate the close connection between them ([T4], [T5]). Since then many authors have used local systems in the study of continuity, derivability, monotonicity and integrability. Beside monotonicity, we use local systems to investigate complete Riemann integrals (Kurzweil-Henstock integrals). We give Perron and descriptive type definitions for these integrals, and show that they are contained in Denjoy type integrals (in the wide sense).

In Chapter V we extend the classical result of Nina Bary, that a continuous function on an interval is the sum of three superpositions of absolutely continuous functions ([Ba]). We also show the close relationship between Nina Bary's wrinkled functions and Foran's condition (M).

The present monograph can be used as a textbook for a special course on mathematical analysis, addressed to graduate students in mathematics. By its systematic nature and by the great attention given to examples, this monograph should be appropriate for all students who want to deepen their knowledge and understanding of one of the most important chapters of mathematical analysis.

In writing this monograph, the author had the benefit of very useful discussions with Professor J. Marik and Professor Clifford E. Weil, during a ten-month stay at Michigan State University, East Lansing, 1991-1992, and also, during the same visit with Professor Paul Humke, who was the first to suggest writing this monograph, after an attempt to write a survey article on a similar topic.

The anonymous reviewers of the manuscript provided significant help.

Professor Solomon Marcus, who guided me in my first steps in real analysis as well as in my Ph.D. thesis, helped me in many respects in preparing this monograph.

The Springer Publishing House provided useful help in improving the final presentation.

To all these persons, to Springer Publishing House and last but not least to my wife Gabriela Ene, who took care of technical aspects related to English and printing, I expend my warmest thanks.

Vasile Ene

Contents

1 Preliminaries **1**

1.1 Notations . . . 1
1.2 The δ–decomposition of a set . . . 2
1.3 Notions related to Hausdorff Measures. Conditions $f.l.$ and $\sigma.f.l.$ . . . 2
1.4 Oscillations . . . 4
1.5 Borel sets F_σ, G_δ; Borel Functions; Analytic sets . . . 8
1.6 Densities; First Category sets . . . 9
1.7 The Baire Category Theorem; Romanovski's Lemma . . . 10
1.8 Vitali's Covering Theorem . . . 11
1.9 The generalized properties PG, $[PG]$, $P_1; P_2G$ . . . 11
1.10 Extreme derivatives . . . 13
1.11 Approximate continuity and derivability . . . 14
1.12 Sharp derivatives $D^{\#}F$ . . . 15
1.13 Local systems; examples . . . 16
1.14 $\mathcal{S}$-open sets . . . 19
1.15 Semicontinuity; $\mathcal{S}$-semicontinuity . . . 21

2 Classes of functions **25**

2.1 Darboux conditions $\mathcal{D}$, $\mathcal{D}_-$, $\mathcal{D}_+$ . . . 25
2.2 Baire conditions $\mathcal{B}_1$, $\underline{\mathrm{B}}_1$, $\overline{\mathcal{B}}_1$ . . . 27
2.3 Conditions $\mathcal{C}_i$; $\mathcal{C}_i^*$; $[\mathcal{C}_iG]$; $[\mathcal{C}G]$ . . . 31
2.4 Conditions internal, internal*, $\mathcal{Z}_i$, uCM . . . 33
2.5 Conditions $\mathcal{D}_-\mathcal{B}_1$, $\mathcal{D}\mathcal{B}_1$ . . . 36
2.6 Conditions $\mathcal{B}_1$, $w\mathcal{B}_1$, $\mathcal{S}_2^-$ and $\mathcal{S}_2^+$-local systems . . . 39
2.7 Conditions VB, $\underline{\mathrm{VB}}$, VBG . . . 41
2.8 Conditions VB^*, $\underline{\mathrm{VB}}^*$, VB^*G . . . 44
2.9 Conditions monotone* and VB^* . . . 47
2.10 Conditions VB^*, VB^*G and $\mathcal{D}$, $\mathcal{D}_-$, $[CG]$, $[\mathcal{C}_iG]$, lower internal, internal 50
2.11 Conditions AC, ACG, . . . 51
2.12 Conditions AC^*, $\underline{\mathrm{AC}}^*$, AC^*G, $\underline{\mathrm{AC}}^*G$ . . . 54
2.13 Conditions L, $\underline{\mathrm{L}}$, LG, $\underline{\mathrm{L}}\mathrm{G}$ . . . 58
2.14 Summability and conditions VB and AC . . . 60
2.15 Differentiability and conditions VBG, VB^*G . . . 62
2.16 A fundamental lemma for monotonicity . . . 65
2.17 Krzyzewski's lemma and Foran's lemma . . . 70
2.18 Conditions (N), T_1, T_2, (S), $(+)$, $(-)$ . . . 71
2.19 Conditions wS, wN . . . 78

2.20 Condition $(\overline{N})$ . . . 78
2.21 Conditions N^∞, $N^{+\infty}$, $N^{-\infty}$ . . . 79
2.22 Conditions M^*, $\overline{M}^*$ . . . 82
2.23 Conditions (M), $\overline{M}$, N_g^∞, $N_g^{+\infty}$ . . . 84
2.24 Derivation bases . . . 87
2.25 Conditions $AC_{D\#}$, AC_{D^o}, AC_D . . . 88
2.26 Condition $Y_{D\#}$, Y_{D^o}, Y_D . . . 89
2.27 Characterizations of $AC^*G \cap \mathcal{C}$, $AC^*G \cap \mathcal{C}_i$, AC and $\underline{AC}$ . . . 90
2.28 Conditions AC_n, AC_ω, AC_∞, $\mathcal{F}$ . . . 93
2.29 Conditions VB_n, VB_ω, VB_∞, $\mathcal{B}$ . . . 96
2.30 Variations V_n, V_ω, V_∞ and the Banach Indicatrix . . . 100
2.31 Conditions S_o, wS_o and AC_∞, VB_∞, (N) . . . 103
2.32 Conditions L_n, L_ω, L_∞, $\mathcal{L}$ . . . 104
2.33 Conditions ΛZ, $\Lambda\overline{Z}$, $f.l.$, $\sigma.f.l.$ . . . 108
2.34 Conditions SAC_n, SAC_ω, SAC_∞, $S\mathcal{F}$ . . . 111
2.35 Conditions SVB_n, SVB_ω, SVB_∞,; $S\mathcal{B}$ . . . 114
2.36 Conditions DW_n, DW_ω, DW_∞, DW^* . . . 116
2.37 Conditions E_n, E_ω, E_∞, $\mathcal{E}$ . . . 118
2.38 Conditions $\mathcal{S}AC$, $\mathcal{S}ACG$, $\mathcal{S}VB$, $\mathcal{S}VBG$, $\mathcal{S}Y$ . . . 121

3 Finite representations for continuous functions **127**
3.1 quasi-derivable $\subsetneq AC^*; DW^*G{+}AC^*; DW^*G$ and approximately quasi-derivable $\subsetneq AC; DW_1G + AC; DW_1G$ . . . 127
3.2 $\mathcal{C} \subsetneq DW_1 + DW_1$ on a perfect nowhere dense set . . . 130
3.3 Wrinkled functions (W) and condition (M) . . . 131
3.4 $\mathcal{C}$ = quasi-derivable + quasi-derivable . . . 134
3.5 $\mathcal{C} = AC^*;\ DW_1G + AC^*;\ DW_1G + AC^*;\ DW_1G$ . . . 135

4 Monotonicity **141**
4.1 Monotonicity and conditions $(-)$, $VB_\omega G$, $\mathcal{D}_-\mathcal{B}_1$ . . . 141
4.2 Monotonicity and conditions $\underline{\text{M}}$,uCM,$\underline{\text{AC}}$,C_i, $\mathcal{C}_i^*$, $\mathcal{D}_i$ . . . 142
4.3 Monotonicity and conditions $N^{-\infty}$; N^∞ . . . 145
4.4 Local monotonicity . . . 149
4.5 $\mathcal{S}$-derivatives and the Mean Value Theorem . . . 149
4.6 Relative monotonicity . . . 151
4.7 An application of Corollary 4.4.1 . . . 151
4.8 A general monotonicity theorem . . . 152
4.9 Monotonicity in terms of extreme derivatives . . . 158

5 Integrals **161**
5.1 Descriptive and Perron type definitions for the Lebesgue integral . . . 161
5.2 Ward type definitions for the Lebesgue integral . . . 166
5.3 Henstock variational definitions for the Lebesgue integral . . . 167
5.4 Riemann type definitions for the Lebesgue integral (The McShane integral) . . . 169
5.5 Theorems of Marcinkiewicz type for the Lebesgue integral . . . 172
5.6 Bounded Riemann# sums and locally small Riemann# sums . . . 173
5.7 Descriptive and Perron type definitions for the $\mathcal{D}^*$-integral . . . 174

5.8 An improvement of the Hake Theorem 180
5.9 An improvement of the Looman-Alexandroff Theorem. The equivalence of the $\mathcal{D}^*$-integral and the $(\mathcal{P}_{j,k})$-integral 184
5.10 Ward type definitions for the $\mathcal{D}^*$-integral 186
5.11 Henstock Variational definitions for the $\mathcal{D}^*$- integral 187
5.12 The Kurzweil-Henstock integral 188
5.13 Cauchy and Harnak extensions of the $\mathcal{D}^*$ - integral 189
5.14 A theorem of Marcinkiewicz type for the $\mathcal{D}^*$- integral 190
5.15 Bounded Riemann sums and locally small Riemann sums 192
5.16 Riemann type integrals and local systems 193
5.17 The $< LPG >$ and $< LDG >$ integrals 197
5.18 The chain rule for the derivative of a composite function 200
5.19 The chain rule for the approximate derivative of a composite function . 202
5.20 Change of variable formula for the Lebesgue integral 204
5.21 Change of variable formula for the Denjoy* integral 205
5.22 Change of variable formula for the $< LDG >$ integral 206
5.23 Integrals of Foran type 207
5.24 Integrals which extend both, Foran's integral and Iseki's integral 210

6 Examples **213**
6.1 The Cantor ternary set, a perfect nowhere dense set 213
6.2 The Cantor ternary function φ 214
6.3 A real bounded $\mathcal{S}_o^+$ closed set which is not of F_σ-type 214
6.4 An $\mathcal{S}_o^+$ lower semicontinuous function which is not $\overline{\mathcal{B}}_1$ 215
6.5 A function $F \in \mathcal{C}_i$, $F \notin \mathcal{C}_i^*$ 215
6.6 A function $F \in \mathcal{D}$, $F \in [\mathcal{C}_i^* G]$, $F \notin [\mathcal{C}G]$ 215
6.7 A function $F \in \mathcal{D}\mathcal{B}_1$, $F \notin [\mathcal{C}_i G]$ 216
6.8 A function $F \in uCM$; $F \notin \ell CM$ 216
6.9 A function concerning conditions $\mathcal{D}_+$, $\mathcal{D}_-$, CM, sCM, lower internal . 217
6.10 A function concerning conditions: $\mathcal{D}_-$, $\mathcal{D}$, internal, $\underline{\mathrm{B}}_1$, $\overline{\mathcal{B}}_1$, $\mathcal{B}_1$, $w\mathcal{B}_1$, $[VBG]$, $(-)$, T_1, T_2 (Bruckner) 217
6.11 A function concerning conditions: $\overline{\mathcal{B}}_1$, $\underline{\mathrm{B}}_1$, $\mathcal{D}_-$, $\mathcal{D}_+$, lower internal, internal, internal* (Dirichlet) 218
6.12 A function concerning conditions: $\mathcal{D}$, $\mathcal{D}_-$, $\mathcal{B}_1$, $\mathcal{C}_i$, $\mathcal{C}_i^*$, lower internal, internal*, VB, VB^*G, $N^{-\infty}$ 219
6.13 A function $F \in \mathcal{D}$, $F \in \underline{\mathrm{B}}_1 \setminus \overline{\mathcal{B}}_1$; $-F \in \overline{\mathcal{Z}}_i \setminus \mathcal{C}_i$, $-F \in \mathcal{D}_-\overline{\mathcal{B}}_1 \setminus \overline{\mathcal{Z}}_i$ 219
6.14 A function $F \in \underline{\mathrm{B}}_1 \setminus \overline{\mathcal{B}}_1$, $F \in$ lower internal, $F \notin \mathcal{D}_-$ 220
6.15 A function $F \in sCM$, $F \notin$ internal* 220
6.16 A function $F \in AC^*G \setminus AC$, $F \in \mathcal{C}_i^* \setminus \mathcal{D}$, $F \in sCM\setminus$ internal* 220
6.17 A function $F \in (D.C.)$, $F \in \mathcal{B}_1$, $F \notin m_2$, $F \notin \mathcal{D}$ 221
6.18 A function $F \in (+) \cap (-)$; $F \notin \mathcal{D}\mathcal{B}_1 T_2$ 221
6.19 A function $G \in \mathcal{D}$, $G \notin \underline{\mathrm{B}}_1$, $G \notin \overline{\mathcal{B}}_1$, $G'_{ap}(x)$ exists *n.e.*, $G'_{ap}(x) \geq 0$ *a.e.* (Preiss) 222
6.20 A function $H \in \mathcal{D}$, $H \notin \overline{\mathcal{B}}_1$, $H \notin \underline{\mathrm{B}}_1$, $H'_{ap}(x)$ exists on $(0,1)$ (Preiss) . . 222
6.21 A function $F \in \mathcal{D}\mathcal{B}_1$, $F(x) = 0$ *a.e.*, F is not identically zero (Croft) . . 223
6.22 A function $F \in \mathcal{D}$, $F \in [\mathcal{C}G]$, $F \in [VBG]$, $F \notin VB^*G$, $F \notin \mathcal{C}$ (Bruckner) 223
6.23 A function $F \in AC^*$, $F \notin VB^*$ 224

6.24 A function $F \in \mathcal{C}$, $F \in T_1$, $F \in VBG$, $F \notin VB^*G$ 224
6.25 A function $F \in [bAC^*G] \cap VB^*G \cap N^{-\infty}$ $F \notin$ lower internal 225
6.26 A function $F \in \mathcal{C} \cap (S) \cap LG$, $F \notin AC^*G$, $F'(x)$ does not exist on a set of positive measure, $F(x) + x \in LG$, $F(x) + x \notin T_1$ 225
6.27 A function $F \in (S) \cap \mathcal{C}$ such that the sum of F and any linear nonconstant function does not satisfy (N) (Mazurkiewicz) 226
6.28 A function $F \in (M)$, $F \notin T_2$ 227
6.29 Functions concerning conditions (M), AC, T_1, T_2, (S), (N), L, L_2G, VBG, $\mathcal{SF}$, quasi-derivable 229
6.30 A function $G \in N^{\infty}$, $F \notin (M)$, $F \notin (+)$ 237
6.31 Functions concerning conditions (S), (N), (M), T_1, T_2, ACG, AC_n, SAC_n, VB_2, VBG, SVB, $\mathcal{F}$, $\mathcal{SF}$ 238
6.32 A function $F \in$ lower semicontinuous, $F \in AC_2$, $F \notin \underline{AC}$ 244
6.33 A function $F_n \in L_{n+1}$ on a perfect set, $F_n \in VB_n$ on no portion of this set, $F_n \in L_{n+1}G$, $F_n \notin AC_nG$ on $[0,1]$ 245
6.34 Functions $F \in L_2G$, $G_s \in (N)$, $G_s' = F'$ *a.e.*, $G_s - F$ is not identically zero, $F \notin SACG$ 247
6.35 A function $F \in \underline{L}_2$, $F \notin T_2$, $F \notin \mathcal{B}$ 250
6.36 A function $F \in VB_2$ on C, $V_2(F;C) \leq 1$ 252
6.37 A function $F_p \in L_{2^p}$, $F_p \notin AC_{2^p-1}$, $F_p \in VB_2$ on C; $V_2(F_p;C) \leq 1$ . . . 252
6.38 A function $G \in VB_2$, $G \notin AC_n$ on C, $G \in \mathcal{F}$ on $[0,1]$ 254
6.39 A function $F_1 \in VB_2$ on C, $V_2(F_1;[0,x] \cap C) = \varphi(x)$ (G. Ene) 254
6.40 A function $F_q \in (N)$ on $[0,1]$, $F_q \notin VB_n$ on C, $F_q \in VB_\omega$ on C (G. Ene) 255
6.41 A function $G_1 \in L_2G$, $G_1 \in \mathcal{F}$, $G_1 \notin SVBG$, $G_1 \notin SACG$, $(G_1)'_{ap}$ does not exist on a set of positive measure 256
6.42 A function $F \in SACG$, $F \notin \mathcal{F}$, $F \notin ACG$ 258
6.43 A function $F \in DW_1$, $F \notin DW^*$ 261
6.44 A function $F \in AC^*; DW_1G$, $F \notin AC^*; DW^*G$ 261
6.45 Functions $F_1, F_2 \in \mathcal{C} \cap AC^*; DW^*G$, F_1, F_2 are derivable *a.e.*, $F_1' = F_2'$ *a.e.*, F_1 and F_2 do not differ by a constant 262
6.46 Functions $F_1, F_2 \in \mathcal{C} \cap AC^*; DW_1G$, F_1, F_2 are approximately derivable *a.e.*, $F_1 + F_2 \notin$ quasi-derivable 262
6.47 Functions $F_1, F_2 \in \mathcal{E} \cap \mathcal{B}$, $F_1, F_2 \notin \mathcal{F}$, $F_1 + F_2 \notin \mathcal{E}$ 264
6.48 A function $G_n \in E_{n+1}$, $G_n \notin E_n$, $G_n \in L_{n^2+2n+1}$, $G_n \notin VB_{n^2+2n}$ 265
6.49 Functions concerning conditions $\underline{L}$,$\underline{E}$,$\overline{\mathcal{F}}$, VB_2G, $\mathcal{B}$, E_1G 268
6.50 A function $F \in \mathcal{E} \cap VB_\omega G$, $F \notin \mathcal{B}$ 270
6.51 A function $F \in (N)$, $F \notin \Lambda Z$ (Foran) 271
6.52 A function $F \in AC \circ \Lambda Z$, $F \notin \Lambda Z$ (Foran) 272
6.53 A function $H \in AC + \Lambda Z$, $H \notin \Lambda Z$ (Foran) 272
6.54 A function $G \in AC \cdot \Lambda Z$, $G \notin \Lambda Z$ (Foran) 272
6.55 A function $F_1 \in AC$, $F_1 \notin L_n$, $F_1 \notin \mathcal{L}$ 273
6.56 Functions $F_1 \in AC_2G$, $F_2 \in \Lambda Z$, $F_1 + F_2 \notin (M)$, $F_1' = -F_2'$ *a.e.* 273
6.57 A function $F \in \Lambda Z$, $F \notin [\mathcal{E}]$ 274
6.58 Functions $F_1 \in (S)$, $F_1 \in AC \circ \sigma.f.l.$, $F_1 \notin \sigma.f.l.$, $F_2 \in L$, $F_1 + F_2 \notin T_2$ 276
6.59 Functions $G_1 \in \sigma.f.l.$, $G_2 \in AC$, $G_1 + G_2 \notin \sigma.f.l.$ 279
6.60 Functions $H_1 \in \sigma.f.l.$, $H_2 \in AC$, $H_1 \cdot H_2 \notin \sigma.f.l.$ 280

6.61 A function $F \in \sigma.f.l.$, $F \in T_1$, $F \notin \mathcal{B}$, F is nowhere approximately derivable, (Foran) 280
6.62 A function $G \in \sigma.f.l.$, $G \in T_1$, G is nowhere derivable, $G'_{ap}(x) = 0$ *a.e.*, $G \notin W$, $G \in W^*$ (Foran) 284
6.63 A function $F \in W$ on a perfect nowhere dense set of positive measure, with each level set perfect, F is nowhere approximately derivable 287
6.64 A function $G_1 \in DW_1 \cap \mathcal{C}$, G_1 is not approximately derivable *a.e.* on a set of positive measure 291
6.65 A function $F \in \mathcal{C}$, F is quasi-derivable, $F \notin AC \circ AC + AC$ 291
6.66 Examples concerning the chain rule for the approximate derivative of a composite function 291

Bibliography **293**

Index **305**

Introduction

Beyond advanced calculus, the prerequisite for understanding this book is the basic theory of functions of a real variable including Lebesgue integral.

We shall be dealing with real functions of a real variable and throughout the book we state explicitly the exact domains of definition and the ranges of them.

We overlap with other books very few, but frequently we shall need results of the classical books of Saks [S3], Natanson [N], Bruckner [Br2] and Thomson [T5], which we take over without proofs (see Chapter 1). On occasion, when several theorems have similar proofs, we prove only one or two of these theorems. Where we do not give a complete proof, however, we provide references.

Chapters 3, 4, 5 and 6 are dependent of Chapters 1 and 2. However, it is not necessary to read all the sections of them, since we give detailed references in all the proofs of what is needed. A reader interested in a particular result from a certain page of the book is not expected to read the entire material until that page, because he is sent back very accurate to the exact notions which he needs.

All results contain short historical remarks, unless they are due to the author himself.

Numerous examples will be needed, so we have gathered them in Chapter 6. We considered appropriate to illustrate some of these examples with pictures, although pictures does not replace proofs (however the author's proofs are independent from pictures).

We have also included a long list of bibliography in order to mention background reading, to give credit for original discoveries or to indicate directions for further study, but we make no claims of the completeness of this list.

We tried to make an index as rich as possible, so that it might be useful. If the reader is interested in a notion or a class of functions, he should look up the Contents or the Index for it.

We think the present book will serve the following purposes: to report on some recent advances in the theory of real functions; to serve as a textbook for a course in the subject; to serve as a reference work for persons studying real analysis independently; and to stimulate further research in this exciting field.

Chapter 1

Preliminaries

1.1 Notations

For convenience, if P is a property for functions on a certain domain, we shall also use P to denote the class of all functions having property P. If a function F satisfies both properties P and P', we denote this by $F \in P \cap P'$. A property P for a function F is said to hold n.e. (nearly everywhere), if it holds, except on a countable set of points. A property P for a function F is said to hold a.e. (almost everywhere), if it holds, except on a null set of points. Let $\mathcal{A}_1 + \mathcal{A}_2 = \{F + G : F \in \mathcal{A}_1, G \in \mathcal{A}_2\}$, where $\mathcal{A}_1$ and $\mathcal{A}_2$ are two classes of real functions defined on the same domain. Let $\mathcal{A}_1 \cdot \mathcal{A}_2 = \{F \cdot G : F \in \mathcal{A}_1, G \in \mathcal{A}_2\}$. Let $\mathcal{C}$ denote the class of all continuous functions on a certain domain. Any set of the form $A \cap (a, b) \neq \emptyset$ is said to be a portion of the set A. We denote by $\overline{A}$ the closure of the set A and by $int(A)$ the interior of the set A.

Definition 1.1.1 *Let $P \subseteq [a, b]$. Then $int(P) = \cup_{i\geq 1}(c_i, d_i)$, where $\{(c_i, d_i)\}_i$ is a sequence of nonoverlapping maximal open intervals. $int(P)$ is uniquely represented in this way. We say that (c_i, d_i), $i \geq 1$ are the components of P.*

A set $P \subset \mathbb{R}$ has compact components if $(c, d) \subset P$ implies that $[c, d] \subset P$. (a, b) is said to be an interval contiguous to the closed set $P \neq \emptyset$, if $a, b \in P$ and $(a, b) \cap P = \emptyset$.

Definition 1.1.2 *Let $P \subseteq [a, b]$, $K_P : [a, b] \mapsto \mathbb{R}$, $K_P(x) = 1$, $x \in P$ and $K_P(x) = 0$, $x \notin P$. K_P is called the characteristic function of P.*

Definition 1.1.3 *Let $F : [a, b] \mapsto \mathbb{R}$, and let P be a closed subset of P, $c = inf(P)$, $d = sup(P)$. Let $\{(c_k, d_k)\}_{k\geq 1}$ be the intervals contiguous to P and $F_P : [c, d] \mapsto \mathbb{R}$, defined as follows: $F_P(x) = F(x)$, $x \in P$ and F_P is linear on each $[c_k, d_k]$.*

We denote by $F_{/P}$ the restriction of the function F to P. We denote by $B(F; X)$ the graph of the function F on the set X. We denote by $f \circ g$ the composition of the functions f and g. Given a real set X we shall denote by $|X|$ the outer Lebesgue measure of X. If X is Lebesgue measurable we denote by $m(X)$ the Lebesgue measure of X. We denote by $i = \overline{1, n}$ (respectively $i = \overline{1, \infty}$) the fact that $i = 1, 2, \ldots, n$ (respectively $i = 1, 2, \ldots$).

1.2 The δ–decomposition of a set

Definition 1.2.1 *([T4], p.104.) Let δ be a positive function and let X be a set of real numbers. By a δ-decomposition of X we shall mean a sequence of sets $\{X_n\}_n$, which is a relabelling of the countable collection $Y_{mj} = \{x \in X : \delta(x) > 1/m\} \cap [j/m, (j+1)/m]$, $m = 1, 2, \ldots$, $j = 0, \pm 1, \pm 2, \ldots$.*

Remark 1.2.1 *([T4],p.104; [T5].pp.32-33). The key features of a δ–decomposition are:*

(i) $\cup_{n=1}^{\infty} X_n = X$;

(ii) If x and y belong to some set X_n then $| x - y | < \min\{\delta(x), \delta(y)\}$;

(iii) If $x \in X \cap \overline{X}_n$ and $y \in (x - \delta(x), x + \delta(x)) \cap X_n$ then again one must have $| x - y | < \{\delta(x), \delta(y)\}$.

1.3 Notions related to Hausdorff Measures. Conditions $f.l.$ and $\sigma.f.l.$

Definition 1.3.1 *([Le3]) Let P be a subset of $[a, b]$. Let $n \geq 1$ be a natural number and let $\beta > 0$ be a real number. We denote by:*

- $\lambda_n^\beta = inf\{\sum_{i=1}^n | J_i |^\beta : \{J_i\}_{i=1}^n$ *is a set of intervals which cover* $P\}$;
- $\lambda_\infty^\beta = inf\{\sum_{i=1}^\infty | J_i |^\beta : \{J_i\}_{i=1}^\infty$ *is a sequence of open intervals which cover* $P\}$;
- $\lambda_\omega^\beta = inf\{\lambda_n^\beta(P) : n = \overline{1, \infty}\}$.

Let $\lambda_n(P) = \lambda_n^1(P)$; $\lambda_\omega(P) = \lambda_\omega^1(P)$; $\lambda_\infty(P) = \lambda_\infty^1(P)$.

Remark 1.3.1 *([Le3]). Let $P \subseteq [a, b]$, $\beta > 0$ and let $n \geq 1$ be a natural number.*

(i) $0 \leq \lambda_n^\beta(P) \leq \lambda_n^\beta(Q)$, *whenever* $P \subseteq Q \subseteq [a, b]$;

(ii) $0 \leq \lambda_\infty^\beta(P) \leq \lambda_\infty^\beta(Q)$, *whenever* $P \subseteq Q \subseteq [a, b]$;

(iii) $0 \leq \lambda_\omega^\beta(P) \leq \lambda_\omega^\beta(Q)$, *whenever* $P \subseteq Q \subseteq [a, b]$;

(iv) $\lambda_\infty^\beta(P) \leq \lambda_\omega^\beta(P) \leq \lambda_{n+1}^\beta(P) \leq \lambda_n^\beta$. *Note that there exist sets P such that all inequalities are strict;*

(v) $\lambda_\infty^\beta(\cup_{i=1}^\infty P_i) \leq \sum_{i=1}^\infty \lambda_\infty^\beta(P_i)$, *where* $P_i \subset [a, b]$, $i = \overline{1, \infty}$. *Note that this inequality does not hold if ∞ is replaced by ω or by a natural number n.*

(vi) $\lambda_\infty(P) = | P |$.

Theorem 1.3.1 *Let P, Q be bounded real sets, $m, n \geq 1$ natural numbers, and let $a, b, \beta > 0$. Then we have:*

(i) $\lambda_{mn}^\beta(P + Q) \leq 2^{\beta - 1}(n \cdot \lambda_m^\beta(P) + m \cdot \lambda_n^\beta(Q))$, *for* $\beta > 1$;

(ii) $\lambda^{\beta}_{mn}(P+Q) \le n \cdot \lambda^{\beta}_{m}(P) + m \cdot \lambda^{\beta}_{n}(Q)$, *for* $0 < \beta \le 1$;

(iii) $\lambda^{\beta}_{mn}(P \cdot Q) \le 2^{\beta-1}(b^{\beta} \cdot n \cdot \lambda^{\beta}_{m}(P) + a^{\beta} \cdot m \cdot \lambda^{\beta}_{n}(P))$, *for* $\beta \ge 1$, $P \subset [0,a]$ *and* $Q \subset [0,b]$;

(iv) $\lambda_{mn}(P \cdot Q) \le b^{\beta} \cdot n \cdot \lambda^{\beta}_{m}(P) + a^{\beta} \cdot m \cdot \lambda^{\beta}_{n}(Q)$, *for* $0 < \beta \le 1$, $P \subset [0,a]$ *and* $Q \subset [0,b]$;

(v) $\lambda^{\beta}_{n}(P) = \lambda^{\beta}_{n}(\overline{P})$;

(vi) $\lambda^{\beta}_{\omega}(P) = \lambda^{\beta}_{\omega}(\overline{P}) = \lambda^{\beta}_{\infty}(\overline{P})$.

Proof. (i) For $\epsilon > 0$, let $J_1, J_2, \ldots, J_m$ and $J'_1, J'_2, \ldots, J'_n$ be closed intervals such that $\lambda^{\beta}_{m}(P) \le \sum_{i=1}^{m} \mid J_i \mid^{\beta} \le \lambda^{\beta}_{m}(P) + \epsilon/(n \cdot 2^{\beta})$, $P \subset \cup_{i=1}^{m} J_i$ and $+\lambda^{\beta}_{n}(Q) \le \sum_{k=1}^{n} \mid J'_k \mid^{\beta} \le \lambda^{\beta}_{n}(Q) + \epsilon/(m \cdot 2^{\beta})$, $Q \subset \cup_{k=1}^{n} J'_k$. Then $P+Q \subset \cup_{i=1}^{m} \cup_{k=1}^{n} (J_i + J'_k)$ and $\lambda^{\beta}_{mn}(P+Q) \le \sum_{i=1}^{m} \sum_{k=1}^{n} \mid J_i + J'_k \mid^{\beta} = \sum_{i=1}^{m} \sum_{k=1}^{n} (\mid J_i \mid + \mid J'_k \mid)^{\beta} \le 2^{\beta-1}(\sum_{i=1}^{m} \sum_{k=1}^{n} \mid J'_k \mid^{\beta} + \sum_{k=1}^{n} \sum_{i=1}^{m} \mid J_i \mid^{\beta}) \le 2^{\beta-1} \cdot m(\lambda^{\beta}_{n}(Q) + \epsilon/(m \cdot 2^{\beta})) + 2^{\beta-1} \cdot n(\lambda^{\beta}_{m}(P) + \epsilon/(n \cdot 2^{\beta})) = 2^{\beta-1} \cdot (m \cdot \lambda^{\beta}_{n}(Q) + n \cdot \lambda^{\beta}_{m}(P)) + \epsilon$. Since ϵ is arbitrary, we obtain (i) (we have used the inequality: $(\mid x \mid + \mid y \mid)^{\beta} \le 2^{\beta-1} \cdot (\mid x \mid^{\beta} + \mid y \mid^{\beta})$, $\beta \ge 0$).

(ii) Using the fact that $\mid x+y \mid^{\beta} \le \mid x \mid^{\beta} + \mid y \mid^{\beta}$, $0 < \beta \le 1$, the proof is similar to that of (i)

(iii) First we show that

(1) $$\mid J \cdot J' \mid^{\beta} \le 2^{\beta-1} \cdot b^{\beta} \cdot \mid J \mid^{\beta} + 2^{\beta-1} \cdot a^{\beta} \cdot \mid J' \mid^{\beta},$$

whenever J and J' are closed intervals with $J \subset [0,a]$, $J' \subset [0,b]$, $\beta \ge 1$. Let's denote $J = [c,d]$ and $J' = [c',d']$. Then $\mid J \cdot J' \mid = (dd' - cc')^{\beta} = (d'(d-c) + c(d'-c'))^{\beta} \le 2^{\beta-1} \cdot (d'(d-c))^{\beta} + 2^{\beta-1} \cdot (c(d'-c'))^{\beta} \le 2^{\beta-1} \cdot b^{\beta} \mid J \mid^{\beta} + 2^{\beta-1} \cdot a^{\beta} \cdot \mid J' \mid^{\beta}$, so we have (1). For $\epsilon > 0$, let $J_1, J_2, \ldots, J_m$ and $J'_1, J'_2, \ldots, J'_n$ be closed intervals with the following properties: $\lambda^{\beta}_{m}(P) \le \sum_{i=1}^{m} \mid J_i \mid^{\beta} \le \lambda^{\beta}_{m}(P) + \epsilon/(n \cdot b^{\beta} \cdot 2^{\beta})$, $\lambda^{\beta}_{n}(Q) \le \sum_{k=1}^{m} \mid J'_k \mid^{\beta} \le \epsilon/(m \cdot a^{\beta} \cdot 2^{\beta})$, $P \subset \cup_{i=1}^{m} J_i$ and $Q \subset \cup_{k=1}^{n} J'_k$. Then $P \cdot Q \subset \cup_{i=1}^{m} \cup_{k=1}^{n} J_i \cdot J'_k$, and by (1), $\lambda^{\beta}_{mn}(P \cdot Q) \le \sum_{i=1}^{m} \sum_{k=1}^{n} \mid J_i \cdot J'_k \mid^{\beta} \le 2^{\beta-1} \cdot b^{\beta} \cdot \sum_{k=1}^{n} \sum_{i=1}^{m} \mid J_i \mid^{\beta} + 2^{\beta-1} \cdot a^{\beta} \cdot \sum_{i=1}^{m} \sum_{k=1}^{n} \mid J'_k \mid^{\beta} \le 2^{\beta-1} \cdot (b^{\beta} \cdot n \cdot \lambda^{\beta}_{m}(P) + a^{\beta} \cdot m \cdot \lambda^{\beta}_{n}(Q)) + \epsilon$. Since ϵ is arbitrary, we have (iii).

(iv) The proof is similar to that of (iii), but using

(2) $$\mid J \cdot J' \mid \le b^{\beta} \cdot \mid J \mid^{\beta} + a^{\beta} \cdot \mid J' \mid^{\beta}, \ 0 < \beta \le 1$$

(v) By Remark 1.3.1, $\lambda^{\beta}_{n}(P) \le \lambda^{\beta}_{n}(\overline{P})$. For $\epsilon > 0$ there exists a finite set of n closed intervals $J_1, J_2, \ldots, J_n$, such that $P \subset \cup_{i=1}^{n} J_i$ and $\lambda^{\beta}_{n}(P) \le \sum_{i=1}^{n} \mid J_i \mid^{\beta} \le \lambda^{\beta}_{n}(P) + \epsilon$. Since $\overline{P} \subset \cup_{i=1}^{n} J_i$, it follows that $\lambda^{\beta}_{n}(P) \le \lambda^{\beta}_{n}(\overline{P}) \le \sum_{i=1}^{n} \mid J_i \mid^{\beta} \le \lambda^{\beta}_{n}(P) + \epsilon$. Since ϵ is arbitrary, it follows (v).

(vi) By (v) we have $\lambda_{\omega}(\overline{P}) = \inf\{\lambda^{\beta}_{n}(\overline{P}) : n = \overline{1,\infty}\} = \inf\{\lambda^{\beta}_{n}(P) : n = \overline{1,\infty}\} = \lambda^{\beta}_{\omega}(P)$. By Remark 1.3.1, $\lambda^{\beta}_{\infty}(\overline{P}) \le \lambda^{\beta}_{\omega}(\overline{P}) \le \lambda^{\beta}_{n}(\overline{P})$, for each natural number $n \ge 1$. For $\epsilon > 0$ there exists a sequence $\{J_i\}_i$ of open intervals, such that $\overline{P} \subset \cup_{i=1}^{\infty} J_i$ and $\lambda^{\beta}_{\infty}(\overline{P}) \le \sum_{i=1}^{\infty} \mid J_i \mid^{\beta} < \lambda^{\beta}_{\infty}(\overline{P}) + \epsilon$. Since $\overline{P}$ is a compact set, there exists a natural number n such that $\overline{P} \subset \cup_{i=1}^{n} J_i$. It follows that $\lambda^{\beta}_{\infty}(\overline{P}) \le \lambda^{\beta}_{\omega}(\overline{P}) \le \lambda^{\beta}_{n}(\overline{P}) \le \sum_{i=1}^{n} \mid J_i \mid^{\beta} \le \sum_{i=1}^{\infty} \mid J_i \mid^{\beta} < \lambda^{\beta}_{\infty}(\overline{P}) + \epsilon$. Since ϵ is arbitrary, we have $\lambda^{\beta}_{\omega}(\overline{P}) = \lambda^{\beta}_{\infty}(\overline{P})$.

Definition 1.3.2 *([S3]). Let* $X \subset \mathbb{R}\times\mathbb{R}$ *and let* $\beta > 0$. *For each* $\epsilon > 0$ *let* $\lambda_\epsilon^\beta(X) = inf\{\sum_{i=1}^\infty (diam(X_i))^\beta : \{X_i\}, i = \overline{1,\infty}$ *is a partition of* X, $diam(X_i) < \epsilon\}$. *Let* $\Lambda^\beta(X) = \lim_{\epsilon\to 0}(\Lambda_\epsilon^\beta(X))$ *and* $\Lambda(X) = \Lambda^1(X)$. *The number* $\Lambda(X)$ *is called the outer length of* X. *If* $\Lambda(X) \neq +\infty$ *then* X *is said to be of finite length (short f.l.). Let* $F : [a,b] \mapsto \mathbb{R}$, $P \subseteq [a,b]$. F *is said to be f.l. (of finite length) on* P, *if* $\Lambda(B(F;P)) \neq +\infty$.

Definition 1.3.3 *Let* $X \subset \mathbb{R}\times\mathbb{R}$. X *is said to have* σ*-finite length (short* $\sigma.f.l.$*) if there exists a partition* $\{X_i\}_i$ *of* X, $X = \cup_{i=1}^\infty X_i$, *such that* $\Lambda(X_i) < +\infty$ *for each* i. *Let* $F : [a,b] \mapsto \mathbb{R}$, $P \subset [a,b]$. F *is said to be* $\sigma.f.l.$ *on* P *if* $B(F;P)$ *has* $\sigma.f.l.$.

Theorem 1.3.2 (Gross) *([Gr], [F1]). Let* $E \subset \mathbb{R}\times\mathbb{R}$, $\Lambda(E) \neq +\infty$ *and let* $(E)_n = \{c :$ *the line* $y = c$ *meets* E *in* n *or more points* $\}$. *Then* $|(E)_n|\cdot n \le \Lambda(E)$.

Proof. Let E^i be the set of points (x,y) in E such that E contains n points (x_k,y) with $x_{k+1} \ge x_k + 1/i$, for $k = \overline{1,n-1}$. Since $(E)_n = \cup_{i=1}^\infty (E^i)_n$, it suffices to show that $|(E^i)_n|\cdot n \le \Lambda(E^i)$. Consider any countable covering of E^i by convex sets A_j, $diam(A_j) < 1/i$. Since each point of $(E^i)_n$ is at in least n of the sets $\mathrm{Pr}_y(A_j)$, it follows that $|(E^i)_n|\cdot n < \sum|\mathrm{Pr}_y(A_j)| \le \sum diam(A_j)$. Thus $|(E^i)_n|\cdot n \le \Lambda(E_i)$.

1.4 Oscillations

Definition 1.4.1 *([S2]). Let* P *be a bounded real set and let* $F : P \mapsto \mathbb{R}$, $x_o \in P$. *We define:*

- $\mathcal{O}(F;P) = \sup\{|F(y) - F(x)| : x,y \in P\}$ *the oscillation of* F *on* P;
- $\mathcal{O}(F;x_o) = \lim_{\delta\to 0}\mathcal{O}(F;[x_o-\delta,x_o+\delta]\cap P)$ *the oscillation of* F *at* x_o;
- $\mathcal{O}_-(F;P) = \inf\{F(y) - F(x) : x,y \in P,\ x \le y\}$;
- $\mathcal{O}_+(F;P) = \sup\{F(y) - F(x) : x,y \in P,\ x \le y\}$.

Definition 1.4.2 *Let* P *be a bounded set,* $F : P \mapsto \mathbb{R}$ *and let* $\emptyset \neq X \subseteq Y \subseteq P$, $x_o \in P$. *We define:*

- $\Omega(F;Y\wedge X) = \sup\{|F(y) - F(x)| : x \in X,\ y \in Y\}$;
- $\Omega_-(F;Y\wedge X) = \inf\{F(y) - F(x) : x \le y,\ x,y \in Y,\ \{x,y\}\cap X \neq \emptyset\}$;
- $\Omega_+(F;Y\wedge X) = \sup\{F(y) - F(x) : x \le y,\ x,y \in Y,\ \{x,y\}\cap X \neq \emptyset\}$;
- $\Omega(F;x_o) = \lim_{\delta\to 0}\Omega(F;P\cap[x_o-\delta,x_o+\delta]\wedge\{x_o\})$.

Remark 1.4.1 *Let* $F : [a,b] \mapsto \mathbb{R}$, $P \subseteq [a,b]$. *We have:*

(i) $\mathcal{O}_-(F;P) = -\mathcal{O}_+(-F;P)$;

(ii) $\mathcal{O}(F;I) \le 2\cdot\mathcal{O}(|F|;I)$, *where* $I \subseteq [a,b]$ *is an interval;*

(iii) $\mathcal{O}(F;P) = \max\{\mathcal{O}_+(F;P); |\mathcal{O}_-(F;P)|\}$;

(iv) $\Omega_-(F;Y\wedge X) = -\Omega_+(-F;Y\wedge X)$;

(v) $\Omega(F;Y\wedge X) = \max\{\Omega_+(F;Y\wedge X); |\,\Omega_-(F;Y\wedge X)\,|\}$;

(vi) $\mathcal{O}(F;P) = \Omega(F;P\wedge P)$ *and* $\mathcal{O}_+(F;P) = \Omega_+(F;P\wedge P)$.

Proposition 1.4.1 *Let* $F,G:[a,b]\mapsto \mathbb{R}$. *Let* $I\subseteq[a,b]$ *be an interval and let* P *be a nonempty subset of* I. *Then we have:*

(i) $\mathcal{O}(F;P)\le\Omega(F;I\wedge P)\le\mathcal{O}(F;I)$; $\mathcal{O}_+(F;P)\le\Omega_+(F;I\wedge P)\le\mathcal{O}_+(F;I)$ *and* $\mathcal{O}(F;I)\le\mathcal{O}(F;P)+2\Omega(F;I\wedge P)\le 3\Omega(F;I\wedge P)$;

(ii) *If* $\lambda>0$ *and* $\Omega(F;I\wedge P)\in(0,+\infty)$ *(resp.* $\Omega_+(F;I\wedge P)\in(0,+\infty)$*) then there exist* $c.d\in I$, $\{c,d\}\cap P\ne\emptyset$, $c<d$, *such that* $\Omega(F;I\wedge P)\le(1+\lambda)\,|\,F(d)-F(c)\,|\le(1+\lambda)\Omega(F;[c,d]\wedge(P\cap[c,d]))$ *(resp.* $\Omega_+(F;I\wedge P)\le(1+\lambda)(F(d)-F(c))\le(1+\lambda)\Omega_+(F;[c,d]\wedge(P\cap[c,d]))$*)*;

(iii) *If* $x,y\in I$, $\{x,y\}\cap P\ne\emptyset$, $x<y$ *then* $|\,F(y)-F(x)\,|\le\Omega(F;[x,y]\wedge(P\cap[x,y]))\le\Omega(F;I\wedge P)$ *and* $\Omega_-(F;I\wedge P)\le\Omega_-(F;[x,y]\wedge(P\cap[x,y]))\le F(y)-F(x)\le\Omega_+(F;[x,y]\wedge(P\cap[c,y]))\le\Omega_+(F;I\wedge P)$;

(iv) *If* $\alpha\in\mathbb{R}$ *then* $\Omega(\alpha F;I\wedge P)=|\,\alpha\,|\,\Omega(F;I\wedge P)$;

(v) *If* $\alpha>0$ *then* $\Omega_+(\alpha F;I\wedge P)=\alpha\Omega_+(F;I\wedge P)$;

(vi) *If* $\alpha,\beta>0$, $|\,F(x)\,|\le\alpha$, $|\,G(x)\,|\le\beta$, *for each* $x\in I$ *then* $\Omega(F\cdot G;I\wedge P)\le\beta\Omega(F;I\wedge P)+\alpha\Omega(G;I\wedge P)$;

(vii) *If* $\alpha,\beta\in\mathbb{R}$ *then* $\Omega(\alpha F+\beta G;I\wedge P)\le|\,\alpha\,|\,\Omega(F;I\wedge P)+|\,\beta\,|\,\Omega(G;I\wedge P)$;

(viii) *If* $\alpha,\beta>0$ *then* $\Omega_+(\alpha F+\beta G;I\wedge P)\le\alpha\Omega_+(F;I\wedge P)+\beta\Omega_+(G;I\wedge P)$;

(ix) *If* $x,y\in P$, $x<y$ *then* $\mathcal{O}(F;[x,y])\le F(x)-F(y)+2\Omega_+(F;[x,y]\wedge(P\cap[x,y]))$;

(x) *If* P *is a closed set and* $x,y\in P$, $x<y$ *then* $\Omega_+(F;[x,y]\wedge\{x,y\})\le\mathcal{O}_+(F;P\cap[x,y])+\sum_{(u,v)\in\mathcal{A}}\Omega_+(F;[u,v]\wedge\{u,v\})$, *where* $\mathcal{A}=\{(u,v):(u,v)$ *is an interval contiguous to* $P\cap[x,y]\}$;

(xi) *If* P *is a closed set and* $x,y\in P$, $x<y$ *then* $\mathcal{O}_+(F;[x,y])\le 2\Omega_+(F;[x,y]\wedge(P\cap[x,y]))+\sum_{(u,v)\in\mathcal{A}}\mathcal{O}_+(F;[u,v])$, *where* $\mathcal{A}=\{(u,v):(u,v)$ *is an interval contiguous to* $P\cap[x,y]\}$;

(xii) *If* P *is a closed set and* $x,y\in P$, $x<y$ *then* $\mathcal{O}_+(F;P\cap[x,y])=\mathcal{O}_+(F_P;[x,y])$.

(xiii) *If* $x,y\in\overline{P}$, $x<y$ *then* $\mathcal{O}(F;[x,y])\le\sup\{|\,F(t)-F(v)\,|: t\in P\cap[x,y],\ v\in(P\cup P_-)\cap[x,y])\}+\sup\{|\,F(t)-F(v)\,|: t\in(P\cup P_+)\cap[x,y],\ v\in P\cap[x,y]\}+2\sum_{(\alpha,\beta)\in\mathcal{A}}\Omega(F;[\alpha,\beta]\wedge\{\alpha,\beta\})$, *where* $\mathcal{A}=\{(\alpha,\beta):(\alpha,\beta)$ *is an interval contiguous to* $\overline{P}\cap[x,y]\}$, $P_+=\{x\in\overline{P}: x$ *is a right accumulation point of* $P\}$ *and* $P_-=\{x\in\overline{P}: x$ *is a left accumulation point of* $P\}$.

Corollary 1.4.1 *Let* $F,G:[a,b]\mapsto\mathbb{R}$, $P\subseteq[a,b]$, $P\ne\emptyset$.

(i) *If* $\lambda > 0$, *P contains more than two points and* $\mathcal{O}(F;P) \in (0,+\infty)$ *(resp.* $\mathcal{O}_+(F;P) \in (0.+\infty)$*) then there exist* $c,d \in P$, $c < d$ *such that* $\mathcal{O}(F;P) \le (1+\lambda) \mid F(d) - F(c) \mid \le (1+\lambda)\mathcal{O}(F;P\cap[c,d])$ *(resp.* $\mathcal{O}_+(F;P) \le (1+\lambda)(F(d)-F(c)) \le (1+\lambda)\mathcal{O}_+(F;P\cap[c,d])$*);*

(ii) *If* $x,y \in P$, $x < y$ *then* $\mid F(y) - F(x) \mid \le \mathcal{O}(F;P\cap[x,y]) \le \mathcal{O}(F;P)$ *and* $\mathcal{O}_-(F;P) \le \mathcal{O}_-(F;P\cap[x,y]) \le F(y)-F(x) \le \mathcal{O}_+(F;P\cap[x,y]) \le \mathcal{O}_+(F;P)$*;*

(iii) $\mathcal{O}(F;P) = \max\{\mid \mathcal{O}_-(F;P) \mid ; \mathcal{O}_+(F;P)\}$*;*

(iv) *If* $\alpha \in \mathbb{R}$ *then* $\mathcal{O}(\alpha F;P) = \mid \alpha \mid \mathcal{O}(F;P)$*;*

(v) *If* $\alpha > 0$ *then* $\mathcal{O}_+(\alpha F;P) = \alpha\mathcal{O}_+(F;P)$*;*

(vi) *Let* $\alpha,\beta \in (0,+\infty)$ *and let* $x,y \in P$, $x < y$. *If* $0 \le F(t) \le \alpha$ *and* $0 \le G(t) \le \beta$ *for each* $t \in P$ *then* $F(y)G(y) - F(x)G(x) = F(y)(G(y)-G(x)) + F(x)(G(y) - G(x))$ *and* $\mathcal{O}_+(F\cdot G;P) \le \beta\mathcal{O}_+(F;P) + \alpha\mathcal{O}_+(G;P)$*;*

(vii) *Let* $\alpha,\beta \in (0,+\infty)$ *and let* $x,y \in P$, $x < y$. *If* $\mid F(t) \mid \le \alpha$ *and* $\mid G(t) \mid \le \beta$, *for each* $t \in P$ *then* $\mid F(y)G(y) - F(x)G(x) \mid \le \beta \mid F(y) - F(x) \mid + \alpha \mid G(y) - G(x) \mid$ *and* $\mathcal{O}(F\cdot G;P) \le \beta\mathcal{O}(F;P) + \alpha\mathcal{O}(G;P)$*;*

(viii) *Let* $\alpha,\beta \in R$ *and let* $x,y \in P$, $x < y$. *Then* $\mid (\alpha F(y) + \beta G(y)) - (\alpha F(x) + \beta G(y)) \mid \le \mid \alpha \mid \cdot \mid F(y) - F(x) \mid + \mid \beta \mid \cdot \mid G(y) - G(x) \mid$ *and* $\mathcal{O}(\alpha F + \beta G;P) \le \mid \alpha \mid \mathcal{O}(F;P) + \mid \beta \mid \mathcal{O}(G;P)$*;*

(ix) *Let* $\alpha,\beta \in (0,+\infty)$ *and let* $x,y \in P$, $x < y$. *Then* $(\alpha F(y) + \beta G(y)) - (\alpha F(x) + \beta G(x)) = \alpha(F(y)-F(x)) + \beta(G(y)-G(x))$ *and* $\mathcal{O}_+(\alpha F + \beta G) \le \alpha\mathcal{O}_+(F;P) + \beta\mathcal{O}_+(G;P)$*;*

(x) $\mathcal{O}(G\circ F;P) = \mathcal{O}(G;F(P))$.

Proposition 1.4.2 *Let P be a bounded real set and let* $F : P \mapsto \mathbb{R}$, $\alpha \in \mathbb{R}$. *Then* $\{x \in P : \mathcal{O}(F;x) \ge \alpha\}$ *is a closed set with respect to P.*

Definition 1.4.3 *Let* $F : [a,b] \mapsto \mathbb{R}$, $P \subseteq [a,b]$, *and let* $n \ge 1$ *be a natural number. We define:*

- $\mathcal{O}^n(F;P) = \inf\{\sum_{i=1}^n \mathcal{O}(F;P_i) : \cup_{i=1}^n P_i = P\}$;
- $\mathcal{O}_+^n(F;P) = \inf\{\sum_{i=1}^n \mathcal{O}_+(F;P_i) : \cup_{i=1}^n P_i = P\}$;
- $\mathcal{O}_-^n(F;P) = \sup\{\sum_{i=1}^n \mathcal{O}_-(F;P_i) : \cup_{i=1}^n P_i = P\}$;
- $\mathcal{O}^\infty(F;P) = \inf\{\sum_{i=1}^\infty \mathcal{O}(F;P_i) : \cup_{i=1}^\infty P_i = P\}$;
- $\mathcal{O}_+^\infty(F;P) = \inf\{\sum_{i=1}^\infty \mathcal{O}_+(F;P_i) : \cup_{i=1}^\infty P_i = P\}$;
- $\mathcal{O}_-^\infty(F;P) = \sup\{\sum_{i=1}^\infty \mathcal{O}_-(F;P_i) : \cup_{i=1}^\infty P_i = P\}$;
- $\mathcal{O}^\omega(F;P) = \inf\{\mathcal{O}^n(F;P_i) : n = \overline{1,\infty}\}$;
- $\mathcal{O}_+^\omega(F;P) = \inf\{\mathcal{O}_+(F;P_i) : n = \overline{1,\infty}\}$;

- $\mathcal{O}_-^\omega(F;P) = \sup\{\mathcal{O}_-(F;P_i) : n = \overline{1,\infty}\}$.

Proposition 1.4.3 *Let $F, G : [a,b] \mapsto \mathbb{R}$ be bounded functions. Let $P \subseteq [a,b]$ and let $m, n \geq 1$ be natural number. Then we have:*

(i) $\mathcal{O}^n(F;P) = \lambda_n(F(P)) = \lambda_n(\overline{F(P)})$;

(ii) $\mathcal{O}^\omega(F;P) = \lambda_\omega(F(P)) = \lambda_\omega(\overline{F(P)}) = \lambda_\infty(\overline{F(P)}) = |\overline{F(P)}|$;

(iii) $\mathcal{O}^\infty(F;P) = \lambda_\infty(F(P)) = |F(P)|$;

(iv) $0 \leq \mathcal{O}^\infty(F;P) \leq \mathcal{O}^\omega(F;P) \leq \mathcal{O}^{n+1}(F;P) \leq \mathcal{O}^n(F;P) \leq \mathcal{O}(F;P)$;

(v) $0 \leq \mathcal{O}_+^\infty(F;P) \leq \mathcal{O}_+^\omega(F;P) \leq \mathcal{O}_+^{n+1}(F;P) \leq \mathcal{O}_+^n(F;P) \leq \mathcal{O}_+(F;P)$;

(vi) $\mathcal{O}_+^n(F;P) + |\mathcal{O}_-^n(F;P)| \geq \mathcal{O}^n(F;P) \geq \max\{\mathcal{O}_+^n(F;P); |\mathcal{O}_-^n(F;P)|\}$;

(vii) $\mathcal{O}_+^\omega(F;P) + |\mathcal{O}_-^\omega(F;P)| \geq \mathcal{O}^\omega(F;P) \geq \max\{\mathcal{O}_+^\omega(F;P); |\mathcal{O}_-^\omega(F;P)|\}$;

(viii) $\mathcal{O}^\infty(F;P) \geq \max\{\mathcal{O}_+^\infty(F;P); |\mathcal{O}_-^\infty(F;P)|\}$;

(ix) $\mathcal{O}^{mn}(F;P) \leq n\mathcal{O}_+^m(F;P) + m\,|\mathcal{O}_-^n(F;P)|$;

(x) $\mathcal{O}_+^{mn}(F+G;P) \leq n\mathcal{O}_+^m(F;P) + m\mathcal{O}_+^n(G;P)$;

(xi) $\mathcal{O}^{mn}(F+G;P) \leq n\mathcal{O}^m(F;P) + m\mathcal{O}^n(G;P)$;

(xii) *If $0 \leq F(x) \leq \alpha$ and $0 \leq G(x) \leq \beta$, for each $x \in P$ then $\mathcal{O}_+^{mn}(F \cdot G;P) \leq \beta n\mathcal{O}_+^m(F;P) + \alpha m\mathcal{O}_+^n(G;P)$;*

(xiii) *If $|F(x)| \leq \alpha$ and $|G(x)| \leq \beta$, for each $x \in P$ then $\mathcal{O}^{mn}(F \cdot G;P) \leq \beta n\mathcal{O}^m(F;P) + \alpha m\mathcal{O}^n(G;P)$;*

(xiv) $\mathcal{O}^n(G \circ F;P) = \mathcal{O}^n(G;F(P))$.

Proof. (i) For $\epsilon > 0$ there exist n closed intervals $J_1, J_2, \ldots, J_n$ such that $F(P) \subseteq \cup_{i=1}^n J_i$ and $\lambda_n(F(P)) + \epsilon > \sum_{i=1}^n |J_i|$. Let $P_i = \{x \in P : F(x) \in J_i\}$. Then $P = \cup_{i=1}^n P_i$ and $\mathcal{O}(F;P_i) \leq |J_i|$. It follows that $\mathcal{O}^n(F;P) \leq \sum_{i=1}^n |J_i| < \lambda_n(F(P)) + \epsilon$, hence $\mathcal{O}^n(F;P) \leq \lambda_n(F(P)) \leq \lambda_n(\overline{F(P)})$. For $\epsilon > 0$ there exist n sets $P_1, P_2, \ldots, P_n$ such that $P = \cup_{i=1}^n P_i$ and $\sum_{i=1}^n \mathcal{O}(F;P_i) < \mathcal{O}^n(F;P) + \epsilon$. Let $J_i = [\inf(P_i), \sup(P_i)]$. Then $F(P) \subseteq \cup_{i=1}^n J_i$ and $|J_i| = \mathcal{O}(F;P_i)$. Hence $\overline{F(P)} \subseteq \cup_{i=1}^n J_i$ and $\sum_{i=1}^n |J_i| \leq \mathcal{O}^n(F;P) + \epsilon$. Therefore $\lambda_n(\overline{F(P)}) \leq \mathcal{O}^n(F;P)$.

(ii) $\mathcal{O}^\omega(F;P) = \lambda_\omega(F(P)) = \lambda_\omega(\overline{F(P)})$ follows by (i). We have $|\overline{F(P)}| = \lambda_\infty(\overline{F(P)}) \leq \lambda_\omega(\overline{F(P)})$ (see Remark 1.3.1). Since $\overline{F(P)}$ is a compact set, it follows that $\lambda_\infty(\overline{F(P)}) = \lambda_\omega(\overline{F(P)})$.

(iii)The proof is similar to that of (i).

(iv) and (v) are evident.

(vi) By Corollary 1.4.1 we have $\mathcal{O}^n(F;P) = \inf\{\sum_{i=1}^n \mathcal{O}(F;P_i) : \cup_{i=1}^n P_i = P\} = \inf\{\sum_{i=1}^n \max\{\mathcal{O}_+(F;P_i); |\mathcal{O}_-(F;P_i)|\} : \cup_{i=1}^n P_i = P\} \leq \inf\{\sum_{i=1}^n(\mathcal{O}_+(F;P_i) + |\mathcal{O}_-(F;P_i)|) : \cup_{i=1}^n P_i = P\} \leq \mathcal{O}_+^n(F;P) + |\mathcal{O}_-^n(F;P)|$.

We show that $\mathcal{O}_+^n(F;P) \leq \mathcal{O}^n(F;P)$. For $\epsilon > 0$ there exist $P_1, P_2, \ldots, P_n$ such that $\cup_{i=1}^n P_i = P$ and $\sum_{i=1}^n \mathcal{O}(F;P_i) < \mathcal{O}^n(F;P) + \epsilon$. Since $\mathcal{O}_+(F;P_i) \leq \mathcal{O}(F;P_i)$, it follows that $\mathcal{O}_+^n(F;P) \leq \mathcal{O}^n(F;P)$. Similarly $|\mathcal{O}_-^n(F;P)| \leq \mathcal{O}^n(F;P)$.

(vii) follows by (vi), and the proof of (viii) is similar to that of (vi).

(ix) For $\epsilon > 0$ there exist $P_1, P_2, \ldots, P_m$ and $P_1', P_2', \ldots, P_n'$ such that $P = \cup_{i=1}^m P_i = \cup_{j=1}^n P_j'$, $\sum_{i=1}^m \mathcal{O}_+(F;P_i) \leq \mathcal{O}_+^m(F;P) + \epsilon/(2n)$ and $\sum_{j=1}^n \mathcal{O}_-(F;P_j') \geq \mathcal{O}_-^n(F;P) - \epsilon/(2m)$. By Corollary 1.4.1, we have $\mathcal{O}^{mn}(F;P) \leq \sum_{i=1}^m \sum_{j=1}^n \mathcal{O}(F;P_i \cap P_j') \leq \sum_{i=1}^m \sum_{j=1}^n (\mathcal{O}_+(F;P_i \cap P_j') + |\mathcal{O}_-(F;P_i \cap P_j')|) \leq \sum_{i=1}^m \sum_{j=1}^n (\mathcal{O}_+(F;P_i) + |\mathcal{O}_-(F;P_j')|) \leq \sum_{i=1}^m \sum_{j=1}^n \mathcal{O}_+(F;P_i) + \sum_{i=1}^m \sum_{j=1}^n |\mathcal{O}_-(F;P_j')| \leq n(\mathcal{O}_+^m(F;P) + \epsilon/(2n)) + m(|\mathcal{O}_-^n(F;P)| + \epsilon/(2m)) = n\mathcal{O}_+^m(F;P) + m|\mathcal{O}_-^n(F;P)| + \epsilon$. Hence we have (ix).

(x) For $\epsilon > 0$ there exist $P_1, P_2, \ldots, P_m$ and $P_1', P_2', \ldots, P_n'$ such that $P = \cup_{i=1}^m P_i = \cup_{j=1}^n P_j'$, $\mathcal{O}_+^m(F;P) + \epsilon/(2n) > \sum_{i=1}^m \mathcal{O}_+(F;P_i)$ and $\mathcal{O}_+^n(G;P) + \epsilon/(2m) > \sum_{j=1}^n \mathcal{O}_+(G;P_j')$. By Corollary 1.4.1, it follows that $\mathcal{O}_+^{mn}(F+G;P) \leq \sum_{i=1}^m \sum_{j=1}^n \mathcal{O}_+(F+G;P_i \cap P_j') \leq \sum_{i=1}^m \sum_{i=1}^j (\mathcal{O}_+(F;P_i \cap P_j') + \mathcal{O}_+(G;P_i \cap P_j')) \leq \sum_{i=1}^m \sum_{j=1}^n (\mathcal{O}_+(F;P_i) + \mathcal{O}_+(G;P_j')) \leq n\mathcal{O}_+^m(F;P) + m\mathcal{O}_+^n(G;P) + \epsilon$, so we have (x).

(xi) The proof is similar to that of (ix).

(xii) As in the proof of (ix), we obtain: $\mathcal{O}_+^{mn}(F \cdot G;P) \leq \sum_{i=1}^m \sum_{j=1}^n \mathcal{O}_+(F \cdot G;P_i \cap P_j') \leq \sum_{i=1}^m \sum_{j=1}^n (\beta\mathcal{O}_+(F;P_i \cap P_j') + \alpha\mathcal{O}_+(G;P_i \cap P_j')) \leq \beta \sum_{i=1}^m \sum_{j=1}^n \mathcal{O}_+(F;P_i) + \alpha \sum_{i=1}^m \sum_{j=1}^n \mathcal{O}_+(G;P_j') \leq \beta n\mathcal{O}_+^m(F;P) + \alpha m\mathcal{O}_+^n(G;P) + \epsilon(\alpha+\beta)/2$.

(xiii) The proof is similar to that of (xii).

(xiv) By (i) we have $\mathcal{O}^n(G \circ F;P) = \lambda_n(G \circ F(P)) = \lambda_n(G(F(P))) = \mathcal{O}_n(G;F(P))$.

1.5 Borel sets F_σ, G_δ; Borel Functions; Analytic sets

Definition 1.5.1 *([Ku], p.27). Let $\mathcal{A}$ be a family of real sets. The family $\mathcal{A}$ of Borel sets is the smallest family of sets which satisfy the following conditions:*

(i) Any closed set belong to $\mathcal{A}$;

(ii) If $X \in \mathcal{A}$ then $\mathbb{R} \setminus X \in \mathcal{A}$;

(iii) If $X_n \in \mathcal{A}$, $n = \overline{1,\infty}$ the $\cup_{n=1}^\infty X_n \in \mathcal{A}$.

Definition 1.5.2 *([Ku], p.26). Let P be a real set. We say that:*

- *P is of F_σ-type, whenever $P = \cup_{n=1}^\infty P_n$, where each P_n is closed;*
- *P is of G_δ-type, whenever $P = \cap_{n=1}^\infty P_n$, where each P_n is open.*

Proposition 1.5.1 *([Ku], pp.26-27, [N], p.139).*

(i) Any set of F_σ-type or G_δ-type is a Borel set;

(ii) Any closed set is of F_σ-type and of G_δ-type;

(iii) Any open set is of G_δ-type and of F_σ-type;

(iv) The difference of two closed sets is of F_σ-type;

(v) The union of countable many sets of F_σ-type is a set of F_σ-type;

(vi) The intersection of a finite number of F_σ-type sets is a set of F_σ-type;

(vii) The intersection of countable many G_δ-type sets is a set of G_δ-type;

(viii) The union of a finite number of G_δ-type sets is a set of G_δ-type;

(ix) Let $P \subset [a,b]$. P is a sets of F_σ-type if and only if $[a,b] \setminus P$ is a set of G_δ-type.

Proposition 1.5.2 *([Oxt], p.31). Let $F : [a,b] \mapsto \mathbb{R}$. Then the set $\{x : F$ is discontinuous at $x\}$ is of F_σ-type.*

Theorem 1.5.1 *([F11],p.197). Given any measurable set E and $\epsilon > 0$*

(i) there is an open set G such that $E \subset G$ and $\mid G \setminus E \mid < \epsilon$;

(ii) there is a G_δ-set A such that $E \subset A$ and $\mid A \setminus E \mid = 0$;

(iii) there is a closed set $P \subset E$ such that $\mid E \setminus P \mid < \epsilon$;

(iv) there is an F_σ-set H such that $H \subset E$ and $\mid E \setminus H \mid = 0$;

(v) if $\mid E \mid < +\infty$, then there exists a compact set $K \subset E$ such that $\mid E \setminus K \mid < \epsilon$.

Definition 1.5.3 *Let $F : [a,b] \mapsto \mathbb{R}$. F is said to be a Borel function if the set $\{x : F(x) \in (\alpha, \beta)\}$ is a Borel set, whenever $\alpha < \beta$ are real numbers.*

Definition 1.5.4 *([Ku], pp.360, 386). A real set P is said to be analytic if there exists a continuous function $F : [a,b] \mapsto \mathbb{R}$ and a Borel set $Q \subset [a,b]$, such that $F(Q) = P$.*

Proposition 1.5.3 *([Ku], p.391, 387, 365).*

(i) Any analytic set is measurable;

(ii) Any uncountable set contains a nonempty perfect set;

(iii) The image of a Borel set under a Borel function is an analytic set.

1.6 Densities; First Category sets

Definition 1.6.1 *([T5], p.22). Let A be a real set and let x_o be a real number. For the interior right density of A at x_o (upper and lower) we write*

- $\underline{d}^i_+(A;x_o) = \liminf_{h\searrow 0} \frac{|A\cap(x_o,x_o+h)|^i}{h}$ *and*
- $\overline{d}^i_+(A;x_o) = \limsup_{h\searrow 0} \frac{|A\cap(x_o,x_o+h)|^i}{h}$,

with the obvious notations for the exterior right densities, the left versions and the bilateral versions ($\mid X \mid^i$ and $\mid X \mid^e$ are the interior, respectively exterior measure of X). If $\underline{d}^i_+(A;x_o) = \underline{d}^i_-(A;x_o) = 1$ then x_o is said to be a point of density for A, and we write $d(A;x_o) = 1$.

Theorem 1.6.1 (Lebesque's Density Theorem) *([Br2], p.18). Let A be a measurable real set. Then $d(A;x) = 1$, for almost every $x \in A$ and $d(A;x) = 0$, for almost every $x \notin A$.*

Definition 1.6.2 *([S3], p.41). Let P and A be real sets, $P \subset A$. P is said to be nowhere dense on A, if no portion of A is contained in $\overline{P}$.*
P is said to be everywhere dense on A if $\overline{P} \supset A$.

Definition 1.6.3 *([S3], p.41). Let P and A be real sets, $P \subset A$. P is said to be of first category with respect to A, if it is the countable union of nowhere dense sets on A. P is said to be of second category with respect to A, if it is not of first category on A. P is said to be residual with respect to A,if $B \setminus A$ is of first category with respect to A.*

Theorem 1.6.2 (Baire) *([Oxt], p.2). Let $P \subset [a,b]$. If P is of first category then the set $[a,b] \setminus P$ is everywhere dense on $[a,b]$.*

Theorem 1.6.3 *([Oxt], p.3).*

(i) Any subset of a set of first category is of first category;

(ii) The union of any countable family of first category sets is of first category;

(iii) If P is a perfect real set then any countable subset of P is of first category on P.

1.7 The Baire Category Theorem; Romanovski's Lemma

Theorem 1.7.1 (Baire's Category Theorem) *([S3], p.54). Let $P \subseteq [a,b]$, $P \neq \emptyset$. If P is of G_δ-type then it is of second category, i.e., if $P = \cup_{i=1}^{\infty} P_n$ then there exists an open interval (c,d) such that $(c,d) \cap P \neq \emptyset$ and $(c,d) \cap P \subset \overline{P_n}$, for some natural number n.*

Lemma 1.7.1 (Romanovski's Lemma) *([Rom], [G5]). Let $\mathcal{A}$ be a family of open intervals in (a,b) and suppose that $\mathcal{A}$ has the following properties:*

(i) If (α,β) and (β,γ) belong to $\mathcal{A}$ then (α,γ) belongs to $\mathcal{A}$;

(ii) If (α,β) belongs to $\mathcal{A}$ then every open interval in (α,β) belongs to $\mathcal{A}$;

(iii) If (α,β) belongs to $\mathcal{A}$ for every interval $[\alpha,\beta] \subset (c,d)$ then $(c,d) \in \mathcal{A}$;

(iv) If all open intervals in (a,b) contiguous to the perfect set $E \subset [a,b]$ belong to $\mathcal{A}$, then there exists an interval I in $\mathcal{A}$ such that $I \cap E \neq \emptyset$.

Then $\mathcal{A}$ contains the interval (a,b).

Proof. By applying condition (iv) to the set $[a,b]$, we find that $\mathcal{A}$ is nonempty. Let $H = (a,b) \setminus (\cup_{I \in \mathcal{A}} I)$ then $\overline{H} \setminus H \subset \{a,b\}$.Let $(a,b) \setminus H = \cup_{k \geq 1}(c_k, d_k)$. We first prove that each interval $(c_k, d_k) \in \mathcal{A}$. To this end, fix k and let $[\alpha, \beta] \subset (c_k, d_k)$. For each t in $[\alpha, \beta]$, there exists an interval I_t in $\mathcal{A}$ that contains t. The collection $\{I_t : t \in [\alpha, \beta]\}$ is an open cover of $[\alpha, \beta]$, and since $[\alpha, \beta]$ is compact, there exists a finite subcover $\{I_{t_i} : i = \overline{1, N}\}$. Let $s_j : j = \overline{0, n}\}$ be the set that contains $\{\alpha, \beta\}$ and all of the endpoints of the intervals I_{t_i} that belong to $[\alpha, \beta]$, and assume that the points are in increasing order. By condition (ii), each of the intervals (s_{j-1}, s_j), for $1 \leq j \leq N$ belongs to $\mathcal{A}$. By repeated application of condition (i), we find that $(\alpha, \beta) \in \mathcal{A}$. By condition (iii), we conclude that (c_k, d_k) belongs to $\mathcal{A}$. Now condition (i) implies that the set $\overline{H}$ is perfect. In addition each of the intervals contiguous to $\overline{H}$ in (a,b) belongs to $\mathcal{A}$. By condition (iv), the set H must be empty. Repeating the argument above, we find that (a,b) belongs to $\mathcal{A}$.

1.8 Vitali's Covering Theorem

Definition 1.8.1 *Let $E \subseteq [a,b]$ and let $\mathcal{A}$ be a family of nondegenerate closed intervals. If for every $x \in E$ and every $\epsilon > 0$, there exists a closed interval $I_{x,\epsilon} \in \mathcal{A}$, such that $x \in I_{x,\epsilon}$, $| I_{x,\epsilon} | < \epsilon$, then E is said to be covered by $\mathcal{A}$ in the sense of Vitali.*

Theorem 1.8.1 (Vitali's Covering Theorem) *([N], p.83). Let $E \subseteq [a,b]$ be covered by a family $\mathcal{A}$ of closed intervals, in the sense of Vitali. For every $\epsilon > 0$, there exists a finite set of pairwise disjoint closed intervals $I_1, I_2, \ldots, I_n \in \mathcal{A}$, such that $| E - (\cup_{k=1}^{n} I_k) | < \epsilon$.*

1.9 The generalized properties PG, $[PG]$, $P_1; P_2G$

Definition 1.9.1 *A property P for a function F is said to be hereditary, if F is P on S, whenever $S \subset E$ and $F \in P$ on E.*
A property P for a function F is said to be strong, if $F \in P$ on $\overline{E}$, whenever $F \in P$ on E.

Definition 1.9.2 *Let $F : [a,b] \mapsto \mathbb{R}$ and let $Q \subseteq [a,b]$. F will be said to be PG on Q, if Q can be written as the union of a countable collection of sets Q_i, over each of which F satisfying property P. If in addition the sets Q_i are supposed to be closed, F is said to be $[PG]$ on Q.*
F is said to be $P_1; P_2G$ on Q, if Q can be written as the union of a countable collection of sets Q_i, over each of which F satisfying property P_1 or property P_2.

Remark 1.9.1 *$P \subset PG$, $[PG] \subset PG$ and $P_1 \subset P_1G \subset P_1; P_2G$ on a set Q.*

Theorem 1.9.1 *Let $F : [a,b] \mapsto \mathbb{R}$ and let Q be a closed subset of $[a,b]$.*

(i) Whenever P is a hereditary strong property the following are equivalent:

a) $F \in PG$ on Q;

b) Each closed subset S of Q contains a portion on which $F \in P$.

c) $F \in PG$ on Z, whenever $Z \subset Q$ and $|Z| = 0$.

(ii) Whenever P is a hereditary property the following are equivalent:

a) $F \in [PG]$ on Q;

b) Each closed subset S of Q contains a portion on which F is P;

(iii) Whenever P is a hereditary property, and P_1 is a property of F on Q which implies that P is strong, the following are equivalent:

a) $F \in P_1 \cap PG$ on Q;

b) $F \in P_1$ on Q, and each closed subset S of Q contains a portion on which $F \in P$;

c) $F \in P_1$ on Q and $F \in PG$ on Z, whenever $Z \subset Q$ and $|Z| = 0$.

Proof. We prove only (iii) (the proofs of (i) and (ii) are similar).

a) $\Rightarrow$ *b)* Let S be a closed subset of Q. Since $F \in PG$ on Q, there exists a collection of sets $\{Q_i\}$, $i = \overline{1,\infty}$, $Q_i \subset Q$ such that $\cup_{i=1}^{\infty} Q_i = Q$ and $F \in P$ on each Q_i. Let $S_i = S \cap Q_i$. Since P is hereditary, $F \in P$ on each S_i. Since $F \in P_1$ on Q, it follows that $F \in P$ on $\overline{S}_i$. By Theorem 1.7.1, there exists a portion $\emptyset \neq (c,d) \cap S \subset \overline{S}_i$ for some i. Since P is hereditary, $F \in P$ on $(c,d) \cap S$.

b) $\Rightarrow$ *a)* By b) it follows that there exists an open interval I such that $I \cap Q \neq \emptyset$ and $F \in P$ on $I \cap Q$. Since $I \cap Q$ is a set of F_σ-type (see Proposition 1.5.1) and P is hereditary, it follows that $F \in [PG]$ on $I \cap Q$. Let$\{I_n\}_n$ be the sequence of all open intervals with rational endpoints such that $F \in [PG]$ on each $I_n \cap Q$. Let $B = \cup_{n=1}^{\infty}(I_n \cap Q)$ and let $H = Q \setminus B$. Then $F \in [PG]$ on B, and we need only to prove that H is empty. Suppose on the contrary that $H \neq \emptyset$. Since H is closed, there exists by hypothesis an open interval J, such that $H \cap J \neq \emptyset$ and $F \in P$ on $H \cap J$. Since H is of F_σ-type (see Proposition 1.5.1) and P is hereditary, it follows that $F \in [PG]$ on $H \cap J$. We may suppose without loss of generality that the endpoints of J are rational. Since $Q \cap J = (H \cap J) \cup (B \cap J)$, it follows that $F \in [PG]$ on $Q \cap J$, a contradiction.

a) $\Rightarrow$ *c)* Let $Z \subset Q$, $|Z| = 0$. Since $F \in PG$ on Q, it follows that there exists a sequence of sets $\{Q_i\}_i$, $Q_i \subset Q$, such that $\cup_{i \geq 1} Q_i = Q$ and $F \in P$ on each Q_i. Let $Z_i = Q_i \cap Z$. Since P is hereditary, $F \in P$ on each Z_i, hence $F \in PG$ on Z.

c) $\Rightarrow$ *b)* Let S be a closed subset of Q. Let $Z \subset S$ be a G_δ-set such that $|Z| = 0$ and $\overline{Z} = S$ (this is possible: let $Z_1 = \{x \in S :$ x is rational or x is an endpoint of an interval contiguous to $S\} = \{x_1, x_2, \ldots\}$; let $G_j = \cup_{i=1}^{\infty}(x_i - 1/(2^{j+1}), x_i + 1/(2^{j+1}))$, $j \geq 1$; let $Z = \cap_{j=1}^{\infty} G_j$; then $Z_1 \subset Z$, $|Z| = 0$ and $\overline{Z}_i = S$, hence $\overline{Z} = S$.) Since $F \in PG$ on Z, it follows that there exists a sequence of sets $\{Z_i\}_i$, such that $Z = \cup_i Z_i$ and F is P on each Z_i. Since F is P_1 it follows that $F \in P$ on $\overline{Z}_i$. By Theorem 1.7.1, there exists an open interval I such that $\emptyset \neq I \cap Z \subset \overline{Z}_i$ for some i. Since P is hereditary, $F \in P$ pn $\overline{I \cap Z}$. But $I \cap S = I \cap \overline{Z} \subset \overline{I \cap Z}$. Since P is hereditary, $F \in P$ on $I \cap S$.

Remark 1.9.2 *The technique used in the proof of Theorem 1.9.1,a) $\Rightarrow$ b) belongs to Saks (see [S3], pp. 233- 234).*

Lemma 1.9.1 *Let $F : [a,b] \mapsto \mathbb{R}$ and let P be a hereditary property such that $F \in [PG]$ on $[a,b]$. Then there exists a sequence $\{I_n\}$ of intervals, whose union is dense in $[a,b]$, such that $F \in P$on each I_n.*

Proof. The proof follows by applying Theorem 1.7.1 infinitely many times.

1.10 Extreme derivatives

Definition 1.10.1 *([Br2], p.52). Let x_o be a real number, and let F be a function defined in a neighborhood of x_o. We define the upper right Dini derivative $D^+F(x_o)$ of F at x_o by:*

$$D^+F(x_o) = \limsup_{h\searrow 0} \frac{F(x_o+h)-F(x_o)}{h}$$

We define the other three unilateral derivatives D_+F, D^-F and D_-F analogously. We denote by $\overline{D}F(x_o)$ (or by $\overline{F}'(x_o)$) the upper (bilateral) derivative of F at x_o:

$$\overline{D}F(x_o) = \limsup_{h\to 0} \frac{F(x_o+h)-F(x_o)}{h}.$$

We denote by $\underline{D}F(x_o)$ (or by $\underline{F}'(x_o)$) the lower (bilateral) derivative of F at x_o:

$$\underline{D}F(x_o) = \liminf_{h\to 0} \frac{F(x_o+h)-F(x_o)}{h}.$$

We shall also write F'_+ and F'_- for the right and left derivatives of F if they exist. Thus $F'_+(x_o)$ is the common value of $D^+F(x_o)$ and $D_+F(x_o)$ if they are equal.

Lemma 1.10.1 *([McS2], p.190). Let $F, G : [a,b] \mapsto \mathbb{R}$. We have*

(i) $\overline{D}(F+G)(x) \le \overline{D}F(x) + \overline{D}G(x)$;

(ii) $\underline{D}(F+G)(x) \ge \underline{D}F(x) + \underline{D}G(x)$;

(iii) $\overline{D}(F+G)(x) \ge \overline{D}F(x) + \underline{D}G(x)$;

(iv) $\underline{D}(F+G)(x) \le \underline{D}F(x) + \overline{D}G(x)$;

(v) $\overline{D}(F-G)(x) \le \overline{D}F(x) - \underline{D}G(x)$;

(vi) $\overline{D}(F-G)(x) \ge \overline{D}F(x) - \overline{D}G(x) \ge \underline{D}(F-G)(x)$;

(vii) $\underline{D}(F-G)(x) \ge \underline{D}F(x) - \overline{D}G(x)$;

(viii) $\overline{D}(F-G)(x) \ge \underline{D}F(x) - \underline{D}G(x) \ge \underline{D}(F-G)(x)$;

(ix) If $x < b$ we may replace in (i)-(viii) $\overline{D}$ and $\underline{D}$ by D^+ and D_+;

(x) If $x > a$ we may replace in (i)-(viii) $\overline{D}$ and $\underline{D}$ by D^- and D_-.

Theorem 1.10.1 *([S3], p.261). Let $F : [a,b] \mapsto \mathbb{R}$. Then the following sets are at most countable:*

(i) $\{x : x$ *is a point of strictly maximum or strictly minimum for* $F\}$;

(ii) $\{x : \limsup_{t\to x} F(t) > \limsup_{t\searrow x} F(t)$ *or* $\liminf_{t\to x} F(t) < \liminf_{t\searrow x} F(t)\}$;

(iii) $\{x : D^+F(x) < D_-F(x) \text{ or } D^-F(x) < D_+F(x)\}$.

Theorem 1.10.2 *([S3], p.113; [Br2], p.54). Let $F : [a,b] \mapsto \mathbb{R}$ be a Borel function. Then the Dini derivatives of F are also Borel functions.*

Corollary 1.10.1 *Let $F : [a,b] \mapsto \mathbb{R}$ be a Borel function and let α be a real number. Then $\{x : F'(x) \le \alpha\}$ is a Borel set.*

Theorem 1.10.3 (Hájek) *([Br2], p.57). Let $F : [a,b] \mapsto \mathbb{R}$. Then $\overline{D}F$ is in Baire class two.*

Corollary 1.10.2 *Let $F : [a,b] \mapsto \mathbb{R}$ and let α be a real number. Then $\{x : F'(x) \le \alpha\}$ is a measurable set.*

Theorem 1.10.4 *([S3], p.226). Let $F : [a,b] \mapsto \mathbb{R}$ and let $P = \{x : \overline{D}^+F(x) < M$ and $\underline{D}^-F(x) > M$ for some positive number $M\}$. Then $|F(P)| \le M \mid P \mid$.*

Remark 1.10.1 *The proof of Theorem 1.10.4 is based on Theorem 1.8.1 (Vitali's Covering Theorem). An improvement of this theorem is due to Denjoy (see [S3], p.271): If $|D^+F(x)| \le M$ at every point x of a set E then $|F(E)| \le M \mid E \mid$.*

Lemma 1.10.2 *Let $F : [a,b] \mapsto \mathbb{R}$ and let P be a closed subset of $[a,b]$. If $x_o \in P$ is a bilateral accumulation point for P and $(F_{/P})'(x_o)$ exists (finite or infinite) then $(F_P)'(x_o) = (F_{/P})'(x_o)$ (see Definition 1.1.3).*

Theorem 1.10.5 *([N], p.214). Let Z be a subset of $[a,b]$, $|Z| = 0$. Then there exists a continuous, increasing function $F : [a,b] \mapsto \mathbb{R}$ such that $F'(x) = +\infty$, for each $x \in Z$.*

1.11 Approximate continuity and derivability

Definition 1.11.1 *([Br2], p.18). Let $F : [a,b] \mapsto \mathbb{R}$, $x_o \in [a,b]$. F is said to be approximately continuous at x_o provided there is a set E such that $d(E; x_o) = 1$ and $F_{/E}$ is continuous at x_o. If F is approximately continuous at every point $x \in [a,b]$, we say simply that F is approximately continuous on $[a,b]$.*

Theorem 1.11.1 *([Br2], p.19). Let $F : [a,b] \mapsto \mathbb{R}$. The function F is measurable if and only if F is approximately continuous a.e.*

Theorem 1.11.2 *([Br2], pp.20-24). Let $F, G, F_n : [a,b] \mapsto \mathbb{R}$ and $x_o \in [a,b]$.*

(i) *If F and G are approximately continuous at x_o then the same is true for the functions $F+G$, $F-G$ and $F \cdot G$. If $G(x_o) \ne 0$ then F/G is also approximately continuous at x_o.*

(ii) *If F is approximately continuous at x_o and G is continuous at $F(x_o)$ then $G \circ F$ is approximately continuous at x_o.*

(iii) *If F_n are approximately continuous on $[a,b]$ and $F_n \mapsto F\,[unif]$ then F is approximately continuous on $[a,b]$.*

Definition 1.11.2 *([Br2], p.149). Let $F : [a,b] \mapsto \mathbb{R}$, $x_o \in [a,b]$. If there exists a set A_{x_o} so that*

$$d(A_{x_o}; x_o) = 1 \text{ and } \lim_{x \to x_o, x \in A_{x_o}} \frac{F(x) - F(x_o)}{x - x_o} = L$$

then we call this limit $L = F'_{ap}(x_o)$, the approximate derivative of F at x_o. If F is approximately derivable at each point $x_o \in [a,b]$, then we say that F is approximately derivable on [a,b]. Similarly we can define the extreme approximate derivatives.

Theorem 1.11.3 *Let P be a measurable subset of [a,b], and let $F : P \mapsto \mathbb{R}$. Let $E = \{x : F'_{ap}(x) = 0\}$. Then $| F(E) |= 0$.*

Proof. The proof follows by Lemma 9.2 of [S3] (p.290).

Theorem 1.11.4 *([S3], p.299). Let P be a measurable subset of $[a,b]$, $| P |> 0$, and let $F : P \mapsto \mathbb{R}$ be a measurable function. Then the extreme approximate derivatives of F are measurable.*

Corollary 1.11.1 *Let P be a measurable subset of $[a,b]$, $| P |> 0$, and let $F : P \mapsto \mathbb{R}$ be a measurable function. Let $E = \{x : F$ is approximately derivable at $x\}$. Then E is measurable.*

1.12 Sharp derivatives $D^{\#}F$

Definition 1.12.1 (Peano) *([T4], I, pp. 73 -74). Let $F : [a,b] \mapsto \mathbb{R}$, $x_o \in [a,b]$. We define the lower and upper sharp derivatives as:*

$$\underline{D}^{\#}F(x_o) = \liminf_{\substack{(y,z) \to (x_o,x_o) \\ y \neq z}} \frac{F(z) - F(y)}{z - y} \quad \text{and}$$

$$\overline{D}^{\#}F(x_o) = \limsup_{\substack{(y,z) \to (x_o,x_o) \\ y \neq z}} \frac{F(z) - F(y)}{z - y}.$$

If $\underline{D}^{\#}F(x_o) = \overline{D}^{\#}F(x_o)$ then we denote their common value by $D^{\#}F(x_o)$, and call it the sharp derivative of F at x_o.

Remark 1.12.1 *Let $F : [a,b] \mapsto \mathbb{R}$, $x_o \in [a,b]$.*

(i) $\underline{D}^{\#}F(x_o) \leq \underline{F}'(x_o) \leq \overline{F}'(x_o) \leq \overline{D}^{\#}F(x_o)$.

(ii) We may replace in Lemma 1.10.1 (i)-(viii), $\overline{D}$ and $\underline{D}$ by $\overline{D}^{\#}$ and $\underline{D}^{\#}$.

1.13 Local systems; examples

Definition 1.13.1 *([T5], p.3). The family $\mathcal{S} = \{\mathcal{S}(x) : x \in \mathbb{R}\}$ is said to be a local system of sets provided it has the following properties:*

(i) $\{x\} \notin \mathcal{S}(x)$;

(ii) if $S \in \mathcal{S}(x)$ *then* $x \in S$;

(iii) if $S_1 \in \mathcal{S}(x)$ *and* $S_2 \supset S_1$ *then* $S_2 \in \mathcal{S}(x)$;

(iv) if $S \in \mathcal{S}(x)$ *and* $\delta > 0$ *then* $S \cap (x - \delta, x + \delta) \in \mathcal{S}(x)$.

The system $\mathcal{S}$ is bilateral provided, every set $S \in \mathcal{S}(x)$ contains points on either side of x. The system $\mathcal{S}$ is bilaterally c-dense, provided for every set $S \in \mathcal{S}(x)$ and any $\delta > 0$, the sets $S \cap (x, x+\delta)$ and $S \cap (x-\delta, x)$ have power c. Let $\mathcal{S}_-(x) = \mathcal{S}(x) \cap (-\infty, x]$ and $\mathcal{S}_+(x) = \mathcal{S}(x) \cap [x, +\infty)$.

Definition 1.13.2 *([T5], p.117). Let $\mathcal{S}$ be a local system. A function $F : [a,b] \mapsto \mathbb{R}$ is said to be $\mathcal{S}$-increasing at a point x provided $\sigma_x = \{y : y = x$ or $(F(y) - F(x))/(y-x) \geq 0\} \in \mathcal{S}(x)$. If " $\geq$ " is replaced by " $>$ ", " $\leq$ " or " $<$ ", we obtain the conditions strictly $\mathcal{S}$-increasing, $\mathcal{S}$-decreasing, strictly $\mathcal{S}$-decreasing.*

Definition 1.13.3 *([T5], p.3). Let $\mathcal{S}$ be a local system, and let $F : [a,b] \mapsto \mathbb{R}$. The extreme limits relative to $\mathcal{S}$ at a point x are defined as:*

- $\mathcal{S} - \limsup_{y \to x} F(y) = \inf\{c : \{t : t = x \text{ or } F(t) < c\} \in \mathcal{S}(x)\}$ *and*
- $\mathcal{S} - \liminf_{y \to x} F(y) = \sup\{c : \{t : t = x \text{ or } F(t) > c\} \in \mathcal{S}(x)\}$

Definition 1.13.4 *([T4], II, p. 281). Let $\mathcal{S}$ be a local system and let $F : [a,b] \mapsto \mathbb{R}$. We denote by*

- $\mathcal{S} - \underline{D}F(x) = \sup\{c \in \mathbb{R} : \{x\} \cup \{y : \frac{F(y)-F(x)}{y-x} > c\} \in \mathcal{S}(x)\}$ *and*
- $\mathcal{S} - \overline{D}F(x) = \inf\{c \in \mathbb{R} : \{x\} \cup \{y : \frac{F(y)-F(x)}{y-x} < c\} \in \mathcal{S}(x)\}$

the extreme $\mathcal{S}$-derivatives of F at a point x. An exact $\mathcal{S}$-derivative of F at x, if it exists, is any number $c \in \overline{\mathbb{R}}$ such that, for any neighborhood U of c, the set of points $\{y : y = x$ or $(F(y) - F(x))/(y-x)) \in U\} \in \mathcal{S}(x)$. In this case we write $\mathcal{S} - DF(x) = c$, with the warning that the number c need not to be unique, nor have an immediate relation with the two extreme $\mathcal{S}$-derivatives. The set of all $\mathcal{S}$- derivatives of a function F at a point x will be denoted by $\mathcal{S}$-$DF(x)$. Clearly, if F is derivable at a point x then F is also $\mathcal{S}$- derivable at x, and the derivatives are equal.

Definition 1.13.5 *([T4], p.229; [BrOT], p.101). A local system $\mathcal{S} = \{\mathcal{S}(x) : x \in \mathbb{R}\}$ will be said to satisfy the intersection conditions listed below, if corresponding to any choice $\{\sigma_x : x \in \mathbb{R}\}$ from $\mathcal{S}$, there must be a positive function δ such that whenever $x, y \in \mathbb{R}$ and $0 < y - x < \min\{\delta(x), \delta(y)\}$, the two sets σ_x and σ_y must intersect in the asserted fashion:*

(1) intersection condition (I.C.): $\sigma_x \cap \sigma_y \cap [x,y] \neq \emptyset$;

(2) *external intersection condition (E.I.C.):*
$\sigma_x \cap \sigma_y \cap (y, 2y - x) \neq \emptyset$ *and* $\sigma_x \cap \sigma_y \cap (2x - y, x) \neq \emptyset$;

(3) *external intersection condition with parameter m (E.I.C.[m]):*
$\sigma_x \cap \sigma_y \cap (y, (m+1)y - mx) \neq \emptyset$ *and* $\sigma_x \cap \sigma_y \cap ((m+1)x - my, x) \neq \emptyset$;

(4) $\sigma_x \cap \sigma_y \cap (-\infty, x] \neq \emptyset$ *and* $\sigma_x \cap \sigma_y \cap [y, +\infty) \neq \emptyset$;

(5) $\sigma_x \cap \sigma_y \cap (-\infty, x] \neq \emptyset$;

(6) $\sigma_x \cap \sigma_y \cap [y, +\infty) \neq \emptyset$.

Definition 1.13.6 *([T5]). Let $\mathcal{S} = \{\mathcal{S}(x) : x \in \mathbb{R}\}$ be a local system. We will say that $\mathcal{S}$ is filtering at a point x, if $S_1 \cap S_2 \in \mathcal{S}(x)$, whenever S_1 and S_2 belong to $\mathcal{S}(x)$.*

Lemma 1.13.1 *Let $F : [a,b] \mapsto \mathbb{R}$ and let $P \subseteq (a,b)$, $P \neq \emptyset$ be a G_δ-set. Let $\mathcal{S}$ be a local system with intersection condition (4). Let $A \subset \{x \in P : F$ is strictly $\mathcal{S}$-increasing at $x\}$ and $B \subset \{x \in P : F$ is strictly $\mathcal{S}$-decreasing at $x\}$. Suppose that $P = \overline{A}$ (resp. $P = \overline{B}$). Then we have:*

(i) B (resp. A) is of first category with respect to P;

(ii) If $E = P \setminus (A \cup B)$ is countable then B (resp. A) is nowhere dense in P.

Proof. Let $\sigma_x = \{y : y = x$ or $(F(y) - F(x))/(y - x) \geq 0$ for $y \neq x\} \cap (a,b)$, for $x \in A$, and let $\sigma_x = \{y : y = x$ or $(F(y) - F(x))/(y - x) < 0$ for $y \neq x\} \cap (a,b)$, for $x \in B$. Let $\delta(x)$ be given by (4), for $x \in A \cup B$. We may suppose without loss of generality that $\delta(x) < 1$. If $x, y \in A \cup B$ and $|x - y| < \min\{\delta(x), \delta(y)\}$ then $\sigma_x \cap \sigma_y \cap (-\infty, x) \neq \emptyset$ and $\sigma_x \cap \sigma_y \cap (y, +\infty) \neq \emptyset$. Suppose that $P = \overline{A}$ (the second part follows similarly).

(i) Let $G_n = \cap_{x \in A}(x - (\delta(x))/n, x + (\delta(x))/n)$ and $H = P \cap (\cap_{n=1}^{\infty} G_n)$. Then H is a dense G_δ-set in P, and $A \subset H \subset P$. We prove that $B \cap H = \emptyset$. Suppose on the contrary that $B \cap H \neq \emptyset$. Let $y \in B \cap H$ and let n be a natural number such that $1/n < \delta(y)$. Then $y \in G_n$, hence there exists $x \in A$ such that $y \in (x - (\delta(x))/n, x + (\delta(x))/n)$. Since $(\delta(x))/n < 1/n < \delta(y)$, it follows that $|y - x| < \min\{\delta(x), \delta(y)\}$. Suppose for example that $x < y$ (the case $x > y$ is similar). Then we have two situations:
1) $F(x) \leq F(y)$. Let $z \in (-\infty, x) \cap \sigma_x \cap \sigma_y \neq \emptyset$. Then $F(z) \leq F(x)$ and $F(z) > F(y)$, a contradiction.
2) $F(x) > F(y)$. Let $z \in (y, +\infty) \cap \sigma_x \cap \sigma_y \neq \emptyset$. Then $F(z) \geq F(x)$ and $F(z) < F(y)$, a contradiction. It follows that $B \cap H = \emptyset$, hence $B \subset P \setminus H$, which is a set of first category with respect to P.

(ii) Suppose on the contrary that $\emptyset \neq (c,d) \cap P \subset \overline{B}$. Let

- $A_{mn} = \{x \in (c,d) \cap A \cap [m/n, (m+1)/n] : \delta(x) \in (1/n, 1/(n-1))\}$ and
- $B_{mn} = \{x \in (c,d) \cap B \cap [m/n, (m+1)/n] : \delta(x) \in (1/n, 1/(n-1))\}$,

where $n = \overline{2, +\infty}$, $m = \overline{0, n-1}$. Then $(c,d) \cap P = (\cup_{m,n}(A_{mn} \cap B_{mn})) \cup (E \cap (c,d))$. By Theorem 1.7.1, there exists an open interval $(\alpha, \beta) \subset (c,d)$ such that either 1) $\emptyset \neq (\alpha, \beta) \cap P \subset \overline{A}_{mn}$ or 2) $\emptyset \neq (\alpha, \beta) \cap P \subset \overline{B}_{mn}$, for some m and n.
1) Let $y \in (\alpha, \beta) \cap B$ and let $x \in (y - \delta(y), y + \delta(y)) \cap A_{mn} \cap (\alpha, \beta)$. Then $|y - x|$

$< \min\{\delta(x), \delta(y)\}$, a contradiction (as in (i), 1) and 2)).
2) Let $x \in (\alpha, \beta) \cap A$ and let $y \in (x - \delta(x), x + \delta(x)) \cap B_{mn} \cap (\alpha, \beta)$. Then $| x - y | < \min\{\delta(x), \delta(y)\}$, a contradiction (as in (i), 1) and 2)).

Definition 1.13.7 *([T5], p.7). Let $\mathcal{S}$ be a local system. Then by the dual of $\mathcal{S}$, denoted by $\mathcal{S}^*$, we mean the system defined so that for $x \in \mathbb{R}$, $S \in \mathcal{S}^*(x)$ if and only if $x \in S$ and $(R \setminus S) \cup \{x\} \notin \mathcal{S}(x)$.*

Proposition 1.13.1 *([T5], p.9). Let $\mathcal{S}$ be a local system and let $\mathcal{S}^*$ be its dual. Then for any function $F : [a, b] \mapsto \mathbb{R}$ the extreme limits must satisfy $\mathcal{S} - \limsup_{y \to x} F(y) = \mathcal{S}^* - \liminf_{y \to x} F(y)$ and $\mathcal{S} - \liminf_{y \to x} F(y) = \mathcal{S}^* - \limsup_{y \to x} F(y)$, at every point x.*

Examples ([T5], pp. 18,22,72).

1) Let $\mathcal{S}_o = \{\mathcal{S}_o(x) : x \in \mathbb{R}\}$ denote the local system defined at each point x as $\mathcal{S}_o(x) = \{S : S$ contains an open interval about the point $x\}$. We can define right and left versions of this, by writing: $\mathcal{S}_o^+(x) = \{U : U$ is a right neighborhood of $x\}$ and $\mathcal{S}_o^-(x) = \{U : U$ is a left neighborhood of $x\}$.

2) Let $\mathcal{S}_\infty = \{\mathcal{S}_\infty(x) : x \in \mathbb{R}\}$ denote the local system defined at each point as $\mathcal{S}_\infty(x) = \{S : S$ contains x and has x as an accumulation point$\}$. We can define right and left versions of this, by writing: $\mathcal{S}_\infty^+(x) = \{S : S$ contains x and has x as a right accumulation point $\}$, and $\mathcal{S}_\infty^-(x) = \{S : S$ contains x and has x as a left accumulation point $\}$.

3) Let $\mathcal{S}_q = \{\mathcal{S}_q(x) : x \in \mathbb{R}\}$ denote the local system defined at each point x as $\mathcal{S}_q(x) = \{S : x \in S$ and there exists a $\delta > 0$ such that $(x - \delta, x + \delta) \setminus S$ is of first category $\}$. We can define right and left versions of this, by writing: $\mathcal{S}_q^+(x) = \{S : x \in S$ and there exists a $\delta > 0)$ such that $(x, x + \delta) \setminus S$ is of first category $\}$, and $\mathcal{S}_q^-(x) = \{S : x \in S$ and there exists a $\delta > 0$ such that $(x - \delta, x) \setminus S$ is of first category $\}$.

4) Let $\mathcal{S}_c = \{\mathcal{S}_x(x) : x \in \mathbb{R}\}$ denote the local system defined at each point x as $\mathcal{S}_c(x) = \{S : x \in S$ and $S \cap U$ is of power c, whenever U is a neighborhood of $x\}$. We can define right and left versions of this, by writing $\mathcal{S}_c^+(x) = \{S : x \in S$ and $S \cap U$ is of power c, whenever U is a right neighborhood of $x\}$, and $\mathcal{S}_c^-(x) = \{S : x \in S$ and $S \cap U$ is of power c, whenever U is a left neighborhood of $x\}$.

5) Let $\mathcal{S}_2 = \{\mathcal{S}_2(x) : x \in \mathbb{R}\}$ denote the local system defined at each point as $\mathcal{S}_2(x) = \{S : x \in S$ and for all $\delta > 0$ both of the sets $S \cap (x, x + \delta)$ and $S \cap (x - \delta, x)$ have positive exterior measure $\}$. We can define right and left versions of this, by writing: $\mathcal{S}_2^+(x) = \{S : x \in S$ and for all $\delta > 0$ the set $S \cap (x, x + \delta)$ has positive exterior measure $\}$ and $\mathcal{S}_2^-(x) = \{S : x \in S$ and for all $\delta > 0$ the set $S \cap (x - \delta, x)$ has positive exterior measure $\}$.

6) Let $\mathcal{S}_{ap} = \{\mathcal{S}_{ap}(x) : x \in \mathbb{R}\}$ denote the local system defined at each point x as $\mathcal{S}_{ap}(x) = \{S : x \in S$ and $\underline{d}^i(S; x) = 1\}$. We can define right and left versions of this, by writing: $\mathcal{S}_{ap}^+(x) = \{S : x \in S$ and $\underline{d}^i_+(S; x) = 1\}$ and $\mathcal{S}_{ap}^-(x) = \{S : x \in S$ and $\underline{d}^i_-(S; x) = 1\}$.

7) Let $\mathcal{S}_3 = \{\mathcal{S}_3(x) : x \in \mathbb{R}\}$ denote the local system defined at each point x as $\mathcal{S}_3(x) = \{S : x \in S$ and for each sequence $\{I_n\}$ of closed intervals converging to x and not containing x, with $| I_n \cap S |^e = 0$, $n = \overline{1, \infty}$, we have $\lim_{n \to +\infty}(| I_n |^e)/(dist(x; I_n)) = 0\}$. We can define right and left versions of this, by writing: $\mathcal{S}_3^+(x)$

(if the intervals I_n are at the right side of x), and $\mathcal{S}_3^-(x)$ (if the intervals I_n are at the left side of x).

Remark 1.13.1 *([T5]).*

(i) $\mathcal{S}_o^* = \mathcal{S}_\infty$; $\mathcal{S}_\infty^* = \mathcal{S}_o$; $(\mathcal{S}_o^+)^* = \mathcal{S}_\infty^+$; $(\mathcal{S}_\infty^+)^* = \mathcal{S}_o^+$;

(ii) $\limsup_{y\searrow x} F(y) = \mathcal{S}_o^+ - \limsup_{y\to x} F(y) = \mathcal{S}_\infty^+ - \limsup_{y\to x} F(y)$ *and* $\liminf_{y\searrow x} F(y) = \mathcal{S}_o^+ - \liminf_{y\to x} F(y) = \mathcal{S}_\infty^+ - \liminf_{y\to x F(y)}$;

(iii) Following Zahorski's classes M_o, M_1, M_2, M_3, M_5 (see [Z1]), the definitions of $\mathcal{S}_\infty$, $\mathcal{S}_c$, $\mathcal{S}_2$, $\mathcal{S}_3$, $\mathcal{S}_{ap}$ respectively seem very natural. The $\mathcal{S}_q$, $\mathcal{S}_q^+$ and $\mathcal{S}_q^-$ local systems are related to the qualitative differentiation and qualitative continuity of Marcus (see [Ma1], [Ma4]).

1.14 $\mathcal{S}$-open sets

Definition 1.14.1 *Let $\mathcal{S} = \{\mathcal{S}(x) : x \in \mathbb{R}\}$ be a local system and let $\emptyset \neq E \subset \mathbb{R}$. E is said to be $\mathcal{S}$-open if $E \in \mathcal{S}(x)$, for $x \in E$. E is said to be $\mathcal{S}$-closed if $\mathbb{R} \setminus E$ is $\mathcal{S}$-open. $\emptyset$ is supposed to be $\mathcal{S}$-open. Clearly, if $E \subset \mathbb{R}$ is an open set then E is $\mathcal{S}$-open.*

Proposition 1.14.1 *Let $E \subset \mathbb{R}$. The following assertions are equivalent:*

(i) E is $\mathcal{S}_o^+$-open;

(ii) E is $\mathcal{S}_q^+$-open.

Proof. (i) ⇒ (ii) is evident.

(ii) ⇒ (i) Suppose that E is not $\mathcal{S}_o^+$-open. Then there exists $x_o \in E$ such that $E \notin \mathcal{S}_o^+(x_o)$. Since E is $\mathcal{S}_q^+$-open there is a $\delta > 0$ such that $E \cap [x_o, x_o + \delta)$ is residual with respect to $[x_o, x_o + \delta)$. Because $E \notin \mathcal{S}_o^+(x_o)$, it follows that $[x_o, x_o + \delta) \not\subset E$. Hence there exists $x_1 \in (x_o, x_o + \delta)$, such that $x_1 \notin E$. Then $E \notin \mathcal{S}_q^+(x_1)$, so $E \cap [x_1, x_o + \delta)$ is not residual with respect to $[x_1, x_o + \delta)$. But $E \cap [x_o, x_o + \delta)$ is residual with respect to $[x_o, x_o + \delta)$, hence $E \cap [x_1, x_o + \delta)$ is residual with respect to $[x_1, x_o + \delta)$, a contradiction.

Corollary 1.14.1 *Let $E \subset \mathbb{R}$. The following assertions are equivalent:*

(i) E is open (resp. closed);

(ii) E is simultaneously $\mathcal{S}_o^+$ and $\mathcal{S}_o^-$ open (resp. closed);

(iii) E is simultaneously $\mathcal{S}_q^+$ and $\mathcal{S}_q^-$ open(resp. closed).

Proposition 1.14.2 *Let $E \subset \mathbb{R}$. The following assertions are equivalent:*

(i) E is $\mathcal{S}_o^+$ open;

(ii) $E \subset \cup_{i\geq 1}[a_i, b_i)$, where $\{(a_i, b_i)\}_{i\geq 1}$ are the components of $int(E)$.

Proof. (i) $\Rightarrow$ (ii) Let $x \in E$. Then there exists $\delta > 0$ such that $[x, x+\delta) \subset E$. It follows that $(x, x+\delta) \subset int(E)$. Hence there exists an i such that $(x, x+\delta) \subset (a_i, b_i)$. Now $x \in [a_i, b_i)$, hence $E \subset \cup_{i\geq 1}[a_i, b_i)$.

(ii) $\Rightarrow$ (i) Let $x \in E$. Then there exists an i such that $x \in [a_i, b_i)$. But $(a_i, b_i) \subset E$, hence $[x, b_i) \subset E$. It follows that E is $\mathcal{S}_o^+$ open.

Proposition 1.14.3 *Let $E \subset \mathbb{R}$. The following assertions are equivalent:*

(i) E is $\mathcal{S}_o^+$ closed;

(ii) If $x_n \to x_o$, $x_n \geq x_o$ and $x_n \in E$ then $x_o \in E$.

Proof. (i) $\Rightarrow$ (ii) Let $x_n \to x_o$, $x_n \geq x_o$, $x_n \in E$ and suppose that $x_o \notin E$. Then $x_o \in \mathbb{R}\setminus E$. Since $\mathbb{R}\setminus E$ is $\mathcal{S}_o^+$ open, there exists a $\delta > 0$ such that $[x_o, x_o+\delta) \subset \mathbb{R}\setminus E$. It follows that there exists n_o such that $x_{n_o} \in (x_o, x_o+\delta) \subset \mathbb{R}\setminus E$, hence $x_{n_o} \notin E$, a contradiction.

(ii) $\Rightarrow$ (i) Suppose that E is not $\mathcal{S}_o^+$ closed. Then $\mathbb{R}\setminus E$ is not $\mathcal{S}_o^+$ open. It follows that there exists $x_o \in \mathbb{R}\setminus E$ such that, for each $\delta > 0$ the set $[x_o, x_o+\delta)$ is not contained in $\mathbb{R}\setminus E$. Then there exists x_δ such that $x_\delta \in [x_o, x_o+\delta) \cap E$. For $\delta = 1/n$, there exists $x_n \in [x_o, x_o + 1/n) \cap E$, $n = \overline{1, \infty}$. Then $x_n \to x_o$, and by (ii), $x_o \in E$, a contradiction.

Proposition 1.14.4 *Let $E \subset \mathbb{R}$. The following assertions are equivalent:*

(i) E is $\mathcal{S}_\infty^+$ (resp. $\mathcal{S}_c^+$; $\mathcal{S}_q^+$) closed;

(ii) If $(u, v)\setminus E$ is empty (resp. at most countable; of first category) then $u \in E$.

Proof. We show for example the part with $\mathcal{S}_\infty^+$ (the other parts follow similarly).

(i) $\Rightarrow$ (ii) Let $(u, v) \subset E$. Suppose that $u \notin E$. Then $u \in \mathbb{R}\setminus E$. Since $\mathbb{R}\setminus E$ is $\mathcal{S}_\infty^+$ open, it follows that $(u, v) \cap (\mathbb{R}\setminus E) \neq \emptyset$, a contradiction.

(ii) $\Rightarrow$ (i) Suppose that E is not $\mathcal{S}_\infty^+$ closed. Then $\mathbb{R}\setminus E$ is not $\mathcal{S}_\infty^+$ open and there exists $x_o \in \mathbb{R}\setminus E$ such that $\mathbb{R}\setminus E \notin \mathcal{S}_\infty^+(x_o)$. It follows that there exists $\delta > 0$ such that $(x_o, x_o+\delta) \cap (\mathbb{R}\setminus E) = \emptyset$, hence $(x_o, x_o+\delta) \subset E$. By (ii), $x_o \in E$, a contradiction.

Proposition 1.14.5 *Let E be a real measurable set. The following assertion are equivalent:*

(i) E is $\mathcal{S}_2^+$ closed;

(ii) If $|(u, v) \cap E| = v - u$ then $u \in E$.

Proof. The proof is similar to that of Proposition 1.14.4.

Remark 1.14.1 *Let $\mathcal{S}$ be a local system.*

(i) If $\{E_i\}_{i\in I}$ is a collection of $\mathcal{S}$ open sets then $\cup_i E_i$ is $\mathcal{S}$ open.

(ii) There exists a local system $\mathcal{S}$ and some $\mathcal{S}$ open sets E_1 and E_2, such that $E_1 \cap E_2$ is not $\mathcal{S}$ open (see for example $\mathcal{S}_\infty$);

(iii) Any $\mathcal{S}_o^+$ open set is an F_σ-set (see Proposition 1.14.2);

(iv) *Any* $\mathcal{S}_o^+$ *closed set is a* G_δ*-set (see Proposition 1.14.2);*

(v) *There exists a bounded* $\mathcal{S}_o^+$ *closed set which is not of* F_σ*-type (see Section 6.3.);*

(vi) *If* $\mathcal{S}$ *is a local system which has right and left versions, and* $E \subseteq [a,b)$ *is* $\mathcal{S}^+$ *open then* $[a,b) \setminus E$ *is* $\mathcal{S}^+$ *closed.*

Theorem 1.14.1 (O'Malley) *([O1], [FrH]). Let* E *be a real bounded* F_σ*-set. If* E *is* $\mathcal{S}_{ap}^+$ *open then there exists* c *in* $\mathbb{R} \setminus E$ *such that* $E \in \mathcal{S}_{ap}^-(c)$.

Remark 1.14.2 *The existence of a point* c *as in Theorem 1.14.1 is called the density property of O'Malley. Freilig and Humke showed in [FrH] that Theorem 1.14.1 remains true if* E *is a* G_δ*-set, but not if* E *is an* $F_{\sigma\delta}$*-set.*

1.15 Semicontinuity; $\mathcal{S}$-semicontinuity

Definition 1.15.1 *Let* E *be a real bounded set and* $F : [a,b] \mapsto \overline{\mathbb{R}}$. *Let* $x_o \in E$. F *is said to be lower left semicontinuous at* x_o*, if* x_o *is a left isolated point in* E*, or if* $\liminf_{x \nearrow x_o, x\in E} F(x) \geq F(x_o)$*, whenever* x_o *is a left accumulation point of* E*. Similarly we define: lower right semicontinuity, upper left semicontinuity and upper right semicontinuity.* F *is said to be lower semicontinuous ar a point* x_o*, if* F *is simultaneously left and right lower semicontinuous (i.e., if* x_o *is an isolated point in* E*, or if* $\liminf_{x\to x_o, x\in E} F(x_o) \geq F(x_o)$*, whenever* x_o *is an accumulation point of* E*). Similarly we can define upper semicontinuity.*
F *is continuous at* x_o*, if* F *is both, lower and upper semicontinuous at* x_o *(i.e., if* x_o *is an isolated point in* E*, or if* $\lim_{x\to x_o, x\in E} F(x) = F(x_o)$*, whenever* x_o *is an accumulation point of* E*).*
F *is said to be lower left semicontinuous (resp. lower right semicontinuous, etc.) on a set* $A \subseteq E$*, if it is so at each point* $x \in A$.

Definition 1.15.2 *([T5], p. 71). Let* $\mathcal{S} = \{\mathcal{S}(x) : x \in \mathbb{R}\}$ *be a local system, and let* $F : [a,b] \mapsto \overline{\mathbb{R}}$, $x_o \in (a,b)$. F *is said to be* $\mathcal{S}$ *lower semicontinuous at* x_o*, if for every real* $\alpha < F(x_o)$ *we have* $\{t \in (a,b) : F(t) > \alpha\} \in \mathcal{S}(x_o)$. F *is said to be* $\mathcal{S}$ *upper semicontinuous at* x_o*, if* $-F$ *is* $\mathcal{S}$ *lower semicontinuous at* x_o. F *is said to be* $\mathcal{S}$ *continuous at* x_o*, if for every* $\epsilon > 0$ *we have:* $\{t : | F(x_o) - F(t) | < \epsilon\} \in \mathcal{S}(x_o)$*, whenever* $F(x_o) \in \mathbb{R}$*;* $\{t : F(t) > \alpha\} \in \mathcal{S}(x_o)$*, whenever* $F(x_o) = +\infty$, $\alpha \in \mathbb{R}$*; and* $\{t : F(t) < \alpha\} \in \mathcal{S}(x_o)$*, whenever* $F(x_o) = -\infty$, $\alpha \in \mathbb{R}$.

Remark 1.15.1 *If* $F : [a,b] \mapsto \overline{\mathbb{R}}$ *is* $\mathcal{S}$ *continuous at* x_o *then it is both,* $\mathcal{S}$ *upper and* $\mathcal{S}$ *lower semicontinuous at* x_o. *If* $\mathcal{S}$ *is filtering then the converse is also true.*

Proposition 1.15.1 *Let* $F : [a,b] \mapsto \overline{\mathbb{R}}$.

(i) *([T5], p. 71). The following assertions are equivalent:*

a) F *is* $\mathcal{S}$ *continuous on* $[a,b]$*;*

b) *If* $u, v, w \in \mathbb{R}$, $u < v$ *then* $F^{-1}((u,v))$, $F^{-1}((w,+\infty])$ *and* $F^{-1}([-\infty,w))$ *are* $\mathcal{S}$ *open;*

(ii) The following assertions are equivalent:

a) F is $\mathcal{S}$ continuous on $A = \{x : F(x) \neq \pm\infty\}$;

b) $F^{-1}((u,v))$ is $\mathcal{S}$ open, whenever $-\infty < u < v < +\infty$

Proposition 1.15.2 *([T5], p. 71). Let $F : (a,b) \mapsto \overline{\mathbb{R}}$. The following assertions are equivalent:*

(i) F is $\mathcal{S}$ lower semicontinuous on (a,b);

(ii) $\{t \in (a,b) : F(t) > \alpha\}$ is $\mathcal{S}$ open, whenever $\alpha \in \mathbb{R}$.

Definition 1.15.3 *Let $\mathcal{S}$ be a local system which has left and right versions, $F : [a,b] \mapsto \overline{\mathbb{R}}$, and let $x_o \in [a,b]$. F is said to be $\mathcal{S}^+$ lower semicontinuous at $x_o \in [a,b)$ if $\{t \in [a,b) : F(t) > \alpha\} \in \mathcal{S}^+(x_o)$, whenever $\alpha < F(x_o)$. By definition F is considered to be $\mathcal{S}^+$ lower semicontinuous at b. F is said to be $\mathcal{S}^+$ upper semicontinuous at x_o, if $-F$ is $\mathcal{S}^+$ lower semicontinuous at x_o.*
F is said to be $\mathcal{S}^+$ continuous at $x_o \in [a,b)$ if: $\{t \in [a,b) : \mid F(x_o) - F(t) \mid < \epsilon\} \in \mathcal{S}^+(x_o)$, whenever $F(x_o) \in \mathbb{R}$ and $\epsilon > 0$; $\{t \in [a,b) : F(t) > \alpha\} \in \mathcal{S}^+(x_o)$, whenever $F(x_o) = +\infty$ and $\alpha \in \mathbb{R}$; $\{t \in [a,b) : F(t) < \alpha\} \in \mathcal{S}^+(x_o)$, whenever $F(x_o) = -\infty$ and $\alpha \in \mathbb{R}$. By definition F is considered to be $\mathcal{S}^+$ continuous at b.
Similarly we can define $\mathcal{S}^-$ lower semicontinuity, etc.
F is said to be $\mathcal{S}^+$ lower semicontinuous (resp. $\mathcal{S}^+$ upper semicontinuous, etc.) on a set $A \subseteq [a,b]$, if F is so at each point $x \in A$.

Remark 1.15.2 *A function $F : [a,b] \mapsto \overline{\mathbb{R}}$ is lower left semicontinuous on $[a,b]$ if and only if F is $\mathcal{S}_o^-$ lower semicontinuous on $[a,b]$.*

Proposition 1.15.3 *Let $\mathcal{S}$ be a local system which has left and right versions, and let $F : [a,b] \mapsto \overline{\mathbb{R}}$. The following assertions are equivalent:*

(i) F is $\mathcal{S}^+$ lower semicontinuous on $[a,b]$;

(ii) $\{t \in [a,b) : F(t) > \alpha\}$ is $\mathcal{S}^+$ open, whenever $\alpha \in \mathbb{R}$;

(iii) $\{t \in [a,b) : F(t) \leq \alpha\}$ is $\mathcal{S}^+$ closed, whenever $\alpha \in \mathbb{R}$.

Proof. (i) $\Leftrightarrow$ (ii) follows by definitions; for (ii) $\Leftrightarrow$ (iii) see Remark 1.14.1.

Corollary 1.15.1 *Let $F : [a,b] \mapsto \overline{\mathbb{R}}$. The following assertions are equivalent:*

(i) F is lower semicontinuous on $[a,b]$;

(ii) $\{t \in [a,b) : F(t) > \alpha\}$ is $\mathcal{S}_o^+$ open, and $\{t \in (a,b] : F(t) > \alpha\}$ is $\mathcal{S}_o^-$ open, whenever $\alpha \in \mathbb{R}$;

(iii) $\{t \in [a,b] : F(t) \leq \alpha\}$ is closed, whenever $\alpha \in \mathbb{R}$.

Proof. (i) $\Leftrightarrow$ (ii) See Remark 1.15.2 and Proposition 1.15.3;
(i) $\Leftrightarrow$ (iii) See [N] (II, p.152).

Corollary 1.15.2 *Let $F : [a,b] \mapsto \overline{\mathbb{R}}$. The following assertions are equivalent:*

(i) $F(x) \leq \limsup_{y \searrow x} F(y)$, $x \in [a,b)$;

(ii) F *is* S^+_∞ *lower semicontinuous on* $[a,b]$;

(iii) $\{t \in [a,b) : F(t) > \alpha\}$ *is* S^+_∞ *open, whenever* $\alpha \in \mathbb{R}$;

(iv) $\{t \in [a,b) : F(t) \leq \alpha\}$ *is* S^+_∞ *closed, whenever* $\alpha \in \mathbb{R}$.

Proof. (i) $\Leftrightarrow$ (ii) See Remark 1.13.1.
(ii) $\Leftrightarrow$ (iii) $\Leftrightarrow$ (iv) See Proposition 1.15.3.

Proposition 1.15.4 *Let* S *be a local system which has left and right versions, and let* $F : [a,b] \mapsto \overline{\mathbb{R}}$.

(i) The following assertions are equivalent:

a) F *is* S^+ *continuous on* $[a,b]$;

b) The sets $\{t \in [a,b) : \alpha < F(t) < \beta\}$, $\{t \in [a,b) : F(t) > \gamma\}$ *and* $\{t \in [a,b) : F(t) < \gamma\}$ *are* S^+ *open, whenever* $\alpha, \beta, \gamma \in \mathbb{R}$;

(ii) The following assertions are equivalent:

a) F *is* S^+ *continuous on* $A = \{t \in [a,b) : F(t) \neq \pm\infty\}$;

b) The set $\{t \in [a,b) : \alpha < F(t) < \beta\}$ *is* S^+ *open, whenever* $\alpha, \beta \in \mathbb{R}$.

Proposition 1.15.5 *([N], p. 153). Let* $F : [a,b] \mapsto \overline{\mathbb{R}}$. *If* F *is lower semicontinuous and* $F(x) > -\infty$, *for each* $x \in [a,b]$ *then there exists a real number* M *such that* $M < F(x)$, *for each* $x \in [a,b]$.

Lemma 1.15.1 *Let* $F : [a,b] \mapsto \mathbb{R}$. *If* $\underline{D}^{\#}F(x) > \infty$ *everywhere on* $[a,b]$ *then* $\underline{D}^{\#}F(x)$ *is lower semicontinuous on* $[a,b]$.

Proof. Let $f(x) = \underline{D}^{\#}F(x)$, $x \in [a,b]$. Let α be a real number and let $E^\alpha = \{x \in [a,b] : f(x) \leq \alpha\}$. We show that E^α is closed. Suppose that there exists $x_o \in \overline{E^\alpha} \setminus E^\alpha$. Then $f(x_o) > \alpha$. Let $\beta = (f(x_o) + \alpha)/2$. By the definition of $\underline{D}^{\#}F(x)$, it follows that there is a $\delta > 0$ such that $(F(z) - F(y))/(z - y) > \beta$, whenever $y, z \in [a,b]$, $x_o - \delta < y < z < x_o + \delta$. Because x_o is an accumulation point of E^α, there exists $x_1 \in E^\alpha \cap (x_o - \delta, x_o + \delta)$. Let $\delta_1 > 0$ such that $(x_1 - \delta_1, x_1 + \delta_1) \subset (x_o - \delta, x_o + \delta)$. We have $(F(z) - F(y))/(z - y) > \beta$, whenever $y, z \in [a,b]$, $x_1 - \delta_1 < y < z < x_1 + \delta_1$, hence $f(x_1) \geq \beta > \alpha$, a contradiction (since $x_1 \in E^\alpha$), By Corollary 1.15.1, f is lower semicontinuous on $[a,b]$.

Chapter 2

Classes of functions

2.1 Darboux conditions $\mathcal{D}$, $\mathcal{D}_-$, $\mathcal{D}_+$

Let $\mathcal{D}$ denote the class of all Darboux functions defined on an interval.

Definition 2.1.1 *Let $F : [a,b] \mapsto \overline{\mathbb{R}}$. F is said to be $\mathcal{D}_-$ (Darboux$_-$) on $[a,b]$, if from $a \le \alpha < \beta \le b$ and $F(\beta) < F(\alpha)$ it follows that $[F(\beta), F(\alpha)] \subset F([\alpha,\beta])$. Let $\mathcal{D}_+ = \{F : -F \in \mathcal{D}_-\}$ on $[a,b]$.*

Remark 2.1.1 *Let $F : [a,b] \mapsto \overline{\mathbb{R}}$.*

(i) $\mathcal{D} = \mathcal{D}_- \cap \mathcal{D}_+$ on $[a,b]$;

(ii) If F is increasing then $F \in \mathcal{D}_-$ on $[a,b]$;

(iii) If $F \in \mathcal{D}$ and F is increasing then F is continuous on $[a,b]$;

(iv) $\mathcal{C} \subset \mathcal{D} \subset \mathcal{D}_-$ on $[a,b]$ (because if F is increasing and $F \notin \mathcal{C}$ then $F \in \mathcal{D}_-$ and $F \notin \mathcal{D}$);

(v) Some other Darboux properties are studied in [El] and [En2].

Lemma 2.1.1 *Let $F : [a,b] \mapsto \mathbb{R}$.*

(i) If $F(x) \le \liminf_{y \searrow x} F(y)$ for each $x \in [a,b)$ then the following assertions are equivalent:

- *a) $F \in \mathcal{D}_-$ on $[a,b]$;*
- *b) $F(x) \ge \liminf_{y \nearrow x} F(y)$ for each $x \in (a,b]$;*

(ii) If $\limsup_{y \nearrow x} F(y) \le F(x)$ for each $x \in (a,b]$ then the following assertions are equivalent:

- *a) $F \in \mathcal{D}_-$ on $[a,b]$;*
- *b) $F(x) \le \limsup_{y \searrow x} F(y)$ for each $x \in [a,b)$.*

Proof. We show for example the first part.

a) $\Rightarrow$ b) Suppose on the contrary that there exists $x_o \in (a,b]$ such that $F(x_o) < \liminf_{y \nearrow x_o} F(y)$. Then there exist a real number A and $x_1 < x_o$ such that $F(x_o) < A$ and $F(y) > A$, for each $y \in [x_1, x_o]$. Applying the definition of $\mathcal{D}_-$ to x_1 and x_o, we obtain a contradiction.

b) $\Rightarrow$ a) Let $a \leq a' < b' \leq b$ such that $F(a') > F(b')$, and let $\alpha \in (F(b'), F(a'))$. Let $B = \{x \in [a',b'] : F(x) \leq \alpha\}$. Then $b' \in B$, hence $B \neq \emptyset$. Let $c = inf(B)$. Suppose on the contrary that $c \notin B$. Then c is a right accumulation point of B and $F(c) \leq \liminf_{y \searrow c} F(y) \leq \liminf_{y \searrow c, y \in B} F(y) \leq \alpha$, hence $c \in B$, a contradiction. Then $F(c) \leq \alpha$. Since $F(a') > \alpha$, it follows that $a' \neq c$ and $F(x) > \alpha$, $x \in [a', c)$. By hypothesis $F(c) \geq \liminf_{y \nearrow c} F(y) \geq \alpha$. It follows that $F(c) = \alpha$, hence $F \in \mathcal{D}_-$ on $[a,b]$.

Corollary 2.1.1 *Let $F : [a,b] \mapsto \mathbb{R}$.*

(i) If $\limsup_{y \nearrow x} F(y) \leq F(x)$, $x \in (a,b]$ and $F(x) \leq \limsup_{y \searrow x} F(y)$, $x \in [a,b)$ then $F \in \mathcal{D}_-$ on $[a,b]$;

(ii) If $\liminf_{y \nearrow x} F(y) \leq F(x)$, $x \in (a,b]$ and $F(x) \leq \liminf_{y \searrow x} F(y)$, $x \in [a,b)$ then $F \in \mathcal{D}_-$ on $[a,b]$;

(iii) If $\limsup_{y \nearrow x} F(y) \geq F(x)$, $x \in (a,b]$ and $F(x) \geq \limsup_{y \searrow x} F(y)$, $x \in [a,b)$ then $F \in \mathcal{D}_+$ on $[a,b]$;

(iv) If $\liminf_{y \nearrow x} F(y) \geq F(x)$, $x \in (a,b]$ and $F(x) \geq \liminf_{y \searrow x} F(y)$, $x \in [a,b)$ then $F \in \mathcal{D}_+$ on $[a,b]$.

Lemma 2.1.2 *Let $F : [a,b] \mapsto \mathbb{R}$, $F(a) < F(b)$. If $F \in \mathcal{D}_+$ then $\mathcal{O}_+^n(F;[a,b]) \geq F(b) - F(a)$, for each natural number $n \geq 1$.*

Proof. Clearly $\mathcal{O}_+^1(F;[a,b]) \geq F(b) - F(a)$ (this follows without condition $\mathcal{D}_+$). Suppose $n \geq 1$ and let $\{E_k\}$, $k = \overline{1,n}$ such that $[a,b] = \cup_{k=1}^n E_k$. It is sufficient to show

$$(1) \quad F(b) - F(a) \leq \sum_{k=1}^{n} \mathcal{O}_+(F; E_k).$$

We may suppose without loss of generality that $a \in E_1$. Let $x_o = a$ and $M_1 = \sup\{F(x) : x \in E_1\}$. Then $\mathcal{O}_+(F;E_1) \geq M_1 - F(a)$. If $M_1 \geq F(b)$ then we have (1). If $M_1 < F(b)$ then there exists $x_1 \in [a,b)$ such that $F(x_1) = M_1$ (since $F \in \mathcal{D}_+$). We have two possibilities:

a) If $x_1 \notin E_1$ we may suppose that $x_1 \in E_2$. Let $M_2 = \sup\{F(x) : x \in [x_1, b] \cap E_2\}$. Then $\mathcal{O}_+(F;E_2) \geq M_2 - F(x_1) = M_2 - M_1$, hence

$$(2) \quad \mathcal{O}_+(F;E_1) + \mathcal{O}_+(F;E_2) \geq M_2 - F(a).$$

b) If $x_1 \in E_1$, let $\epsilon_1 \in (0, (F(b) - M_1)/2)$. Since $F \in \mathcal{D}_+$, there exists $y_1 \in (x_1, b)$ such that $F(y_1) = M_1 + \epsilon_1$. Let $\epsilon_2 = (\epsilon_1)/(2^1)$. Since $F \in \mathcal{D}_+$. there exists $y_2 \in (x_1, y_1)$ such that $F(y_2) = M_1 + \epsilon_2$. Continuing we obtain a decreasing sequence $\{y_k\}_k$, such that $x_1 < x_k < y_{k-1}$ and $F(y_k) = M_1 + (\epsilon_1)/(2^{k-1})$. Then $F(y_k) \to M_1$, $k \to \infty$. Let $x_1' = \lim_{k\to\infty} y_k$. Then $y_k \notin E_1$, for each k. We may suppose that

E_2 contains infinitely many points y_k. Let $M_2 = \sup\{F(x) : x \in [x_1', b] \cap E\}$. Then $\mathcal{O}_+(F; E_2) \geq M_2 - M_1 > 0$, hence

$$(3) \quad \mathcal{O}_+(F; E_1) + \mathcal{O}_+(F; E_2) \geq M_2 - F(a).$$

If $M_2 \geq F(b)$ then by (2) and (3) we have (1). If $M_2 < F(b)$ we continue as above until we find M_k, $k \leq n$ such that $M_k \geq F(b)$, hence we have (1).

Corollary 2.1.2 *Let $F : [a, b] \mapsto \mathbb{R}$ and let I be a closed subinterval of $[a, b]$. If $F \in \mathcal{D}_+$ and $\mathcal{O}_+(F; I) \neq +\infty$ then $\mathcal{O}_+^\omega(F; I) = \mathcal{O}_+(F; I)$.*

Proof. By Proposition 1.4.3, $\mathcal{O}_+^\omega(F; I) \leq \mathcal{O}_+(F; I)$. Suppose on the contrary that the inequality is strict. Then there exists a positive integer n such that $\mathcal{O}_+^\omega(F; \mathcal{O}) \leq \mathcal{O}_+^n(F; I) < \mathcal{O}_+(F; I)$. By the definition of $\mathcal{O}_+(F; I)$, there exists $c, d \in I$, $c < d$ such that $\mathcal{O}_+(F; I) \geq F(d) - F(c) > \mathcal{O}_+^n(F; I)$. By Lemma 2.1.2, $\mathcal{O}_+^n(F; I) \geq \mathcal{O}_+^n(F; [c, d]) \geq F(d) - F(c) > \mathcal{O}_+^n(F; I)$, a contradiction.

2.2 Baire conditions $\mathcal{B}_1$, $\underline{\mathcal{B}}_1$, $\overline{\mathcal{B}}_1$

Definition 2.2.1 *Let $F : [a, b] \mapsto \overline{\mathbb{R}}$, $P = \overline{P} \subseteq [a, b]$. F will be said to be $\mathcal{B}_1$ (in Baire class one) on P, if there exists a sequence of continuous functions $F_n : [a, b] \mapsto \mathbb{R}$ such that $F(x) = \lim_{n\to\infty} F_n(x)$, $x \in [a, b]$. F will be said to be $\underline{\mathcal{B}}_1$ (resp. $\overline{\mathcal{B}}_1$) on P, if the set $\{x \in P : F(x) > \alpha\}$ (resp. $\{x \in P : F(x) < \alpha\}$ is of F_σ-type for each $\alpha \in \mathbb{R}$.*

Theorem 2.2.1 (Lebesgue-Baire) *([N], p.139). Let $F : [a, b] \mapsto \mathbb{R}$ and let P be a closed subset of $[a, b]$. The following assertions are equivalent:*

(i) $F \in \mathcal{B}_1$ on P;

(ii) $F \in \underline{\mathcal{B}}_1 \cap \overline{\mathcal{B}}_1$ on P;

(iii) Each nonempty perfect set $Q \subset P$ contains at least one point x such that $F_{/Q}$ is continuous at x.

Proof. The proof of this theorem can be found in [N] (p. 139). The typical proof of *(iii)* $\Rightarrow$ *(ii)* uses transfinite numbers. There is however the following proof, due to Gordon, which avoids the use of transfinite numbers (see [G8], p. 116):
It is sufficient to consider $P = [a, b]$. The first step is to prove the following: if r and s are arbitrary real numbers with $r < s$ then there exist disjoint F_σ-sets A and B such that $[a, b] = A \cup B$, $A \subset \{x \in [a, b] : F(x) > r\}$ and $B \subset \{x \in [a, b] : F(x) < s\}$. Let $\mathcal{P}$ be the collection of all open intervals in (a, b) that have such a decomposition. We will verify that $\mathcal{P}$ satisfies the four conditions of Lemma 1.7.1. It will then follow that (a, b) has the required decomposition, and hence $[a, b]$ as well. Suppose that (α, β) and (β, γ) belong to $\mathcal{P}$. Let $(\alpha, \beta) = C_r \cup C_s$ and $(\beta, \gamma) = D_r \cup D_s$ be the corresponding decompositions. Suppose that $F(\beta) > r$; the case $F(\beta) < s$ is similar. Then $(\alpha, \gamma) = (C_r \cup D_r \cup \{\beta\}) \cup (C_s \cup D_s)$ is an appropriate decomposition of (α, γ). Hence $(\alpha, \gamma) \in \mathcal{P}$. If $(u, v) \subset (\alpha, \beta)$ then $(u, v) = (C_r \cap (u, v)) \cup (C_s \cap (u, v))$ is an appropriate decomposition of (u, v) (see Proposition 1.5.1). Hence $(u, v) \in \mathcal{P}$. This shows that $\mathcal{P}$ satisfies conditions (i) and (ii) of Lemma 1.7.1. Now suppose that

$(\alpha, \beta) \in \mathcal{P}$ for every interval $[\alpha, \beta] \subset (c, d)$. Let u be the midpoint of (c, d). We will prove that $(u, d) \in \mathcal{P}$. The proof that $(c, u) \in \mathcal{P}$ is similar. Then $(c, d) \in \mathcal{P}$ (by condition (i) in Lemma 1.7.1). Let $\{c_n\}$ be an increasing sequence in (u, d), and let $c_o = u$. For each n, let $(c_{n-1}, c_n) = A_n \cup B_n$ be an appropriate decomposition of (c_{n-1}, c_n). Define $P_r = \{n : F(c_n) > r\}$ and $P_s = \{n : n \notin P_r \text{ and } F(c_n) < s\}$. By Proposition 1.5.1 (v), $(u, d) = (\cup_{n=1}^{\infty} A_n \cup \{c_n : n \in P_r\}) \cup (\cup_{n=1}^{\infty} B_n \cup \{c_n : n \in P_s\})$ is an appropriate decomposition of (u, d). This shows that $\mathcal{P}$ satisfies condition (iii) in Lemma 1.7.1. Finally, suppose that all of the intervals contiguous to the perfect set $E \subset [a, b]$ belong to $\mathcal{P}$. By hypothesis, there exists a point $z \in E$ such that $F_{/E}$ is continuous at z. Suppose for the sake of definiteness that $F(z) < s$. Now there exists an interval $[c, d]$ such that $c, d \in E$, $E \cap (c, d) \neq \emptyset$, $x \in [c, d]$ and $F(x) < s$, for all $x \in E \cap [c, d]$. Let $[c, d] \setminus E = \cup_{n=1}^{\infty} (c_n, d_n)$ and let $(c_n, d_n) = A_n \cup B_n$ be an appropriate decomposition of $(c_n, d_n$ for each n. Since $E \cap (c, d)$ is an F_σ-set, $(c, d) = (\cup_{n=1}^{\infty} A_n) \cup (\cup_{n=1}^{\infty} B_n \cup (E \cap (c, d)))$ is an appropriate decomposition of (c, d). Hence $\mathcal{P}$ satisfies condition (iv) in Lemma 1.7.1. Now let r be an arbitrary real number and let $\{s_n\}$ be a decreasing sequence of real numbers that converges to r. By the above result, for each n there exist disjoint F_σ-sets A_n and B_n, such that $[a, b] = A_n \cup B_n$, $A_n \subset \{x \in [a, b] : F(x) > r\}$ and $B_n \subset \{x \in [a, b] : F(x) < s_n\}$. It is easy to verify that $\{x \in [a, b] : F(x) > r\} = \cup_{n=1}^{\infty} A_n$, and is therefore an F_σ-set, hence $F \in \underline{\mathcal{B}}_1$. Similarly $F \in \overline{\mathcal{B}}_1$.

Remark 2.2.1 *Let P be a closed subset of $[a, b]$, $c = \inf(P)$, $d = \sup(P)$.*

(i) Theorem 2.2.1, (i),(ii) remains true if $F : [a, b] \mapsto \overline{\mathbb{R}}$.

(ii) $\mathcal{B}_1 = \overline{\mathcal{B}}_1 \cap \underline{\mathcal{B}}_1$, and $\mathcal{B}_1$ is a linear space on $[a, b]$ (see Theorem 2.2.1).

(iii) All functions in $\mathcal{C}$, $\mathcal{B}$, $\underline{\mathcal{B}}_1$, $\overline{\mathcal{B}}_1$ are Borel functions (see [Ku], [Ga]).

(iv) If $F : [a, b] \mapsto \mathbb{R}$, $F \in \mathcal{B}_1$ then $F_P \in \mathcal{B}_1$ on $[c, d]$ (see Theorem 2.2.1).

(v) If $F : [a, b] \mapsto \mathbb{R}$, $F \in \mathcal{B}_1$ and P is perfect then the set $\{x : F_{/P}$ is discontinuous at $x\}$ is of first category on P (see [Oxt], p.32).

(vi) If $F : [a, b] \mapsto \mathbb{R}$ and $A = \{x \in P : F_{/P}$ is discontinuous at $x\}$ is a countable set then $F \in \mathcal{B}_1$ on P (see Theorem 2.2.1).

(vii) Let $F, F_n : [a, b] \mapsto \mathbb{R}$, $F_n \to F\,[unif]$, $F_n \in \mathcal{B}_1$ on P. Then $F \in \mathcal{B}_1$ (see [Ku], [Ga]).

Theorem 2.2.2 (Preiss) *([Pr1]). Let $F : [a, b] \mapsto \overline{\mathbb{R}}$. The following assertions are equivalent:*

(i) $F \in \mathcal{B}_1$ on $[a, b]$;

(ii) For each closed subset P of $[a, b]$, and for any real numbers $\alpha < \beta$, at most one of the sets $\{x \in P : F(x) \geq \beta\}$ and $\{x \in P : F(x) \leq \alpha\}$ is dense in P;

(iii) For each closed subset P of $[a, b]$, there exists at most one real number p (depending on P) such that $\overline{\{x \in P : F(x) < p\}} = \overline{\{x \in P : F(x) > p\}} = P$;

(iv) For each closed subset P of $[a,b]$, and for any real numbers $\alpha < \beta$, at most one of the sets $\{x \in P : F(x) > \beta\}$ and $\{x \in P : F(x) < \alpha\}$ is dense in P.

Proof. $(i) \Rightarrow (ii)$. Let P be a perfect subset of $[a,b]$, and let $\alpha < \beta$. By Remark 2.2.1 (i), it follows that $A = \{x \in P : F(x) \leq \alpha\}$ and $B = \{x \in P : F(x) \geq \beta\}$ are G_δ- sets. If $\overline{A} = \overline{B} = P$ then A and B are residual with respect to P, hence $A \cap B$ is residual with respect to P, a contradiction. It follows that at most one of the sets A or B is dense in P.

$(ii) \Rightarrow (i)$ Let

$$\overline{arctan}(x) = \begin{cases} \arctan(x), & x \in (-\infty, +\infty) \\ -\pi/2, & x = -\infty \\ \pi/2, & x = +\infty \end{cases}$$

and let $g : [a,b] \mapsto [-1,1]$, $g(x) = (2/\pi) \cdot \overline{arctan}(F(x))$. Let P be a perfect subset of $[a,b]$. By hypothesis, for $-1 < \alpha < \beta < 1$, it follows that at most one of the sets $\{x \in P : g(x) \leq \alpha\} = \{x \in P : F(x) \leq \tan\alpha\}$ or $\{x \in P : g(x) \geq \beta\} = \{x \in P : F(x) \geq \tan\beta\}$ is dense in P. Suppose on the contrary that $\mathcal{O}(g_{/P}; \{x\}) > 0$, for each $x \in P$. Let $P_n = \{x \in P : \mathcal{O}(g_{/P}; \{x\}) \geq 1/n\}$. Then $P = \cup_{n=1}^{\infty} P_n$. By Theorem 1.7.1, there exist a positive integer n_o and $[c,d] \cap P \subset \overline{P}_{n_o}$, such that $(c,d) \cap P \neq \emptyset$, $c, d \in P$. Let $m = \inf\{g(x) : x \in [c,d] \cap P\}$ and $M = \sup\{g(x) : x \in [c,d] \cap P\}$. Let k be a positive integer such that $(M-m)/k < 1/(3n_o)$. Let $Q_i = \{x \in [c,d] \cap P : M + i(m-m)/k \leq g(x) \leq m + (i+1)(M-m)/k\}$, $i = \overline{0, k-1}$. By Theorem 1.7.1 it follows that there exists $i_o \in \{0, 1, \ldots, k-1\}$ and $c_1, d_1 \in P$, $(c_1, d_1) \cap P \neq \emptyset$, $(c_1, d_1) \subset (c,d)$, such that $[c_1, d_1] \cap P \subset \overline{Q}_{i_o}$. Let $\alpha_o = m + i_o(M-m)/k$ and $\beta_o = m + (i_o+1)(M-m)/k$. Then

$$(1) \quad \overline{\{x \in [c_1, d_1] \cap P : g(x) \leq \beta_o\}} = \overline{\{x \in [c_1, d_1] \cap P : g(x) \geq \alpha_o\}} = [c_1, d_1] \cap P.$$

Let $\beta_1 = \beta_o + (M-m)/k$ and $\alpha_1 = \alpha_o - (M-m)/k$. Since $\mathcal{O}(F_{/P}; \{x\}) \geq 1/(n_o)$ on Q_{i_o}, it follows that $\overline{\{x \in [c_1, d_1] \cap P : g(x) \geq \beta_1\} \cup \{x \in [c_1, d_1] \cap P : g(x) \leq \alpha_1\}} = \overline{\{x \in [c_1, d_1] \cap P : g(x) \geq \beta_1\}} \cup \overline{\{x \in [c_1, d_1] \cap P : g(x) \leq \alpha_1\}} = [c_1, d_1] \cap P$. By Theorem 1.7.1, it follows that there exists for example

$$(2) \quad [c_2, d_2] \cap P \subset \overline{\{x \in [c_1, d_1] \cap P : g(x) \geq \beta_1\}},$$

where $(c_2, d_2) \cap P \neq \emptyset$, $[c_2, d_2] \subset [c_1, d_1]$. By (1) and (2) we obtain a contradiction. Thus there exists $x_o \in P$ such that $\mathcal{O}(g_{/P}; \{x_o\}) = 0$. By Theorem 2.2.1, it follows that $g \in \mathcal{B}_1$ on $[a,b]$. Hence $F(x) = \overline{\tan}((\pi/2) \cdot g(x)) \in \mathcal{B}_1$ on $[a,b]$ (where $\overline{\tan}x = \tan x, x \in (-\pi/2, \pi/2)$; $\overline{\tan}(-\pi/2) = -\infty$; $\overline{\tan}(\pi/2) = +\infty$).

$(ii) \Rightarrow (iii)$ Suppose that F does not satisfy (iii). Then there exist a closed set $P \subset [a,b]$ and real numbers $\alpha < \beta$ such that $\overline{\{x \in P : F(x) > \alpha\}} = \overline{\{x \in P : F(x) < \alpha\}} = \overline{\{x \in P : F(x) > \beta\}} = \overline{\{x \in P : F(x) < \beta\}} = P$. Hence $\overline{\{x \in P : F(x) \leq \alpha\}} = \overline{\{x \in P : F(x) \geq \beta\}} = P$. This contradicts (ii).

$(iii) \Rightarrow (iv)$ Suppose that F does not satisfy (iv). Then there exist a closed subset P of $[a,b]$ and $\alpha < \beta$ such that $\overline{\{x \in P : F(x) > \alpha\}} = \overline{\{x \in P : F(x) < \alpha\}} = P$. Since $\{x \in P : F(x) > \beta\} \subset \{x \in P : F(x) > \alpha\}$ and $\{x \in P : F(x) < \alpha\} \subset \{x \in P : F(x) < \beta\}$, we have $\overline{\{x \in P : F(x) < \alpha\}} = \overline{\{x \in P : F(x) > \alpha\}} = \overline{\{x \in P : F(x) < \beta\}} = \overline{\{x \in P : F(x) > \beta\}} = P$, and this contradicts (iii).

$(iv) \Rightarrow (ii)$ Suppose that F does not satisfy (ii). Then there exist a closed subset

P of $[a,b]$ and $\alpha < \beta$ such that $\overline{\{x \in P : F(x) \geq \beta\}} = \overline{\{x \in P : F(x) \leq \alpha\}} = P$. Let $\alpha < \alpha_1 < \beta_1 < \beta$. The $\overline{\{x \in P : F(x) > \beta_1\}} = \overline{\{x \in P : F(x) < \alpha_1\}} = P$. This contradicts (iv).

Theorem 2.2.3 *Let $F : [a,b] \mapsto \mathbb{R}$.*

(i) If F is $\mathcal{S}_o^+$ lower semicontinuous on $[a,b)$ then $F \in \underline{\mathcal{B}}_1$ on $[a,b]$;

(ii) If F is lower semicontinuous on [a,b] then $F \in \mathcal{B}_1$ on $[a,b]$.

Proof. (i) By Proposition 1.15.3, the set $\{x \in [a,b) : F(x) > \alpha\}$ is $\mathcal{S}_o^+$ open. By Remark 1.14.1, (iii), the set $\{x \in [a,b) : F(x) > \alpha\}$ is of F_σ-type, hence $F \in \underline{\mathcal{B}}_1$ on $[a,b]$.

By Corollary 1.15.1, (ii) and Proposition 1.14.2, $F \in \underline{\mathcal{B}}_1$on $[a,b]$. Let $\alpha \in \mathbb{R}$. It follows that $\{x \in [a,b] : F(x) < \alpha\} = \cup_{n=1}^{\infty}\{x \in [a,b] : F(x) \leq \alpha - 1/n\}$. By Corollary 1.15.1, (iii), this set is of F_σ-type, hence $F \in \overline{\mathcal{B}}_1$ on $[a,b]$. By Theorem 2.2.1, $F \in \mathcal{B}_1$ on $[a,b]$.

Remark 2.2.2 *There exists a function $F : [a,b] \mapsto \mathbb{R}$, such that F is $\mathcal{S}_o^+$ lower semicontinuous on $[a,b]$, but F is not $\overline{\mathcal{B}}_1$ (see Section 6.4), so by Remark 2.1.1, (ii), F is not $\mathcal{B}_1$ on $[a,b]$. It follows that we cannot replace in Theorem 2.2.3, (i) condition $\underline{\mathcal{B}}_1$ by $\mathcal{B}_1$.*

Theorem 2.2.4 *Let $F : [a,b] \mapsto \mathbb{R}$ and let $\mathcal{S} = \{\mathcal{S}(x) : x \in \mathbb{R}\}$ be a local system with intersection condition (4), such that $\mathcal{S}\,DF(x)$ exists (finite or infinite) at each point $x \in [a,b]$. Then $\mathcal{S}\,DF(x)$ is $\mathcal{B}_1$ of $[a,b]$.*

Proof. Let $f(x) = \mathcal{S}\,DF(x)$. Suppose that $f \notin \mathcal{B}_1$. By Theorem 2.2.2, (ii), there exist a closed subset P of $[a,b]$ and real numbers $\alpha < \beta$, such that $\overline{\{x \in P : f(x) > \alpha\}} = \overline{\{x \in P : f(x) > \beta\}} = \overline{\{x \in P : f(x) < \alpha\}} = \overline{\{x \in P : f(x) < \beta\}} = P$. Applying Lemma 1.13.1, (i) to $F(x) - \alpha x$ and $F(x) - \beta x$, it follows that $\{x \in P : f(x) > \alpha\}$, $\{x \in P : f(x) > \beta\}$, $\{x \in P : f(x) < \alpha\}$ and $\{x \in P : f(x) < \beta\}$ are first category sets with respect to P. Hence $\{x \in P : f(x) = \alpha\}$ and $\{x \in P : f(x) = \beta\}$ are residual sets with respect to P, a contradiction.

Remark 2.2.3 *Theorem 2.2.4 was first obtained by Preiss, but only for an $\mathcal{S}_{ap}$ local system: If $F : [a,b] \mapsto \mathbb{R}$ has a finite or infinite approximate derivative everywhere then $F'_{ap} \notin \mathcal{B}_1$ on (a,b) (see [Pr1]). Note that F is not necessarily in $\mathcal{B}_1$ (see Section 6.20.).*

Theorem 2.2.5 (Thomson) *([T5], p. 74). Let $\mathcal{S}$ be a local system satisfying an intersection condition of the form $\sigma_x \cap \sigma_y \neq \emptyset$ and let $F : [a,b] \mapsto \overline{\mathbb{R}}$. If F is simultaneously $\mathcal{S}$ lower and $\mathcal{S}$ upper semicontinuous then $F \in \mathcal{B}_1$.*

Remark 2.2.4 *The proof of Theorem 2.2.5 is based on Theorem 2.2.2.*

2.3 Conditions C_i; C_i^*; $[C_iG]$; $[CG]$

Definition 2.3.1 *Let $P \subseteq [a,b]$, $x_o \in P$ and $F : P \mapsto \mathbb{R}$. F is said to be C_i at x_o if $\limsup_{x\nearrow x_o, x\in P} F(x) \le F(x_o)$, whenever x_o is a left accumulation point for P, and $F(x_o) \le \liminf_{x\searrow x_o, x\in P} F(x)$, whenever x_o is a right accumulation point for P. Let $C_d = \{F : -F \in C_i\}$ at x_o. F is said to be C_i^* at x_o if $\lim_{x\nearrow x_o, x\in P} F(x)$ exists and is finite, with $\lim_{x\nearrow x_o, x\in P} F(x) \le F(x_o)$, whenever x_o is a left accumulation point for P, and if $\lim_{x\searrow x_o, x\in P} F(x)$ exists and is finite, with $F(x_o) \le \lim_{x\searrow x_o, x\in P} F(x)$, whenever x_o is right accumulation point for P. Let $C_d^* = \{F : -F \in C_i^*\}$ at x_o. F is said to be C_i (resp. C_d, C_i^*, C_d^*) on P, if F is so at each point $x \in P$. We define conditions C_iG, C_i^*G, $[CG]$, etc., using Definition 1.9.2*

Proposition 2.3.1 *Let $P \subseteq [a,b]$, $x_o \in P$ and $F : P \mapsto \mathbb{R}$.*

(i) $C_i^* \cap C_d^* = C_i \cap C_d = C$ on P.

(ii) $C_i^* \subset C_i$ on P.

(iii) If F is bounded and increasing on P then $F \in C_i^*$ on P.

(iv) C_i is a real semi-linear space on P.

(v) C_i^* is a real semi-linear space on P.

(vi) The following assertions are equivalent:

(1) $F \in C_i$ at x_o;

(2) $\Omega_-(F; x_o) = 0$;

(3) For every $\epsilon > 0$ there exists a $\delta > 0$ such that $\Omega_-(F; P \cap (x-\delta, x+\delta) \wedge \{x_o\}) > -\epsilon$;

(4) For $\epsilon > 0$ there exists a $\delta > 0$ such that $F(y) - \epsilon \le F(x_o) \le F(z) + \epsilon$, whenever $y \in (x_o - \delta, x_o] \cap P$ and $z \in [x_o, x_o + \delta) \cap P$.

(vii) The following assertions are equivalent:

(1) $F \in C_i^*$ at x_o;

(2) $\mathcal{O}_-(F; x_o) = 0$;

(3) For $\epsilon > 0$ there exists a $\delta > 0$ such that $\mathcal{O}_-(F; P \cap (x-\delta, x+\delta)) > \epsilon$;

(4) $F \in C_i$ at x_o, and for $\epsilon > 0$ there exists a $\delta > 0$ such that $| F(z) - F(y) | < \epsilon$, whenever $y, z \in (x_o - \delta, x_o + \delta) \cap P$ and $y < z < x_o$ or $x_o < y < z$.

(viii) If $F : [a,b] \mapsto \mathbb{R}$ then the following assertions are equivalent:

(1) $F \in C_i$ on $[a,b]$;

(2) F is S_o^+ lower semicontinuous on $[a,b)$ and F is S_o^- upper semicontinuous on $(a,b]$.

Proof. (ii) See Section 6.5.

(vii) (3) $\Rightarrow$ (1) Suppose for example that x_o is a right accumulation point for P. Let $\epsilon > 0$. Then there is a $\delta > 0$ such that $\mathcal{O}_-(F; P \cap (x_o - \delta, c_o + \delta)) > -\epsilon$. It follows that $\liminf_{x \searrow x_o, x \in P} F(x) \geq F(x_o)$ and $\limsup_{x \searrow x_o, x \in P} F(x) \neq +\infty$ (because, if $\limsup_{x \searrow x_o, x \in P} F(x) = +\infty$ and $x_1 \in (x_o, x_o + \delta) \cap P$ then there exists $y_1 \in (x_o, x_1)$ such that $F(y_1) > F(x_1) + 2\epsilon$, a contradiction). Let $\ell_o = \liminf_{x \searrow x_o, x \in P} F(x)$ and $L_o = \limsup_{x \searrow x_o, x \in P} F(x)$. Then $\ell_o = L_o$ (indeed: suppose that $\ell_o < L_o$; for $\epsilon_o = (L_o - \ell_o)/2$ let $\delta_o > 0$ be given by (3); then there exists $x_1 \in (x_o, x_o + \delta_o) \cap P$, with $F(x_1) < \ell_o + \epsilon_o/2$ and $y_1 \in (x_o, x_1) \cap P$, with $F(y_1) > L_o - \epsilon_o/2$; it follows that $F(x_1) - F(y_1) < \ell_o + \epsilon_o/2 - L_o + \epsilon_o/2 = -\epsilon_o$, a contradiction).

The other statements are evident.

Proposition 2.3.2 *Let $F : [a,b] \mapsto \mathbb{R}$ and let P be a closed subset of $[a,b]$, $c = \inf(P)$, $d = \sup(P)$.*

(i) $F \in \mathcal{C}_i$ on P if and only if $F_P \in \mathcal{C}_i$ on $[c,d]$;

(ii) If F is increasing on the closure of each interval contiguous to P, and $F \in \mathcal{C}_i$ on P then $F \in \mathcal{C}_i$ on $[c,d]$.

Proof. (i) Suppose that $F \in \mathcal{C}_i$ on P. Then F_P (see Definition 1.1.3) is $\mathcal{C}_i$ on $[a,b] \setminus P$. Let $x_o \in P \cap (c,d)$. If x_o is the left endpoint of some interval contiguous to P, then there is nothing to prove. If x_o is the left endpoint of no interval contiguous to P then x_o is a left accumulation point of P. Since $F \in \mathcal{C}_i$ on P, for $\epsilon > 0$ there is a $\delta > 0$ such that $F(u) - \epsilon \leq F(x_o)$, for each $u \in (x_o - \delta, x_o] \cap P$ (see Proposition 2.3.1, (vi)). Let $x_1 = \inf(P \cap (x_o - \delta, x_o])$ and let $\delta_1 = x_0 - x_1$. It follows that $F_P(u) - \epsilon \leq F_P(x_o) = F(x_o)$, $u \in (x_o - \delta_1, x_o]$. Hence $\limsup_{x \nearrow x_o} F_P(x) \leq F_P(x_o)$. Similarly $F_P(x_o) \leq \liminf_{x \searrow x_o} F_P(x)$. Thus F_P is $\mathcal{C}_i$ on P. The converse is obvious.

(ii) The proof is similar to that of (i).

Proposition 2.3.3 *Let $F : [a,b] \mapsto \mathbb{R}$. If $F \in \mathcal{C}_i$ on $[a,b]$ then $F \in \mathcal{C}$\.]. on $[a,b]$, hence $F \in \mathcal{B}_1$ on $[a,b]$.*

Proof. By Theorem 1.10.1, (ii), $\liminf_{t \to x} F(t) = \liminf_{t \searrow x} F(t)$ *n.e.* on $[a,b]$. By hypothesis $F(x) \leq \liminf_{t \searrow x} F(t)$, hence $F(x) \leq \liminf_{t \to x} F(t)$ *n.e.* on $[a,b]$. By Theorem 1.3.1, (ii), $\limsup_{t \to x} F(t) = \limsup_{t \nearrow x} F(t)$ *n.e.* on $[a,b]$. By hypothesis $\limsup_{t \nearrow x} F(t) \leq F(x)$, hence $\limsup_{t \to x} F(t) \leq F(x)$ *n.e.* on $[a,b]$. It follows that $F \in \mathcal{C}$ *n.e.* on $[a,b]$. By Remark 2.2.1, (vi), $F \in \mathcal{B}_1$ on $[a,b]$.

Proposition 2.3.4 *Let $F_n, F : [a,b] \mapsto \mathbb{R}$, $n \geq 1$. If $F_n \in \mathcal{C}_i$ and $F_n \to F$ [unif] then $F \in \mathcal{C}_i$ on $[a,b]$.*

Proof. Let $\epsilon > 0$ and choose N so large that if $n \geq N$ then $| F_n(x) - F(x) | < \epsilon/3$, for all $x \in [a,b]$. Let $x_o \in (a,b)$. Since $F_N \in \mathcal{C}_i$ at x_o, it follows that there exists a $\delta > 0$ such that $a < x_o - \delta < x_o + \delta < b$ and $F_n(u) - \epsilon/3 < F_N(x_o) < F_N(v) + \epsilon/3$, for each $u \in (x_o - \delta, x_o]$ and $v \in [x_o, x_o + \delta)$. We obtain that $F(x_o) < F_n(x_o) + \epsilon/3 < F_N(v) + 2\epsilon/3 < F(v) + \epsilon$ and $F(x_o) > F_N(x_o) - \epsilon/3 > F_N(u) - 2\epsilon/3 > F(u) - \epsilon$. By Proposition 2.3.1, (vi), $F \in \mathcal{C}_i$ at x_o.

Lemma 2.3.1 *Let $F : [a,b] \mapsto \mathbb{R}$ and let P be a closed subset of $[a,b]$. The following assertions are equivalent:*

(i) $F \in [\mathcal{C}_i G]$ *(resp.* $[\mathcal{C}G]$*) on* P;

(ii) Each closed subset $S \subset P$ *contains a portion* $(c,d) \cap S \neq \emptyset$ *such that* $F \in \mathcal{C}_i$ *(resp.* $F \in \mathcal{C}$*) on* $(c,d) \cap S$.

Proof. See Theorem 1.9.1, (ii).

Proposition 2.3.5 *Let* $F : [a,b] \mapsto \mathbb{R}$ *and let* P *be a closed subset of* $[a,b]$.

(i) $[\mathcal{C}G] = [\mathcal{C}_i G] \cap [\mathcal{C}_d G]$ *on* P;

(ii) $[\mathcal{C}_i G]$ *is a semi-linear space on* P;

(iii) $[\mathcal{C}G]$ *is a linear space on* P;

(iv) $[\mathcal{C}G] \subseteq [\mathcal{C}_i G] \subseteq \mathcal{B}_1$ *on* P;

(v) $\mathcal{C} \subset [\mathcal{C}G] \cap \mathcal{D} \subset [\mathcal{C}_i^* G] \cap \mathcal{D} \subseteq [\mathcal{C}_i G] \cap \mathcal{D} \subset \mathcal{D}\mathcal{B}_1 \subset \mathcal{D}$ *on* $[a,b]$.

Proof. (i), (ii) and (iii) are evident.

(iv) Let $F \in [\mathcal{C}_i G]$ on P, and let S be a closed subset of P. We may suppose that S is perfect. By Lemma 2.3.1, there exist $c, d \in S$ such that $(c,d) \cap S \neq \emptyset$ and $F \in \mathcal{C}_i$ on $Q = [c,d] \cap S$. By Proposition 2.3.2, (i), $F_Q \in \mathcal{C}_i$ on $[c,d]$. By Proposition 2.3.3, $F_Q \in \mathcal{C}$ *n.e.* on $[c,d]$, hence $F_{/Q} \in \mathcal{C}$ *n.e.*. But Q is uncountable, hence $F_{/Q}$ contains uncountably many points x at which $F_{/Q} \in \mathcal{C}$. By Theorem 2.2.1, (i), (iii), $F \in \mathcal{B}_1$ on P.

(v) See Sections 6.22.1., 6.6, 6.7 and 6.10.1.

Remark 2.3.1 *(i) Condition* $[\mathcal{C}G]$ *was defined by Ellis in [El]. In [O2], O'Malley gave the following definition: A function* $F : [a,b] \mapsto \mathbb{R}$ *is said to be* $\mathcal{B}_1^*$ *on* $[a,b]$*, if for every perfect subset* P *of* $[a,b]$ *there exists a portion* $(c,d) \cap P$ *such that* $F_{/(c,d)\cap P}$ *is* $\mathcal{C}$. *By Lemma 2.3.1, it follows that* $[\mathcal{C}G] = \mathcal{B}_1^*$ *on* $[a,b]$.

(ii) In [Ki], Kirchheim noticed (without proof) that in the definition of $\mathcal{B}_1^*$*, the assumption that* P *is perfect can be dropped.*
Open question: Is the following assertion true: $F \in \mathcal{B}_1^*$ *on* $[a,b]$ *if and only if the sets* $\{x \cap \overline{P} : F(x) \le \alpha\}$ *and* $\{x \in P : F(x) \ge \beta\}$ *are dense in* P*, whenever* $P \subseteq [a,b]$ *and* $-\infty < \alpha < \beta < +\infty$ *?*

(iii) For more information on this topic (often in metric spaces) see : [CL], [O2], [JR], [LuMZ], [Ros].

2.4 Conditions internal, internal*, $\mathcal{Z}_i$, uCM

Definition 2.4.1 (Garg) *([Ga]). Let* $F : [a,b] \mapsto \overline{\mathbb{R}}$. F *is said to be lower internal if* $F(x) \le \limsup_{y \searrow x} F(y)$*, for* $a \le x < b$ *and* $F(x) \ge \liminf_{y \nearrow x} F(y)$*, for* $a < x \le b$. F *is said to be upper internal if* $-F$ *is lower internal.* F *is said to be internal if it is simultaneously upper and lower internal.*

Proposition 2.4.1 *Let* $F : [a,b] \mapsto \overline{\mathbb{R}}$. *The following assertions are equivalent:*

(i) F *is lower internal on* $[a,b]$;

(ii) F is S^+_∞ lower semicontinuous on $[a,b)$, and F is S^-_∞ semicontinuous on $(a,b]$.

Proof. See Remark 1.13.1, (ii).

Proposition 2.4.2 *Let $F : [a,b] \mapsto \overline{\mathbb{R}}$ and $\alpha \in \mathbb{R}$. The following assertions are equivalent:*

(i) F is lower internal;

(ii) $\{x \in [a,b) : F(x) > \alpha\}$ is S^+_∞ open, and $\{x \in (a,b] : F(x) < \alpha\}$ is S^-_∞ open.

(iii) $\{x \in [a,b) : F(x) \le \alpha\}$ is S^+_∞ closed, and $\{x \in (a,b] : F(x) \ge \alpha\}$ is S^-_∞ closed.

Proof. See Corollary 1.15.2

Corollary 2.4.1 *Let $F : [a,b] \mapsto \overline{\mathbb{R}}$ and $\alpha \in \mathbb{R}$. The following assertions are equivalent:*

(i) F is internal;

(ii) $\{x \in [a,b) : F(x) > \alpha\}$ and $\{x \in [a,b) : F(x) < \alpha\}$ are S^+_∞ open sets; and $\{x \in (a,b] : F(x) < \alpha\}$ and $\{x \in (a,b] : F(x) > \alpha\}$ are S^-_∞ open sets;

(iii) $\{x \in [a,b) : F(x) \le \alpha\}$ and $\{x \in [a,b) : F(x) \ge \alpha\}$ are S^+_∞ closed sets; and $\{x \in (a,b] : F(x) \ge \alpha\}$ and $\{x \in (a,b] : F(x) \le \alpha\}$ are S^-_∞ closed sets.

Proposition 2.4.3 *Let $F : [a,b] \mapsto \mathbb{R}$ and let P be a closed subset of $[a,b]$, $c = \inf(P)$, $d = \sup(P)$. Let $\{(c_k,d_k)\}_{k\ge 1}$ be the intervals contiguous to P. If F is lower internal on $[a,b]$ and $\sum_{k\ge1} \Omega_-(F;[c_k,d_k]\wedge\{c_k,d_k\}) > -\infty$ then F_P is lower internal on $[c,d]$.*

Proof. F_P is defined in Definition 1.1.3. Let $x_o \in P$ be a right accumulation point of P. Suppose that $F_P(x_o) > \limsup_{x\searrow x_o} F_P(x)$. Let $\alpha \in (\limsup_{x\searrow x_o} F_P(x), F_P(x_o))$. Then there exists a $\delta > 0$ such that

(1) $\quad F_P(x) < \alpha,$ *for each* $x \in (x_o, x_o+\delta)$.

Let $x_1 \in (x_o, x_o+\delta) \cap P$ such that

(2) $$\sum_{(c_k,d_k)\subseteq(x_o,x_1)} \Omega_-(F;[c_k,d_k]\wedge\{c_k,d_k\}) > -\frac{F(x_o)-\alpha}{2}$$

(this is possible by hypothesis). Let $x \in (x_o,x_1)$. If $x \in P$ then by (1), $F(x) < \alpha$. If $x \notin P$ then there exists a positive integer n such that $x \in (c_n,d_n) \subset (x_o,x_1)$. By (2), $F(d_n) - F(x) > -(F(x_o)-\alpha)/2$, hence $F(x) < F(d_n) + (F(x_o)-\alpha)/2 < \alpha + (F(x_o)-\alpha)/2 = (\alpha+F(x_o))/2$. It follows that $F(x) < \alpha$, for all $x \in [x_o,x_1]$, hence $\limsup_{x\searrow x_o} F(x) \le \alpha < F(x_o)$. This contradicts the fact that F is lower internal.

Definition 2.4.2 (Zygmund) *([S3], p.203). Let $F : [a,b] \mapsto \mathbb{R}$. F is said to be $\overline{\mathcal{Z}}_i$ (resp. $\underline{\mathcal{Z}}_i$) on $[a,b]$, if $F(x) \ge \limsup_{y\nearrow x} F(y)$ (resp. $F(x) \ge \liminf_{y\nearrow x} F(y)$), for $x \in (a,b]$ and $F(x) \le \limsup_{y\searrow x} F(y)$ (resp. $F(x) \le \liminf_{y\searrow x} F(y)$) for $x \in [a,b]$. F is said to be $\overline{\mathcal{Z}}_d$ (resp. $\underline{\mathcal{Z}}_d$) if $-F$ is $\overline{\mathcal{Z}}_i$ (resp. $\underline{\mathcal{Z}}_i$).*

Proposition 2.4.4 *Let $F, F_n : [a,b] \mapsto \mathbb{R}$, $n \geq 1$. If F_n are lower internal (resp. $\overline{\mathcal{Z}}_i$) and $F_n \to F$ [unif] then F is lower internal (resp. $\overline{\mathcal{Z}}_i$).*

Proof. The proof is evident.

Definition 2.4.3 (Preiss) *Let $F : [a,b] \mapsto \mathbb{R}$. F is said to be lower internal*, if $F(x+) \geq F(x)$, whenever $x \in (a,b]$ and $F(x+)$ exists, and $F(x-) \leq F(x)$, whenever $x \in [a,b)$ and $F(x-)$ exists. F is said to be upper internal* if $-F$ is lower internal*. F is said to be internal* if it is simultaneously upper and lower internal*.*

Remark 2.4.1 *If $F(x+)$ exists and is finite for each $x \in [a,b)$, $F(x-)$ exists and is finite for each $x \in (a,b]$, and F is lower internal* (resp. internal*) then F is $\mathcal{C}_i^*$ (resp. $\mathcal{C}$) on $[a,b]$.*

Definition 2.4.4 (C.M.Lee) *Let $F : [a,b] \mapsto \mathbb{R}$. F is said to be uCM if it is increasing on $[c,d] \subseteq [a,b]$, whenever it is so on (c,d). F is said to be ℓCM if $-F$ is uCM. Let $CM = \ell CM \cap uCM$ and $sCM = \{F : F(x) + \lambda x \in CM$ for each $\lambda \in \mathbb{R}\}$.*

Theorem 2.4.1 *Let $F : [a,b] \mapsto \mathbb{R}$, $F \in uCM$ and let $h : [a,b] \mapsto \mathbb{R}$, h is increasing and continuous on $[a,b]$. Then $F - h$ is uCM on $[a,b]$.*

Proof. Let $g = F - h$ and let $(c,d) \subseteq [a,b]$ such that g is increasing on (c,d). Then $F = g + h$ is increasing on (c,d). Since $F \in uCM$, F is increasing on $[c,d]$. Suppose that there exists $x_1 \in (c,d)$ such that $g(x_1) > g(d)$. Let $\epsilon = g(x_1) - g(d)$. Since h is continuous, it follows that there exists $\delta \in (0, d - x_1)$ such that $h(x) > h(d) - \epsilon$, for each $x \in (d-\delta, d)$. Since $g(x) > g(x_1)$, for each $x \in (d-\delta, d)$, it follows that $F(x) = g(x) + h(x) > F(d)$. This contradicts the fact that F is increasing on $[c,d]$. Hence g is increasing on $[c,d]$, and so $g \in uCM$ on $[a,b]$.

Remark 2.4.2 *The fact that h is increasing in Theorem 2.4.1 is essential (see Section 6.9.).*

Corollary 2.4.2 *Let$F : [a,b] \mapsto \mathbb{R}$. The following assertions are equivalent:*

(i) $F \in sCM$ on $[a,b]$;

(ii) $F(x) + \lambda x$ and $\lambda x - F(x)$ are uCM, for each $\lambda \in [0, +\infty)$.

Proof. $(i) \Rightarrow (ii)$ This is evident.

$(ii) \Rightarrow (i)$ If $\lambda = 0$ then $F \in CM$ on $[a,b]$. If $\lambda < 0$ then by Theorem 2.4.1, $F(x) + \lambda x$ is uCM. By hypothesis $-F(x) - \lambda x$ is uCM, hence $F(x) + \lambda x$ is CM. If $\lambda > 0$ then, by hypothesis $F(x) + \lambda x$ is uCM. By Theorem 2.4.1, $-F(x) - \lambda x$ is uCM, hence $F(x) + \lambda x$ is CM.

Theorem 2.4.2 *Let $F : [a,b] \mapsto \mathbb{R}$. Then we have:*

(i) $\mathcal{C}_i$ + lower internal = lower internal;

(ii) $\mathcal{C}$ + internal = internal;

(iii) $\mathcal{C}_i^*$ + lower internal = lower internal*;*

(iv) $\mathcal{C}$ + *internal** = *internal**;

(v) $\mathcal{C}_i \subset \overline{\mathcal{Z}}_i \subset \mathcal{D}_-\overline{\mathcal{B}}_1$;

(vi) $\mathcal{C}_i^* \subset \mathcal{C}_i \subset \mathcal{D}_-\mathcal{B}_1 \subset \mathcal{D}_- \subset$ *lower internal* $\subset$ *lower internal** $\subset uCM$;

(vii) $\mathcal{C} \subset \mathcal{D}\mathcal{B}_1 \subset \mathcal{D} \subset$ *internal* $\subset$ *internal** $\subset sCM \subset CM \subset uCM$;

(viii) $\mathcal{D}_-\mathcal{B}_1 \subset$ *lower internal*$^* \cap \mathcal{B}_1 \subset uCM \cap \mathcal{B}_1$;

(ix) $\mathcal{D}\mathcal{B}_1 \subset$ *internal*$^* \cap \mathcal{B}_1 \subset sCM \cap \mathcal{B}_1$.

Proof. (i)-(iv) follow by definitions.

(v) $\mathcal{C}_i \subset \overline{\mathcal{Z}}_i$ follows by Section 6.13., and $\overline{\mathcal{Z}}_i \subset \overline{\mathcal{B}}_1$ follows by Theorem 2.2.3, (i). We show that $\overline{\mathcal{Z}}_i \subseteq \mathcal{D}$. Let $\alpha, \beta \in [a,b]$, $\alpha < \beta$, $F(\alpha) > F(\beta)$ and let $r \in (F(\beta), F(\alpha))$. Let $E = \{x \in [\alpha,\beta] : F(x) \geq r\}$. Since $\alpha \in E$, $E \neq \emptyset$. Let $c = \sup(E)$. If $c \in E$ then $F(c) \geq r$. If $c \notin E$ then c is a left accumulation point for E and $F(c) \geq \limsup_{x \nearrow c} F(x) \geq \limsup_{x \nearrow c, x \in E} F(x) \geq r$. But $F(\beta) < r$, hence $c \neq \beta$. It follows that for each $x \in (c,\beta)$, $F(x) < r$. We have $F(c) \leq \limsup_{x \searrow c} F(x) \leq r$. It follows that $F(c) = r$, hence $F \in \mathcal{D}_-$ on $[a,b]$. The fact that $\overline{\mathcal{Z}}_i \subset \mathcal{D}_-\overline{\mathcal{B}}_1$ follows by Section 6.13.

(vi) For $\mathcal{C}_i^* \subset \mathcal{C}_i$ see Proposition 2.3.1, (ii); $\mathcal{C}_i \subset \mathcal{D}_-\mathcal{B}_1$ follows by (v), Proposition 2.3.3 and Section 6.12.; $\mathcal{D}_-\mathcal{B}_1 \subset \mathcal{D}_-$ follows by Section 6.10.1; $\mathcal{D}_- \subset$ *lower internal* follows by definitions and Section 6.14.; lower *internal* $\subset$ *lower internal** follows by Section 6.12.

We show that lower *internal** $\subset uCM$. Suppose that F is lower *internal** and let $(c,d) \subseteq (a,b)$ such that F is increasing on $[c,d]$. Suppose for example that $F(c) > F(x_1)$, for some $x_1 \in (c,d)$. Then $F(x) < F(x_1)$, for each $x \in (c, x_1]$. Since F is lower *internal**, it follows that $F(c) \leq \lim_{x \searrow c} F(x) \leq F(x_1)$ (this limit exists because F is increasing on (c,d)), a contradiction.

lower *internal** $\subset uCM$ follows by Section 6.16.

(vii) $\mathcal{C} \subset \mathcal{D}\mathcal{B}_1$ follows by Section 6.12.; $\mathcal{D}\mathcal{B}_1 \subset \mathcal{D}$ follows by Section 6.10.1.; $\mathcal{D} \subset$ *internal* follows by (vi), Remark 2.1.1, (i) and Section 6.11.; *internal* $\subset$ *internal** follows by Section 6.11.; *internal** $\subset sCM$ follows by (vi), (iv) and Corollary 2.4.2 and Section 6.15. or Section 6.16.; $sCM \subset CM$ follows by Section 6.9.; $CM \subset uCM$ follows by Section 6.8.

(viii) $\mathcal{D}_-\mathcal{B}_1 \subset$ *lower internal*$^* \cap \mathcal{B}_1$ follows by Section 6.12.; *lower internal*$^* \cap \mathcal{B}_1 \subset uCM \cap \mathcal{B}_1$ follows by (vi) and Section 6.16.

(ix) $\mathcal{D}\mathcal{B}_1 \subset \mathcal{B}_1$ follows by(vii) and Section 6.12.; *internal*$^* \cap \mathcal{B}_1 \subset sCM \cap \mathcal{B}_1$ follows by (vii) and Section 6.16.

2.5 Conditions $\mathcal{D}_-\mathcal{B}_1$, $\mathcal{D}\mathcal{B}_1$

Theorem 2.5.1 *Let* $F : [a,b] \mapsto \overline{\mathbb{R}}$, $F \in \mathcal{B}_1$. *The following assertions are equivalent:*

(i) F *is* $\mathcal{D}_-$ *on* $[a,b]$;

(ii) F *is* $\mathcal{S}_c^+$ *lower semicontinuous on* $[a,b)$ *and* $\mathcal{S}_c^-$ *upper semicontinuous on* $(a,b]$;

(iii) F *is* $\mathcal{S}_\infty^+$ *lower semicontinuous on* $[a,b)$ *and* $\mathcal{S}_\infty^-$ *upper semicontinuous on* $(a,b]$;

(iv) *F is lower internal on* $[a,b]$;

(v) $\{x \in [a,b) : F(x) > r\}$ *is* S_c^+ *open (resp.* S_∞^+ *open) and* $\{x \in (a,b] : F(x) < r\}$ *is* S_c^- *open (resp.* S_∞^- *open), whenever* $r \in \mathbb{R}$;

(vi) $\{x \in [a,b) : F(x) \le r\}$ *is* S_c^+ *closed (resp.* S_∞^+ *closed) and* $\{x \in (a,b] : F(x) \ge r\}$ *is* S_c^- *closed), whenever* $r \in \mathbb{R}$.

Proof. $(i) \Rightarrow (ii)$ Let $x_o \in (a,b]$ (for $x_o \in [a,b)$ the proof is similar).
(I) Suppose that $F(x_o) \in \mathbb{R}$ and that there exists $\epsilon > 0$ such that $\{x : F(x) < F(x_o)+\epsilon\} \notin S_c^-(x_o)$. Then there exists $\delta > 0$ such that $A = (x_o) - \delta, x_o) \cap \{x : F(x) < F(x_o)+\epsilon\}$ is at most countable. It follows that there exists $x_1 \in (x_o - \delta, x_o)$ such that $F(x_1) \ge F(x_o)+\epsilon$. Since $F \in \mathcal{D}_-$, for each $y \in (F(x_o), F(x_o)+\epsilon)$ there exists $x_y \in (x_1, x_o)$ such that $F(x_y) = y$, hence A is uncountable, a contradiction.
(II) Suppose that $F(x_o) = -\infty$ and that there exists $r \in \mathbb{R}$ such that $\{x : F(x) < r\} \notin S_c^-(x_o)$. Then there exists $\delta > 0$ such that $A = (x_o - \delta, x_o) \cap \{x : F(x) < r\}$ is at most countable. It follows that there exists $x_1 \in (x_o - \delta, x_o)$ such that $F(x_1) \ge r$. Since $F \in \mathcal{D}_-$, for $y \in (-\infty, r)$ there exists $x_y \in (x_1, x_o)$ such that $F(x_y) < r$, hence A is uncountable, a contradiction.

$(ii) \Rightarrow (iii)$ is evident and $(iii) \Leftrightarrow (iv)$ follows by Proposition 2.4.1.

$(iv) \Rightarrow (i)$ Suppose that $F \notin \mathcal{D}_-$. Then there exists an interval $[c,d]$, $F(c) > F(d)$ and $r \in (F(d), F(c))$ such that $r \notin F([c,d])$. Let $A = \{x \in [c,d] : F(x) > r\} = \{x \in [c,d] : F(x) \ge r\}$ and $B = \{x \in [c,d] : F(x) < r\} = \{x \in [c,d] : F(x) \le r\}$. Then $d \in A$, $c \in B$ and $[c,d] = A \cup B$. Let $P = \overline{A} \cap \overline{B} = Fr(A) = Fr(B) = \overline{A} \setminus int(A) = \overline{B} \setminus int(B)$. Since $[c,d]$ is a convex set it follows that $P \ne \emptyset$ and $[c,d] = P \cup int(A) \cup int(B)$. Clearly $c, d \in P$. Let (u,v) be an interval contiguous to P. Then $(u,v) \subset int(A) \cup int(B)$. Since (u,v) is a convex set, it follows that either $(u,v) \subset int(A)$ or $(u,v) \subset int(B)$. We have three cases:
(I) Suppose P is finite. Then $P \cap (c,d) \ne \emptyset$ (because if $P \cap (c,d) = \emptyset$ then (c,d) is an interval contiguous to P; since F is *lower internal* and $F(c) < r$, it follows that $(c,d) \subset int(A)$; since F is *lower internal*, $F(d) \ge r$, a contradiction). Let $P \cap (c,d) = \{c_1, c_2, \ldots, c_n\}$. Then $F(c_1) \ge r$. But $c_i \in P \subset \overline{B}$, hence $(c_1, c_2) \subset int(B)$. Since F is *lower internal*, $F(c_1) \le r$. It follows that $F(c_1) = r$, a contradiction. Hence P cannot be finite.
(II) Suppose that P is countable infinite. Since P is closed it follows that $P' = \{x \in P : x$ is an accumulation point of $P\}$ is also closed and at most countable. By Theorem 1.7.1, P' contains an isolated point x_o. Thus near x_o there are only isolated points of P. Among the later, let $x_1 < x_2 < x_3 < x_4$ be four consecutive points. Suppose that $(x_1, x_2) \subset int(A)$. Since $x_2 \in P \subset \overline{B}$, $(x_2, x_3) \subset int(B)$. Since F is *lower internal*, it follows that $r \le F(x_2) \le r$, hence $F(x_2) = r$, a contradiction. Suppose that $(x_1, x_2) \subset int(B)$. Since $x_2 \in P \subset \overline{A}$, it follows that $(x_2, x_3) \subset int(A)$ and $(x_3, x_4) \subset int(B)$, hence $F(x_3) = r$, a contradiction. It follows that P cannot be countable.
(III) Suppose that P is uncountable. Then we may suppose without loss of generality that P is perfect. Let $E_1 = A \cap P$ and $E_2 = B \cap P$. We have three situations:
1) $P \not\subset \overline{E}_2$. Then there exist $c_1, d_1 \in P$ such that $(c_1, d_1) \cap P \ne \emptyset$ and $[c_1, d_1] \cap E_2 = \emptyset$. It follows that $[c_1, d_1] \cap P \subset E_1 \subset A$. We show that $(c_1, d_1) \subset A$. Clearly $(c_1, d_1) \subset ((c_1, d_1) \cap P) \cup int(B)$. Let $x \in (c_1, d_1)$ and suppose that $x \notin A$. Then

$x \notin int(A)$. Let (a_1, b_1) be the component of $int(B)$ which contains x. It follows that $a_1, b_1 \in P$ and $F(y) < r$, for each $y \in (a_1, b_1)$. Since F is *lower internal*, $F(a_1) < r$, a contradiction (because $a_1 \in [c_1, d_1] \cap P \subset A$. hence $F(a_1) > r$). Thus $(c_1, d_1) \subset A$, hence $(c_1, d_1) \subset int(A)$, a contradiction (because $(c_1, d_1) \cap P \neq \emptyset$). So $P \not\subset \overline{E}_2$ is impossible.
2) $P \not\subset \overline{E}_1$. Similarly to 1), we obtain that this is impossible.
3) $P = \overline{E}_1 = \overline{E}_2$. Since $F \in \mathcal{B}_1$, by Theorem 2.2.1, $E_1 = \{x \in P : F(x) > r\} = \{x \in P : F(x) \geq r\}$ and $E_2 = \{x \in P : F(x) < r\} = \{x \in P : F(x) \leq r\}$ are G_δ-sets. It follows that E_1 and E_2 are residual sets with respect to P, hence $E_1 \cap E_2$ is residual with respect to P, a contradiction (because $E_1 \cap E_2 = \emptyset$).

That (v) and (vi) are equivalent with (ii) and (iii) follow by Proposition 1.15.3.

Remark 2.5.1 *Consider the following theorem of [Br2] (p. 8): Let $F : [a,b] \mapsto \mathbb{R}$, $F \in \mathcal{B}_1$. The following assertions are equivalent:*

(i) *$F \in \mathcal{D}$ on $[a,b]$;*

(ii) *(Young) For each x there exist sequences $x_n \nearrow x$ and $y_n \searrow x$ such that $F(x) = \lim_{n\to\infty} F(x_n) = \lim_{n\to\infty} F(y_n)$;*

(iii) *(Sen; Masera) For each x we have $F(x) \in [\liminf_{z\nearrow x} F(z), \limsup_{z\nearrow x} F(z)] \cap [\liminf_{z\searrow x} F(z), \limsup_{z\searrow x} F(z)]$ (i.e., F is internal);*

(iv) *(Neugebauer) For each $r \in \mathbb{R}$ the sets $\{x : F(x) \leq r\}$ and $\{x : F(x) \geq r\}$ have compact components (i.e., these sets are simultaneously $\mathcal{S}^+_\infty$ and $\mathcal{S}^-_\infty$ closed (see Proposition 1.14.4*

(v) *(Kuratowski; Sierpinski; Choquet) The graph of F is connected;*

(vi) *(Maximoff) For each x there exists a perfect set P (depending on x) such that x is a bilateral point of accumulation of P and $F_{/P}$ is $\mathcal{C}$ at x;*

(vii) *(Zahorski) For each $r \in \mathbb{R}$ the sets $\{x : F(x) < r\}$ and $\{x : F(x) > r\}$ are bilaterally c-dense in themselves (i.e., they are simultaneously $\mathcal{S}^+_c$ open and $\mathcal{S}^-_c$ open);*

(viii) *(Zahorski)For each $r \in \mathbb{R}$ the sets $\{x : F(x) < r\}$and $\{x : F(x) > r\}$ are bilaterally dense in themselves (i.e., they are simultaneously $\mathcal{S}^+_\infty$ open and $\mathcal{S}^-_\infty$ open).*

This theorem (except (v) and (vi)) follows also by Theorem 2.5.1. Moreover, it is true for functions $F : [a,b] \mapsto \overline{\mathbb{R}}$.

Remark 2.5.2 *In Theorem 2.5.1, the fact that $F \in \mathcal{B}_1$ is essential (see Sections 6.10.2 and 6.11.).*

Remark 2.5.3 *It is well known that if a continuous function $F : [a,b] \mapsto \mathbb{R}$ is constant on a dense subset of $[a,b]$ then F is constant on $[a,b]$. But there exists a $\mathcal{DB}_1$ function $F : [a,b] \mapsto \mathbb{R}$ such that F is constant a.e. on $[a,b]$ and F is not constant on all of $[a,b]$ (see Section 6.21.).*

Corollary 2.5.1 *Let $F_n, F : [a,b] \mapsto \mathbb{R}$. Then we have:*

(i) $\mathcal{C}_i + \mathcal{D}_-\mathcal{B}_1 = \mathcal{D}_-\mathcal{B}_1$;

(ii) $\mathcal{C} + \mathcal{D}\mathcal{B}_1 = \mathcal{D}\mathcal{B}_1$;

(iii) $\mathcal{C}_i + (\mathcal{D}_- \cap [\mathcal{C}_i G]) = \mathcal{D}_- \cap [\mathcal{C}_i G]$ *and* $\mathcal{C}_i^* + (\mathcal{D}_- \cap [\mathcal{C}_i^* G]) = \mathcal{D}_- \cap [\mathcal{C}_i^* G]$

(iv) $\mathcal{C} + (\mathcal{D} \cap [\mathcal{C}G]) = \mathcal{D} \cap [\mathcal{C}G]$;

(v) If $F_n \in \mathcal{D}_-\mathcal{B}_1$ *and* $F_n \to F$ *[unif] then* $F \in \mathcal{D}_-\mathcal{B}_1$;

(vi) If $F_n \in \mathcal{D}\mathcal{B}_1$ *and* $F_n \to F$ *[unif] then* $F \in \mathcal{D}\mathcal{B}_1$;

(vii) $\mathcal{D}\mathcal{B}_1 \circ \mathcal{C} = \mathcal{C} \circ \mathcal{D}\mathcal{B}_1 = \mathcal{D}\mathcal{B}_1$, *for functions defined on intervals.*

Proof. (i) Let $f, g : [a,b] \mapsto \mathbb{R}$, $f \in \mathcal{C}_i$, $g \in \mathcal{D}_-\mathcal{B}_1$ and $F = f + g$. By Proposition 2.3.3, $f \in \mathcal{B}_1$. By Remark 2.2.1, (ii), $F \in \mathcal{B}_1$. Since $g \in \mathcal{D}_-\mathcal{B}_1$, by Theorem 2.5.1, g is *lower internal*. By Theorem 2.4.2,(i), F is *lower internal*, and by Theorem 2.5.1,$F \in \mathcal{D}_-\mathcal{B}_1$.

(ii) and (iii) follow by (i);(iv) follows by (ii);

(v) By Remark 2.2.1, (vii), $F \in \mathcal{B}_1$. By Theorem 2.4.2, (vi), F_n is *lower internal*. By Proposition 2.4.4,F is *lower internal*, and by Theorem 2.5.1, $F \in \mathcal{D}_-\mathcal{B}_1$.

(vi) follows by (v); for (vii) see [Br2] (p. 16).

2.6 Conditions $\mathcal{B}_1$, $w\mathcal{B}_1$, $\mathcal{S}_2^-$ and $\mathcal{S}_2^+$-local systems

Definition 2.6.1 (O'Malley) *A function* $F : [a,b] \mapsto \overline{\mathbb{R}}$ *is said to be* $w\mathcal{B}_1$ *(wide* $\mathcal{B}_1$*), if for* $-\infty < u < v < +\infty$ *and for every open interval I, the sets* $\{x : F(x) \le u\}$ *and* $\{x : F(x) \ge v\}$ *are not simultaneously dense in* $I \cap \overline{F^{-1}((u,v))}$*, where* $I \cap \overline{F^{-1}((u,v))} \ne \emptyset$.

Remark 2.6.1 $\mathcal{B}_1 \subset w\mathcal{B}_1$ *(see Theorem 2.2.1 and Section 6.10.1.).*

Theorem 2.6.1 (O'Malley) *([O4]). Let* $F : [a,b] \mapsto \overline{\mathbb{R}}$, $F \in \mathcal{D}$*. The following assertions are equivalent:*

(i) $F \in w\mathcal{B}_1$ *on* $[a,b]$;

(ii) For $-\infty < u < v < +\infty$ *and for every open interval I with* $I \cap \overline{F^{-1}((u,v))} \ne \emptyset$*, there exists an open subinterval J of I with* $F \cap \overline{F^{-1}((u,v))} \ne \emptyset$*, such that either* $J \subset F^{-1}((u,+\infty])$ *or* $J \subset F^{-1}([-\infty,v))$.

Proof. $(i) \Rightarrow (ii)$ Since $\{x : F(x) \le u\}$ and $\{x : F(x) \ge v\}$ are not simultaneously dense in $I \cap \overline{F^{-1}((u,v))}$, there exists an open subinterval J of I with $E = J \cap \overline{F^{-1}((u,v))} \ne \emptyset$, and either $F(x) > u$ or $F(x) < v$ on E. Then $U = J \setminus \overline{F^{-1}((u,v))}$ is an open set and each component of U has at least one endpoint in $\overline{F^{-1}((u,v))}$. Suppose for example that $F(x) > u$ on E. We show that $F(x) > u$ on J. Suppose that there is $x_1 \in J_1$ such that $F(x_1) \le u$. Then $x_1 \in U$. Let (r,s) be the component of U which contains x_1. It follows that (r,s) has at least one endpoint in E. Suppose for example that $s \in E$. Then $F(s) > u$. Let $y \in (u, \min\{v; F(s)\})$. Since $F \in \mathcal{D}$, there exists $x_y \in (x_1, s)$ such that $F(x_y) = y$. Hence $x_y \in E$, a contradiction.

$(ii) \Rightarrow (i)$ This is evident.

Theorem 2.6.2 *Let $F : [a,b] \mapsto \overline{\mathbb{R}}$ and $\alpha, \beta \in \mathbb{R}$. If F is measurable and $F \in \mathcal{D} \cap w\mathcal{B}_1$ then the following assertions are equivalent:*

(i) F is simultaneously $\mathcal{S}_2^+$ and $\mathcal{S}_2^-$ lower semicontinuous and $\mathcal{S}_2^+$ and $\mathcal{S}_2^-$ upper semicontinuous on $[a,b]$;

(ii) F is $\mathcal{S}_2^+$ continuous on $\{x \in [a,b) : F(x) \neq \pm\infty\}$ and F is $\mathcal{S}_2^-$ continuous on $\{x \in (a,b] : F(x) \neq \pm\infty\}$;

(iii) $\{x \in [a,b) : F(x) < \alpha\}$ and $\{x \in [a,b) : F(x) > \alpha\}$ are $\mathcal{S}_2^+$ open sets; and $\{x \in (a,b] : F(x) > \alpha\}$ and $\{x \in (a,b] : F(x) < \alpha\}$ are $\mathcal{S}_2^-$ open sets;

(iv) $\{x \in [a,b) : \alpha < F(x) < \beta\}$ is $\mathcal{S}_2^+$ open and $\{x \in (a,b] : \alpha < F(x) < \beta\}$ is $\mathcal{S}_2^-$ open.

Proof. *(i)* $\Leftrightarrow$ *(iii)* See Proposition 1.15.3.

(ii) $\Leftrightarrow$ *(iv)* See Proposition 1.15.4, (ii).

(iii) $\Rightarrow$ *(iv)*Suppose (for example) on the contrary that the set $E = \{x \in [a,b) : \alpha < F(x) < \beta\}$ is $\mathcal{S}_2^+$ open for no $\alpha, \beta \in \mathbb{R}$. Then there exists $x_o \in E$ and $\delta > 0$ such that $\mid E \cap (x_o, x_o + \delta) \mid = 0$. Clearly $F(x_o) \in (u,v)$. Let $x_1 \in (x_o, x_o + \delta)$ such that $x_1 \notin E$. Suppose for example that $F(x_1) \geq \beta$. Let $y \in (F(x_o), \beta)$. Since $F \in \mathcal{D}$, there exists $x_y \in (x_o, x_1)$ such that $F(x_y) = y$. It follows that $E \cap (x_o, x_1)$ is uncountable. Hence $(x_o, x_1) \cap \overline{E} \neq \emptyset$. By Theorem 2.6.1, there exists an open interval $J \subset (x_o, x_1)$, with $J \cap \overline{E} \neq \emptyset$, such that $F(x) > \alpha$ on J, for example. Since $\mid E \cap J \mid = 0$ it follows that

(1) $\quad F(x) \geq \beta$ *a.e. on J.*

Since $J \cap \overline{E} \neq \emptyset$ it follows that $J \cap E \neq \emptyset$. Let $x_2 \in J \cap E$. Then $x_2 \in \{x \in [a,b) : F(x) < \beta\}$ which is an $\mathcal{S}_2^+$ open set. This contradicts (1).

(iv) $\Rightarrow$ *(iii)* Let α be a real number and $E = \{x \in [a,b) : F(x) > \alpha\}$. We show that E is $\mathcal{S}_2^+$ open. Let $E_o = \{x \in [a,b) : \alpha < F(x) < +\infty\}$ and $E_{+\infty} = \{x \in [a,b) : F(x) = +\infty\}$. Then $E_o = \cup_{n=1}^{\infty}\{x \in [a,b) : \alpha < F(x) < \alpha + n\}$ is $\mathcal{S}_2^+$ open. Let $x_o \in E$. If $x_o \in E_o$ then $E_o \in \mathcal{S}_2^+(x_o)$. If $x_o \in E_{+\infty}$ then $F(x_o) = +\infty$. Suppose that $E \notin \mathcal{S}_2^+(x_o)$. Then there exists $\delta > 0$ such that $\mid E \cap (x_o, x_o + \delta) \mid = 0$. Let $x_1 \in (x_o, x_o + \delta)$ such that $F(x_1) \leq \alpha$. Since $F \in \mathcal{D}$, for each $y \in (\alpha, +\infty)$ there exists $x_y \in (x_o, x_1)$ such that $F(x_y) = y$. Because $x_y \in E_o$, it follows that $0 < \mid (x_o, x_1) \cap E \mid < \mid (x_o, x_1) \cap E \mid < \mid (x_o, x_o + \delta) \cap E \mid$, a contradiction.

Remark 2.6.2 *(i) in [O4], O'Malley denotes condition (iii) of Theorem 2.6.2 by m_2 (for measurable functions). Note that Zahorski introduced in [Z1] a class of functions M_2. In fact we have $M_2 = m_2 \cap \mathcal{B}_1$ and $M_2 \subset \mathcal{DB}_1$ (see [Br2], p. 87, or Theorem 2.5.1).*

(ii) Condition (iv) of Theorem 2.6.2 is in fact the Denjoy Clarkson Property (short D.C.) (see [O4], [Mu]).

Corollary 2.6.1 *Let $F : [a,b] \mapsto \overline{\mathbb{R}}$, $F \in \mathcal{B}_1$ and let $\alpha, \beta \in \mathbb{R}$. The following assertions are equivalent:*

(i) F is simultaneously $\mathcal{S}_2^+$ and $\mathcal{S}_2^-$ lower semicontinuous and $\mathcal{S}_2^+$ and $\mathcal{S}_2^-$ upper semicontinuous;

(ii) $F \in \mathcal{D}$ *on* $[a,b]$, F *is* S_2^+ *continuous on* $\{x \in [a,b) : F(x) \neq \pm\infty\}$, *and* F *is* S_2^- *continuous on* $\{x \in (a,b] : F(x) \neq \pm\infty\}$;

(iii) $\{x \in [a,b) : F(x) < \alpha\}$ *and* $\{x \in [a,b) : F(x) > \alpha\}$ *are* S_2^+ *open sets, and* $\{x \in (a,b] : F(x) < \alpha\}$ *and* $\{x \in (a,b] : F(x) > \alpha\}$ *are* S_2^- *open sets;*

(iv) $F \in \mathcal{D}$ *on* $[a,b]$, $\{x \in [a,b) : \alpha < F(x) < \beta\}$ *is* S_2^+ *open and* $\{x \in (a,b] : \alpha < F(x) < \beta\}$ *is* S_2^- *open.*

Proof. See Theorem 2.6.2.

Remark 2.6.3 *In [Mu], Mukhopadhyay proved the equivalence of (iii) and (iv) in Corollary 2.6.1 for finite functions.*

2.7 Conditions VB, $\underline{VB}$, VBG

Definition 2.7.1 *Let* $F : [a,b] \mapsto \mathbb{R}$ *and let* $P, Q \subseteq [a,b]$ *such that* $\{(x,y) \in P \times Q : x < y\} \neq \emptyset$. F *is said to be* $VB(P \wedge Q)$ *(resp.* $\overline{VB}(P \wedge Q)$*) if there exists* $M \in (0,+\infty)$ *such that*

$$(1) \quad \sum_{k=1}^{n} | F(b_k) - F(a_k) | < M \quad (resp.$$

$$(2) \quad \sum_{k=1}^{n} (F(b_k) - F(a_k)) \; < \; M),$$

whenever $\{[a_k, b_k]\}$, $k = \overline{1,n}$ *is a finite set of nonoverlapping closed intervals with* $a_k \in P$ *and* $b_k \in Q$. F *is said to be* $\underline{VB}(P \wedge Q)$ *if* $-F$ *is* $\overline{VB}(P \wedge Q)$, *i.e.,*

$$(3) \quad \sum_{k=1}^{n} (F(b_k) - F(a_k)) \; > \; -M.$$

Clearly $VB(P \wedge Q) = \overline{VB}(P \wedge Q) \cap \underline{VB}(P \wedge Q)$. *For* $P \subseteq Q \subseteq [a,b]$ *we define* $VB(P;Q) = VB(P \wedge Q) \cap VB(Q \wedge P)$ *and* $\underline{VB}(P;Q) = \underline{VB}(P \wedge Q) \cap \underline{VB}(Q \wedge P)$. *If* $P = Q$ *then* $VB(P;P) = VB$ *on* P *(the class of functions with bounded variation). Let* $\underline{VB}(P;P) = \underline{VB}$ *on P. We define conditions* VBG *and* $[VBG]$ *on P, using Definition 1.9.2.*

Remark 2.7.1 *(i) In the definition of* VB *(resp.* $\overline{VB}$, $\underline{VB}$*) we can replace (1) (resp. (2), (3)) by*

$$(1') \quad \sum_{k=1}^{n} \mathcal{O}(G; [a_k, b_k] \cap P) << \quad (resp.$$

$$(2') \quad \sum_{k=1}^{n} \mathcal{O}_+(F; [a_k, b_k] \cap P) \; < \; M,$$

$$(3') \quad \sum_{k=1}^{n} \mathcal{O}_-(F; [a_k, b_k] \cap P) \; > \; -M)$$

(see Corollary 1.4.1, (i), (ii)).

(ii) *In the definition of* VB *and* $\underline{VB}$ *we may replace condition "*$a_k, b_k \in P$*" by* $[a_k, b_k] \cap P \neq \emptyset$*" (see Corollary 1.4.1, (i), (ii)).*

(iii) $\{[a_k, b_k]\}_k$ *may be supposed to be an infinite sequence of nonoverlapping closed intervals, in Definition 2.7.1.*

Remark 2.7.2 *Let* $P \subseteq Q \subseteq [a,b]$. *Then*

(i) $VB(P;Q)$, $VB(P \wedge Q)$ *and* $VB(Q \wedge P)$ *are real linear spaces;*

(ii) $\underline{VB}(P;Q)$, $\underline{VB}(P \wedge Q)$ *and* $\underline{VB}(Q \wedge P)$ *are real semi-linear spaces;*

(iii) VBG *and* $[VBG]$ *are real linear spaces on* P. *Moreover, for bounded functions on* P, VBG *and* $[VBG]$ *are real algebras;*

(iv) *If* $F \in VB$ *on* P *and* $F_{/\overline{P}} \in \mathcal{C}$ *then* $F \in VB$ *on* $\overline{P}$.

Definition 2.7.2 *Let* $F : [a,b] \mapsto \mathbb{R}$, $P \subseteq [a,b]$. *Let* $V(F;P) = \sup\{\sum_{k\geq 1} \mid F(b_k) - F(a_k) \mid : \{[a_k, b_k]\}_k$ *is a sequence of nonoverlapping closed intervals,* $a_k, b_k \in P\}$. *If* $F \in VB$ *on* P *then* $V(F;P) = \inf\{M : M$ *is given by the fact that* $F \in VB$ *on* $P\}$. *Let* $\overline{V}(F;P) = \sup\{\sum_{k\geq 1}(F(b_k) - F(a_k)) : \{[a_k, b_k]\}_k$ *is a sequence of nonoverlapping closed intervals,* $a_k, b_k \in P\}$. *If* $F \in \overline{VB}$ *then* $\overline{V}(F;P) = \inf\{M : M$ *is given by the fact that* $F \in \overline{VB}$ *on* $P\}$.

Lemma 2.7.1 *([S3], p.221). Let* $F : [a,b] \mapsto \mathbb{R}$, $P \subseteq [a,b]$.

(i) F *is bounded and increasing on* P *if and only if* F *coincide on* P *with a function which is bounded and increasing on* $[a,b]$.

(ii) *If* F *is bounded and increasing on* P *then* F *is* VB *on* P. *Moreover, if* F *is increasing on* $\overline{P}$ *then* F *is* VB *on* $\overline{P}$.

Theorem 2.7.1 *Let* $F : [a,b] \mapsto \mathbb{R}$, $P \subseteq [a,b]$, $c = \inf(P)$, $d = sup(P)$. *The following assertions are equivalent:*

(i) $F \in VB$ *on* P;

(ii) *There exists* $M \in (0, +\infty)$ *such that* $\sum_{i=1}^{n-1} \mid F(x_i) - F(x_{i-1}) \mid < M$, *whenever* $c = x_o < x_1 < \ldots < x_{n-1} < x_n = d$, $x_1, x_2, \ldots, x_{n-1} \in P$;

(iii) $F \in \underline{VB}$ *on* $P \cup \{c, d\}$;

(iv) F *is bounded and* $\underline{VB}$ *on* P;

(v) *There exist* $F_1, F_2 : [a,b] \mapsto \mathbb{R}$, F_1, F_2 *increasing on* $[a,b]$, *such that* $F = F_1 - F_2$ *on* P;

(vi) *There exists* $f : [a,b] \mapsto \mathbb{R}$, $f \in VB$ *on* $[a,b]$ *such that* $F = f$ *on* P.

Proof. (*i*) $\Rightarrow$ (*ii*) Let $M \in (0, +\infty)$ be given by the fact that $F \in VB$ on P, and let $x_o \in P$. For each $x \in P$ we have $| F(x) - F(x_o) | < M$, hence F is bounded on P. Since $F(c), F(d) \in \mathbb{R}$, it follows that F is bounded on $P \cup \{c, d\}$. Let $r \in (0, +\infty)$ such that $| F(x) | < r$, for each $x \in P \cup \{c, d\}$, and let $x = x_o < x_1 < \ldots < x_{n-1} < x_n = d$, $x_1, x_2, \ldots, x_{n-1} \in P$. Then $\sum_{i=0}^{n-1} | F(x_{i+1}) - F(x_i) | = | F(x_i) - F(x_o) | + \sum_{i=1}^{n-2} | F(x_{i+1}) - F(x_i) | + | F(x_n) - F(x_{n-1}) | < 2r + M + 2r = 4r + M$.

(*ii*) $\Rightarrow$ (*i*) and (*ii*) $\Rightarrow$ (*iii*) are evident.

(*iii*) $\Rightarrow$ (*iv*) Let $M \in (0, +\infty)$ be given by the fact that $F \in \underline{VB}$ on $P \cup \{c, d\}$, and let $x \in P$. Then $-M < F(x) - F(c)$ and $-M < F(d) - F(x)$. It follows that $F(c) - M \le F(x) \le M + F(d)$, for each $x \in P$, hence F is bounded on P.

(*iv*) $\Rightarrow$ (*ii*) Let $M \in (0, +\infty)$ be given by the fact that F is $\underline{VB}$ on P, and let $r \in (0, +\infty)$ such that $| F(x) | < r$ for each $x \in P \cup \{c, d\}$. Let $c = x_o < x_1 < \ldots < x_{n-1} < x_n = d$, $x_1, x_2, \ldots, x_{n-1} \in P$. Let $A = \{i \in \{1, 2, \ldots, n-2\} : F(x_{i+1}) - F(x_i) < 0\}$. It follows that $\sum_{i=0}^{n-1} | F(x_{i+1}) - F(x_i) | = | F(x_1) - F(x_o) | + | F(x_n) - F(x_{n-1}) | + \sum_{i=1}^{n-2} | F(x_{i+1}) - F(x_i) | \le 4r + \sum_{i=1}^{n-2} | F(x_{i+1}) - F(x_i) | = 4r + \sum_{i=2}^{n-2} (F(x_{i+1}) - F(x_i)) - 2 \sum_{i \in A} (F(x_{i+1}) - F(x_i)) \le 4r + F(x_{n-1}) - F(x_2) + 2m < 6r + 2M$.

(*i*) $\Rightarrow$ (*v*) Let $F_1(x) = V(F; [a, x] \cap P)$ and $G = F_1 - F$. Then F_1 is increasing on $[a, b]$ and G is bounded and increasing on P. By Lemma 2.7.1, there exists a function $F_2 : [a, b] \mapsto \mathbb{R}$, increasing on $[a, b]$, such that $F_2 = G$ on P.

(*v*) $\Rightarrow$ (*vi*) follows by Lemma 2.7.1; (*vi*) $\Rightarrow$ (*i*) is evident.

Corollary 2.7.1 *Let $P \subseteq [a, b]$. Then $\underline{VB} \subseteq VBG$ on P.*

Proof. If F is bounded on P then by Theorem 2.7.1, $F \in VB \subset VBG$ on P. If F is not bounded on P then let $P_n = \{x \in P : -n \le F(x) \le n\}$, for each positive integer n. It follows that $P = \cup_{n=1}^{\infty} P_n$, and F is bounded and $\underline{VB}$ on each P_n. By Theorem 2.7.1, (i), (iv), $F \in VB$ on each P_n, hence $F \in VBG$ on P.

Proposition 2.7.1 *Let $F : [a, b] \mapsto \mathbb{R}$, $P \subseteq [a, b]$, $c = \inf(P)$, $d = \sup(P)$. Let (c_k, d_k), $k \ge 1$ be the intervals contiguous to$\overline{P}$. Let $P_- = \{x \in \overline{P} : x$ is a left accumulation point of $P\}$ and $P_+ = \{x \in \overline{P} : x$ is a right accumulation point of $P\}$. The following assertions are equivalent:*

(*i*) $F \in VB$ *on* $\overline{P}$;

(*ii*) $F \in \overline{VB}$ *on* $\overline{P}$;

(*iii*) $F \in VB(P \wedge (P \cup P_-)) \cap VB((P \cup P_-) \wedge P) \cap VB(P \wedge (P \cup P_+)) \cap VB((P \cup P_+) \wedge P)$;

(*iv*) $F \in \overline{VB}(P \wedge (P \cup P_-)) \cap \overline{VB}((P \cup P_-) \wedge P) \cap \overline{VB}(P \wedge (P \cup P_+)) \cap \overline{VB}((P \cup P_+) \wedge P)$;

(*v*) $F \in \overline{VB}(P \wedge (P \cup P_-)) \cap \overline{VB}((P \cup P_+) \wedge P)$ *and* $\sum_{k \ge 1} \mathcal{O}_+(F; [c_k, d_k]) < +\infty$.

Proof. For (*i*) $\Leftrightarrow$ (*ii*) see Theorem 2.7.1, (i), (iii); (*i*) $\Leftrightarrow$ (*iii*), (*ii*) $\Leftrightarrow$ (*iv*) and (*i*) $\Rightarrow$ (*v*) follow by definitions.

(*v*) $\Rightarrow$ (*iv*) We show for example that $F \in \overline{VB}(P \wedge (P \cup P_+))$. Let $\{[a_i, b_i]\}$, $i = \overline{1, n}$ be a finite set of nonoverlapping closed intervals, with $a_i \in P$ and $b_i \in P \cup P_+$. We consider only the nontrivial case $b_i \in P_+$, $i = \overline{1, n}$. Then there exists a natural number k_i such that $b_i = d_{k_i}$, $i = \overline{1, n}$. Clearly $c_{k_i} \in P \cup P_-$ and $a_i \le c_{k_i} < d_{k_i} = b_i$. Let

$r = \sum_{k\geq 1} \mathcal{O}_+(F; \{c_k, d_k\})$ and let M be given by the fact that $F \in \overline{VB}(P \wedge (P \cup P_-))$. We obtain that $\sum_{i=1}^n (F(b_i) - F(a_i)) = \sum_{i=1}^n (F(d_{k_i}) - F(c_{k_i}) + F(c_{k_i}) - F(a_i)) < r + M < +\infty$, hence $F \in \overline{VB}(P \wedge (P \cup P_+))$.

Corollary 2.7.2 *Let $F : [a,b] \mapsto \mathbb{R}$, $P \subseteq [a,b]$, $c = \inf(P)$, $d = \sup(P)$. Let (c_k, d_k), $k \geq 1$ be the intervals contiguous to $\overline{P}$. We have:*

(i) $F \in \overline{VB}(P \wedge (P \cup P_-)) \cup \overline{VB}((P \cup P_+) \wedge P)$ and $\sum_{k\geq 1} V(F; [c_k, d_k]) < +\infty$ then $F \in VB$ on $[c,d]$;

(ii) $F \in VB$ on $\overline{P}$ if and only if $F_{\overline{P}} \in VB$ on $[c,d]$;

(iii) If $F \in VB$ on $\overline{P}$ and F is increasing on each $[c_k, d_k]$ then $F \in VB$ on $[c,d]$.

Proof.(i) We have $\sum_{k\geq 1} \mathcal{O}_+(F; \{c_k, d_k\}) \leq \sum_{k\geq 1} \mathcal{O}(F; [c_k, d_k]) < \sum_{k\geq 1} V(F; [c_k, d_k]) < +\infty$. That $F \in VB$ on $\overline{P}$ follows by Proposition 2.7.1, (v), (i). Let $\{[a_i, b_i]\}$, $i = \overline{1,n}$ be a finite set of nonoverlapping closed subintervals of $[c,d]$. Let $A = \{i : (a_i, b_i) \cap P \neq \emptyset\}$, $B = \{1, 2, \ldots, n\} \setminus A$, $a_i' = \inf([a_i, b_i] \cap \overline{P})$ and $b_i' = \sup([a_i, b_i] \cap \overline{P})$, $i \in A$. Then $\sum_{i=1}^n \mid F(b_i) - F(a_i) \mid \leq \sum_{i\in A} (\mid F(b_i') - F(a_i') \mid + \mid F(b_i) - F(b_i') \mid + \mid F(a_i') - F(a_i) \mid) + \sum_{i\in B} \mid F(b_i) - F(a_i) \mid \leq V(F; \overline{P}) + \sum_{k\geq 1} V(F; [c_k, d_k]) < +\infty$. Hence $F \in VB$ on $[c,d]$.

(ii) and (iii) follow by definitions (for F_P see Definition 1.1.3).

Theorem 2.7.2 *([N], pp. 205, 218). Let $F : [a,b] \mapsto \mathbb{R}$.*

a) If F is increasing then:

(i) $\{x : F$ is discontinuous at $x\}$ is at most countable;

(ii) $F(x^-)$ exists and $F(x^-) \leq F(x)$ for each $x \in (a,b]$; $F(x^+)$ exists and $F(x) \leq F(x^+)$, for each $x \in [a,b)$.

b) If $F \in VB$ then:

(i) $\{x : F$ is discontinuous at $x\}$ is at most countable;

(ii) for each $x \in (a,b]$, $F(x^-)$ exists; for each $x \in [a,b)$, $F(x^+)$ exists.

2.8 Conditions VB^*, $\underline{VB}^*$, VB^*G

Definition 2.8.1 *Let $F : [a,b] \mapsto \mathbb{R}$, $P \subseteq [a,b]$. F is said to be VB^* (see [S3], p. 228) (resp. $\overline{VB}^*$) on P, if there exists $M \in (0, +\infty$ such that $\sum_{k=1}^n \mathcal{O}(F; [a_k, b_k]) < M$ (resp. $\sum_{k=1}^n \Omega_+(F; [a_k, b_k] \wedge (P \cap [a_k, b_k])) < M$), whenever $\{[a_k, b_k]\}$, $k = \overline{1,n}$, is a finite set of nonoverlapping closed intervals with $a_k, b_k \in P$. F is said to be $\underline{VB}^*$ on P, if $-F \in \overline{VB}^*$ on P, i.e., $\sum_{k=1}^n \Omega_-(F; [a_k, b_k] \wedge (P \cap [a_k, b_k])) > -M$. We define conditions VB^*G and $[VB^*G]$ on P, using Definition 1.9.2.*
Let $V^*(F; P) = \sup\{\sum_{k=1}^n \mathcal{O}(F; [a_k, b_k]) : \{[a_k, b_k]\}$, $k = \overline{1,n}$ is a finite set of nonoverlapping closed intervals, with $a_k, b_k \in P\}$. If $F \in VB^*$ on P then $V^*(F; P) = \inf\{M : M$ is given by the fact that $F \in VB^*$ on $P\}$.

Remark 2.8.1 *In Definition 2.8.1, $\{[a_k, b_k]\}_k$ may be suppose to be an infinite sequence of nonoverlapping closed intervals.*

Proposition 2.8.1 *Let $F : [a,b] \mapsto \mathbb{R}$, $P \subseteq [a,b]$, $c = \inf(P)$, $d = \sup(P)$.*

(i) $VB^* = \overline{VB}^* \cap \underline{VB}^*$ *on* P;

(ii) $VB^* = VB$ *on* $[a,b]$;

(iii) $VB^* \subsetneq VB$ *on* P, *hence* $VB^*G \subset VBG$ *on* P;

(iv) $VB^* \subsetneq \overline{VB}^* \subsetneq \overline{VB}$ *on* P;

(v) VB^* *is a real algebra on* P;

(vi) VB^*G *is a real space on* P. *Moreover,* VB^*G *is a real algebra on* P *for bounded functions on* $[c,d]$;

(vii) $\overline{VB}^*$ *is a semi-linear space on* P. *Moreover,* $\overline{VB}^* \cdot \overline{VB}^* = \overline{VB}^*$ *on* P *for bounded positive functions on* $[c,d]$.

Proof. See Proposition 1.4.1, Corollary 1.4.1 and Section 6.24.

Proposition 2.8.2 *Let $F : [a,b] \mapsto \mathbb{R}$, $P \subseteq [a,b]$ and let I be the smallest interval containing P. The following assertions are equivalent:*

(i) $F \in \overline{VB}^*$ *on* P;

(ii) $F \in \overline{VB}(P; I)$;

(iii) *There exists a positive real number M such that $\sum_{k=1}^{n}(F(x_k) - F(a_k)) < M$ and $\sum_{k=1}^{n}(F(b_k) - F(x_k)) < M$, whenever $\{[a_k, b_k]\}$, $k = \overline{1,n}$ is a finite set of nonoverlapping closed intervals, with $a_k, b_k \in P$ and $x_k \in [a_k, b_k]$.*

Proof. (i) $\Rightarrow$ (ii) We show for example that $F \in \overline{VB}(P \wedge I)$. Let $\{[a_k, b_k]\}$, $k = \overline{1,n}$ be a finite set of nonoverlapping closed intervals, $a_k \in P$, $b_k \in I$ and $b_k < a_{k+1}$, $k = \overline{1, b-1}$. Let $[c,d]$ be the smallest closed interval containing I and let $a_{n+1} \in P \cap [b_n, d] \neq \emptyset$. Let M be given by the fact that $F \in \overline{VB}^*$ on P. We have $\sum_{k=1}^{n}(F(b_k) - F(a_k)) \leq \sum_{k=1}^{n} \Omega_+(F; [a_k, b_{k+1}] \wedge (P \cap [a_k, b_k])) < M$.

(ii) $\Rightarrow$ (iii) The proof is evident.

(iii) $\Rightarrow$ (i) We may suppose that $r_k = \Omega_+(F; [a_k, b_k] \wedge (P \cap [a_k, b_k])) > 0$, whenever $[a_k, b_k]$, $k = \overline{1,n}$ are as in (i). Then there exist $x_k, y_k \in [a_k, b_k]$, $x_k < y_k$ such that at least one of them belongs to P and $r_k/2 < F(y_k) - F(x_k)$. We consider only the case when all $x_k \in P$ (the other situations are similar). Clearly $[x_k, b_k]$, $k = \overline{1,n}$ are nonoverlapping closed intervals, with $x_k, b_k \in P$. By (iii) we have $(1/2)\sum_{k=1}^{n} r_k < \sum_{k=1}^{n}(F(y_k) - F(x_k)) < M$.

Theorem 2.8.1 *Let $F : [a,b] \mapsto \mathbb{R}$, $P \subseteq [a,b]$, $c = \inf(P)$, $d = \sup(P)$. Let $\{(c_k, d_k)\}_k$ be the intervals contiguous to P. The following assertions are equivalent:*

(i) $F \in VB^*$ *on* P;

(ii) *(<u>Saks</u>, [S3], p.229).* $F \in VB^*$ *on* $\overline{P}$ *(hence VB^* is a hereditary strong property);*

(iii) $F \in \overline{VB}^*$ *on* $P \cup \{c, d\}$;

(iv) $F \in \overline{VB}^*$ *and F is bounded on P;*

(v) $F \in VB \cap \overline{VB}^*$ *on P;*

(vi) $F \in \underline{VB} \cap \overline{VB}^*$ *on P;*

(vii) *(Saks, [S3], p. 232).* $F \in VB$ *on* $\overline{P}$ *and* $\sum_{k\geq 1} \mathcal{O}(F;[c_k,d_k]) < +\infty$;

(viii) $F \in \overline{VB}(P\wedge(P\cup P_-)) \cap \overline{VB}((P\cup P_+)\wedge P)$ *and* $\sum_{k\geq 1} \Omega_+(F;[c_k,d_k]\wedge\{c_k,d_k\}) < +\infty$;

(ix) *There exists a positive real number M such that* $\sum_{k=1}^n \mathcal{O}(F;[a_k,b_k]) < M$, *whenever* $\{[a_k,b_k]\}$, $k=\overline{1,n}$ *is a finite set of nonoverlapping closed subintervals of* $[c,d]$, *with* $[a_k,b_k]\cap P \neq \emptyset$;

(x) $F \in VB(P;[c,d])$;

(xi) $F \in VB(P\wedge[c,d])$ *(resp.* $F \in VB([c,d]\wedge P)$*)*;

(xii) $F \in \overline{VB}(P;[c,d])$;

(xiii) $F \in \overline{VB}(P\cup\{c,d\};[c,d])$.

Proof. For $(i) \Leftrightarrow (ii)$ see [S3] (p.229); $(ii) \Rightarrow (iii)$ is evident.

$(iii) \Rightarrow (iv)$ Let M be given by the fact that $F \in \overline{VB}^*$ on $P\cup\{c,d\}$. Then $\Omega_+(F;[c,d]\wedge(P\cup\{c,d\}) < M$. It follows that $F(d)-M < F(x) < F(c)+M$, for each $x\in[c,d]$. Hence F is bounded on P.

$(iv) \Rightarrow (v)$ See Proposition 2.8.1, (iv) and Theorem 2.7.1, (i), (iv).

$(v) \Rightarrow (vi)$ This is evident.

$(vi) \Rightarrow (i)$ Let $\{[a_k,b_k]\}$, $k=\overline{1,n}$ be a finite set of nonoverlapping closed intervals, $a_k,b_k\in P$. Let M_1, M_2 be given by the facts that $F\in\underline{VB}$ on P and $F\in\overline{VB}^*$ on P. By Proposition 1.4.1, (ix) we have $\mathcal{O}(F;[a_k,b_k]) < F(a_k)-F(b_k)+2\Omega_+(F;[a_k,b_k]\wedge(P\cap[a_k,b_k])$, so $\sum_{k=1}^n \mathcal{O}(F;[a_k,b_k]) < M_1+M_2$. It follows that $F\in VB^*$ on P.

$(ii) \Rightarrow (vii) \Rightarrow (viii)$ are evident.

$(viii) \Rightarrow (iii)$ By Proposition 2.7.1, (v), $F\in VB$ on $\overline{P}$, and let M_1 be given by this fact. Let $M_2 = \sum_{k\geq 1}\Omega_+(F;[c_k,d_k]\wedge\{c_k,d_k\})$. Let $\{[a_k,b_k]\}$, $k=\overline{1,n}$ be a finite set of nonoverlapping closed intervals, $a_k\in P\cup\{c,d\}$, $b_k\in[c,d]$. Let $r_k = \sup([a_k,b_k]\cap\overline{P})$. Then $r_k\in\overline{P}$, $k=\overline{1,n}$. It follows that $\sum_{k=1}^n(F(b_k)-F(a_k)) = \sum_{k=1}^n(F(r_k)-F(a_k)) + \sum_{k=1}^n(F(b_k)-F(r_k)) < M_1+M_2$. Hence $F \in \overline{VB}((P\cup\{c,d\})\wedge[c,d])$. Similarly $F\in\overline{VB}([c,d]\wedge(P\cup\{c,d\})$, hence $F\in\overline{VB}(P\cup\{c,d\};[c,d])$. By Proposition 2.8.2, $F\in\overline{VB}^*$ on $P\cup\{c,d\}$.

$(ii) \Rightarrow (ix)$ Let M be given by the fact that $F\in VB^*$ on $\overline{P}$. Let $\{[a_k,b_k]\}$, $k=\overline{1,n}$ be a finite set of nonoverlapping closed intervals, $[a_k,b_k]\cap P\neq\emptyset$. We may suppose without loss of generality that $a_k\in P$, $k=\overline{1,n}$ and $a_{k+1}\geq b_k$, $k=\overline{1,n-1}$. Let $a_{n+1}=d$. Then $\sum_{k=1}^n\mathcal{O}(F;[a_k,b_k]) \leq \sum_{k=1}^n\mathcal{O}(F;[a_k,a_{k+1}]) < M$.

$(ix) \Rightarrow (x) \Rightarrow (xi)$ are evident.

$(xi) \Rightarrow (xii)$ Suppose that $F\in VB(P\wedge[c,d]) \subseteq \overline{VB}(P\wedge[c,d])$. We have to show that $F\in\overline{VB}([c,d]\wedge P)$. Let M be given by the fact that $F\in VB(P\wedge[c,d])$. Let $\{[a_k,b_k]\}$, $k=\overline{1,n}$ be a finite set of nonoverlapping closed intervals, with $a_k\in[c,d]$, $b_k\in P$, $k=\overline{1,n}$ and $b_k\leq a_{k+1}$, $k=\overline{1,n-1}$. If $c=a_1$ then $\sum_{k=1}^n(F(b_k)-F(a_k)) + \sum_{k=1}^n(F(a_k)-F(b_{k-1})) + F(d)-F(b_n) = F(d)-F(c)$, hence

$\sum_{k=1}^{n}(F(b_k) - F(a_k)) < M + \mid F(d) - F(c) \mid$. If $c < a_1$ then let $b_o \in [c, d_1] \cap P \neq \emptyset$. It follows that $\sum_{k=1}^{n}(F(b_k) - F(a_k)) + \sum_{k=1}^{n}(F(a_k) - F(b_{k-1})) = F(b_n) - F(b_o)$, hence $\sum_{k=1}^{n}(F(b_k) - F(a_k)) < 2M$.

$(xii) \Rightarrow (xiii) \Rightarrow (iii)$ follow by Proposition 2.8.2.

Corollary 2.8.1 *Let $F : [a,b] \mapsto \mathbb{R}$, $P \subseteq [a,b]$, $c = \inf(P)$, $d = \sup(P)$.*

(i) $\underline{VB}^* \subseteq VB^*G$ on P;

(ii) If $F \in VB^*$ on P then F is bounded on $[c,d]$.

Proof. (i) If F is bounded on P then the proof follows by Theorem 2.8.1, (i), (iv). Suppose that F is not bounded on P. Let $P_n = \{x \in P : -n \leq F(x) \leq n\}$, $n \geq 1$. Then $P = \cup_{n=1}^{\infty} P_n$ and F is bounded on each P_n, hence $F \in VB^*$ on each P_n. It follows that $F \in VB^*G$ on P.

(ii) See Theorem 2.8.1, (i), (ii).

Theorem 2.8.2 *Let $F : [a,b] \mapsto \mathbb{R}$ and let P be a closed subset of $[a,b]$. The following assertions are equivalent:*

(i) $F \in VB^*G$ on P;

(ii) For each perfect subset $S \subset P$ there exists a portion $S \cap (c,d)$ such that $F \in VB^*$ on $S \cap (c,d)$;

(iii) $F \in VB^*G$ on each $Z \subset P$, whenever $\mid Z \mid = 0$.

Proof. See Theorem 2.8.1, (i), (ii) and Theorem 1.9.1.

2.9 Conditions monotone* and VB^*

Definition 2.9.1 *Let $F : [a,b] \mapsto \mathbb{R}$, $P, Q \subseteq [a,b]$ such that the set $\{(x,y) \in P \times Q : x < y\}$ is nonempty. F is said to be $I(P \wedge Q)$ (resp. $sI(P \wedge Q)$) if $F(x_1) \leq F(x_2)$ (resp. $F(x_1) < F(x_2)$), whenever $x_1 < x_2$, $x_1 \in P$, $x_2 \in Q$. F is said to be $D(P \wedge Q)$ (resp. $sD(P \wedge Q)$) if $F(x_1) \geq F(x_2)$ (resp. $F(x_1) > F(x_2)$). Let $M(P \wedge Q) = I(P \wedge Q) \cup D(P \wedge Q)$.*
If $P \subseteq Q \subseteq [a,b]$ then we define: $I(P;Q) = I(P \wedge Q) \cap I(Q \wedge P)$, $D(P;Q) = D(P \wedge Q) \cap D(Q \wedge P)$ and $M(P;Q) = M(P \wedge Q) \cap M(Q \wedge P)$. Similarly we obtain $sI(P;Q)$, $sD(P;Q)$ and $sM(P;Q)$.
Let $c = \inf(P)$, $d = \sup(P)$. If $F \in I(P;[c,d])$ then we denote this by "$F \in$ increasing on P". Similarly we define decreasing*, monotone*, etc. (these conditions were introduced by Krzyzewski in [Kr1]).*
*We define increasing*G, strictly increasing*G, etc, using Definition 1.9.2.*

Proposition 2.9.1 *Let $F : [a,b] \mapsto \mathbb{R}$, $P \subseteq Q \subseteq [a,b]$, $c = \inf(P)$, $d = \sup(P)$.*

(i) $M(P;P)$ = monotone on P, $sI(P;P)$ = strictly increasing on P, etc.

(ii) If $F \in VB(P;Q)$ then there exist $F_1, F_2 : [a,b] \mapsto \mathbb{R}$ such that $F = F_1 - F_2$ and $F_1, F_2 \in I(P;Q)$;

(iii) *If F is bounded on Q and $F \in I(P;Q)$ then $F \in VB(P;Q)$, hence increasing* $\subset VB^*$ on P;*

(iv) *If F is decreasing* on P and $F \in C_i$ on $\overline{P}$ then F is decreasing* on $\overline{P}$.*

Proof. (i) This is evident.

(ii) Let $F_1(x) = \sup\{\sum_{k=1}^{n} | F(b_k) - F(a_k) | : \{[a_k, b_k]\},\ k = \overline{1,n}$ is a finite set of nonoverlapping closed intervals, with $\{a_k, b_k\} \cap P \neq \emptyset$ and $a_k, b_k \in [a, x] \cap Q\}$. Let $x, y \in Q$, $x < y$ such that $\{x, y\} \cap P \neq \emptyset$. Then $F_1(y) - F_1(x) \geq | F(y) - F(x) | \geq F(y) - F(x)$, so F_1 is $I(P;Q)$. Let $F_2(x) = F_1(x) - F_2(x)$. Then F_2 is $I(P;Q)$.

(iii) The first part is evident. For the second part see Theorem 2.8.1, (i), (x).

(iv) Let $c \leq x < y \leq d$, $y \in \overline{P}$. Suppose that y is a left accumulation point of P, and let $z \in P$, $x < z < y$. Then $F(x) \geq F(z) \geq F(y)$. Suppose that y is a right accumulation point of P. Since $F \in C_i$ on $\overline{P}$, for $\epsilon > 0$ there exists $z \in P$, $z \geq y$ such that $F(y) \leq F(z) + \epsilon \leq F(x) + \epsilon$, hence $F(y) \leq F(x)$. It follows that F is *decreasing** on $\overline{P}$.

Remark 2.9.1 *If in Proposition 2.9.1, (ii) we put $P = Q$ then we obtain the Jordan Decomposition Theorem.*

Corollary 2.9.1 *Let $F : [a,b] \mapsto \mathbb{R}$, $P \subseteq [a,b]$, $c = inf(P)$, $d = sup(P)$. If $F \in VB^*$ on P then there exist $F_1, F_2 : [c,d] \mapsto \mathbb{R}$ such that F_1, F_2 are increasing* on $\overline{P}$ and $F = F_1 - F_2$ on $[c,d]$.*

Proof. By Theorem 2.8.1, (i), (ii), (x), $F \in VB(\overline{P};[c,d])$. The proof follows now by Proposition 2.9.1, (ii).

Lemma 2.9.1 *Let $F : [a,b] \mapsto \mathbb{R}$, $F(a) < F(b)$ such that $\limsup_{y \nearrow x} F(y) \geq F(x)$, $x \in (a,b]$ and $F(x) \geq \limsup_{y \searrow x} F(y)$, $x \in [a,b)$.*

(i) *If $b_o = \inf\{x : F(x) = F(b)\}$ then $F(b_o) = F(b)$;*

(ii) *If $x_y = \inf\{x \in (a, b_o) : F(x) = y\}$, $y \in (F(a), F(b))$ then $F(x_y) = y$;*

(iii) *There exists a set $A \subseteq [a, b_o]$, $a, b_o \in A$ such that $F(A) = [F(a), F(b)]$ and $F \in sI([a, b_o] \wedge A)$;*

(iv) *If $A_+ = \{x : x$ is a right accumulation point of $A\}$ then $F(A \cup A_+) = [F(a), F(b)]$ and $F \in I([a, b_o] \wedge (A \cup A_+))$.*

Proof. By Corollary 2.1.1, (iii), $F \in \mathcal{D}_+$ on $[a,b]$.

(i) Suppose on the contrary that $F(b_o) \neq F(b)$. Then b_o is a right accumulation point of the set $\{x : F(x) = F(b)\}$, hence $F(b_o) \geq \limsup_{y \searrow b_o} F(y) \geq F(b)$. It follows that $F(b_o) > F(b)$. Since $F(a) < F(b)$ and $F \in \mathcal{D}_+$ on $[a,b]$, there exists $c \in (a, b_o)$ such that $F(c) = F(b)$. This contradicts the definition of b_o.

(ii) Since $F \in \mathcal{D}_+$ on $[a,b]$, the set $\{x \in (a, b_o) : F(x) = y\}$ is nonempty. Suppose on the contrary that $F(x_y) \neq y$. Then x_y is a right accumulation point of the set $\{x \in (a, b_o) : F(x) = y\}$, hence $F(x_y) \geq \limsup_{x \searrow x_y} F(x) \geq y$. It follows that $F(x_y) > y$. But $F(a) < y$. This contradicts the definition of x_y.

(iii) Let $A = \{x_y : y \in [F(a), F(b)]\}$, where x_y is defined as in (ii). Clearly $a, b_o \in A$. Since $F(x_y) = y$, $F(A) = [F(a), F(b)]$. Let $a \leq x_1 < x_2 \leq b_o$, $x_2 \in A$.

Then $x_2 = x_{F(x_2)}$ and $F(x_1) \neq F(x_2)$. Suppose on the contrary that $F(x_2) < F(x_1)$. Since $F(x_2) > F(a)$ and $F \in \mathcal{D}_+$, there exists $c \in (a, x_1)$ such that $F(c) = F(x_2)$, a contradiction (since $x_2 = x_{F(x_2)}$). Thus $F(x_1) < F(x_2)$ and $F \in sI([a, b_o] \wedge A)$.

(iv) Let $x_o \in A_+$, $x_o \neq b_o$. We show that $F(x_o) = \inf\{F(x) : x \in A,\ x > x_o\}$. But F is strictly increasing and bounded on A (see (iii)). It follows that the above infimum is finite and belongs to $[F(a), F(b)]$. Since $F \in sI([a, b_o] \wedge A)$, $F(x_o) < F(x)$ for each $x \in A$, $x > x_o$. Hence $F(x_o) \leq \inf\{F(x) : x \in A,\ x > x_o\}$. But $F(x_o) \geq \limsup_{x \searrow x_o} F(x) \geq \limsup_{x \searrow x_o, x \in A} F(x) = \inf\{F(x) : x \in A,\ x > x_o\}$ (since F is strictly increasing and bounded on A). It follows that $F(x_o) = \inf\{F(x) : x \in A,\ x > x_o\}$, so $F(x_o) \in [F(a), F(b)]$ and $F(A \cup A_+) = [F(a), F(b)]$. Let $a < x_1 < x_2 \leq b_o$, $x_2 \in A_+$. Since $F \in sI([a.b_o] \wedge A)$, $F(x_1) < F(x)$ for each $x \in A$, $x > x_2$, hence $F(x_1) \leq \inf\{F(x) : x \in A,\ x > x_2\} = F(x_2)$. Thus $F \in I([a, b_o] \wedge (A \cup A_+))$.

Corollary 2.9.2 *Let $F : [a, b] \mapsto \mathbb{R}$, $F(a) < F(b)$, $F \in \mathcal{C}_d$.*

(i) If $a_o = \sup\{x \in [a, b] : F(x) = F(a)\}$ then $F(a_o) = F(a)$ and $F(x) > F(a_o)$ for each $x \in (a_o, b]$;

(ii) If $b_o = \inf\{x \in [a_o, b] : F(x) = F(b)\}$ then $a_o < b_o$, $F(b_o) = F(b)$ and $F(x) < F(b_o)$ for each $x \in [a_o, b_o)$;

(iii) $F([a_o, b_o]) = [F(a_o), F(b_o)]$;

(iv) There exists a set $A \subset [a_o, b_o]$ such that $a_o, b_o \in A$, $F(A) = [F(a), F(b)]$ and $F \in sI([a_o, b_o] \wedge A)$;

(v) $F \in I([a_o, b_o] \wedge (A \cup A_+))$ and $F(A \cup A_+) = [F(a), F(b)]$, where $A_+ = \{x : x$ is a right accumulation point of $A\}$;

(vi) There exists a set $B \subset [a_o, b_o]$ such that $a_o, b_o \in B$, $F(B) = [F(a), F(b)]$ and $F \in sI(B \wedge [a_o, b_o])$;

(vii) $F \in I((B \cup B_-) \wedge [a_o, b_o])$ and $F(B \cup B_-) = [F(a), F(b)]$, where $B_- = \{x : x$ is a left accumulation point of $B\}$.

Proof. (i) By Lemma 2.9.1, (i), (ii), $F(a_o) = F(a)$ and $F(x) > F(a_o)$ for each $x \in (a_o, b]$.

(ii) By (i), $F(a_o) = F(a) < F(b)$. By Lemma 2.9.1, (i), (iii), applied on the interval $[a_o, b]$, it follows that $F(b_o) = F(b)$ and $F(x) < F(b)$, $x \in [a_o, b_o)$.

(iii) By Theorem 2.4.2, (vi), $F \in \mathcal{D}_+$ on $[a_o, b_o]$, hence $F([a_o, b_o]) \supseteq [F(a), F(b)]$. By (i) and (ii), $F([a_o, b_o]) \subseteq [F(a_o), F(b_o)] = [F(a), F(b)]$, hence $F([a_o, b_o]) = [F(a_o), F(b_o)] = [F(a), F(b)]$.

(iv)-(vii) follow by Lemma 2.9.1 (applied on the interval $[a_o, b_o]$).

Lemma 2.9.2 *Let $g : [a, b] \mapsto [c, d]$, $f : [c, d] \mapsto \mathbb{R}$, $A \subseteq [a, b]$ and $g(A) = B$. If f is strictly-increasingG on B and $f \circ g$ is strictly-increasingG on A then g is strictly-increasingG on A.*

Proof. Since $f \circ g$ is *strictly − increasingG* on A, there exists a sequence of sets $\{A_n\}_n$ such that $A = \cup_{n=1}^{\infty} A_n$ and $f \circ g$ is strictly increasing on each A_n. Because f is *strictly – increasingG* on B, there exists a sequence of sets $\{B_k\}_k$ such that $B = \cup_{k=1}^{\infty} B_k$ and f is strictly increasing on each B_k. Let $A_{nk} = A_n \cap g^{-1}(B_k)$ and $B_{nk} = g(A_{nk})$. Then $B_{nk} \subset B_k$, $f \circ g$ is strictly increasing on each A_{nk} and f is strictly increasing on each B_{nk}. Suppose that g is not strictly increasing on some A_{nk}. Then there exist $x_1, x_2 \in A_{nk}$, $x_1 < x_2$ such that $g(x_1) \geq g(x_2)$. But $g(x_1), g(x_2) \in B_{nk}$, hence $f(g(x_1)) \geq f(g(x_2))$. It follows that $f \circ g$ is not strictly increasing on A_{nk}, a contradiction.

2.10 Conditions VB^*, VB^*G and $\mathcal{D}$, $\mathcal{D}_-$, $[CG]$, $[C_iG]$, lower internal, internal

Theorem 2.10.1 *Let $F : [a,b] \mapsto \mathbb{R}$ and let P be an uncountable subset of $[a,b]$. If F is VB^*G on P then F is continuous n.e. on P.*

Proof. (the proof is similar to that of [Br2], p. 196). Since $F \in VB^*G$ on P, there exists a sequence of sets Q_i, such that $P = \cup_i Q_i$ and $F \in VB^*$ on each Q_i. By Theorem 2.8.1, (i), (ii), $F \in VB^*$ on $\overline{Q}_i$. Let $E_n = \{x \in [a,b] : \mathcal{O}(F;x) \geq 1/n\}$. Then E_n is closed for each n (see Proposition 1.4.2). Suppose on the contrary that $E_n \cap (\cup \overline{Q}_i)$ is uncountable. Then there exists a natural number i_o such that $E_n \cap \overline{Q}_{i_o}$ is uncountable. Let A be a nonempty perfect subset of $E_n \cap \overline{Q}_{i_o}$. Clearly $F_{/A}$ is VB^*. Since $A \subseteq E_n$, $\mathcal{O}(F;x) \geq 1/n$ for all $x \in A$. Thus the oscillation of F on any interval determined by two bilateral accumulation points of A is at least $1/n$. Since A is perfect, we can choose as many such intervals as we like and we can make them pairwise disjoint. It follows that $F \notin VB^*$ on A, a contradiction. Thus the set of points of discontinuity of F is at most countable.

Theorem 2.10.2 *Let $F; [a,b] \mapsto \mathbb{R}$, $P \subseteq [a,b]$. If F is lower internal on $[a,b]$ and $F \in VB^*$ on P then $F \in C_i$ on $\overline{P}$.*

Proof. Let x_o be a left accumulation point of $\overline{P}$. Suppose that $d = \limsup_{x \nearrow x_o, x \in \overline{P}} F(x) > F(x_o)$. Let $F(x_o) < c < d$. It follows that there exists $\{x_i\}$, $i = \overline{1,\infty}$, $x_i \in \overline{P}$, $x_i \nearrow x_o$ such that $F(x_i) > c$. Since $F \in VB^*$ on P, by Theorem 2.8.1, (i), (ii), $F \in VB^*$ on $\overline{P}$. Let M be a constant given by this fact. It follows that $\sum_{i=1}^{\infty} \mathcal{O}(F;[x_i, x_{i+1}]) < M$. Let $\epsilon = (c - F(x_o))/2$. Then there exists a positive integer i_o such that $\sum_{i=i_o}^{\infty} \mathcal{O}(F;[x_i, x_{i+1}]) < \epsilon$. Let $x \in [x_{i_o}, x_o)$. Then there exists a positive integer $i \geq i_o$ such that $x \in [x_i, x_{i+1}]$. Since $F(x_1) > c$ and $\epsilon > \mathcal{O}(F;[x_i, x_{i+1}]) > | F(x_i) - F(x) | > F(x_i) - F(x)$, it follows that $F(x) > F(x_i) - \epsilon > c - \epsilon = (c + F(x_o))/2$. Hence $F(x) > (F(x_o) + c)/2$, for each $x \in [x_{i_o}, x_o)$. It follows that $\limsup_{x \nearrow x_o} F(x) \geq (F(x_o) + c)/2 > F(x_o)$, a contradiction (because F is *lower internal* on $[a,b]$).

Theorem 2.10.3 *Let $F : [a,b] \mapsto \mathbb{R}$, $P \subseteq [a,b]$.*

(i) $VB \subset VB^*G \subset \mathcal{B}_1$ on $[a,b]$;

(ii) $VB \cap$ (lower internal) $= VB \cap \mathcal{D}_- \subset C_i$ on $[a,b]$;

(iii) If F is internal on $[a,b]$ and $F \in VB^*$ on P then $F_{/\overline{P}} \in C$;

(iv) If $F \in \mathcal{D}$ on $[a,b]$ and $F \in VB^*$ on P then $F_{/\overline{P}} \in \mathcal{C}$;

(v) $VB^*G \cap (\text{lower internal}) = VB^*G \cap \mathcal{D}_- \subseteq [\mathcal{C}_i G]$ on $[a,b]$.

(vi) $VB^*G \cap \text{internal} = VB^*G \cap \mathcal{D} \subset [\mathcal{C}G]$ on $[a,b]$.

Proof. (i) $VB \subset VB^*G$ on $[a,b]$ follows by Proposition 2.8.1, (ii) and Section 6.12.; $VB^*G \subset \mathcal{B}_1$ on $[a,b]$ follows by Theorem 2.10.1, Remark 2.2.1, (vi) and Section 6.22.2.

(ii) $VB \cap (\text{lower internal}) = VB \cap \mathcal{D}_-$ on $[a,b]$ follows by (i) and Theorem 2.5.1, (i), (iv); $VB \cap \mathcal{D}_- \subset \mathcal{C}_i$ on $[a,b]$ follows by Theorem 2.10.2 and Section 6.12.

(iii) This follows by Theorem 2.10.2 and Proposition 2.3.1, (i).

(iv) This follows by (iii) and Theorem 2.4.2, (vii).

(v) $VB^*G \cap (\text{lower internal}) = VB^*G \cap \mathcal{D}_-$ on $[a,b]$ follows by (i) and Theorem 2.5.1, (i), (iv); $VB^*G \cap (\text{lower internal}) \subseteq [\mathcal{C}_i G]$ follows by Theorem 2.10.2.

(vi) $VB^*G \cap \text{internal} = VB^*G \cap \mathcal{D}$ on $[a,b]$ follows by (v) and Remark 2.1.1, (i); $VB^*G \cap \mathcal{D} \subset [\mathcal{C}G]$ follows by (iv) and Section 6.22.2.

Remark 2.10.1 *In Theorem 2.10.3, (iii) and (iv), condition VB^* cannot be replaced by VB (see Section 6.22.2.).*

2.11 Conditions AC, $\underline{AC}$, ACG, $\underline{ACG}$

Definition 2.11.1 *Let $F : [a,b] \mapsto \mathbb{R}$, $P, Q \subseteq [a,b]$ such that $\{(x,y) \in P \times Q : x < y\} \neq \emptyset$. F is said to be $AC(P \wedge Q)$ (resp. $\overline{AC}(P \wedge Q)$) if for every $\epsilon > 0$ there exists a $\delta > 0$ such that*

$$(1) \quad \sum_{k=1}^{n} | F(b_k) - F(a_k) | < \epsilon \ \ (\text{resp.}$$

$$(2) \quad \sum_{k=1}^{n} (F(b_k) - F(a_k)) < \epsilon),$$

whenever $\{[a_k, b_k]\}$, $k = \overline{1,n}$ is a finite set of nonoverlapping closed intervals with $a_k \in P$, $b_k \in Q$. F is said to be $\underline{AC}(P \wedge Q)$ if $-F$ is $\overline{AC}(P \wedge Q)$, i.e.,

$$(3) \quad \sum_{k=1}^{n} (F(b_k) - F(a_k)) > -\epsilon.$$

Clearly $AC(P \wedge Q) = \overline{AC}(P \wedge Q) \cap \underline{AC}(P \wedge Q)$.
For $P \subseteq Q \subseteq [a,b]$, let $AC(P;Q) = AC(P \wedge Q) \cap AC(Q \wedge P)$ and $\underline{AC}(P;Q) = \underline{AC}(P \wedge Q) \cap \underline{AC}(Q \wedge P)$.
If $P = Q$ then $AC(P;P) = AC$ on P (the class of absolutely continuous functions).
Let $\underline{AC}(P;P) = \underline{AC}$ on P.
We define conditions ACG, $[ACG]$, $\underline{ACG}$, etc. on P, using Definition 1.9.2.

Remark 2.11.1 *(i) Conditions $\overline{AC}$ and $\underline{AC}$ were defined by Ridder in [R1] (p. 236).*

(ii) In the definition of AC (resp. $\overline{AC}$, $\underline{AC}$) we can replace (1) (resp. (2), (3)) by

$$(1)' \quad \sum_{k=1}^{n} \mathcal{O}(F;[a_k,b_k]\cap P) < \epsilon \ (resp.$$

$$(2)' \quad \sum_{k=1}^{n} \mathcal{O}_+(F;[a_k,b_k]\cap P) < \epsilon,$$

$$(3)' \quad \sum_{k=1}^{n} \mathcal{O}_-(F;[a_k,b_k]\cap P) > -\epsilon)$$

(see Corollary 1.4.1, (i), (ii)).

(iii) $\{[a_k,b_k]\}_k$ may be supposed to be an infinite sequence of nonoverlapping closed intervals in Definition 2.11.1.

Theorem 2.11.1 *Let $F, F_n : [a,b] \mapsto \mathbb{R}$, $P \subseteq Q \subseteq [a,b]$, $c = \inf(P)$, $d = \sup(P)$ and let (c_k,d_k), $k \geq 1$ be the intervals contiguous to P. Let $P_+ = \{x \in \overline{P} : x$ is a right accumulation point of $P\}$, $P_- = \{x \in \overline{P} : x$ is a left accumulation point of $P\}$.*

(i) $AC(P \wedge Q)$, $AC(Q \wedge P)$ and $AC(P;Q)$ are real algebras;

(ii) ACG and $[ACG]$ are real linear spaces on P; Moreover, for bounded functions ACG and $[ACG]$ are real algebras;

(iii) $\underline{AC}(P \wedge Q)$, $\underline{AC}(Q \wedge P)$ and $\underline{AC}(P;Q)$ are real semi-linear spaces;

(iv) $I(P;Q) \subseteq \underline{AC}(P;Q)$;

(v) $AC \subset VB$ on P, hence $ACG \subset VBG$ on P;

(vi) $\overline{AC} \subseteq VB$ on $\overline{P}$, hence $[\overline{ACG}] \subseteq [VBG]$ on $\overline{P}$;

(vii) $\overline{AC} \subseteq VBG$ on P, hence $\overline{ACG} \subseteq VBG$ on P;

(viii) $F \in AC$ on $\overline{P}$ if and only if $F \in AC$ on P and $F_{/\overline{P}} \in \mathcal{C}$;

(ix) $ACG \cap \mathcal{C} \subseteq [ACG] \subseteq ACG$ on $[a,b]$;

(x) $ACG \cap [\mathcal{C}G] = [ACG]$ on $[a,b]$;

(xi) $F \in AC$ on $\overline{P}$ if and only if
$F \in AC(P\wedge(P\cup P_-)) \cap AC((P\cup P_-)\wedge P) \cap AC(P\wedge(P\cup P_+)) \cap AC((P\cup P_+)\wedge P)$;

(xii) $F \in \overline{AC}$ on $\overline{P}$ if and only if
$F \in \overline{AC}(P\wedge(P\cup P_-)) \cap \overline{AC}((P\cup P_-)\wedge P) \cap \overline{AC}(P\wedge(P\cup P_+)) \cap \overline{AC}((P\cup P_+)\wedge P)$;

(xiii) $F \in AC$ on $\overline{P}$ if and only if
$F \in AC(P\wedge(P\cup P_-)) \cap AC((P\cup P_+)\wedge P)$ and $\sum_{k\geq 1} | F(d_k) - F(c_k) | < +\infty$;

(xiv) $F \in \overline{AC}$ on $\overline{P}$ if and only if
$F \in \overline{AC}(P\wedge(P\cup P_-)) \cap \overline{AC}((P\cup P_+)\wedge P)$ and $\sum_{k\geq 1} \mathcal{O}_+(G;\{c_k,d_k\}) < +\infty$;

(xv) If $F \in \overline{AC}(P \wedge (P \cup P_-)) \cap \overline{AC}((P \cup P_+) \wedge P)$, $F \in \overline{AC}$ on each $[c_k, d_k]$ and $\sum_{k\leq 1} V(F;[c_k,d_k]) \leq +\infty$ then $F \in \overline{AC}$ on $[c,d]$;

(xvi) If $F \in \overline{AC}(P \wedge (P \cup P_-)) \cap \overline{AC}((P \cup P_+) \wedge P)$ and F is decreasing on each $[c_k, d_k]$ then $F \in \overline{AC}$ on $[c,d]$;

(xvii) $F \in \overline{AC}$ on $\overline{P}$ if and only if $F_{\overline{P}} \in \overline{AC}$ on $[c,d]$;

(xviii) $F \in AC$ on $\overline{P}$ if and only if $F_{\overline{P}} \in AC$ on $[c,d]$;

(xix) If $F_{\overline{P}} \in C_i$ and $F \in \underline{AC}$ on P then $F \in \underline{AC}(P \wedge (P \cup P_-)) \cap \underline{AC}((P \cup P_+) \wedge P)$;

(xx) If $F_{\overline{P}} \in C_i$ and $F \in \overline{AC}$ on P then $F \in \overline{AC}((P \cup P_-) \wedge P) \cap \overline{AC}(P \wedge (P \cup P_+))$;

(xxi) If $F \in \underline{AC}$ on P then $F_{/P} \in C_i^* \subset C_i$;

(xxii) If $0 \leq F_n(x) - F(x)$ on P, $F_n - F$ is increasing on P, $F_n \to F$ $[unif]$ and $F_n \in \underline{AC}$ on P then $F \in \underline{AC}$ on P.

Proof. (i) - (iv) follow by definitions.

(v) and (vi) For $\epsilon = 1$ let $\delta > 0$ be given by the fact that $F \in AC$ on P (resp. $F \in \overline{AC}$ on $\overline{P}$). Let $\{[a_i, b_i]\}$, $i = \overline{1,p}$ be a finite set of nonoverlapping closed intervals with endpoints in $\overline{P}$, numbered from left to right, such that $b_i - a_i < \delta$ for each i, and $\overline{P} \subset \cup_{i=1}^n [a_i, b_i]$. It follows that $V(F;P) \leq \sum_{i=1}^n V(F; P \cup \{a_i, b_i\}) \cap [a_i, b_i]) + \sum_{i=1}^{n-1} \mid F(a_{i+1}) - F(b_i) \mid$ (resp. $V(F;\overline{P}) \leq \sum_{i=1}^n V(F;\overline{P} \cap [a_i,b_i]) + \sum_{i=1}^{n-1} \mid F(a_{i+1} - F(a_i) \mid$, because by Theorem 2.7.1, (i), (iii), $F \in VB$ on $[a_i, b_i] \cap \overline{P}$). Hence $F \in VB$ on P (respective $\overline{P}$). $AC \subset VB$ and $ACG \subset VBG$ follow by Section 6.2.

(vii) For $\epsilon = 1$ there is a $\delta > 0$ such that $F \in \overline{VB}$ on $P \cap I$ with constant -1, whenever I is an interval with $\mid I \mid < \delta$, $P \cap I \neq \emptyset$. Then $[c,d]$ is the union of a finite set of nonoverlapping intervals $J_1, J_2, \ldots, J_p$, with $\mid J_i \mid < \delta$, $i = \overline{1,p}$. By Corollary 2.7.1, $F \in VBG$ on each $P \cap J_i$, hence $F \in VBG$ on P.

(viii), (xi) and (xii) follow by definitions; for (ix) and (x) see (viii).

(xiii) Let $F \in AC$ on $\overline{P}$. By (v), $F \in VB$ on $\overline{P}$. Let M be a constant given by this fact . Then $\sum_{k\geq 1} \mid F(d_k) - F(c_k) \mid < M < +\infty$. The other part follows by (xi). We show the converse. Let $\epsilon > 0$. For $\epsilon/2$ let $\delta > 0$ be given by the fact that $F \in AC(P \wedge (P \cup P_-)) \cap AC((P \cup P_+) \wedge P)$. Let p be a natural number such that $\sum_{k\geq p+1} \mid F(d_k) - F(c_k) \mid < \epsilon/2$. Let $\mu = \inf\{\delta, d_1 - c_1; d_2 - c_2; \ldots; d_p - c_p\}$. Let $\{[a_i, b_i]\}$, $i = \overline{1,n}$ be a finite set of nonoverlapping closed intervals with endpoints in $\overline{P}$ and let $\sum_{i=1}^n (b_i - a_i) < \mu$. We show that $F \in AC(P \wedge (P \cup P_+))$. Suppose that $a_i \in P$ and $b_i \in P \cup P_+$, $i = \overline{1,n}$. Then there exists a natural number k_i such that $b_i = d_{k_i}$. Clearly $c_{k_i} \in P \cup P_-$ and $a_i \leq c_{k_i} < d_{k_1} = b_i$. Then $\sum_{i=1}^n \mid F(b_i) - F(a_i) \mid \leq \sum_{i=1}^n \mid F(d_{k_i}) - F(c_{k_i}) \mid + \sum_{k=1}^n \mid F(c_{k_i}) - F(a_i) \mid < \epsilon$. Similarly $F \in AC((P \cup P_-) \wedge P)$. By (ii), $F \in AC$ on $\overline{P}$.

(xiv) The proof is similar to that of (xiii), using (vi) and (xii) instead of (v) and (xi).

(xv) Let $\epsilon > 0$. For $\epsilon/5$ let $\delta > 0$ be given by the fact that $F \in \overline{AC}$ on $\overline{P}$ (see (xiv)). Let p be a natural number such that $\sum_{k\geq p+1} V(F;[c_k,d_k]) < \epsilon/5$. For $\epsilon/(5p)$, let $\delta_k > 0$ be given by the fact that $F \in \overline{AC}$ on $[c_k, d_k]$, $k = \overline{1,p}$. Let $\mu = \inf\{\delta; \delta_1; \delta_2; \ldots; \delta_p; d_1 - c_1; d_2 - c_2; \ldots; d_p - c_p\}$. Let $\{[a_i, b_i]\}$, $i = \overline{1,n}$ be a finite set of nonoverlapping closed subintervals of $[c,d]$, with $\sum_{i=1}^n (b_i - c_i) < \mu$.

Let's define $A = \{i : (a_i, b_i) \cap P \neq \emptyset\}$, $B = \{i : (a_i, b_i) \subset \cup_{k \geq p+1}(c_k, d_k)\}$ and $C = \{i : (a_i, b_i) \subset \cup_{k=1}^{n}[c_k, d_k]\}$. Clearly $A \cup B \cup C = \{1, 2, ..., n\}$. For $i \in A$ let $\alpha_i = \inf(P \cap [a_i, b_i])$, $\beta_i = \sup(P \cap [a_i, b_i])$. Let $\mathcal{P} = \{(a_i, \alpha_i) : i \in A\} \cup \{(\beta_i, b_i) : i \in A\}$, $\mathcal{P}_1 = \{(u, v) : (u, v) \subset \cup_{k \geq p+1}(c_k, d_k)\}$ and let $\mathcal{P}_2 = \{(u, v) : (u, v) \subset \cup_{k=1}^{n}(c_k, d_k)\}$. Then we have $\sum_{i \in A}(F(b_i) - F(a_i)) = \sum_{i \in A}(F(\beta_i) - F(\alpha_i)) + \sum_{(u,v) \in \mathcal{P}_1}(F(v) - F(u)) + \sum_{(u,v) \in \mathcal{P}_2}(F(v) - F(u)) < (3\epsilon)/5$. Similarly we obtain that $\sum_{i \in B}(F(b_i) - F(a_i)) < \epsilon/5$ and $\sum_{i \in C}(F(b_i) - F(a_i)) < \epsilon/5$. Therefore $\sum_{i=1}^{n}(F(b_i) - F(a_i)) < \epsilon$, and $F \in \overline{AC}$ on $\overline{P}$. By (vi), $F \in VB$ on $\overline{P}$. Let M be a constant given by this fact. We have $\sum_{k \geq 1} V(F; [c_k, d_k]) = \sum_{k \geq 1}(F(c_k) - F(d_k)) < M$. By (xv), $F \in \overline{AC}$ on $[c, d]$.

(xvii) - (xxii) follow now easily.

Remark 2.11.2 *In Theorem 2.11.1, (xvii), condition "$\overline{AC}$ on $\overline{P}$" cannot be replaced by "$\overline{AC}(P \wedge (P \cup P_-)) \cap \overline{AC}((P \cup P_+) \wedge P)$" (see Section 6.10.1.).*

2.12 Conditions AC^*, $\underline{AC}^*$, AC^*G, $\underline{AC}^*G$

Definition 2.12.1 *Let $F : [a, b] \mapsto \mathbb{R}$, $P \subseteq [a, b]$, $c = \inf(P)$, $d = \sup(P)$. F is said to be AC^* (resp. $\overline{AC}^*$) on P, if for each $\epsilon > 0$ there exists a $\delta > 0$ such that $\sum_{k=1}^{n} \mathcal{O}(F; [a_k, b_k]) < \epsilon$ (resp. $\sum_{k=1}^{n} \Omega_+(F; [a_k, b_k] \wedge (P \cap [a_k, b_k])) < \epsilon$), whenever $\{[a_k, b_k]\}$, $k = \overline{1, n}$ is a finite set of nonoverlapping closed intervals, with $a_k, b_k \in P$ and $\sum_{k=1}^{n}(b_k - a_k) < \delta$. F is said to be $\underline{AC}^*$ on P is $-F$ is $\overline{AC}^*$ on P, i.e., $\sum_{k=1}^{n} \Omega_-(F; [a_k, b_k] \wedge (P \cap [a_k, b_k]) > -\epsilon$. If in addition F is bounded on $[c, d]$ then we obtain the conditions bAC^*, $b\overline{AC}^*$, $b\underline{AC}^*$ on P.*

We define the classes AC^*G, $\underline{AC}^*G$, $[AC^*G]$, etc. on P, using Definition 1.9.2.

Remark 2.12.1 *(i) In [S3] (p. 231), Saks denoted bAC^* by AC^*.*

(ii) Condition $\underline{AC}^*$ was introduced by Ridder in [R1].

(iii) $\{[a_k, b_k]\}_k$ may be supposed to be an infinite sequence of nonoverlapping closed intervals, in Definition 2.12.1.

Lemma 2.12.1 *Let $F : [a, b] \mapsto \mathbb{R}$, $P \subseteq [a, b]$. If $F \in AC^*$ on P then there exists $\{I_k\}$, $k = \overline{1, p}$ a finite set of nonoverlapping closed intervals with endpoints in $\overline{P}$ such that $\overline{P} \subseteq \cup_{k=1}^{p} I_k$ and $F \in bAC^*$ on each $P \cap I_k$.*

Proof. For $\epsilon = 1$ let δ be given by the fact that $F \in AC^*$ on P. Let $\{I_k\}$, $k = \overline{1, p}$ be a finite set of nonoverlapping closed intervals with endpoints in $\overline{P}$ such that $\overline{P} \subseteq \cup_{k=1}^{p} I_k$ and $| I_k | < \delta$, $k = \overline{1, p}$. Then $F \in AC^* \cap VB^*$ on each $P \cap I_k$. By Corollary 2.8.1, (ii), F is bounded on each I_k.

Lemma 2.12.2 *Let $F : [a, b] \mapsto \mathbb{R}$, $P \subseteq [a, b]$ and let I be the smallest interval containing P. The following assertions are equivalent:*

(i) $F \in \overline{AC}^*$ on P;

(ii) $F \in \overline{AC}(P; I)$ on P;

(iii) (Ridder) For every $\epsilon > 0$ there is a $\delta > 0$ such that $\sum_{k=1}^{p}(F(x_k) - F(a_k)) < \epsilon$ and $\sum_{k=1}^{p}(F(b_k) - F(x_k)) < \epsilon$, whenever $\{[a_k, b_k]\}$, $k = \overline{1,p}$ is a finite set of nonoverlapping closed intervals, such that $a_k, b_k \in P$, $\sum_{k=1}^{p}(b_k - a_k) < \delta$ and $x_k \in [a_k, b_k]$, $k = \overline{1,p}$.

Proof. The proof is similar to that of Proposition 2.8.2.

Lemma 2.12.3 *Let $F : [a,b] \mapsto \mathbb{R}$, $P \subseteq [a,b]$ and let I be the smallest interval containing P. The following assertions are equivalent:*

(i) $F \in AC^*$ on P;

(ii) $F \in \overline{AC^*} \cap \underline{AC^*}$ on P;

(iii) $F \in AC(P; I)$.

Proof. *(i)* ⇔ *(ii)* See Proposition 1.4.1, (i) and Remark 1.4.1, (v).
(i) ⇔ *(iii)* See Lemma 2.12.2.

Proposition 2.12.1 *Let $F : [a,b] \mapsto \mathbb{R}$, $P \subseteq [a,b]$, $c = \inf(P)$, $d = \sup(P)$.*

(i) $AC^* \subseteq AC$ on P, hence $AC^*G \subseteq ACG$ on P;

(ii) $AC^* \subseteq \overline{AC^*} \subseteq \overline{AC}$ on P, hence $AC^*G \subseteq \overline{AC^*}G \subseteq \overline{AC}G$ on P;

(iii) $AC^* = AC$ and $\overline{AC^*} = \overline{AC}$ on $[a,b]$;

(iv) $AC^* \subseteq \overline{AC^*} \subseteq VB^*G$ on P, hence $AC^*G \subseteq \overline{AC^*}G \subseteq VB^*G$ on P;

(v) $b\overline{AC^*} = VB^* \cap \overline{AC^*}$ on P, hence $bAC^* = VB^* \cap AC^*$ on P;

(vi) $b\overline{AC^*}G = \overline{AC^*}G$ on P, hence $bAC^*G = AC^*G$ on P;

(vii) AC^* is a real algebra on P;

(viii) AC^*G is a real linear space on P. Moreover, for bounded functions on $[c,d]$, AC^*G is a real algebra on P;

(ix) bAC^* is a real linear space on P;

(x) $\overline{AC^*}$ and $\overline{AC^*}G$ are real semi- linear spaces on P;

(xi) $[AC^*G] = [bAC^*G]$ on $\overline{P}$ and $[\overline{AC^*}G] = [b\overline{AC^*}G]$ on $\overline{P}$.

Proof. (iv) For $\epsilon = 1$ let $\delta > 0$ be given by the fact that $F \in \overline{AC^*}$ on P. Let $\{I_k\}$, $k = \overline{1,p}$ be a finite set of nonoverlapping closed intervals with endpoints in $\overline{P}$, such that $\overline{P} \subseteq \cup_{k=1}^{p} I_k$ and $| I_k |< \delta$, $k = \overline{1,p}$. Then $F \in \overline{VB^*}$ on each $P \cap I_k$. By Corollary 2.8.1, (i), $F \in VB^*G$ on each $P \cap I_k$, hence $F \in VB^*G$ on P. The other parts are evident.

(v) $VB^* \cap \overline{AC^*} \subseteq b\overline{AC^*}$ on P follows by Corollary 2.8.1, (ii). Let $F \in b\overline{AC^*}$on P. Similarly to (iv), $F \in \overline{VB^*}$ on each $P \cap I_k$. By Theorem 2.8.1, (i), (iv), $F \in VB^*$on each $P \cap I_k$. Let $M > 0$ such that $| F(x) |\leq M$ on $[c,d]$. Then $V^*(F;P) \leq \sum_{k=1}^{p} V^*(F; P \cap I_k) + 2Mp < +\infty$, hence $F \in VB^*$ on P. It follows that $b\overline{AC^*} \subseteq VB^* \cap \overline{AC^*}$ on P.

(xi) We have $[AC^*G] \subseteq AC^*G \subseteq VB^*G = [VB^*G]$ on $\overline{P}$ and $bAC^* = VB^* \cap AC^*$. It follows that $[AC^*G] = [bAC^*G]$ on $\overline{P}$.
We have $[\overline{AC}^*G] \subseteq \overline{AC}^*G \subseteq VB^*G = [VB^*G]$ on $\overline{P}$ and $b\overline{AC}^* = VB^* \cap \overline{AC}^*$. It follows that $[\overline{AC}^*G] = [b\overline{AC}^*G]$ on $\overline{P}$.

The other items are evident.

Theorem 2.12.1 *Let $F : [a,b] \mapsto \mathbb{R}$, $P \subseteq [a,b]$, $c = \inf(P)$, $d = \sup(P)$ and let $\{(c_k,d_k)\}_k$ be the intervals contiguous to $\overline{P}$. Let $P_+ = \{x \in \overline{P} : x$ is a right accumulation point of $P\}$ and $P_- = \{x \in \overline{P} : x$ is a left accumulation point of $P\}$. The following assertions are equivalent:*

(i) $F \in bAC^*$ *on* $\overline{P}$;

(ii) $F \in AC \cap VB^*$ *on* $\overline{P}$;

(iii) $F \in AC$ *on* $\overline{P}$ *and* $\sum_{k\geq 1} \Omega(F;[c_k,d_k] \wedge \{c_k,d_k\}) < +\infty$;

(iv) $F \in AC(P \wedge (P \cup P_-)) \cap AC((P \cup P_+) \wedge P)$ *and* $\sum_{k\geq 1} \Omega(F;[c_k,d_k] \wedge \{c_k,d_k\}) < +\infty$;

(v) $F \in AC(P \wedge (P \cup P_-)) \cap AC((P \cup P_+) \wedge P)$ *and* $F \in VB^*$ *on* P;

(vi) $F \in bAC^*$ *on* P *and* $F_{/\overline{P}} \in \mathcal{C}$.

Proof. *(i)* ⇒ *(ii)* By Proposition 2.12.1, (i), (v), $bAC^* = AC^* \cap VB^* \subseteq AC \cap VB^*$ on $\overline{P}$.

(ii) ⇒ *(iii)* Since $F \in VB^*$ on P, it follows that $\sum_{k\geq 1} \mathcal{O}(F;[c_k,d_k]) < +\infty$. Now the proof follows by Proposition 1.4.1, (i).

(iii) ⇒ *(iv)* See Theorem 2.11.1,(xi).

(iv) ⇒ *(i)* Using Proposition 1.4.1, (xiii) we can see easily that $F \in AC^*$on $\overline{P}$. By Proposition 2.12.1, (i) and Theorem 2.11.1,(v), $F \in VB$ on $\overline{P}$. By Proposition 1.4.1, (i) and Theorem 2.8.1, (ii), (vii), it follows that $F \in VB^*$ on $\overline{P}$. By Corollary 2.8.1,(ii), F is bounded on $[c,d]$.

(ii) ⇒ *(v)* See Theorem 2.11.1, (xi).

(v) ⇒ *(iv)* The proof is similar to that of *(ii)* ⇒ *(iii)*.

(i) ⇒ *(vi)* This follows by definitions.

(vi) ⇒ *(ii)* By Proposition 2.12.1, (v), $F \in VB^* \cap AC^*$ on P. By Theorem 2.8.1, (i), (ii), $F \in VB^*$ on $\overline{P}$. Since $AC^* \subseteq AC$ on P and $F_{/\overline{P}} \in \mathcal{C}$, by Theorem 2.11.1, (viii), $F \in AC$ on $\overline{P}$.

Remark 2.12.2 *By Theorem 2.12.1, (i), (ii), $bAC^* \subseteq VB^*$. But $AC^* \not\subseteq VB^*$ (see Section 6.23.).*

Theorem 2.12.2 *Let $F : [a,b] \mapsto \mathbb{R}$, $P \subseteq [a,b]$, $c = \inf(P)$, $d = \sup(P)$ and let $\{(c_k,d_k)\}_k$ be the intervals contiguous to $\overline{P}$. Let $P_+ = \{x \in \overline{P} : x$ is a right accumulation point of $P\}$ and $P_- = \{x \in \overline{P} : x$ is a left accumulation point of $P\}$. The following assertions are equivalent:*

(i) $F \in b\overline{AC}^*$ *on* $\overline{P}$;

(ii) $F \in \overline{AC} \cap VB^*$ *on* $\overline{P}$;

(iii) $F \in \overline{AC}$ *on* $\overline{P}$ *and* $\sum_{k\geq 1} \Omega_+(F;[c_k,d_k] \wedge \{c_k,d_k\}) < +\infty$;

(iv) $F \in \overline{AC}(P \wedge (P \cup P_-)) \cap \overline{AC}((P \cup P_+) \wedge P)$ *and* $F \in VB^*$ *on* P;

(v) $F \in \overline{AC}(P \wedge (P \cup P_-)) \cap \overline{AC}((P \cup P_+) \wedge P)$ *and* $\sum_{k\geq 1} \Omega_+(F;[c_k,d_k] \wedge \{c_k,d_k\}) < +\infty$;

(vi) $F \in b\overline{AC}^*$ *on* P *and* $F \in C_i^*$ *at each point of* $\overline{P}$;

(vii) $F \in b\overline{AC}^*$ *on* P *and* $F \in C_i$ *at each point of* $\overline{P}$.

Proof. The proof is similar to that of Theorem 2.12.1.

Theorem 2.12.3 *Let* $F : [a,b] \mapsto \mathbb{R}$, $P \subseteq [a,b]$. *The following assertions are equivalent:*

(i) $F \in AC^*$ *on* $\overline{P}$ *and* $F \in C$ *at each point* $x \in \overline{P}$;

(ii) *For every* $\epsilon > 0$ *there is a* $\delta > 0$ *such that* $\sum_{k=1}^n O(F;[a_k,b_k]) < \epsilon$, *whenever* $\{[a_k,b_k]\}$, $k = \overline{1,n}$ *is a finite set of nonoverlapping closed intervals with* $[a_k,b_k] \cap P \neq \emptyset$ *and* $\sum_{k=1}^n (b_k - a_k) < \delta$;

(iii) *(Seng,[Se])* $F \in AC(P;[a,b])$.

Proof.$(i) \Rightarrow (ii)$ Let $c = \inf(P)$, $d = \sup(P)$ and let $\{(c_k,d_k)\}_k$ be the intervals contiguous to $\overline{P}$. For $\epsilon > 0$ let δ be given by the fact that $F \in AC^*$ on $\overline{P}$. Let N be a positive integer such that $\sum_{k\geq N+1}(d_k - c_k) < \delta$. Then

$$(1) \qquad \sum_{k\geq N+1} O(F;[c_k,d_k]) < \epsilon.$$

Since $F \in C$ at each $x \in \overline{P}$, there exists $\mu > 0$ such that $\mu < (d_k - c_k)/2$, $k = \overline{1,N}$ and

$$(2) \qquad \sum_{k=1}^{N} (O(F;[c_k,c_k+\mu]) + O(F;[d_k-\mu,d_k])) < \epsilon.$$

Let $\delta_1 = \min\{\delta,\mu\}$ and let $\{[a_k,b_k]\}$, $k = \overline{1,n}$ be a finite set of nonoverlapping closed intervals, such that $[a_k,b_k] \cap P \neq \emptyset$ and $\sum_{k=1}^n (b_k - a_k) < \delta_1$. Let $\alpha_k = \inf(P \cap [a_k,b_k])$ and $\beta_k = \sup(P \cap [a_k,b_k])$. By (1) and (2), since $F \in AC^*$ on $\overline{P}$, it follows that $\sum_{k=1}^n O(F;[a_k,b_k]) \leq \sum_{k=1}^n (O(F;[a_k,\alpha_k]) + O(F;[\alpha_k,\beta_k]) + O(F;[\beta_k,b_k])) < 3\epsilon$.

$(ii) \Rightarrow (i)$ For $\epsilon > 0$ let δ be given by (ii). Let $x \in \overline{P}$ and let $x_1 \in P$ such that $| x_1 - x_o | < \delta/8$. Then $[x_o - \delta/8, x_o + \delta/8] \subset [x_1 - \delta/4, x_1 + \delta/4]$. It follows that $O(F;[x_o - \delta/8, x_o + \delta/8]) \leq O(F;[x_1 - \delta/4, x_1 + \delta/4]) < \epsilon$, hence $F \in C$ at x_o. That $F \in AC^*$ on P follows by definitions. Let $\{[a_k,b_k]\}$, $k = \overline{1,n}$ be a finite set of nonoverlapping closed intervals with endpoints in $\overline{P}$ and $\sum_{k=1}^n (b_k - a_k) < \delta/2$. We consider only the case $[a_k,b_k] \cap P = \emptyset$, $k = \overline{1,n}$ (the other cases follow by (ii)). Then there exists $\{J_k\}$, $k = \overline{1,n}$ be a finite set of nonoverlapping closed intervals such that $[a_k,b_k] \subseteq J_k$, $J_k \cap P \neq \emptyset$ and $\sum_{k=1}^n | J_k | < \delta$. It follows that $\sum_{k=1}^n O(F;[a_k,b_k]) \leq \sum_{k=1}^n O(F;J_k) < \epsilon$. Thus $F \in AC^*$ on $\overline{P}$.

$(ii) \Rightarrow (iii)$ This is evident.

$(iii) \Rightarrow (ii)$ Let ϵ, δ and $\{[a_k,b_k]\}$ be as in (ii). Let $t_k \in [a_k,b_k] \cap P$. then $\sum_{k=1}^n \Omega(F;[a_k,b_k] \wedge \{t_k\}) = \sum_{k=1}^n \max\{| F(y_k) - F(t_k) |;| F(t_k) - F(x_k) |: a_k \leq x_k \leq t_k \leq y_k < b_k\} < \epsilon$. Now by Proposition 1.4.1, (i) it follows that $\sum_{k=1}^n O(F;[a_k,b_k]) \leq 2\sum_{k=1}^n \Omega(F;[a_k,b_k] \wedge \{t_k\}) < 2\epsilon$.

Lemma 2.12.4 *Let $F : [a,b] \mapsto \mathbb{R}$ and let P be a closed subset of $[a,b]$. The following assertions are equivalent:*

(i) $F \in [AC^*G]$ *(resp.* $[\overline{AC}^*G]$*) on* P*;*

(ii) For each perfect subset $S \subset P$ *there exists a portion* $S \cap (c,d)$ *such that* $F \in bAC^*$ *(resp.* $b\overline{AC}^*$*) on* $S \cap [c,d]$

Proof. See Theorem 1.9.1 and Proposition 2.12.1, (xi).

Lemma 2.12.5 *Let $F : [a,b] \mapsto \mathbb{R}$ and let P be a closed subset of $[a,b]$. The following assertions are equivalent:*

(i) $F \in AC^*G \cap \mathcal{C}$ *(resp.* $\overline{AC}^*G \cap \mathcal{C}_i$*) on* P*;*

(ii) $F \in \mathcal{C}$ *(resp.* $\mathcal{C}_i$*) on* P *and for each perfect subset* $S \subset P$*, there exists a portion* $S \cap [c,d]$ *such that* $F \in bAC^*$ *(resp.* $b\overline{AC}^*$*) on* $S \cap [c,d]$*;*

(iii) $F \in \mathcal{C}$ *(resp.* $\mathcal{C}_i$*) on* P *and* $F \in AC^*G$ *(resp.* $\overline{AC}^*G$*) on* Z*, whenever* $Z \subset P$ *and* $|Z| = 0$*.*

Proof. See Theorem 1.9.1 and Theorem 2.12.1, (i), (vi) (resp. Theorem 2.12.2, (i),(vii)).

2.13 Conditions L, $\underline{L}$, LG, $\underline{L}G$

Definition 2.13.1 *Let* $F : [a,b] \mapsto \mathbb{R}$ *and let* $P, Q \subseteq [a,b]$ *such that* $\{(x,y) \in P \times Q : x < y\} \neq \emptyset$*. F is said to be* $L(P \wedge Q)$ *(resp.* $\overline{L}(P \wedge Q)$*) if there exists* $M \in (0,+\infty)$ *such that* $|F(y) - F(x)| \leq M(y-x)$ *(resp.* $F(y) - F(x) \leq M(y-x)$*), whenever* $x < y$, $x \in P$, $y \in Q$*. F is said to be* $\underline{L}(P \wedge Q)$ *if* $-F \in \overline{L}(P \wedge Q)$*, i.e.,* $F(y) - F(x) \geq -M(y-x)$*. Clearly* $L(P \wedge Q) = \overline{L}(P \wedge Q) \cap \underline{L}(P \wedge Q)$*.*
For $P \subseteq Q \subseteq [a,b]$ *let* $L(P;Q) = L(P \wedge Q) \cap L(Q \wedge P)$ *and* $\underline{L}(P;Q) = \underline{L}(P \wedge Q) \cap \underline{L}(Q \wedge P)$*. If* $P = Q$ *then* $L(P;P) = L$ *on* P *(the class of Lipschitz functions). Let* $\underline{L}(P;P) = \underline{L}$ *on* P*.*
We define the classes LG, $\underline{L}G$, $[LG]$*, etc. on* P*, using Definition 1.9.2.*

Remark 2.13.1 *In [Kr1], Krzyzewski introduced condition* L^* *on* P*, and we have* $L^* = L(P;[a,b])$*.*

Proposition 2.13.1 *Let* $F : [a,b] \mapsto \mathbb{R}$, $P \subseteq Q \subseteq [a,b]$, $c = \inf(P)$ *and* $d = \sup(P)$*. Let* $P_+ = \{x \in \overline{P} : x$ *is a right accumulation point of* $P\}$ *and* $P_- = \{x \in \overline{P} : x$ *is a left accumulation point of* $P\}$*. We have:*

(i) $\underline{L}(P;Q) \subseteq \underline{AC}(P;Q)$*, hence* $L(P;Q) \subseteq AC(P;Q)$*;*

(ii) $\underline{L} \subseteq \underline{AC}$*, hence* $L \subset AC$ *on* P*;*

(iii) $F \in \underline{L}$ *on* $\overline{P}$ *if and only if*
$F \in \underline{L}(P \wedge (P \cup P_-)) \cap \underline{L}((P \cup P_-) \wedge P) \cap \underline{L}(P \wedge (P \cup P_+)) \cap \underline{L}((P \cup P_+) \wedge P)$*;*

(iv) $F \in L$ *on* $\overline{P}$ *if and only if*
$F \in L(P \wedge (P \cup P_-)) \cap L((P \cup P_-) \wedge P) \cap L(P \wedge (P \cup P_+)) \cap L((P \cup P_+) \wedge P)$*;*

(v) *$\underline{L}(P;[c,d]) \subseteq \underline{AC}^* \cap VB^* = b\underline{AC}^*$ on P, hence $L(P;[c,d]) \subsetneq AC^* \cap VB^* = bAC^*$;*

(vi) *If $F \in \overline{L}(P;[c,d])$ on P with a constant M, and F is lower internal on $[a,b]$ then $F \in \overline{L}(\overline{P};[c,d])$ with the same constant M.*

Proof. (i)-(iv) The proofs follow by definitions and Section 6.29. (the functions F_{PC} and G_{PC}).

(v) $\underline{L}(P;[c,d]) \subseteq \underline{AC}^*$ on P follows by definitions. Let M be a constant given by the fact that $F \in \underline{L}(P;[c,d])$ on P. Then $G(x) = F(x) - Mx$ is *increasing** on P. By Proposition 2.9.1, (iii), $G \in VB^*$ on P, hence $F \in VB^*$ on P.

(vi) Let $G(x) = F(x) - Mx$. By (v), $F \in VB^*$ on P, hence $F \in \mathcal{C}_i$ on $\overline{P}$ (see Theorem 2.10.2). It follows that $G \in \mathcal{C}_i$ on $\overline{P}$ and G is *decreasing** on P. By Proposition 2.9.1, (iv), G is *decreasing** on $\overline{P}$, hence $F \in \overline{L}(\overline{P};[c,d])$ with constant M.

Lemma 2.13.1 (Nina Bary) *Let $F : [a,b] \mapsto \mathbb{R}$, $P \subseteq [a,b]$, $M \in (0,+\infty)$. If $F \in \mathcal{D}$ on $[a,b]$, $| F'(x) |< M$, $x \in P$ and $| F([a,b] \setminus P) |= 0$ then $F \in L$ with constant M on $[a,b]$.*

Proof. Let $x_1, x_2 \in [a,b]$, $x_1 < x_2$. Then $F([x_1,x_2]) = F([x_1,x_2] \cap P) \cup F([x_1,x_2] \setminus P)$ and $| F([x_1,x_2]) | = | F([x_1,x_2] \cap P) |$. By Theorem 1.10.4, $| F([x_1,x_2] \cap P) | \leq M | [x_1,x_2] \cap P | \leq M(x_2 - x_1)$. So $| F([x_1,x_2]) | \leq M(x_2 - x_1)$. Since $F \in \mathcal{D}$, $F([x_1,x_2])$ is an interval, hence $| F(x_1) - F(x_2) | \leq M(x_2 - x_1)$, and this completes the proof.

Theorem 2.13.1 *Let $F : [a,b] \mapsto \mathbb{R}$. The following assertions are equivalent:*

(i) *$F \in \underline{L}$ on $[a,b]$;*

(ii) *$\underline{D}^{\#}F(x) > -\infty$ on $[a,b]$;*

(iii) *$\underline{F}'(x) > -\infty$ and there exists a real number M such that $\underline{F}'(x) > M$ n.e. on $[a,b]$.*

Proof. *(i)* ⇒ *(ii)* This is obvious.

(ii) ⇒ *(iii)* By Lemma 1.15.1, $\underline{D}^{\#}F(x)$ is *lower semicontinuous* on $[a,b]$. By Proposition 1.15.5 and Remark 1.12.1, we have (iii).

(iii) ⇒ *(i)* Let $G(x) = F(x) - Mx$ on $[a,b]$. Then $\underline{G}'(x) > -\infty$ and $\underline{G}'(x) > 0$ *n.e.* on $[a,b]$. It follows that $E = \{x : \underline{G}'(x) \leq 0\}$ is countable. Suppose on the contrary that G is not increasing on $[a,b]$. Then there exist $x_1, x_2 \in [a,b]$, $x_1 < x_2$ such that $G(x_1) > G(x_2)$. Let $y_o \in (G(x_2), G(x_1)) \setminus G(E)$ and let $A = \{x \in [x_1,x_2] : G(x) \leq y_o\}$. Then $A \neq \emptyset$. Let $x_o = \sup(A)$. If $x_o \in A$ then $G(x_o) \geq y_o$. If $x_o \notin A$ then $G(x_o) \geq y_o$ (because $\underline{G}'(x_o) > -\infty$). Hence $x_o \in (x_1,x_2)$ and $G(x) < y_o$ on (x_o,x_2). Since $\underline{G}'(x_o) > -\infty$ it follows that $G(x_o) = y_o$, hence $\overline{D}^+G(x_o) \leq 0$, a contradiction (because $x_o \notin E$). So G is increasing, hence $F \in \underline{L}$ on $[a,b]$.

Corollary 2.13.1 *Let $F : [a,b] \mapsto \mathbb{R}$. The following assertions are equivalent:*

(i) *$F \in L$ on $[a,b]$;*

(ii) *$-\infty < \underline{D}^{\#}F(x) \leq \overline{D}^{\#}F(x) < +\infty$ on $[a,b]$;*

(iii) $-\infty < \underline{F}'(x) \leq \overline{F}'(x) < +\infty$ *on* $[a,b]$ *and there exists* $M \in (0,+\infty)$ *such that* $\max\{|\underline{F}'(x)|; \overline{F}'(x)\} < M$ *n.e. on* $[a,b]$.

Theorem 2.13.2 *Let* $F : [a,b] \mapsto \mathbb{R}$, $P \subseteq [a,b]$ *and let* M *be a real number. The following assertions are equivalent:*

(i) $D^+F(x) < +\infty$ *(resp.* $\overline{D}F(x) < M$*) on* P*;*

(ii) There exist a sequence of sets $\{P_n\}_n$*, with* $P = \cup_n P_n$*, and a sequence* $\{J_n\}_n$ *of open sets such that* $P_n \subset J_n$ *and* $F \in \overline{L}(P_n \wedge J_n)$ *(resp.* $F \in \overline{L}(P_n; J_n)$*;* $F \in \overline{L}(P_n; J_n)$ *with a constant less that* M*).*

Proof. *(i)* $\Rightarrow$ *(ii)* Let $E_n = \{x \in P : (F(t) - F(x))/(t-x) \leq n,\ o < t - x < 1/n\}$, $n = \overline{1,\infty}$. Let $P_n^i = [i/(2n),(i+1)/(2n)] \cap E_n$. Then $P = \cup_{n,i} P_n^i$. Let J_n^i be an open interval such that $P_n^i \subset J_n^i$ and $|J_n^i| < 3/(4n)$. Let $x \in P_n^i$, $y \in J_n^i$, $x < y$. Then $F(y) - F(x) < n(y-x)$, hence $\overline{L}(P_n^i \wedge J_n^i)$. The other parts follow similarly.

(ii) $\Rightarrow$ *(i)* This is evident.

Corollary 2.13.2 (Denjoy) *([S3], pp. 237, 234). Let* $F : [a,b] \mapsto \mathbb{R}$, $P \subseteq [a,b]$*. If* $D^+F(x) < +\infty$ *(resp.* $\overline{D}F(x) < +\infty$*) on* P *then* $F \in VBG$ *(resp.* VB^*G*) on* P.

Proof. By Theorem 2.13.2, $F \in \overline{L}(P_n \wedge J_n) \subseteq \overline{AC} \subseteq VBG$ on P_n (resp. $F \in \overline{L}(P_n; J_n) \subseteq \overline{L}(P_n \wedge [\inf(P_n), \sup(P_n)]) \subseteq VB^*$) on P_n, hence $F \in VBG$ (resp. VB^*G) on P.

Corollary 2.13.3 (Denjoy) *([S3], p.235). Let* $F : [a,b] \mapsto \mathbb{R}$, $P \subseteq [a,b]$*. If* $-\infty < D_+F(x) \leq D^+F(x) < +\infty$ *on* P *then* $F \in AC^*G$ *on* P

Proof. By Theorem 2.13.2, $F \in L(P_n \wedge J_n) \subseteq AC(P_n \wedge [\inf(P_n), \sup(P_n)]) \subseteq AC^*$ on P_n, hence $F \in AC^*G$ on P.

2.14 Summability and conditions VB and AC

Definition 2.14.1 *([N]). A function* $f : E \mapsto \overline{\mathbb{R}}$*, measurable on the real measurable set* E *is said to be summable if* f *is Lebesgue integrable on* E*.*

Definition 2.14.2 *([N]). A function* $f : [a,b] \mapsto \mathbb{R}$ *is said to be singular, if* $f \in VB$ *on* $[a,b]$ *and* $f'(x) = 0$ *a.e..*

Theorem 2.14.1 *([N], I, pp.205, 212). Let* $F : [a,b] \mapsto \mathbb{R}$ *be an increasing function. Then we have:*

(i) F *is derivable a.e.;*

(ii) F' *is summable and* $(\mathcal{L})\int_a^b F'(t)dt \leq F(b) - F(a)$.

Theorem 2.14.2 *([N], I, p. 252-255, II, p. 166). Let* $f : [a,b] \mapsto \overline{\mathbb{R}}$ *be a summable function. Then we have:*

(i) $F(x) = (\mathcal{L})\int_a^x f(t)dt$ *is AC on* $[a,b]$ *and* $F'(x) = f(x)$ *a.e.;*

(ii) For every $\epsilon > 0$ *there exists a function* $u : [a,b] \mapsto (-\infty + \infty]$ *such that:*

> *a)* u *is lower semicontinuous;*
>
> *b)* $u(x) \geq f(x)$*;*
>
> *c)* u *is summable and* $(\mathcal{L}) \int_a^b u(t)dt \leq \epsilon + (\mathcal{L}) \int_a^b f(t)dt$.

Theorem 2.14.3 *([N], I, p. 219). Let* $F : [a,b] \mapsto \mathbb{R}$, $F \in VB$*. Then we have:*

(i) F *is derivable a.e.;*

(ii) F' *is summable;*

(iii) There exists $F_1, F_2 : [a,b] \mapsto \mathbb{R}$ *such that* $F_1 \in AC$, F_2 *is singular,* $F_2(a) = 0$ *and* $F = F_1 + F_2$ *(this representation is unique).*

Theorem 2.14.4 *([S3], p. 227). Let* $F : [a,b] \mapsto \mathbb{R}$*, and let* P *be a measurable subset of* $[a,b]$*. If* F *is derivable on* P *then* $| F(P) | \leq (\mathcal{L}) \int_P | F'(t) | dt$.

Theorem 2.14.5 *Let* $F : [a,b] \mapsto \mathbb{R}$, $F \in \underline{AC}$*. Then we have:*

(i) F *is derivable a.e.;*

(ii) (<u>Ridder</u>) If $F'(x) \geq 0$ *a.e. then* F *is increasing on* $[a,b]$*;*

(iii) F' *is summable and* $(\mathcal{L}) \int_a^x F'(t)dt \leq F(x) - F(a)$*;*

(iv) There exist $F_1, F_2 : [a,b] \mapsto \mathbb{R}$ *such that* $F_1 \in AC$, F_2 *is singular and increasing,* $F_2(a) = 0$ *and* $F = F_1 + F_2$ *(this representation is unique).*

Proof. (i) By Theorem 2.11.1, $F \in VB$ on $[a,b]$. By Theorem 2.14.3, F is derivable *a.e.*.

(ii) It is sufficient to show that $F(b) > F(a)$. Let $A = \{x : F'(x) \geq 0\}$. By (i), A is measurable and $| A |= b - a$. Let $\epsilon > 0$. If $x \in A$ then there exists $\delta_x > 0$ such that for each $h \in (0, \delta_x)$ we have $(F(x+h) - F(x))/h > -\epsilon$. The closed intervals $[x, x+h]$ cover the set A in the Vitali sense. By Theorem 1.8.1, we can choose a finite set of pairwise disjoint closed intervals $[x_i, x_i + h_i]$, $i = \overline{1,n}$ such that $b - a - \sum_{i=1}^n h_i < \delta$, where δ depends on ϵ, and is given by the fact that $F \in \underline{AC}$ on $[a,b]$. It follows that $F(x_1) - F(a) + \sum_{k=1}^{n-1} (F(x_{k+1}) - F(x_k + h_k)) + F(b) - F(x_n + h_n) > -\epsilon$. Hence $F(b) - F(a) > -\epsilon(b-a)$. Since ϵ is arbitrarily chosen, $F(b) \geq F(a)$.

(iii) By Theorem 2.11.1, $F \in VB$ on $[a,b]$. By Theorem 2.14.3, (ii), F is summable. Let $F_1(x) = (\mathcal{L}) \int_a^x F'(t)dt$ and $F_2 = F - F_1$. By Theorem 2.14.2, (i), $F_1 \in AC \subseteq \underline{AC}$ on $[a,b]$. But $F_2'(x) = F'(x) - F_1'(x)$ *a.e.*. By (ii), F_2 is increasing on $[a,b]$. It follows that $(\mathcal{L}) \int_a^x F'(t)dt \leq F(x) - F(a)$.

(iv) By the proof of (iii) it remains to show that the representation $F = F_1 + F_2$ is unique. Suppose that there exists another representation $F = G_1 + G_2$ which satisfies (iv). Then $F_1 - G_1 = G_2 - F_2$. But $F_1 - G_1 \in AC$ (see Theorem 2.11.1, (i)) and $(F_1 - G_1)' = 0$ *a.e.*. By (ii) it follows that $F_1 - G_1 = 0$. Hence $F_1 = G_1$ and $F_2 = G_2$ on $[a,b]$.

Corollary 2.14.1 *([N], I, p. 246). Let* $F : [a,b] \mapsto \mathbb{R}$, $F \in AC$*. If* $F'(x) = 0$ *a.e. then* F *is constant.*

Corollary 2.14.2 *([N], I, p. 255). Let $F : [a,b] \mapsto \mathbb{R}$, $F \in AC$. Then we have:*

(i) F is derivable a.e;

(ii) F is summable and $(\mathcal{L}) \int_a^x F'(t)dt = F(x) - F(a)$.

Corollary 2.14.3 *Let $F : [a,b] \mapsto \mathbb{R}$ and let P be a closed subset of $[a,b]$, $c = \inf(P)$, $d = \sup(P)$.*

(i) $F \in \underline{AC}$ on P if and only if $F = F_1 + F_2$ on $[c,d]$, $F_1 \in AC$ on P, F_2 is increasing and singular on $[c,d]$;

(ii) $F \in b\underline{AC}^*$ on P if and only if $F = F_1 + F_2$ on $[c,d]$, $F_1 \in bAC^*$ on P, F_2 is increasing and singular on $[c,d]$.

Proof. (i) Let $F \in \underline{AC}$ on P. By Theorem 2.11.1, (xvii), $F_P \in \underline{AC}$ on $[c,d]$. By Theorem 2.14.5, (iv), $F_P = f_1 + f_2$, $f_1 \in AC$ on $[c,d]$, f_2 is increasing and singular on $[c,d]$. Let $F_2 = f_2$ on $[c,d]$, and let $F_1 = F - F_2$ on $[c,d]$. Since $F_1 = f_1$ on P it follows that $F_1 \in AC$ on P. The converse is evident.

(ii) By Theorem 2.12.2, i), (ii), $b\underline{AC}^* = AC \cap VB^*$ on P. By (i), $F = F_1 + F_2$ on $[c,d]$, $F_1 \in AC$ on P, F_2 is increasing and singular on $[c,d]$. Since $F \in VB^*$ on $[c,d]$, by Proposition 2.8.1, (v), $F_1 = F - F_2 \in VB^*$ on P. By Theorem 2.12.1, $F_1 \in bAC^*$ on P. The converse is evident.

Theorem 2.14.6 (Zahorski; Tolstoff) *([Br2], p. 124, [Z2], [To1]). Let $Z \subset [a,b]$ be a set of G_δ-type and of measure zero. For $\epsilon > 0$ there exists $F : [a,b] \mapsto \mathbb{R}$, $F(a) = 0$, $F(b) < \epsilon$, such that F is AC and increasing on $[a,b]$, $F'(x) = +\infty$, for all $x \in Z$ and $F'(x) > 0$, for all $x \in [a,b] \setminus Z$.*

Lemma 2.14.1 *Let $f : [a,b] \mapsto \mathbb{R}$ be a summable function, $x_o \in [a,b]$ and $F(x) = (\mathcal{L}) \int_a^x f(t)dt$. If f is lower semicontinuous at x_o then $\underline{D}^{\#} F(x_o) \geq f(x_o)$.*

Proof. We may assume that $f(x_o) > -\infty$ (otherwise there is nothing to prove). For any real number $r < f(x_o)$, there exists a $\delta > 0$ such that $f(x) > r$, for all $x \in [a,b]$ for which $| x - x_o | < \delta$. Let $y, z \in [a,b]$ such that $x_o - \delta < y < z < x_o + \delta$. Then we have $(\mathcal{L}) \int_y^x f(t)dt \geq r(z-y)$, hence $(F(z) - F(y))/(z-y) \geq r$. It follows that $\underline{D}^{\#} F(x_o) \geq r$. Since r is arbitrary, it follows that $\underline{D}^{\#} F(x_o) \geq f(x_o)$.

Remark 2.14.1 *Particularly Lemma 2.14.1 is true for $\underline{D}F(x_o)$, and this was shown in [S3] (p. 107).*

2.15 Differentiability and conditions VBG, VB^*G

Theorem 2.15.1 *([S3], p. 230). Let $F : [a,b] \mapsto \mathbb{R}$, $P \subseteq [a,b]$, $F \in VB^*G$ on P. Then we have:*

(i) F is derivable a.e. on P;

(ii) $|F(N)| = \Lambda(B(F;N)) = 0$, where $N = \{x \in P : F'(x)$ does not exist finite or infinite$\}$.

Remark 2.15.1 *Particularly Theorem 2.15.1 is true for functions AC^*G or $\underline{AC}^*G$ (because $AC^*G \subseteq \underline{AC}^*G \subseteq VB^*G$, see Proposition 2.12.1, (iv)).*

Corollary 2.15.1 *([S3], p. 236). Let $F : [a,b] \mapsto \mathbb{R}$. Then the set $\{x : F'(x) = +\infty\}$ is of measure zero.*

Proof. See Theorem 2.15.1 and Corollary 2.13.2.

Remark 2.15.2 *There is the following generalization of Corollary 2.15.1, due to Denjoy (see [S3], p. 227)): Let $F : [a,b] \mapsto \mathbb{R}$. Then the set $\{x : \lim_{h\searrow 0} \mid F(x+h) - F(x) \mid /h = +\infty\}$ is of measure zero.*

Lemma 2.15.1 *Let $F : [a,b] \mapsto \mathbb{R}$, $P = \overline{P} \subseteq [a,b]$, $c = \inf(P)$, $d = \sup(P)$, and let $\{(c_k, d_k)\}_k$ be the intervals contiguous to P. Let $R_1 = \{x \in P : F'(x)$ exists finite or infinite $\}$ and $R_2 = \{x \in P : F_P'(x)$ exists finite or infinite $\}$. Let $T = (R_1 \setminus R_2) \cup (R_2 \setminus R_1)$ and let $Q \subseteq P$. If $\sum_{k\geq 1} \Omega_-(F;[c_k,d_k] \wedge \{c_k,d_k\}) > \infty$ then we have:*

(i) $F \in VB^*G$ on Q if and only if $F_P \in VB^*G$ on Q;

(ii) $\mid F(T) \mid = \mid T \mid = 0$ and $F'(x) = F_P'(x)$ a.e. on R_1.

Proof. For F_P see Definition 1.1.3.

(i) Let $F \in VB^*G$ on Q.We may suppose that $F \in VB^*$ on Q. Let $x, y \in Q$, $x < y$. Since $\mathcal{O}(F;[x,y]) \geq \mathcal{O}(F;[x,y] \cap P) = \mathcal{O}(F_P;[x,y])$, it follows that $F_P \in VB^*$ on Q. Conversely, let $F_P \in VB^*$ on Q. By Theorem 2.8.1, (i), (ii), $F_P \in VB^*$ on $\overline{Q}$. Let $x, y \in \overline{Q}$, $x < y$. By Proposition 1.4.1, (x) and Corollary 1.4.1, (iii), since $\mathcal{O}(F;[x,y] \cap P) = \mathcal{O}(F_P;[x,y])$, we have:

$$(1) \quad \Omega_-(F;[x,y] \wedge \{x,y\}) \geq -\mathcal{O}(F;[x,y]) + \sum_{(u,v)\in\mathcal{A}} \Omega_-(F;[u,v] \wedge \{u,v\}),$$

where $\mathcal{A} = \{(u,v) : (u,v)$ is an interval contiguous to $[x,y] \cap P\}$. Since $F_P \in VB^* \subseteq VB$ on $\overline{Q}$, it follows that $F \in VB$ on $\overline{Q}$. Let $a' = \inf(Q)$, $b' = \sup(Q)$ and let $\{(a_i,b_i)\}_i$ be the intervals contiguous to $\overline{Q}$. Let $\mathcal{A}_i = \{(u,v) : (u,v)$ is an interval contiguous to $[a_i,b_i] \cap P\}$ and let M be given by the fact that $F_P \in VB^*$ on $\overline{Q}$. By (1), $\sum_{i\geq 1} \Omega_-(F;[a_i,b_i] \wedge \{a_i,b_i\}) \geq -\sum_{i\geq 1} \mathcal{O}(F_P;[a_i,b_i]) + \sum_{i\geq 1} \sum_{(u,v)\in\mathcal{A}_i} \Omega_-(F;[u,v] \wedge \{u,v\}) > -M + \sum_{k\geq 1} \Omega_-(F;[c_k,d_k] \wedge \{c_k,d_k\}) > -\infty$. By Theorem 2.8.1, (viii), $F \in VB^*$ on $\overline{Q}$.

(ii) Let $T_1 = R_2 \setminus R_1$ and $T_2 = R_1 \setminus R_2$. By Corollary 2.13.2, $F_P \in VB^*G$ on T_1. By (i), $F \in VB^*G$ on T_1. Since $F'(x)$ does not exist (finite or infinite) on T_1, by Theorem 2.15.1, $\mid F(T_1) \mid = \Lambda(B(F;T_1)) = 0$. Similarly we obtain that $\mid F(T_2) \mid = \Lambda(B(F;T_2)) = 0$. It follows that $\mid R_1 \mid = \mid R_2 \mid = \mid R_1 \cap R_2 \mid$. Since $F'(x) = F_P'(x)$ on $(R_1 \cap R_2) \setminus A$, where $A = \{x \in P : x$ is isolated in $P\}$ and A is countable, it follows that $F'(x) = F_P'(x)$ *a.e.* on R_1.

Remark 2.15.3 *(i) Lemma 2.15.1, (ii) is an extension of a theorem of Saks (see [S2], p. 138), because we may put $\sum_{k\geq 1} \Omega_-(F;[c_k,d_k] \wedge \{c_k,d_k\}) > -\infty$ instead of $\sum_{k\geq 1} \mathcal{O}_-(F;[c_k,d_k]) > -\infty$.*

(ii) Particularly Lemma 2.15.1 remains true if "$\sum_{k\geq 1}\Omega_-(F;[c_k,d_k]\wedge\{c_k,d_k\}) > -\infty$" is replaced by "$F(c_k)\leq F(d_k)$ and $F((c_k,d_k))\subseteq[F(c_k),F(d_k)]$" (because, if $x\in[c_k,d_k]$ then $F(d_k)-F(x)\geq 0$ and $F(x)-F(c_k)\geq 0$, hence $\Omega_-(F;[c_k,d_k]\wedge\{c_k,d_k\})=0$).

Lemma 2.15.2 *Let $F:[a,b]\mapsto \mathbb{R}$, $P=\overline{P}\subseteq[a,b]$, $c=\inf(P)$, $d=\sup(P)$. If $F\in VB^*$ on P then there exists a set $Q\subset P$ such that $|F(Q)|=|Q|=0$ and $F'(x)=F'_P(x)$ on $P\setminus Q$.*

Proof. Let $E=\{x\in P: F'(x)$ does not exist finite or infinite$\}$, $E_1=\{x\in P: F'_P(x)$ does not exist finite or infinite$\}$ and $E_2=\{x\in P: F'(x)$ and $F'_P(x)$ exist and $F'(x)\neq F'_P(x)\}$. Since $F_P(x)=F(x)$ on P, it follows that $F'_P(x)=F'(x)$ at all points $x\in P$, except perhaps the endpoints of the intervals contiguous to P. Hence E_2 is countable. By Theorem 2.15.1, (ii), $|F(E)|=|E|=0$. Since $VB^*\subseteq VB$, by Corollary 2.7.2, it follows that $F_P\in VB=VB^*$ on $[c,d]$. By Theorem 2.15.1, (ii), $|F_P(E_1)|=|E_1|=0$. Clearly $|F(E_2)|=0$. Let $Q=E\cup E_1\cup E_2$. Then $|F(Q)|=|Q|=0$ and $F'(x)=F'_P(x)$ on $P\setminus Q$.

Theorem 2.15.2 (Saks) *([S3], p. 222). Let $F:[a,b]\mapsto\mathbb{R}$, $P\subseteq[a,b]$. If F is measurable on P and VB on a set $P_1\subseteq P$ then:*

(i) there exists a VB function $G:[a,b]\mapsto\mathbb{R}$ which coincide with F on P_1. Moreover, $P_2=\{x\in P: F(x)=G(x)\}$ is measurable and $P_1\subseteq P_2\subseteq P$;

(ii) F is approximately derivable a.e. on P_1.

Theorem 2.15.3 (Denjoy-Khintchine) *([S3], p. 222). Let $F:[a,b]\mapsto\mathbb{R}$ and let P be a measurable subset of $[a,b]$. If F is measurable and VBG on P then F is approximately derivable a.e. on P.*

Corollary 2.15.2 *Let $g:[a,b]\mapsto[c,d]$, $f:[c,d]\mapsto\mathbb{R}$, $A\subset[a,b]$, $|A|>0$ and $g(A)=B$. If g is measurable on $[a,b]$, f is approximately derivable on B, $f'_{ap}(y)>0$ for each $y\in B$, $f\circ g$ is approximately derivable on A, and $(f\circ g)'_{ap}(x)>0$ for each $x\in A$ then g is approximately derivable a.e. on A.*

Proof. It is easy to prove that $f\circ g\in$ *strictly − increasingG* on A and $f\in$ *strictly increasingG* on B. By Lemma 2.9.2, g is *strictly increasingG* on A. Then $g\in VBG$ on A. Since g is measurable, by Theorem 2.15.3 it follows that g is approximately derivable *a.e.* on A.

Remark 2.15.4 *(i) Particularly, Theorem 2.15.3 is true for functions ACG or $\underline{ACG}$ (because $ACG\subseteq\underline{ACG}\subseteq VBG$, see Theorem 2.11.1, (vii)).*

(ii) In Theorem 2.15.3, VBG cannot be replaced by L_2G, AC_2G or VB_2G (these classes will be defined in Sections 2.32., 2.28., 2.29.), see Section 6.29. (the functions G_{PP}, H_{PP} or K_{PP}).

Theorem 2.15.4 (Denjoy) *([S3], p.239). Let $F:[a,b]\mapsto\mathbb{R}$, $P\subseteq[a,b]$. If two extreme approximate derivatives on the same side are finite for F n.e. on P, then $F\in ACG$ on P.*

Theorem 2.15.5 (Denjoy) *([S3], p. 237). Let $F : [a,b] \mapsto \mathbb{R}$, $P \subseteq [a,b]$. If $\overline{D}_{ap}F(x) < +\infty$ n.e. on P or $\underline{D}_{ap}F(x) > -\infty$ n.e. on P then $F \in VBG$ on P.*

Corollary 2.15.3 *Let $F : [a,b] \mapsto \mathbb{R}$ and let P be a measurable subset of $[a,b]$. If F is measurable on P then $\{x \in P : F'_{ap}(x) = +\infty\}$ is a null set.*

Proof. See Theorem 2.15.5 and Theorem 2.15.3.

Theorem 2.15.6 (Thomson) *([T5], p. 147). Let $\mathcal{S}$ be a local system satisfying intersection condition (I.C.), and let $F : [a,b] \mapsto \mathbb{R}$ be a measurable function. If $\mathcal{S}\,DF(x)$ exists everywhere on a set E then F is approximately derivable a.e. on E and $F'_{ap}(x) = \mathcal{S}\,DF(x)$ a.e. on E.*

2.16 A fundamental lemma for monotonicity

Lemma 2.16.1 *Let $F : [a,b] \mapsto \mathbb{R}$, $F(a) < F(b)$ and let $P = \{x : F'(x) \geq 0\}$. If $F \in \mathcal{C}_d$ on $[a,b]$ and $\mid F(P) \mid = 0$, then there exist $E \subset [F(a), F(b)]$ and $K \subset [F(a), F(b)]$ such that:*

(i) $\mid E \mid = 0$ and $\mid K \mid > (F(b) - F(a))/2$;

(ii) E and K are compact sets;

(iii) $F(E) = K$;

(iv) F is continuous and strictly increasing on E.

Proof. By Theorem 2.4.2, (vi), $F \in \mathcal{D}_+$ on $[a,b]$. Let $a_o = \sup\{x \in [a,b] : F(x) = F(a)\}$ and $b_o = \inf\{x \in [a_o, b] : F(x) = F(b)\}$. By Corollary 2.9.2, (i)-(iii), $F(a_o) = F(a)$, $F(b_o) = F(b)$ and $F([a_o, b_o]) = [F(a_o), F(b_o)]$. By Corollary 2.9.2, (iv), there exists a set $A \subset [a_o, b_o]$, $a_o, b_o \in A$, such that $F \in sI([a_o, b_o] \wedge A)$ and

(1) $F(A) = [F(a_o), F(b_o)]$.

We show that $\overline{A}$ is nowhere dense. Suppose that there exists an interval $[c,d]$, $c, d \in A$, such that $\overline{A} \supset [c,d]$. It follows that $[c,d] \subset A \cup A_+$. By Corollary 2.9.2, (v), F is increasing on $[c,d]$, $F(x) < F(d)$ and $F([c,d]) = [F(c), F(d)]$. By Theorem 2.15.1, $\mid F(P) \mid = F(d) - F(c)$. This contradicts the fact that $\mid F(P) \mid = 0$. Hence $\overline{A}$ is nowhere dense.
Let $\{(c_k, d_k)\}$, $k = \overline{1,\infty}$ be the intervals contiguous to $\overline{A}$ ($k = \overline{1,\infty}$ because $\overline{A}$ is nowhere dense). Let $A_o = \overline{A} \setminus (\cup_{k=1}^{\infty}\{c_k\})$. Then $A_o \subset A \cup A_+$ and $F \in I([a_o, b_o] \wedge A_o)$ (see Corollary 2.9.2, (v)). Let $A_1 = \{x \in A_o : \underline{D}^+F(x) = \infty\}$.
We show that

(2) $\mid F(A_1) \mid = F(b) - F(a) = F(b_o) - F(a_o)$.

Since $F \in I([a_o, b_o] \wedge A_o)$ it follows that $\underline{D}^-F(x) \geq 0$ for each $x \in A_o$. Let $\underline{D}^+F(x) > -\infty$ for each $x \in B_o$, hence $\underline{F}'(x) > -\infty$ for each $x \in B_o$. By Corollary 2.13.2, $F \in VB^*G$ on B_o. By Theorem 2.15.1, since $\mid F(P) \mid = 0$ it follows that $\mid F(B_o) \mid = 0$.

Since $F(A) = [F(a), F(b)]$ we have (2).
Let $e \in (0,1)$. Let N be a positive integer such that

$$(3) \quad \sum_{k=N}^{\infty} (d_k - c_k) < \frac{b_o - a_o}{2}.$$

We shall construct a cover in the Vitali sense for the set $F(A_1)$. Let $x \in A$. Then $\underline{D}^+ F(x) = -\infty$. It follows that there exist

(4) $k(x) \geq N$ *and* $\alpha(x) \in [c_{k(x)}, d_{k(x)})$ *such that* $F(\alpha(x)) < F(x)$.

Since $F \in I([a_o, b_o] \wedge A_o)$, we have

(5) $F(x) < F(d_{k(x)})$.

Let $0 < \epsilon(x) < \min\{F(x) - F(\alpha(x)); (F(b) - F(a))/4\}$ and let $J_{x,\epsilon(x)} = [F(x) - \epsilon(x), F(x)]$. Then $\{J_{x,\epsilon(x)}\}$, $x \in A_1$, $\epsilon(x)$ is a cover in the Vitali sense for the set $F(A_1)$. Since $F(d_{k(x)}) \geq F(x) - \epsilon(x) > F(\alpha(x))$ and $F \in \mathcal{D}_+$, it follows that $\{t \in [\alpha(x), d_{k(x)}] : F(t) = F(x) - \epsilon(x)\} \neq \emptyset$.
Let $a_{x,\epsilon(x)} = \sup\{t \in [\alpha(x), d_{k(x)}] : F(t) = F(x) - \epsilon(x)\}$. By Corollary 2.9.2, (i), $F(a_{x,\epsilon(x)}) = F(x) - \epsilon(x)$. Since $F \in \mathcal{D}_+$, $\{t \in [a_{x,\epsilon(x)}, d_{k(x)}] : F(t) = F(x)\} \neq \emptyset$.
Let $b_{x,\epsilon(x)} = \inf\{t \in [a_{x,\epsilon(x)}, d_{k(x)}] : F(t) = F(x)\}$. By Corollary 2.9.2, (ii), $F(b_{x,\epsilon(x)}) = F(x)$ and $a_{x,\epsilon(x)} < b_{x,\epsilon(x)}$. By Corollary 2.9.2, (iii), it follows that $F([a_{x,\epsilon(x)}, b_{x,\epsilon(x)}]) = [F(a_{x,\epsilon(x)}), F(b_{x,\epsilon(x)})] = J_{x,\epsilon(x)}$. By Theorem 1.8.1, there exist a positive integer n, $x_i \in A_1$ and $0 < \epsilon(x_i) < \min\{F(x_i) - F(\alpha(x_i)); (F(b) - F(a))/4\}$, $i = \overline{1,n}$ such that $\sum_{i=1}^n \mid J_{x_i,\epsilon(x_i)} \mid > (F(b) - F(a)) \cdot (1 - e)$ and

(6) $\{J_{x_i,\epsilon(x_i)}\}$, $i = \overline{1,n}$ are pairwise disjoint.

We may suppose without loss of generality that

(7) $F(x_1) < F(x_2) < \ldots < F(x_n)$.

It follows that there exist a positive integer $q \leq n$ and a set $\{r_1, r_2, \ldots, r_q\} \subseteq \{1, 2, \ldots, n\}$, $r_1 < r_2 < \ldots < r_q = n$, $r_o = 0$, such that

$$(8) \quad F(x_{r_o+1}) \leq F(x_{r_1}) \leq F(d_{k(x_{r_o+1})}) < F(x_{r_1+1}) \leq F(x_{r_2}) \leq F(d_{k(x_{r_1+1})}) < \ldots$$
$$< F(x_{r_{q-1}+1}) \leq F(x_{r_q}) = F(x_n) \leq F(d_{k(x_{r_{q-1}+1})}).$$

We show (8). By (5), $F(x_1) = F(x_{r_o+1}) \leq F(d_{k(x_{r_o+1})})$. Let r_1 be the last positive integer such that $F(x_{r_1}) \leq F(d_{k(x_{r_o+1})})$. It follows that $F(x_{r_o+1}) \leq F(x_{r_1}) \leq F(d_{k(x_{r_o+1})})$. If $r_1 = n$ then let $q = 1$. If $r_1 \neq n$ then by (7), $F(x_{r_1+1}) > F(d_{k(x_{r_o+1})})$. By (5), $F(x_{r_1+1}) \leq F(d_{k(x_{r_1+1})})$. Let r_2 be the last positive integer such that $F(x_{r_2}) \leq F(d_{k(x_{r_1+1})})$. If $r_2 = n$ then let $q = 2$. If $r_2 \neq n$ then we continue until we find a positive integer $q \leq n$ such that $r_q = n$, and we obtain (8).
(I) Let $a_1 = a_{x_{r_o+1},\epsilon(x_{r_o+1})}$ and $b_1 = b_{x_{r_o+1},\epsilon(x_{r_o+1})}$. Then

(9) $[a_1, b_1] \subset [c_{k(x_{r_o+1})}, d_{k(x_{r_o+1})}]$.

Let $a_2 = \sup\{t \in [b_1, d_{k(x_{r_o+1})}] : F(t) = F(x_2) - \epsilon(x_2)\}$ and

$b_2 = \inf\{t \in [a_2, d_{k(x_{r_o+1})}] : F(t) = F(x_2)\}$. Continuing, let

$$a_{r_1} = \sup\{t \in b_{r_1-1}, d_{k(x_{r_o+1})}] : F(t) = F(x_{r_1}) - \epsilon(x_{r_1})\} \text{ and}$$

$$b_{r_1} = \inf\{t \in [a_{r_1}, d_{k(x_{r_o+1})}] : F(t) = F(x_{r_1})\}.$$

By Corollary 2.9.2, (i)-(iii), we have

(10) $F(a_i) = F(x_i) - \epsilon(x_i),\ F(b_i) = F(x_i),\ F([a_i, b_i]) = [F(a_i), F(b_i)]\ i = \overline{1, r_1}.$

By (6), (10) and (9) we have

(11) $a_1 < b_1 < a_2 < b_2 < \ldots < a_{r_1} < b_{r_1}$ and $[a_i, b_i] \subset [c_{k(x_{r_o+1})}, d_{k(x_{r_o+1})}]$.

(II) Let $a_{r_1+1} = a_{x_{r_1+1}, \epsilon(x_{r_1+1})}$ and $b_{r_1+1} = b_{x_{r_1+1}, \epsilon(x_{r_1+1})}$. Then

(12) $[a_{r_1+1}, b_{r_1+1}] \subset [c_{k(x_{r_1+1]})}, d_{k(x_{r_1+1})}]$.

Continuing, we define

$$a_{r_2} = \sup\{t \in [b_{r_2-1}, d_{k(x_{r_1+1})}] : F(t) = F(x_{r_2}) - \epsilon(x_{r_2})\} \text{ and}$$

$$b_{r_2} = \inf\{t \in [a_{r_2}, d_{k(x_{r_1+1})}] : F(t) = F(x_{r_2})\}.$$

Similarly as above, we obtain

(13) $F(a_i) = F(x_i) - \epsilon(x_i),\ F(b_i) = F(x_i),\ F([a_i, b_i]) = [F(a_i), F(b_i)],$

$i = \overline{r_1 + 1, r_2};$

(14) $a_{r_1+1} < b_{r_1+1} < \ldots < a_{r_2} < b_{r_2}$ and $[a_i, b_i] \subset [c_{k(x_{r_1+1})}, d_{k(x_{r_1+1})}],\ i = \overline{r_1 + 1, r_2}.$

By (11) we conclude that $b_{r_1} \le d_{k(x_{r_o+1})}$, and by (8), $F(d_{k(x_{r_o+1})}) < F(d_{k(x_{r_1+1})})$. Since $F \in I([a_o, b_o] \wedge A_o)$, it follows that $d_{k(x_{r_o+1})} < d_{k(x_{r_1+1})}$, hence $d_{k(x_{r_o+1})} \le c_{k(x_{r_1+1})}$. By (14), $c_{k(x_{r_1+1})} \le a_{r_1+1}$. So $b_{r_1} < a_{r_1+1}$.

(III) Continuing we obtain

(15) $a_1 < b_1 < \ldots < a_{r_1} < b_{r_1} < a_{r_1+1} < b_{r_1+1} < \ldots < a_{r_q-1} < b_{r_q-1} < a_{r_q}$

$= a_n < b_{r_q} = b_n;$

(16) $F(a_i) = F(x_i) - \epsilon(x_i),\ F(b_i) = F(x_i),\ F([a_i, b_i]) = [F(a_i), F(b_i)],\ i = \overline{1, n};$

(17) $[a_i, b_i] \subset [c_{k(x_{r_j+1})}, d_{k(x_{r_j+1})}],\ i = \overline{r_j + 1, r_{j+1}},\ j = \overline{0, q-1}.$

By (3), (4) and (17) we have

(18) $\displaystyle\sum_{i=1}^{n}(b_i - a_i) < \frac{b_o - a_o}{2}.$

Moreover, we obtain

(19) $\displaystyle\sum_{i=1}^{n}(F(b_i) - F(a_i)) > (F(b_o) - F(a_o)) \cdot (1 - e)$ and

$$(20) \quad F(b_i) - F(a_i) < \frac{F(b_o) - F(a_o)}{4}, \; i = \overline{1, n}.$$

Let $e_i \in (0,1)$, $i = \overline{1, \infty}$ such that

$$(21) \quad (1 - e_1) \cdot (1 - e_2) \cdot (1 - e_3) \ldots > \frac{1}{2}.$$

Let $e = e_1$ and $n = n_1$, where n_1 is a positive integer. Let $E_1 = \cup_{i_1=1}^{n_1}[a_{i_1}, b_{i_1}]$ and $K_1 = \cup_{i_1=1}^{n_1}[F(a_{i_1}), F(b_{i_1})]$. By (18) and (19), we have

$$(22) \quad | E_1 | < (b_o - a_o)/2 \text{ and } | K_1 | > (F(b_o) - F(a_o)) \cdot (1 - e_1).$$

For each $[a_{i_1}, b_{i_1}]$, $i_1 = \overline{1, n_1}$ we make a construction similar to that for the interval $[a_o, b_o]$. For $e_2 \in (0,1)$, there exist a positive integer n_{i_1} and a set of pairwise disjoint closed intervals $[a_{i_1,i_2}, b_{i_1,i_2}]$, $i_2 = \overline{1, n_{i_1}}$, such that

$$(23) \quad a_{i_1} \leq a_{i_1,1} < b_{i_1,1} < a_{i_1,2} < b_{i_1,2} < \ldots < a_{i_1,n_{i_1}} < b_{i_1,n_{i_1}} \leq b_{i_1};$$

$$(24) \quad F(a_{i_1}) \leq F(a_{i_1,1}) < F(b_{i_1,1}) < F(a_{i_1,2}) < F(b_{i_1,2}) < \ldots$$
$$< F(a_{i_1,n_{i_1}}) < F(b_{i_1,n_{i_1}}) \leq F(b_{i_1});$$

$$(25) \quad \sum_{i_2=1}^{n_{i_1}} (b_{i_1,i_2} - a_{i_1,i_2}) < \frac{b_{i_1} - a_{i_1}}{2};$$

$$(26) \quad F([a_{i_1,i_2}, b_{i_1,i_2}]) = [F(a_{i_1,i_2}), F(b_{i_1,i_2})];$$

$$(27) \quad \sum_{i_2=1}^{n_{i_1}} (F(b_{i_1,i_2}) - F(a_{i_1,i_2})) > (F(b_{i_1}) - F(a_{i_1})) \cdot (1 - e_2);$$

$$(28) \quad F(b_{i_1,i_2}) - F(a_{i_1,i_2}) < \frac{F(b_{i_1}) - F(a_{i_1})}{4} < \frac{F(b_o) - F(a_o)}{4^2}, \; i_2 = \overline{1, n_{i_1}}.$$

Let $E_2 = \cup_{i_1,i_2}[a_{i_1,i_2}, b_{i_1,i_2}]$ and $K_2 = \cup_{i_1,i_2}[F(a_{i_1,i_2}), F(b_{i_1,i_2})]$. By (25) and (28)

$$(29) \quad | K_2 | > \sum_{i_1=1}^{n_1} (F(b_{i_1}) - F(a_{i_1}))(1 - e_2) > (F(b_o) - F(a_o))(1 - e_1)(1 - e_2).$$

Suppose we have constructed the sets E_{p-1} and K_{p-1}, $p \geq 2$. For $e_p \in (0,1)$ there exist a positive integer $n_{i_1,i_2 \ldots, i_{p-1}}$ and a set of pairwise disjoint closed intervals $[a_{i_1,i_2,\ldots,i_p}, b_{i_1,i_2,\ldots,i_p}]$, $i_p = \overline{1, n_{i_1,i_2,\ldots,i_{p-1}}}$, such that

$$(30) \quad a_{i_1,\ldots,i_{p-1}} \leq a_{i_1,\ldots,i_{p-1},1} < b_{i_1,\ldots,i_{p-1},1} < a_{i_1,\ldots,i_{p-1},2} < b_{i_1,\ldots,i_{p-1},2} < \ldots$$
$$< a_{i_1,\ldots,i_{p-1},n_{i_1,\ldots,i_{p-1}}} < b_{i_1,\ldots,i_{p-1},n_{i_1,\ldots,i_{p-1}}} \leq b_{i_1,\ldots,i_{p-1}};$$

$$(31) \quad F(a_{i_1,\ldots,i_{p-1}}) \leq F(a_{i_1,\ldots,i_{p-1},1} < b_{i_1,\ldots,i_{p-1},1}) < F(a_{i_1,\ldots,i_{p-1},2}) < F(b_{i_1,\ldots,i_{p-1},2})$$
$$< \ldots < F(a_{i_1,\ldots,i_{p-1},n_{i_1,\ldots,i_{p-1}}}) < F(b_{i_1,\ldots,i_{p-1},n_{i_1,\ldots,i_{p-1}}}) \leq F(b_{i_1,\ldots,i_{p-1}});$$

(32) $$\sum_{i_p=1}^{n_{i_1,\dots,i_{p-1}}} (b_{i_1,\dots,i_p} - a_{i_1,\dots,i_p}) < \frac{b_{i_1,\dots,i_{p-1}} - a_{i_1,\dots,i_{p-1}}}{2};$$

(33) $$F([a_{i_1,\dots,i_p}, b_{i_1,\dots,i_p}]) = [F(a_{i_1,\dots,i_p}), F(b_{i_1,\dots,i_p})],\ i_p = \overline{1, n_{i_1,\dots,i_{p-1}}};$$

(34) $$\sum_{i_p=1}^{n_{i_1,\dots,i_{p-1}}} (F(b_{i_1,\dots,i_p}) - F(a_{i_1,\dots,i_p})) > (F(b_{i_1,\dots,i_{p-1}}) - F(a_{i_1,\dots,i_{p-1}}))(1 - e_p);$$

(35) $$F(b_{i_1,\dots,i_p}) - F(a_{i_1,\dots,i_p}) < \frac{F(n_{i_1,\dots,i_{p-1}}) - F(a_{i_1,\dots,i_{p-1}})}{4} < \frac{F(b_o) - F(a_o)}{4^p},$$

$$i_p = \overline{1, n_{i_1,\dots,i_{p-1}}}$$

Let $E_p = \cup_{i_1,\dots,i_p}[a_{i_1,\dots,i_p}, b_{i_1,\dots,i_p}]$ and $K_p = [F(a_{i_1,\dots,i_p}), F(b_{i_1,\dots,i_p})]$.

By (32) we have

(36) $$| E_p | < \frac{b_o - a_o}{2^p},$$

and by (34),

(37) $$| K_p | > (F(b_o) - F(a_o))(1 - e_1)\dots(1 - e_p).$$

Let $E = \cap_{p=1}^{\infty} E_p$ and $K = \cap_{p=1}^{\infty} K_p$.

(i) By (36), $|E| = 0$. Since each K_p is measurable, by (37) and (21), the set K is also measurable and $|K| > (F(b_o) - F(a_o))/2 = (F(b) - F(a))/2$.

(ii) Since E_p, K_p, $p \geq 1$ are compact sets, it follows that E and K are also compact sets.

(iii) Let $x_o \in E$. Then there exists a sequence $\{[a_{i_1,\dots,i_p}, b_{i_1,\dots,i_p}]\}$, $p \geq 1$ of closed intervals, each containing x_o. By (36),

$$\{x_o\} = \bigcap_{\substack{i_1,\dots,i_p \\ p\geq 1}} [a_{i_1,\dots,i_p}, b_{i_1,\dots,i_p}].$$

By (33), it follows that

$$F(x_o) = \bigcap_{\substack{i_1,\dots,i_p \\ p\geq 1}} F([a_{i_1,\dots,i_p}, b_{i_1,\dots,i_p}]) = \bigcap_{\substack{i_1,\dots,i_p \\ p\geq 1}} [F(a_{i_1,\dots,i_p}), F(b_{i_1,\dots,i_p})] \in \bigcap_{p\geq 1} K_p = K,$$

hence $F(E) \subseteq K$. Let $y_o \in K$. Then there exists a sequence $\{[a_{i_1,\dots,i_p}, b_{i_1,\dots,i_p}]\}$, $p \geq 1$ of closed intervals such that each interval $[F(a_{i_1,\dots,i_p}), F(b_{i_1,\dots,i_p})]$ contains y_o. By (35), $\{y_o\} = \cap_{\substack{i_1,\dots,i_p \\ p\geq 1}} [F(a_{i_1,\dots,i_p}), F(b_{i_1,\dots,i_p})]$. By (33) and (36), the intersection $\cap_{\substack{i_1,\dots,i_p \\ p\geq 1}} [a_{i_1,\dots,i_p}, b_{i_1,\dots,i_p}]$ degenerates to a point x_o. Hence $F(x_o) = \cap_{\substack{i_1,\dots,i_p \\ p\geq 1}} [F(a_{i_1,\dots,i_p}), F(b_{i_1,\dots,i_p})] = \{y_o\}$ and $F(x_o) = y_o$. It follows that $K \subseteq F(E)$.

(iv) Let $x' < x"$, $x', x" \in E$. Then there exist two sequences of closed intervals $\{[a_{i'_1,\dots,i'_p}, b_{i'_1,\dots,i'_p}]\}$, $p \geq 1$ and $\{[a_{i_1",\dots,i_p"}, b_{i_1",\dots,i_p"}]\}$, $p \geq 1$, such that $x' = \cap_{\substack{i'_1,\dots,i'_p \\ p\geq 1}} [a_{i'_1,\dots,i'_p}, b_{i'_1,\dots,i_p"}]$ and $x" = \cap_{\substack{i_1",\dots,c_p" \\ p\geq 1}} [a_{i_1",\dots,i_p"}, b_{i_1",\dots,i_p"}]$. Since $x' < x"$, there exists $p_o \geq 1$ such that $i'_j = i_j"$, $j = \overline{1, p_o - 1}$ and $i'_{p_o} < i_{p_o}"$. It follows that $x' \in [a_{i'_1,\dots,i'_{p_o}}, b_{i'_1,\dots,i'_{p_o}}]$ and $x" \in [a_{i'_1,\dots,i'_{p_o-1},i_{p_o}"}, b_{i'_1,\dots,i'_{p_o-1},i_{p_o}"}]$. By (31) and (33) it follows that $F(a_{i'_1,\dots,i'_{p_o-1},i'_{p_o}}) < F(x') < F(b_{i'_1,\dots,i'_{p_o-1},i_{p_o}"}) < F(a_{i'_1,\dots,i'_{p_o-1},i_{p_o-1}"}) < F(x") < F(b_{i'_1,\dots,o'_{p_o-1},i_{p_o}"})$. Hence $F(x') < F(x")$. Since F is increasing on E, it follows that $F \in C_i$, hence $F \in C_i \cap C_d = C$ on E.

Corollary 2.16.1 *Let $F : [a,b] \mapsto \mathbb{R}$, $F \in C_d$. If $\mid \{x : F'(x)$ exists, finite or infinite$\} \mid = 0$ then there exists a perfect null set $P \subset [a,b]$, such that $\mid F(P) \mid > 0$, F is strictly increasing and continuous on P.*

Proof. Since F is not decreasing on $[a,b]$ (this is so because otherwise F would be derivable *a.e.* on $[a,b]$, which is a contradiction), it follows that there exists $[c,d] \subset [a,b]$ such that $F(c) < F(d)$. By Lemma 2.16.1, there exists $E \subset [c,d]$ and $K \subset [F(c), f(d)]$ such that $\mid E \mid = 0$, $\mid K \mid > (F(d) - F(c))/2$, E and K are compact sets, $F(E) = K$, and F is strictly increasing and continuous on E. Let $P = \{x : x$ is an accumulation point of $E\}$. Then P is a perfect set, $\mid P \mid = 0$, $\mid F(P) \mid > 0$ and F is strictly increasing and continuous on P.

Remark 2.16.1 *Particularly, Corollary 2.16.1 is true for continuous functions. This result was shown by Foran in [F7].*

2.17 Krzyzewski's lemma and Foran's lemma

Lemma 2.17.1 (Krzyzewski) *([Kr1]). LetF : $[a,b] \mapsto \mathbb{R}$, $P \subseteq [a,b]$. If $F'(x)$ exists, finite or infinite, at each point $x \in P$ and $\mid F(P) \mid = 0$ then $F'(x) = 0$ a.e. on P.*

Proof. Let $B = \{x \in P : \mid F'(x) \mid > 0\}$. Then $B = \cup_{n=1}^{\infty} B_n$, where $B_n = \{t \in B : \mid F(s) - F(t) \mid \geq \mid s - t \mid /n$, for $\mid s - t \mid < 1/n\}$. To prove that B is a null set, it is enough to show that for each fixed n and any interval of length less that $1/n$, $B_n \cap I$ is a null set. Fix B_n and take such an I. Let $A = B_n \cap I$. Now take $\epsilon > 0$ and cover $F(A)$ by a sequence of intervals I_k of total length less than ϵ. Put $A_k = F^{-1}(I_k) \cap A$, and note that the union of the A_k covers A. Thus $\mid A \mid \leq \sum_k \mid A_k \mid \leq \sum_k \sup_{s,t \in A_k} \mid s - t \mid \leq \sum_k n \cdot \sup_{s,t \in A_k} \mid F(s) - F(t) \mid$ (because $A_k \subset I \cap B_n$). Now from the definition of A_k it follows that $\sup_{s,t \in A_k} \mid F(s) - F(t) \mid \leq \mid I_k \mid$ so that $\mid A \mid \leq \sum_k n \mid I_k \mid \leq n\epsilon$. Since ϵ is arbitrary and n is fixed, it follows that $\mid A \mid = 0$.

Remark 2.17.1 *Lemma 2.17.1 was originally proved by Vallée Poussin under the additional hypothesis that $F \in AC$. However the above proof of Lemma 2.12.5 is due to Serrin and Varberg ([SerV]). For more comments on the history of this lemma, see [Go1] and [Go2].*

Lemma 2.17.2 (Foran) *([F9]). Let $P \subseteq [a,b]$ be a measurable set and $F : [a,b] \mapsto \mathbb{R}$ be a measurable function. If F has an approximate derivative (finite or infinite) at each point of a set $E \subseteq P$ and $\mid F(E) \mid = 0$ then $F'_{ap}(x) = 0$ a.e. on E.*

Proof. By Theorem 2.15.5, $F \in VBG$ on E. It follows that $E = \cup_{i=1}^{n} E_n$ and $F \in VB$ on each E_n. By Theorem 2.15.2,(i), for each $n \geq 1$ there exists $F_n : [a,b] \mapsto \mathbb{R}$, $F \in VB$, such that $F_n = F$ on E_n, and $P_n = \{x \in P : F_n(x) = F(x)\}$ is measurable. By Theorem 1.6.1, $d(P_n; x) = 1$ for almost all $x \in E_n$. Then $F'_n(x) = F'_{ap}(x)$ *a.e.* on E_n. Since $\mid F_n(E_n) \mid = 0$ and F_n is derivable *a.e.* on $[a,b]$ (see Theorem 2.14.3), by Lemma 2.17.1, $F'_n(x) = 0$ *a.e.* onE_n. Hence $F'_{ap}(x) = 0$ *a.e.* on E_n. Therefore $F'_{ap}(x) = 0$ *a.e.* on E.

Remark 2.17.2 *Foran asserted in fact that Lemma 2.17.2 is true even if one drops the measurability condition for F (but the proof isn't quite clear, because we do not know if "$F'_n(x) = F'_{ap}(x)$ a.e. on E_n" without assuming that F is measurable).*

2.18 Conditions (N), T_1, T_2, (S), $(+)$, $(-)$

Definition 2.18.1 *([S3], p. 224). Let P be a real set and $F : P \mapsto \mathbb{R}$. F is said to satisfy Lusin's condition (N) on P if $\mid F(Z) \mid = 0$, whenever $Z \subset P$ and $\mid Z \mid = 0$. We can define condition NG on P, using Definition 1.9.2, but $NG = (N)$.*

Definition 2.18.2 *([S3], p. 277). Let P be a real set and $F : P \mapsto \mathbb{R}$. F is said to satisfy Banach's condition T_1 on P, if the set $\{y \in F(P) : F^{-1}(y)$ is infinite$\}$ is of measure zero. F is said to satisfy Banach's condition T_2 on P, if the set $\{y \in F(P) : F^{-1}(y)$ is uncountable$\}$ is of measure zero. We can define conditions T_1G and T_2G on P, using Definition 1.9.2, but $T_2G = T_2$.*

Definition 2.18.3 *([S3], p. 282). Let P be a real set and $F : P \mapsto \mathbb{R}$. F is said to satisfy Banach's condition (S) on P, if for every $\epsilon > 0$ there exists a $\delta > 0$ such that $\mid F(Z) \mid < \epsilon$, whenever $Z \subset P$ and $\mid Z \mid < \delta$. We define condition SG on P, using Definition 1.9.2*

Remark 2.18.1 *If in the definitions of (N) and (S) the set P is measurable then we may suppose without loss of generality that Z is also measurable (because a measurable set is always contained in a G_δ-set having the same outer measure).*

Definition 2.18.4 *Let $F : [a,b] \mapsto \mathbb{R}$, $P = \{x : 0 \le F'(x) \le +\infty\}$ and $N = \{x : 0 \ge F'(x) \ge -\infty\}$. Let $a \le c < d \le b$. We denote by $(+)$ and $(-)$ the following properties:*

(+) If $F(c) < F(d)$ then $\mid F(P \cap [c,d]) \mid \ge F(d) - F(c)$;

(−) If $F(c) > F(d)$ then $\mid F(N) \cap [c,d]) \mid \ge F(c) - F(d)$.

Theorem 2.18.1 (Banach) *([S3], p. 278). Let $F : [a,b] \mapsto \mathbb{R}$, $F \in \mathcal{C}$. The following assertions are equivalent:*

(i) $F \in T_1$ on $[a,b]$;

(ii) $F(\{x : F'(x)$ does not exist, finite or infinite$\})$ is a null set.

Theorem 2.18.2 (Rademacher-Ellis) *Let $P \subseteq [a,b]$ be a measurable set and let $F : P \mapsto \mathbb{R}$ be a measurable function. The following assertions are equivalent:*

(i) $F \in (N)$ on P;

(ii) F transforms every measurable set into a measurable set.

Proof. $(i) \Rightarrow (ii)$ ([El], p.476). Let $A \subset P$ be a measurable set. By Lusin's Theorem (see [S3], p.72), we have $A = (\cup_{n=1}^{\infty} A_n) \cup Z$, where $\mid Z \mid = 0$, $A_n = \overline{A}_n$ and $F_{/A_n} \in \mathcal{C}$. It follows that $F(A_n)$ is closed, hence $F(A_n)$ is measurable. Since $F \in (N)$, $\mid F(Z) \mid = 0$, so $F(A)$ is measurable.

$(ii) \Rightarrow (i)$ ([N], p.249). Suppose that $F \notin (N)$ on P. Then there exists $Z \subset P$, $\mid Z \mid = 0$ such that $\mid F(Z) \mid > 0$. By hypothesis $F(Z)$ is measurable. It follows that $F(Z)$ contains a measurable subset B. Let $Z_1 = F^{-1}(B) \cap Z$. Then $\mid Z_1 \mid = 0$ and $F(Z_1) = B$, a contradiction.

Theorem 2.18.3 *Let $F : [a, b] \mapsto \mathbb{R}$, $F \in \mathcal{D}_-\mathcal{B}_1T_2$. If $F(b) < F(a)$ then for almost all $y \in [F(b), F(a)]$, the set $F^{-1}(y)$ contains an isolated point x_y such that $\overline{F}'(x_y) \le 0$.*

Proof. Let $E_y = F^{-1}(y)$, $Y_1 = \{y : E_y$ is uncountable$\}$ and $Y_2 = \{y \in [F(b), F(a)] :$ there exists $x \in E_y$ such that F attains a strict relative maximum or minimum at $x\}$. Since $F \in T_2$, $| Y_1 |= 0$, and by Theorem 1.10.1, (i), Y_2 is at most countable. Let $y \in (F(b), F(a)) \setminus (Y_1 \cup Y_2)$. We prove that E_y contains an isolated point x_y, for which there exists an open interval $I_y = (a_y, b_y)$ such that, $E_y \cap I_y = \{x_y\}$, $F(x) > y$ for $x \in [a_y, x_y)$, and $F(x) < y$ for $x \in (x_y, b_y]$. Clearly $\overline{F}'(x_y) \le 0$. We have two situations:

(I) $E_y = \{x_y\}$. Since $F \in \mathcal{D}_-$ it follows that $F(x) > y$ for $x \in [a, x_y)$, and $F(x) < y$ for $x \in (x_y, b]$. Hence $I_y = (a, b)$.

(II) E_y contains more than one point. Since $F \in \mathcal{B}_1$, E_y is a set of G_δ-type (see Theorem 2.2.1, (i), (ii)). Since E_y is at most countable, by Theorem 1.7.1, the set $A_y = \{x \in E_y : x$ is an isolated point of $E_y\}$ is infinite when E_y is infinite, and $A_y = E_y$ when E_y is finite. Put $A_y = \{x_1, x_2, \ldots\}$. For each positive integer n let (a_n, b_n) be the maximal open interval such that $E_y \cap (a_n, b_n) = \{x_n\}$. We have two possibilities:

a) Suppose that there exist $x_i < x_j$, $x_i, x_j \in A_y$ such that $(x_i, x_j) \cap E_y = \emptyset$. (If E_y is finite then for each two consecutive points $x_i < x_j$, $x_i, x_j \in A_y$ we have $(x_i, x_j) \cap E_y = \emptyset$.) Let $c < x_i < x_j < d$ such that $(c, d) \cap E_y = \{x_i, x_j\}$. We have two cases:

(i) There exists $z_1 \in (c_i, x_i)$ with $F(z_1) > y$. Then $F(x) > y$ for each $x \in [z_1, x_i)$ (because $F \in \mathcal{D}_-$). Since $y \notin Y_2$, there exists $z_2 \in (x_i, y_j)$ such that $F(z_2) < y$. Let $x_y = x_i$. Then $F(x) = y$ for each $x \in [z_1, x_y)$. Since $F \in \mathcal{D}_-$, $F(x) < y$ for each $x \in (x_y, z_2]$. Hence $I_y = (z_1, z_2]$.

(ii) $F(x) < y$ for each $x \in (c, x_i)$. Then there exists $z_3 \in (x_i, x_j)$ such that $F(z_3) > y$ (because $y \notin Y_2$). Since $F \in \mathcal{D}_-$, $F(x) > y$ for each $x \in [z_3, x_j)$. Since $y \notin Y_2$, there exists $z_4 \in (x_j, d)$ such that $F(z_4) < y$. Let $x_y = x_j$ and $I_y = (z_3, z_4)$.

b) Suppose that $(x_i, x_j) \cap E_y \neq \emptyset$ for every $x_i, x_j \in A_y$, $x_i < x_j$. Then $P = [a, b] \setminus (\cup_{n\ge1}(A_n, b_n))$ is a nonempty perfect subset of $[a, b]$. We have three cases:

(i) $F(a_n) - y < 0$ and $F(b_n) - y < 0$ for some n. Since $F \in \mathcal{D}_-$, $F(x) < y$ for each $y \in (x_n, b_n]$. Since $y \notin Y_2$, there exists $z_n \in (a_n, x_n)$ such that $F(z_n) > y$. Since $F \in \mathcal{D}_-$, $F(x) > y$ for each $x \in [z_n, x_n)$. Put $x_y = x_n$ and $I_y = (z_n, b_n)$.

(ii) $y - F(a_n) < 0$ and $y - F(b_n) < 0$ for some n. Since $F \in \mathcal{D}_-$, $F(x) > y$ for each $x \in (a_n, x_n)$. Since $y \notin Y_2$, there exists $z_n \in (x_n, b_n]$ such that $F(z_n) < y$. Since $F \in \mathcal{D}_-$, $F(x) < y$ for each $x \in (x_n, z_n]$. Put $x_y = x_n$ and $I_y = (a_n, z_n)$.

(iii) $F(a_n) - y$ and $F(b_n) - y$ are of opposite sign. Let $C_y = \{x \in P : F_{/P} \in \mathcal{C}$ at $x\}$. Since $F \in \mathcal{B}_1$, by Theorem 1.6.3, Theorem 1.7.1 and Remark 2.2.1,(v), it follows that C_y is uncountable. Hence $E_y$is uncountable, a contradiction.

Remark 2.18.2 *Particularly, Theorem 2.18.3 is true if $\mathcal{D}_-$ is replaced by $\mathcal{D}$. In this case one obtain a stronger result, due to Bruckner ([Br2], p. 177):*
Let $F : [a, b] \mapsto \mathbb{R}$, $F \in \mathcal{D}\mathcal{B}_1T_2$. Then for almost all $y \in F([a, b])$ the set $F^{-1}(y)$ either consists of a single point or contains a pair of isolated neighbors (i.e., there exist $x_1, x_2 \in E_y$ such that x_1 and x_2 are isolated in E_y, and the interval (x_1, x_2) contains no point of E_y).

Theorem 2.18.4 *Let $F : [a, b] \mapsto \mathbb{R}$, $F \in \mathcal{D}_-\mathcal{B}_1T_2$ and let $N = \{x : F'(x) \le 0\}$. If $F(a) > F(b)$ then N is uncountable, $F(N)$ is measurable and $F \in (-)$ on $[a, b]$.*

Proof. By Remark 2.2.1, (iii), F is a Borel function. By Corollary 1.10.1, N is a Borel set. By Proposition 1.5.3, (iii), (i), $F(N)$ is an analytic set, and is therefore measurable. Let Y_1, Y_2, E_y be the sets defined in the proof of Theorem 2.18.3. For each $y \in (F(b), F(a)) \setminus (Y_1 \cup Y_2)$ select a point x_y such that $\overline{F}'(x_y) \le 0$ and x_y is isolated in E_y (x_y exists, see Theorem 2.18.3). Let X be the set of points selected. By Corollary 2.13.2, $F \in VB^*G$ on X. Let $X_o = \{x \in X : F'(x)$ exists, finite or infinite$\}$. Clearly $X_o \subset N$. Let $X_1 = X \setminus X_o$. By Theorem 2.15.1, $| F(X_1) | = 0$, hence $| F(X) | = | F(X_o) | = F(a) - F(b)$. It follows that $| F(N) | \ge | F(X_o) | = F(a) - F(b)$, hence N is uncountable. Therefore $F \in (-)$ on $[a, b]$.

Remark 2.18.3 *(i) In fact Theorem 2.18.4 is announced by Garg in [Ga] (p. 70), because $\mathcal{D}_-\mathcal{B}_1$ = lower internal $\cap \mathcal{B}_1$ (see Theorem 2.5.1).*

(ii) Theorem 2.18.4 extends the following result of Bruckner ([Br2], p. 179):
Let $F : [a, b] \mapsto \mathbb{R}$, $F \in \mathcal{DB}_1T_2$. Let $P = \{x : +\infty \ge F'(x) \ge 0\}$ and $N = \{x : 0 \ge F'(x) \ge -\infty\}$. Then $P \cup N$ is uncountable and the sets $F(P)$ and $F(N)$ are measurable. If $F(a) < F(b)$ then $| F(P) | \ge F(b) - F(a)$. If $F(a) < F(b)$ then $| F(N) | \ge F(a) - F(b)$.
Particularly, Bruckner's theorem is true if $\mathcal{DB}_1T_2$ is replaced by $\mathcal{C} \cap T_2$, and this result is due to Banach (see [S3], p. 280).

Theorem 2.18.5 *Let $F : [a, b] \mapsto \mathbb{R}$ and let r be a real number. If $F \in \mathcal{D}_-\mathcal{B}_1T_2$ then the sets $\{x \in [a, b) : F(x) > r\}$ and $\{x \in (a, b] : F(x) < r\}$ are either empty or contain an interval.*

Proof. Suppose that $A = \{x \in [a, b) : F(x) > r\} \ne \emptyset$. Let $a_1 \in A$. If $[a_1, b) \subset A$ then there is nothing to prove. Suppose that there exists $b_1 \in (a_1, b)$ such that $F(b_1) \le r$. As in the proof of Theorem 2.18.3, it follows that for $y \in [r, F(a_1)] \setminus (Y_1 \cup Y_2)$ there exist x_y and an open interval $I_y = (a_y, b_y)$ such that $\{x_y\} = E_y \cap I_y$ and $F(x) > y$ for each $x \in (a_y, x_y)$. Hence $[a_y, x_y) \subset A$. For $\{x \in (a, b] : F(x) < r\}$ the proof is similar.

Remark 2.18.4 *In connection with Theorem 2.18.5, we have also the following result due to Bruckner ([Br2], p. 179):*
Let $F : [a, b] \mapsto \mathbb{R}$, $F \in \mathcal{DB}_1T_2$. For every pair of real numbers $c < d$ the set $E_{c,d} = \{x : c < F(x) < d\}$ is either empty or contains an interval.

Theorem 2.18.6 *Let $F : [a, b] \mapsto \mathbb{R}$ and let $H = \{x : F \in \mathcal{C}$ at $x\}$. If $F \in (-)$ on $[a, b]$ then H is a G_δ-set, everywhere dense in $[a, b]$.*

Proof. By Proposition 1.5.2 and Proposition 1.5.1(ix), H is a G_δ-set. Let $J \subset [a, b]$ be an interval. If F is monotone on J then F is continuous *n.e.* on J (see Theorem 2.7.2, a)). Hence $J \cap H \ne \emptyset$. If F is not monotone on J then there exist $x_1, x_2 \in J$, $x_1 < x_2$ such that $F(x_1) > F(x_2)$. Since $F \in (-)$ it follows that $| F(N \cap [x_1, x_2]) | \ge F(x_1) - F(x_2)$, where $N = \{x : F'(x) \le 0\}$. Hence $N \cap [x_1 x_2]$ is uncountable, and $F \in VB^*G$ on $N \cap [x_1, x_2]$ (see Corollary 2.13.2). By Theorem 2.10.1, F is continuous *n.e.* on $N \cap [x_1, x_2]$. Hence $[x_1, x_2]$ contains uncountably many points of continuity. Thus H is everywhere dense in $[a, b]$.

Lemma 2.18.1 (Banach) *Let $P \subseteq [a,b]$ be a closed set and let $F : P \mapsto \mathbb{R}$, $F \in \mathcal{C}$. Then there exists a G_δ-set $A \subset P$ such that F assumes each value $y \in F(P)$ exactly once.*

Proof. The proof is as in [S3] (p. 283), where the set A is shown to be measurable. But clearly A is also a G_δ-set.

Lemma 2.18.2 (Banach) *Let $P \subseteq [a,b]$ be a measurable set and let $F : P \mapsto \mathbb{R}$ be a measurable bounded function. If $F \in (N)$ then we have:*

(i) For every $\epsilon > 0$ there exists a G_δ-set $Q \subset P$ such that $| F(P) \setminus F(Q) |> \epsilon$ and on which F assumes each of its values at most once;

(ii) There exists a Borel set S such that $| F(P)\setminus F(S) |= 0$ and on which F assumes each of its values countable infinitely many times.

Proof. The proof is similar to that of Lemma 7.2. of [S3] (p. 283):

(i) By Lusin's Theorem (see [S3], p. 73), for each $n \geq 1$ there exists a closed set $P_n \subset P$ such that $| P \setminus P_n |< 1/n$ and $F_{/P_n} \in \mathcal{C}$. We may suppose without loss of generality that $P_n \subset P_{n+1}$, $n = \overline{1,\infty}$. Let $H = P \setminus (\cup_{n=1}^{\infty} P_n)$. Then $| H |= 0$, hence $| F(H) |= 0$ (because $F \in (N)$). Since $F(P)$ is bounded we have $| F(P) |=| F(\cup_{n=1}^{\infty} P_n) |= \lim_{n\to\infty} | F(P_n) |$. For $\epsilon > 0$ there exists n_o such that $| F(P) \setminus F(P_{n_o}) |< \epsilon$. By Lemma 2.18.1, for P_{n_o} there exists a G_δ-set $Q \subset P_{n_o}$ such that each value $y \in F(P_{n_o})$ is assumed exactly once by F on Q.

(ii) By (i) it follows that for each positive integer n there exists a G_δ-set $Q_n \subset P$ such that $| F(P) \setminus F(Q_n) |< 1/n$, and on which F assumes each of its values at most once. Let $S = \cup_{n=1}^{\infty} Q_n$. Then $| F(P) \setminus F(S) |= 0$, and for each $y \in F(S)$ the set $\{x \in S : F(x) = y\}$ is at most countable.

Theorem 2.18.7 *Let $P \subseteq [a,b]$ be a measurable set and let $F : P \mapsto \mathbb{R}$ be a measurable function. Then $(N) \subseteq T_2$ on P.*

Proof. We may suppose that F is bounded on P (if not then for each $n \geq 1$ let $P_n = \{c \in P :| F(x) |\leq n\}$; it follows that$F_{/P_n}$ is measurable and bounded). The proof continues as in [S3] (p.284), using Lemma 2.18.2 instead of Lemma 7.2 of [S3] (p. 283).

Remark 2.18.5 *In fact Banach proved Theorem 2.18.7 for continuous functions $F : [a,b] \mapsto \mathbb{R}$ (see[S3], p. 283). Another proof (for continuous functions) was given by Nina Bary in [Ba] (p. 195), using analytic sets. But Theorem 2.18.7 was first proved by Ellis in [El] (using Banach's result and Lusin's Theorem).*

Theorem 2.18.8 *Let $P \subseteq [a,b]$ be a measurable set and let $F : P \mapsto \mathbb{R}$ be a measurable bounded function. Then $(S) = (N) \cap T_1$ on P.*

Proof. $(N) \cap T_1 \subset (S)$ on P is true even if F and P are not supposed to be measurable (see [S3], p. 283). Let $F \in (S)$ on P. By definitions it follows that $F \in (N)$ on P. Suppose that $F \notin T_1$ on P. Then $Y = \{y : s(y) = +\infty\}$ has positive measure and $Y = F(P) \setminus (\cup_{n=1}^{\infty} Y_n)$, where $Y_n = \{y : s(y) \leq n\}$ and $s : \mathbb{R} \mapsto \overline{\mathbb{R}}$, $s(y) =$ the number of roots of the equation $F(x) = y$, $x \in P$. Since $s(y)$ is measurable (see Lemma 2.30.1, (i)), it follows that Y_n is measurable. By Theorem 2.18.2, $F(P)$ is measurable and $| Y |> 0$. Let B be a Borel subset of Y, $| B |> 0$. Since F is measurable, $A = F^{-1}(B)$ is measurable. By Lemma 2.18.2,(i) there exists a G_δ-set $A_1 \subset A$ such that $| F(A_1) |>| B | /2$, and each value $y \in F(A_1)$ is assumed at most once on A_1.Let $E_1 = A \setminus A_1$. Then $F(E_1) = B$. By Lemma 2.18.2, (i), there exists a G_δ-set $A_2 \subset E_1$ such that $| F(A_2) |>| B | /2$ and each value $y \in F(A_2)$ is assumed at most once on A_2. Let $E_2 = E_1 \setminus A_2 = A \setminus (A_1 \cup A_2)$. Continuing we obtain a sequence $\{A_i\}_i$ of pairwise disjoint G_δ-sets, such that $| F(A_i) |>| B | /2$, $i = \overline{1, \infty}$. Let $\epsilon > 0$, $\epsilon <| B | /2$ and let δ be given by the fact that $F \in (S)$ on P. Since A_i are pairwise disjoint and $| P |\leq b - a$, it follows that there exists i_o such that $| A_{i_o} |< \delta$. Then $| F(A_{i_o}) |< \epsilon <| B | /2 <| F(A_{i_o}) |$, a contradiction.

Remark 2.18.6 *In fact Banach proved Theorem 2.18.8 for continuous functions $F : [a, b] \mapsto \mathbb{R}$ (see [S3], p. 284). In [I3], Iseki proved Theorem 2.18.8 for continuous bounded functions $F : P \mapsto \mathbb{R}$ with P a Borel set.*

Theorem 2.18.9 *Let $F : [a, b] \mapsto \mathbb{R}$, $P \subseteq [a, b]$, $c = \inf(P)$, $d = \sup(P)$.*

(i) $VB \subsetneq f.l. \subsetneq T_1 \subset T_1G \subsetneq T_2$ on P.

(ii) $VBG \subsetneq \sigma.f.l. \subsetneq T_1G \subsetneq T_2$ on P.

(iii) If $F \in T_2$ on $[a, b]$ then $F_{\overline{P}} \in T_2$ on $[c, d]$.

(iv) $VB \cap \mathcal{C} \subset VB^*G \cap \mathcal{C} \subset T_1 \subset T_2$ on $[a, b]$.

(v) $ACG \subset SG \subsetneq (N)$ on P.

(vi) $AC \subset AC^*G \subset (S)$, but $ACG \not\subset (S)$ for continuous functions on $[a, b]$.

(vii) (<u>Banach-Zarecki</u>). $VB \cap (N) \cap \mathcal{C} = AC$ on $\overline{P}$.

(viii) $VBG \cap (N) \cap \mathcal{C} = ACG \cap \mathcal{C}$ on $\overline{P}$.

(ix) $VB^*G \cap (N) \cap \mathcal{C} = AC^*G$ on $\overline{P}$.

(x) $(S) \subset SG \subsetneq (N)$ on P;

(xi) $S = S \circ S$ and $(N) = (N) \circ (N)$ on P.

(xii) (<u>Nina Bary</u>). $S = L \circ AC = AC \circ AC$ for continuous functions on $[a, b]$.

(xiii) (<u>Nina Bary</u>). $T_1 = L \circ VB = AC \circ VB$ for continuous functions on $[a, b]$.

(xiv) $\mathcal{D}\_T_1 = \mathcal{C}_i^* \cap T_1$ for bounded functions on $[a, b]$.

(xv) $\mathcal{D}\_T_1 = \mathcal{C}_i \cap T_1 \subset \mathcal{D}\_\mathcal{B}_1T_2$ on $[a, b]$.

(xvi) (<u>Ridder</u>, [R4]). $\mathcal{D} \cap T_1 = \mathcal{C} \cap T_1$ on $[a, b]$.

(xvii) $\mathcal{B}_1T_2 \cap$ lower internal $= \mathcal{D}\_\mathcal{B}_1T_2 \subsetneq (-) \subset$ lower internal on $[a, b]$.

(xviii) $\mathcal{B}_1T_2 \cap$ internal $= \mathcal{D}\mathcal{B}_1T_2 \subset (-) \cap (+) \subset$ internal on $[a, b]$.

(xix) For functions satisfying VB on $[a,b]$ we have:
$(-) = \mathcal{D}_- =$ *lower internal* $=$ *lower internal** $= \mathcal{C}_i^* = \mathcal{C}_i$ *and*
$(+)\cap(-) = \mathcal{D} =$ *internal* $=$ *internal** $= \mathcal{C}$.

Proof. (i) $VB \subseteq f.l.$ and $T_1 \subseteq T_1G \subseteq T_2$ follow by definitions; $f.l. \subseteq T_1$ follows by Theorem 1.3.2; $T_1 \subset T_1G$ follows by Section 6.29 (the function G_{PP}).

(ii) This follows by (i).

(iii) This is evident.

(iv) See [S3] (p. 279), Sections 6.12, Section 6.24 and Section 6.10.1. (or Section 6.26.).

(v) By definitions $AC \subseteq (S)$, hence $ACG \subseteq SG$ on P; $SG \subseteq (N)$ is evident; $ACG \subset SG$ follows by Section 6.31.

(vi) $AC \subset AC^*G$ on $[a,b]$ follows by Section 6.16. We have $AC^*G \subseteq VB^*G \cap ACG \subseteq T_1 \cap (N) = (S)$ (see (iv), (v), Proposition 2.12.1 and Theorem 2.18.8). For $AC^*G \subset (S)$ see Section 6.31, and for $ACG \not\subset (S)$ see Section 6.26.

(vii) See [S3] (p. 227).

(viii) See [S3] (p. 228).

(ix) By Proposition 2.12.1,(vi), Theorem 2.18.9,(vii) and Theorem 2.12.1, (i),(ii), $AC^*G\cap\mathcal{C} = bAC^*G\cap\mathcal{C} \subseteq VB^*G\cap(N)\cap\mathcal{C} \subseteq VB^*G\cap ACG\cap\mathcal{C} = bAC^*G\cap\mathcal{C}$.

(x) This follows by Section 6.29 (the function G_{PP}).

(xi) This is evident.

(xii) and (xiii) These follow by [Ba] (pp. 208, 633), or by [S3] (pp. 287-289).

(xiv) $\mathcal{C}_i^* \cap T_1 \subseteq \mathcal{D}_-T_1$ on $[a,b]$ follows by Theorem 2.4.2, (vi). Let $F \in \mathcal{D}_-T_1$ on $[a,b]$. By Theorem 2.4.2, (vi), F is *lower internal* $\cap\, T_1$ on $[a,b]$.
We show that $F \in \mathcal{C}_i^*$ on $[a,b]$. Let $x_o \in (a,b]$ and suppose that $F(x_o^-)$ does not exist (finite, because F is bounded). Let $\alpha,\beta \in \mathbb{R}$ such that $\liminf_{x\nearrow x_o} F(x) < \alpha < \beta < \limsup_{x\nearrow x_o} F(x)$. Then there exists an increasing sequence $\{x_n\}_n$, $x_n \nearrow x_o$, $x_n \in (a,x_o]$, such that $F(x_{2i-1}) > \beta$ and $F(x_{2i}) < \alpha$, $i \geq 1$. Let $y \in (\alpha,\beta)$. Since $F \in \mathcal{D}_-$ there exists $x_{iy} \in (x_{2i-1}, x_{2i})$, such that $F(x_{iy}) = y$, a contradiction (because $F \in T_1$). It follows that $F(x_o^-)$ exists. Since F is *lower internal*, $F(x_o^-) < F(x_o)$, hence $F \in \mathcal{C}_i^*$ on $[a,b]$.

(xv) By Theorem 2.4.2,(vi), $\mathcal{C}_i \cap T_i \subseteq \mathcal{D}_-T_1$ on $[a,b]$. Let $F \in \mathcal{D}_-T_1$ on $[a,b]$. Suppose that there exists $x_o \in (a,b]$ such that $L_o = \limsup_{x\nearrow x_o} F(x) > F(x_o)$. Let $\alpha,\beta \in \mathbb{R}$ such that $F(x_o) < \alpha < \beta < L_o$, and let $\{x_n\}_n$ be a strictly increasing sequence of real numbers, such that $x_n \nearrow x_o$ and $F(x_n) > \beta$, $n \geq 1$. Since $F \in \mathcal{D}_-$, there exists $z_1 \in (x_1, x_o)$ such that $F(z_1) = \alpha$. Let k_1 be the first natural number such that $x_n \in (z_1, x_o)$ for each $n \geq k_1$. Since $F \in \mathcal{D}_-$ there exists $z_2 \in (x_{k_1}, x_o)$ such that $F(z_2) = \alpha$. Continuing we obtain $x_{k_o} < z_1 < x_{k_1} < z_2 < x_{k_2} < z_3 < \ldots < x_{k_p} \leq z_{p+1} < \ldots$, $k_o = 1$, $p = \overline{0,+\infty}$. Hence $\{x : F(x) = y\}$ is infinite, a contradiction (because $F \in T_1$). So $F \in \mathcal{C}_i$ on $[a,b]$. $\mathcal{C}_i \cap T_1 \subseteq \mathcal{D}_-\mathcal{B}_1T_2$ follows by (i) and Theorem 2.4.2, (vi). $\mathcal{D}_- \subset \mathcal{D}_-\mathcal{B}_1T_2$ follows by Section 6.26.

(xvi) See (xv).

(xvii) $\mathcal{B}_1T_2 \cap$ *lower internal* $= \mathcal{D}_-\mathcal{B}_1T_2$ follows by Theorem 2.5.1, (i),(iv); $\mathcal{D}_-\mathcal{B}_1T_2 \subset (-)$ follows by Theorem 2.18.4 and Section 6.10.
We show that $(-) \subseteq$ *lower internal* on $[a,b]$. Suppose for example that there exists $x_o \in (a,b]$ such that $\liminf_{x\nearrow x_o} F(x) > F(x_o)$. Let $\ell_o = \liminf_{x\nearrow x_o} F(x)$ and $L_o = \limsup_{x\nearrow x_o} F(x)$. Let $\epsilon > 0$, $\epsilon < (\ell_o - F(x_o))/2$. Let $x_1 \in (0,x_o)$ such that $F(x) \in (\ell_o - \epsilon/2, L_o + \epsilon/2)$, whenever $x \in (x_1,x_o)$. Let $x_2 \in (x_1,x_o)$ such that

$F(x_2) > L_o - \epsilon/2$. Since $F(x_2) - F(x_o) \leq | F(x_1, x_o) \cap N) | \leq | F((x_1, x_o)) | < L_o - \ell_o + \epsilon$, where $N = \{x : F'(x) \leq 0\}$. But $F(x_1) - F(x_o) > L_o - \epsilon/2 + 3\epsilon/2$, a contradiction. That $(-) \subset$ *lower internal* follows by Section 6.10.1.

(xviii) See (xvii), Sections 6.18. and 6.10.1.

(xix) By Theorem 2.4.2,(vi), $\mathcal{C}_i^* \subseteq \mathcal{C}_i \subseteq \mathcal{D}_- \subseteq$ *lower internal* $\subseteq$ *lower internal**. By Theorem 2.7.2, b), since $F \in VB \cap$ *lower internal**, it follows that $F \in \mathcal{C}_i^*$ on $[a, b]$. Hence $\mathcal{C}_i^* = \mathcal{C}_i = \mathcal{D}_- =$ *lower internal* $=$ *lower internal** for VB functions on $[a, b]$. Since $VB \subseteq \mathcal{B}_1 T_2$, by (xvii) it follows that $VB \cap \mathcal{D}_- \subseteq (-) \subseteq$ *lower internal* on $[a, b]$. This completes the proof for the first part. Now the second part follows easily.

Remark 2.18.7 *(i) $(N)+(N) \neq (N)$ for continuous functions on $[a, b]$ (see Section 6.31.).*

(ii) $(N) + ($a linear nonconstant function$) \neq (N)$ for continuous functions on $[a, b]$;

(iii) $AC \circ AC + AC \not\subset (N)$ for continuous functions on $[a, b]$ (see Section 6.27.);

(iv) In Theorem 2.18.9, (iv), for $VB^*G \cap C \subset T_1$, the continuity hypothesis is not superfluous (see Section 6.12.). This hypothesis may however be replaced by a weaker one: "the function F has no point of discontinuity other that of the first kind" (see [S3], p. 279).

Theorem 2.18.10 (Nina Bary and D. Menchoff) *([S3], p.289). Let $F : [a, b] \mapsto \mathbb{R}$, $F \in C$. The following assertions are equivalent:*

(i) $F \in AC \circ AC$ on $[a, b]$;

(ii) $F(\{x : F'(x)$ does not exist finite$\})$ has measure zero.

Lemma 2.18.3 (Foran) *Let $F : [a, b] \mapsto \mathbb{R}$ and let P be a perfect subset of $[a, b]$, $| P | = 0$. If $F \in (N) \cap C$ on $[a, b]$ then there exists a strictly increasing function $g : [a, b] \mapsto \mathbb{R}$ such that $g \in AC$ on $[a, b]$, g is linear on each interval contiguous to P, $g^{-1} \in AC$, g^{-1} is linear on each interval contiguous to $g(P)$, and $\Lambda(B(F \circ g^{-1}; g(P))) = 0$.*

Proof. ([F4], p. 35). Since $F(P)$ is a compact set of measure zero, there is a perfect set P' such that $P' \times F(P)$ is of linear measure zero. For example P' can be constructed as follows: Let $Q = F(P)$, $\epsilon_1 = diam(Q)$ and $n_1 = 1$. Since Q is a compact set of measure zero, ϵ_{i+1} and n_{i+1} can be defined inductively so that Q can be covered by n_{i+1} intervals of equal length e_{i+1}, such that $n_{i+1} \cdot e_{i+1} < \epsilon_i / 2^{i+1}$. Let P' be the set of all numbers x of the form $\sum_i a_i \cdot \epsilon_i$, with $a_i = 0$ or 1. Then P' can be covered by 2^{i+1} intervals of length ϵ_{i+1} and hence $P' \times Q$ can be covered by $n_{i+1} \cdot 2^{i+1}$ squares of diameter $\epsilon_{i+1} \cdot \sqrt{2}$. Since $n_{i+1} \cdot 2^{i+1} \cdot \epsilon_{i+1} \cdot \sqrt{2} < \epsilon_i \cdot \sqrt{2}$ and since $\epsilon_i \to 0$ it follows that $P' \times Q$ is of linear measure zero. Since P' is a perfect set, there is a strictly increasing function $g(x)$ such that $g(P) = P'$ and such that $g(x)$ is linear on each interval contiguous to P. Then, since g and g^{-1} are strictly increasing and satisfy condition (N), both g and g^{-1} are AC. Since $B(F \circ g^{-1}; g(P)) \subset P' \times F(P)$, it follows that $\Lambda(B(F \circ g^{-1}; g(P))) = 0$.

2.19 Conditions wS, wN

Definition 2.19.1 (Foran) *Let P be a real set and $F : P \mapsto \mathbb{R}$. F is said to be wN (weak (N)) on P if $| F(Z) |= 0$, whenever Z is a compact subset of P and $| Z |= 0$.*

Definition 2.19.2 (Iseki) *Let P be a real set and $F : P \mapsto \mathbb{R}$. F is said to be wS (weak (S)) on P if for every $\epsilon > 0$ there exists a $\delta > 0$ such that $| F(z) |< \epsilon$, whenever Z is a compact subset of P and $| Z |< \delta$.*

Theorem 2.19.1 *Let $F : [a, b] \mapsto \mathbb{R}$, $P \subseteq [a, b]$.*

(i) $(S) \subseteq wS$ on P;

(ii) If P is measurable then $(S) = wS \cap (N)$ on P;

(iii) If P is of F_σ-type then $wS \subseteq (N)$ on P, hence $(S) = wS$ on P;

(iv) $(N) \subseteq wN$ on P;

(v) $wS \subseteq wN$ on P.

Proof. (i), (iv) and (v) are evident.

(ii) $(S) \subseteq wS \cap (N)$ follows by (i) and Theorem 2.18.9, (x). Let $F \in wS \cap (N)$ on P. For $\epsilon > 0$ let δ be given by the fact that $F \in wS$ on P. Let Z be a measurable subset of P, $| Z |< \delta$. We have two situations:
(I) Z is a set of F_σ-type. Then there exist $Q_1 \subset Q_2 \subset \ldots \subset Q_n \subset \ldots$, compact sets, such that $Z = \cup_{n=1}^\infty Q_n$. But $F(Z) = F(\cup_{n=1}^\infty Q_n) = \cup_{n=1}^\infty F(Q_n)$. Since $\{F(Q_n)\}_n$ is an increasing sequence of sets, it follows that $| F(Z) | = \lim_{n\to\infty} | F(Q_n) |$. But $| Q_n |< \delta$, $n = \overline{1,\infty}$, hence $| F(Q_n) |< \epsilon$, $n = \overline{1,\infty}$. It follows that $| F(Z) |\le \epsilon$, hence $F \in (S)$ on P.
(II) Z is not of F_σ-type. By Theorem 1.5.1 there exists $A \subset Z$ such that A is of F_σ-type and $| Z \setminus A |= 0$. We have $| F(Z) |\le| F(A) | + | F(Z \setminus A) |$. But $| F(A) |\le \epsilon$ (see (I)), and $| F(Z \setminus A) |= 0$ (since $F \in (N)$). Hence $| F(Z) |\le \epsilon$. So $F \in (S)$ on P.

(iii) Let $Z \subset P$, $| Z |= 0$. For $\epsilon > 0$ let δ be given by the fact that $F \in wS$ on P. Then there exists an open set Q such that $Z \subset Q$ and $| Q |< \delta$. It follows that Z is contained in the set $Q \cap P$ which is of F_σ-type (see Proposition 1.5.1, (vi), (iii)). Similarly to (ii), (II) we obtain that $| F(Z) |\le| F(Q \cap P) |< \epsilon$. Hence $| F(Z) |= 0$, so $F \in (N)$ on P.

Remark 2.19.1 *For more information on this topic see [F3] and [I3].*

2.20 Condition $(\overline{N})$

Definition 2.20.1 *Let $F : [a, b] \mapsto \mathbb{R}$, $P \subseteq [a, b]$. F is said to be $(\overline{N})$ on P if $\mathcal{O}_+^\infty(F; Z) = 0$, whenever $Z \subset P$ and $| Z |= 0$. F is said to be $(\underline{N})$ on P if $-F$ is $(\overline{N})$ on P, i.e., $\mathcal{O}_-^\infty(F; Z) = 0$.*
We can define the class $\overline{N}G$, using Definition 1.9.2, but $\overline{N}G = (\overline{N})$ on P.

Lemma 2.20.1 *Let $F : [a, b] \mapsto \mathbb{R}$ and let $E_k \subset [a, b]$, $k = \overline{1,\infty}$. If $F \in (\overline{N})$ on each E_k then $F \in (\overline{N})$ on $E = \cup_{k=1}^\infty E_k$.*

Proof. Let $\epsilon > 0$, $Z \subset E$, $|Z| = 0$ and $Z_k = Z \cap E_k$, $k = \overline{1,\infty}$. Then $|Z_k| = 0$, $k = \overline{1,\infty}$. Since $F \in (\overline{N})$ on E_k, for $\epsilon/2^k$ there exists Z_{ki}, $i = \overline{1,\infty}$ such that $Z_k = \cup_{i=1}^\infty Z_{ki}$ and $0 \le \sum_{k=1}^\infty \mathcal{O}_+(F; Z_{ki}) < \epsilon/2^k$, hence $0 \le \sum_{k=1}^\infty \sum_{i=1}^\infty \mathcal{O}_+(F; Z_{ki}) < \epsilon$. Since $Z = \cup_{k=1}^\infty Z_k = \cup_{k=1}^\infty \cup_{i=1}^\infty Z_{ki}$ and ϵ is arbitrary, it follows that $\mathcal{O}_+^\infty(F; Z) = 0$.

Theorem 2.20.1 *Let $F : [a, b] \mapsto \mathbb{R}$, $P \subseteq [a, b]$, $c = \inf(P)$, $d = \sup(P)$.*

(i) $F \in (N)$ if and only if $\mathcal{O}^\infty(F; Z) = 0$, whenever $Z \subset P$, $|Z| = 0$;

(ii) $(N) \subseteq (\underline{N}) \cap (\overline{N})$ on P;

(iii) $\overline{AC} \subseteq \overline{ACG} \subseteq (\overline{N})$ on P;

(iv) If $F \in (\overline{N})$ on P then $F_{\overline{P}} \in (\overline{N})$ on $[c, d]$.

Proof. (i) Let $F \in (N)$ on P, $\epsilon > 0$ and $Z \subset P$, $|Z| = 0$. Then $|F(Z)| = 0$. It follows that there exists an open set G such that $F(Z) \subset G$ and $|G| < \epsilon$. Then $G = \cup_{i\ge1}(c_i, d_i)$. Let $Z_i = \{x \in Z : F(x) \in (c_i, d_i)\}$. Then $\mathcal{O}(F; Z_i) \le d_i - c_i$ for each i. Hence $0 \le \sum_{i\ge1} \mathcal{O}(F; Z_i) \le \epsilon$. Since ϵ was arbitrary it follows that $\mathcal{O}^\infty(F; Z) = 0$. Conversely, let $\epsilon > 0$ and $Z \subset P$, $|Z| = 0$. Since $\mathcal{O}^\infty(F; Z) = 0$, there exist Z_i, $i = \overline{1,\infty}$ such that $\cup_{i=1}^\infty Z_i = Z$ and $\sum_{i=1}^\infty \mathcal{O}(F; X_i) \le \epsilon$. Let $J_i = [\inf(F(Z_i)), \sup(F(Z_i))]$. Then $|J_i| = \mathcal{O}(F; Z_i)$. Hence $F(Z) \subset \cup_{i=1}^\infty J_i$ and $\sum_{i=1}^\infty |J_i| < \epsilon$. It follows that $|F(Z)| < \epsilon$. Since ϵ was arbitrary it follows that $|F(Z)| = 0$, hence $F \in (N)$ on P.

(ii) See (i) and Corollary 1.4.1, (iii).

(iii) Clearly $\overline{AC} \subseteq \overline{ACG}$ on P. To show that $\overline{ACG} \subseteq (\overline{N})$ on P it is sufficient to show that $\overline{AC} \subseteq (\overline{N})$ on P. Let $Z \subset P$, $|Z| = 0$. For $\epsilon > 0$ let δ be given by the fact that $F \in \overline{AC}$ on P. Let G be an open set such that $Z \subset G$ and $|G| < \delta$. Then $G = \cup_{i\ge1}(a_i, b_i)$, where $\{(a_i, b_i)\}_i$ are nonoverlapping intervals. Then $0 \le \sum_{i\ge1} \mathcal{O}_+(F; (a_i, b_i) \cap Z) < \epsilon$. It follows that $\mathcal{O}_+^\infty(F; Z) = 0$, hence $F \in (\overline{N})$ on P.

(iv) This follows by definitions.

2.21 Conditions N^∞, $N^{+\infty}$, $N^{-\infty}$

Definition 2.21.1 (Saks) *Let $F : [a, b] \mapsto \mathbb{R}$, $P \subseteq [a, b]$. F is said to be $N^{+\infty}$ on P if the set $F(\{x \in P : (F_{/P})'(x) = +\infty\})$ is of measure zero. F is said to be $N^{-\infty}$ on P if $-F$ is $N^{+\infty}$ on P, i.e., the set $F(\{x \in P : (F_{/P})'(x) = -\infty\})$ is of measure zero. Let $N^\infty = N^{-\infty} \cap N^{+\infty}$ on P.*

Theorem 2.21.1 *Let $F : [a, b] \mapsto \mathbb{R}$. Let P be a closed subset of $[a, b]$, $c = \inf(P)$, $d = \sup(P)$ and let $\{(c_k, d_k)\}_k$ be the intervals contiguous to P.*

(i) $F \in N^{-\infty}$ on P if and only if $F_P \in N^{-\infty}$ on $[c, d]$.

(ii) If $\sum_{k\ge1} \Omega_-(F; [c_k, d_k] \wedge \{c_k, d_k\}) > -\infty$ and $F \in N^{-\infty}$ on $[a, b]$ then $F_P \in N^{-\infty}$ on $[c, d]$.

Proof. (i) See Lemma 1.10.2.

(ii) Let $R_1 = \{x \in P : F'(x)$ exists, finite or infinite$\}$, $R_2 = \{x \in P : F_P'(x)$ exists, finite or infinite$\}$ and $T = (R_1 \setminus R_2) \cup (R_2 \setminus R_1)$. By Lemma 2.15.1, (ii), $|F(T)| = |T| = 0$. Let's define $R_1^{-\infty} = \{x \in P : F'(x) = -\infty\}$ and $R_2^{-\infty} = \{x \in P :$

$F'_P(x) = -\infty\}$. Then $R_2^{-\infty} = (R_2^{-\infty} \cap R_1) \cup (R_2^{-\infty} \setminus R_1) \subset R_1^{-\infty} \cup (R_2 \setminus R_1) \cup R_1^{-\infty} \cup T$. Since $F \in N^{-\infty}$ on $[a,b]$, $| F(R_1^{-\infty}) |= 0$. It follows that $| F(R_2^{-\infty}) |= 0$, hence $F_P \in N^{-\infty}$ on $[c,d]$.

Lemma 2.21.1 *Let $F : [a,b] \mapsto \mathbb{R}$ and let P be a closed subset of $[a,b]$.*

(i) $(\underline{N}) \subseteq N^{-\infty}$ on P, hence $(N) \subseteq N^{\infty}$ on P.

(ii) (Saks, Ridder). $(N) = N^{\infty}$ on $[a,b]$, for $\mathcal{D} \cap T_1$ functions.

Proof. (i) Let $F \in (\underline{N})$ on P. Then $F_P \in (\underline{N})$ on $[c,d]$, where $c = \inf(P)$, $d = \sup(P)$. Let $E-\infty = \{x \in [c,d] : F'_P(x) = -\infty\}$. Then $E^{-\infty} = \cup_{k=1}^{\infty} P_k$ and F_P is *strictly decreasing** on P_k. Hence F is strictly decreasing on P_k. By Corollary 2.15.1, $| E^{-\infty} |= 0$, hence $| P_k |= 0$, $k = \overline{1,\infty}$. It follows that $| \mathcal{O}_-^{\infty}(F;P_k) |= \mathcal{O}^{\infty}(F;P_k) = 0$, hence $| F(P_k) |= 0$. Thus $| F(E^{-\infty}) |= 0$ and $F_P \in N^{-\infty}$ on $[c,d]$.

(ii) ([S3], p.131). By Theorem 2.18.9, (xvi), $\mathcal{D} \cap T_1 = \mathcal{C} \cap T_1$. Let $F \in \mathcal{C} \cap T_1 \cap N^{\infty}$ on $[a,b]$. Let $Z \subset [a,b]$, $| Z |= 0$, $Z_1 = \{x \in Z : F'(x)$ exists and is finite$\}$, $Z_2 = \{x \in Z : F'(x)$ exists and is infinite$\}$ and $Z_3 = Z \setminus (Z_1 \cup Z_2)$. Then $Z = Z_1 \cup Z_2 \cup Z_3$ and $| F(Z_2) |= 0$. By Theorem 2.18.1, $| F(Z_3) |= 0$. By Corollary 2.13.3, $F \in AC^*G$ on Z_1. Hence $| F(Z_1) |= 0$. Thus $| F(Z) |= 0$, and $F \in (N)$ on $[a,b]$.

Theorem 2.21.2 *Let $F : [a,b] \mapsto \mathbb{R}$. The following assertions are equivalent:*

(i) $F \in \underline{AC}$ on $[a,b]$;

(ii) $F \in (\underline{N}) \cap (-) \cap VB$ on $[a,b]$;

(iii) $F \in N^{-\infty} \cap (-) \cap VB$ on $[a,b]$;

(iv) $F \in N^{-\infty} \cap (-)$ on $[a,b]$ and F' is summable on $E = \{x : F'(x) \leq 0\}$.

Proof. By Corollary 1.10.2, the set E is measurable.

(i) ⇒ *(ii)* $\underline{AC} \subseteq (\underline{N})$ follows by Theorem 2.20.1,(iii); $\underline{AC} \subseteq VB \cap \mathcal{C}_i^*$ follows by Theorem 2.11.1, (vi), (xxi). By Theorem 2.18.9, (xix), $F \in (-)$ on $[a,b]$.

(ii) ⇒ *(iii)* See Lemma 2.21.1.

(iii) ⇒ *(iv)* By Theorem 2.14.3, F is derivable *a.e.* on $[a,b]$ and F' is summable on E.

(iv) ⇒ *(i)* Let $E^{-\infty} = \{x : F'(x) = -\infty\}$ and $E^{-} = \{x : -\infty < F'() \leq 0\}$. Clearly $E = E^{-\infty} \cup E^{-}$. Let $g : [a,b] \mapsto \mathbb{R}$, $g(x) = F'(x)$ if $x \in E^{-}$, and $g(x) = 0$ if $x \notin E^{-}$. Let $G(x) = (\mathcal{L}) \int_a^x | g(t) | \, dt$, $x \in [a,b]$. Since F' is summable on E, by Theorem 2.14.2, (i), G is AC and increasing on $[a,b]$. By Theorem 2.14.4, for $a \leq c < d \leq b$ with $F(c) > F(d)$, we have

$$(1) \quad F(c) - F(d) \leq | F([c,d] \cap E) | = | F([c,d] \cap E^{-}) | \leq (\mathcal{L}) \int_{[c,d] \cap E^{-}} | F'(t) | \, dt$$

$$= (\mathcal{L}) \int_{[c,d] \cap E^{-}} | g(t) | \, dt \leq G(d) - G(c) = | G(d) - G(c) | .$$

For $\epsilon > 0$ let δ be given by the fact that $G \in AC$ on $[a,b]$. Let $\{[a_k,b_k]\}$, $k = \overline{1,n}$ be a finite set of nonoverlapping closed intervals, such that $\sum_{k=1}^{n}(b_k - a_k) < \delta$. We may suppose that $F(b_k) - F(a_k) < 0$, $k = \overline{1,n}$. By (1), $\sum_{k=1}^{n}(F(b_k) - F(a_k)) \geq -\sum_{k=1}^{n} | G(b_k) - G(a_k) | \geq -\epsilon$. Hence $F \in \underline{AC}$ on $[a,b]$.

Remark 2.21.1 *Theorem 2.21.2 is an extension of a theorem of Saks (see [S3], Theorem 6 of p. 135).*

Corollary 2.21.1 *Let $F : [a,b] \mapsto \mathbb{R}$ and let P be a closed subset of $[a,b]$.*

(i) $\underline{AC} = \underline{AC}G \cap C_i \cap VB = (\underline{N}) \cap C_i \cap VB = N^{-\infty} \cap C_i \cap VB = N^{-\infty} \cap C_i^* \cap VB = N^{-\infty} \cap (-) \cap VB = N^{-\infty} \cap$ *lower internal*$^* \cap VB = N^{-\infty} \cap$ *lower internal* $\cap VB = N^{-\infty} \cap \mathcal{D}_- \cap VB$ *on* $[a,b]$.

(ii) $AC = ACG \cap C \cap VB = (N) \cap C \cap VB = N^{\infty} \cap C \cap VB = N^{\infty} \cap (-) \cap (+) \cap VB = N^{\infty} \cap$ *internal*$^* \cap VB = N^{\infty} \cap$ *internal* $\cap VB = N^{\infty} \cap \mathcal{D} \cap VB = (N) \cap (-) \cap (+) \cap VB = (N) \cap$ *internal*$^* \cap VB = (N) \cap$ *internal* $\cap VB = (N) \cap D \cap VB$ *on* $[a,b]$.

(iii) $\underline{AC} = \underline{AC}G \cap C_i \cap VB = (\underline{N}) \cap C_i \cap VB$ *on* P.

(iv) $AC = ACG \cap C \cap VB = (N) \cap C \cap VB$ *on* P.

Proof. See Theorem 2.21.2 and Theorem 2.18.9, (xix).

Remark 2.21.2 *Corollary 2.21.1, (iv) contains the Banach Zarecki Theorem (see Theorem 2.18.9, (vii)). That $\mathcal{D} \cap VB \cap N^{\infty} = AC$ on $[a,b]$ was shown by Ridder in [R4] (p. 165).*

Corollary 2.21.2 *Let $F : [a,b] \mapsto \mathbb{R}$. Let $\mathcal{P}$ be any of the following classes of functions on $[a,b]$: $C_i^* \cap T_2 \cap N^{-\infty}$, $C_i \cap T_2 \cap N^{-\infty}$, $\mathcal{D}_-B_1T_2 \cap N^{-\infty} =$ lower internal $\cap$ $B_1\, T_2 \cap N^{-\infty}$. Then the following assertions are equivalent:*

(i) $F \in \underline{AC}$ *on* $[a,b]$;

(ii) $F \in \mathcal{P}$ *on* $[a,b]$ *and* F' *is summable on the set* $\{x : F'(x) \leq 0\}$.

Proof. See Theorem 2.12.2, Theorem 2.4.2, (vi) and Theorem 2.18.9, (xvii).

Remark 2.21.3 *Saks showed that a function $F \in \underline{AC} \cap C$ on $[a,b]$ if and only if $F \in C \cap T_2 \cap N^{-\infty}$ on $[a,b]$ and F' is summable on $\{x : F'(x) \leq 0\}$ (see [S3], p. 135).*

Corollary 2.21.3 *Let $F : [a,b] \mapsto \mathbb{R}$ and let $\mathcal{P}$ be any of the following classes of functions on $[a,b]$: $(+) \cap (-) \cap N^{\infty}$, $\mathcal{C} \cap (N)$, $\mathcal{D}B_1 \cap (N)$, $\mathcal{C} \cap T_2 \cap N^{\infty}$, $\mathcal{D}B_1T_2 \cap N^{\infty}$. The following assertions are equivalent:*

(i) $F \in AC$ *on* $[a,b]$;

(ii) $F \in \mathcal{P}$ *on* $[a,b]$ *and* F' *is summable on the set* $\{x : F'(x) \leq 0\}$.

Proof. (*i*) $\Rightarrow$ (*ii*) This is evident. (*ii*) $\Rightarrow$ (*i*) Let $\mathcal{P} = (+) \cap (-) \cap N^{\infty}$. By Theorem 2.21.1, $F \in \underline{AC} \subseteq VB$ on $[a,b]$. By Corollary 2.21.2, $F \in (+) \cap N^{\infty} \cap VB \subset (+) \cap N^{+\infty} \cap VB = \overline{AC}$. It follows that $F \in AC$ on $[a,b]$. The other parts follow by the facts that $\mathcal{C} \cap (N) \subseteq \mathcal{D}B_1 \cap (N) \subseteq \mathcal{D}B_1T_2 \cap N^{\infty} \subseteq (+) \cap (-) \cap N^{\infty}$ and $\mathcal{C} \cap T_2 \cap N^{\infty} \subset \mathcal{D}B_1T_2 \cap N^{\infty}$ (see Theorem 2.18.9).

Remark 2.21.4 *Saks proved Corollary 2.21.3 for $\mathcal{P} = \mathcal{C} \cap (N)$ (see [S3], p. 285).*

2.22   Conditions M^*, $\overline{M}^*$

Definition 2.22.1 *Let $F : [a, b] \mapsto \mathbb{R}$. F is said to be M^* (resp. $\overline{M}^*$) on $[a, b]$ if $F \in AC$ (resp $\overline{AC}$) on P, whenever $P \subseteq [a, b]$ and $F \in VB^*$ on P. F is said to be $\underline{M}^*$ on $[a, b]$ if $-F \in \overline{M}^*$ on $[a, b]$, i.e., $F \in \underline{AC}$ on P. Clearly $M^* = \overline{M}^* \cap \underline{M}^*$ on $[a, b]$.*

Theorem 2.22.1 $\underline{M}^* + \underline{AC}^* G \subseteq N^{-\infty}$ *on $[a, b]$, hence $\underline{M}^* \subseteq N^{-\infty}$ and $\underline{AC}^* G \subseteq N^{-\infty}$ on $[a, b]$.*

Proof. Let $F, G, H : [a, b] \mapsto \mathbb{R}$, $H = F + G$, $F \in \underline{M}^*$, $G \in \underline{AC}^* G \subseteq VB^* G$ on $[a, b]$. Let $E^{-\infty} = \{x : H'(x) = -\infty\}$. By Corollary 2.13.2, $H \in VB^* G$ on $E^{-\infty}$, hence $F \in VB^* G$ on $E^{-\infty}$. It follows that there exists a sequence $\{P_n\}_n$ of closed sets such that $E^{-\infty} \subset \cup_{n=1}^{\infty} P_n$ and $F \in VB^* \cap \underline{AC}$ on each P_n (because $F \in \underline{M}^*$), hence $F \in b\underline{AC}^*$ on P_n (see Theorem 2.12.2). It follows that $F \in \underline{AC}^* G$ on $E^{-\infty}$, hence $H \in \underline{AC}^* G$ on $E^{-\infty}$. By Theorem 2.13.2 and Proposition 2.13.1, $H \in \overline{AC}^* G$ on $E^{-\infty}$. Hence $H \in AC^* G \subseteq (N)$ on $E^{-\infty}$. Since $\mid E^{-\infty} \mid = 0$ (see Corollary 2.15.1), it follows that $\mid H(E^{-\infty}) \mid = 0$. Thus $H \in N^{-\infty}$ on $[a, b]$.

Theorem 2.22.2 *Let $F : [a, b] \mapsto \mathbb{R}$, $F \in$ lower internal. The following assertions are equivalent:*

(i) $F \in N^{-\infty}$ on $[a, b]$;

(ii) $F \in \underline{M}^*$ on $[a, b]$;

(iii) $F \in \underline{AC}$ on P, whenever $P \subseteq [a, b]$ and $F \in \overline{L}(P; [\inf(P), \sup(P)]\}$;

(iv) $F \in \underline{AC}$ on P, whenever $P \subseteq [a, b]$ and $F \in \overline{L}(P; [\inf(P), \sup(P)])$ with a negative constant;

(v) $\mid F(Z) \mid = 0$, whenever Z is a subset of $[a, b]$, with $\mid Z \mid = 0$ and $F \in \overline{L}(Z; [\inf(Z), \sup(Z)])$;

(vi) $\mid F(Z) \mid = 0$, whenever Z is a subset of $[a, b]$, with $\mid Z \mid = 0$ and $F \in \overline{L}(Z; [\inf(Z), \sup(Z)])$, with a negative constant.

Proof. $(i) \Rightarrow (ii)$ Let $P \subseteq [a, b]$ such that $F \in VB^*$ on P. Then $F \in VB^*$ on $\overline{P}$ and $F_{\overline{P}} \in VB$ on $[c, d]$, where $c = \inf(P)$, $d = \sup(P)$. By Theorem 2.10.2, $F \in \mathcal{C}_i$ on $\overline{P}$, hence $F_{\overline{P}} \in \mathcal{C}_i$ on $[c, d]$ (see Proposition 2.3.2). Let $E^{-\infty} = \{x \in \overline{P} : F'(x) = -\infty\}$ and $E_1^{-\infty} = \{x \in \overline{P} : F'(x) = -\infty\}$. By Lemma 2.15.2, there exists a set $Q \subset \overline{P}$ such that

(1) $\mid F(Q) \mid = \mid Q \mid = 0$ *and*

(2) $F'(x) = (F_{\overline{P}})'(x) = 0$ *on* $\overline{P} \setminus Q$.

By (2), $E_1^{-\infty} = (E_1^{-\infty} \cap Q) \cup (E_1^{-\infty} \cap (\overline{P} \setminus Q)) \subset Q \cup E^{-\infty}$. Since $F \in N^{-\infty}$ on $[a, b]$, by (1) we have $\mid F_{\overline{P}}(E_1^{-\infty}) \mid = \mid F(E_1^{-\infty}) \mid \leq \mid F(Q) \mid + \mid F(E^{-\infty}) \mid = 0$. Hence $F_{\overline{P}} \in N^{-\infty}$ on $[c, d]$. By Corollary 2.21.1, (i), it follows that $F_{\overline{P}} \in N^{-\infty} \cap (-) \cap VB = \underline{AC}$ on $[c, d]$. Thus $F \in \underline{AC}$ on P and $F \in \underline{M}^*$ on $[a, b]$.

   $(ii) \Rightarrow (iii)$ See Proposition 2.13.1.
   $(iii) \Rightarrow (iv) \Rightarrow (i)$ and $(v) \Rightarrow (vi) \Rightarrow (i)$ are evident.
   $(iii) \Rightarrow (v)$ Let Z be as in (v). Then $F \in AC \subseteq (N)$ on Z. Hence $\mid F(Z) \mid = 0$.

Remark 2.22.1 *(i) Each of the assertions (ii) – (vi) in Theorem 2.22.2 implies that $F \in N^{-\infty}$ on $[a,b]$, even if F is not supposed to be lower internal.*

(ii) Clearly Theorem 2.22.2 is valid for the subclasses of lower internal functions, such as: $\mathcal{C}_i^*$, $\mathcal{C}_i$, $\mathcal{D}_-$, $\mathcal{D}$, etc.

Corollary 2.22.1 *(i) $M^* = N^\infty$ on $[a,b]$, for internal functions;*

(ii) $\mathcal{C}_i^* \cap N^{-\infty} \cap VB^*G \subset \mathcal{C}_i \cap N^{-\infty} \cap VB^*G \subset \mathcal{D}_- \cap N^{-\infty} \cap VB^*G =$ lower internal $\cap N^{-\infty} \cap VB^*G \subset [b\underline{AC}^*G] \subseteq \underline{AC}^*G$ on $[a,b]$;

(iii) $\mathcal{C} \cap N^\infty \cap VB^*G \subseteq \mathcal{D} \cap N^\infty \cap VB^*G =$ internal $\cap N^\infty \cap VB^*G$ on $[a,b]$;

(iv) Let $\mathcal{P}$ be any of the following classes of functions: $\mathcal{C}_i^*$, $\mathcal{C}_i$. $\mathcal{D}_-$, lower internal. Then $\mathcal{P} \cap [b\underline{AC}^*G] = \mathcal{P} \cap [\underline{AC}^*G] = \mathcal{P} \cap \underline{AC}^*G = \mathcal{P} \cap N^{-\infty} \cap VB^*G$ on $[a,b]$;

(v) Let $\mathcal{P}$ be any of the following classes of functions: $\mathcal{C}$, $\mathcal{D}$, internal. Then $\mathcal{P} \cap [bAC^*G] = \mathcal{P} \cap [AC^*G] = \mathcal{P} \cap AC^*G = \mathcal{P} \cap N^\infty \cap VB^*G$ on $[a,b]$;

(vi) Let $\mathcal{P}_1$ and $\mathcal{P}_2$ be two classes of functions on $[a,b]$. Then $(\mathcal{P}_1 \cap N^{-\infty}) + (\mathcal{P}_2 \cap \underline{AC}^*G) = \mathcal{P}_1 \cap N^{-\infty}$ and $(\mathcal{P}_2 \cap N^{-\infty}) + (\mathcal{P}_1 \cap \underline{AC}^*G) = \mathcal{P}_1 \cap N^{-\infty}$ on $[a,b]$, whenever:

a) $\mathcal{P}_1 =$ lower internal, $\mathcal{P}_2 = \mathcal{C}_i^*$ or $\mathcal{P}_2 = \mathcal{C}_i$;

b) $\mathcal{P}_1 = \mathcal{C}_i$, $\mathcal{P}_2 = \mathcal{C}_i^*$ or $\mathcal{P}_2 = \mathcal{C}_i$;

c) $\mathcal{P}_1 = \mathcal{P}_2 = \mathcal{C}_i^*$;

(vii) Let $\mathcal{P}_1$ and $\mathcal{P}_2$ be two classes of functions on $[a,b]$. Then $(\mathcal{P}_1 \cap N^\infty) + (\mathcal{P}_2 \cap AC^*G) = \mathcal{P}_1 \cap N^\infty$ and $(\mathcal{P}_2 \cap N^\infty) + (\mathcal{P}_1 \cap AC^*G) = \mathcal{P}_1 \cap N^\infty$ on $[a,b]$, whenever $\mathcal{P}_1 =$ internal, or $\mathcal{P}_1 = \mathcal{P}_2 = \mathcal{C}$.

Proof. (i) See Theorem 2.22.2, (i), (ii).

(ii) $\mathcal{C}_i^* \cap N^{-\infty} \cap VB^*G \subset \mathcal{C}_i \cap N^{-\infty} \cap VB^*G \subset \mathcal{D}_- \cap N^{-\infty} \cap VB^*G$ follows by Theorem 2.4.2, (vi) and Section 6.12.; $\mathcal{D}_- \cap N^{-\infty} \cap VB^*G =$ *lower internal* $\cap N^{-\infty} \cap VB^*G$ follows by Theorem 2.10.3, (v). Let $F \in$ *lower internal* $\cap N^{-\infty} \cap VB^*G$ on $[a,b]$. But $VB^*G = [VB^*G]$ on $[a,b]$. It follows that $[a,b] = \cup_{i=1}^\infty P_n$, $P_n = \overline{P}_n$, such that $F \in VB^*$ on each P_n. By Theorem 2.22.2, (i), (ii), $F \in \underline{AC} \cap VB^* = b\underline{AC}^*$ on each P_n. Hence $F \in [b\underline{AC}^*G]$ on $[a,b]$. That the inclusion is strict follows by Section 6.25.

(iii) See (ii).

(iv) We have $[b\underline{AC}^*G] \subseteq [\underline{AC}^*G] \subseteq \underline{AC}^*G \subseteq N^{-\infty} \cap VB^*G$ on $[a,b]$ (see Theorem 2.21.1 and Proposition 2.12.1, (iv)). Now the proof follows by (ii).

(v) The proof is similar to that of (iv).

(vi) See Theorem 2.22.2, Theorem 2.22.1 and Theorem 2.4.2.

(vii) See (vi).

Remark 2.22.2 *Saks showed that $\mathcal{C} \cap AC^*G = \mathcal{C} \cap N^\infty \cap VB^*G$ on $[a,b]$ (see [S2], p. 133).*

Corollary 2.22.2 *Let $F : [a,b] \mapsto \mathbb{R}$ F internal. The following assertions are equivalent:*

(i) $F \in M^*$ on $[a,b]$;

(ii) $F \in AC$ on each subset E of $[a,b]$ on which F is monotone.*

Proof. *(i)* $\Rightarrow$ *(ii)* Suppose that F is *monotone** on a subset E of $[a,b]$. By Proposition 2.9.1, (ii), $F \in VB^*$ on E. But $F \in M^*$, so $F \in AC$ on E.

(ii) $\Rightarrow$ *(i)* By Theorem 2.22.3, (ii), (iv) it follows that $F \in \underline{M}^* \cap \overline{M}^* = M^*$ on $[a,b]$.

Theorem 2.22.3 *$(N^\infty \cap \mathcal{D}) \circ (N^\infty \cap \mathcal{D}) = N^\infty \cap \mathcal{D}$ on $[a,b]$.*

Proof. Let $g : [a,b] \mapsto \mathbb{R}$, $g \in N^\infty \cap \mathcal{D}$, $f : g([a,b]) \mapsto \mathbb{R}$, $f \in N^\infty \cap \mathcal{D}$ and $F = f \circ g$. Then $F \in \mathcal{D} \subseteq$ *lower internal* on $[a,b]$.
We show that $F \in N^{-\infty}$ on $[a,b]$. Let $P \subseteq [a,b]$, $c = \inf(P)$, $d = \sup(P)$, such that $F \in \overline{L}(P;[c,d])$ with a negative constant M. By Proposition 2.13.1, (vi), $F \in \overline{L}(\overline{P};[c,d])$ with the same constant M. It follows that F is *strictly decreasing** $\subseteq VB^* \subseteq VB$ on $\overline{P}$. Clearly $F_{/\overline{P}}$, $g_{/\overline{P}}$ and $f_{/g(\overline{P})}$ are injective functions.
We show that g is *strictly monotone** on $\overline{P}$. Suppose on the contrary that there exist $x_1 < x_2 < x_3$, $x_1, x_3 \in P$ such that $g(x_2)$ does not belong to the open interval with endpoints $g(x_1)$ and $g(x_3)$. Suppose for example that $g(x_2) \notin (g(x_1), g(x_3))$. If $g(x_2) \le g(x_1) < g(x_3)$ then, since $g \in \mathcal{D}$, there exists $x_4 \in [x_2, x_3)$ such that $g(x_4) = g(x_1)$. Hence $F(x_4) = F(x_1)$, a contradiction. If $g(x_1) < g(x_3) \le g(x_2)$ then we obtain again a contradiction. It follows that g is *strictly monotone** $\subseteq VB^*$ on $\overline{P}$ Since $g \in M^*$ on $[a,b]$ (see Corollary 2.22.1, (i)), $g \in AC$ on $\overline{P}$. By Theorem 2.10.3, (iv), $g_{/\overline{P}} \in \mathcal{C}$. Let $Q = g(\overline{P})$. Then Q is a compact set. Clearly f is strictly monotone on Q. Suppose for example that f is strictly increasing on Q. Let $y_1 < y_2 < y_3$, $y_1, y_3 \in Q$. Let $x_1, x_3 \in P$ such that $g(x_1) = y_1$ and $g(x_3) = y_3$. Then $x_1 < x_3$. Since $g \in \mathcal{D}$, there exists $x_2 \in (x_1, x_3)$ such that $g(x_2) = y_2$. Since F is *strictly decreasing** on $\overline{P}$, it follows that $F(x_1) > F(x_2) > F(x_3)$. Hence $f(y_1) < f(y_2) < f(y_3)$, so f is *strictly increasing** $\subseteq VB^*$ on Q. Since $f \in M^*$ on $g([a,b])$ (see Corollary 2.22.1, (i)), $f \in AC$ on Q. Thus $F \in AC \circ AC = (S) \subset (N)$, so $F \in AC$ on $\overline{P}$ (see Theorem 2.18.9, (xii), (vii)). By Theorem 2.22.2,(iv), (i), $F \in N^{-\infty}$ on $[a,b]$. Similarly $F \in N^{+\infty}$ on $[a,b]$. Hence $F \in N^\infty$ on $[a,b]$.

2.23 Conditions (M), $\overline{M}$, N_g^∞, $N_g^{+\infty}$

Definition 2.23.1 *Let $F : [a,b] \mapsto \mathbb{R}$, $P \subseteq [a,b]$. F is said to be (M) (resp. $\overline{M}$) on P if $F \in AC$ (resp. $\overline{AC}$) on Q, whenever $Q = \overline{Q} \subseteq P$ and $F \in VB \cap C$ (resp. $VB \cap C_d$) on Q. F is said to be $\underline{M}$ on P if $-F \in \overline{M}$ on P, i.e., $F \in \underline{AC}$ on Q, whenever $F \in VB \cap C_i$ on Q. Clearly $(M) = \overline{M} \cap \underline{M}$ on P.*

Remark 2.23.1 *Foran introduced condition (M) for continuous functions in [F7] (see Remark 2.7.2, (iv)).*

Definition 2.23.2 (Ridder) *([R4]). Let $F : [a,b] \mapsto \mathbb{R}$. F is said to be $N_g^{-\infty}$ ($N^{-\infty}$ in a modified (in german: geändert) sense) on $[a,b]$ if $F_P \in N^{-\infty}$ on P, whenever P is a closed subset of $[a,b]$. F is said to be $N_g^{+\infty}$ on $[a,b]$ if $-F \in N_g^{-\infty}$ on $[a,b]$, i.e., $F_P \in N^{+\infty}$ on P. Let $N_g^\infty = N_g^{-\infty} \cap N_g^{+\infty}$ on $[a,b]$, i.e., $F_P \in N^\infty$ on P.*

Theorem 2.23.1 *Let $F : [a,b] \mapsto \mathbb{R}$, $F \in [C_iG]$. The following assertions are equivalent:*

(i) $F \in N_g^{-\infty}$ *on* $[a,b]$;

(ii) $F \in \underline{M}$ *on* $[a,b]$;

(iii) $F \in \underline{AC}$ *on P, whenever* $P = \overline{P} \subseteq [a,b]$ *and* $F \in \overline{L} \cap C_i$ *on P;*

(iv) $F \in \underline{AC}$ *on P, whenever* $P = \overline{P} \subseteq [a,b]$ *and* $F \in \overline{L} \cap C_i$ *on P with a negative constant;*

(v) $\mid F(Z) \mid = 0$, *whenever* $Z \subset [a,b]$, $\mid Z \mid = 0$ *and* $F \in \overline{L} \cap C_i$ *on* $\overline{Z}$;

(vi) $\mid F(Z) \mid = 0$, *whenever* $Z \subset [a,b]$, $\mid Z \mid = 0$ *and* $F \in \overline{L} \cap C_i$ *on* $\overline{Z}$ *with a negative constant.*

Proof. *(i)* $\Rightarrow$ *(ii)* Let P be a closed subset of $[a,b]$ such that $F \in VB \cap C_i$ on P. Let $c = \inf(P)$, $d = \sup(P)$. Then $F_P \in VB \cap C_i$ on $[c,d]$ (see Proposition 2.3.2, (i) and Corollary 2.7.2). Since $F \in N_g^{-\infty}$ on $[a,b]$, by Theorem 2.21.1,(i), $F_P \in N^{-\infty}$ on $[c,d]$. By Corollary 2.21.1, (i), $F_P \in \underline{AC}$ on P.

(ii) $\Rightarrow$ *(iii)* Since $\overline{L} \subseteq \overline{AC} \subseteq VB$ on $P = \overline{P} \subseteq [a,b]$, the proof follows by definitions.

(iii) $\Rightarrow$ *(iv)* $\Rightarrow$ *(vi)* and *(iii)* $\Rightarrow$ *(v)* $\Rightarrow$ *(vi)* are evident.

(vi) $\Rightarrow$ *(i)* Let $P = \overline{P} \subseteq [a,b]$, $c = \inf(P)$, $d = \sup(P)$. Let $E^{-\infty} = \{x \in P : (F_P)'(x) = -\infty\}$. Since $F \in [C_iG]$ on $[a,b]$, it follows that $F_p \in [C_iG]$ on $[c,d]$. Then $P = \cup_{k=1}^\infty P_k$, such that $P_k = \overline{P}_k$ and $F \in C_i$ on each P_k. Hence $F_{P_k} \in C_i$ on $[c_k, d_k]$, where $c_k = \inf(P_k)$, $d_k = \sup(P_k)$. Let $E_k^{-\infty} = \{x \in P_k : (F_{P_k})'(x) = -\infty\}$. Then $E^{-\infty} \subset \cup_{k=1}^\infty E_k^{-\infty}$. By Theorem 2.13.2 it follows that $E_k^{-\infty} = \cup_{i=1}^\infty E_{ki}$ such that $F_{P_k} \in \overline{L}(E_{ki}; [c_{ki}, d_{ki}])$ on E_{ki} with a negative constant, where $c_{ki} = \inf(E_{ki})$, $d_{ki} = \sup(E_{ki})$. Then $F_{P_k} \in \overline{L}(\overline{E}_{ki}; [c_{ki}, d_{ki}])$ on $\overline{E}_{ki}$ (this follows by Proposition 2.13.1, (vi)). Hence $F \in \overline{L}$ with a negative constant on $\overline{E}_{ki}$. By Corollary 2.15.1, $\mid E^{-\infty} \mid = 0$, $\mid E_k^{-\infty} \mid = 0$ and $\mid E_{ki} \mid = 0$. By hypothesis it follows that $\mid F(E_{ki}) \mid = 0$, hence $\mid F(E_k^{-\infty}) \mid = 0$. Thus $\mid F(E^{-\infty}) \mid = 0$.

Remark 2.23.2 *Theorem 2.23.1, (ii)-(vi) remains true if F is defined on a compact set Q.*

Corollary 2.23.1 (Foran) *([F7]) Let* $F : Q \mapsto \mathbb{R}$, $Q = \overline{Q} \subseteq [a,b]$, $F \in \mathcal{C}$. *The following assertions are equivalent:*

(i) $F \in (M)$ *on Q;*

(ii) $F \in AC$ *on any subset E of Q on which it is monotone.*

Proof. *(i)* $\Rightarrow$ *(ii)* If F is monotone on E then, since $F \in \mathcal{C}$ on Q, F is monotone on $\overline{E}$, hence $F \in VB$ on $\overline{E}$. Now the proof follows by definitions.
(ii) $\Rightarrow$ *(i)* By Remark 2.23.2, F is $\underline{M}$ and $\overline{M}$ on Q, hence $F \in (M)$ on Q.

Corollary 2.23.2 $N_g^\infty \cap [\mathcal{C}G] = (M) \cap [\mathcal{C}G]$ *on* $[a,b]$.

Theorem 2.23.2 *Let $F : [a,b] \mapsto \mathbb{R}$ and let P be a closed subset of $[a,b]$, $c = \inf(P)$, $d = \sup(P)$.*

(i) $(\underline{N}) \subseteq N_g^{-\infty} \subseteq N^{-\infty}$ on $[a,b]$, hence $(N) \subseteq N_g^{\infty} \subseteq N^{\infty}$ on $[a,b]$.

(ii) $(N) \subset (M) \cap T_2 \subset (M) = N_g^{\infty} \subseteq N^{\infty} = M^*$ for continuous functions on $[a,b]$.

(iii) $(S) = (N) \cap T_1 = (M) \cap T_1 = N^{\infty} \cap T_1 = M^* \cap T_1$ for continuous functions on $[a,b]$.

(iv) $F \in \underline{M}$ if and only if $F_P \in \underline{M}$ on $[c,d]$.

(v) $F \in (M)$ on P if and only if $F_P \in (M)$ on $[c,d]$.

(vi) $([\mathcal{C}_i G] \cap \underline{M}) + [\underline{AC}G] = [\mathcal{C}_i G] \cap \underline{M}$ on $[a,b]$, hence $([\mathcal{C}G] \cap (M)) + [ACG] = [\mathcal{C}G] \cap (M)$ on $[a,b]$.

(vii) $(AC \circ AC) + AC = (S) + AC \subseteq (N) + AC \subseteq (M) + AC = (M)$, for continuous functions on $[a,b]$.

(viii) $(M) \subset (M) \circ L$, for continuous functions on $[a,b]$.

Proof. (i) Let $F \in (\underline{N})$ on $[a,b]$ and let E be a closed subset of $[a,b]$, $I = [\inf(E), \sup(E)]$. Then $F \in (\underline{N})$ on E. By Theorem 2.20.1, $F_E \in (\underline{N}) \subseteq N^{-\infty}$ on I. It follows that $F \in (N_g^{-\infty})$ on $[a,b]$. The other assertions follow by definitions and Theorem 2.20.1.

(ii) $(N) \subset (M) \cap T_2$ follows by Corollary 2.23.2, Theorem 2.18.7 and Section 6.31 or Section 6.29 (the function G_{PP}); $(M) \cap T_2 \subset (M)$ follows by Section 6.28. The other assertions follow by Corollary 2.23.2 and Corollary 2.22.1, (i).

(iii) See (ii), Theorem 2.18.8 and Lemma 2.21.1.

(iv) Let $F \in \underline{M}$ on P and let $E = \overline{E} \subseteq [c,d]$ such that $F_P \in VB \cap \mathcal{C}_i$ on E. Then $F \in VB \cap \mathcal{C}_i$ on $P \cap E$. Since $F \in \underline{M}$ on P, $F \in \underline{AC}$ on $P \cap E$, hence $F_P \in \underline{ACG}$ on E. By Corollary 2.21.1, (iii), $F_P \in \underline{AC}$ on E, hence $F_P \in \underline{M}$ on $[c,d]$. The converse is evident.

(v) See (iv).

(vi) By Theorem 2.11.1, (xxi), $[\underline{AC}G] \subset [\mathcal{C}_i G]$ on $[a,b]$ and by Proposition 2.3.5, (ii), $[\mathcal{C}_i G] + [\mathcal{C}_i G] = [\mathcal{C}_i G]$ on $[a,b]$. Let $F_1 \in \underline{M} \cap [\mathcal{C}_i G]$ and $F_2 \in [\underline{AC}G]$ on $[a,b]$, $F = F_1 + F_2$. Let E be a closed subset of $[a,b]$, such that $F \in VB \cap \mathcal{C}_i$ on E. Then by Theorem 2.11.1, (vi), $F_2 \in [VBG]$ on E. It follows that $F_1 = F - F_2 \in [VBG] \cap [\mathcal{C}_i G]$ on E. Since $F_1 \in \underline{M}$ it follows that $F_1 \in [\underline{AC}G]$ on E. By Corollary 2.21.1, (iii), $F \in \underline{AC}$ on E. Then $F \in \underline{M}$ on $[a,b]$.

(vii) See Theorem 2.18.9 and (vi).

(viii) See Corollary 6.29.1, (ii).

Theorem 2.23.3 *Let $F : [a,b] \mapsto \mathbb{R}$. If $F \in \mathcal{C}_d \cap \overline{M}$ on $[a,b]$ then F is derivable on a set of positive measure. Moreover, if there exist $a \leq c < d \leq b$ such that $F(c) < F(d)$ then $| F(P) | > 0$, where $P = \{x \in [c,d] : F'(x) \geq 0\}$.*

Proof. If F is decreasing on $[a,b]$ then, by Theorem 2.14.1, F is derivable *a.e.* on $[a,b]$. Suppose that F is not decreasing on $[a,b]$. Then there exist $c,d \in [a,b]$, $c < d$ such that $F(c) < F(d)$. Suppose on the contrary that $\mid F(P) \mid = 0$. By Lemma 2.16.1, there exist $E \subset [c,d]$ and $K \subset [F(c), F(d)]$ such that $\mid E \mid = 0$, $\mid K \mid > (F(d) - F(c))/2$, E and K compact sets, $F(E) = K$ and F is strictly increasing on E. By Lemma 2.7.1, $F \in VB$ on E. By hypothesis, $F \in \mathcal{C}_d$ on E. Since $F \in \overline{M}$ on $[a,b]$ it follows that $F \in \overline{AC}$ on E. But $F \in \underline{AC}$ on E (because F is increasing on E), hence $F \in AC \subset (N)$ on E. It follows that $\mid F(E) \mid = 0$, a contradiction. Hence $\mid F(P) \mid > 0$. Let $E^{+\infty} = \{x \in [c,d] : F'(x) = +\infty\}$. By Corollary 2.15.1, $\mid E^{+\infty} \mid = 0$. By Theorem 2.23.1, (i), (ii) and Theorem 2.23.2, (i), $F \in N^{+\infty}$ on $[a,b]$. Then $\mid F(E^{+\infty}) \mid = 0$, hence $\mid F(P \setminus E^{+\infty}) \mid > 0$. By Corollary 2.13.3, $F \in AC^*G \subseteq (N)$ on $P \setminus E^{+\infty}$. It follows that $\mid P \setminus E^{+\infty} \mid > 0$. By Proposition 2.3.3, $F \in \mathcal{B}_1$ on $[a,b]$. By Remark 2.2.1, (iii), F is a Borel function, and by Corollary 1.10.1, P is a Borel set. Hence P is measurable.

Corollary 2.23.3 (Foran) *([F7]). Let $F : [a,b] \mapsto \mathbb{R}$. If $F \in \mathcal{C} \cap (M)$ on $[a,b]$ then F is derivable on a set of positive measure.*

2.24 Derivation bases

Definition 2.24.1 *([T4], pp. 99-101). Let $P \subseteq [a,b]$ and let $\delta : P \mapsto (0,+\infty)$.*

(i) Let $\beta_\delta^\#[P] = \{([y,z];x)\ [y,z] \subset (x-\delta(x), x+\delta(x))$ and $x \in P\}$ and $D^\#[P] = \{\beta_\delta^\#[P] : \delta : P \mapsto (0,+\infty)\}$. $D^\#[P]$ is called the sharp derivation basis on P. If δ are constant functions then we obtain the uniform sharp derivation basis $U^\#[P]$ on P.

(ii) Let $\beta_\delta^\circ[P] = \{([y,z];x) : x \in [y,z] \subset (x-\delta(x), x+\delta(x))$ and $x \in P\}$ and $D^\circ[P] = \{\beta_\delta^\circ[P] : \delta : P \mapsto (0,+\infty)\}$. $D^\circ[P]$ is called the ordinary derivation basis on P. If δ are constant functions then we obtain the uniform ordinary derivation basis $U^\circ[P]$ on P.

(iii) Let $\beta_\delta[P] = \{([y,z];x) : x \in P$ and either $y = z$, $z < x+\delta(x)$ or $z = x$, $y > x-\delta(x)\}$ and $D[P] = \{\beta_\delta[P] : \delta : P \mapsto (0,+\infty)\}$. $D[P]$ is called the derivation basis on P. If δ are constant functions then we obtain the uniform derivation basis $U[P]$ on P.

Lemma 2.24.1 *([H4], p. 83). Let $[a,b] \mapsto (0,+\infty)$. Then there exist a partition $a = x_o < x_1 < x_2 < \ldots < x_n = b$ and $t_i \in [x_{i-1}, x_i] \subset (t_i - \delta(t_i), t_i + \delta(t_i))$, $i = \overline{1,n}$.*

Proof. If no such partition exists then no such partition of $[a,(a+b)/2]$ exists, or no such partition of $[(a+b)/2, b]$ exists, or both. Then we can obtain a sequence $\{[a_i,b_i]\}_i$ of closed intervals with $[a_{i+1}, b_{i+1}] \subset [a_i,b_i]$, $b_{i-1} - a_{i-1} = (b_i - a_i)/2$, $i = \overline{1,\infty}$, which therefore contain a single common point t, and after a certain stage, every interval of the sequence lies in $(t - \delta(t), t + \delta(t))$. The intervals themselves are then their own partitions, a contradiction.

2.25 Conditions $AC_{D\#}$, AC_{D°, AC_D

Definition 2.25.1 *Let $F : [a,b] \mapsto \mathbb{R}$, $P \subseteq [a,b]$. F is said to be $AC_{D\#}$ (resp. $\overline{AC}_{D\#}$) on P, if for every $\epsilon > 0$ there exist $\mu > 0$ and $\delta : P \mapsto (0,+\infty)$ such that $\sum_{i=1}^{n} \mid F(d_i) - F(c_i) \mid < \epsilon$ (resp. $\sum_{i=1}^{n}(F(d_i) - F(c_i)) < \epsilon$), whenever $[c_i, d_i]$, $i = \overline{1,n}$ are nonoverlapping closed intervals with $\sum_{i=1}^{n}(d_i - c_i) < \mu$ and $([c_i, d_i]; t_i) \in \beta_\delta^\#[P]$. F is said to be $\underline{AC}_{D\#}$ on P if $-F \in \overline{AC}_{D\#}$ on P, i.e., $\sum_{i=1}^{n}(F(d_i) - F(c_i)) > -\epsilon$. Clearly $AC_{D\#} = \overline{AC}_{D\#} \cap \underline{AC}_{D\#}$ on P. If we put D^0 and β_δ° instead of $D^\#$ and $\beta_\delta^\#$, we obtain conditions $AC_{D_\circ}$, $\overline{AC}_{D^\circ}$, $\underline{AC}_{D^\circ}$ on P. If we put D and β_δ instead of $D^\#$ and $\beta_\delta^\#$, we obtain conditions AC_D, $\overline{AC}_D$, $\underline{AC}_D$ on P.*

Remark 2.25.1 *(i) $AC_{U^\circ} \subseteq AC_{D^\circ} \subseteq AC_D$ on P, hence $AC_{U^\circ}G \subseteq AC_{D^\circ}G \subseteq AC_DG$ on P.*

(ii) $\overline{AC}_{U^\circ} \subseteq \overline{AC}_{D^\circ} \subseteq \overline{AC}_D$ on P, hence $\overline{AC}_{U^\circ}G \subseteq \overline{AC}_{D^\circ}G \subseteq \overline{AC}_DG$ on P.

(iii) $AC_{D\#} \subseteq AC_{D^\circ}$ and $\overline{AC}_{D\#} \subseteq \overline{AC}_{D^\circ}$ on P.

(iv) AC_{D° was defined by Gordon in [G2].

Lemma 2.25.1 *Let $F : [a,b] \mapsto \mathbb{R}$, $P \subseteq [a,b]$. The following assertions are equivalent:*

(i) $F \in AC_{D\#}$ (resp. $F \in \underline{AC}_{D\#}$) on P;

(ii) For $\epsilon > 0$ there exist $\mu > 0$ and $\delta : P \mapsto (0,+\infty)$ such that $\sum_{i=1}^{n} V(F;[c_i,d_i]) < \epsilon$ (resp. $\sum_{i=1}^{n} \underline{V}(F;[c_i,d_i]) > -\epsilon$), whenever $[c_i,d_i]$, $i = \overline{1,n}$ are nonoverlapping closed intervals with $\sum_{i=1}^{n}(d_i - c_i) < \mu$ and $([c_i,d_i];t_i) \in \beta_\delta^\circ[P]$.

Proof. $(i) \Rightarrow (ii)$ For $\epsilon > 0$ let μ and δ be given by the fact that $F \in AC_{D\#}$ (resp. $F \in \underline{AC}_{D\#}$) on P. Let $([c_i,d_i];t_i) \in \beta_\delta^\#[P]$, $i = \overline{1,n}$, such that $[c_i,d_i]$, $i = \overline{1,n}$ are nonoverlapping closed intervals with $\sum_{i=1}^{n}(d_i - c_i)\mu$. Let$\{[c_{ij},d_{ij}]\}$, $j = \overline{1,k_i}$ be a finite set of nonoverlapping closed intervals contained in $[c_i,d_i]$. It follows that $([c_{ij},d_{ij}];t_i) \in \beta_\delta^\#$, $i = \overline{1,n}$, $j = \overline{1,k_i}$. Since $\sum_{i=1}^{n}\sum_{j=1}^{k_i}(d_{ij} - c_{ij}) < \mu$, we have $\sum_{i=1}^{n}\sum_{j=1}^{k_i} \mid F(d_{ij}) - F(c_{ij}) \mid < \epsilon$ (resp. $\sum_{i=1}^{n}\sum_{j=1}^{k_i}(F(d_{ij}) - F(c_{ij})) > -\epsilon$). Hence $\sum_{i=1}^{n} V(F;[c_i,d_i]) \le \epsilon$ (resp. $\sum_{i=1}^{n} \underline{V}(F;[c_i,d_i]) > -\epsilon$).

$(ii) \Rightarrow (i)$ For $\epsilon > 0$ let μ and δ be given by (ii). Let $\{[c_i,d_i]\}$, $i = \overline{1,n}$ be nonoverlapping closed intervals with $\sum_{i=1}^{n}(d_i - c_i) < \mu$ and $([c_i,d_i];t_i) \in \beta_\delta^\#[P]$. Then $\sum_{i=1}^{n} \mid F(d_i) - F(c_i) \mid \le \sum_{i=1}^{n} V(F;[c_i,d_i]) < \epsilon$ (resp. $\sum_{i=1}^{n}(F(d_i) - F(c_i)) \ge \sum_{i=1}^{n} \underline{V}(F;[c_i,d_i]) > -\epsilon$).

Proposition 2.25.1 *Let $F : [a,b] \mapsto \mathbb{R}$, $P \subseteq [a,b]$.*

(i) $\underline{AC}(P;[a,b]) = \underline{AC}_{U^\circ}$ on P;

(ii) $AC(P;[a,b]) = AC_{U^\circ}$ on P.

Proof. (i) For $\epsilon > 0$ let $\mu > 0$ be given by the fact that $F \in \underline{AC}(P;[a,b])$. Let $\{[a_i,b_i]\}$, $i = \overline{1,n}$ be a finite set of nonoverlapping closed intervals such that $\{a_i,b_i\} \cap P \ne \emptyset$ and $\sum_{i=1}^{n}(b_i - a_i) < \mu$. Let $\delta : P \mapsto (0,+\infty)$, $\delta(x) = \mu$. Let $t_i \in \{a_i,b_i\} \cap P$. Then $(a_i,b_i) \subset (t_i - \mu, t_i + \mu)$ and $\sum_{i=1}^{n}(F(b_i) - F(a_i)) > -\epsilon - \epsilon = -2\epsilon$.

Conversely, for each $\epsilon > 0$, let μ and δ be given by the fact that $F \in \underline{AC}_{U^\circ}$ on P. Let $\delta_1 = \min\{\mu, \delta\}$ and let $[a_i, b_i]$, $i = \overline{1,n}$ be nonoverlapping closed intervals such that $a_i \in P$, $b_i \in [a,b]$ and $\sum_{i=1}^n (b_i - a_i) < \delta_1$. Then $[a_i, b_i] \subset (a_i - \delta, a_i + \delta)$, hence $\sum_{i=1}^n (F(b_i) - F(a_i)) > -\epsilon$, so $F \in \underline{AC}(P \wedge [a,b])$. Similarly it follows that $F \in \underline{AC}([a,b] \wedge P)$.

(ii) The proof is similar to that of (i).

2.26 Condition $Y_{D\#}$, Y_{D^o}, Y_D

Definition 2.26.1 *Let $F : [a,b] \mapsto \mathbb{R}$. F is said to be $Y_{D\#}$ (resp. $\overline{Y}_{D\#}$) on $[a,b]$ if for each $Z \subset [a,b]$, $| Z | = 0$, and for every $\epsilon > 0$, there exists $\delta : Z \mapsto (0, +\infty)$, such that $\sum_{i=1}^n | F(d_i) - F(c_i) | < \epsilon$ (resp. $\sum_{i=1}^n (F(d_i) - F(c_i)) < \epsilon$), whenever $[c_i, d_i]$, $i = \overline{1,n}$ are nonoverlapping closed intervals and $([c_i, d_i]; t_i) \in \beta_\delta^\#[Z]$. F is said to be $\underline{Y}_{D\#}$ on $[a,b]$ if $-F \in \overline{Y}_{D\#}$ on $[a,b]$, i.e., $\sum_{i=1}^n (F(d_i) - F(c_i)) > -\epsilon$. Clearly $Y_{D\#} = \overline{Y}_{D\#} \cap \underline{Y}_{D\#}$ on $[a,b]$. If we put D^0 and β_δ^o instead of $D^\#$ and $\beta_\delta^\#$, we obtain conditions Y_{D^o}, $\overline{Y}_{D^o}$, $\underline{Y}_{D^o}$ on $[a,b]$. If we put D and β_δ instead of $D^\#$ and $\beta_\delta^\#$, we obtain conditions Y_D, $\overline{Y}_D$, $\underline{Y}_D$ on $[a,b]$.*

Remark 2.26.1 *(i) $Y_{D^o} = Y_D$ and $\overline{Y}_{D^o} = \overline{Y}_D$ on $[a,b]$.*

(ii) $Y_{D\#} \subseteq Y_{D^o}$ and $\overline{Y}_{D\#} \subseteq \overline{Y}_{D^o}$ on $[a,b]$.

(iii) Y_{D^o} was defined by Lee Peng Yee in [L2], but he called it "the strong Lusin condition". This condition also appears in Gordon's Lemma 2 of [G2]. A generalization of Y_{D^o} was given by Kurzweil and Jarnik in [KJ2] (see Definition 3.6., p. 124).

Lemma 2.26.1 *Let $F : [a,b] \mapsto \mathbb{R}$. The following assertions are equivalent:*

(i) $F \in Y_{D\#}$ (resp. $F \in \underline{Y}_{D\#}$) on $[a,b]$;

(ii) For each $Z \subset [a,b]$, $| Z | = 0$ and for every $\epsilon > 0$ there exists $\delta : Z \mapsto (0, +\infty)$, such that $\sum_{i=1}^n V(F; [c_i, d_i]) < \epsilon$ (resp. $\sum_{i=1}^n \underline{V}(F; [c_i, d_i]) > -\epsilon$), whenever $[c_i, d_i]$, $i = \overline{1,n}$ are nonoverlapping closed intervals and $([c_i, d_i]; t_i) \in \beta_\delta^\#[Z]$.

Proof. The proof is similar to that of Lemma 2.25.1.

Lemma 2.26.2 *Let $F : [a,b] \mapsto \mathbb{R}$. If $F \in \underline{Y}_D$ then $| F(\{x : F'_+(x) = -\infty\}) | = 0$ and $F \in \mathcal{C}_i$ on $[a,b]$.*

Proof. Applying the definition of $\overline{Y}_D$ for $Z = \{x\}$, $x \in [a,b]$, it follows that $F \in \mathcal{C}_i$ on $[a,b]$. Let $R^{-\infty} = \{x : F'_+(x) = -\infty\}$. Then $R^{-\infty} = \cup_{n=1}^\infty R_n$ and $F \in \overline{L}(R_n \wedge [c_n, d_n])$ on R_n with a negative constant, where $c_n = \inf(R_n)$, $d_n = \sup(R_n)$. But $| R^{-\infty} | = 0$ (see Remark 2.15.2), hence $| R_n | = 0$, $n = \overline{1,\infty}$. Fix n and let $\epsilon > 0$. Let $\delta_n : R_n \mapsto (0, +\infty)$ be given by the fact that $F \in \underline{Y}_D$ on $[a,b]$. For $t \in R_n \setminus \{d_n\}$, let $t^* \in [t, \min\{d_n, t + \delta_n(t)\})$. Then $F(t) > F(t^*)$. Since $F \in \mathcal{C}_i \subseteq$ *right lower semicontinuous* we obtain that $\lim_{t^* \searrow t} F(t^*) = F(t)$. It follows that $\{[F(t^*), F(t)]\}_{t,t^*}$ is a Vitali cover of the set $F(R_n \setminus \{d_n\})$. By Theorem 1.8.1, there exists a finite set of closed intervals $[t_i, t_i^*]$, $i = \overline{1,p}$ such that $[F(t_i^*), F(t_i)], i = \overline{1,p}$

are pairwise disjoint intervals and $|F(R_n)| < \sum_{i=1}^{n}(F(t_i) - F(t_i^*)) < \epsilon$. Suppose that $t_1 < t_2 < \ldots < t_p$. Since $F \in \overline{L}(R_n \wedge [c_n, d_n])$. on R_n, it follows that $t_1 < t_1^* < t_2 < t_2^* < \ldots < t_p < t_p^*$. Since $F \in \underline{Y}_D$ we have $\sum_{i=1}^{n}(F(t_i) - F(t_i^*)) < \epsilon$. Hence $|F(R_n)| < 2\epsilon$. It follows that $|F(R_n)| = 0$ and $|F(R^{-\infty})| = 0$, so $|F(\{x : F'_+(x) = -\infty\})| = 0$.

Theorem 2.26.1 *Let $F : [a,b] \mapsto \mathbb{R}$.*

(i) $\underline{Y}_{D^\circ} \subseteq VB^*G \cap \mathcal{C}_i \cap N^{-\infty}$ on $[a,b]$.

(ii) $\underline{Y}_{D\#} \subseteq VB \cap \mathcal{C}_i \cap N^{-\infty}$ on $[a,b]$.

Proof. (i) By Lemma 2.26.2, (i), $F \in \mathcal{C}_i \cap N^{-\infty}$ on $[a,b]$. We show that $F \in VB^*G$ on $[a,b]$. By Theorem 2.8.2, it is sufficient to show that $F \in VB^*G$ on each $Z \subset [a,b]$, whenever $|Z| = 0$. For $\epsilon > 0$ let $\delta : Z \mapsto (0, +\infty)$ be given by the fact that $F \in \underline{Y}_{D^\circ}$ on $[a,b]$. Let $Z_n = \{x \in Z : \delta(x) \geq 1/n\}$, $n = \overline{1,\infty}$. Then $Z = \cup_{n=1}^{\infty} Z_n$. Let $Z_{ni} \cap [i/n, (i+1)/n)$. Fix n and i such that $Z_{ni} \neq \emptyset$. We show that $F \in \underline{VB}^*$ on Z_{ni}. Let $[c_k, dk]$, $k = \overline{1,n}$ be nonoverlapping closed intervals with endpoints in Z_{ni}. Let $x_k \in [c_k, d_k]$. Then $([c_k, x_k]; c_k)$ and $([x_k, d_k]; d_k)$ belong to $\beta_\delta^\circ[Z_{ni}]$, hence $\sum_{k=1}^{n}(F(x_k) - F(c_k)) > -\epsilon$ and $\sum_{k=1}^{n}(F(d_k) - F(x_k)) > -\epsilon$. By Proposition 2.8.2, (i), (iii), $F \in \underline{VB}^*$ on Z_{ni}. By Corollary 2.8.1, (i), $F \in VB^*G$ on Z_{ni}, hence $F \in VB^*G$ on Z.

(ii) By Remark 2.26.1, (ii), we need only to prove that $F \in VB$ on $[a,b]$. Let $\epsilon > 0$ and $t \in [a,b]$. By Lemma 2.26.1, for$Z = \{t\}$, there exists $\delta(t) > 0$ such that $\underline{V}(F; [u,v]) > -\epsilon$, whenever $[u,v] \subset (t - \delta(t), t + \delta(t))$, hence $F \in \underline{VB}$ on $[u,v]$. By Theorem 2.7.1, (i), (iv), $F \in VB$ on $[u,v]$. By Lemma 2.24.1, there exist a partition $a = x_o < x_1 < \ldots < x_n = b$ and $t_i \in [x_{i-1}, x_i]$, $i = \overline{1,n}$ such that $[x_{i-1}, x_i] \subset (t_i - \delta(t_i), t_i + \delta(t_i))$, $i = \overline{1,n}$. It follows that $F \in VB$ on $[x_{i-1}, x_i]$, $i = \overline{1,n}$, hence $F \in VB$ on $[a,b]$.

2.27 Characterizations of $AC^*G \cap \mathcal{C}$, $AC^*G \cap \mathcal{C}_i$, AC and $\underline{AC}$

Lemma 2.27.1 *Let $F : [a,b] \mapsto \mathbb{R}$. Then $\underline{AC}_{D^\circ}G \subseteq \underline{Y}_{D^\circ}$ and $\underline{AC}_{D\#}G \subseteq \underline{Y}_{D\#}$ on $[a,b]$.*

Proof. Let $Z \subset [a,b]$, $|Z| = 0$. Then $Z = \cup_{n=1}^{\infty} Z_n$, where the sets Z_n, $n = \overline{1,\infty}$ are pairwise disjoint and $F \in \underline{AC}_{D^\circ}$ (resp. $F \in \underline{AC}_{D\#}$) on each Z_n. Let $\epsilon > 0$. For each n there exist $\delta_n : Z_n \mapsto (0, +\infty)$ and a positive number μ_n such that $\sum_{i=1}^{s_n}(F(d_{ni}) - F(c_{ni})) > -\epsilon/2^n$ (resp. $\sum_{i=1}^{s_n} \underline{V}(F; [c_{ni}, d_{ni}]) > -\epsilon/2^n$), whenever $[c_{ni}, d_{ni}]$, $i = \overline{1,s_n}$ are nonoverlapping closed intervals with $\sum_{i=1}^{s_n}(d_{ni} - c_{ni}) < \mu_n$ and $([c_{ni}, d_{ni}]; t_{ni}) \in \beta_{\delta_n}^\circ[Z_n]$ (for the second part see Lemma 2.25.1). For each n choose an open set U_n such that $Z_n \subset U_n$ and $|U_n| < \mu_n$. Let $\delta : Z \mapsto (0, +\infty)$, $\delta(t) = \min\{\delta_n(t); d(t; U_n)\}$, $t \in Z_n$. Let $[c_j, d_j]$, $j = \overline{1,m}$ be nonoverlapping closed intervals such that $([c_j, d_j]; t_j) \in \beta_\delta^\circ[Z]$. Let $A_n = \{j \in \{1, 2, \ldots, m\} : t_j \in Z_n\}$. Then $\sum_{j=1}^{m}(F(d_i) - F(c_i)) = \sum_{n=1}^{\infty}\sum_{j \in A_N}(F(d_i) - F(c_i)) > \sum_{n=1}^{\infty}(-\epsilon/2^n) > -\epsilon$ (resp. $\sum_{j=1}^{m} \underline{V}(F; [c_j, d_j]) > -\epsilon$). Hence $F \in \underline{Y}_{D^\circ}$ (resp., by Lemma 2.26.1, $F \in \underline{Y}_{D\#}$) on $[a,b]$.

Theorem 2.27.1 *Let $F : [a,b] \mapsto \mathbb{R}$. The following assertions are equivalent:*

(i) $F \in \underline{AC}^*G \cap C_i$ *on* $[a, b]$;

(ii) $F \in C_i$ *on* $[a, b]$, *and for each perfect subset* S *of* $[a, b]$ *there exists a portion* $S \cap (c, d)$ *such that* $F \in b\underline{AC}^*$ *on* $S \cap (c, d)$;

(iii) $F \in C_i$ *on* $[a, b]$, *and* $F \in \underline{AC}^*G$ *on* Z, *whenever* $Z \subset [a, b]$, $|Z| = 0$;

(iv) There exists a sequence $\{P_n\}_n$ *of closed sets such that* $F \in \underline{AC}(P; [a, b])$ *and* $\cup_n P_n = [a, b]$;

(v) $F \in \underline{AC}_{U^\circ}G$ *on* $[a, b]$;

(vi) $F \in \underline{AC}_{D^\circ}G$ *on* $[a, b]$;

(vii) $F \in \underline{Y}_{D^\circ}$ *on* $[a, b]$;

(viii) $F \in C_i \cap VB^*G \cap N^{-\infty}$ *on* $[a, b]$.

Proof. *(i)* $\Leftrightarrow$ *(ii)* $\Leftrightarrow$ *(iii)* follow by Lemma 2.12.5.

(i) $\Rightarrow$ *(iv)* By Proposition 2.12.1, (vi), $\underline{AC}^*G \cap C_i = b\underline{AC}^*G \cap C_i$ on $[a, b]$. It follows that $[a, b] = \cup_{n=1}^{\infty} P_n$ and $F \in b\underline{AC}^*$ on each P_n. For a fixed n we show that $F \in \underline{AC}(P_n; [a, b])$. By Theorem 2.12.2, (i), (vii), $F \in b\underline{AC}^*$ on $\overline{P}_n$. Let $\{(c_k, d_k)\}$, $k = \overline{1, \infty}$ be the intervals contiguous to $\overline{P}_n$ (if we have a finite number of such intervals, there is nothing to prove). For $\epsilon > 0$ let δ be given by the fact that $F \in b\underline{AC}^*$ on $\overline{P}_n$. Let N be a positive integer such that $\sum_{k=N+1}^{\infty}(d_k - c_k) < \delta$. Then

$$(1) \qquad \sum_{k=N+1}^{\infty} \Omega(F; [c_k, d_k] \wedge \{c_k, d_k\}) > -\epsilon.$$

Let $\mu > 0$ such that $\mu < (d_k - c_k)/2$, $k = \overline{1, N}$ and

$$(2) \qquad \sum_{k=1}^{N} (\Omega_-(F; [c_k, c_k + \mu] \wedge \{c_k\}) + \Omega_-(F; [d_k - \mu, d_k] \wedge \{d_k\})) > -\epsilon$$

(this is possible because $F \in C_i$ at each point $\overline{P}_n$). Let $\delta_1 = \min\{\delta, \mu\}$. Let $\{[a_i, b_i]\}$, $i = \overline{1, m}$ be a finite set of nonoverlapping closed intervals such that $a_i \in \overline{P}_n$, $b_i \in [a, b]$ and $\sum_{i=1}^{m}(b_i - a_i) < \delta$. If $b_i \in \overline{P}_n$ then let $b_i' = b_i$, and if $b_i \notin \overline{P}_n$ then let $b_i' = \sup([a_i, b_i] \cap \overline{P}_n)$. By (1) and (2), since $F \in b\underline{AC}^*$ on $\overline{P}_n$, we have $\sum_{i=1}^{m}(F(b_i) - F(a_i)) = \sum_{i=1}^{m}(F(b_i) - F(b_i') + F(b_i') - F(a_i)) > -\epsilon - 2\epsilon = -3\epsilon$. Hence $F \in \underline{AC}(\overline{P}_n \wedge [a, b])$. Similarly we obtain that $F \in \underline{AC}([a, b] \wedge \overline{P}_n)$. It follows that $F \in \underline{AC}(\overline{P}_n; [a, b])$.

(iv) $\Leftrightarrow$ *(v)* follows by Proposition 2.25.1; *(v)* $\Rightarrow$ *(vi)* is evident; *(vi)* $\Rightarrow$ *(vii)* follows by Lemma 2.27.1, (i); *(vii)* $\Rightarrow$ *(viii)* follows by Theorem 2.26.1, (i); *(viii)* $\Rightarrow$ *(i)* follows by Corollary 2.22.1, (ii).

Corollary 2.27.1 *Let* $F : [a, b] \mapsto \mathbb{R}$. *The following assertions are equivalent:*

(i) $F \in AC^*G \cap C$ *on* $[a, b]$;

(ii) $F \in C$ *on* $[a, b]$, *and for each perfect subset* S *of* $[a, b]$ *there exists a portion* $S \cap (c, d)$ *such that* $F \in bAC^*$ *on* $S \cap (c, d)$;

(iii) $F \in C$ *on* $[a,b]$, *and* $F \in AC^*G$ *on* Z, *whenever* $Z \subset [a,b]$, $|Z| = 0$;

(iv) *There exists a sequence* $\{P_n\}_n$ *of closed sets such that* $F \in AC(P_n; [a,b])$ *and* $\cup_n P_n = [a,b]$;

(v) $F \in AC_{U^\circ}G$ *on* $[a,b]$;

(vi) $F \in AC_{D^\circ}G$ *on* $[a,b]$;

(vii) $F \in Y_{D^\circ}$ *on* $[a,b]$;

(viii) $F \in C \cap VB^*G \cap N^\infty$ *on* $[a,b]$.

Theorem 2.27.2 *Let* $F : [a,b] \mapsto \mathbb{R}$. *The following assertions are equivalent:*

(i) $F \in \underline{AC}$ *on* $[a,b]$;

(ii) $F \in C_i$ *on* $[a,b]$ *and* $F \in \underline{AC}$ *on* Z, *whenever* $Z \subset [a,b]$, $|Z| = 0$;

(iii) $F \in VB \cap N^{-\infty} \cap C_i$ *on* $[a,b]$;

(iv) $F \in \underline{AC}_{D^\circ}$ *on* $[a,b]$;

(v) $F \in \underline{AC}_{U\#}$ *on* $[a,b]$;

(vi) $F \in \underline{AC}_{U\#}G$ *on* $[a,b]$;

(vii) $F \in \underline{AC}_{D\#}$ *on* $[a,b]$;

(viii) $F \in \underline{AC}_{D\#}G$ *on* $[a,b]$;

(ix) $F \in \underline{Y}_{D\#}$ *on* $[a,b]$.

Proof. *(i)* ⇒ *(ii)* See Theorem 2.11.1, (xxi).

(ii) ⇒ *(i)* Let Z be the set of all rational numbers od $[a,b]$. By Theorem 2.11.1, (xix), $F \in \underline{AC}(Z \wedge (Z \cup Z_-)) \cap \underline{AC}((Z \cup Z_+) \wedge Z)$. Since $Z_+ = [a,b)$ and $Z_- = (a,b]$, $f \in \underline{AC}$ on $[a,b]$.

(i) ⇔ *(iii)* See Corollary 2.21.1, (i).

(i) ⇒ *(iv)* This follows by definitions.

(iv) ⇒ *(i)* For $\epsilon > 0$ let μ and δ be given by the fact that $F \in AC_{D^\circ}$ on $[a,b]$. Let $[c_k,d_k]$, $k = \overline{1,n}$ be nonoverlapping closed intervals with $\sum_{k=1}^n (d_k - c_k) < \mu$. By Lemma 2.24.1 there exist a partition $c_k = x_{k,0} < x_{k,1} < \ldots < x_{k,n_k} = d_k$, and $t_{k,i} \in [x_{k,i-1}, x_{k,i}] \subset (t_{k,i} - \delta(t_{k,i}), t_{k,i} = \delta(t_{k,i}))$. It follows that $\sum_{k=1}^n (F(d_k) - F(c_k) = \sum_{k=1}^n \sum_{i=1}^{n_k} (F(x_{k,i} - F(x_{k,i-1})) > -\epsilon$, hence $F \in \underline{AC}$ on $[a,b]$.

(i) ⇒ *(v)* ⇒ *(vi)* ⇒ *(viii)* and *(v)* ⇒ *(vii)* ⇒ *(viii)* follow by definitions; *(viii)* ⇒ *(ix)* follows by Lemma 2.27.1; *(ix)* ⇒ *(iii)* follows by Theorem 2.26.1, (ii).

Corollary 2.27.2 *Let* $F : [a,b] \mapsto \mathbb{R}$. *The following assertions are equivalent:*

(i) $F \in AC$ *on* $[a,b]$;

(ii) $F \in C$ *on* $[a,b]$, *and* $F \in AC$ *on* Z, *whenever* $Z \subset [a,b]$, $|Z| = 0$;

(iii) $F \in VB \cap N^\infty \cap AC$ *on* $[a,b]$;

(iv) $F \in AC_{D^\circ}$ on $[a,b]$;

(v) $F \in AC_{U\#}$ on $[a,b]$;

(vi) $F \in AC_{U\#}G$ on $[a,b]$;

(vii) $F \in AC_{D\#}$ on $[a,b]$;

(viii) $F \in AC_{D\#}G$ on $[a,b]$;

(ix) $F \in Y_{D\#}$ on $[a,b]$.

Remark 2.27.1 *For other characterizations of $\underline{AC}$ and AC, see Theorem 4.2.3 and Corollary 4.2.5*

2.28 Conditions AC_n, AC_ω, AC_∞, $\mathcal{F}$

Definition 2.28.1 *Let $F : [a,b] \mapsto \mathbb{R}$, $P \subseteq [a,b]$ and let n be a positive integer. F is said to be AC_n (resp. $\overline{AC}_n$) on P, if for every $\epsilon > 0$ there exists $\delta > 0$ such that $\sum_{k=1}^p \mathcal{O}^n(F; P \cap I_k) < \epsilon$ (resp. $\sum_{k=1}^p \mathcal{O}_+^n(F; P \cap I_k) < \epsilon$), whenever $\{I_k\}$, $k = \overline{1,p}$ is a finite set of nonoverlapping closed intervals with $I_k \cap P \neq \emptyset$ and $\sum_{k=1}^p | I_k |< \delta$. F is said to be $\underline{AC}_n$ on P if $-F \in \overline{AC}_n$ on P, i.e., $\sum_{k=1}^p \mathcal{O}_-^n(F; P \cap I_k) > -\epsilon$. If $\mathcal{O}^n$, $\mathcal{O}_+^n$, $\mathcal{O}_-^n$ are replaced by $\mathcal{O}^\omega$, $\mathcal{O}_+^\omega$, $\mathcal{O}_-^\omega$, we obtain conditions AC_ω, $\overline{AC}_\omega$, $\underline{AC}_\omega$ on P. If $\mathcal{O}^n$, $\mathcal{O}_+^n$, $\mathcal{O}_-^n$ are replaced by $\mathcal{O}^\infty$, $\mathcal{O}_+^\infty$, $\mathcal{O}_-^\infty$, we obtain conditions AC_∞, $\overline{AC}_\infty$, $\underline{AC}_\infty$ on P.*
We define conditions AC_nG, $AC_\omega G$, $AC_\infty G$, $[AC_nG]$, etc. on P, using Definition 1.9.2.

Remark 2.28.1 *(i) In Definition 2.28.1, $\{I_k\}_k$ may be supposed to be an infinite sequence of nonoverlapping closed intervals, and if P is a closed set then the intervals I_k may be supposed to have endpoints in P (because if $c_k = \inf(I_k)$, $d_k = \sup(I_k)$, $I_k \cap \neq \emptyset$ then $c_k, d_k \in P$).*

(ii) In 1975 Foran introduced condition $A(n)$ (see [F2]), which is equivalent to AC_n: Given a natural number $n \geq 1$ and $\epsilon > 0$, there exists a $\delta > 0$ such that, if $\{I_k\}$, $k = \overline{1,p}$ is a finite set of nonoverlapping closed intervals with $P \cap I_k \neq \emptyset$ and $\sum_{k=1}^p | I_k |< \delta$ then there exist interval J_{ki}, $i = \overline{1,n}$ such that $B(F; P \cap (\cup_{i=1}^p I_k)) \subset \cup_{k=1}^p \cup_{i=1}^n (I_k \times J_{ki})$ and $\sum_{k=1}^p \sum_{i=1}^n | J_{ki} |< \epsilon$.

(iii) In the definition of AC_n, "$\mathcal{O}^n(F; P \cap I_k)$" may be replaced by "$\lambda_n(F(P \cap I_k))$" (see Proposition 1.4.3, (i)). This definition of AC_n was given by C.M.Lee in [Le3].

(iv) In the definition of AC_ω, "$\mathcal{O}_\omega(F; P \cap I_k)$" may be replaced by one of the following: "$\lambda_\omega(F(P \cap I_k))$", "$\lambda_\omega(\overline{F(P \cap I_k)})$", "$\lambda_\infty(\overline{F(P \cap I_k)})$", "$| \overline{F(P \cap I_k)} |$" (see Proposition 1.4.3, (ii)).

(v) In the definition of AC_∞, "$\mathcal{O}^\infty(F; P \cap I_k)$" may be replaced by "$\lambda_\infty(F(P \cap I_k))$" or "$| F(P \cap I_k) |$" (see Proposition 1.4.3, (iii)).

Proposition 2.28.1 *Let $F : [a,b] \mapsto \mathbb{R}$, $P \subseteq [a,b]$ and let n be a positive integer. The following assertions are equivalent:*

(i) $F \in AC_n$ *on* P;

(ii) For every $\epsilon > 0$ there exists a $\delta > 0$ with the following property: if $\{I_k\}$, $k = \overline{1,s}$ are nonoverlapping intervals with each $P \cap I_k \neq \emptyset$ and $\sum_{k=1}^{s} | I_k | < \delta$, then for each k there exist P_{kj}, $j = \overline{1,n}$ such that $P \cap I_k = \cup_{j=1}^{n} P_{kj}$ and $\sum_{k=1}^{s} \sum_{j=1}^{n} \mathcal{O}(F; P_{kj}) < \epsilon$.

Definition 2.28.2 *Let $F : [a,b] \mapsto \mathbb{R}$, $P \subseteq [a,b]$. F is said to be $\underline{\mathcal{F}}$ on P, if $P = \cup_{k=1}^{\infty} P_k$ and if there exist positive integers n_k, $k = \overline{1,\infty}$ such that $F \in \underline{AC}_{n_k}$ on P_k. F is said to be $\overline{\mathcal{F}}$ on P if $-F \in \underline{\mathcal{F}}$ on P, i.e., $F \in \overline{AC}_{n_k}$ on P_k. F is said to be in Foran's class $\mathcal{F}$ on P, if $F \in AC_{n_k}$ on P_k (see [F2]). If in addition $P_k = \overline{P}_k$, $k = \overline{1,\infty}$ then we obtain the classes $[\underline{\mathcal{F}}]$, $[\overline{\mathcal{F}}]$, $[\mathcal{F}]$ on P. Clearly $[\underline{\mathcal{F}}] \subseteq \underline{\mathcal{F}}$ and $[\mathcal{F}] \subset \mathcal{F}$.*

Theorem 2.28.1 *Let $F, G : [a,b] \mapsto \mathbb{R}$, $P \subseteq [a,b]$ and let m and n be positive integers.*

(i) $\underline{AC}_\omega \cap \mathcal{D}_- = \underline{AC}$ *on* $[a,b]$.

(ii) $AC_\infty \cap \mathcal{D} = AC$ *on* $[a,b]$.

(iii) $\underline{AC}_1 = \underline{AC}$ *and* $AC_1 = AC$ *on* P.

(iv) $AC_n \subseteq \underline{AC}_n \cap \overline{AC}_n$; $AC_\omega \subseteq \underline{AC}_\omega \cap \overline{AC}_\omega$; $AC_\infty \subseteq \underline{AC}_\infty \cap \overline{AC}_\infty$ *on* P.

(v) $\underline{AC}_n \subseteq \underline{AC}_{n+1} \subseteq \underline{AC}_\omega \subseteq \underline{AC}_\infty$ *on* P.

(vi) $AC_n \subset AC_{n+1} \subseteq AC_\omega \subseteq AC_\infty \subseteq (S) \subset (N)$ *on* P.

(vii) $\overline{AC}_m \cap \underline{AC}_n \subseteq AC_{mn}$ *on* P;

(viii) $\overline{AC}_m + \overline{AC}_n \subseteq \overline{AC}_{mn}$ *on* P.

(ix) (Foran, [F2]). $AC_m + AC_n \subseteq AC_{mn}$ *on* P.

(x) $\overline{AC}_m \cdot \overline{AC}_n \subseteq AC_{mn}$ *on P for bounded positive functions.*

(xi) (Foran, [F2]). $AC_m \cdot AC_n \subseteq AC_{mn}$ *on P for bounded functions.*

(xii) (Foran, [F2]). If $F_{/\overline{P}} \in \mathcal{C}$ *and* $F \in AC_n$ *on* P *then* $F \in AC_n$ *on* $\overline{P}$.

(xiii) If $F_{/\overline{P}} \in \mathcal{C}$ *and* $F \in AC_\omega$ *on* P *then* $F \in AC_\omega$ *on* $\overline{P}$.

(xiv) $(M) \subset AC_2 G \circ (M)$ *on* P.

Proof. (i) $\underline{AC} \subseteq \underline{AC}_\omega \cap \mathcal{D}_-$ follows by definitions. For the converse see Corollary 2.1.2.

(ii) Let I be a closed subinterval of $[a,b]$. By Proposition 1.4.3, (iii), $\mathcal{O}^\infty(F;I) = | F(I) |$. Since $F \in \mathcal{D}$ it follows that $F(I)$ is an interval, hence $| F(I) | = \mathcal{O}(F;I)$. Thus $\mathcal{O}^\infty(F;I) = \mathcal{O}(F;I)$. Now the proof follows by definitions.

(iii) See Remark 2.11.1, (ii).

(iv) See Proposition 1.4.3, (vi), (vii), (viii).

(v) See Proposition 1.4.3,(v).

(vi) $AC_n \subset AC_{n+1} \subsetneq AC_\omega \subsetneq AC_\infty$ follows by Proposition 1.4.3, (iv) and Section 6.33. We show that $AC_\infty \subsetneq (S)$. For $\epsilon > 0$ let δ be given by the fact that $F \in AC_\infty$ on P. Let $E \subset P$, $\mid E \mid < \delta/2$. It follows that there exists $\{I_k\}_k$ a sequence of nonoverlapping closed intervals such that $E \subset \cup_{k=1}^{\infty} I_k$ and $\sum_{k=1}^{\infty} \mathcal{O}^\infty(F; E \cap I_k) < \epsilon$. By Proposition 1.4.3, (iii), $\mid F(E) \mid \leq \sum_{k=1}^{\infty} \mid F(E \cap I_k) \mid = \sum_{k=1}^{\infty} \mathcal{O}^\infty(F; E \cap I_k) < \epsilon$. Hence $F \in (S)$ on P. For $(S) \subset (N)$ see Theorem 2.18.9, (x).

(vii) Let $\epsilon > 0$, $\epsilon_1 = \epsilon/(2n)$, $\epsilon_2 = \epsilon/(2m)$. For ϵ_1 and ϵ_2 let δ_1 and δ_2 be given by the facts that $F \in \overline{AC}_m$ and $F \in \underline{AC}_n$. Let $\delta_o = \min\{\delta_1, \delta_2\}$. Let $\{I_k\}$, $k = \overline{1,p}$ be a finite set of nonoverlapping closed intervals such that $P \cap I_k \neq \emptyset$ and $\sum_{k=1}^{p} \mid I_k \mid < \delta_o$. By Proposition 1.4.3, (ix), we have $\sum_{k=1}^{p} \mathcal{O}^{mn}(F; P \cap I_k) \leq n \sum_{k=1}^{p} \mathcal{O}_+^m(F; P \cap I_k) + m \sum_{k=1}^{p} \mid \mathcal{O}_-^n(F; P \cap I_k) \mid \leq n \cdot \epsilon/(2n) + m \cdot \epsilon/(2m) = \epsilon$.

(viii) Let $\epsilon > 0$, $\epsilon_1 = \epsilon/(2n)$, $\epsilon_2 = \epsilon/(2m)$, $F \in \overline{AC}_m$ and $G \in \overline{AC}_n$ on P. Let δ_1 and δ_2 be given by the facts that $F \in \overline{AC}_m$ on P and $G \in \overline{AC}_n$ on P. Let $\delta_o = \min\{\delta_1, \delta_2\}$. Let $\{I_k\}$, $k = \overline{1,p}$ be a finite set of nonoverlapping closed intervals such that $P \cap I_k \neq \emptyset$ and $\sum_{k=1}^{p} \mid I_k \mid < \delta_o$. By Proposition 1.4.3, (x) it follows that $\sum_{k=1}^{p} \mathcal{O}_+^{mn}(F + G; P \cap I_k) \leq n \sum_{k=1}^{p} \mathcal{O}_+^m(F; P \cap I_k) + m \sum_{k=1}^{p} \mathcal{O}_+^n(G; P \cap I_k) < n \cdot \epsilon/(2n) + m \cdot \epsilon/(2m) = \epsilon$.

(ix), (x), (xi) Using Proposition 1.4.3, (xi), (xii), (xiii), the proofs are similar to that of (viii).

(xii) Since $F_{/\overline{P}} \in \mathcal{C}$ it follows that $F(\overline{P} \cap I) = \overline{F(P \cap I)}$, whenever I is a closed subinterval of $[a, b]$. By Theorem 1.3.1 it follows that $\lambda_n(F(\overline{P} \cap I)) = \lambda_n(\overline{F(P \cap I)}) = \lambda_n(F(P \cap I))$. By Remark 2.28.1, (iii) we have (xii).

(xiii) The proof is similar to that of(xii),using Remark 2.28.1,(iv).

(xiv) See Corollary 6.29.1, (i).

Remark 2.28.2 *In Theorem 2.28.1, (i), we cannot give up condition $\mathcal{D}_-$ (see Section 6.32).*

Corollary 2.28.1 *Let $P \subseteq [a, b]$ and let m and n be positive integers.*

(i) $\mathcal{F} = \underline{\mathcal{F}} \cap \overline{\mathcal{F}}$ *on P.*

(ii) $\underline{ACG} \subsetneq \underline{AC_n}G \subsetneq \underline{AC_{n+1}}G \subsetneq \underline{\mathcal{F}} \subsetneq \underline{AC_\omega}G \subsetneq \underline{AC_\omega}G$ *on P.*

(iii) $ACG \subset AC_nG \subset AC_{n+1}G \subsetneq \mathcal{F} \subsetneq AC_\omega G \subsetneq AC_\infty G \subsetneq (N)$ *on P (see Section 6.33.).*

(iv) $\overline{AC}_m G \cap \underline{AC}_n G \subseteq AC_{mn}G$ *on P.*

(v) $\overline{AC}_m G + \overline{AC}_n G \subseteq \overline{AC}_{mn} G$ *on P.*

(vi) $AC_m G + AC_n G \subseteq AC_{mn} G$ *on P.*

(vii) $\overline{AC}_n G \cdot \overline{AC}_n G \subseteq \overline{AC}_{mn} G$ *on P, for bounded positive functions;*

(viii) $AC_m G \cdot AC_n G \subseteq AC_{mn} G$ *on P, for bounded functions.*

(ix) $\overline{\mathcal{F}} + \overline{\mathcal{F}} = \overline{\mathcal{F}}$ *on P.*

(x) (Foran, [F2]). $\mathcal{F} + \mathcal{F} = \mathcal{F}$ *on P.*

(xi) $\overline{\mathcal{F}} \cdot \overline{\mathcal{F}} = \overline{\mathcal{F}}$ *on P, for bounded positive functions.*

(xii) (Foran, [F2]). $\mathcal{F} \cdot \mathcal{F} = \mathcal{F}$ *on P, for bounded functions.*

Remark 2.28.3 $AC_2G \subsetneq \mathcal{F} \not\subset$ *approximately derivable a.e., for continuous functions on* $[a,b]$ *(see Section 6.41.). Hence the Denjoy-Khintchine Theorem (Theorem 2.15.3) is not valid if* VBG *is replaced by* AC_2G *or by* $\mathcal{F}$.

Lemma 2.28.1 *Let* $F : [a,b] \mapsto \mathbb{R}$, ; $P = \overline{P} \subseteq [a,b]$ *and let n be a positive integer. The following assertions are equivalent:*

(i) $F \in [\underline{AC}_nG]$ *(resp.* $[AC_nG]$*) on P;*

(ii) Each closed subset S of P contains a portion $S \cap (c,d)$ *such that* $F \in \underline{AC}_n$ *(resp.* $F \in AC_n$*) on* $S \cap (c,d)$.

Proof. See Theorem 1.9.1.

Lemma 2.28.2 *Let* $F : [a,b] \mapsto \mathbb{R}$, $P = \overline{P} \subseteq [a,b]$. *The following assertions are equivalent:*

(i) $F \in [\underline{\mathcal{F}}]$ *(resp.* $[\mathcal{F}]$*) on P;*

(ii) Each closed subset S of P contains a portion $S \cap (c,d)$ *such that* $F \in \underline{AC}_n$ *(resp.* $F \in AC_n$*) on* $S \cap (c,d)$ *for some positive integer n.*

Proof. See Theorem 1.9.1.

Remark 2.28.4 *(i) Particularly, Lemma 2.28.1 is true for functions* $AC_nG \cap C \subseteq [AC_nG]$.

(ii) Particularly, Lemma 2.28.2 is true for functions $\mathcal{F} \cap C \subseteq [\mathcal{F}]$ *(see Theorem 2.28.1,(xii)).*

2.29 Conditions VB_n, VB_ω, VB_∞, $\mathcal{B}$

Definition 2.29.1 *Let* $F : [a,b] \mapsto \mathbb{R}$, $P \subseteq [a,b]$ *and let n be a positive integer. F is said to be* VB_n *(resp.* $\overline{VB}_n$*) on P, if there exists a number* $M \in (0,+\infty)$ *such that* $\sum_{k=1}^{p} \mathcal{O}^n(F; P \cap I_k) < M$ *(resp.* $\sum_{k=1}^{p} \mathcal{O}^n_+(F; P \cap I_k) < M$*), whenever* $\{I_k\}$, $k = \overline{1,p}$ *is a finite set of nonoverlapping closed intervals with* $P \cap I_k \neq \emptyset$. *F is said to be* $\underline{VB}_n$ *on P, i.e.,* $\sum_{k=1}^{p} \mathcal{O}^n_-(F; P \cap I_k) > -M$. *If* $\mathcal{O}^n$, $\mathcal{O}^n_+$, $\mathcal{O}^n_-$ *are replaced by* $\mathcal{O}^\omega$, $\mathcal{O}^\omega_+$, $\mathcal{O}^\omega_-$, *we obtain conditions* VB_ω, $\overline{VB}_\omega$, $\underline{VB}_\omega$ *on P. If* $\mathcal{O}^n$, $\mathcal{O}^n_+$, $\mathcal{O}^n_-$ *are replaced by* $\mathcal{O}^\infty$, $\mathcal{O}^\infty_+$, $\mathcal{O}^\infty_-$, *we obtain conditions* VB_∞, $\overline{VB}_\infty$, $\underline{VB}_\infty$.
We define conditions VB_nG, $VB_\omega G$, $VB_\infty G$, $[VB_nG]$, *etc. on P, using Definition 1.9.2. Clearly* $[VB_nG] \subseteq VB_nG$, *etc.*

Remark 2.29.1 *(i) In Definition 2.29.1,* $\{I_k\}_k$ *may be supposed to be an infinite sequence of nonoverlapping closed intervals. If P is a closed set then the intervals may be supposed to have endpoints in P (because if* $c_k = \inf(I_k)$, $d_k = \sup(I_k)$, $I_k \cap P \neq \emptyset$ *then* c_k, $d_k \in P$*).*

(ii) *In 1975, Foran introduced a condition called $B(n)$ (see [F2]), which is equivalent to VB_n:*
Given a natural number n, there exists a number $M \in (0, +\infty)$ such that, if $\{I_k\}$, $k = \overline{1,p}$ is a finite set of nonoverlapping closed intervals with $I_k \cap P \neq \emptyset$ then there exists a finite set of intervals $\{J_{ki}\}$, $i = \overline{1,n}$ such that $B(F; P \cap (\cup_{k=1}^{p} I_k)) \subset \cup_{k=1}^{p} \cup_{i=1}^{n} (I_k \times J_{ki})$ and $\sum_{k=1}^{p} \sum_{i=1}^{n} \mid J_{ki} \mid < M$.

(iii) *In the definition of VB_n, "$\mathcal{O}^n(F; P \cap I_k)$" may be replaced by "$\lambda_n(F(P \cap I_k))$" (see Proposition 1.4.3, (i)).*

(iv) *In the definition of VB_ω, "$\mathcal{O}^\omega(F; P \cap I_k)$" may be replaced by one of the following: "$\lambda_\omega(F(P \cap I_k))$", "$\lambda_\omega(\overline{F(P \cap I_k)})$", $\lambda_\infty(\overline{F(P \cap I_k)})$", "$\mid \overline{F(P \cap I_k)} \mid$" (see Proposition 1.4.3, (ii).*

(v) *In the definition of VB_∞, "$\mathcal{O}^\infty(F; P \cap I_k)$" may be replaced by "$\lambda_\infty(F(P \cap I_k))$" or "$\mid F(P \cap I_k) \mid$" (see Proposition 1.4.3, (iii)).*

Definition 2.29.2 *Let $F : [a, b] \mapsto \mathbb{R}$, $P \subseteq [a, b]$. F is said to be $\overline{\mathcal{B}}$ on P, if $P = \cup_{k=1}^{\infty} P_k$ and there exist natural numbers n_k, $k = \overline{1,\infty}$ such that $F \in \overline{VB}_{n_k}$ on P_k. F is said to be $\underline{\mathcal{B}}$ on P if $-f \in \overline{\mathcal{B}}$ on P, i.e., $F \in \underline{VB}_{n_k}$ on P_k. F is said to be in Foran's class $\mathcal{B}$ on P, if $F \in VB_{n_k}$ on P_k. If in addition P_k ar supposed to be closed sets then we obtain the classes $[\overline{\mathcal{B}}]$, $[\underline{\mathcal{B}}]$, $[\mathcal{B}]$ on P. Clearly $[\underline{\mathcal{B}}] \subseteq \underline{\mathcal{B}}$ and $[\mathcal{B}] \subseteq \mathcal{B}$.*

Theorem 2.29.1 *Let $F : [a, b] \mapsto \mathbb{R}$, $P \subseteq [a, b]$ and let m and n be positive integers.*

(i) $\overline{VB}_\omega \cap \mathcal{D}_+ \subseteq \overline{VB} = VB$ *on* $[a, b]$.

(ii) $VB_\omega \cap \mathcal{D} \subseteq VB$ *on* $[a, b]$.

(iii) $VB_1 = VB$ *(Foran, [F2]) and* $\underline{VB}_1 = \underline{VB}$ *on* P.

(iv) $VB_n \subseteq \underline{VB}_n \cap \overline{VB}_n$; $VB_\omega \subseteq \underline{VB}_\omega \cap \overline{VB}_\omega$; $VB_\infty \subseteq \underline{VB}_\infty \cap \overline{VB}_\infty$.

(v) $\underline{VB}_n \subseteq \underline{VB}_{n+1} \subseteq \underline{VB}_\omega \subseteq \underline{VB}_\infty$ *on* P.

(vi) $VB_n \subseteq VB_{n+1} \subset VB_\omega \subseteq VB_\infty$ *on* P.

(vii) $\overline{VB}_m \cap \underline{VB}_n \subseteq VB_{mn}$ *on* P.

(viii) $\overline{VB}_m + \overline{VB}_n \subseteq \overline{VB}_{mn}$ *on* P.

(ix) *(Foran, [F2]).* $VB_m + VB_n \subseteq VB_{mn}$ *on* P.

(x) $\overline{VB}_m \cdot \overline{VB}_n \subseteq \overline{VB}_{mn}$ *on P for bounded functions.*

(xi) *(Foran, [F2]).* $VB_m \cdot VB_n \subseteq VB_{mn}$ *on P for bounded functions.*

(xii) *(Foran, [F2]). If* $F_{/\overline{P}} \in \mathcal{C}$ *and* $F \in VB_n$ *on P then* $F \in VB_n$ *on* $\overline{P}$.

(xiii) *If* $F_{/\overline{P}} \in \mathcal{C}$ *and* $F \in VB_\omega$ *on P then* $F \in VB_\omega$ *on* $\overline{P}$.

(xiv) $\overline{AC}_n \subseteq \overline{VB}_n$, $\overline{AC}_\omega \subseteq \overline{VB}_\omega$, $\overline{AC}_\infty \subseteq \overline{VB}_\infty$ *on P for bounded functions.*

(xv) $AC_n \subseteq VB_n$ *(Foran, [F2]),* $AC_\omega \subseteq VB_\omega$, $AC_\infty \subseteq VB_\infty$ *on P for bounded functions.*

(xvi) (Foran, [F2]). $VB_n \subseteq f.l. \subseteq VB_\infty$ *on P.*

Proof. Similarly to the proof of Theorem 2.28.1, we obtain (i) (see also Proposition 2.7.1, (ii), (iii), (v), (vi) (see Section 6.40.), (vii),..., (xiii).

(xiv) Let $F \in \overline{AC}_n$ on P, and for $\epsilon > 0$ let δ be given by this fact. Let J be the minimal closed interval which contains P. Then $J = \cup_{i=1}^p J_i$, where $J_1, J_2, \ldots, J_p$ are nonoverlapping closed intervals with $|J_i| < \delta$, $i = \overline{1,p}$. Let $\{I_k\}$, $k = \overline{1,m}$ be a finite set of nonoverlapping closed intervals such that $P \cap I_k \neq \emptyset$. Since F is bounded, there exists a positive number M such that $|F(x)| \leq M$, $x \in P$. We have $\sum_{k=1}^m \mathcal{O}_+^n(F; P\cap I_k) = \sum_{i=1}^p \sum_{k,\, I_k \subset J_i} \mathcal{O}_+^n(; P \cap I_k) + \sum_{k,\, I_k \subset J} \mathcal{O}_+^n(F; P\cap I_k) \leq p + 2p \cdot 2M = p(1+4M)$ (because there exist at most $2p$ intervals I_k which are contained in no J_i). It follows that $F \in \overline{VB}_n$ on P, hence $\overline{AC}_n \subseteq \overline{VB}_n$ on P. The other parts follow similarly.

(xv) The proof is similar to that of (xiv).

(xvi) For $\epsilon > 0$ let $\{I_k\}_k$ be a sequence of nonoverlapping closed intervals such that $P \subset \cup_{k=1}^\infty I_k$, $|I_k| < \epsilon$ and $\sum_{k=1}^\infty |I_k| - |P| < \epsilon$. By Remark 2.29.1, (ii), there exist intervals J_{ki}, $i = \overline{1,n}$ such that $B(F;P) \subset \cup_{k=1}^\infty \cup_{i=1}^n (I_k \times J_{ki})$ and $\sum_{k=1}^\infty \sum_{i=1}^n |J_{ki}| < +\infty$, where M is given by the fact that $F \in VB_n$ on P. Each interval J_{ki} can be written as the union of at most $[|J_{ki}| / |I_k|] + 1$ intervals J_{kip} of length $|I_k|$ (where $[x]$ is the greatest integer less than x). But then $\sum_{k,i,p} diam(I_k \times J_{kip}) \leq \sum_{k,i,p}(|I_k| + |J_{kip}|) \leq \sum_{k,i} 2 \cdot (|J_{ki}| / |I_k| + 1) \cdot |I_k| \leq \sum_{k,i} 2 \cdot (|J_{ki}| + |I_k|) \leq 2(M + n(|P| + \epsilon))$. Since ϵ is arbitrary, it follows that $\Lambda(B(F;P)) \leq 2M + 2n|P| < +\infty$, hence $F \in f.l.$ on P.

We show that $f.l. \subseteq VB_\infty$ on P. Let $[a_k, b_k]$, $k = \overline{1,p}$ be nonoverlapping closed intervals, $[a_k, b_k] \cap P \neq \emptyset$. We may suppose that $b_i < a_{i+1}$, $i = \overline{1, p-1}$. Since $F \in f.l.$, it follows that there exists $M \in (0, +\infty)$ such that $\Lambda(B(F;P)) < M$. But $\Lambda = \Lambda_1$ is a Carathéodory outer measure (see [S3], p.53). Therefore $\sum_{i=odd} \lambda_\infty(F; P \cap [a_i, b_i]) \leq \sum_{i=odd} \lambda(B(F; P \cap [a_i, b_i]) < \lambda(B(F;P)) < M$. A similar result follows for $i = even$. Hence $\sum_{i=1}^n \lambda_\infty(F; P \cap [a_i, b_i]) < 2M$. Thus $F \in VB_\infty$ on P.

Corollary 2.29.1 *Let* $P \subseteq [a,b]$ *and let m and n be positive integers.*

(i) $\mathcal{B} = \underline{\mathcal{B}} \cap \overline{\mathcal{B}}$ *on P.*

(ii) $\underline{VB}G \subseteq \underline{VB}_nG \subseteq \underline{VB}_{n+1}G \subseteq \underline{\mathcal{B}} \subseteq \underline{VB}_\omega G \subseteq \underline{VB}_\infty G$ *on P.*

(iii) $VBG \subseteq VB_nG \subseteq VB_{n+1}G \subseteq \mathcal{B} \subseteq VB_\omega G \subseteq VB_\infty G$ *on P.*

(iv) $\overline{VB}_mG \cap \underline{VB}_nG \subseteq VB_{mn}G$ *on P.*

(v) $\overline{VB}_MG + \overline{VB}_nG \subseteq \overline{VB}_{mn}G$ *on P.*

(vi) $VB_mG + VB_nG \subseteq VB_{mn}G$ *on P.*

(vii) $\overline{VB}_mG \cdot \overline{VB}_n \subseteq \overline{VB}_{mn}G$ *on P for bounded positive functions.*

(viii) $VB_mG \cdot VB_nG \subseteq VB_{mn}G$ *on P for bounded functions.*

(ix) $\overline{\mathcal{B}} + \overline{\mathcal{B}} = \overline{\mathcal{B}}$ *on P.*

(x) (Foran, [F2]). $\mathcal{B} + \mathcal{B} = \mathcal{B}$ on P.

(xi) $\overline{\mathcal{B}} \cdot \overline{\mathcal{B}} = \overline{\mathcal{B}}$ on P for bounded positive functions.

(xii) (Foran, [F2]). $\mathcal{B} \cdot \mathcal{B} = \mathcal{B}$ on P for bounded functions.

(xiii) $\underline{AC_n}G \subsetneq \underline{VB_n}G$; $\underline{AC_\omega}G \subsetneq \underline{VB_\omega}G$; $\underline{AC_\infty}G \subsetneq \underline{VB_\infty}G$ on P.

(xiv) $AC_nG \subsetneq VB_nG$; $AC_\omega G \subsetneq VB_\omega G$; $AC_\infty G \subsetneq VB_\infty G$ on P.

(xv) $\underline{\mathcal{F}} \subseteq \underline{\mathcal{B}}$ on P.

(xvi) $\mathcal{F} \subset \mathcal{B} \subset \sigma.f.l. \subset T_2$ on P (see Section 6.2., Section 6.58. and Section 6.61.or Remark 3.1.1, (vi)).

Remark 2.29.2 *(i) The Banach-Zarecki Theorem (see Theorem 2.18.9, (vii))is not valid if AC and VB are replaced by AC_n and VB_n, $n \geq 2$ (see Section 6.31).*

(ii) The Banach-Zarecki Theorem is not valid if AC is replaced by AC_n for some $n \geq 2$, VB is replaced by VB_2, and (N) is replaced by $\mathcal{F}$ (see Section 6.38.). Moreover this theorem is no longer valid even if AC is replaced by AC_2, VB by VB_2, and (N) by L_4 (see Section 6.37).

(iii) $\mathcal{F} \not\subset$ approximately derivable a.e., for continuous functions on $[a, b]$ (see Section 6.41.), hence $ACG \subset \mathcal{F}$.

Lemma 2.29.1 *Let $F : [a, b] \mapsto \mathbb{R}$, $P = \overline{P} \subseteq [a, b]$, and let n be a positive integer. The following assertions are equivalent:*

(i) $F \in [\underline{VB_n}G]$ (resp. $[VB_nG]$) on P;

(ii) Each closed subset S of P contains a portion $S \cap (c, d)$ such that $F \in \underline{VB_n}$ (resp. VB_n) on $S \cap (c, d)$ (see Theorem 1.9.1).

Lemma 2.29.2 *Let $F : [a, b] \mapsto \mathbb{R}$, $P = \overline{P} \subseteq [a, b]$. The following assertions are equivalent:*

(i) $F \in [\underline{\mathcal{B}}]$ (resp. $[\mathcal{B}]$) on P;

(ii) Each closed subset S of P contains a portion $S \cap (c, d)$ such that $F \in \underline{VB_n}$ (resp. VB_n) on $S \cap (c, d)$ for some n (see Theorem 1.9.1).

Remark 2.29.3 *(i) Particularly Lemma 2.29.1 is true for functions $VB_nG \cap C \subseteq [VB_nG]$ (see Theorem 2.29.1, (xii)).*

(ii) Particularly Lemma 2.29.2 is true for functions $\mathcal{B} \cap C \subseteq [\mathcal{B}]$ (see Theorem 2.29.1, (xii)).

Lemma 2.29.3 *Let $F : [a, b] \mapsto \mathbb{R}$, $F \in (-)$ on $[a, b]$. If $F \in VB_\omega G$ on $H = \{x : F$ is continuous at $x\}$ then there exists a sequence $\{I_n\}_n$ of closed intervals whose union is dense in $[a, b]$, such that $F \in VB \cap C_i$ on each I_n.*

Proof. Since $F \in VB_\omega G$ on H, there exists a finite or infinite sequence of sets H_n such that $H = \cup_n H_n$ and $F \in VB_\omega$ on H_n. By Theorem 2.18.6, H is a G_δ-set, everywhere dense in $[a, b]$. By Theorem 1.7.1, there exist a positive integer p and an interval $[c, d]$ such that $H \cap (c, d) \neq \emptyset$ and $H \cap [c, d] \subset \overline{H_p}$. Let $I \subset [c, d]$ be an interval. Since F is continuous on H it follows that $\overline{F(I \cap H)} = \overline{F(I \cap H_p)}$ (indeed: it is sufficient to show that $F(I \cap H) \subset \overline{F(I \cap H_p)}$; let $z_o \in F(I \cap H)$; then $z_o = F(x_o)$, $x_o \in I \cap H$; since F is continuous at x_o, for $\epsilon > 0$ there is a $\delta > 0$ such that $F(x) \in (z - \epsilon, z_o + \epsilon)$, for all $x \in (x_o - \delta, x_o + \delta)$.; let $x \in (x_o - \delta, x_o + \delta) \cap (I \cap H_p)$; then $(z_o - \epsilon, z_o + \epsilon) \cap F(I \cap H_p) \neq \emptyset$, hence $z_o \in \overline{F(I \cap H_p)}$). By definitions it follows that $F \in VB_\omega$ on $[c, d] \cap H$. We show that $F \in VB$ on $[c, d]$. Suppose that $F \notin VB$ on $[c, d]$. Then there exist $c = a_o < a_1 < \ldots < a_n < a_{n+1} = d$ such that

$$(1) \quad \sum_{i=0}^{n} | F(a_{i+1}) - F(a_i) | > 4M + | F(d) - F(c) |,$$

where M is given by the fact that $F \in VB_\omega$ on $[c, d] \cap H$. Let $A = \{i : F(a_{i+1}) < F(a_i)\}$. Since $\sum_{i=0}^{n}(F(a_{i+1}) - F(a_i)) = F(d) - F(c)$, by (1) it follows that

$$(2) \quad \sum_{i \in A}(F(a_i) - F(a_{i+1}) \geq 2M.$$

Since $F \in (-)$ on $[a, b]$, $F(a_i) - F(a_{i+1}) \leq | F(N \cap [a_i, a_{i+1}]) |$ for each $i \in A$, where $N = \{x : F'(x) \leq 0\}$. By Theorem 1.10.4, the set $\{x \in [a, b] : F'(x) = 0\}$ maps into a set of measure zero. It follows that $N_i = \{x \in [a_i, a_{i+1}] : -\infty \leq F'(x) < 0\}$ is uncountable for each $i \in A$. By Corollary 2.13.2, $F \in VB^*G$ on N_i. Let $N_i' = \{x \in N_i : F$ is continuous at $x\}$. Clearly $N_i' \subset H$. By Theorem 2.10.1 we have

$$(3) \quad F(a_i) - F(a_{i+1}) \leq | F(N_i' \cap [a_i, a_{i+1}]) | \leq | F(H \cap [a_i, a_{i+1}]) | \leq | \overline{F(H \cap [a_i, a_{i+1}])} |$$

By (2) and (3) we obtain $\sum_{i \in A} | \overline{F(H \cap [a_i, a_{i+1}])} | \geq 2M$. This contradicts the fact that $F \in VB_\omega$ on $[c, d]$. Since $F \in (-)$, by Theorem 2.18.9, (xix), $F \in C_i$ on $[c, d]$. The argument we have just given applies equally well to any subinterval of $[a, b]$. The conclusion of our lemma follows by repeated applications of this process.

Corollary 2.29.2 (Bruckner) *([Br2]). Let $F : [a, b] \mapsto \mathbb{R}$, $F \in \mathcal{DB}_1$ on $[a, b]$. If $F \in VBG$ on $[a, b]$ then there exists a sequence $\{I_n\}_n$ of intervals whose union is dense in $[a, b]$, such that $F \in VB \cap C$ on each I_n.*

2.30 Variations V_n, V_ω, V_∞ and the Banach Indicatrix Function

Definition 2.30.1 *Let $F : [a, b] \mapsto \mathbb{R}$, $P \subseteq [a, b]$ and let n be a positive integer.*

- *Let $V_n(F; P) = \sup\{\sum_{k=1}^{n} \lambda_n(F(P \cap I_k)) : \{I_k\}, k = \overline{1, n}$ is a finite set of nonoverlapping closed intervals, with $I_k \cap P \neq \emptyset\}$.*
- *Let $V_\omega(F; P) = \sup\{\sum_{k=1}^{n} | \overline{(F(P \cap I_k)} | : \{I_k\}, k = \overline{1, n}$ is a finite set of nonoverlapping closed intervals, with $I_k \cap P \neq \emptyset\}$.*
- *Let $V_\infty(F; P) = \sup\{\sum_{k=1}^{n} | F(P \cap I_k | : \{I_k\}, k = \overline{1, n}$ is a finite set of nonoverlapping closed intervals, with $I_k \cap P \neq \emptyset\}$.*

Remark 2.30.1 *Let $F : [a,b] \mapsto \mathbb{R}$, $P \subseteq [a,b]$ and let n be a positive integer.*

- *$F \in VB_n$ on P if and only if $V_n(F;P) \neq +\infty$ and then $V_n(F;P) = \inf\{M : M$ is given by the fact that $F \in VB_n$ on $P\}$.*
- *$F \in VB_\omega$ on P if and only if $V_\omega(F;P) \neq +\infty$ and then $V_\omega(F;P) = \inf\{M : M$ is given by the fact that $F \in VB_\omega$ on $P\}$.*
- *$F \in VB_\infty$ on P if and only if $V_\infty(F;P) \neq +\infty$ and then $V_\infty(F;P) = \inf\{M : M$ is given by the fact that $F \in VB_\infty$ on $P\}$.*

Remark 2.30.2 *It is possible to calculate such variations (see Section 6.39.).*

Definition 2.30.2 *Let $P \subseteq [a,b]$, $F : P \mapsto \mathbb{R}$ and $s : \mathbb{R} \mapsto \overline{\mathbb{R}}$, $s(y) =$ the number of roots of the equation $F(x) = y$, $x \in P$. $s(y)$ is called the Banach indicatrix function.*

Lemma 2.30.1 *Let $P \subseteq [a,b]$ be a measurable set and let $F : P \mapsto \mathbb{R}$ be a bounded measurable function, $m = \inf(F(P))$, $M = \sup(F(P))$. If $F(A)$ is measurable, whenever A is a measurable subset of P then $s(y)$ is measurable and $(\mathcal{L})\int_m^M s(y)dy = V_\infty(F;P) = \sup\{\sum_{k\geq 1} \mid F(P_k) \mid : \{P_k\}_k$ is a finite or infinite set of measurable, pairwise disjoint subsets of P, with $\cup_{k\geq 1} P_k = P\}$.*
Moreover, if $F \in VB_\infty$ on P then $\Phi(X) = V_\infty(F;X)$ is an additive set function, where Φ is defined on all measurable subsets Z of P.

Proof. Let $\{P_k\}_k$ be a finite or infinite set of measurable, pairwise disjoint subsets of P, with $\cup_{k\geq 1} P_k = P$. Then

(1) $\displaystyle\sum_{k\geq 1} K_{F(P_k)}(y) \leq s(y)$, *for each* $y \in [m, M]$.

For each positive integer n, let $I_1^n = [a, a+(b-a)/2^n]$ and $I_k^n = (a+(k-1)(b-a)/2^n, a+k(b-a)/2^n]$, $n = \overline{2, 2^n}$. Let $s_n(y) = \sum_{k=1}^{2^n} K_{F(P\cap I_k^n)}(y)$. But $F(P \cap I_k^n)$ is measurable by hypothesis, hence $s_n(y)$ is a positive measurable function. Clearly $\{s_n(y)\}_n$ is increasing. We show that $s_n(y) \to s(y)$, $n \to \infty$. Let $s^*(y) = \lim_{n\to\infty} s_n(y)$. Then $s^*(y)$ is a positive measurable function. By (1), $s_n(y) \leq s(y)$, hence $s^*(y) \leq s(y)$. For each y let $q(y)$ be a positive integer such that $q(y) \leq s(y)$. Then there exist q(y) distinct roots $x_1 < x_2 < \ldots < x_{q(y)}$ of the equation $F(x) = y$, where $x \in P$. Let $n(y)$ be a positive integer such that $(b-a)/2^{n(y)} < \min\{x_{i+1} - x_i : i = \overline{1, q(y)-1}\}$. It follows that there exist $k_1 < k_2 < \ldots < k_{q(y)}$ such that $x_i \in P \cap I_{k_i}^{n(y)}$, $i = \overline{1, q(y)}$. Hence $K_{F(P\cap I_{k_i}^{n(y)})}(y) = 1$ and $s_{n(y)}(y) \geq q(y)$. If $q(y) = s(y) < +\infty$ then $q(y) = s(y) \geq q(y)$, hence $s_{n(y)}(y) = s(y) = q(y)$. If $s(y) = +\infty$ then $q(y)$ can be taken arbitrarily large, hence $s^*(y) = +\infty$. It follows that $s(y) = s^*(y)$ and $\lim_{n\to\infty} s_n(y) = s(y)$. By Theorem Beppo-Levi, we have $\lim_{n\to\infty}(\mathcal{L})\int_m^M s_n(y)dy = (\mathcal{L})\int_m^M s(y)dy$. By (1) it follows that $\sum_{k\geq 1}(\mathcal{L})\int_m^M K_{F(P_k)}(y)dy \leq (\mathcal{L})\int_m^M s(y)dy$.
We prove the second part. Let $\{X_i\}_i$ be a sequence of measurable, pairwise disjoint subsets of P, and let $X = \cup_{i=1}^\infty X_i$. From the first part it follows that $\sum_{i=1}^\infty V_\infty(F;X_i) = \sum_{i=1}^\infty(\mathcal{L})\int_m^M (s_{/X_i})(y)dy = (\mathcal{L})\int_m^M (s_{/X_i})(y)dy = (\mathcal{L})\int_m^M (s_{/X})(y)dy = V_\infty(F;X)$, hence $\Phi(X) = \sum_{i=1}^\infty \Phi(X_i)$.

Corollary 2.30.1 *Let $P \subseteq [a,b]$ be a measurable set and let $F : P \mapsto \mathbb{R}$ be a bounded measurable function. If $F \in (N) \cap VB_\infty$ on P then the Banach indicatrix $s(y)$ is measurable and $(\mathcal{L}) \int_{\mathbb{R}} s(y)dy = V_\infty(F;P) = \sup\{\sum_{k\geq 1} \mid F(P_k) \mid : \{P_k\}_k$ is a finite or infinite collection of pairwise disjoint measurable subsets of P, with $\cup_{k\geq 1} P_k = P\} \neq +\infty$ and $\Phi(X) = V_\infty(F;X)$ is an additive set function and $\Phi \in AC$ on P, where Φ is defined on all measurable subsets X of P.*

Proof. By Theorem 2.18.2 and Lemma 2.30.1, Φ is additive. Let $X \subset P$, $\mid X \mid = 0$. Let $\{X_k\}_k$ be a finite or infinite collection of measurable pairwise disjoint subsets of X, with $X = \cup_{k\geq 1} X_k$. Since $F \in (N)$, $\sum_{k\geq 1} \mid F(X_k) \mid = 0$. By Lemma 2.30.1, $\Phi(X) = 0$, hence $\Phi \in AC$ on P (see [S3], p. 30).

Corollary 2.30.2 (Iseki) *([I3], p. 16; [I5], pp. 38- 39). Let $P \subseteq [a,b]$ be a Borel (resp. analytic) set and let $F : P \mapsto \mathbb{R}$. If F is bounded and continuous on P then the Banach indicatrix $s(y)$ is measurable and $(\mathcal{L}) \int_{\mathbb{R}} s(y)dy = V_\infty(F;P) = \sup\{\sum_{k\geq 1} \mid F(P_k) \mid : \{P_k\}_k$ is a finite or infinite collection of pairwise disjoint Borel (resp. analytic) sets, with $\cup_{k\geq 1} P_k\}$.*

Proof. The proof is similar to that of Lemma 2.30.1, because a continuous image of a Borel (resp. analytic) set is always a measurable set (see [Ku], p. 249).

Remark 2.30.3 *Lemma 2.30.1 and Corollary 2.30.2 may be regarded as generalizations of a well known Banach theorem (see [S3], p. 280).*

Corollary 2.30.3 *Let $P \subseteq [a,b]$ and let $F : P \mapsto \mathbb{R}$ be a bounded function, $F \in VB_\infty$ on P.*

(i) If P is a measurable set and F is a measurable function satisfying (N) on P then $F \in T_1$ on P.

(ii) If F is a Borel (resp. analytic) set and $F \in C$ on P then $F \in T_1$ on P.

Proof. (i) By Corollary 2.30.1 it follows that $s(y)$ is Lebesgue integrable, hence $s(y)$ is finite *a.e.*. Therefore $F \in T_1$ on P.

(ii) The proof is as in (i), using Corollary 2.30.2.

Lemma 2.30.2 *([En1]). Let $P \subseteq [a,b]$, $F : P \mapsto \mathbb{R}$ and let n be a positive integer. Let $\{[a_i,b_i]\}_i$ be a sequence of nonoverlapping closed intervals such that $b_i \leq a_{i=1}$, $P \cap [a_i,b_i] \neq \emptyset$ and $F \in VB_n$ on each $P \cap [a_i,b_i]$. Then $\sum_{i=1}^\infty V_n(F;P \cap [a_i,b_i]) \leq V_n(F;P \cap [a_1,\beta])$, where $\beta = \sup\{b_1,b_2,\ldots\}$.*

Proof. Let $V_i = V_n(F;P \cap [a_i,b_i])$. Then there exists a sequence $\{I_k^i\}_k$ of nonoverlapping closed intervals, with $I_k^i \cap P \neq \emptyset$, $I_k^i \subset [a_i,b_i]$, such that for any intervals J_{km}^i, $m = \overline{1,n}$, with $B(F;P \cap (\cup_{k\geq 1} I_k^i)) \subset \cup_{k\geq 1} \cup_{m=1}^n (I_k^i \times J_{km}^i)$ we have

$$(1) \quad \sum_{k\geq 1} \sum_{m=1}^{n} \mid J_{km}^i \mid \geq V_i.$$

Let $M > V_n(F;P \cap [a_1,\beta])$. Then there exist intervals J_{km}^i, $k = \overline{1,\infty}$, $i = \overline{1,\infty}$, $m = \overline{1,n}$, such that $B(F;P \cap (\cup_{k,i} I_k^i)) \subset \cup_{k,i} \cup_{m=1}^n (I_k^i \times J_{km}^i)$ and $\sum_{k,i} \sum_{m=1}^n \mid J_{km}^i \mid < M$. By (1) it follows that $\sum_i V_i < M$, which implies our assertion.

2.31 Conditions S_o, wS_o and AC_∞, VB_∞, (N)

Definition 2.31.1 (Iseki) *([I3]). Let $F : [a,b] \mapsto \mathbb{R}$, $P \subseteq [a,b]$. F is said to be S_o on P, if for every $\epsilon > 0$ there exists a $\delta > 0$ such that $\sum_{i=1}^n \mid F(P_i) \mid < \epsilon$, whenever P_i, $i = \overline{1,n}$ are measurable, pairwise disjoint subsets of P, with $\sum_{i=1}^n \mid P_i \mid < \delta$. If in addition P_i, $i = \overline{1,n}$ are supposed to be compact sets, we obtain condition wS_o (weak S_o).*

Remark 2.31.1 *Let $P \subseteq [a,b]$.*

(i) $S_o \subseteq (S) \subseteq (N)$ on P.

(ii) $S_o \subseteq wS_o \subseteq wS$ on P.

Theorem 2.31.1 *Let $P \subseteq [a,b]$ be a measurable set and let $F : P \mapsto \mathbb{R}$ be a bounded measurable function. The following assertions are equivalent:*

(i) $F \in AC_\infty$ on P;

(ii) $F \in wS_o \cap (N)$ on P;

(iii) $F \in S_o$ on P;

(iv) $F \in VB_\infty \cap (N)$ on P.

Proof. *(i)* $\Rightarrow$ *(ii)* For $\epsilon > 0$ let δ be given by the fact that $F \in AC_\omega$ on P. Let P_k, $k = \overline{1,n}$ be a finite set of pairwise disjoint compact subsets of P, such that $\sum_{k=1}^n \mid P_k \mid < \delta/2$. For each P_k there exists a finite set of nonoverlapping closed intervals I_{kj}, $j = \overline{1,p}$, with endpoints in P_k (this is possible since P_k is compact), such that $P_k \subset \cup_{k=1}^n \cup_{j=1}^p I_{kj}$ and $\sum_{k=1}^n \sum_{j=1}^p \mid I_{kj} \mid < \delta$. Then $\sum_{k=1}^n \mid F(P_k) \mid \le \sum_{k=1}^n \sum_{j=1}^p \mid F(I_{kj} \cap P) \mid < \epsilon$, hence $F \in wS_o$ on P. By Theorem 2.28.1, (vi), $F \in (N)$ on P.

(ii) $\Rightarrow$ *(iii)* The proof is similar to that of Iseki ([I3], Theorem 14). Let $\epsilon > 0$. For $\epsilon/2$ let δ be given by the fact that $F \in wS_o$ on P. Let Q be a measurable subset of P. By Theorem 1.5.1, there exists an F_σ-type set A such that $A \subset Q$ and $\mid Q \setminus A \mid = 0$. Since $F \in (N)$ on P we have $\mid F(Q \setminus A) \mid = 0$. By Remark 2.31.1, (ii), $F \in wS_o$ on P. Thus $\mid F(Q) \mid = \mid F(A) \cup F(Q \setminus A) \mid \le \mid F(A) \mid + \mid F(Q \setminus A) \mid = \mid F(A) \mid$. Since $A \subset Q$ it follows that $\mid F(Q) \mid = \mid F(A) \mid$. The set A can be expressed as the limit of an infinite, increasing sequence of compact sets. Hence, for $\epsilon > 0$ there exists a compact subset A_ϵ of A, such that $\mid F(Q) \mid = \mid F(A) \mid < \mid F(A_\epsilon) \mid + \epsilon$. Let P_i, $i = \overline{1,n}$ be a finite set of measurable pairwise disjoint subsets of P, such that $\sum_{i=1}^n \mid P_i \mid < \delta$. Then as above, there exists a compact set $Q_i \subset P_i$, such that $\mid F(P_i) \mid \le \mid F(Q_i) \mid + \epsilon/(2n)$, $i = \overline{1,n}$. It follows that $\sum_{i=1}^n \mid Q_i \mid < \delta$ and $\sum_{i=1}^n \mid F(P_i) \mid \le \sum_{i=1}^n \mid F(Q_i) \mid + \epsilon/2 < \epsilon/2 + \epsilon/2 = \epsilon$, hence $F \in S_o$ on P.

(iii) $\Rightarrow$ *(i)* For $\epsilon > 0$ let $\delta > 0$ be given by the fact that $F \in S_o$ on P. Let $\{I_k\}$, $k = \overline{1,n}$ be a finite set of nonoverlapping closed intervals such that $P \cap I_k \neq \emptyset$ and $\sum_{k=1}^n \mid I_k \mid < \delta$. Let $P_k = P \cap I_k$. Then $\sum_{k=1}^n \mid P_k \mid < \delta$ and $\sum_{k=1}^n \mid F(P_k \mid < \epsilon$, hence $F \in AC_\infty$ on P.

(iii) $\Rightarrow$ *(iv)* Let $F \in S_o$ on P. By Remark 2.31.1, (i), $F \in (N)$ on P. For $\epsilon = 1$ let $\delta > 0$ be given by the fact that $F \in S_o$ on P. Let $\{P_k\}$, $k = \overline{1,p}$ be a finite set of

measurable, pairwise disjoint subsets of P, with $P = \cup_{k=1}^{p} P_k$, $diam(P_k) < \delta$, $k = \overline{1,p}$. By Lemma 2.30.1, $\Phi(P) = \sum_{k=1}^{p} \Phi(P_k) \leq p$, hence $F \in VB_\infty$ on P.

$(iv) \Rightarrow (iii)$ Let $F \in VB_\infty \cap (N)$ on P. By Corollary 2.30.1, $\Phi \in AC$ on P. By a well known theorem of Saks (see [S3], p. 31), it follows that for every $\epsilon > 0$ there exists a $\delta > 0$ such that for every measurable set $X \subset P$, $\Phi(X) < \epsilon$, whenever $\mid X \mid < \delta$. Let $\{P_k\}$, $k = \overline{1,p}$ be a finite collection of measurable, pairwise disjoint subsets of P, with $\sum_{k=1}^{p} \mid P_k \mid < \delta$. It follows that $\sum_{k=1}^{p} \mid F(P_k) \mid \leq \sum_{k=1}^{p} \Phi(P_k) = \Phi(\cup_{k=1}^{p} P_k) < \epsilon$, hence $F \in S_o$ on P.

Corollary 2.31.1 *Let $F : [a,b] \mapsto \mathbb{R}$, $P \subseteq [a,b]$.*

(i) If P is a set of F_σ-type then $AC_\infty = wS_o = S_o$ on P, for bounded measurable functions;

(ii) $AC = AC_\infty = S_o = wS_o$ on $[a,b]$ for continuous functions;

(iii) $S_o \subset (S) \subset (N)$ and $wS_o \subset wS$ on $[a,b]$, for continuous functions;

(iv) $VB \cap (N) = VB_\infty \cap (N) = AC_\infty = AC$ on $[a,b]$, for Darboux functions.

Proof. (i) This follows by Remark 2.31.1, (ii), Theorem 2.19.1, (iii) and Theorem 2.31.1.

(ii) See (i) and Theorem 2.28.1, (ii).

(iii) By Theorem 2.18.9, (vi), $AC \subset (S)$ for continuous functions on $[a,b]$. Now the proof follows by (ii) and Remark 2.31.1.

(iv) See Theorem 2.28.1, (ii), Theorem 2.29.1, (ii) and Theorem 2.31.1.

Remark 2.31.2 *Corollary 2.31.1, (iv) contains the Banach-Zarecki Theorem (see Theorem 2.18.9, (vii)), and Theorem 2.31.1, (i) $\Leftrightarrow$ (iv) is a generalization of it.*

2.32 Conditions L_n, L_ω, L_∞, $\mathcal{L}$

Definition 2.32.1 *Let $F : [a,b] \mapsto \mathbb{R}$, $P \subseteq [a,b]$ and let n be a positive integer. F is said to be L_n (resp. $\overline{L}_n$) on P, if there exists a number $M \in (0,+\infty)$ such that $\mathcal{O}^n(F; P \cap I) \leq M \cdot diam(P \cap I)$ (resp. $\mathcal{O}^n_+(F; P \cap I) \leq M \cdot diam(P \cap I)$), whenever I is a closed interval with $P \cap I \neq \emptyset$. F is said to be $\underline{L}_n$ on P if $-F \in \overline{L}_n$ on P, i.e., $\mathcal{O}^n_-(F; P \cap I) \geq -M \cdot diam(P \cap I)$. If $\mathcal{O}^n$, $\mathcal{O}^n_+$, $\mathcal{O}^n_-$ are replaced by $\mathcal{O}^\omega$, $\mathcal{O}^\omega_+$, $\mathcal{O}^\omega_-$, we obtain the conditions L_ω, $\overline{L}_\omega$, $\underline{L}\omega$ on P. If $\mathcal{O}^n$, $\mathcal{O}^n_+$, $\mathcal{O}^n_-$ are replaced by $\mathcal{O}^\infty$, $\mathcal{O}^\infty_+$, $\mathcal{O}^\infty_-$, we obtain the conditions L_∞, $\overline{L}_\infty$, $\underline{L}\infty$ on P.*
We can define conditions L_nG, $\overline{L}_nG$, $[L_nG]$, etc. on P, using Definition 1.9.2.

Remark 2.32.1 *(i) If in Definition 2.32.1, P is a closed set then the interval I may be supposed to have endpoints in P, hence $diam(P \cap I) = \mid I \mid$ (because, if $c = \inf(P)$, $d = \sup(P)$ and $P \cap I \neq \emptyset$ then $c, d \in P$).*

(ii) In the definition of L_n, "$\mathcal{O}^n(F; P \cap I)$" may be replaced by "$\lambda_n(F(P \cap I))$" (see Proposition 1.4.3, (i)).

(iii) In the definition of L_ω, "$\mathcal{O}^\omega(F; P \cap I)$" may be replaced by one of the following: "$\lambda_\omega(F(P \cap I))$", "$\lambda_\omega(\overline{F(P \cap I)})$", "$\lambda_\infty(\overline{F(P \cap I)})$", "$\mid \overline{F(P \cap I)} \mid$" (see Proposition 1.4.3, (ii)).

(iv) *In the definition of* L_∞, "$\mathcal{O}^\infty(F; P \cap I)$ *may be replaced by* "$\lambda_\infty(F(P \cap I))$" *or by* "$| F(P \cap I) |$" *(see Proposition 1.4.3, (iii)).*

Definition 2.32.2 *Let* $F : [a, b] \mapsto \mathbb{R}$, $P \subseteq [a, b]$. *F is said to be* $\underline{\mathcal{L}}$ *on P, if* $P = \cup_{k=1}^\infty P_k$ *and if there exist natural numbers* n_k, $k = \overline{1, \infty}$ *such that* $F \in \underline{L}_{n_k}$ *on* P_k. *F is said to be* $\overline{\mathcal{L}}$ *on P, if* $-F \in \underline{\mathcal{L}}$ *on P, i.e.,* $F \in \overline{L}_{n_k}$ *on* P_k. *F is said to* $\mathcal{L}$ *on P if* $F \in L_{n_k}$ *on* P_k. *If in addition each* P_k *is closed then we obtain the classes* $[\underline{\mathcal{L}}]$, $[\overline{\mathcal{L}}]$, $[\mathcal{L}]$ *on P. Clearly* $[\underline{\mathcal{L}}] \subseteq \underline{\mathcal{L}}$ *and* $[\mathcal{L}] \subseteq \mathcal{L}$ *on P.*

Lemma 2.32.1 *Let* $F : [a, b] \mapsto \mathbb{R}$, $P \subseteq [a, b]$ *and let n be a positive integer. If F is bounded on P then the following assertions are equivalent:*

(i) $F \in L_n$ *(resp.* $\overline{L}_n$, L_ω, $\overline{L}_\omega$, L_∞, $\overline{L}_\infty$*) on P;*

(ii) *For* $\epsilon > 0$ *there is a* $\delta > 0$ *such that* $\sum_{k=1}^\infty \mathcal{O}^n(F; P \cap I_k) < \epsilon$ *(resp.* $\sum_{k=1}^\infty \mathcal{O}_+^n(F; P \cap I_k) < \epsilon$, $\sum_{k=1}^\infty \mathcal{O}^\omega(F; P \cap I_k) < \epsilon$, $\sum_{k=1}^\infty \mathcal{O}_+^\omega(F; P \cap I_k) < \epsilon$, $\sum_{k=1}^\infty \mathcal{O}^\infty(F; P \cap I_k) < \epsilon$, $\sum_{k=1}^\infty \mathcal{O}_+^\infty(F; P \cap I_k) < \epsilon$*), whenever* $\{I_k\}_k$ *is a sequence of closed intervals with* $P \cap I_k \neq \emptyset$ *and* $\sum_{k=1}^\infty | I_k | < \delta$.

Proof. We prove the first part (the others are similar).

(i) $\Rightarrow$ *(ii)* Let $\epsilon > 0$ and let M be given by the fact that $F \in L_n$ on P. Let $\delta = \epsilon/M$ and let $\{I_k\}_k$ be a sequence of closed intervals, with $P \cap I_k \neq \emptyset$ and $\sum_{k=1}^\infty | I_k | < \delta$. Then $\sum_{k=1}^\infty \mathcal{O}^n(F; P \cap I_k) \leq M \cdot \sum_{k=1}^\infty | I_k | < M \cdot \delta == \epsilon$.

(ii) $\Rightarrow$ *(i)* Suppose that $F \notin L_n$ on P. Then there exist $a_i, b_i \in P$, $a_i < b_i$, $i = \overline{1, \infty}$ such that

(1) $\quad \mathcal{O}^n(F; P \cap [a_i, b_i]) > 2^i (b_i - a_i)$.

For $\epsilon > 0$ let δ be given by (ii). Let i_o be the positive integer for which $\delta \cdot i_o \leq \epsilon < \delta \cdot (i_o + 1)$. By (1), since F is bounded on P, it follows that $b_i - a_i \to 0$, $i \to \infty$. Then there exists a positive integer $j > i_o$ such that $b_j - a_j < \delta/2$. Let $k_o \geq 2$ be the positive integer for which

(2) $\quad (b_j - a_j) \cdot k_o \leq \delta < (b_j - a_j) \cdot (k_o + 1)$.

By (1) and (2) we have $k_o \cdot \mathcal{O}^n(F; P \cap [a_j, b_j]) > k_o \cdot 2^j \cdot (b_j - a_j) > (k_o/(k_o + 1)) \cdot (k_o + 1) \cdot (b_j - a_j) \cdot 2^j > (2/3) \cdot \delta \cdot 2^j > \delta \cot 2^{j-1} \geq 2^{i_o} \cdot \delta \geq (i_o + 1) \cdot \delta > \epsilon$. This contradicts (ii).

Lemma 2.32.2 *Let* $P \subseteq [a, b]$, $F : P \mapsto \mathbb{R}$, $G : F(P) \mapsto \mathbb{R}$ *and let m and n be positive integers.*

(i) *If* $G \in L_m$ *on* $F(P)$ *with a constant* $M > 0$ *then* $\lambda_{mn}(G \circ F(P)) \leq M \cdot \lambda_m(F(P))$.

(ii) *In (i), n or m or both may be replaced by* ω *or by* ∞.

Proof. (i) For $\epsilon > 0$ there exist P_i, $i = \overline{1, n}$ such that $\cup_{i=1}^n P_i = P$ and $\lambda_n(F(P)) = \mathcal{O}^n(F; P) \leq \sum_{i=1}^n \mathcal{O}(F; P_i) + \epsilon$ (see Proposition 1.4.3, (i)). Since $G \in L_m$ it follows that $\mathcal{O}^m(G; F(P)) < (M + \epsilon) \cdot diam(F(P_i)) = (M + \epsilon) \cdot \mathcal{O}(F; P)$. By Proposition 1.4.3, (xiv), $\mathcal{O}^m(G; F(P_i)) = \mathcal{O}^m(G \circ F; P_i)$. It follows that there exist P_{ij}, $j = \overline{1, m}$ such that $\cup_{j=1}^m P_{ij} = P_i$ and $\sum_{j=1}^m \mathcal{O}(G \circ F; P_{ij}) < (M + \epsilon) \cdot \mathcal{O}(F; P_i)$, $i = \overline{1, n}$. We obtain that $P = \cup_{i=1}^n \cup_{j=1}^m P_{ij}$ and $\lambda_{mn}(G \circ F(P)) = \mathcal{O}^{mn}(G \circ F; P) \leq \sum_{i=1}^n \sum_{j=1}^m \mathcal{O}(G \circ F; P_{ij}) < (M + \epsilon) \cdot \sum_{i=1}^n \mathcal{O}(F; P_i) < (M + \epsilon) \cdot (\mathcal{O}^n(F; P) + \epsilon)$. Since ϵ is arbitrary, we have (i).

(ii) The proofs are similar to that of (i).

Theorem 2.32.1 *Let $F : [a,b] \mapsto \mathbb{R}$, $P \subseteq [a,b]$ and let m and n be positive integers.*

(i) $\overline{L}_\omega \cap \mathcal{D}_+ = \overline{L}$ *on* $[a,b]$.

(ii) $L_\infty \cap \mathcal{D} = L$ *on* $[a,b]$.

(iii) $L_1 = L$ *and* $\overline{L}_1 = \overline{L}$ *on* P.

(iv) $L_n \subseteq \overline{L}_n \cap \underline{L}_n$, $L_\omega \subseteq \overline{L}_\omega \cap \underline{L}_\omega$, $L_\infty \subseteq \overline{L}_\infty \cap \underline{L}_\infty$ *on* P.

(v) $\underline{L}_n \subseteq \underline{L}_{n+1} \subseteq \underline{L}_\omega \subseteq \underline{L}_\infty$ *on* P.

(vi) $L_n \subseteq L_{n+1} \subseteq L_\omega \subseteq L_\infty$ *on* P.

(vii) $\overline{L}_m \cap \underline{L}_n \subseteq L_{mn}$ *on* P.

(viii) $\overline{L}_m + \overline{L}_n \subseteq \overline{L}_{mn}$ *on* P.

(ix) $L_m + L_n \subseteq L_{mn}$ *on* P.

(x) $\overline{L}_m \cdot \overline{L}_n \subseteq \overline{L}_{mn}$ *on* P *for bounded positive functions.*

(xi) $L_m \cdot L_n \subseteq L_{mn}$ *on* P *for bounded functions.*

(xii) $F_{/\overline{P}} \in C$ *and* $F \in L_n$ *on* P *then* $F \in L_n$ *on* $\overline{P}$.

(xiii) $F_{/\overline{P}} \in C$ *and* $F \in L_\omega$ *on* P *then* $F \in L_\omega$ *on* $\overline{P}$.

(xiv) $\overline{L}_n \subseteq \overline{AC}_n$, $\overline{L}_\omega \subseteq \overline{AC}_\omega$, $\overline{L}_\infty \subseteq \overline{AC}_\infty$ *on* P.

(xv) $L_n \subset AC_n$, $L_\omega \subseteq AC_\omega$, $L_\infty \subseteq AC_\infty$ *on* P.

(xvi) $L_m \circ L_n \subseteq L_{mn}$ *on* P.

(xvii) $L_m \circ AC_n \subseteq AC_{mn}$ *on* P.

(xviii) $L_m \circ VB_n \subseteq VB_{mn}$ *on* P.

Proof. (i) - (xiii) follow similarly to Theorem 2.28.1. (xiv) and (xv) follow by Lemma 2.32.1 and Section 6.55.

(xvi) Let $F \in L_n$ on P with a constant M_1 and let $G \in L_m$ on $F(P)$ with a constant M_2. Let I be a subinterval of $[a,b]$, $I \cap P \neq \emptyset$. By Lemma 2.32.2, we have $\lambda_{mn}(G \circ F(P \cap I)) \leq M_2 \cdot \lambda_m(F(P \cap I))$Ł $\leq M_1 \cdot M_2 \cdot diam(P \cap I)$. It follows that $G \circ F \in L_{mn}$ on P.

(xvii) Let $F \in AC_n$ on P and let $G \in L_m$ on $F(P)$ with a constant M. Let $\epsilon > 0$. For ϵ/M let δ be given by the fact that $F \in AC_n$ on P. Let $\{I_k\}$, $k = \overline{1,p}$ be a finite set of nonoverlapping closed intervals such that $P \cap I_k \neq \emptyset$ and $\sum_{k=1}^{p} \mid I_k \mid < \delta$. By Lemma 2.32.2, we have $\sum_{k=1}^{p} \lambda_{mn}(G \circ F(P \cap I_k)) < \sum_{k=1}^{p} \lambda_n(F(P \cap I_k)) < M \cdot (\epsilon/M) = \epsilon$, hence $G \circ F \in AC_{mn}$ on P (see Remark 2.28.1, (iii)).

(xviii) The proof is similar to that of (xvii).

Remark 2.32.2 *We have* $\underline{L}_1 \subseteq \underline{AC}_1 \subseteq \underline{VB}_1 = VB \subseteq VB_n \subseteq \mathcal{B} \subseteq T_2$ *for continuous functions on a closed set, but* $\underline{L}_2 \subseteq \underline{AC}_2 \subseteq \underline{VB}_2 \not\subset T_2$ *(see Section 6.35.).*

Corollary 2.32.1 *Let $P \subseteq [a,b]$ and let m and n be positive integers.*

(i) $\mathcal{L} = \underline{\mathcal{L}} \cap \overline{\mathcal{L}}$ *on* P.

(ii) $\underline{L}G \subseteq \underline{L}_nG \subseteq \underline{L}_{n+1}G \subseteq \underline{\mathcal{L}} \subseteq \underline{L}_\omega G \subseteq \underline{L}_\infty G$ *on* P.

(iii) $LG \subseteq L_nG \subseteq L_{n+1}G \subseteq \mathcal{L} \subseteq L_\omega G \subseteq L_\infty G$ *on* P.

(iv) $LG \subseteq L_nG \subset AC_nG \subseteq \mathcal{F} \subset (N)$ *and* $\mathcal{L} \subset \mathcal{F}$ *on* P *(see Section 6.55.).*

(v) $\overline{L}_mG \cap \underline{L}_nG \subseteq L_{mn}G$ *on* P.

(vi) $\overline{L}_mG + \overline{L}_nG \subseteq \overline{L}_{mn}G$ *on* P.

(vii) $L_mG + L_nG \subseteq L_{mn}G$ *on* P.

(viii) $\overline{L}_mG \cdot \overline{L}_nG \subseteq \overline{L}_{mn}G$ *on* P *for bounded positive functions.*

(ix) $L_mG \cdot L_nG \subseteq L_{mn}G$ *on* P *for bounded functions.*

(x) $\overline{\mathcal{L}} + \overline{\mathcal{L}} = \overline{\mathcal{L}}$ *on* P.

(xi) $\mathcal{L} + \mathcal{L} = \mathcal{L}$ *on* P.

(xii) $\overline{\mathcal{L}} \cdot \overline{\mathcal{L}} = \overline{\mathcal{L}}$ *on* P *for bounded positive functions.*

(xiii) $\mathcal{L} \cdot \mathcal{L} = \mathcal{L}$ *on* P *for bounded functions.*

(xiv) $\mathcal{L} \circ \mathcal{L} = \mathcal{L}$ *on* P.

(xv) $\mathcal{L} \circ \mathcal{F} = \mathcal{F}$ *on* P.

(xvi) $\mathcal{L} \circ \mathcal{B} = \mathcal{B}$ *on* P.

Lemma 2.32.3 *Let $F : [a,b] \mapsto \mathbb{R}$, $P = \overline{P} \subseteq [a,b]$, and let n be a positive integer. The following assertions are equivalent:*

(i) $F \in [\underline{L}_nG]$ *(resp.* $F \in [L_nG]$*) on* P;

(ii) Each closed subset S of P contains a portion $S \cap (c,d)$ such that $F \in \underline{L}_n$ (resp. $F \in L_n$) on $S \cap (c,d)$ (see Theorem 1.9.1).

Lemma 2.32.4 *Let $F : [a,b] \mapsto \mathbb{R}$, $P = \overline{P} \subseteq [a,b]$. The following assertions are equivalent:*

(i) $F \in [\underline{\mathcal{L}}]$ *(resp.* $F \in [\mathcal{L}]$*) on* P;

(ii) Each closed subset S of P contains a portion $S \cap (c,d)$ such that $F \in \underline{L}_n$ (resp. $F \in L_n$) on $S \cap (c,d)$, for some positive integer n (see Theorem 1.9.1).

Remark 2.32.3 *(i) Particularly, Lemma 2.32.3 is true for functions* $L_nG \cap C \subseteq [L_nG]$.

(ii) Particularly, Lemma 2.32.4 is true for functions $\mathcal{L} \cap C \subseteq [\mathcal{L}]$ *(see Theorem 2.32.1, (xii)).*

2.33 Conditions ΛZ, $\Lambda\overline{Z}$, *f.l.*, $\sigma.f.l.$

Definition 2.33.1 *Let $F : [a,b] \mapsto \mathbb{R}$, $P \subseteq [a,b]$. F is said to be $\Lambda\overline{Z}$ on P, if for every $Z \subset P$ with $| Z |= 0$, and for every $\epsilon > 0$ there exists $\{Z_k\}_k$ a sequence of subsets of Z such that $Z = \cup_{k=1}^{\infty} Z_k$, $\sum_{k=1}^{\infty} diam(Z_k) < \epsilon/2$ and $\sum_{k=1}^{\infty} \mathcal{O}_+(F; Z_k) < \epsilon/2$. F is said to be $\Lambda\underline{Z}$ on P if $-F \in \Lambda\overline{Z}$ on P, i.e., $\sum_{k=1}^{\infty} \mathcal{O}_-(F; Z_k) > -\epsilon/2$.*
(Foran, [F4]) F is said to be ΛZ on P if $\Lambda(B(F;Z)) = 0$.

Lemma 2.33.1 *Let $F : [a,b] \mapsto \mathbb{R}$, $P \subseteq [a,b]$ and let n be a positive integer. The following assertions are equivalent:*

- *(i) $F \in \Lambda Z$ (resp. $F \in \Lambda\overline{Z}$) on P;*
- *(ii) For $\epsilon > 0$ and $Z \subset P$, $| Z |= 0$, there exists a sequence $\{Z_k\}_k$ of subsets of Z such that $Z = \cup_{k=1}^{\infty} Z_k$, $\sum_{k=1}^{\infty} diam(Z_k) < \epsilon/2$ and $\sum_{k=1}^{\infty} \mathcal{O}^n(F; Z_k) < \epsilon/2$ (resp. $\sum_{k=1}^{\infty} \mathcal{O}^n_+(F; Z_k) < \epsilon/2$).*

Proof. We prove the first part.

$(i) \Rightarrow (ii)$ Let $F \in \Lambda Z$ on P and let $Z \subset P$ with $| Z |= 0$. Then $\lambda(B(F;Z)) = 0$, and for each $\epsilon > 0$ we have $\Lambda^1_\epsilon(B(F;Z)) = 0$. It follows that there exists a sequence of sets $\{D_k\}_k$, such that $D_k \subset \mathbb{R} \times \mathbb{R}$, $B(F;Z) \subset \cup_{k=1}^{\infty} D_k$ and $\sum_{k=1}^{\infty} diam(D_k) < \epsilon/2$. Let $Z_k = \{x \in Z : (x, F(x)) \in D_k\}$. Then $Z = \cup_{k=1}^{\infty} Z_k$, $\sum_{k=1}^{\infty} diam(Z_k) \leq \sum_{k=1}^{\infty} diam(D_k) < \epsilon/2$, therefore by Proposition 1.4.3, (iv), $\sum_{k=1}^{\infty} \mathcal{O}^n(F; Z_k) \leq \sum_{k=1}^{\infty} \mathcal{O}(F; Z_k) \leq \sum_{k=1}^{\infty} diam(D_k) < \epsilon/2$.

$(ii) \Rightarrow (i)$ Let $Z \subset P$, $| Z |= 0$ and let $\epsilon > 0$. For ϵ/n there exists a sequence of sets $\{Z_k\}_k$ such that $Z = \cup_{k=1}^{\infty} Z_k$, $\sum_{k=1}^{\infty} diam(Z_k) < \epsilon/(2n)$ and $\sum_{k=1}^{\infty} \mathcal{O}^n(F; Z_k) < \epsilon/(2n)$. It follows that for each k there exist sets I_{kj}, $j = \overline{1,n}$ such that $Z_k \subset \cup_{j=1}^{n} Z_{kj}$ and $\sum_{k=1}^{\infty}\sum_{j=1}^{n} \mathcal{O}(F; Z_{kj}) < \epsilon/(2n)$. Let's define $D_{kj} = [\inf(Z_k), \sup(Z_k)] \times [\inf(F(Z_{kj})), \sup(F(Z_{kj}))]$, $k = \overline{1,\infty}$, $j = \overline{1,n}$. It follows that $\Lambda^1_\epsilon(B(F;Z)) \leq \sum_{k=1}^{\infty}\sum_{j=1}^{n} diam(D_{kj}) \leq n \cdot \sum_{k=1}^{\infty}(diam(Z_k) + \mathcal{O}^n(F; Z_k)) \leq \epsilon$. Hence $\Lambda(B(F;Z)) = 0$.

We prove the second part.

$(i) \Rightarrow (ii)$ See Proposition 1.4.3, (v).

$(ii) \Rightarrow (i)$ Let $Z \subset P$, $| Z |= 0$ and let $\epsilon > 0$. For ϵ/n there exists a sequence of sets $\{Z_k\}_k$ such that $Z = \cup_{k=1}^{\infty} Z_k$, $\sum_{k=1}^{\infty} diam(Z_k) < \epsilon/(2n)$ and $\sum_{k=1}^{\infty} \mathcal{O}^n_+(F; Z_k) < \epsilon/(2n)$. It follows that for each k there exist sets Z_{kj}, $j = \overline{1,n}$ such that $Z_k \subset \cup_{j=1}^{n} Z_{kj}$ and $\sum_{k=1}^{\infty}\sum_{j=1}^{n} \mathcal{O}_+(F; Z_{kj}) < \epsilon/(2n)$. We have $\sum_{k=1}^{\infty}\sum_{j=1}^{n} diam(Z_k) < \epsilon/2$.

Lemma 2.33.2 *Let $F : [a,b] \mapsto \mathbb{R}$, $P \subseteq [a,b]$ and let n be a positive integer. The following assertions are equivalent:*

- *(i) $F \in f.l.$ on P;*
- *(ii) There exists $M \in (0, +\infty)$ such that for each $\epsilon > 0$ we have $\sum_{k=1}^{\infty} diam(P_k) < M$ and $\sum_{k=1}^{\infty} \mathcal{O}^n(F; P_k) < M$, whenever $\{P_k\}_k$ is a sequence of subsets of P, with $P = \cup_{k=1}^{\infty} P_k$ and $diam(P_k) < \epsilon$.*

Proof. $(i) \Rightarrow (ii)$ Let $\Lambda(B(F;P)) = r \neq +\infty$. Then for every $\epsilon > 0$ we have $\Lambda^1_\epsilon(B(F;P)) \leq r$. For $M = r + 1$ there exists a sequence of sets $\{D_k\}_k$ such that $D_k \subset \mathbb{R} \times \mathbb{R}$, $B(F;P) \subset \cup_{k=1}^{\infty} D_k$, $diam(D_k) < \epsilon$ and $\sum_{k=1}^{\infty} diam(D_k) < M$. Let $P_k = \{x \in P : (x, F(x)) \in D_k\}$. Then $P = \cup_{k=1}^{\infty} P_k$, $diam(P_k) < diam(D_k) < \epsilon$

and $\mathcal{O}^n(F;P_k) \le \mathcal{O}(F;P_k) \le \mathcal{O}(F;P_k) < diam(D_k) < \epsilon$ (see Proposition 1.4.3, (iv)). Hence $\sum_{k=1}^{\infty} diam(P_k) < M$ and $\sum_{k=1}^{\infty} \mathcal{O}^n(F;P_k) < M$.

(*ii*) $\Rightarrow$ (*i*) Since $\mathcal{O}^n(F;P_k) < \epsilon$ it follows that there exist sets P_{kj}, $j = \overline{1,n}$ such that $P_k = \cup_{j=1}^{n} P_{kj}$ and $\mathcal{O}(F;P_{kj}) < \mathcal{O}^n(F;P_k) < \epsilon$. Let $D_{kj} = [\inf(P_k), \sup(P_k)] \times [\inf(F(P_{kj})), \sup(F(P_{kj}))]$, $k = \overline{1,\infty}$, $j = \overline{1,n}$. Then $diam(D_{kj} < \sqrt{2}\epsilon$, hence we obtain that $\Lambda^1_{\sqrt{2}\epsilon}(B(F;P)) \le \sum_{k=1}^{\infty}\sum_{j=1}^{n} diam(D_{kj}) \le \sum_{k=1}^{\infty}\sum_{j=1}^{n}(diam(P_k) + \mathcal{O}(F;P_{kj})) \le n\sum_{k=1}^{\infty} diam(P_k) + \sum_{k=1}^{\infty}\sum_{j=1}^{n} \mathcal{O}^n(F;P_k) \le nM + nM = 2nM$. It follows that $\Lambda(B(F;p)) \le 2nM$.

Theorem 2.33.1 (Foran) *Let $F : [a,b] \mapsto \mathbb{R}$. If F is a Baire function and $F \in \Lambda Z$ then $F \in \sigma.f.l.$ on $[a,b]$.*

Proof. The Baire functions are characterized by the fact that their graphs are Borel sets (see [Ku], p. 300). Davies showed that every analytic set of positive linear measure contains a closed subset of finite non-zero linear measure (see [Da]). Consequently, if F is a Baire function and $B(F)$ the graph of F is not $\sigma.f.l.$ then there exists a closed set $E_o \subset B(F)$, such that $0 < \Lambda(E_o) < +\infty$. Suppose that for each γ, with $0 \le \gamma < \alpha < \omega_1$, a closed set $E_\gamma \subset B(F)$ has been chosen such that $0 < \Lambda(E_\gamma) < +\infty$, and such that for every $\beta < \gamma$, $E_\gamma \cap E_\beta = \emptyset$. Then, since $B(F)\setminus(\cup_{\gamma<\alpha}E_\gamma)$ is a Borel set and $B(F)$ is not $\sigma.f.l.$, it follows that $\Lambda(B(F))\setminus(\cup_{\gamma<\alpha}E_\gamma) = +\infty$, and that there exists a closed set $E_\alpha \subset B(F) \setminus (\cup_{\gamma<\alpha}E_\gamma)$, such that $0 < \Lambda(E_\alpha) < +\infty$. In this manner one chooses an uncountable collection of sets $\{E_\alpha\}_{\alpha<\omega_1}$, which are closed, pairwise disjoint, contained in $B(F)$ and satisfy $0 < \Lambda(E_\alpha) < +\infty$. Let $E'_\alpha = \mathrm{Pr}_x(E_\alpha)$. Since the line is not the union of uncountably many pairwise disjoint, measurable sets of positive measure (if it were, some interval $[k, k+1]$ would be the union of uncountably many measurable sets of measure greater than $1/n$), it follows that one of the sets E'_α is of measure 0. But $\mid E'_\alpha \mid = 0$ and $\Lambda(E_\alpha) > 0$ contradicts the fact that $F \in \Lambda Z$, and thus the theorem is proved.

Theorem 2.33.2 *Let $P \subseteq [a,b]$ and let n be a positive integer.*

(*i*) $\Lambda Z \subsetneq \Lambda\overline{Z} \cap \Lambda\underline{Z}$ *on* P.

(*ii*) $AC \subsetneq AC_n \subsetneq \Lambda Z$ *on* P.

(*iii*) $\overline{L}_n + \Lambda\overline{Z} = \Lambda\overline{Z}$ *on* P.

(*iv*) $L_n + \Lambda Z = \Lambda Z$ *on* P.

(*v*) $\Lambda\overline{Z} \subsetneq (\overline{N})$ *on* P.

(*vi*) $\Lambda Z \subset (N)$ *on* P.

(*vii*) $\overline{L}_n \cdot \Lambda\overline{Z} = \Lambda\overline{Z}$ *on* P *for bounded positive functions.*

(*viii*) $L_n \cdot \Lambda Z = \Lambda Z$ *on* P *for bounded functions.*

(*ix*) $L_n \circ \Lambda Z = \Lambda Z$ *on* P.

(*x*) $\Lambda Z \subset AC \circ \Lambda Z$ *on* $[a,b]$ *for continuous functions.*

(*xi*) $\Lambda Z \subset AC + \Lambda Z$ *on* $[a,b]$ *for continuous functions.*

(xii) $\Lambda Z \subset AC \cdot \Lambda Z$ *on* $[a,b]$ *for continuous functions.*

(xiii) $L_n + f.l. = f.l.$ *on* P.

(xiv) $L_n \cdot f.l. = f.l.$ *on* P.

(xv) $L_n \circ f.l. = f.l.$ *on* P.

Proof. (i) This follows by Corollary 1.4.1, (iii).

(ii) This follows by definitions and Lemma 2.33.1.

(iii) Let $F \in \Lambda\overline{Z}$ and $G \in \overline{L}_n$ on P. Let M be given by the later fact. Let $Z \subset P$, $|Z| = 0$ and let $\epsilon > 0$. Then there exist Z_k, $k = \overline{1,\infty}$, such that $Z = \cup_{k=1}^{\infty} Z_k$, $\sum_{k=1}^{\infty} diam(Z_k) < \epsilon/2(n+m)$ and $\sum_{k=1}^{\infty} \mathcal{O}_+(F;Z_k) < \epsilon/2(n+m)$. By Proposition 1.4.3, (xi), $\sum_{k=1}^{\infty} \mathcal{O}_+^n(F+G;Z_k) < n\sum_{k=1}^{\infty} \mathcal{O}_+(F;Z_k) + \sum_{k=1}^{\infty} \mathcal{O}_+^n(G;Z_k) \le n\epsilon/2(n+m) + M\epsilon/2(n+M) = \epsilon/2$.

(iv), (vii) and (viii) follow similarly to (iii); (v) follows by definitions; (vi) follows by definitions and Section 6.51.

(ix) Let $F \in \Lambda Z$ on P, and let $G \in L_n$ on P with a constant M. Let $\epsilon > 0$ and let $Z \subset P$, $|Z| = 0$. For $\epsilon/2(M+1)$ there exists a sequence $\{Z_k\}_k$ of subsets of Z, such that $Z = \cup_{k=1}^{\infty} Z_k$, $\sum_{k=1}^{\infty} diam(Z_k) < \epsilon 2(M+1)$ and $\sum_{k=1}^{\infty} \mathcal{O}(F;Z_k) < \epsilon/2(M+1)$.By Proposition 1.4.3, (xiv), $\sum_{k=1}^{\infty} \mathcal{O}^n(G\circ F;Z_k) = \sum_{k=1}^{\infty} \mathcal{O}^n(G;F(Z_k)) \le M\sum_{k=1}^{\infty} diam(F(Z_k)) = M\sum_{k=1}^{\infty} \mathcal{O}(F;Z_k) \le M\epsilon/2(M+1) < \epsilon/2$. By Lemma 2.33.1, $G\circ F \in \Lambda Z$ on P.

(x) follows by Section 6.52.; (xi) follows by Section 6.53.; (xii) follows by Section 6.54.

(xiii) Let $F \in f.l.$ on P. Then there exists $M \in (0,+\infty)$ such that for every $\epsilon > 0$, $\sum_{k=1}^{\infty} diam(P_k) < M$ and $\sum_{k=1}^{\infty} \mathcal{O}^n(F;P_k) < M$, whenever $\{P_k\}_k$ is a sequence of subsets of P with $P = \cup_{k=1}^{\infty}$ and $diam(P_k) < \epsilon$ (see Lemma 2.33.2). Let $G \in L_n$ on P with a constant M_1. By Proposition 1.4.3, (xi), it follows that $\sum_{k=1}^{\infty} \mathcal{O}^n(G+F;P_k) \le \sum_{k=1}^{\infty} \mathcal{O}^n(G;P) + n\sum_{k=1}^{\infty} \mathcal{O}(F;P_k) \le M_1 \cdot \sum_{k=1}^{\infty} diam(P_k) + n\cdot M \le M_1\cdot M + n\cdot M$, and by Lemma 2.33.2, we have that $G+F \in f.l.$.

(xiv) and (xv) follow similarly to (xiii).

Remark 2.33.1 *$AC + \Lambda Z \not\subset (N)$ on $[a,b]$ for continuous functions (see Section 6.53.).*

Corollary 2.33.1 *Let $P \subseteq [a,b]$.*

(i) $ACG \subset \mathcal{F} \subset \Lambda Z$ *on* P.

(ii) $\overline{\mathcal{L}} + \Lambda\overline{Z} = \Lambda\overline{Z}$ *on* P.

(iii) $\mathcal{L} + \Lambda Z = \Lambda Z$ *on* P.

(iv) $\overline{\mathcal{L}} \cdot \Lambda\overline{Z} = \Lambda\overline{Z}$ *on P for bounded positive functions.*

(v) $\mathcal{L} \cdot \Lambda Z = \Lambda Z$ *on P for bounded functions.*

(vi) $\mathcal{L} \circ \Lambda Z = \Lambda Z$ *on* P.

(vii) $\mathcal{L} + \sigma.f.l. = \sigma.f.l.$ *on* P.

(viii) $\mathcal{L} \cdot \sigma.f.l. = \sigma.f.l.$ *on* P *for bounded functions.*

(ix) $\mathcal{L} \circ \sigma.f.l. = \sigma.f.l.$ *on* P.

(x) $\sigma.f.l. \subset AC \circ \sigma.f.l.$ *on* $[a,b]$ *for continuous functions.*

(xi) $\sigma.f.l. \subset AC + \sigma.f.l.$ *on* $[a,b]$ *for continuous functions.*

(xii) $\sigma.f.l. \subset AC \cdot \sigma.f.l.$ *on* $[a,b]$ *for continuous functions.*

(xiii) $(M) \subset L_2G \circ (M)$ *on* $[a,b]$ *for continuous functions.*

(xiv) $LG \cap (S) \subset LG$ *on* $[a,b]$ *for continuous functions.*

Proof. (i) $ACG \subset \mathcal{F}$ follows by Corollary 2.28.1, (iii). By Theorem 2.33.2, (ii), $\mathcal{F} \subseteq \Lambda Z$. By Corollary 2.28.1, (1), it follows that $AC + \mathcal{F} = \mathcal{F}$. But by Theorem 2.33.2, (xi), $\Lambda Z \subset AC + \Lambda Z$, hence $\mathcal{F} \subset \lambda Z$.

(ii) - (ix) follow by Theorem 2.33.2; (x) follows by Section 6.58.; (xi) follows by Section 6.59.; (xii) follows by Section 6.60.; (xiii) follows by Corollary 6.29.1; (xiv) follows by Section 6.26.

Remark 2.33.2 *(i) The question is if in Corollary 2.33.1, (xiii),* L_2G *can be replaced by* LG *or* ACG*?*

(ii) $L_2G \not\subset ACG$ *(see Section 6.29, the function* G_{PP}*) and* $AC \not\subset \mathcal{L}$ *(because by Theorem 2.33.2, (xi) and Corollary 2.33.1, (iii),* $AC + \Lambda Z \supset \Lambda Z = \Lambda Z + \mathcal{L}$*, hence* $AC \setminus \mathcal{L} \neq \emptyset$*).*

(iii) $AC_2G + \Lambda Z \not\subset (M)$ *(see Section 6.56.), hence* $\mathcal{F} + \Lambda Z \not\subset (M)$.

2.34 Conditions SAC_n, SAC_ω, SAC_∞, $S\mathcal{F}$

Definition 2.34.1 (Iseki) *([I4]). Let* $W \subseteq [a,b]$*.* W *is said to be a sparse figure if there exist a natural number* p *and* $a_1 < b_1 < a_2 < b_2 < \ldots < a_p < b_p$*, such that* $W = \cup_{k=1}^{p}[a_k, b_k]$ *and* $\max\{b_1 - a_1, b_2 - a_2, \ldots, b_p - a_p\} < \min\{a_2 - b_1, a_3 - b_3, \ldots, a_p - b_{p-1}\}$*. Let* $P \subseteq [a,b]$*.* W *is said to be a sparse figure for* P *if* $[a_k, b_k] \cap P \neq \emptyset$*,* $k = \overline{1,p}$*. The intervals* $[a_k, b_k]$ *are called the components of* W*.*

Lemma 2.34.1 *Let* $\{I_k\}$*,* $k = \overline{1,n}$ *be a finite set of nonoverlapping closed nondegenerate subintervals of* $[a,b]$*. Then there exist three sparse figures* W_1*,* W_2*,* W_3 *whose interiors are pairwise disjoint, such that* $\cup_{k=1}^{n} I_k = W_1 \cup W_2 \cup W_3$*.*

Proof. Let $d = \inf\{| I_k | : k = \overline{1,n}\}$. Let $n_k = [| I_k | /d]$ (where $[x]$ is the greatest integer less than x), $k = \overline{1,n}$. If $| I_k | = d \cdot n_k$ then I_k is the union of n_k closed intervals with length d. If $| I_k | = d \cdot n_k$ then I_k is the union of $n_k + 1$ nonoverlapping closed intervals with length d, except the last, which has the length less than d. Let $J_1, J_2, \ldots, J_m$ be the set of all new created closed intervals, numbered from the left to the right. We define $W_1 = \cup_{i=0}^{[(m-1)/3]} J_{3i+1}$, $W_2 = \cup_{i=0}^{[(m-2)/3]} J_{3i+2}$, $W_3 = \cup_{i=1}^{[m/3]} J_{3i}$.

Definition 2.34.2 *Let $F : [a,b] \mapsto \mathbb{R}$, $P \subseteq [a,b]$ and let n be a positive integer. F is said to be SAC_n on P if for $\epsilon > 0$ there exists a $\delta > 0$ such that $\sum_{k=1}^{p} \mathcal{O}^n(F; [a_k,b_k] \cap P) < \epsilon$, whenever $W = \cup_{k=1}^{p}[a_k,b_k]$ is a sparse figure for P, with $\mid W \mid < \delta$. Let $SAC_1 = SAC$ on P (this condition was introduced by Iseki in [I3], but he called it "sparsely continuous").*
F is said to be SAC_ω (resp. SAC_∞) on P if $\sum_{k=1}^{p} \mid \overline{F([a_k,b_k] \cap P)} \mid < \epsilon$ (resp. $\sum_{k=1}^{p} \mid F([a_k,b_k] \cap P) \mid < \epsilon$).
We define the classes SAC_nG and $[SAC_nG]$ on P, using Definition 1.9.2.
We can also define the following conditions: $S\underline{AC}_n$, $S\overline{AC}_n$, $S\underline{AC}_nG$, $S\underline{AC}_\omega$, etc.

Remark 2.34.1 *In the definition of SAC_n, "$\mathcal{O}^n(F; [a_k,b_k] \cap P)$" may be replaced by $\lambda_n(F([a_k,b_k] \cap P))$ (see Proposition 1.4.3,(i)).*

Proposition 2.34.1 *Let $F : [a,b] \mapsto \mathbb{R}$, $P \subseteq [a,b]$ and let n be a positive integer. The following assertions are equivalent:*

(i) $F \in SAC_n$ on P;

(ii) For every $\epsilon > 0$ there exists a $\delta > 0$ with the following property: if $W = \cup_{k=1}^{s}[a_k,b_k]$ is a sparse figure for P with $\mid W \mid < \delta$ then for each k there exist P_{kj}, $j = \overline{1,n}$ such that $P \cap [a_k,b_k] = \cup_{j=1}^{n} P_{kj}$ and $\sum_{k=1}^{s}\sum_{j=1}^{n} \mathcal{O}(F; P_{kj} < \epsilon$.

Lemma 2.34.2 *Let $F : [a,b] \mapsto \mathbb{R}$, $P \subseteq [a,b]$.*

(i) $AC_\omega = SAC_\omega$ on P.

(ii) $AC_\infty = SAC_\infty$ on P.

Proof. (i) $AC_\omega \subseteq SAC_\omega$ on P is evident. Conversely, for $\epsilon > 0$ let $\delta > 0$ be given by the fact that $F \in SAC_\omega$ on P. Let $\{I_j\}$, $j = \overline{1,p}$ be a sequence of nonoverlapping closed intervals, such that $I_j \cap P \neq \emptyset$ and $\sum_{j=1}^{p} \mid I_j \mid < \delta$. By Lemma 2.34.1, there exist three sparse figures W_1, W_2, W_3 whose interiors are pairwise disjoint, such that $\cup_{j=1}^{p} I_j = W_1 \cup W_2 \cup W_3$. Clearly $\mid W_i \mid < \delta$, $i = 1,2,3$. Let $\mathcal{W}_i = \{K : K$ is a component of $W_i\}$, $i = 1,2,3$ and let $\mathcal{W} = \mathcal{W}_1 \cup \mathcal{W}_2 \cup \mathcal{W}_3$. It follows that $\sum_{j=1}^{p} \mid \overline{F(I_j \cap P)} \mid \leq \sum_{j=1}^{p} \sum_{K \in \mathcal{W} \cap I_j} \mid \overline{F(K \cap P)} \mid \leq \sum_{i=1}^{3} \sum_{K \in \mathcal{W}_i} \mid \overline{F(K \cap P)} \mid < 3\epsilon$ (because $\overline{A \cup B} = \overline{A} \cup \overline{B}$).

(ii) The proof is similar to that of (i).

Definition 2.34.3 *Let $F : [a,b] \mapsto \mathbb{R}$, $P \subseteq [a,b]$. F is said to be $S\mathcal{F}$ on P, if $P = \cup_{k=1}^{\infty} P_k$ and if there exists a sequence $\{n_k\}_k$ of positive integers such that $F \in SAC_{n_k}$ on P_k. If in addition the sets P_k are supposed to be closed then we obtain the class $[S\mathcal{F}]$ on P. Clearly $[S\mathcal{F}] \subseteq S\mathcal{F}$. Similarly we can define the classes $S\underline{\mathcal{F}}$, $[S\underline{\mathcal{F}}]$, etc.*

Theorem 2.34.1 *Let $F : [a,b] \mapsto \mathbb{R}$, $P \subseteq [a,b]$, and let m and n be positive integers.*

(i) $SAC = AC$ on $[a,b]$.

(ii) $AC_n \subseteq SAC_n \subseteq SAC_{n+1} \subseteq AC_\omega \subseteq AC_\infty \subseteq (S) \subset (N)$ on P.

(iii) $SAC_m + SAC_n \subseteq SAC_{mn}$ on P.

(iv) $SAC_m \cdot SAC_n \subseteq SAC_{mn}$ *on P for bounded functions.*

(v) If $F_{/\overline{P}} \in \mathcal{C}$ *and* $F \in SAC_n$ *on P then* $F \in SAC_n$ *on* $\overline{P}$.

Proof. (i) Clearly $AC \subset SAC$ on $[a,b]$. Conversely, let $\epsilon > 0$. For $\epsilon/3$ let $\delta > 0$ be given by the fact that $F \in SAC$ on $[a,b]$. Let $\{I_k\}$, $k = \overline{1,p}$ be a finite set of nonoverlapping closed intervals, such that $\sum_{k=1}^p \mid I_k \mid < \delta$. By Lemma 2.34.1, there exist three sparse figures W_1, W_2, W_3 whose interiors are pairwise disjoint, such that $\cup_{k=1}^p I_k = W_1 \cup W_2 \cup W_3$. Let $\mathcal{W}_i = \{K : K \text{ is a component of } W_i\}$, $i = 1,2,3$. We have $\sum_{k=1}^p \mathcal{O}(F;I_k) \le \sum_{k=1}^p \sum_{i=1}^3 \sum_{K\in\mathcal{W}_i} \mathcal{O}(F;K) = \sum_{i=1}^3 \sum_{K\in\mathcal{W}_i} \mathcal{O}(F;K) < \epsilon/3 + \epsilon/3 + \epsilon/3 = \epsilon$, hence $F \in AC$ on $[a,b]$.

(ii) $AC_n \subseteq SAC_n \subseteq SAC_{n+1}$ follow by definitions. $SAC_{n+1} \subseteq SAC_\omega = AC_\omega$ follow by definitions and Lemma 2.34.2. $AC_\omega \subseteq AC_\infty \subseteq (S) \subset (N)$ follow by Theorem 2.28.1, (vi).

(iii), (iv), (v) The proofs are similar to those of Theorem 2.28.1, (ix), (xi), (xii).

Corollary 2.34.1 *Let* $F : [a,b] \mapsto \mathbb{R}$, $P \subseteq [a,b]$ *and let m and n be positive integers.*

(i) $ACG \subseteq SACG$ *on P. Moreover,* $ACG \subset SACG$ *for continuous functions on* $[a,b]$.

(ii) $SAC_nG \subseteq SAC_{n+1}G \subseteq \mathcal{SF} \subset SG \subset (N)$ *on P.*

(iii) $\mathcal{F} \subseteq \mathcal{SF}$ *on P. Moreover,* $\mathcal{F} \subset \mathcal{SF}$ *for continuous functions on* $[a,b]$.

(iv) $SAC_mG + SAC_nG \subseteq SAC_{mn}G$ *on P.*

(v) $SAC_m \cdot SAC_nG \subseteq SAC_{mn}G$ *on P for bounded functions.*

(vi) $\mathcal{SF} + \mathcal{SF} = \mathcal{SF}$ *on P.*

(vii) $\mathcal{SF} \cdot \mathcal{SF} = \mathcal{SF}$ *on P for bounded functions.*

Proof. (i) See Theorem 2.34.1, (ii) and Section 6.42.

(ii) See Theorem 2.34.1, (ii) and Section 6.31 or Section 6.29 (the functions K_{CP} or H_{CP}).

(iii) See Theorem 2.34.1, (ii) and Section 6.42.

(iv) - (vii) follow by Theorem 2.34.1.

Remark 2.34.2 *(i) For continuous functions on* $[a,b]$ *we have:* $SACG \not\subset \mathcal{F}$ *(see Section 6.42), and* $\mathcal{F} \not\subset SACG$ *(see Section 6.41). Hence* $SACG \subset \mathcal{SF}$ *and* $\mathcal{F} \subset \mathcal{SF}$.

(ii) In Corollary 2.34.1, (ii), we cannot replace SG by (S) (see Section 6.29, the function G_{PP}*).*

Lemma 2.34.3 *Let* $F : [a,b] \mapsto \mathbb{R}$, $P = \overline{P} \subseteq [a,b]$ *and let n be a positive integer. The following assertions are equivalent:*

(i) $F \in [SAC_nG]$ *on P;*

(ii) Each closed subset S of P contains a portion $S \cap (c,d)$ *such that* $F \in SAC_n$ *on* $S \cap (c,d)$ *(see Theorem 1.9.1).*

Lemma 2.34.4 *Ket $F : [a,b] \mapsto \mathbb{R}$, $P = \overline{P} \subseteq [a,b]$. The following assertions are equivalent:*

(i) $F \in [S\mathcal{F}]$ on P;

(ii) Each closed subset S of P contains a portion $S \cap (c,d)$ such that $F \in SAC_n$ on $S \cap (c,d)$, for some positive integer n (see Theorem 1.9.1).

Remark 2.34.3 *(i) Particularly, Lemma 2.34.3 is true for functions $SAC_n \cap C \subseteq [SAC_nG]$.*

(ii) Particularly, Lemma 2.34.4 is true for functions $S\mathcal{F} \cap C \subseteq [S\mathcal{F}]$ (see Theorem 2.34.1, (v)).

2.35 Conditions SVB_n, SVB_ω, SVB_∞,; $S\mathcal{B}$

Definition 2.35.1 *Let $F : [a,b] \mapsto \mathbb{R}$, $P \subseteq [a,b]$ and let n be a positive integer. F is said to be SVB_n on P if there exists $M \in (0,+\infty)$, such that $\sum_{k=1}^{p} \mathcal{O}^n(F;[a_k,b_k]\cap P) < M$, whenever $W = \cup_{k=1}^{p}[a_k,b_k]$ is a sparse figure for P (see Definition 2.34.1). Let $SVB_1 = SVB$ on P.*
F is said to be SVB_ω (resp. SVB_∞) on P if $\sum_{k=1}^{p} \mid \overline{F(P\cap[a_k,b_k])} \mid < M$ (resp. $\sum_{k=1}^{p} \mid F(P\cap[a_k,b_k]) \mid < M$).
We define the classes SVB_nG and $[SVB_nG]$ on P, using Definition 1.9.2.
Similarly we can define the classes $S\underline{VB}_n$, $S\overline{VB}_n$, $S\underline{VB}_nG$, $S\underline{VB}_\omega$, etc.

Remark 2.35.1 *In the definition of SVB_n, "$\mathcal{O}^n(F;P\cap[a_k,b_k])$" may be replaced by "$\lambda_n(F(P\cap[a_k,b_k]))$" (see Proposition 1.4.3, (i).*

Lemma 2.35.1 *Let $F : [a,b] \mapsto \mathbb{R}$, $P \subseteq [a,b]$.*

(i) $VB_\omega = SVB_\omega$ on P.

(ii) $VB_\infty = SVB_\infty$ on P.

Proof. The proof is similar to that of Lemma 2.34.3.

Definition 2.35.2 *Let $F : [a,b] \mapsto \mathbb{R}$, $P \subseteq [a,b]$. F is said to be $S\mathcal{B}$ on P if $P = \cup_{k=1}^{\infty} P_k$ and if there exists a sequence $\{n_k\}_k$ of positive integers such that $F \in SVB_{n_k}$ on P_k. If in addition P_k are supposed to be closed sets then we obtain the class $[S\mathcal{B}]$ on P. Clearly $[S\mathcal{B}] \subseteq S\mathcal{B}$. Similarly we define the classes $S\underline{\mathcal{B}}$, $[S\underline{\mathcal{B}}]$, etc.*

Theorem 2.35.1 *Let $F : [a,b] \mapsto \mathbb{R}$, $P \subseteq [a,b]$ and let m and n be positive integers.*

(i) $SVB = VB$ on $[a,b]$.

(ii) $VB_n \subseteq SVB_n \subseteq SVB_{n+1} \subseteq VB_\omega \subseteq VB_\infty$ on P.

(iii) $SVB_m + SVB_n \subseteq SVB_{mn}$ on P.

(iv) $SVB_m \cdot SVB_n \subseteq SVB_{mn}$ on P for bounded functions.

(v) If $F_{/\overline{P}} \in C$ and $F \in SVB_n$ on P then $F \in SVB_n$ on $\overline{P}$.

Proof. The proof is similar to that of Theorem 2.34.1 (using Theorem 2.29.1 instead of Theorem 2.28.1).

Remark 2.35.2 *Is it true that* $SAC_n \subseteq SVB_n \subseteq f.l.$ *on* $P \subseteq [a, b]$?

Corollary 2.35.1 *Let* $P \subseteq [a, b]$ *and let* m *and* n *be positive integers.*

(i) $VBG \subseteq SVBG$ *on* P.

(ii) $SVB_nG \subseteq SVB_{n+1}G \subseteq SB$ *on* P.

(iii) $B \subseteq SB$ *on* P.

(iv) $SVB_mG + SVB_nG \subseteq SVB_{mn}G$ *on* P.

(v) $SVB_mG \cdot SVB_nG \subseteq SVB_{mn}G$ *on* P *for bounded functions.*

(vi) $SB + SB = SB$ *on* P.

(vii) $SB \cdot SB = SB$ *on* P *for bounded functions.*

Remark 2.35.3   *(i)* *The Banach-Zarecki Theorem (see Theorem 2.18.9, (vii)) is no longer true if* VB *is replaced by* SVB *and* AC *by* SAC_n *(see Section 6.31).*

(ii) $SVB \cap C \cap L_2 \not\subseteq SAC$ *(see Section 6.41).*

Lemma 2.35.2 *Let* $F : [a, b] \mapsto \mathbb{R}$, $P = \overline{P} \subseteq [a, b]$ *and let* n *be a positive integer. The following assertions are equivalent:*

(i) $F \in [SVB_nG]$ *on* P;

(ii) *Each closed subset of* P *contains a portion* $S \cap (c, d)$ *such that* $F \in SVB_n$ *on* $S \cap (c, d)$ *(see Theorem 1.9.1).*

Lemma 2.35.3 *Let* $F : [a, b] \mapsto \mathbb{R}$, $P = \overline{P} \subseteq [a, b]$. *The following assertions are equivalent:*

(i) $F \in [SB]$ *on* P;

(ii) *Each closed subset* S *of* P *contains a portion* $S \cap (c, d)$ *such that* $F \in SVB_n$ *on* $S \cap (c, d)$, *for some positive integer* n *(see Theorem 1.9.1).*

Remark 2.35.4   *(i)* *Particularly Lemma 2.35.2 is true for functions* $SVB_nG \cap C \subseteq [SVB_nG]$.

(ii) *Particularly Lemma 2.35.3 is true for functions* $SB \cap C \subseteq [SB]$ *(see Theorem 2.35.1, (v)).*

2.36 Conditions DW_n, DW_ω, DW_∞, DW^*

Definition 2.36.1 *Let $F : [a,b] \mapsto \mathbb{R}$, $P \subseteq [a,b]$, and let n be a positive integer. F is said to be DW_n (resp. $\overline{DW}_n$) on P, if for every $\epsilon > 0$ there exists a sequence $\{I_k\}_k$ of nonoverlapping closed intervals which cover P, such that $\sum_{k\geq 1} \mathcal{O}^n(F; P\cap I_k) < \epsilon$ (resp. $\sum_{k\geq 1} \mathcal{O}^n_+(F; P\cap I_k) < \epsilon$). F is said to be $\underline{DW}_n$ on P if $-F \in \overline{DW}_n$ on P, i.e., $\sum_{k\geq 1} \mathcal{O}^n_-(F; P\cap I_k) > -\epsilon$. If $\mathcal{O}^n$, $\mathcal{O}^n_+$, $\mathcal{O}^n_-$ are replaced by $\mathcal{O}^\omega$, $\mathcal{O}^\omega_+$, $\mathcal{O}^\omega_-$ we obtain the conditions DW_ω, $\overline{DW}_\omega$, $\underline{DW}_\omega$ on P. If $\mathcal{O}^n$, $\mathcal{O}^n_+$, $\mathcal{O}^n_-$ are replaced by $\mathcal{O}^\infty$, $\mathcal{O}^\infty_+$, $\mathcal{O}^\infty_-$ we obtain the conditions DW_∞, $\overline{DW}_\infty$, $\underline{DW}_\infty$ on P.*
We define the classes DW_nG, $[DW_nG]$, $(AC^* \cup DW_1)G$, etc., using Definition 1.9.2. Clearly $[DW_nG] \subseteq DW_nG$, etc.

Remark 2.36.1 *(i) Condition DW_1 (dwindle) was introduced by Iseki in [I2], and condition DW_n, $n \geq 2$ was introduced by C.M. Lee in [Le3].*

(ii) In the definition of DW_n, "$\mathcal{O}^n(F; P\cap I_k)$" may be replaced by "$\lambda_n(F(P\cap I_k))$" (see Proposition 1.4.3, (i)).

(iii) In the definition of DW_ω, "$\mathcal{O}^\omega(F; P\cap I_k)$" may be replaced be one of the following: "$\lambda_\omega(F(P\cap I_k))$", "$\lambda_\omega(\overline{F(P\cap I_k)})$", "$\lambda_\infty(\overline{F(P\cap I_k)})$", "$|\ \overline{F(P\cap I_k)}\ |$" (see Proposition 1.4.3, (ii)).

(iv) In the definition of DW_∞, "$\mathcal{O}^\infty(F; P\cap I_k)$ " may be replaced by "$\lambda_\infty(F(P\cap I_k))$" or by "$|\ F(P\cap I_k)\ |$" (see Proposition 1.4.3, (iii)).

Definition 2.36.2 *Let $F : [a,b] \mapsto \mathbb{R}$, $P \subseteq [a,b]$. F is said to be DW^* on P if for every $\epsilon > 0$ there exists a sequence $\{I_k\}_k$ of open intervals which cover P, such that $\sum_{k\geq 1} \mathcal{O}(F; I_k) < \epsilon$. We define condition DW^*G on P, using Definition 1.9.2.*

Lemma 2.36.1 *Let $[a,b] \subset \cup_{k=1}^\infty (\alpha_k, \beta_k)$. Then there exist a partition $a = a_o < a_1 < \ldots < a_p = b$ and p different positive integers $k_1, k_2, \ldots, k_p$ such that $[a_{i-1}, a_i] \subset (\alpha_{k_i}, \beta_{k_i})$, $i = \overline{1,p}$.*

Proof. By the Heine-Borel Covering Theorem, there exists a positive integer m such that $[a,b] \subset \cup_{k=1}^m (\alpha_k, \beta_k)$. Let $k_1 < m$ such that $a \in (\alpha_{k_1}, \beta_{k_1})$. If $b \in (\alpha_{k_1}, \beta_{k_1})$ then let $p = 1$ and $a_1 = b$. If $b \notin (\alpha_{k_1}, \beta_{k_1})$ then there exists $k_2 < m$ such that $\beta_{k_1} \in (\alpha_{k_2}, \beta_{k_2})$. If $b \in (\alpha_{k_2}, \beta_{k_2})$ then $p = 2$, $a_1 \in (\alpha_{k_1}, \beta_{k_1})$ and $a_2 = b$. Continuing in this manner, we obtain the conclusion of our lemma.

Lemma 2.36.2 *Let $F : [a,b] \mapsto \mathbb{R}$, $F \in \mathcal{C}$. Let $P \subseteq [a,b]$ and $\epsilon > 0$. If $F \in DW^*$ on P then there exists a sequence $\{I_k\}_k$ of nonoverlapping closed intervals which cover P and $\sum_{k=1}^\infty \mathcal{O}(F; I_k) < \epsilon$.*

Proof. Since $F \in DW^*$ on P, there exists a sequence $\{J_i\}_i$ of open intervals such that $P =\subset \cup_{i=1}^\infty J_i$ and $\sum_{i=1}^\infty \mathcal{O}(F; J_i) < \epsilon/2$. Let $G = \cup_{i=1}^\infty J_i$. Let (a_i, b_i), $j \geq 1$ be the components of the open set G. Let $[\alpha_j, \beta_j] \subset [a_j, b_j]$ such that $\mathcal{O}(F; [a_j, \alpha_j]) + \mathcal{O}(F; [\beta_j, b_j]) < \epsilon/2^{j+1}$ (this is possible because $F \in \mathcal{C}$). Let $A_j = \{i : J_i \subset (a_j, b_j)\}$. It follows that $[\alpha_j, \beta_j] \subset \cup_{i\in A_j} J_i$. By Lemma 2.36.1, there exists a partition of each interval $[\alpha_j, \beta_j]$ such that each of its components is contained in one single interval J_i, $i \in A_j$. Then the required sequence $\{I_k\}_k$ will consist of all components of the partition of $[\alpha_j, \beta_j]$ and the intervals $[a_j, \alpha_j]$ and $[\beta_j, b_j]$, $j \geq 1$.

Lemma 2.36.3 *Let $F : [a,b] \mapsto \mathbb{R}$, $F \in \mathcal{C}$ and let $P = \overline{P} \subset [a,b]$, $|P| = 0$. If $F \in AC^*$ on P then $F \in DW^*$ on P.*

Proof. By Theorem 2.12.3, (ii), for $\epsilon > 0$ let δ be given by the fact that $F \in AC^*$ on P and $F \in \mathcal{C}$ at each point $x \in P$. Since P is a compact null set, there exists a finite set of nonoverlapping open intervals $\{(a_k, b_k)\}$, $k = \overline{1,p}$ such that $P \subset \cup_{k=1}^{p}(a_k, b_k)$ and $\sum_{k=1}^{p}(b_k - a_k) < \delta$. Then $\sum_{k=1}^{p} \mathcal{O}(F; [a_k, b_k]) < \epsilon$. It follows that $F \in DW^*$ on P.

Theorem 2.36.1 *Let $F : [a,b] \mapsto \mathbb{R}$, $P \subseteq [a,b]$ and let n be a positive integer.*

(i) $DW_n \subseteq \underline{DW}_n \cap \overline{DW}_n$, $DW_\omega \subseteq \underline{DW}_\omega \cap \overline{DW}_\omega$, $DW_\infty \subseteq \underline{DW}_\infty \cap \overline{DW}_\infty$ *on P.*

(ii) $\underline{SW}_n \subseteq \underline{DW}_{n+1} \subseteq \underline{DW}_\omega \subseteq \underline{DW}_\infty$, *hence* $\underline{DW}_n G \subseteq \underline{DW}_{n+1} G \subseteq \underline{DW}_\omega G \subseteq \underline{DW}_\infty G$ *on P.*

(iii) $DW_n \subseteq DW_{n+1} \subseteq DW_\omega \subseteq DW_\infty$, *hence* $DW_n G \subseteq DW_{n+1} G \subseteq DW_\omega G \subseteq DW_\infty G$ *on P.*

(iv) $\underline{DW}_\omega \cap \mathcal{D}_- = \underline{DW}_1$ *on* $[a,b]$;

(v) If $F \in DW_\infty G$ *on P then* $|F(P)| = 0$, *hence* $F \in (N) \subset (M)$ *on P.*

(vi) If $F \in DW_\infty G \cap \mathcal{D}$ *on* $[a,b]$ *then F is constant on* $[a,b]$.

(vii) $DW^*G = DW^*$ *on P.*

(viii) $DW^* \subset DW_1$ *on P.*

(ix) AC^*; $DW^*G \subset AC^*$; $DW_1 G$ *on P.*

Proof. (i) See Proposition 1.4.3, (vi), (vii), (viii).

(ii) See Proposition 1.4.3, (v).

(iii) See Proposition 1.4.3, (iv).

$\underline{DW}_1 \subseteq \underline{DW}_\omega \cap \mathcal{D}_-$ follows by definitions. For the converse see Corollary 2.1.2.

(v) It is sufficient to show that if DW_∞ on P then $|F(P)| = 0$. For $\epsilon > 0$ there exists a sequence $\{I_k\}_k$ of nonoverlapping closed intervals which cover P such that $\sum_{k\geq 1} |F(P \cap I_k)| < \epsilon$. It follows that $|F(P)| \leq \sum_{k\geq 1} |F(P \cap I_k)| < \epsilon$, hence $|F(P)| = 0$.

(vi) If $F \in \mathcal{D}$ on $[a,b]$ then $F([a,b])$ is an interval. By (v) it follows that $|F([a,b])|$, hence F is constant on $[a,b]$.

(vii) Suppose that $F \in DW^*G$ on P. Then there exists a sequence $\{P_k\}_k$ of sets, such that $P = \cup_{k\geq 1} P_k$ and $F \in DW^*$ on each P_k. Let $\epsilon > 0$. For $\epsilon/2^k$ there exists a sequence $\{I_{kj}\}_j$ of open intervals, such that $P_k \subset \cup_{j\geq 1} I_{kj}$ and $\sum_{j\geq 1} \mathcal{O}(F; I_{kj}) < \epsilon/2^k$. Since $P \subset \cup_{k\geq 1} \cup_{j\geq 1} I_{kj}$ and $\sum_{k\geq 1} \sum_{j\geq 1} \mathcal{O}(F; I_{kj}) < \epsilon$, it follows that $F \in DW^*$ on P. The converse is evident.

(viii) See Lemma 2.36.2 and Section 6.43.

(ix) See (viii) and Section 6.44.

2.37 Conditions E_n, E_ω, E_∞, $\mathcal{E}$

Definition 2.37.1 *Let $F : [a,b] \mapsto \mathbb{R}$, $P \subseteq [a,b]$ and let n be a positive integer. F is said to be E_n (resp. $\overline{E}_n$, $\underline{E}_n$, E_ω, $\overline{E}_\omega$, $\underline{E}_\omega$, E_∞, $\overline{E}_\infty$, $\underline{E}_\infty$) on P, if F is DW_n (resp. $\overline{DW}_n$, $\underline{DW}_n$, DW_ω, $\overline{DW}_\omega$, $\underline{DW}_\omega$, DW_∞, $\overline{DW}_\infty$, $\underline{DW}_\infty$) on Z, whenever $Z \subset P$, $|Z| = 0$.*
We define the conditions E_nG, $[E_nG]$, etc., using Definition 1.9.2.

Definition 2.37.2 *Let $F : [a,b] \mapsto \mathbb{R}$, $P \subseteq [a,b]$. F is said to be $\mathcal{E}$ (resp. $\overline{\mathcal{E}}$) on P, if $P = \cup_{k=1}^{\infty} P_k$ and if there exists a sequence $\{n_k\}_k$ of positive integers, such that $F \in E_{n_k}$ (resp. $\overline{E}_{n_k}$) on P_k. F is said to be $\underline{\mathcal{E}}$ on P if $-F$ is $\overline{\mathcal{E}}$ on P, i.e., $F \in \underline{E}_{n_k}$ on P_k. If in addition P_k are supposed to be closed sets then we obtain the classes $[\mathcal{E}]$, $[\overline{\mathcal{E}}]$, $[\underline{\mathcal{E}}]$ on P. Clearly $[\mathcal{E}] \subseteq \mathcal{E}$ and $[\overline{\mathcal{E}}] \subseteq \overline{\mathcal{E}}$ on P.*

Lemma 2.37.1 *Let $F : [a,b] \mapsto \mathbb{R}$, $P \subseteq [a,b]$ and let n be a positive integer. The following assertions are equivalent:*

(i) $F \in \overline{E}_n$ (resp. $F \in E_n$) on P;

(ii) For every $\epsilon > 0$, $\delta > 0$ and $Z \subset P$, with $|Z| = 0$, there exists a sequence $\{I_k\}_k$ of nonoverlapping closed intervals such that $Z \subset \cup_{k=1}^{\infty} I_k$, $\sum_{k=1}^{\infty} |I_k| < \delta$ and $\sum_{k=1}^{\infty} \mathcal{O}_+^n(F; Z \cap I_k) < \epsilon$ (resp. $\sum_{k=1}^{\infty} \mathcal{O}^n(F; Z \cap I_k) < \epsilon$).

Proof. $(i) \Rightarrow (ii)$ We prove only the first part. Let $F \in \overline{E}_n$ on P. Let $\epsilon > 0$, $\delta > 0$ and $Z \subset P$, with $|Z| = 0$. Then there exists a sequence $\{J_i\}_i$ of nonoverlapping closed intervals such that $Z \subset \cup_{i=1}^{\infty} J_i$ and $\sum_{i=1}^{\infty} |J_i| < \delta$. Clearly $|Z \cap J_i| = 0$, $i = \overline{1,\infty}$, hence $F \in \overline{DW}_n$ on each $Z \cap J_i$. For $\epsilon/2^i$ there exists a sequence $\{I_k^i\}_k$ of nonoverlapping closed intervals such that $I_k^i \subset J_i$ and $\sum_{k=1}^{\infty} \mathcal{O}_+^n(F; Z \cap I_k^i) < \epsilon/2^i$, $i = \overline{1,\infty}$. Clearly $\{I_k^i\}_k$ is a sequence of nonoverlapping closed intervals which cover Z, $\sum_{i=1}^{\infty}\sum_{k=1}^{\infty} |I_k^i| < \delta$ and $\sum_{i=1}^{\infty}\sum_{k=1}^{\infty} \mathcal{O}_+^n(F; Z \cap I_k^i) < \epsilon$.
$(ii) \Rightarrow (ii)$ This is evident.

Remark 2.37.1 *Lemma 2.37.1 remains true if n is replaced by ω or by ∞.*

Theorem 2.37.1 *Let $F : [a,b] \mapsto \mathbb{R}$, $P \subseteq [a,b]$, and let m and n be positive integers.*

(i) $\overline{SAC}_n \subseteq \overline{E}_n$, $\overline{SAC}_\omega \subseteq \overline{E}_\omega$, $\overline{SAC}_\infty \subseteq \overline{E}_\infty$ on P.

(ii) $SAC_n \subseteq E_n$, $SAC_\omega \subseteq E_\omega$, $SAC_\infty \subseteq E_\infty$ on P.

(iii) $E_n \subseteq \overline{E}_n \cap \underline{E}_n$, $E_\omega \subseteq \overline{E}_\omega \cap \underline{E}_\omega$, $E_\infty \subseteq \overline{E}_\infty \cap \underline{E}_\infty$ on P.

(iv) $\overline{E}_n \subseteq \overline{E}_{n+1} \subseteq \overline{E}_\omega \subseteq \overline{E}_\infty = (\overline{N})$ on P.

(v) $E_n \subset E_{n+1} \subseteq E_\omega \subseteq E_\infty = (N)$ on P.

(vi) $\underline{AC}_m \cap \overline{E}_n \subseteq E_{mn}$ on P.

(vii) $\overline{AC}_m + \overline{E}_n \subseteq \overline{E}_{mn}$ on P.

(viii) $AC_m + E_n \subseteq E_{mn}$ on P.

(ix) $\overline{AC}_m \cdot \overline{E}_n \subseteq \overline{E}_{mn}$ *on P for bounded positive functions.*

(x) $AC_m \cdot E_n \subseteq E_{mn}$ *on P for bounded functions.*

(xi) $L_m \circ E_n \subseteq E_{mn}$, $L_m \circ E_\omega \subseteq E_\omega$, $L_m \circ E_\infty \subseteq E_\infty$ *on P.*

Proof. (i) We show for example the first part. For $\epsilon > 0$ let δ be given by the fact that $F \in \overline{SAC}_n$ on P. Let $Z \subset P$, with $| Z |= 0$. Then there exists a sequence $\{I_k\}_k$ of nonoverlapping closed intervals which cover Z, such that $\sum_{k=1}^\infty | I_k |< \delta$. For $\epsilon/2$ let δ_1 be given by the fact that $F \in \overline{SAC}_n$ on P. Let n_1 be a positive integer such that $\sum_{k=n_1+1}^\infty | I_k |< \delta_1$. For $\epsilon/2^2$ let δ_2 be given by the fact that $F \in \overline{SAC}_n$ on P. Let n_2 be a positive integer such that $n_2 > n_1$ and $\sum_{k=n_2+1}^\infty | I_k |< \delta_2$. Continuing, for $\epsilon/2^p$ let δ_p be given by the fact that $F \in \overline{SAC}_n$ on P. Let n_p be a positive integer such that $n_p > n_{p-1}$ and $\sum_{k=n_p+1}^\infty | I_k |< \delta_p$. Applying Lemma 2.34.1 to the intervals I_k, $k = \overline{n_p+1, n_{p+1}}$, $p = \overline{0, +\infty}$, $n_o = 1$, $\delta_o = \delta$, it follows that there exist three sparse figure $W_{p,1}$, $W_{p,2}$, $W_{p,3}$, whose interiors are pairwise disjoint, such that $\cup_{k=n_p+1}^{n_{p+1}} I_k = W_{p,1} \cup W_{P,2} \cup W_{p,3}$. But $\sum_{k=n_p+1}^{n_{p+1}} | I_k |< \delta_p$, hence $| W_{p,i} |< \delta_p$, $i = 1, 2, 3$. Let $\mathcal{W}_{p,i} = \{K : K$ is a component of $W_{p,i}\}$, $i = 1, 2, 3$. It follows that $\sum_{K \in \mathcal{W}_{p,i}} \mathcal{O}_+^n(F; P \cap K) < \epsilon/2^p$ and $Z \subset \cup_{k=1}^\infty I_k = \cup_{p=0}^\infty \cup_{i=1}^3 W_{p,i}$. Therefore we obtain $\sum_{p=0}^\infty \sum_{i=1}^3 \sum_{K \in \mathcal{W}_{p,i}} \mathcal{O}_+^n(F; P \cap K) < 3\epsilon$, so $F \in \overline{E}_n$ on P.

(ii) The proof is similar to that of (i).

(iii) See Proposition 1.4.3, (vi), (vii), (viii).

(iv) These follow by definitions and Proposition 1.4.3, (v).

(v) These follow be definitions, Proposition 1.4.3, (iv) and Section 6.48.

(vi) Let $F \in \underline{AC}_m \cap \overline{E}_n$ on P. Let $\epsilon > 0$ and $Z \subset P$, with $| Z |= 0$. For $\epsilon/(2n)$ let δ be given by the fact that $F \in \underline{AC}_m$ on P. Since $F \in \overline{E}_n$ on P, by Lemma 2.37.1, there exists a sequence $\{I_k\}_k$ of nonoverlapping closed intervals which cover Z, such that $\sum_{k=1}^\infty | I_k |< \delta$ and $\sum_{k=1}^\infty \mathcal{O}_+^n(F; Z \cap I_k) < \epsilon/(2m)$. Since $F \in \underline{AC}_m$ on P we have $\sum_{k=1}^\infty | \mathcal{O}_-^m(F; Z \cap I_k) |< \epsilon/(2n)$. By Proposition 1.4.3, (ix), it follows that $\sum_{k=1}^\infty \mathcal{O}^{mn}(F; Z \cap I_k) \le m \sum_{k=1}^\infty \mathcal{O}_+^n(F; Z \cap I_k) + n \sum_{k=1}^\infty | \mathcal{O}_-^m(F; Z \cap I_k) |< M\epsilon/(2m) + +n\epsilon/(2n) = \epsilon$. Hence $F \in E_{mn}$ on P.

(vii) Let $F_1 \in \overline{AC}_m$ on P, $F_2 \in \overline{E}_n$ on P and $\epsilon > 0$. For $\epsilon/(2n)$ let δ be given by the fact that $F_1 \in \overline{AC}_m$ on P. Let $Z \subset P$, $| Z |= 0$. By Lemma 2.37.1, for $\epsilon/(2m)$, δ and Z, there exists a sequence $\{I_k\}_k$ of nonoverlapping closed intervals which cover Z, such that $\sum_{k=1}^\infty | I_k |< \delta$ and $\sum_{k=1}^\infty \mathcal{O}_+^n(F_2; Z \cap I_k) < \epsilon/(2m)$. Since $F_1 \in \overline{AC}_m$ on P, we have $\sum_{k=1}^\infty \mathcal{O}_+^m(F_1; Z \cap I_k) < \epsilon/(2n)$. By Proposition 1.4.3, (x), it follows that $\sum_{k=1}^\infty \mathcal{O}_+^{mn}(F_1 + FG_2; Z \cap I_k) < n \sum_{k=1}^\infty \mathcal{O}_+^m(F_1; Z \cap I_k) + m \sum_{k=1}^\infty \mathcal{O}_+^n(F_2; Z \cap I_k) < n\epsilon(2n) + m\epsilon/(2m) = \epsilon$. Hence $F_1 + F_2 \in \overline{E}_{mn}$ on P.

(viii), (ix), (x) The proofs are similar to that of (vii), using Proposition 1.4.3, (xi), (xii), (xiii).

(xi) Let $F_1 \in E_n$ on P, and let $F_2 \in L_m$ on $F(P)$ with a constant M. Let $\epsilon > 0$ and $Z \subset P$, $| Z |= 0$. For ϵ/M there exists a sequence $\{I_k\}_k$ of nonoverlapping closed intervals which cover Z, such that $\sum_{k=1}^\infty \lambda_n(F_1(Z \cap I_k)) < \epsilon/M$. By Lemma 2.32.2, (i), it follows that $\sum_{k=1}^\infty \lambda_{mn}(F_2 \circ F_1(Z \cap I_k)) < M \sum_{k=1}^\infty \lambda_n(F_1(Z \cap I_k)) < M\epsilon/M = \epsilon$, hence $F_2 \circ F_1 \in E_{mn}$ on P. The other parts follow similarly, using Lemma 2.32.2, (ii).

Remark 2.37.2 *Is it true that* $\overline{L}_m \circ \overline{E}_n \subseteq \overline{E}_{mn}$ *on* $P \subseteq [a, b]$*?*

Corollary 2.37.1 *Let*$P \subseteq [a, b]$ *and let m and n be positive integers.*

(i) $\underline{E}_nG \subseteq \underline{E}_{n+1}G \subseteq \underline{\mathcal{E}} \subseteq \underline{E}_\omega G \subseteq \underline{E}_\infty G$ *on* P.

(ii) $E_nG \subseteq E_{n+1}G \subseteq \mathcal{E} \subseteq E_\omega G \subseteq E_\infty G$ *on* P.

(iii) $\underline{AC}_mG \cap \overline{E}_nG \subseteq E_{mn}G$ *on* P.

(iv) $\overline{AC}_mG + \overline{E}_nG \subseteq \overline{E}_{mn}$ *on* P.

(v) $AC_mG + E_nG \subseteq E_{mn}G$ *on* P.

(vi) $\overline{AC}_mG \cdot \overline{E}_nG \subseteq \overline{E}_{mn}$ *on* P *for bounded positive functions.*

(vii) $AC_mG \cdot E_nG \subseteq E_{mn}G$ *on* P *for bounded functions.*

(viii) $\underline{\mathcal{F}} + \underline{\mathcal{E}} = \underline{\mathcal{E}}$ *on* P.

(ix) $\mathcal{F} + \mathcal{E} = \mathcal{E}$ *on* P.

(x) $\overline{\mathcal{F}} \cdot \overline{\mathcal{E}} = \overline{\mathcal{E}}$ *on* P *for bounded positive functions.*

(xi) $\mathcal{F} \cdot \mathcal{E} = \mathcal{E}$ *on* P *for bounded functions.*

(xii) $L_mG \circ E_nG \subseteq E_{mn}G$ *on* P.

(xiii) $\mathcal{L} \circ \mathcal{E} = \mathcal{E}$ *on* P.

(xiv) $\mathcal{E} \subseteq \underline{\mathcal{E}} \cap \overline{\mathcal{E}}$ *on* P.

(xv) $\underline{\mathcal{F}} \cap \overline{\mathcal{E}} \subseteq \mathcal{E}$ *on* P.

(xvi) $\underline{\mathcal{F}} \subset \underline{\mathcal{E}} \subseteq \Lambda\underline{Z} \subseteq (\underline{N}) \subseteq \underline{M}$ *on* P.

(xvii) $\mathcal{L} \subset \mathcal{F} \subset \mathcal{E} \subset \Lambda Z \subset (N) \subset (M)$ *on* P.

(xviii) $AC^*;\, DW^*G \subset AC^*;\, DW_1G \subseteq E_1G \cap T_1 \subseteq \mathcal{E} \cap T_1 \subset (S)$ *on* $[a,b]$ *for continuous functions.*

Proof. (i) - (xvi) follow by definitions, Theorem 2.37.1 and Section 6.49.

(xvii) $\mathcal{L} \subset \mathcal{F}$ follows by Section 6.55. $\mathcal{F} \subset \mathcal{E}$ follows by Section 6.47. By Remark 2.33.2, (iii), $\mathcal{F} + \Lambda Z \not\subset (M)$, but by (ix), $\mathcal{F} + \mathcal{E} = \mathcal{E} \subset (M)$, hence $\mathcal{E} \subset \Lambda Z$. $\Lambda Z \subset (N)$ follows by Theorem 2.33.2, (vi). $(N) \subset (M)$ follows by Theorem 2.23.2, (ii).

(xviii) $AC^*;\, DW^*G \subset AC^*;\, DW_1G$ follows by Theorem 2.36.1, (ix). $AC^*;\, DW_1G \subseteq E_1G$ follows by definitions. By Theorem 2.36.1, (iii), (v), Theorem 2.15.1 and Theorem 2.18.1 we obtain that $AC^*;\, DW_1G \subseteq T_1$ for continuous functions on $[a,b]$. $E_1 \cap T_1 \subseteq \mathcal{E} \cap T_1$ is evident. $\mathcal{E} \cap T_1 \subset (S)$ follows by (xvii), Theorem 2.18.9, (xix) and Section 6.27.

Remark 2.37.3 *(i)* $\mathcal{E} + \mathcal{E} \neq \mathcal{E}$ *(see Section 6.47.).*

(ii) $\mathcal{E} \cap \mathcal{B} \neq \emptyset$ *(because* $\mathcal{F} \subset \mathcal{E} \cap \mathcal{B}$*), but* $\mathcal{B} \not\subset \mathcal{E}$ *(see Section 6.2.) and* $\mathcal{E} \not\subset \mathcal{B}$ *(see Section 6.50.).*

Lemma 2.37.2 *Let $F : [a,b] \mapsto \mathbb{R}$, $P = \overline{P} \subseteq [a,b]$, and let n be a positive integer. The following assertions are equivalent:*

(i) $F \in [\underline{E}_n G]$ (resp. $F \in [E_n G]$) on P;

(ii) Each closed subset S of P contains a portion $S \cap (c,d)$ such that $F \in \underline{E}_n$ (resp. $F \in E_n$) on $S \cap (c,d)$ (see Theorem 1.9.1).

Lemma 2.37.3 *Let $F : [a,b] \mapsto \mathbb{R}$, $P = \overline{P} \subseteq [a,b]$. The following assertions are equivalent:*

(i) $F \in [\underline{\mathcal{E}}]$ (resp. $F \in [\mathcal{E}]$) on P;

(ii) Each closed subset S of P contains a portion $S \cap (c,d)$ such that $F \in \underline{E}_n$ (resp. $F \in E_n$) on $S \cap (c,d)$, for some positive integer n (see Theorem 1.9.1).

Lemma 2.37.4 *Let $F : [a,b] \mapsto \mathbb{R}$, $P \subseteq [a,b]$, and let n be a positive integer. The following assertions are equivalent:*

(i) $F \in E_n$ on P;

(ii) For every $Z \subset P$, with $\mid Z \mid = 0$, and for every $\epsilon > 0$ there exists a sequence $\{I_k\}_k$ of nonoverlapping closed intervals which cover Z, and for each positive integer k there exist intervals J_{ki}, $i = \overline{1,n}$ such that $B(F;Z) \subset \cup_{k=1}^{\infty} \cup_{i=1}^{n} (I_k \times J_{ki})$ and $\sum_{k=1}^{\infty} \sum_{i=1}^{n} (\mid I_k \mid + \mid J_{ki} \mid) < \epsilon$.

Proof. (i) $\Rightarrow$ (ii) Let $Z \subset P$, $\mid Z \mid = 0$ and $\epsilon > 0$. It follows that there exists a sequence $\{K_i\}_i$ of nonoverlapping closed intervals which cover Z, such that $\sum_{i=1}^{\infty} \mid K_i \mid < \epsilon/(2n)$. Since $\mid Z \cap K_i \mid = 0$, $F \in DW_n$ on $Z \cap K_i$. Hence for $\epsilon/(2^{i+2})$ there exists a sequence $\{K_{ij}\}_j$ of nonoverlapping closed intervals which cover $Z \cap K_i$ and $\sum_{j=1}^{\infty} \lambda_n(F(K_{ij} \cap Z)) < \epsilon/(2^{i+2})$. Clearly $Z \subset \cup_{i=1}^{\infty} \cup_{j=1}^{\infty} K_{ij}$ and $\sum_{i=1}^{\infty} \sum_{j=1}^{\infty} \mid K_{ij} \mid < \epsilon/2n$. Since $\sum_{j=1}^{\infty} \lambda_n(F(K_{ij} \cap Z)) < \epsilon/(n \cdot 2^{i+2})$, it follows that there exists a finite set of intervals J_{ijk}, $k = \overline{1,n}$ such that $\sum_{j=1}^{\infty} \sum_{k=1}^{n} \mid J_{ijk} \mid < \epsilon/(2^{i+1})$ and $\sum_{k=1}^{n} \sum_{i=1}^{\infty} \sum_{j=1}^{\infty} (\mid K_{ij} \mid + \mid J_{ijk} \mid) < \epsilon/2 + \epsilon/2 = \epsilon$, hence $F \in DW_n$ on Z. It follows that $F \in E_n$ on P.

(ii) $\Rightarrow$ (i) Let $Z \subset P$, $\mid Z \mid = 0$, $\epsilon > 0$, and let $\{I_k\}_k$ and J_{ki}, $i = \overline{1,n}$ be as in (ii). Then $\lambda_n(F(I_k \cap P)) \leq \sum_{i=1}^{n} \mid J_{ki} \mid$, hence $\sum_{k=1}^{\infty} \lambda_n(F(I_k \cap P)) < \epsilon$. It follows that $F \in DW_n$ on Z, hence $F \in E_n$ on P.

Remark 2.37.4 *In [E4] the author introduced condition E_n by (ii) of Lemma 2.37.4.*

2.38 Conditions SAC, $SACG$, SVB, $SVBG$, SY

Definition 2.38.1 *Let $X \subset \mathbb{R}$, $\mathcal{S} = \{\mathcal{S}(x) : x \in \mathbb{R}\}$ be a local system and let $\mathcal{S}[X] = \{\mathcal{S}(x) : x \in X\}$. Given a set-value function $\Delta : X \mapsto \mathcal{S}[X]$, a finite collection of interval-point pairs $\Pi = \{(I_j, x_j) : j = \overline{1,n}\}$ is called a Δ- fine partial partition of X, if $\{I_j\}$, $j = \overline{1,n}$ are nonoverlapping closed intervals with endpoints in $\Delta(x_j)$, $x_j \in I_j$. x_j is said to be the tag of (I_j, x_j). Let $\Pi_p(X;\Delta) = \{\Pi : \Pi$ is a Δ-fine partial partition of $X\}$. If $X = [a,b]$ and $\cup_{j=1}^{n} I_j = [a,b]$ then Π is said to be a Δ-fine partition of $[a,b]$. Let $\Pi([a,b];\Delta) = \{\Pi : \Pi$ is a Δ-fine partition of $[a,b]\}$.*

Theorem 2.38.1 *Given a set-value function $\Delta : [a,b] \mapsto \mathcal{S}[[a,b]]$, if $\mathcal{S}$ has the intersection condition (I.C.) then $\Pi([a,b];\Delta)$ is nonempty.*

Proof. (the proof is similar to that of Gordon in [G5], p. 155). Let $\mathcal{H} = \{(u,v) \subset (a,b) : [s,t] \subset [u,v]$ implies that $\Pi([s,t];\Delta) \neq \emptyset\}$. we show that $\mathcal{H}$ satisfies the four conditions of Lemma 1.7.1. Clearly $\mathcal{H}$ satisfies the first two conditions. Suppose that $(\alpha,\beta) \subset \mathcal{H}$ for each interval $(\alpha,\beta) \subset (c,d)$. Choose $c_1 \in \Delta(c) \cap (c,d)$ and $d_1 \in \Delta(d) \cap (c,d)$. Let $\Pi_1 \in \Pi([c_1,d_1];\Delta)$. Then $\Pi = \Pi_1 \cup \{([c,c_1];c),([d_1,d];d)\} \in \Pi([c,d];\Delta)$, so we have (iii).
We show (iv). Suppose that E is a perfect subset of $[a,b]$, and that each interval contiguous to $E \cup \{a,b\}$ belongs to $\mathcal{H}$. Since $\mathcal{S}$ satisfies (I.C.), it follows that there exists $\delta : [a,b] \mapsto (0,+\infty)$ such that $\Delta(x) \cap \Delta(y) \cap [x,y] \neq \emptyset$, whenever $0 < y - x < \min\{\delta(x),\delta(y)\}$. For each positive integer n, let $A_n = \{x \in E : \delta(x) \geq 1\}$, and for each integer i we define $A_{ni} = A_n \cap [(i-1)/n, i/n]$. Then $E = \cup_{n,i} A_{ni} = \cup_{n,i} \overline{A}_{ni}$. By Theorem 1.7.1, there exists an interval (u,v), $u,v \in E$ and some set A_{ni} such that $E \cap (u,v) \neq \emptyset$ and $E \cap [u,v] = \overline{A}_{ni} \cap [u,v]$. Let $[c,d] \subset [u,v]$. We will consider one of the several cases, the others are similar. Suppose that $c \in E$ and $d \notin E$. Choose an interval (s,t) contiguous to E in (u,v) such that $d \in (s,t)$. Since $c,s \in E$, there exists an integer $p \geq n$ such that $c,s \in A_p$. Since A_{ni} is dense in $E \cap [u,v]$, there exist $c_1 \in [c, c+1/p] \cap A_{ni}$ and $s_1 \in [s - 1/p, s] \cap A_{ni}$, with $c_1 < s_1$. Now $[c,d] = [c,c_1] \cup [c_1,s_1] \cup [s_1,s] \cup [s,d]$, so it is sufficient to prove that each of these intervals has a Δ-fine partition. Since $[s,d] \subset [s,t]$ and $(s,t) \in \mathcal{H}$, the interval $[s,d]$ has a Δ-fine partition. Since $c,c_1 \in A_p$ and $c_1 - c \leq 1/p \leq \min\{\delta(c),\delta(c_1)\}$, it follows that there exists $y \in \Delta(c) \cap \Delta(c_1) \cap [c,c_1] \neq \emptyset$. Let $\Pi = \{(c,y];c),([y,c_1];c_1)\} \in \Pi([c,c_1],\Delta)$. Similarly the intervals $[c_1,s_1]$ and $[s_1,s]$ have Δ- fine partitions. Hence $[c,d]$ has a Δ-fine partition. We conclude that $(u,v) \in \mathcal{H}$ and this shows that $\mathcal{H}$ satisfies (iv).

Definition 2.38.2 *Let $\mathcal{S}$ be a local system and let $F : [a,b] \mapsto \mathbb{R}$, $P \subseteq [a,b]$. F is said to be $\mathcal{S}Y$ (resp. $\mathcal{S}\underline{Y}$) on P, if for every measurable null subset Z of P and for every $\epsilon > 0$, there exists a set-value function $\Delta : Z \mapsto \mathcal{S}[Z]$ such that for any partial partition $\Pi = \{([c_i,d_i];x_o) : i = \overline{1,n}\} \in \Pi_p(Z;\Delta)$ we have $|\sum_{i=1}^n (F(d_i) - F(c_i))| < \epsilon$ (resp. $\sum_{i=1}^n (F(d_i) - F(c_i)) > -\epsilon$). F is said to be $\mathcal{S}\overline{Y}$ on P if $-F \in \mathcal{S}\underline{Y}$ on P, i.e., $\sum_{i=1}^n (F(d_i) - F(c_i)) < \epsilon$.*

Definition 2.38.3 *Let $F : [a,b] \mapsto \mathbb{R}$ and $P \subseteq [a,b]$. F is said to be $\mathcal{S}AC$ (resp. $\mathcal{S}\underline{AC}$) on P, if for every $\epsilon > 0$ there exists $\delta > 0$ and a set-value function $\Delta : P \mapsto \mathcal{S}[P]$, such that $|\sum_{i=1}^n (F(d_i) - F(c_i))| < \epsilon$ (resp. $\sum_{i=1}^n (F(d_i) - F(c_i)) > -\epsilon$), whenever $\{([c_i,d_i];x_i) : i = \overline{1,n}\} \in \Pi_p(P;\Delta)$ and $\sum_{i=1}^n (d_i - c_i) < \delta$. F is said to be $\mathcal{S}\overline{AC}$ on P if $-F$ is $\mathcal{S}\underline{AC}$ on P, i.e., $\sum_{i=1}^n (F(d_i) - F(c_i)) < \epsilon$. We define the conditions $\mathcal{S}ACG$, $\mathcal{S}\underline{ACG}$, $[\mathcal{S}ACG]$, etc., using Definition 1.9.2.*

Remark 2.38.1 *If $F \in \mathcal{S}ACG$ on P then F is $\mathcal{S}$-continuous on P.*

Lemma 2.38.1 *Let $F : [a,b] \mapsto \mathbb{R}$, $P \subseteq [a,b]$ and let $\mathcal{S}$ be a local system.*

(i) $\mathcal{S}\underline{ACG} \subseteq \mathcal{S}\underline{Y}$ on P.

(ii) $\mathcal{S}ACG \subseteq \mathcal{S}Y$ on P.

(iii) If $\mathcal{S}\underline{D}F(x) > -\infty$ for each $x \in [a,b]$ then $F \in \mathcal{S}\underline{ACG}$ on $[a,b]$.

Proof. (i) Let $Z \subset P, |Z| = 0$. Then there exists a sequence $\{Z_k\}_k$ of pairwise disjoint sets, such that $Z = \cup_{n=1}^{\infty} Z_n$ and $F \in \mathcal{S}\underline{AC}$ on each Z_k. Let $\epsilon > 0$. We define $\Delta : Z \mapsto \mathcal{S}[Z]$ as follows: for $\epsilon/2^k$ let $\eta_k > 0$ and $\Delta_k : Z_k \mapsto \mathcal{S}[Z_k]$ be given by the fact that $F \in \mathcal{S}\underline{AC}$ on Z_k. Since $|Z_k| = 0$, there exists an open set G_k such that $Z_k \subset G_k$ and $|G_k| < \eta_k$. Let $\delta_k(x) = \min\{\eta_k, d(x; [a,b] \setminus G_k)\}$, $x \in Z_k$. Now we define $\Delta(x) = \Delta_k(x) \cap (x - \delta_k(x), x + \delta_k(x))$, $x \in Z_k$, $k = \overline{1, \infty}$. Let $\{([c_i, d_i]; x_i) : i = \overline{1, m}\} \in \Pi_p(Z; \Delta)$ and let $I_k = \{i : x_i \in Z_k\}$. Clearly $\sum_{i \in I_k}(d_i - c_i) < \eta_k$. It follows that $\sum_{i=1}^{m}(F(d_i) - F(c_i)) = \sum_k \sum_{i \in I_k}(F(d_i) - F(c_i)) > \sum_k \epsilon/2^k = -\epsilon$. Hence $F \in \mathcal{S}\underline{Y}$ on P.

(ii) The proof is similar to that of (i).

(iii) For each positive integer n let $P_n = \{x \in [a,b] : \mathcal{S}\underline{D}F(x) > -n\}$. Clearly $[a,b] = \cup_{n=1}^{\infty} P_n$. Let $\Delta : [a,b] \mapsto \mathcal{S}[[a,b]]$ be a set-value function defined as follows: for each $x \in P_n$ let $\Delta(x) = \{y : \text{either } y = x \text{ or } (F(y) - F(x))/(y - x) > -n\} \in \mathcal{S}(x)$. Fix some n. Let $\epsilon > 0$ and $\delta = \epsilon/n$. Let $\{([a_j, b_j]; x_j) : j = \overline{1, m}\} \in \Pi_p(P_n; \Delta)$ such that $\sum_{j=1}^{m}(b_j - a_j) < \delta$. It follows that $\sum_{j=1}^{m}(F(b_j - a_j)) > -n \cdot \sum_{j=1}^{m}(b_j - a_j) > -\epsilon$. Therefore $F \in \mathcal{S}\underline{AC}$ on P_n.

Lemma 2.38.2 *Let $\mathcal{S}$ be a local system and let $F : [a,b] \mapsto \mathbb{R}$, $F \in \mathcal{S}\underline{Y}$. If $\mathcal{S}\underline{D}F(x) \geq 0$ a.e. then F is increasing on $[a,b]$.*

Proof. Let $A = \{x : \mathcal{S}\underline{D}F(x) \geq 0\}$ and $Z = [a,b] \setminus A$. Then $|Z| = 0$. For $\epsilon > 0$ let $\Delta : [a,b] \mapsto \mathcal{S}[[a,b]]$ be a set-value function defined as follows: for $t \in A$ let $\Delta(t) = \{x : \text{either } x = t \text{ or } (F(t) - F(x))/(t - x) > -\epsilon\}$. Since $F \in \mathcal{S}\underline{Y}$ on $[a,b]$, there exists $\Delta_1 : Z \mapsto \mathcal{S}[Z]$ such that $\sum_{i=1}^{n}((F(d_i) - F(c_i)) > -\epsilon$, whenever $\Pi = \{([c_i, d_i]; x_i) : i = \overline{1, n}\} \in \Pi_p(Z; \Delta_1)$. For each $t \in Z$ let $\Delta(t) = \Delta_1(t)$. Let $\Pi = \{([a_i, b_i]; z_i) : i = \overline{1, m}\} \in \Pi([a,b]; \Delta)$. Let Π_Z be the set formed of all members of Π with tags in Z. Let $\Pi' = \Pi \setminus \Pi_Z$. Then $F(b) - F(a) = \sum_{i=1}^{n}(F(b_i) - F(a_i)) = \sum_{i \in \Pi_Z}(F(b_i) - F(a_i)) + \sum_{i \in \Pi'}(F(b_i) - F(a_i)) > \epsilon - (b - a) \cdot \epsilon$. Since ϵ is arbitrary, $F(b) \geq F(a)$. Similarly it follows that $F(y) \geq F(x)$, whenever $a \leq x < y \leq b$, hence F is increasing on $[a,b]$.

Lemma 2.38.3 *Let $\mathcal{S}$ be a bilateral, filtering local system which has the intersection condition (I.C.), and let $F : [a,b] \mapsto \mathbb{R}$. If $F \in \mathcal{S}Y$ then $F \in (N)$ on $[a,b]$.*

Proof. We shall use Fu's technique of [Fu] (p. 313). Suppose that $F \notin (N)$. Then there exists $Z \subset [a,b]$, $|Z| = 0$ such that $|F(Z)| > 0$. We may suppose that $|F(Z)| \neq +\infty$ (because if $|F(Z)| = +\infty$ then $|F(Z_n)| \to +\infty$, where $Z_n = \{x \in Z : -n \leq F(x) \leq n\}$ for each positive integer n; hence there exists n_o such that $|F(Z_{n_o})| \in (0, 2n_o)$, , so we can use Z_{n_o} instead of Z). Let $\epsilon = (1/4) \mid F(Z) \mid$. Since $F \in \mathcal{S}Y$, there exists a set-value function $\Delta : Z \mapsto \mathcal{S}[Z]$ such that $\sum_{i=1}^{m} | F(b_i) - F(a_i) | < \epsilon$, whenever $\{([a_i, b_i]; x_i) : i = \overline{1, m}\} \in \Pi_p(Z; \Delta)$. We may suppose that $[a,b] = [0,1]$. Since $\mathcal{S}$ has the intersection condition (I.C.), there exists $\delta : Z \mapsto (0, +\infty)$ such that $\Delta(x) \cap \Delta(y) \cap [x,y] \neq \emptyset$, whenever $|x - y| < \min\{\delta(x), \delta(y)\}$. Let $X_m = \{x \in Z : \delta(x) > 1/m\} = \cup_{j=0}^{m-1} Y_{mj}$, where $Y_{mj} = X_m \cap [j/m, (j+1)/m]$. Since $Z = \cup_{m=1}^{\infty} X_m$, it follows that $F(Z) = \cup_{m=1}^{\infty} F(X_m)$. Hence $|F(X)| > (1/m) \mid F(Z) \mid$, for some m. But $|F(X_m)| \leq \sum_{j=0}^{m-1} |F(Y_{mj})|$. Let $A_m = \{j : |F(Y_{mj}| > 0\}$. Then there exist $a_j, b_j \in Y_{mj}$ such that $|F(b_j) - F(a_j)| > (1/2) \mid F(Y_{mj}) \mid$, for some $j \in A_m$. Since $0 < b_j - a_j \leq 1/m < \min\{\delta(a_j), \delta(b_j)\}$, it follows that there exists $c_j \in \Delta(a_j) \cap \Delta(b_j) \cap [a_j, b_j] \neq \emptyset$. Then $\{([a_j, c_j]; a_j) : j \in A_m\} \in \Pi_p(Z; \Delta)$. It follows that $\sum_{j\in A_m}(|F(c_j)-F(a_j)|+|F(b_j)-F(c_j)|) \ge \sum_{j\in A_m}|F(b_j)-F(a_j)|> (1/2)\cdot\sum_{j\in A_m}|F(Y_{mj})| = (1/2)\cdot\sum_{j=0}^{m-1}|F(Y_{mj})| \ge |F(X_m)| > (1/2)|F(Z)| = 2\epsilon$, a contradiction.

Definition 2.38.4 *Let $\mathcal{S}$ be a local system, $F:[a,b]\mapsto \mathbb{R}$ and $P\subseteq[a,b]$. F is said to be $\mathcal{S}VB$ on P if there exist a set-value function $\Delta : P\mapsto \mathcal{S}[P]$ and $M\in(0,+\infty$ (depending on Δ) such that $\sum_{i=1}^{n}|F(b_i)-F(a_i)|<M$, whenever $\{([a_i,b_i];x_i) : i=\overline{1,n}\}\in\Pi_p(P;\Delta)$. We define the condition $\mathcal{S}VBG$ on P, using Definition 1.9.2.*

Theorem 2.38.2 *Let $\mathcal{S}$ be a local system and $P\subseteq[a,b]$. Then $\mathcal{S}AC\subseteq\mathcal{S}VB$ on P, hence $\mathcal{S}ACG\subseteq\mathcal{S}VBG$ on P.*

Proof. We shall use Gordon's technique of [G5] (p. 161). Let $F:[a,b]\mapsto\mathbb{R}$, $F\in\mathcal{S}AC$ on P. For $\epsilon=1$ let $\Delta_1 : P\mapsto\mathcal{S}[P]$ and $\delta>0$ be given by this fact. Let N be the first positive integer such that $(b-a)/N<\delta$. For each $i=\overline{1,N}$ let $J_i=[a-(i-1)\cdot(b-a)/N, a+i(b-a)/N]$. Let $\Delta(x)=\Delta_1(x)\cap(x-\delta/2,x+\delta/2)$, for each $x\in P$. Let $\Pi=\{([c_j,d_j];x_j) : j=\overline{1,n}\}\in\Pi_p(P;\Delta)$. Let Π_i be the subset of Π that has intervals in J_i and let $\Pi_o=\Pi\setminus(\cup_{i=1}^{N}\Pi_i)$. Then Π_o contains at most $N-1$ intervals and $|F(d_j)-F(c_j)|<1$, whenever $([c_j,d_j];x_j)\in\Pi_o$. It follows that $\sum_{j=1}^{N}|F(b_j)-F(a_j)|<\sum_{j\in\Pi_o}|F(b_j)-F(a_j)|+\sum_{i=1}^{N}\sum_{j\in\Pi_i}|F(b_j)-F(a_j)|<(N-1)+N=2N-1$, hence $F\in\mathcal{S}VB$ on P.

Definition 2.38.5 (Thomson) *([T5], p. 25). Suppose that at every point x of a set $X\subset R$ there has been given a set E_x with the property that $x\in E_x$, and x is also an accumulation point of E_x. Such a set we shall call a path leading to x. We define relative to the system of paths $E=\{E_x\}_{x\in X}$ a local system $\mathcal{S}_E$ so that at each point $x\in X$, $\mathcal{S}_E(x)=\{A$: there is a $\delta>0$ such that $A\supset E_x\cap(x-\delta,x+\delta)\}$.*
We say that E has the internal intersection condition (I.I.C.) if there exists $\delta : X\mapsto(0,+\infty)$ such that $E_x\cap E_y\cap(x,y)\neq\emptyset$, whenever $0<y-x<\min\{\delta(x),\delta(y)\}$.

Remark 2.38.2 *Let $F:[a,b]\mapsto\mathbb{R}$, $X\subseteq[a,b]$. F is $\mathcal{S}_E$ continuous (short E-continuous) at $x\in X$ if and only if $\lim_{y\in E_x,y\to x}F(y)=F(x)$ (see Definition 1.15.2 and Definition 2.38.5).*

Lemma 2.38.4 *Let $F:[a,b]\mapsto\mathbb{R}$, $P\subseteq[a,b]$ and let $E=\{E_x\}_{x\in\overline{P}}$ be a system of paths which has intersection condition (I.I.C.).If $F\in\mathcal{S}_EVB$ on P and F is $\mathcal{S}_E$ continuous at each point $x\in\overline{P}$ then there exists an F_σ-type set Q such that $P\subseteq Q\subseteq\overline{P}$ and $F\in[CG]\cap[VBG]$ on Q.*

Proof. By condition (I.I.C.), it follows that there exists $\delta_o:\overline{P}\mapsto(0,+\infty)$ such that $E_x\cap E_y\cap(x,y)\neq\emptyset$, whenever $0<y-x<\min\{\delta_o(x),\delta_o(y)\}$. Since $F\in\mathcal{S}_EVB$ on P, there exist $\delta_1 : P\mapsto(0,+\infty)$ and $M\in(0,+\infty)$ (depending on δ_1), such that $\sum_{i=1}^{n}|F(b_i)-F(a_i)|<M$, whenever $\{([a_i,b_i];x_i) : i=\overline{1,n}\}\in\Pi_p(P;\delta_1)$. Let $\delta(x)=\min\{\delta_o(x),\delta_1(x)\}$, $x\in P$. For each positive integer n let $P_n=\{x\in P : \delta(x)\ge 1/n\}$, and for each integer i let $P_{ni}=P_n\cap[(i-1)/n,i/n]$.
We show that $F_{/\overline{P}_{ni}}\in C$, whenever $\overline{P}_{ni}$ is nonempty. Fix n and i and let x_o be a left accumulation point of P_{ni}. Suppose on the contrary that there exist a strictly increasing sequence $\{x_k\}_k\subset P_{ni}$ converging to x_o and $\epsilon_o>0$ such that $F(x_k)>F(x_o)+\epsilon_o$, $k=\overline{1,\infty}$. Since F is $\mathcal{S}_E$ continuous at x_o,there exists $0<\alpha_o\le\min\{1/n,\delta_o(x_o)\}$, such

that $\mid F(x) - F(x_o) \mid < \epsilon_o/8$, for each $x \in (x_o - \alpha_o, x_o + \alpha_o) \cap E_{x_o}$. We may suppose without loss of generality that $x_o - \alpha_o < x_1 < x_o$. Clearly $x_o \in ((i-1)/n, i/n]$, hence $0 < x_o - x_1 < \alpha_o < \min\{\delta_o(x_o), \delta_o(x_1)\}$. It follows that there exists $y_1 \in E_{x_o} \cap E_{x_1} \cap (x_1, x_o) \neq \emptyset$. We have $([x_1, y_1]; x_1) \in \Pi_p(P;\delta)$ and $F(x_1) - F(y_1) > (7/8)\epsilon_o$. Let $x_{k_1} > y_1$ and let $y_2 \in E_{x_{k_1}} \cap E_{x_o} \cap (x_{k_1}, x_o) \neq \emptyset$. It follows that $([x_{k_1}, y_2]; x_{k_1}) \in \Pi_p(P;\delta)$ and $F(x_{k_1}) - F(y_2) > (7/8)\epsilon_o$. Continuing we obtain $x_{k_s} > y_s$ and s_{s+1} such that $([x_{k_s}, y_{s+1}]; x_{k_s}) \in \Pi_p(P;\delta)$ and $F(x_{k_s}) - F(y_{s+1}) > (7/8)\epsilon_o$. Let q be a positive integer with $q \cdot (7/8)\epsilon_o > M$. It follows that $\sum_{j=1}^{q}(F(x_{k_j}) - F(y_{j+1})) > q \cdot (7/8)\epsilon_o > M$, a contradiction. Hence $F_{/\overline{P}_{ni}}$ is left continuous. Similarly we can show the right continuity at x_o, so $F \in [CG]$ on $Q = \cup_{n,i}\overline{P}_{ni}$.
We show that $F \in VB$ on each nonempty set P_{ni}. Let $\{[a_j, b_j]\}$, $j = \overline{1,m}$ be a finite set of nonoverlapping closed intervals, with $a_j, b_j \in P_{ni}$. Let $z_j \in E_{a_j} \cap E_{b_j} \cap (a_j, b_j) \neq \emptyset$. Then $\sum_{j=1}^{m} \mid F(b_j) - F(a_j) \mid \leq \sum_{j=1}^{m}(\mid F(b_j) - F(y_j) \mid + \mid F(y_j) - F(a_j) \mid) < 2M$, hence $F \in VB$ on P_{ni}. Since $F_{/\overline{P}_{ni}} \in \mathcal{C}$, it follows that $F \in VB$ on $\overline{P}_{ni}$, hence $F \in [VBG]$ on Q.

Lemma 2.38.5 *Let $F : [a,b] \mapsto \mathbb{R}$ and let P be a closed subset of $[a,b]$. Let $E = \{E_x\}_{x \in P}$ be a system of paths which has intersection condition (I.I.C.). If $F \in \mathcal{S}_E Y$ on P then $F \in [CG] \cap [VBG]$ on P.*

Proof. If we apply the definition of $\mathcal{S}_E Y$ to $Z = \{x\}$, $x \in P$, then it follows that F is $\mathcal{S}_E$ continuous at x. Let Q be a perfect subset of P and let Z be a G_δ-type subset of Q such that $\mid Z \mid = 0$ and $\overline{Z} = Q$. For $\epsilon = 1$ let $\delta : P \mapsto (0, +\infty)$ such that $\mid \sum_{i=1}^{m}(F(b_i) - F(a_i)) \mid < 1$, whenever $\{([a_i, b_i]; x_i) : i = \overline{1,m}\} \in \Pi_p(Z;\delta)$. We may suppose without loss of generality that $E_x \cap E_y \cap (x,y) \neq \emptyset$, whenever $0 < y - x < \min\{\delta(x), \delta(y)\}$. For each positive integer n let $Z_n = \{x \in Z : \delta(x) \geq 1/n\}$, and for each integer i let $Z_{ni} = Z_n \cap [(i-1)/n, i/n]$. Then $\sum_{j=1}^{q} \mid F(d_i) - F(c_i) \mid < 2$, whenever $\{([c_j, d_j]; y_j) : j = \overline{1,q}\} \in \Pi_p(Z_{ni};\delta)$, hence $F \in \mathcal{S}_E VB$ on Z_{ni}. As in the proof of Lemma 2.38.4 (for $M = 2$), we obtain that $F_{/\overline{Z}_{ni}} \in \mathcal{C} \cap VB$ on $\overline{Z}_{ni}$. Since Z is a G_δ-set, by Theorem 1.7.1, it follows that there exists an open interval I such that $\emptyset \neq I \cap Z \subset \overline{Z}_{ni}$ for some n and i. But $\emptyset \neq I \cap Q = I \cap \overline{Z} \subset \overline{I \cap Z} \subset \overline{Z}_{ni}$. It follows that $F_{/I \cap Q} \in \mathcal{C} \cap VB$. By Theorem 1.9.1, $F \in [CG] \cap [VBG]$ on P.

Corollary 2.38.1 *Let $E = \{E_x\}_{x \in [a,b]}$ be a system of paths which has intersection condition (I.I.C.). Then $\mathcal{S}_E Y \subseteq [ACG]$ on $[a,b]$.*

Proof. Let $F : [a,b] \mapsto \mathbb{R}$, $F \in \mathcal{S}_E Y$. By Lemma 2.38.5, $F \in [CG] \cap [VBG]$ on $[a,b]$. By Lemma 2.38.3, $F \in (N)$ on $[a,b]$. It follows that $F \in [CG] \cap [VBG] \cap (N) = [ACG]$ on $[a,b]$ (see Theorem 2.18.9, (vii)).

Chapter 3

Finite representations for continuous functions

3.1 quasi-derivable $\subseteq AC^*; DW^*G{+}AC^*; DW^*G$ and approximately quasi-derivable $\subseteq AC; DW_1G + AC; DW_1G$

Definition 3.1.1 (Nina Bary) *Let $F : [a,b] \mapsto \mathbb{R}$. F is said to be quasi-derivable (resp. approximately quasi-derivable) on $[a,b]$, if on each subinterval of $[a,b]$, $F'(x)$ (resp. $F'_{ap}(x)$) exists and is finite at every point x of a set which has positive measure.*

Remark 3.1.1 *Let $F : [a,b] \mapsto \mathbb{R}$.*

(i) *quasi – derivable $\subset$ approximately quasi – derivable on $[a,b]$ (see Section 6.62 or Section 6.46, the function $F_1 + F_2$).*

(ii) *derivable a.e. $\subset$ quasi – derivable on $[a,b]$ (see Section 6.29, the function G_{PP}).*

(iii) *approximately derivable a.e. $\subset$ approximately quasi – derivable on $[a,b]$ (see Section 6.29, the function G_{PP}).*

(iv) *$VB \subseteq$ quasi – derivable on $[a,b]$ (see (ii) and Theorem 2.14.3).*

(v) *$\mathcal{D}_- \cap [\mathcal{B}] \subset$ quasi – derivable on $[a,b]$. (Indeed, let $[c,d] \subseteq [a,b]$. Then $F \in [\mathcal{B}]$ on $[c,d]$. By Lemma 2.29.2, there exist a positive integer n and $[c_1,d_1] \subset [c,d]$ such that $F \in \underline{VB}_n$ on $[c_1,d_1]$. Since $F \in \mathcal{D}_-$ on $[a,b]$, by Theorem 2.29.1, (i), $F \in VB$ on $[c_1,d_1]$. By Theorem 2.14.3, (i), F is derivable a.e. on $[c_1,d_1]$, hence F is quasi-derivable on $[a,b]$. That the inclusion is strict, follows by Section 6.29, the function G_{PP}.)*

(vi) *$\mathcal{B} \subsetneq$ quasi – derivable on $[a,b]$ for continuous functions (see (v)). But $\mathcal{B} \subseteq \sigma.f.l.$ (see Corollary 2.29.1) and $\sigma.f.l. \cap T_1 \not\subset$ approximately quasi – derivable, for continuous functions on $[a,b]$ (see Section 6.61). Hence $\mathcal{B} \subset \sigma.f.l.$ on $[a,b]$.*

(vii) *$\overline{M} \cap C_d \subset$ quasi – derivable on $[a,b]$ (see Theorem 2.23.3 and Section 6.2.). In fact, Banach showed first that $(N) \subsetneq$ quasi – derivable for continuous functions, and in 1980, Foran improved this result showing that $(M) \subseteq$ quasi – derivable for continuous functions on $[a,b]$ (see [F7]).*

Lemma 3.1.1 (Nina Bary) *Let $\{P_m\}_m$ be an increasing sequence of perfect subsets of $[a,b]$, and let $I_{n_1n_2\ldots n_m}$, $n_i=\overline{1,\infty}$, $i=\overline{1,m}$ be the intervals contiguous to P_m, such that $I_{n_1}\supset I_{n_1n_2}\supset\ldots\supset I_{n_1n_2\ldots n_m}\supset\ldots$. Let $a>0$ and let $\{f_m\}_m$ be a sequence of continuous functions, such that $f_m(x)=0$ on P_m and $\mid f_m(x)\mid < a/2^{n_1+n_2+\ldots+n_m}$, $x\in I_{n_1n_2\ldots n_m}$. Let $R_m(x)=\sum_{i=m}^{\infty}a_if_i(x)$, where $\{a_i\}_i$ is a sequence of real numbers with $\mid a_i\mid\le 1$. Then we have:*

(i) $R_m(x)=0,\ x\in P_m$;

(ii) $\sum_{(n_1,\ldots,n_m)}\mathcal{O}(R_m;I_{n_1\ldots n_m}) < a/2^{m-2}$;

(iii) $R_m\in AC^*$ *on* P_m.

Proof. (i) This is evident.

(ii) By Remark 1.4.1, $\mathcal{O}(\mid f_{m+k}\mid;I_{n_1\ldots n_m}) < a/2^{n_1+n_m+\ldots+n_{m+k}}$. It follows that $\sum_{(n_1,\ldots,n_m)}\mathcal{O}(R_m;I_{n_1\ldots n_m}) < 2\sum_{(n_1,\ldots,n_m)}a/2^{n_1+\ldots n_{m+k}} < a/2^{m-2}$.

(iii) See (i), (ii) and Theorem 2.12.1, (iii).

Construction 3.1.1 (Nina Bary) *Let $F:[a,b]\mapsto\mathbb{R}$ and $F\in\mathcal{C}$. Let A be a measurable subset of $[a,b]$ such that $\mid A\cap[c,d]\mid>0$, whenever $[c,d]\subset[a,b]$ and F is derivable on A. Let $r>0$ and let $F_1,F_2:[a,b]\mapsto\mathbb{R}$, $F=F_1+F_2$ be defined as below. The procedure of construction F_1 and F_2 has been used by Nina Bary (see [Ba], pp. 222 - 228). We shall call F_1 and F_2 the Bary functions associated to F, A and r.*

Step 1. Let $P_1=\overline{P_1}\subset A$, $\mid P_1\mid>(1/2)\mid A\mid$. Let $I_{n_1}=(a_{n_1},b_{n_1})$, $n_1=\overline{1,\infty}$ be the intervals contiguous to $P_1\cup\{a,b\}$. Since $F\in\mathcal{C}$ there exist a positive integer j_{n_1} and $a_{n_1}=d_0^{n_1}<d_1^{n_1}<\ldots<d_{j_{n_1}}^{n_1}=b_{n_1}$, such that

(1) $\mathcal{O}(F;[d_1^{n_1},d_{i+1}^{n_1}]) < r/2^{n_1}$, $y=\overline{0,j_{n_1}-1}$

Let $Q_i^{n_1}=\overline{Q_i^{n_1}}\subset A\cap[d_i^{n_1},d_{i+1}^{n_1}]$, such that $\mid Q_i^{n_1}\mid\ge(1/2)\mid A\cap[d_i^{n_1},d_{i+1}^{n_1}]\mid$, $i=\overline{0,j_{n_1}-1}$. Let $P^{n_1}=\cup_{i=1}^{j_{n_1}-1}Q_1^{n_1}$ and let $P_2=P_1\cup(\cup_{n_1=1}^{\infty}P^{n_1})\subset A$. Then $\mid P_2\mid>(3/4)\mid A\mid$ (because $\mid P_2\setminus P_1\mid\ge(1/2)\mid A\setminus P_1\mid$, hence $\mid P_2\mid=\mid P_1\mid+\mid P_2\setminus P_1\mid\ge(1/2)\mid A\mid+(1/2)\mid A\setminus P_1\mid\ge(1/2)\mid A\mid+(1/4)\mid A\mid=(3/4)\mid A\mid$). Let $I_{n_1n_2}=(a_{n_1n_2},b_{n_1n_2})\subset I_{n_1}$ be the intervals contiguous to $P_2\cup\{a,b\}$. Let $f_1:[a,b]\mapsto\mathbb{R}$, $f_1(x)=F(x)$, $x\in P_1$ and $f_1(x)=F(d_i^{n_1})+(F(d_{i+1}^{n_1})-F(d_i^{n_1}))/\mid Q_i^{n_1}\mid\cdot(\mathcal{L})\int_{d_i^{n_1}}^{x}K_{Q_i^{n_1}}(t)dt$, $x\in[d_i^{n_1},d_{n+1}^{n_1}]$, $n_1=\overline{1,\infty}$, $i=\overline{0,j_{n_1}-1}$. Then $f_1\in\mathcal{C}$ on $[a,b]$; $f_1\in L$ on I_{n_1}; f_1 is constant on each $I_{n_1n_2}$; f_1 is monotone on each $[d_i^{n_1},d_{i+1}^{n_1}]$. Let $h_1:[a,b]\mapsto\mathbb{R}$, $h_1=F-f_1$. By (1), $\mid h_1(x)\mid\le r/2^{n_1}$ on I_{n_1} and $h_1(x)=0$ on P_1, hence $\mid h_1(x)\mid\le r/2$ on $[a,b]$ and $F=f_1+h_1$. By Theorem 2.12.1, (i), (iii), $h_1\in AC^*$ on P_1. By Corollary 2.13.3, $F\in AC^*G$ on A, hence $F\in AC^*G$ on P_2. Then $f_1=F-h_1\in AC^*G$ onP_1. Hence $f_1\in AC^*G$ on $[a,b]$ (because $f_1\in L\subseteq AC^*$ on I_{n_1}). It follows that $h_1=F-f_1\in AC^*G$ on P_2. Since f_1 is constant on $I_{n_1n_2}$, we obtain that $\{x\in I_{n_1n_2}: h_1 \text{ is derivable at } x\}=A\cap I_{n_1n_2}$.

Step i ($i\ge 2$). We repeat the construction of Step 1, using h_{i-1} and P_i instead of F and P_1. Therefore we obtain: $P_{i+1}=\overline{P_{i+1}}\supset P_i$; $P_{i+1}\subset A$; $\mid P_{i+1}\cap I_{n_1\ldots n_i}\mid\ge(1/2)\mid A\cap I_{n_1\ldots n_i}\mid$; $\mid P_{i+1}\mid>(1-1/2^{i+1})\cdot\mid A\mid$ (because $\mid P_{i+1}\mid=\mid P_i\mid+\mid P_{i+1}\setminus P_i\mid>(1-1/2^i)\cdot\mid A\mid+(1/2)\mid A\setminus P_i\mid>(1-1/2^i)\cdot\mid A\mid+(1/2^{i+1})\cdot\mid A\mid=(1-1/2^{i+1})\cdot\mid A\mid$); $\{I_{n_1\ldots n_{i+1}}\}_{n_1,\ldots,n_{i+1}}$ are the intervals contiguous to $P_{i+1}\cup\{a,b\}$; $f_i:[a,b]\mapsto\mathbb{R}$, $f_i\in\mathcal{C}\cap AC^*G$ on $[a,b]$; $f_i=h_{i-1}$ on P_i; f_i is constant on each $I_{n_1\ldots n_{i+1}}$;

$h_i = h_{i-1} - f_i$, $| h_i(x) | < r/2^{n_1+\dots+n_i}$ on $I_{n_1\dots n_i}$; $h_i(x) = 0$ on P_i, hence $| h_i(x) | < r/2^i$ on $[a,b]$; $F = f_1 + \dots + f_i + h_i$. It follows that $F = \sum_{i=1}^{\infty} f_i$ (since $| h_i(x) | < r/2^i$). Let $F_1, F_2 : [a,b] \mapsto \mathbb{R}$, $F_1(x) = \sum_{i=1}^{\infty} f_{2i-1}(x)$ and $F_2(x) = \sum_{i=1}^{\infty} f_{2i}(x)$. Let $R_m(x) = \sum_{i=m}^{\infty}(h_i(x) - h_{i+1}(x))$ and $Q = \cup_{i=1}^{\infty} P_i$. Then $| Q |=| A |$ and

$$(2) \quad F_1(x) = \sum_{i=1}^{k} f_{2i-1}(x) + R_{2k}(x);$$

$$(3) \quad F_2(x) = \sum_{i=1}^{k} f_{2i}(x) + R_{2k+1}(x).$$

By Lemma 3.1.1 we have

$$(4) \quad R_i(x) = 0 \text{ on } P_i;$$

$$(5) \quad \sum_{(n_1,\dots,n_i)} \mathcal{O}(R_i; I_{n_1\dots n_i}) < r/2^{i-2};$$

$$(6) \quad R_i \in AC^* \text{ on } P_i.$$

Since $\sum_{i=1}^{k} f_{2i-1}$ is constant on $I_{n_1\dots n_{2k}}$, it follows that

$$(7) \quad \mathcal{O}(F_1; I_{n_1\dots n_{2k}}) = \mathcal{O}(R_{2k}; I_{n_1\dots n_{2k}}).$$

Since $\sum_{i=1}^{k} f_{2i}$ is constant on $I_{n_1\dots n_{2k+1}}$, it follows that

$$(8) \quad \mathcal{O}(F_2; I_{n_1\dots n_{2k+1}}) = \mathcal{O}(R_{2k+1}; I_{n_1\dots n_{2k+1}}).$$

Lemma 3.1.2 *Let F, F_1, $F_2 : [a,b] \mapsto \mathbb{R}$ and Q be defined as in Construction 3.1.1. Then we have:*

(i) $F_1, F_2 \in \mathcal{C}$ and $F = F_1 + F_2$ on $[a,b]$;

(ii) $| F_2(x) | \le r$ on $[a,b]$;

(iii) $F_1, F_2 \in AC^*G$ on Q;

(iv) F_1, F_2 are derivable a.e. on Q;

(v) $F_1, F_2 \in DW^*$ on $[a,b] \setminus Q$.

Proof. (i) and (ii) are evident.

(iii) Since $f_i \in AC^*G$ on $[a,b]$, by (2) and (6), it follows that $F_1 \in AC^*G$ on P_{2k}, hence $F_1 \in AC^*G$ on Q. By (3) and (6), $F_2 \in AC^*G$ on P_{2k+1}, hence $F_2 \in AC^*G$ on Q.

(iv) See (iii) and Remark 2.15.1.

(v) Let $\epsilon > 0$ and let k be a positive integer such that $r/2^{2k} < \epsilon$. Then $[a,b] \setminus Q \subset \cup_{(n_1,\dots,n_{2k+2})} I_{n_1\dots n_{2k+2}}$ and $\sum_{(n_1,\dots,n_{2k+2})} \mathcal{O}(F; I_{n_1\dots n_{2k+2}}) < r/2^{2k} < \epsilon$ (see (5) and (7)). Hence $F_1 \in DW^*$ on $[a,b] \setminus Q$. Similarly $F_2 \in DW^*$ on $[a,b] \setminus Q$.

Remark 3.1.2 *a) If in the Construction 3.1.1 the function F is supposed to be approximately derivable then:*

(i) $F_1, F_2 \in \mathcal{C}$ *and* $F = F_1 + F_2$ *on* $[a, b]$;

(ii) $|F(x)| < r$ *on* $[a, b]$;

(iii) $F_1, F_2 \in ACG$ *on* Q;

(iv) F_1, F_2 *are approximately derivable a.e. on* Q;

(v) $F_1, F_2 \in DW_1$ *on* $[a, b] \setminus Q$.

(Indeed, by Theorem 2.15.4, $h_i \in ACG$ *on* P_{i+1}*, and the proof continues analogously to that of Lemma 3.1.1).*

b) It is possible to obtain two functions G_1, G_2 *with the same properties as* F_1, F_2 *in Lemma 3.1.1, and in addition* $G_2(a) = G_2(b) = 0$. *(Indeed, let* $H(x) = F_2(a) + (F_2(b) - F_2(a))/ \mid P_1 \mid \cdot (\mathcal{L}) \int_a^x K_{F_1}(t)dt$, $G_1 = F_1 + H$, $G_2 = F_2 - H$. *Now the proof follows by the following facts:* $\mid H \mid < 2r$ *on* $[a, b]$*;* H *is constant on* I_{n_1}*;* $H \in L$ *on* $[a, b]$.

Theorem 3.1.1 *Let* $F : [a, b] \mapsto \mathbb{R}$.

(i) quasi – derivable $\subseteq AC^*; DW^*G + AC^*; DW^*G$ *on* $[a, b]$ *for continuous functions. Moreover, if* $F \in \mathcal{C}$ *and* F *is derivable a.e. then there exist* $F_1, F_2 \in \mathcal{C}$, F_1, F_2 *derivable a.e., such that* $F = F_1 + F_2$ *and* $F_1, F_2 \in AC^*; DW^*G$ *on* $[a, b]$.

(ii) approximately quasi – derivable $\subseteq AC; DW_1G + AC; DW_1G$ *on* $[a, b]$ *for continuous functions. Moreover, if* $F \in \mathcal{C}$ *and* F *is approximately derivable a.e. then there exist* $F_1, F_2 \in \mathcal{C}$, F_1, F_2 *approximately derivable a.e., such that* $F = F_1 + F_2$ *and* $F_1, F_2 \in AC; DW_1G$ *on* $[a, b]$.

Proof. Let $A = \{x : F \text{ is derivable (resp. approximately derivable) at } x\}$. Applying Construction 3.1.1 and Lemma 3.1.2 (resp. Remark 3.1.2) to F and A, we have (i) (resp. (ii)).

Remark 3.1.3 *(i) We do not know if in Theorem 3.1.1, (i) the inclusion is strict. But we do know that quasi – derivable* $\subseteq AC^*; DW^*G + AC^*; DW^*G \subset AC^*; DW_1G + AC^*; DW_1G \subset (S) + (S)$ *(see Section 6.46 and Corollary 2.37.1, (xviii)). The fact that quasi – derivable* $\subseteq (S) + (S) = (AC \circ AC) + (AC \circ AC)$ *was shown by Nina Bary in [Ba]. However quasi – derivable* $\supset (AC \circ AC) + AC$ *(see (Section 6.65).*

(ii) There exist $F_1, F_2 : [a, b] \mapsto \mathbb{R}$, $F_1, F_2 \in \mathcal{C} \cap AC^*; DW^*G \subseteq (S)$, F_1, F_2 *derivable a.e.,* $F_1' = F_2'$ *a.e., but* F_1 *and* F_2 *do not differ by a constant (see Section 6.45)*

3.2 $\mathcal{C} \subseteq DW_1 + DW_1$ on a perfect nowhere dense set

Theorem 3.2.1 *Let* P *be a perfect nowhere dense subset of* $[a, b]$*, and let* $F : P \mapsto \mathbb{R}$. *Then* $\mathcal{C} \subseteq DW_1 + DW_1$ *for continuous functions on* P.

Proof. Let $A = [a, b] \setminus P$. By Lemma 3.1.2, there exist a set Q of F_σ-type, $Q \subseteq A$, $\mid Q \mid = \mid A \mid$, and two continuous functions $f, g \in DW^*$ on $[a, b] \setminus Q \supseteq P$. Let $F_1 = f_{/P}$ and $F_2 = g_{/P}$. Then $F = F_1 + F_2$ and $F_1, F_2 \in DW_1 \cap \mathcal{C}$ on P.

3.3 Wrinkled functions (W) and condition (M)

Definition 3.3.1 (Nina Bary) *Let $P \subseteq [a,b]$ be a measurable set and let $F : P \mapsto \mathbb{R}$. F is said to be (W) (wrinkled) on P, if for every measurable subset A of P, $|A| > 0$, there exists $B \subset A$, such that $|B| = 0$, $F(B)$ is measurable, $|F(B)| > 0$ and F is monotone on P. (Without loss of generality, A may be supposed to be perfect, because any measurable set is the union of an F_σ-set and a null set (see Theorem 1.5.1)).*

Remark 3.3.1 *(i) If $F \in (W)$ on P then $F \in (W)$ on E, whenever E is a measurable subset of P and $|E| > 0$.*

(ii) There exists a continuous wrinkled function defined on an interval (see Section 6.63). Nina Bary constructed first such an example in 1929 (see [Ba], p. 241 - 248).

Theorem 3.3.1 *Let P be a measurable subset of $[a,b]$, $|P| > 0$ and let $F : P \mapsto \mathbb{R}$. If $F \in (W) \cap \mathcal{C}$ and $F = F_1 + F_2$, $F_1 \in (M) \cap \mathcal{C}$ then*

(i) $F_2 \in \mathcal{C}$ on P;

(ii) F_2 is approximately derivable on no set of positive measure.

Proof. (i) This is obvious.

(ii) Suppose on the contrary that F_2 is approximately derivable on a measurable set $E \subseteq P$, $|E| > 0$. By Theorem 2.15.4, $F_2 \in ACG$ on E. Since $F \in (W)$, there exists $E_1 \subset E$, $|E_1| = 0$ such that $F(E_1)$ is measurable, $|F(E_1)| > 0$ and F is monotone on E_1. Then $F_1 = F - F_2 \in VBG$ on E_1. Since $F_1 \in (M)$ it follows that $F_1 \in ACG$ on E_1. Hence $F = F_1 + F_2 \in ACG \subseteq (N)$ on E_1. Then $|F(E_1)| = 0$, a contradiction.

Corollary 3.3.1 *Let P be a measurable subset of $[a,b]$, $|P| > 0$, and let $F : P \mapsto \mathbb{R}$. If $F \in (W) \cap \mathcal{C}$ then F is approximately derivable on no set of positive measure.*

Proof. Since $F(x) = 0 + F(x)$ on P, the proof follows by Theorem 3.3.1.

Corollary 3.3.2 *Let $F : [a,b] \mapsto \mathbb{R}$.*

(i) $(W) \cap ((M) + approximately\ quasi-derivable) = \emptyset$, for continuous functions on $[a,b]$.

(ii) $(W) \cap ((M) + quasi-derivable) = \emptyset$, for continuous functions on $[a,b]$.

(iii) $(W) \cap ((M) + (M)) = \emptyset$, for continuous functions on $[a,b]$.

(iv) $(W) \cap (AC \circ AC + AC \circ AC + AC) = \emptyset$, for continuous functions on $[a,b]$.

Proof. (i) See Theorem 3.3.1.

(ii) See (i) and Remark 3.1.1, (i).

(iii) See (ii) and Remark 3.1.1, (vii).

(iv) We have $AC \circ AC + AC \circ AC + AC = (S) + (S) + AC \subseteq (M) + (M) + AC \subseteq (M) + (M)$ (see Theorem 2.18.9, (xii) and Theorem 2.23.2, (ii), (vii)). Now the proof follows by (iii).

Remark 3.3.2 *Let P be a nowhere dense subset of* $[a,b]$, $|P|>0$, *and let* $F: P\mapsto \mathbb{R}$.

(i) $(W) \subsetneq DW_1 + DW_1 \subsetneq (N)+(N) \subsetneq (M)+(M)$, *for continuous functions on P (see Theorem 3.2.1 and Theorem 2.36.1, (iii), (v)). This means that in Corollary 3.3.2, (iii), the interval* $[a,b]$ *cannot be replaced by a perfect nowhere dense set of positive measure.*

(ii) There exists a function in $(W)\cap\mathcal{C}$ *on a set of positive measure, which is not approximately derivable a.e. on this set (see Section 6.63).*

(iii) There exists a function in $DW_1\cap\mathcal{C}$ *on a set of positive measure which is not approximately derivable a.e. (see Section 6.64).*

Theorem 3.3.2 *Let P be a measurable subset of* $[a,b]$, $|P|>0$, *and let* $F:P\mapsto\mathbb{R}$, $F\in\mathcal{C}$. *The following assertions are equivalent:*

(i) $F\in(W)$ *on P;*

(ii) $F\notin(M)$ *on Q, whenever* $Q=\overline{Q}\subset P$ *and* $|Q|>0$.

Proof. *(i)* $\Rightarrow$ *(ii)* Let $Q=\overline{Q}\subset P$, $|Q|>0$. By (i), it follows that there exists $E\subseteq Q$, $|E|=0$, such that F is monotone on E, $F(E)$ is measurable and $|F(E)|>0$. Hence $F\notin(M)$ on Q.

(ii) $\Rightarrow$ *(i)* Let A be a perfect subset of P, $|A|>0$. Since $F\notin(M)$ on A, by Corollary 2.23.1 there exists $B\subset A$ such that F is monotone on B and $F\notin AC$ on B. Since $F\in\mathcal{C}$, F is monotone on $\overline{B}$, and $F(\overline{B})$ is a compact set. Hence $F(\overline{B})$ is measurable. Suppose on the contrary that $|F(\overline{B})|=0$. By Theorem 2.18.9, (vii), $F\in AC$ on $\overline{B}$, hence $F\in AC$ on B, a contradiction. It follows that $|F(\overline{B})|>0$. Suppose on the contrary that $|\overline{B}|>0$. Since F is monotone on $\overline{P}$, it follows that F is approximately derivable on a measurable set $E\subset\overline{B}$ and $|E|=|\overline{B}|$. By Theorem 2.15.4, $F\in ACG$ on E. It follows that there exists $Q=\overline{Q}\subset E$, $|Q|>0$, such that $F\in ACG\subset(M)$ on Q, a contradiction.

Theorem 3.3.3 *Let P be a measurable subset of* $[a,b]$ *and* $F:P\mapsto\mathbb{R}$, $F\in\mathcal{C}$. *The following assertions are equivalent:*

(i) $F\in(W)$ *on P;*

(ii) F is strictly monotone on some perfect set $B\subset A$, *with* $|B|=0$ *and* $|F(B)|>0$, *whenever A is a measurable subset of P,* $|A|>0$.

Proof. *(i)* $\Rightarrow$ *(ii)* Let A be a measurable subset of P, $|A|>0$. Since $F\in(W)$, there exists $B_1\subset A$, $|B_1|=0$, such that $F_{/B_1}$ is monotone, $F(B_1)$ is measurable and $|F(B_1)|>0$. Suppose on the contrary that $|B_1|>0$. Since $F\in\mathcal{C}$, $F_{/\overline{B}_1}$ is monotone, hence F is approximately derivable *a.e.* on $\overline{B}_1$, a contradiction (see Corollary 3.3.1). Let $H=\{y\in F(\overline{B}_1): F^{-1}(y)\cap\overline{B}_1$ contains more than one point$\}$. Then H is countable. Suppose that $H=\{y_1,y_2,\ldots\}$. Let $\epsilon<|F(B_1)|/4$, $a_n=\inf(\overline{B}_1\cap F^{-1}(y_n))$, $b_n=\sup(\overline{B}_1\cap F^{-1}(y_n))$. Since $F\in\mathcal{C}$, there exists $\delta_n>0$ such that $F(\overline{B}_1\cap(a_n-\delta_n,b_n+\delta_n))\subset(y_n-\epsilon/2^{n+1},y+\epsilon/2^{n+1})$. Let $G=\cup_{n=1}^{\infty}(a_n-\delta,b_n+\delta)$. Then $|F(\overline{B}_1\cap G)|>\epsilon$. Let B be the set of accumulation points of the closed set

$\overline{B}_1 \setminus G$. Then B is a perfect subset of A, $| B |= 0$, $F_{/B}$ is strictly monotone, $F(B)$ is a compact set (because $F \in \mathcal{C}$ and $| F(B) |> (3/4) | F(B_1) |> 0$.

(*ii*) $\Rightarrow$ (*i*) This is evident.

Lemma 3.3.1 *Let P be a measurable subset of $[a,b]$, $| P |> 0$, and let $F : P \mapsto \mathbb{R}$, $F \in \mathcal{C}$. Let $E = \{x \in P : F$ is approximately derivable at x and $F'_{ap}(x) > 0\}$. If E has positive measure then there exists a perfect subset A of E, $| A |> 0$ such that $F_{/A}$ is strictly increasing.*

Proof. That E is measurable follows by Corollary 1.11.1. Let $E_n = \{x \in E :| \{x : 1/n \leq (F(t) - F(x))/(t - x), | t - x |< h\} | > (3/4) \cdot 2h$, whenever $h \in (0, 1/n)\}$, $n = \overline{1,\infty}$. Let $E_{ni} \cap [i/n, (i+1)/n]$, for each integer i. Then $E = \cup_{n,i} E_{ni}$, and there exist p and j such that $| E_{pj} |> 0$. If $x < y$, $x, y \in E_{pj}$ then $F(y) - F(x) \geq (1/p)(y - x)$. Since $F \in \mathcal{C}$, it follows that $F(y) - F(x) \geq (1/p)(y - x)$, for $x < y$, $x, y \in \overline{E}_{pj}$. Let A be a perfect subset of $\overline{E}_{pj}$, which has positive measure. Then F is strictly increasing on A.

Theorem 3.3.4 *Let P be a perfect subset of $[a,b]$, $| P |> 0$ and let $F : P \mapsto \mathbb{R}$, $F \in (W) \cap \mathcal{C}$. Let $g : P \mapsto Q$, $f : Q \mapsto \mathbb{R}$, $g(P) = Q$, $f, g \in \mathcal{C}$, such that $F = f \circ g$. Let $E = \{x \in Q : f$ is approximately derivable at $x\}$. If $| f(Q \setminus E) |= 0$ then $g \in (W)$ on P.*

Proof. We shall use Nina Bary's technique of [Ba] (p. 238). Let A be a perfect subset of P, $| A |> 0$. Since $F \in (W)$, by Theorem 3.3.3, there exists a perfect subset B of A, $| B |> 0$, such that $F_{/B}$ is strictly monotone and $| F(B) |> 0$. Let $B' = g(B)$. By Theorem 2.15.4, $F \in ACG$ on E. Since $| f(Q \setminus E) |= 0$, it follows that $f \in (N)$ on Q, hence $| B' |> 0$ (because, if $| B' |= 0$ then $| F(B) |=| f(B') |= 0$, a contradiction). Let $E_o = \{x \in E : f'_{ap}(x) = 0\}$. By Theorem 1.11.3, $| f(E_o) |= 0$, hence $B' \setminus E_o$ is measurable and $| B' \setminus E_o |> 0$ (because, if $| B' \setminus E_o |= 0$ then $| F(B) |= 0$, a contradiction). It follows that there exists a measurable set $E_1 \subset E \cap B'$, $| E_1 |> 0$ where f'_{ap} does not change the sign. Suppose that $f'_{ap}(x) > 0$ on E_1. By Lemma 3.3.1, there exists a perfect set $H \subset E$, $| H |> 0$ such that $f_{/H}$ is strictly increasing. Let $H_1 = g^{-1}(H)$. Since F is strictly monotone on H_1, it follows that g is strictly monotone on H_1, $| H_1 |= 0$ and $| g(H_1) |=| H |> 0$. Hence $g \in (W)$ on P.

Corollary 3.3.3 (Nina Bary) *Let P be a perfect subset of $[a,b]$, $| P > 0$ and let $F : P \mapsto \mathbb{R}$, $F \in (W) \cap \mathcal{C}$. Let $g : P \mapsto Q$, $f : Q \mapsto \mathbb{R}$, $g(P) = Q$, $f, g \in \mathcal{C}$ such that $F = f \circ g$. If $f \in AC$ on Q then $g \in (W)$ on P.*

Proof. Let $E = \{x \in Q : f$ is approximately derivable at $x\}$. Since $f \in AC$ on Q, it follows that $| Q \setminus E |= 0$. But $AC \subset (N)$, hence $| f(Q \setminus E) |= 0$. By Theorem 3.3.4, $g \in (W)$ on P.

Theorem 3.3.5 *Let P be a perfect subset of $[a,b]$, $| P > 0$ and $F : P \mapsto \mathbb{R}$, $F \in \mathcal{C}$. Let $g : P \mapsto Q$, $f : Q \mapsto \mathbb{R}$, $g(P) = Q$, $f, g \in \mathcal{C}$, such that $F = f \circ g$. Let $E = \{x \in P : g$ is approximately derivable at $x\}$. If $F \in (W)$ on P and $| g(P \setminus E) |= 0$ then $f \in (W)$ on P.*

Proof. Let A be a subset of Q, $| A |> 0$. Let $A_1 = g^{-1}(A)$. Then A_1 is a closed subset of P. But $g \in ACG$ on E, hence $g \in (N)$ on P. It follows that $| A_1 |> 0$. Let

$E_o = \{x \in P : g'_{ap}(x) = 0\}$. Since $| A |> 0$, $| g(E_o) |= 0$ and $| g(P \setminus E) |= 0$, it follows that there exists $E_1 \subset A_1 \cap E$, $| E_1 |> 0$ where f'_{ap} does not change the sign. Suppose that $f'_{ap}(x) > 0$ on E_1. By Lemma 3.3.1, there exists a perfect subset H of E_1, $| H |> 0$, such that $g_{/H}$ is strictly increasing. $F \in (W)$ on P implies that there exists a perfect subset B of H such that $| B |= 0$, $F_{/B}$ is strictly monotone and $| F(B) |> 0$. Let $B' = g(B) \subset A$. Since $g \in ACG$ on E it follows that $| B' |= 0$. Since $g_{/B}$ is strictly increasing and $F_{/B}$ is strictly monotone, it follows that $f_{/B'}$ is strictly monotone and $| f(B') |=| f(g(B)) |=| F(B) |> 0$, so $f \in (W)$ on Q.

Lemma 3.3.2 *Let $f : [a,b] \mapsto \mathbb{R}$, $f \in \mathcal{C}$ and let Q be a positive measurable subset of $[a,b]$. Let $g : [a,b] \mapsto \mathbb{R}$ be a derivable function. If $f \in (W)$ on Q then $f + g \in (W)$ on Q.*

Proof. Let E be a positive measurable subset of Q, and let E_1 be a perfect subset of E, $| E_1 |= 0$, such that $| f(E_1) |> 0$ and $f_{/E_1}$ is monotone. Suppose for example that $f_{/E_1}$ is increasing on E_1. Then f_{E_1} (see Definition 1.1.3) is increasing on $[c,d]$, where $c = \inf(E_1)$ and $d = \sup(E_1)$. Let $E_{+\infty} = \{x \in [c,d] : (f_{E_1})'(x) = +\infty\}$. Then $E_{+\infty} \subset E_1$ and $| f_{E_1}(E_{+\infty}) |> 0$. Let $G = f_{E_1} + g$ on $[c,d]$. Then $E_{+\infty} = \{x \in [c,d] : G'(x) = +\infty\}$. By Corollary 2.13.3, $g \in AC^*G$ on $[c,d]$. Hence $G \in VB^*G$ on $[c,d]$. Suppose on the contrary that $| G(E_{+\infty}) |= 0$. By Corollary 2.27.1, (i), (viii), $G \in AC^*G$ on $[c,d]$. Then $f_{E_1} \in AC$, a contradiction. Hence $| G(E_{+\infty}) |> 0$. But G is *strictly increasing** on $E_{+\infty}$. It follows that $E_{+\infty} = \cup_{i=1}^{\infty} P_i$ and G is *strictly increasing** on each P_i. Since $| G(E_{+\infty}) |> 0$, there exists i_o such that G is *increasing** on P_{i_o} and $| G(P_{i_o}) |> 0$. It follows that G is increasing on $\overline{P}_{i_o}$ and $| G(\overline{P}_{i_o}) |> 0$ (clearly $G(\overline{P}_{i_o})$ is measurable and $| \overline{P}_{i_o} |= 0$). Hence $G \in (W)$ on P.

Definition 3.3.2 *Let $F : [a,b] \mapsto \mathbb{R}$. F is said to be W^* on $[a,b]$, if for every interval $I \subset [a,b]$ there exists a perfect subset P of I, $| P |= 0$ such that $F_{/P}$ is monotone and $| F(P) |> 0$.*

Remark 3.3.3 *(i) $(W) \subset W^*$ for continuous on an interval (see Section 6.62).*

(ii) Any typical continuous function is W^* on $[a,b]$. (Indeed: it is known that a typical continuous function on $[a,b]$ has a finite or infinite derivative at no point of $[a,b]$ (see Corollary 2 of [Br2], p. 213). By Remark 2.16.1, $F \in W^*$ on $[a,b]$. A study of typical continuous functions can be found in [Br2].

(iii) Open question: Is (W) typical for continuous functions on an interval?

3.4 $\mathcal{C}$ = quasi-derivable + quasi-derivable

Theorem 3.4.1 (Nina Bary) *$\mathcal{C}$ = quasi-derivable + quasi-derivable, for continuous functions on $[0,1]$.*

Proof. Let $F : [0,1] \mapsto \mathbb{R}$, $F \in \mathcal{C}$. Let $\{P_i\}_i$ and $\{S_i\}_i$ be defined as in Section 6.46. Let $I_{n_1...n_i}$ be the intervals contiguous to P_i, $I_{n_1...n_i} \subset I_{n_1...n_{i-1}}$. Let $f_1 : [0,1] \mapsto \mathbb{R}$, $f_1 \in \mathcal{C}$, such that $f_1(x) = F(x)$ on P_1; $f_1(x)$ is a polynomial on each I_{n_1}; and $| F(x) - f_1(x) | < 1/2^{n_1}$, $x \in I_{n_1}$. Let $h_1 = F - f_1$. Then $h_1(x) = 0$ on P_1 and $| h_1(x) |< 1/2^{n_1}$, $x \in \overline{I}_{n_1}$. Continuing, we obtain $\{f_i\}_i$ and $\{h_i\}_i$ with the following properties: $h_i = h_{i-1} - f_i$; $h_i(x) = 0$ on P_i; $| h_i(x) |< 1/2^{n_1+\ldots+n_i}$, $x \in I_{n_1\ldots n_i}$, so $| h_i(x) |< 1/2^i$ on $[0,1]$; $f_i(x) = h_{i-1}(x)$ on P_i; $f_i(x)$ is a polynomial on each $I_{n_1\ldots n_i}$; $F = f_1 + f_2 + \ldots + f_i + h_i$. It follows that $F(x) = \sum_{i=1}^{\infty} f_i(x)$ on $[0,1]$. We define $F_1, F_2 : [0,1] \mapsto \mathbb{R}$ as follows: $F_1(x) = \sum_{i=1}^{\infty} f_{2i-1}(x)$, $F_2(x) = \sum_{i=1}^{\infty} f_{2i}(x)$. Then $F_1, F_2 \in \mathcal{C}$ and $F = F_1 + F_2$ on $[0,1]$. Then $F_1(x) = \sum_{i=1}^{k} f_{2i-1}(x) + R_{2k}(x)$ and $F_2(x) = \sum_{i=1}^{k} f_{2i}(x) + R_{2k+1}(x)$, where $R_i = (h_i - h_{i-1}) + (h_{i+2} - h_{i+3}) + \ldots$. By Lemma 3.1.1, $R_i(x) = 0$ on P_i and $R_i \in AC^*$ on P_i, hence $R_i'(x) = 0$ *a.e.* on P_i. Since f_i is a polynomial on each $I_{n_1\ldots n_i}$, it follows that $\sum_{i=1}^{k} f_{2i-1}$ is derivable on each $I_{n_1\ldots n_{2k-1}}$. Hence $\sum_{i=1}^{k} f_{2i-1}$ is derivable on $P_{2k} \setminus P_{2k-1}$. Since $R_{2k}'(x) = 0$ on P_{2k}, it follows that F_1 is derivable *a.e.* on $P_{2k} \setminus P_{2k-1}$.Hence F_1 is derivable *a.e.* on $A_1 = \cup_{k=1}^{\infty}(P_{2k} \setminus P_{2k-1})$. Similarly F_2 is derivable *a.e.* on $A_2 = \cup_{k=1}^{\infty}(P_{2k-1} \setminus P_{2k})$. It follows that $| (c,d) \cap A_1 |> 0$ and $| (c,d) \cap A_2 |> 0$, whenever $0 \le c < d \le 1$. Hence F_1 and F_2 are quasi-derivable on $[0,1]$.

Remark 3.4.1 *In Theorem 3.4.1 we cannot give up the condition quasi- derivable for either of the functions F_1 or F_2, so F cannot be written as the sum of an (M) function and a quasi-derivable function (see Corollary 3.3.2, (ii)).*

3.5 $\mathcal{C} = AC^*; DW_1G + AC^*; DW_1G + AC^*; DW_1G$

Remark 3.5.1 *By Theorem 2.18.9, $AC \circ AC \circ \ldots \circ AC = AC \circ AC = (S)$, for continuous functions on $[a,b]$. By Corollary 3.3.2, (iv), $(AC \circ AC) + (AC \circ AC) + AC \subset \mathcal{C}$ for continuous functions on $[a,b]$. In [Ba], Nina Bary showed that $\mathcal{C} = (AC \circ AC) + (AC \circ AC) + (AC \circ AC) = (S) + (S) + (S)$, for continuous functions on $[a,b]$. In what follows we improve this result showing that $\mathcal{C} = AC^*; DW_1G + AC^*; DW_1G + AC^*; DW_1G$, for continuous functions on $[a,b]$ (by Corollary 2.37.1, (xviii) we have that $AC^*; DW_1G \subset (S)$, for continuous functions on $[a,b]$).*

Lemma 3.5.1 *Let P be a perfect nowhere dense subset of $[a,b]$ and $r > 0$. Let $F : [a,b] \mapsto \mathbb{R}$, $F \in \mathcal{C}$. Then there exist $F_1, F_2 : [a,b] \mapsto \mathbb{R}$ such that:*

(i) $F = F_1 + F_2$ *and* $F_1, F_2 \in \mathcal{C}$ *on* $[a,b]$;

(ii) $F_1, F_2 \in DW_1$ *on* P;

(iii) $| F_2(x) |< r$ *on* $[a,b]$;

(iv) $F_2(a) = F_2(b) = 0$.

Proof. Let $A = [a,b] \setminus P$. By Remark 3.1.2, b), there exist $Q \subset A$ and $G_1, G_2 : [a,b] \mapsto \mathbb{R}$ such that: $G_1, G_2 \in \mathcal{C}$ and $F_{P\cup\{a,b\}} = G_1 + G_2$ (see Definition 1.1.3); $| G_2(x) |< r$ on $[a,b]$; $G_1, G_2 \in DW^*$ on $[a,b] \setminus Q \supset P$; $G_2(a) = G_2(b) = 0$. Let $F_2 = G_2$ and $F_1 = F - G_2$ on $[a,b]$. Then $F_1 = G_1$ on P, hence $F_1, F_2 \in DW_1$ on P. The other statements are evident.

Lemma 3.5.2 *Let P be a perfect nowhere dense subset of $[a,b]$ and $r > 0$. Let $F : [a,b] \mapsto \mathbb{R}$, $F \in \mathcal{C}$. Then there exists a perfect nowhere dense set $Q \supset P$ such that $Q \setminus P$ is a perfect nowhere dense set of positive measure in each interval contiguous to $P \cup \{a,b\}$, and there exist $F_1, F_2 : [a,b] \mapsto \mathbb{R}$ such that :*

(i) $F = F_1 + F_2$ *and* $F_1, F_2 \in \mathcal{C}$ *on* $[a, b]$;

(ii) $F_1, F_2 \in DW_1$ *on* P;

(iii) $F_1 \in AC = AC^*$ *on each interval contiguous to* P;

(iv) F_1 *is constant on each interval contiguous to* $Q \cup \{a, b\}$;

(v) $F_2(a) = F_2(b) = 0$;

(vi) $| F_2(x) | < r$ *on* $[a, b]$.

Proof. We shall use Nina Bary's technique of [Ba] (pp. 609 - 611). By Lemma 3.5.1, for $r/2$ there exist $G_1, G_2 : [a, b] \mapsto \mathbb{R}$, such that: $F = G_1 + G_2$ and $G_1, G_2 \in \mathcal{C}$ on $[a, b]$; $G_1, G_2 \in DW_1$ on P; $| G_2(x) | < r/2$ on $[a, b]$; $G_2(a) = G_2(b) = 0$. Let $I_{n_1} = (a_{n_1}, b_{n_1})$, $n_1 = \overline{1, \infty}$ be the intervals contiguous to $P \cup \{a, b\}$. Since $F \in \mathcal{C}$, for each n_1 there exists j_{n_1} and $a_{n_1} = d_0^{n_1} < d_1^{n_1} < \ldots < d_{j_{n_1}+1}^{n_1} = b_{n_1}$, such that

(1) $\quad \mathcal{O}(F; [d_i^{n_1}, d_{i+1}^{n_1}]) < r/2, \quad i = \overline{0, j_{n_1}}$.

Let $Q_i^{n_1} \subset [d_i^{n_1}, d_{i+1}^{n_1}]$ be a perfect nowhere dense set of positive measure. Let $Q^{n_1} = \cup_{i=0}^{j_{n_1}} Q_i^{n_1}$ and $Q = P \cup (\cup_{n_1=1}^{\infty} Q^{n_1})$. Let $I_{n_1 n_2}$, $n_1 = \overline{1, \infty}$, $n_2 = \overline{1, \infty}$ be the intervals contiguous to $Q \cup \{a, b\}$, $I_{n_1 n_2} \subset I_{n_1}$. Let $F_1 : [a, b] \mapsto \mathbb{R}$ be defined as follows: $F_1(x) = G_1(x)$, $x \in P \cup \{a, b\}$; $F_1(x) = F(x)$, $x = d_i^{n_1}$, $i = \overline{1, j_{n_1}}$, $n_1 = \overline{1, \infty}$; $F_1(x) = F(d_i^{n_1}) + (F_1(d_{i+1}^{n_1}) - F(d_i^{n_1})) / | Q_i^{n_1} | \cdot (\mathcal{L}) \int_{d_i^{n_1}}^{x} K_{Q_i^{n_1}}(t)dt$, $x \in [d_i^{n_1}, d_{i+1}^{n_1}]$, $i = \overline{0, j_{n_1}}$, $n_1 = \overline{1, \infty}$. Let $F_2 = F - F_1$ on $[a, b]$. Clearly $F_2 = G_2$ on $P \cup \{a, b\}$.

(i) - (v) These statements are obvious.

(vi) We have

(2) $\quad | F_2(x) | = | G_2(x) | < r/2, \ x \in P \cup \{a, b\}$.

Since $F_1(d_i^{n_1}) = F(d_i^{n_1})$, $i = \overline{1, j_{n_1}}$, it follows that

(3) $\quad F_2(d_i^{n_1}) = 0$.

But F_1 is monotone on each $[d_i^{n_1}, d_{i+1}^{n_1}]$, $i = \overline{0, j_{n_1}}$. By (1) and (3), $| F_2 | < r/2$, $x \in [d_i^{n_1}, d_{i+1}^{n_1}]$, $i = \overline{1, j_{n_1} - 1}$. By (1) and (2), $| F_2(x) | < r, x \in [d_0^{n_1}, d_1^{n_1}] \cup [d_{j_{n_1}}^{n_1}, d_{j_{n_1}+1}^{n_1}]$. Hence $| F_2(x) | < r$ on $\overline{I}_{n_1}$. By (2), $| F_2(x) | < r$ on $[a, b]$.

Theorem 3.5.1 *Let* $F : [a, b] \mapsto \mathbb{R}$. *Then* $(\mathcal{C}) = AC^*; DW_1G + AC^*; DW_1G + AC^*; DW_1G$ *for continuous functions on* $[a, b]$.

Proof. We shall use Nina Bary's technique of [Ba] (pp. 611 - 621). Let $P_1 \subset [a, b]$ be a perfect nowhere dense set, $a, b \in P_1$, $| P_1 | > 0$. Let I_{n_1}, $n_1 \overline{1, \infty}$ be the intervals contiguous to P_1. Let $f_1 : [a, b] \mapsto \mathbb{R}$, $f_1(x) = (\mathcal{L}) \int_a^x K_{P_1}(t)dt$. Then $f_1 \in AC = AC^*$ on $[a, b]$, hence $f_1 \in AC^*$ on P_1 and f_1 is constant on each I_{n_1}. Let $S_1 = P_1$. Applying Lemma 3.5.2 to $F - f_1$, P_1 and $r = 1$, it follows that there exists a perfect nowhere dense set P_2, $P_2 \supset P_1$, such that:

1) $P^{n_1} = (P_2 \setminus P_1) \cap I_{n_1}$ is a perfect nowhere dense set of positive measure;

2) There exist $u_1, v_1 : [a, b] \mapsto \mathbb{R}$ such that: $u_1, v_1 \in \mathcal{C}$ on $[a, b]$; $F - f_1 = u_1 + v_1$ on $[a, b]$; $u_1, v_1 \in DW_1$ on S_1; $u_1 \in AC = AC^*$ on I_{n_1}; u_1 is constant on $I_{n_1 n_2}$; $v_1(a) = v_1(b) = 0$; $| v_1(x) | < 1$ on $[a, b]$ ($I_{n_1 n_2}$, $n_1, n_2 = \overline{1, \infty}$ are the intervals

contiguous to P_2, $I_{n_1n_2} \subset I_{n_1}$.) Then $F = f_1 + u_1 + v_1$ and $P_2 = P_1 \cup S_2$, where $S_2 = \cup_{n_1=1}^{\infty} P^{n_1}$.

Step 1. Applying Lemma 3.5.2 to $v_{1/\overline{I}_{n_1}}$, P^{n_1} and $r = 1/2^{n_1}$, it follows that there exists a perfect nowhere dense set $Q^{n_1} \subset \overline{I}_{n_1}$, $Q_{n_1} \supset P_{n_1}$ such that:

1) $P^{n_1n_2} = (Qn_1 \setminus P^{n_1}) \cap I_{n_1n_2}$ is a perfect nowhere dense set of positive measure;

2) There exist $g_{n_1}, h_{n_1} : \overline{I}_{n_1} \mapsto \mathbb{R}$ such that : $g_{n_1}, h_{n_1} \in \mathcal{C}$ on $\overline{I}_{n_1}$; $v_1 = g_{n_1} + h_{n_1}$ on $\overline{I}_{n_1}$; $g_{n_1}, h_{n_1} \in DW_1$ on P^{n_1}; $g_{n_1} \in AC = AC^*$ on $I_{n_1n_2}$; g_{n_1} is constant on $I_{n_1n_2n_3}$; $h_{n_1}(x) = 0$ at the endpoints of I_{n_1}, $| h_{n_1}(x) |< 1/2^{n_1}$ on $\overline{I}_{n_1}$. $I_{n_1n_2n_3}$, $n_1, n_2, n_3 = \overline{1,\infty}$ are the intervals contiguous to P_3, $I_{n_1n_2n_3} \subset I_{n_1n_2}$, $P_3 = P_2 \cup S_3$, $S_3 = \cup_{n_1,n_2} P^{n_1n_2}$.

a) Let $v_2 : [a,b] \mapsto \mathbb{R}$, $v_3(x) = v_1(x)$, $x \in P_1$ and $v_2(x) = g_{n_1}(x)$, $x \in \overline{I}_{n_1}$. Then: $v_2 \in \mathcal{C}$ on $[a,b]$; $v_2 \in AC = AC^*$ on $I_{n_1n_2}$; v_2 is constant on $I_{n_1n_2n_3}$; $v_2 \in DW_1G$ on S_2.

b) Let $w_1 = v_1 - v_2$ on $[a,b]$. Then: $w_1 \in \mathcal{C}$ on $[a,b]$; $w_1(x) = 0$ on P_1; $w_1 = h_{n_1}$ on $\overline{I}_{n_1}$, hence $| w_1(x) |< 1/2^{n_1}$ on $\overline{I}_{n_1}$; $| w_1(x) |< 1/2$ on $[a,b]$; $w_1 \in AC^*$ on P_1 (see Theorem 2.12.1, (i), (iii)); $w_1 \in DW_1G$ on S_2.

c) Let $u_2 = u_1$ on $[a,b]$. Then: $u_2 \in \mathcal{C}$ on $[a,b]$; $u_2 \in AC = AC^*$ on I_{n_1}; u_2 is constant on $I_{n_1n_2}$, hence $u_2 \in AC^*G$ on S_2.

d) Let $f_2 = f_1 + w_1$ on $[a,b]$. Then: $f_2 \in \mathcal{C}$; $f_2 = f_1$ on P_1; $f_2 \in DW_1G$ on S_2.

It follows that $F = f_2 + u_2 + v_2$ on $[a,b]$.

Step 2. Applying Lemma 3.5.2 to $f_{2/\overline{I}_{n_1n_2}}$, $P^{n_1n_2}$ and $r = 1/2^{n_1+n_2}$, it follows that there exists a perfect nowhere dense set $Q^{n_1n_2} \subset \overline{I}_{n_1n_2}$, $Q^{n_1n_2} \supset P^{n_1n_2}$ such that:

1) $P^{n_1n_2n_3} = (Q^{n_1n_2} \setminus P^{n_1n_2}) \cap I_{n_1n_2n_3}$ is a perfect nowhere dense set of positive measure;

2) There exist $g_{n_1n_2}, h_{n_1n_2} : \overline{I}_{n_1n_2} \mapsto \mathbb{R}$ such that: $g_{n_1n_2}, h_{n_1n_2} \in \mathcal{C}$ on $\overline{I}_{n_1n_2}$; $f_2 = g_{n_1n_2} + h_{n_1n_2}$ on $\overline{I}_{n_1n_2}$; $g_{n_1n_2}, h_{n_1n_2} \in DW_1$ on $P^{n_1n_2}$; $g_{n_1n_2} \in AC = AC^*$ on $I_{n_1n_2n_3}$; $g_{n_1n_2}$ is constant on $I_{n_1n_2n_3n_4}$; $h_{n_1n_2}(x) = 0$ at the endpoints of $I_{n_1n_2}$; $| h_{n_1n_2}(x) |< 1/2^{n_1+n_2}$ on $\overline{I}_{n_1n_2}$. $I_{n_1n_2n_3n_4}$, $n_1n_2n_3n_4 = \overline{1,\infty}$ are the intervals contiguous to P_4, $I_{n_1n_2n_3n_4} \subset I_{n_1n_2n_3}$, $P_4 = P_3 \cup S_4$ and $S_4 = \cup_{(n_1,n_2,n_3)} P^{n_1n_2n_3}$.

a) Let $f_3 : [a,b] \mapsto \mathbb{R}$, $f_3 = f_2$ on P_2, $f_3 = g_{n_1n_2}$ on $\overline{I}_{n_1n_2}$. Then: $f_3 \in \mathcal{C}$ on $[a,b]$; $f_3 \in AC = AC^*$ on $I_{n_1n_2n_3}$; f_3 is constant on $I_{n_1n_2n_3n_4}$; $f_3 \in DW_1G$ on S_3.

b) Let $w_2 = f_2 - f_3$ on $[a,b]$. Then: $w_2 \in \mathcal{C}$ on $[a,b]$; $w_2(x) = 0$ on P_2; $w_2 = h_{n_1n_2}$ on $\overline{I}_{n_1n_2}$, hence $| w_2(x) |< 1/2^{n_1+n_2}$ on $\overline{I}_{n_1n_2}$; $| w_2(x) |< 1/2^2$ on $[a,b]$; $w_2 \in AC^*$ on P_2 (see Theorem 2.12.1, (i), (iii)); $w_2 \in DW_1G$ on S_3.

c) Let $v_3 = v_2$ on $[a,b]$. Then: $v_3 \in \mathcal{C}$ on $[a,b]$; $v_3 \in AC = AC^*$ on $I_{n_1n_2}$; v_3 is constant on $I_{n_1n_2n_3}$, hence $v_3 \in AC^*G$ on S_3.

d) Let $u_3 = u_2 + w_2$ on $[a,b]$. Then: $u_3 \in \mathcal{C}$ on $[a,b]$; $u_3 = u_2$ on P_2; $u_3 \in DW_1G$ on S_3.

It follows that $F = f_3 + u_3 + v_3$ on $[a,b]$.

Step 3. Applying Lemma 3.5.2 to $u_{3/\overline{I}_{n_1n_2n_3}}$, $P^{n_1n_2n_3}$ and $r = 1/2^{n_1+n_2+n_3}$, it follows that there exists a perfect nowhere dense set $Q^{n_1n_2n_3} \subset \overline{I}_{n_1n_2n_3}$, $Qn_1n_2n_3 \supset P^{n_1n_2n_3}$ such that:

1)$P^{n_1n_2n_3n_4} = (Q^{n_1n_2n_3} \setminus P^{n_1n_2n_3}) \cap I_{n_1n_2n_3n_4}$ is a perfect nowhere dense set of positive measure.

2) There exist $g_{n_1n_2n_3}, h_{n_1n_2n_3} : \overline{I}_{n_1n_2n_3} \mapsto \mathbb{R}$ such that $g_{n_1n_2n_3}, h_{n_1n_2n_3} \in \mathcal{C}$ on $\overline{I}_{n_1n_2n_3}$; $u_3 = g_{n_1n_2n_3} + h_{n_1n_2n_3}$ on $\overline{I}_{n_1n_2n_3}$; $g_{n_1n_2n_3}, h_{n_1n_2n_3} \in DW_1$ on $P^{n_1n_2n_3}$; $g_{n_1n_2n_3} \in AC = AC^*$ on $I_{n_1n_2n_3n_4}$; $g_{n_1n_2n_3}$ is constant on $I_{n_1n_2n_3n_4n_5}$; $h_{n_1n_2n_3}(x) = 0$ at the endpoints of $I_{n_1n_2n_3}$; $| h_{n_1n_2n_3}(x) |< 1/2^{n_1+n_2+n_3}$ on $\overline{I}_{n_1n_2n_3}$. $I_{n_1n_2n_3n_4n_5}$, $n_i = \overline{1,\infty}$, $i = \overline{1,5}$ are the

intervals contiguous to P_5; $I_{n_1n_2n_3n_4n_5} \subset I_{n_1n_2n_3n_4}$; $P_5 = P_4 \cup S_5$; $S_5 = \cup_{(n_1,\dots,n_5)} P^{n_1\dots n_5}$.
a) Let $u_4 : [a,b] \mapsto \mathbb{R}$, $u_4 = u_3$ on P_3, $u_4 = g_{n_1n_2n_3}$, $x \in \overline{I}_{n_1n_2n_3}$. Then: $u_4 \in \mathcal{C}$ on $[a,b]$; $u_4 \in AC = AC^*$ on $I_{n_1\dots n_4}$; u_4 is constant on $I_{n_1\dots n_5}$; $u_4 \in DW_1G$ on S_4.
b) Let $w_3 = u_3 - u_4$ on $[a,b]$. Then: $w_3 \in \mathcal{C}$ on $[a,b]$; $w_3(x) = 0$ on P_3; $w_3 = h_{n_1n_2n_3}$ on $\overline{I}_{n_1n_2n_3}$, hence $| w_3(x) | < 1/2^{n_1+n_2+n_3}$ on $\overline{I}_{n_1n_2n_3}$; $| w_3(x) | < 1/2^3$ on $[a,b]$; $w_3 \in AC^*$ (see Theorem 2.12.1, (i), (iii)); $w_3 \in DW_1G$ on S_4.
c) Let $f_4 = f_3$ on $[a,b]$. Then: $f_4 \in \mathcal{C}$ on $[a,b]$; $f_4 \in AC = AC^*$ on $I_{n_1n_2n_3}$; f_4 is constant on $I_{n_1\dots n_4}$, hence $f_4 \in AC^*G$ on S_4.
d) Let $v_4 = v_3 + w_3$ on $[a,b]$. Then $v_4 \in \mathcal{C}$ on $[a,b]$; $v_4 = v_3$ on P_3; $v_4 \in DW_1G$ on S_4.
It follows that $F = f_4 + u_4 + v_4$ on $[a,b]$.
By induction on k we obtain a sequence of perfect nowhere dense sets $\{P_k\}_k$, $P_k = P_{k-1} \cup S_k$, $S_k = \cup_{(n_1,\dots,n_{k-1})} P^{n_1\dots n_{k-1}}$, where $P^{n_1\dots n_{k-1}}$ is a perfect nowhere dense set of positive measure contained in $\overline{I}_{n_1\dots n_{k-1}}$ (where $I_{n_1\dots n_k}$ is an interval contiguous to P_k, $I_{n_1\dots n_k} \subset I_{n_1\dots n_{k-1}}$). We also obtain four sequences of continuous functions on $[a,b]$, $\{f_k\}_k$, $\{u_k\}_k$, $\{v_k\}_k$ and $\{w_k\}_k$ such that

(1) $\quad F = f_k + u_k + v_k$ *on* $[a,b]$.

If $k = 3p+1$ then $f_{k+1} = f_k + w_k$, $u_{k+1} = u_k$, $v_{k+1} = v_k - w_k$ on $[a,b]$ and

(2) $\quad f_k \in AC^*$ *on* $I_{n_1\dots n_{k-1}}$, *hence* $f_k \in AC^*$ *on* S_k; f_k *is constant on* $I_{n_1\dots n_k}$;

(3) $\quad u_k \in DW_1G$ *on* S_k; $u_k \in AC^*$ *on* $I_{n_1\dots n_k}$; u_k *is constant on* $I_{n_1\dots n_{k+1}}$;

(4) $\quad v_k \in DW_1G$ *on* S_k;

If $k = 3p+2$ then $f_{k+1} = f_k - w_k$; $u_{k+1} = u_k + w_k$; $v_{k+1} = v_k$ on $[a,b]$ and

(5) $\quad f_k \in DW_1G$ *on* S_k;

(6) $\quad u_k \in AC^*$ *on* $I_{n_1\dots n_{k-1}}$; u_k *is constant on* $I_{n_1\dots n_k}$; $u_k \in AC^*G$ *on* S_k;

(7) $\quad v_k \in AC^*$ *on* $I_{n_1\dots n_k}$; v_k *is constant on* $I_{n_1\dots n_{k+1}}$; $v_k \in DW_1G$ *on* S_k.

If $k = 3p+3$ then $f_{k+1} = f_k$; $u_{k+1} = u_k - w_k$; $v_{k+1} = v_k + w_k$ on $[a,b]$ and

(8) $\quad f_k \in AC^*$ *on* $I_{n_1\dots n_k}$; f_k *is constant on* $I_{n_1\dots n_{k+1}}$; $f_k \in DW_1G$ *on* S_k;

(9) $\quad u_k \in DW_1G$ *on* S_k;

(10) $\quad v_k \in AC^*$ *on* $I_{n_1\dots n_{k-1}}$; v_k *is constant on* $I_{n_1\dots n_k}$; $v_k \in AC^*G$ *on* S_k.

For $k = \overline{1,\infty}$ we have:

(11) $\quad w_k(x) = 0$ *on* P_k; $w_k \in DW_1G$ *on* S_{k+1}; $| w_k(x) | < 1/2^{n_1+\dots+n_k}$ *on* $\overline{I}_{n_1\dots n_k}$;

$| w_k(x) | < 1/2^k$ *on* $[a,b]$; $w_k \in AC^*$ *on* P_k.

It follows that

(12) $\quad f_{k+1} = f_k$; $u_{k+1} = u_k$; $v_{k+1} = v_k$ *on* P_k.

Since $| f_{k+1} - f_k | \leq | w_k | \leq 1/2^k$, $| u_{k+1} - u_k | < | w_k | < 1/2^k$ and $| v_{k+1} - v_k | < | w_k | < 1/2^k$ on $[a,b]$, it follows that there exist $f, u, v : [a,b] \mapsto \mathbb{R}$, $f, u, v \in \mathcal{C}$, such that $f_k \to f$ $[unif]$, $u_k \to u$ $[unif]$ and $v_k \to v$ $[unif]$ on $[a,b]$. By (1), $F = f + u + v$ on $[a,b]$. We have:

$$(13) \quad f(x) = f_k(x) + \sum_{i=k}^{\infty} a_i w_i(x) = f_k(x) + R_k^1(x), \ a_i \in \{-1,0,1\};$$

$$(14) \quad u(x) = u_k(x) + \sum_{i=k}^{\infty} b_i w_i(x) = u_k(x) + R_k^2(x), \ b_i \in \{-1,0,1\};$$

$$(15) \quad v(x) = v_k(x) + \sum_{i=k}^{\infty} c_i w_i(x) = v_k(x) + R_k^3(x), \ c_i \in \{-1,0,1\};$$

Then R_k^i, $i = 1,2,3$ satisfy the conditions of Lemma 3.1.1. Hence

$$(16) \quad R_k^i(x) = 0 \ on \ P_k; \ R_k^i \in AC^*, \ i = 1,2,3 \ and$$

$$(17) \quad \sum_{(n_1,\dots,n_k)} \mathcal{O}(R_k^i; I_{n_1\dots n_k}) < 1/2^{k-2}, \ i = 1,2,3.$$

By (13), (14), (15) and (16) we have

$$(18) \quad f = f_k, \ u = u_k \ and \ v = v_k \ on \ P_k.$$

If $k = 3p+1$ then by (3), (4), (18), $u, v \in DW_1G$ on S_k, and by (2), (13), (16), $f \in AC^*G$ on S_k.
If $k = 3p+2$ then by (5), (7), (18), $f, v \in DW_1G$ on S_k, and by (6), (14), (16), $u \in AC^*G$ on S_k.
If $k = 3p+3$ then by (8), (9), (18), $f, u \in DW_1G$ on S_k, and by (10), (15), (16), $v \in AC^*G$ on S_k.
Let $A_i = \cup_{p=1}^{\infty} S_{3p+1}$, $i = 1,2,3$. It follows that $A_1 \cup A_2 \cup A_3 = \cup_{k=1}^{\infty} P_k$. Let $E = [a,b] \setminus (A_1 \cup A_2 \cup A_3)$. Then we have:

$$f \in AC^*G \ on \ A_1; \ u, v \in DW_1G \ on \ A_1;$$

$$u \in AC^*G \ on \ A_2; \ f, v \in DW_1G \ on \ A_2;$$

$$v \in AC^*G \ on \ A_3; \ f, u \in DW_1G \ on \ A_3;$$

We show that $f \in DW^*$ on E. Let $\epsilon > 0$ and let $k = 3p+1$ such that $1/2^{k-1} < \epsilon/2$. Then by (2), (13), (17) we have that $\sum_{(n_1,\dots,n_k)} \mathcal{O}(f; I_{n_1\dots n_k}) = \sum_{(n_1,\dots,n_k)} \mathcal{O}(R_k^1; I_{n_1\dots n_k}) < 1/2^{k-2} < \epsilon$. Since $E \subset \cup_{(n_1,\dots,n_k)} I_{n_1\dots n_k}$ we obtain that $f \in DW^*$ on E. Similarly, $u, v \in DW^* \subseteq DW_1$ on E. Therefore $f, u, v \in AC^*; DW_1G$ on $[a,b]$.

Chapter 4

Monotonicity

4.1 Monotonicity and conditions $(-)$, $VB_\omega G$, $\mathcal{D}_-\mathcal{B}_1$

Definition 4.1.1 *Let $F : [a,b] \mapsto \mathbb{R}$. F is said to satisfy Bruckner's condition B_i on $[a,b]$, if F is increasing on each closed subinterval of $[a,b]$ on which it is VB. F is said to be B_d on $[a,b]$, if $-F \in B_i$ on $[a,b]$.*

Theorem 4.1.1 *Let $F : [a,b] \mapsto \mathbb{R}$ be a function satisfying the following conditions:*

(i) $F \in (-)$ on $[a,b]$;

(ii) $F \in VB_\omega G$ on $H = \{x : F$ is continuous at $x\}$;

(iii) $F \in B_i$ on each compact interval contained in $A = \{x : F \in C_i$ at $x\}$.

Then F is increasing on $[a,b]$.

Proof. By (i), (ii) and Lemma 2.29.3, there exists a sequence $\{I_n\}_n$ of closed intervals whose union is dense in $[a,b]$, such that $F \in VB \cap C_i$ on each I_n. By (iii), F is increasing on each I_n. Let J_n be the maximal open interval such that $J_n \supset I_n$ and F is increasing on J_n. Let $Q = [a,b] \setminus (\cup_{n=1}^{\infty} J_n)$. Suppose on the contrary that $Q \neq (a,b)$. Then Q is a closed set. We show that Q is perfect. If Q contains an isolated point $x_o \in (a,b)$ then F is increasing on some interval J_n, having x_o as a left endpoint, and also on some J_m, having x_o as a right endpoint. By Theorem 2.4.2, (vi) and Theorem 2.18.9, since $F \in (-)$, it follows that $F \in uCM$ on $[a,b]$, hence F is increasing on $\overline{J}_n$ and on $\overline{J}_m$. It follows that F is increasing on $J_n \cup \{x_o\} \cup J_m$, a contradiction. Hence Q is perfect (perhaps without a and b). Let $H_1 = H \cap Q$. Then H_1 is a G_δ-set. We show that H_1 is everywhere dense in Q. Let J be an open subinterval of $[a,b]$ containing points of Q. Let $x_o \in Q \cap J$. Since $x_o \in Q$, F cannot be increasing on all of J. It follows that J contains points $z_1 < z_2$ such that $F(z_1) > F(z_2)$. Let $N = \{x \in [z_1, z_2] : -\infty \leq F'(x) \leq 0\}$. By Theorem 1.10.4, the set $\{x : F'(x) = 0\}$ maps into a set of measure zero. Let $N' = \{x \in [z_1, z_2 \ : -\infty \leq F'(x) \leq 0\}$. Since $F \in (-)$ on $[a,b]$, N' is uncountable. By Corollary 2.13.2, $F \in VB^*G$ on N'. Let $N" = \{x \in N' : F$ is continuous at $x\}$. By Theorem 2.10.1, since $F \in (-)$ on $[a,b]$, we have $F(z_1) - F(z_2) \leq \mid F(N") \mid \leq \mid F(H_1 \cap [z_1, z_2]) \mid$. It follows that $[z_1, z_2] \subset J$ and $H_1 \cap [z_1, z_2]$ contains uncountably many points of continuity. Hence H_1 is everywhere dense in Q. Now the proof continues analogously to that of Lemma 2.29.3 (the set H in the proof of Lemma 2.29.3 is replaced by H_1). Thus we obtain that $F \in VB$ on J. Let $c = \inf(Q \cap J)$, $d = \sup(Q \cap J)$. Then $F \in VB$ on $[c, d]$. By (iii), F is increasing on $[c, d]$. This contradicts the fact that $[c, d]$ contains infinitely many points of Q. Hence $Q = \{a, b\}$ and F is increasing on (a, b). Since $F \in uCM$, F is increasing on $[a, b]$.

Corollary 4.1.1 *Let $F : [a, b] \mapsto \mathbb{R}$ be a function satisfying the following conditions:*

(i) $F \in \mathcal{D}_-\mathcal{B}_1$ (resp. $\mathcal{DB}_1$) on $[a, b]$;

(ii) $F \in VBG$ on $a, b]$;

(iii) $F \in B_i$ on each compact interval contained in $A = \{x : F$ is C_i (resp. C) at $x\}$.

Then F is increasing (resp. C and increasing) on $[a, b]$.

Proof. Since $F \in VBG$, by Theorem 2.18.9, (ii), $F \in T_2$ on $[a, b]$. By Theorem 2.18.9, (xvii), $F \in (-)$ on $[a, b]$. By Corollary 2.29.1, (iii), $F \in VB_\omega G$ on $[a, b]$. By Theorem 4.1.1, F is increasing on $[a, b]$. The continuity for the second part follows by Theorem 2.18.9, (xix).

Remark 4.1.1 *The second part of Corollary 4.1.1 is in fact Bruckner's reduction theorem (see [Br2], pp. 176 - 184), and the first part is due to Garg (see [Ga]). For many applications of Bruckner's theorem see [Br2] (pp. 184 - 190).*

4.2 Monotonicity and conditions $\underline{M}$, uCM, $\underline{AC}$, C_i, C_i^*, $\mathcal{D}_i$

Theorem 4.2.1 *Let $F : [a, b] \mapsto \mathbb{R}$. If $F \in C_i \cap \underline{M}$ on $[a, b]$ and $F'(x) \geq 0$ a.e. where F is derivable, then F is increasing on $[a, b]$.*

Proof. Suppose on the contrary that there exist $x_1, x_2 \in [a, b]$, $x_1 < x_2$ such that $F(x_1) > F(x_2)$. Let $N = \{x \in [x_1, x_2] : 0 \geq F'(x) \geq -\infty\}$, $N_o = \{x \in N : F'(x) = 0\}$, $E_- = \{x \in N : 0 > F'(x) > -\infty\}$ and $E_{-\infty} = \{x \in N : F'(x) = -\infty\}$. Then $N = N_o \cup E_- \cup E_{-\infty}$, $| F(N_o) | = 0$ (see Theorem 1.10.4) and $| E_- | = | E_{-\infty} | = 0$ (by hypothesis). By Corollary 2.13.3, $F \in AC^*G$ on E_-. By Theorem 2.18.9, (vi), (x), $F \in (N)$ on E_-, hence $| F(E_-) | = 0$. By Theorem 2.23.2, (i) and Theorem 2.23.1, (i), (ii), $F \in N^{-\infty}$ on $[x_1, x_2]$, hence $| F(E_{-\infty}) | = 0$. Thus $| F(N) | = 0$, a contradiction (see Theorem 2.23.3).

Corollary 4.2.1 *Let $F : [a, b] \mapsto \mathbb{R}$. If $F \in (M) \cap C$ on $[a, b]$ and $F'(x) \geq 0$ a.e. where F is derivable, then F is AC and increasing on $[a, b]$.*

Proof. Since $C = C_i \cap C_d$ and $(M) = \overline{M} \cap \underline{M}$, the proof follows by Theorem 4.2.1.

Remark 4.2.1 *If in Corollary 4.2.1 we put (N) instead of (M) then we obtain a theorem of Nina Bary (see [Ba], p. 199).*

Corollary 4.2.2 *Let $F_1, F_2 : [a, b] \mapsto \mathbb{R}$.*

(i) (Foran). *If* $F_1 \in \mathcal{C}\cap(N)$ *on* $[a,b]$, F_1, F_2 *are approximately derivable a.e. (resp. derivable a.e.) and* $(F_1)'_{ap} = (F_2)'_{ap}$ *a.e. (resp.* $F_1' = F_2'$ *a.e.) then* $F_2 - F_1$ *is constant.*

(ii) *If* $F_1 \in \mathcal{C}\cap\mathcal{F}$, $F_2 \in \mathcal{C}\cap\mathcal{E}$ *on* $[a,b]$, F_1, F_2 *are approximately derivable a.e. (resp. derivable a.e.) and* $(F_1)'_{ap} = (F_2)'_{ap}$ *a.e. (resp.* $F_1' = F_2'$ *a.e.) then* $F_2 - F_1$ *is constant.*

Proof. (i) Let $F = F_2 - F_1$. By Theorem 2.23.2, (ii), (vi), $F \in (M)\cap\mathcal{C}$ on $[a,b]$. But $F'(x) = 0$ a.e. where $F'(x)$ exists, so by Corollary 4.2.1, F is constant on $[a,b]$.

(ii) Let $F = F_2 - F_1$. By Corollary 2.37.1, $F \in \mathcal{E} \subset (N) \subset (M)$. But $F'(x) = 0$ a.e. where $F'(x)$ exists, so by Corollary 4.2.1, F is constant on $[a,b]$.

Remark 4.2.2 *(i) In Corollary 4.2.2, (i), we cannot replace the class ACG by* L_2G, AC_2G *or by* $\mathcal{F}$ *(see Section 6.34).*

(ii) In Corollary 4.2.2, (i), we cannot replace ACG by $\mathcal{E}$, *and* (N) *by* $\mathcal{E}$ *(see Section 6.47).*

(iii) In Corollary 4.2.2, (ii), we cannot replace $\mathcal{F}$ *by* $\mathcal{E}$ *(see Section 6.47).*

Theorem 4.2.2 *Let* $F : [a,b] \mapsto \mathbb{R}$, $F \in uCM \cap [C_iG]$ *(resp.* $[C_i^*G]$*) on* $[a,b]$. *Let* $A = int\{x : F \text{ is } C_i \text{ (resp. } C_i^*\text{) at } x\}$, *and let* $F \in \underline{M}$ *(resp.* (M)*) on each closed interval contained in A. If* $F'(x) \geq 0$ *a.e. where F is derivable on A then F is increasing on* $[a,b]$.

Proof. We proof for example the first part. Since $F \in [C_iG]$ on $[a,b]$, applying Theorem 1.7.1 infinitely many times, it follows that A is a dense open subset of $[a,b]$. Then F is increasing on the closure of any component of A. (Indeed: let J be a component of A and let $[c,d] \subset J$; by Theorem 4.2.1, F is increasing on $[c,d]$; since $[c,d]$ is arbitrary chosen, it follows that F is increasing on J; since $F \in uCM$ on $[a,b]$, F is increasing on $\overline{J}$.) Suppose on the contrary that $A \neq (a,b)$. Then the set $Q = [a,b] \setminus A$ is perfect, if necessary without a and b (because Q is obviously closed, and if $x_o \in (a,b)\cap Q$ is isolated in Q, then F is increasing on some $\overline{I}_1$, where I_1 is a component of A, having x_o as a right endpoint, and on some $\overline{I}_2$, where I_2 is a component of A, having x_o as a left endpoint; then F is increasing on $I_1 \cup \{x_o\} \cup I_2$, hence $F \in C_i$ on $I_1\cup\{x_o\}\cup I_2$ (see Proposition 2.3.1, (ii), (iii)); it follows that $x_o \in A$, a contradiction.) By Theorem 1.7.1, there exist $c, d \in Q$, $c < d$, such that $(c,d)\cap Q \neq \emptyset$ and $F_{/P} \in C_i$, where $P = [c,d]\cap Q$. By Proposition 2.3.2, (ii), $F \in C_i$ on $[c,d]$, hence $(c,d) \subset A$, a contradiction. Thus $A = (a,b)$. Let I be a closed subinterval of (a,b). By Theorem 4.2.1, F is increasing on I. Since I was arbitrary, F is increasing on (a,b). Since $F \in uCM$ on $[a,b]$, it follows that F is increasing on $[a,b]$.

Corollary 4.2.3 *Let* $F : [a,b] \mapsto \mathbb{R}$, $F \in \mathcal{D}_- \cap [C_iG]$. *Let* $A = int\{x : F \text{ is } C_i \text{ at } x\}$, *and let* $F \in \underline{M}$ *on each closed interval contained in A. If* $F'(x) \geq 0$ *a.e. where F is derivable on A then F is increasing on* $[a,b]$.

Proof. The proof follows by Theorem 4.2.2 and by the fact that $\mathcal{D}_- \subset uCM$.

Corollary 4.2.4 *Let $F : [a,b] \mapsto \mathbb{R}$, $F \in \mathcal{D} \cap [CG]$ on $[a,b]$. Let $A = int\{x : F \in C$ at $x\}$, and let $F \in \underline{M}$ (resp. (M)) on each closed interval contained in A. If $F'(x) \geq 0$ a.e. where F is derivable on A then F is continuous and increasing (resp. AC and increasing) on $[a,b]$.*

Proof. The proof is similar to that of Theorem 4.2.2, using the fact that $\mathcal{D} \cap VB \subseteq C$ on a closed interval (see Theorem 2.18.9, (xix)) and $\mathcal{D} \subset uCM$. That $F \in AC$ follows by the definition of (M).

Theorem 4.2.3 *Let $F : [a,b] \mapsto \mathbb{R}$, $A = int\{x : F \in C_i$ (resp. $F \in C$) ar $x\}$ and $N = \{x \in A : F$ is derivable at x and $F'(x) < 0\}$. The following assertions are equivalent:*

(i) $F \in \underline{AC}$ (resp. $F \in AC$) on $[a,b]$;

(ii) $F \in \mathcal{D}_- \cap [C_iG]$ (resp. $F \in \mathcal{D} \cap [CG]$) on $[a,b]$, $F \in \underline{M}$ (resp. $F \in (M)$) on each closed interval contained in A, and F' is summable on N.

Proof. $(i) \Rightarrow (ii)$ By Theorem 2.11.1, (vi) (resp. (v)), $\underline{AC} \subseteq VB$ (resp. $AC \subseteq VB$). It follows that $F \in \underline{M}$ (resp. $F \in (M)$) on $[a,b]$, hence $F \in \underline{M}$ (resp. $F \in (M)$) on A. By Theorem 2.14.3, F' is summable on $[a,b]$, hence F' is summable on N. By Theorem 2.11.1 and Theorem 2.4.2, (vi), $\underline{AC} \subset C_i \subset \mathcal{D}_- \cap [C_iG]$ (resp. $AC \subset C \subset \mathcal{D} \cap [CG]$) on $[a,b]$.

$(ii) \Rightarrow (i)$ Let $g, G : [a,b] \mapsto \mathbb{R}$, $g(x) = F'(x)$, $x \in N$, $g(x) = 0$, $x \notin N$; $G(x) = (\mathcal{L}) \int_a^x g(t)dt$. Then G is AC and decreasing on $[a,b]$. Let $H : [a,b] \mapsto \mathbb{R}$, $H = F - G$. By Corollary 2.5.1, (iii) (resp. (iv)), $H \in \mathcal{D}_- \cap [C_iG]$ (resp. $H \in \mathcal{D} \cap [CG]$) on $[a,b]$. Since $G \in C$, it follows that $int\{x : H \in C_i$ (resp. $H \in C$) at $x\} = A$. Let $[c,d] \subset A$. By Theorem 2.23.2, (vi), $H \in \underline{M}$ (resp. $H \in (M)$) on $[c,d]$. Let x be any point of A at which both F and G are derivable. Then H is derivable at x and $H'(x) = F'(x) - G'(x)$. If $x \in N$ then $H'(x) = 0$. If $x \notin N$ then $F'(x) \geq 0$ and $G'(x) \leq 0$, hence $H'(x) \geq 0$. Consequently $H'(x) \geq 0$ *a.e.* where $H'(x)$ exists on A. By Corollary 4.2.3 (resp. Corollary 4.2.4, H is increasing (resp. increasing and AC) on $[a,b]$. Since $F = H + G$, it follows that $F \in \underline{AC}$ (resp. AC) on $[a,b]$ (see Theorem 2.11.1, (i), (iii)).

Remark 4.2.3 *Theorem 4.2.3 generalizes Theorem 7.7 of [S3] (p. 285) and Theorem 1 of [FlF1] (p.216).*

Corollary 4.2.5 *Let $F : [a,b] \mapsto \mathbb{R}$, $A = int\{x : F \in C$ at $x\}$ and $N = \{x \in A : F$ is derivable at x and $F'(x) < 0\}$. The following assertions are equivalent:*

(i) $F \in \underline{AC} \cap C$ on $[a,b]$;

(ii) $F \in \mathcal{D} \cap [CG]$ on $[a,b]$, $F \in \underline{M}$ on each closed interval contained in A, and F' is summable on N.

Proof. The proof is similar to that of Theorem 4.2.3, using Corollary 4.2.4 instead of Corollary 4.2.3.

Corollary 4.2.6 *Let $F, G, H : [a,b] \mapsto \mathbb{R}$ such that $H = F + G$, $F \in \mathcal{D} \cap [CG]$ on $[a,b]$ and $F \in (M)$ on each closed interval contained in $A = int\{x : F \in C$ at $x\}$, $G \in \underline{AC}$ on $[a,b]$. If $H'(x) \geq 0$ a.e. where H is derivable then $F \in AC$ and H is increasing on $[a,b]$.*

Proof. By Theorem 2.14.5, $G = g_1 + g_2$, $g_1 \in AC$ on $[a,b]$, g_2 is increasing on $[a,b]$ and $g_2'(x) = 0$ *a.e.* on $[a,b]$. Let $F_1 = F + g_1$. Then $H = F_1 + g_2$, $F_1 \in \underline{M}$ on each closed interval contained in A, and $A = int\{x : F_1 \in \mathcal{C} \text{ at } x\}$. By Corollary 2.5.1, (iv), $F_1 \in \mathcal{D} \cap [\mathcal{C}G]$ on $[a,b]$. Also $F_1'(x) \geq 0$ *a.e.* where $F_1'(x)$ exists and is finite on A. By Corollary 4.2.4, $F_1 \in AC$ on $[a,b]$ and F_1 is increasing on $[a,b]$. Hence $F = F_1 - g_1 \in AC$ and H is increasing on $[a,b]$.

4.3 Monotonicity and conditions $N^{-\infty}$; N^{∞}

Definition 4.3.1 *Let $F : [a,b] \mapsto \mathbb{R}$ and let $c, d \in [a,b]$, $c < d$. We denote by $(+)'$ and $(-)'$ the following properties:*

$(+)'$ *if $F(c) < F(d)$ then $\mid F(P \cap [c,d]) \mid > 0$;*

$(-)'$ *if $F(d) < F(c)$ then $\mid F(N \cap [c,d]) \mid > 0$,*

where $P = \{x : 0 \leq F'(x) \leq +\infty\}$ and $N = \{x : 0 \geq F'(x) \geq -\infty\}$.

Remark 4.3.1 *(i) $\mathcal{C}_i \cap \underline{M} \subseteq (-)'$ on $[a,b]$ (see Theorem 2.23.3).*

(ii) $(-) \subseteq (-)'$ on $[a,b]$.

Theorem 4.3.1 *Let $F : [a,b] \mapsto \mathbb{R}$. The following assertions are equivalent:*

(i) F is increasing on $[a,b]$;

(ii) $F \in (-)' \cap N^{-\infty}$ on $[a,b]$ and $F'(x) \geq 0$ a.e. where F is derivable on $[a,b]$.

Proof. *(i)* $\Rightarrow$ *(ii)* This is evident.

(ii) $\Rightarrow$ *(i)* Suppose on the contrary that there exist $c, d \in [a,b]$, $c < d$ such that $F(c) > F(d)$. Let $N = \{x \in [c,d] : 0 \geq F'(x) \geq -\infty\}$, $N_o = \{x \in [c,d] : F'(x) = 0\}$, $E_- = \{x \in [c,d] : 0 > F'(x) > -\infty\}$ and $E_{-\infty} = \{x \in [c,d] : F'(x) = -\infty\}$. Then $N = N_o \cup E_- \cup E_{-\infty}$. By Theorem 1.10.4, $\mid F(N_o) \mid = 0$ and by hypothesis we have $\mid E_- \mid = \mid E_{-\infty} \mid = 0$. Since $F \in N^{-\infty}$, $\mid F(E_{-\infty}) \mid = 0$. By Corollary 2.13.3, $F \in AC^*G$ on E_-. It follows that $F \in (N)$ on E_-, hence $\mid F(E_-) \mid = 0$. Thus $\mid F(N) \mid = 0$, which contradicts condition $(-)'$.

Remark 4.3.2 *We cannot give up condition $(-)'$ in Theorem 4.3.1 (see Section 6.30, the function $-F$).*

Corollary 4.3.1 *Let $F : [a,b] \mapsto \mathbb{R}$. The following assertions are equivalent:*

(i) F is increasing on $[a,b]$;

(ii) $F \in \mathcal{D}_-\mathcal{B}_1T_2 \cap N^{-\infty}$ on $[a,b]$ and $F'(x) \geq 0$ a.e. where F is derivable on $[a,b]$.

Proof. *(i)* $\Rightarrow$ *(ii)* This is evident.

(ii) $\Rightarrow$ *(i)* By Theorem 2.18.9, (xvii), and Remark 4.3.1, $F \in (-) \subseteq (-)'$ on $[a,b]$. Now the proof follows by Theorem 4.3.1.

Corollary 4.3.2 *Let $F : [a,b] \mapsto \mathbb{R}$, $F \in \mathcal{D}\_\mathcal{B}_1T_2$ and let $E = \{x : -\infty \leq F'(x) < 0\}$. If $\mid F(x) \mid = 0$ then F is increasing on $[a,b]$.*

Proof. Clearly $F \in N^{-\infty}$ on $[a,b]$. By Lemma 2.17.1 (applied to the set E), it follows that $F'(x) = 0$ *a.e.* on E, *so* $\mid E \mid = 0$. Hence $F'(x) \geq 0$ *a.e.* where $F'(x)$ exists. By Corollary 4.3.1, F is increasing on $[a,b]$.

Remark 4.3.3 *C.M. Lee proved Corollary 4.3.2 for $\mathcal{DB}_1T_2$ functions (see [Le2]).*

Corollary 4.3.3 *Let $F : [a,b] \mapsto \mathbb{R}$, $F \in \mathcal{D}\_\mathcal{B}_1 \cap (N)$ on $[a,b]$. If $F'(x) \geq 0$ a.e. where F is derivable then F is increasing on $[a,b]$.*

Proof. By Theorem 2.18.9, $F \in T_2$ on $[a,b]$, hence $F \in \mathcal{D}\_\mathcal{B}_1T_2$ on $[a,b]$. Since $F'(x) \geq 0$ *a.e.* where F is derivable, the set $E = \{x : -\infty \leq F'(x) < 0\}$ has measure 0. Since $F \in (N)$, it follows that $\mid F(E) \mid = 0$. Now Corollary 4.3.2 completes the proof.

Corollary 4.3.4 (C.M. Lee) *Let $F : [a,b] \mapsto \mathbb{R}$, $F \in \mathcal{DB}_1 \cap (N)$. If $F'(x) \geq 0$ a.e. where F is derivable then F is increasing and AC on $[a,b]$.*

Proof. Since $\mathcal{D} \subset \mathcal{D}_-$, by Corollary 4.3.3, F is increasing on $[a,b]$, hence $F \in VB$ on $[a,b]$. But $VB \cap \mathcal{D} \subset \mathcal{C}$ (see Theorem 2.18.9, (xx)). By Theorem 2.18.9, (vii), $F \in AC$ on $[a,b]$.

Corollary 4.3.5 *Let $F : [a,b] \mapsto \mathbb{R}$. The following assertions are equivalent:*

(i) F is increasing on $[a,b]$;

(ii) $F \in [\mathcal{C}_iG] \cap N^{-\infty} \cap uCM \cap T_2$ on $[a,b]$ and $F'(x) \geq 0$ a.e. where F is derivable.

Proof. $(i) \Rightarrow (ii)$ This is evident.

$(ii) \Rightarrow (i)$ By Lemma 1.9.1, since $F \in [\mathcal{C}_iG]$ on $[a,b]$, there exists a sequence of intervals I_n whose union is dense in $[a,b]$, such that $F \in \mathcal{C}_i$ on each I_n. Let J_n be a closed subinterval of I_n. By Theorem 2.4.2, (vi), $\mathcal{C}_i \subset \mathcal{D}\_\mathcal{B}_1$, and by Corollary 4.3.1, F is increasing on J_n. Since J_n is arbitrarily chosen, it follows that F is increasing on I_n.The intervals I_n can be chosen to be maximal open intervals of monotonicity of F. We show that in fact there exists only one such maximal open interval, namely (a,b). Suppose on the contrary that there is more than one such maximal open interval, and let $Q = [a,b] \setminus (\cup_{n\geq 1} I_n)$. Since $F \in uCM$ on $[a,b]$, F is increasing on each $\overline{I}_n$, hence Q is a perfect subset of $[a,b]$ (perhaps without a and b). By Theorem 1.7.1, there exist $c,d \in Q$, $c < d$ such that $Q \cap (c,d) \neq \emptyset$ and $F_{/Q\cap[c,d]} \in \mathcal{C}_i$. By Proposition 2.3.2, (i), $F \in \mathcal{C}_i$ on $[c,d]$. As above it follows that F is increasing on $[c,d]$, a contradiction. Hence F is increasing on (a,b). Since $F \in uCM$ on $[a,b]$, F is increasing on $[a,b]$.

Remark 4.3.4 *Particularly Corollary 4.3.5 is valid for $[\underline{AC}G] \cap uCM$ functions, and this result is due to C.M. Lee (see [Le1]).*

Theorem 4.3.2 *Let $F, G, H : [a,b] \mapsto \mathbb{R}$, $H = F + G$, such that $H'(x) \geq 0$ a.e. where H is derivable.*

(i) If $F \in N^{\infty} \cap (-)$ and $G \in \underline{AC}$ then $F \in \underline{AC}$ and H is increasing on $[a,b]$.

(ii) *If* $F \in (-) \cap (+) \cap N^{\infty}$ *and* $G \in \underline{AC}$ *then* $F \in AC$ *and H is increasing on* $[a, b]$.

(iii) *If* $F \in \mathcal{D}_{-}\mathcal{B}_1 T_2 \cap N^{-\infty}$ *and* $G \in \underline{AC}$ *then* $F \in \underline{AC}$ *and H is increasing on* $[a, b]$.

Proof. (i) By Theorem 2.11.1, (vi), $G \in VB$ on $[a, b]$. By Theorem 2.14.3, G is derivable *a.e.* on $[a, b]$ and G' is summable. Let $N = \{x : F'(x) \leq 0\}$. By Corollary 1.10.2, N is measurable. It follows that H is derivable *a.e.* on N and $H'(x) \geq 0$ *a.e.* on N. Then $G'(x) \geq -F'(x) \geq 0$ *a.e.* on N, hence F' is summable on N. By Theorem 2.21.2, $F \in \underline{AC}$ on $[a, b]$. Clearly $H \in \underline{AC}$ on $[a, b]$, and by Theorem 2.14.5, (ii), H is increasing on $[a, b]$.

(ii) By (i), $F \in \underline{AC}$ and H is increasing on $[a, b]$. By Theorem 2.11.1, (vi), $F \in VB$, and by Corollary 2.21.1, (i), $F \in \overline{AC}$ on $[a, b]$. Hence $F \in AC$ on $[a, b]$.

(iii) By Theorem 2.18.9, (xvii), $\mathcal{D}_{-}\mathcal{B}_1 T_2 \subset (-)$ on $[a, b]$. Now the proof follows by (i).

Theorem 4.3.3 *Let* $F, G, H : [a, b] \mapsto \mathbb{R}$, $H = F + G$ *such that* $F \in \mathcal{D}_{-}\mathcal{B}_1 T_2 \cap N^{-\infty}$, $G \in [\underline{AC}^* G]$, $H \in uCM$. *If* $H'(x) \geq 0$ *a.e. where H is derivable then H is increasing and* $F \in [\underline{AC}^* G]$ *on* $[a, b]$.

Proof. By Lemma 1.9.1, there exists a sequence of closed intervals I_n whose union is dense in $[a, b]$ such that $G \in \underline{AC}^* = \underline{AC}$ on I_n. By Theorem 2.18.9, (xvii), $F \in (-) \cap N^{-\infty}$ on $[a, b]$. By Theorem 4.3.2, H is increasing on I_n. It follows that there exists a maximal nonempty open set E such that H is increasing on each component of E. Suppose on the contrary that $E \neq (a, b)$. Let $Q = \overline{[a, b] \setminus E}$. Since $H \in uCM$, H is increasing on the closure of each component of E, hence Q is a perfect set. It follows that Q is nowhere dense in $[a, b]$ (indeed: suppose that Q contains an interval $[a_1, b_1]$; by Theorem 1.7.1, there exists $[a_2, b_2] \subset [a_1, b_1]$ such that $G \in AC^* = AC$ on $[a_2, b_2]$; similarly as above it follows that H is increasing on $[a_2, b_2]$, a contradiction). By Proposition 2.12.1, (iv), $G \in VB^*G = [VB^*G]$ on $[a, b]$. By Theorem 1.7.1, there exist $c, d \in Q$, $c < d$, such that $(c, d) \cap Q \neq \emptyset$ and $G \in \underline{AC}^* \cap VB^*$ on $P = [c, d] \cap Q$. Let $\{[c_k, d_k]\}_k$ be the intervals I_n contained in $[c, d]$ (we have infinitely many such intervals because Q is nowhere dense in $[a, b]$). Since $G \in VB^*$ on P, by Theorem 2.8.1, (vii), $\sum_{k=1}^{\infty} \mathcal{O}(G; [c_k, d_k]) < +\infty$. Since $F = H - G$ and H is an increasing function on each $[c_k, d_k]$, it follows that $\Omega_{-}(F; [c_k, d_k] \wedge \{c_k, d_k\}) \geq -\mathcal{O}(G; [c_k, d_k])$, hence we obtain that $\sum_{k=1}^{\infty} \Omega_{-}(F; [c_k, d_k] \wedge \{c_k, d_k\}) > -\infty$. By Proposition 2.4.3, F_P is *lower internal* on $[c, d]$. Clearly $F_P \in \mathcal{B}_1 \cap T_2$ on $[c, d]$. By Theorem 2.5.1, (i), (iv), $F_P \in \mathcal{D}_{-}\mathcal{B}_1$ on $[c, d]$. By Theorem 2.21.1, (ii), $F_P \in N^{-\infty}$ on $[c, d]$. Since H is increasing on I_n, it follows that H_P is derivable *a.e.* on $[c_k, d_k]$ and $(H_P)'(x) \geq 0$ *a.e.* on $[c_k, d_k]$. Since H is increasing on $[c_k, d_k]$, by Lemma 2.15.1 and Remark 2.15.3, (ii), $(H_P)'(x) \geq 0$ *a.e.* where $(H_P)'(x)$ exists on P. Since $\underline{AC}^* \subseteq \underline{AC}$ on P, by Theorem 2.11.1, $G_P \in \underline{AC}$ on $[c, d]$. By Theorem 4.3.2, H_P is increasing on $[c, d]$, hence H is increasing on P. Since H is also increasing on each $[c_k, d_k]$, it follows that H is increasing on $[c, d]$, a contradiction. Hence $E = (a, b)$ and H is increasing on (a, b). Since $F \in uCM$ on $[a, b]$, it follows that H is increasing on $[a, b]$. Since $G \in VB^*G$ on $[a, b]$, we obtain that $F \in VB^*G$ on $[a, b]$. By Corollary 2.22.1, (ii), $F \in \underline{AC}^*G$ on $[a, b]$.

Corollary 4.3.6 *Let $F, G, H : [a,b] \mapsto \mathbb{R}$, $H = F + G$ such that $F'(x) \geq 0$ a.e. where H is derivable on $[a,b]$.*

(i) If $F \in \mathcal{D}_-\mathcal{B}_1T_2 \cap N^{\infty}$ and $G \in \underline{AC}^*G \cap C_i$ then $F \in [\underline{AC}^*G]$ and H is increasing on $[a,b]$.

(ii) If $F \in \mathcal{D}_-\mathcal{B}_1T_2 \cap N^{-\infty}$ and $G \in \underline{AC}^*G \cap C_i^*$ then $F \in [\underline{AC}^*G] \cap C_i^*$ and H is increasing on $[a,b]$.

(iii) If $F \in \mathcal{D}\mathcal{B}_1T_2 \cap N^{-\infty}$ and $G \in \underline{AC}^*G \cap C_i^*$ then $F \in [\underline{AC}^*G] \cap C$ and H is C and increasing on $[a,b]$.

(iv) If $F \in \mathcal{D}\mathcal{B}_1T_2 \cap N^{\infty}$ and $G \in \underline{AC}^*G \cap C_i$ then $F \in [AC^*G]$ and H is increasing on $[a,b]$.

(v) If $F \in \mathcal{D}\mathcal{B}_1T_2 \cap N^{\infty}$ and $G \in \underline{AC}^*G \cap C_i^*$ then $F \in AC^*G \cap C$ and H is increasing on $[a,b]$.

(vi) If $F \in C \cap T_2 \cap N^{\infty}$ and $G \in \underline{AC}^*G \cap C_i$ then $F \in AC^*G \cap C$ and H is increasing on $[a,b]$.

(vii) If $F \in \mathcal{D}\mathcal{B}_1 \cap (N)$ and $G \in \underline{AC}^*G \cap C_i^*$ then $F \in AC^*G \cap C$ and H is increasing on $[a,b]$.

(viii) If $F \in \mathcal{D}\mathcal{B}_1 \cap (N)$ and $G \in AC^*G \cap C$ then $F \in AC^*G \cap C$, H is AC and increasing on $[a,b]$.

Proof. (i) See Corollary 2.5.1, (i), Theorem 2.4.2, (vi) and Theorem 4.3.3.

(ii) By (i), H is increasing on $[a,b]$ and $F \in [\underline{AC}^*G]$ on $[a,b]$. It follows that for each $x \in [a,b)$, $H(x+)$ exists and is finite. Since $G \in C_i^*$ on $[a,b]$, $G(x+)$ exists and is finite for each $x \in [a,b)$. Hence $F(x+) = H(x+) - G(x+)$ exists and is finite. Similarly $F(x-)$ exists and is finite for each $x \in (a,b]$. By Theorem 2.4.2, (vi), F is *lower internal* on $[a,b]$, and by Remark 2.4.1, $F \in C_i^*$ on $[a,b]$.

(iii) By (ii), H is increasing on $[a,b]$, $F \in [\underline{AC}^*G]$ on $[a,b]$, $F(x+)$ exists for $x \in [a,b)$ and $F(x-)$ exists for $x \in (a,b]$. By Theorem 2.4.2, (vii), F is *internal* on $[a,b]$. By Remark 2.4.1, $F \in C$ on $[a,b]$.

(iv) By (i), H is increasing on $[a,b]$ and $F \in [\underline{AC}^*G]$ on $[a,b]$. By Proposition 2.12.1, (iv), $F \in VB^*G$ on $[a,b]$. Then $-F \in \mathcal{D}_- \cap N^{-\infty} \cap VB^*G$ on $[a,b]$. By Corollary 2.22.1, (ii), $-F \in [\underline{AC}^*G]$ on $[a,b]$, hence $F \in [\overline{AC}^*G]$ on $[a,b]$. Thus $F \in [AC^*G]$ on $[a,b]$.

(v) By (iv), H is increasing and $F \in [AC^*G]$ on $[a,b]$. Now the proof continues as in (iii).

(vi) See (iv).

(vii) By Theorem 2.18.7 and Lemma 2.21.1, (i), $F \in \mathcal{D}\mathcal{B}_1T_2 \cap N^{\infty}$ on $[a,b]$. Now the proof follows by (v).

(viii) By (vii), H is increasing on $[a,b]$ and $F \in AC^*G \cap C$ on $[a,b]$. It follows that $H \in AC^*G \cap C$ on $[a,b]$. By Theorem 2.18.9, (vi), (vii), (x) since H is increasing on $[a,b]$, it follows that $H \in AC$ on $[a,b]$.

4.4 Local monotonicity

Lemma 4.4.1 (Thomson) *Let $\mathcal{S}$ be a local system satisfying intersection condition (I.C.), and let $F : [a,b] \mapsto \mathbb{R}$. If F is $\mathcal{S}$- increasing on $[a,b]$ then F is increasing on $[a,b]$.*

Proof. Let P be the collection of all x for which there exists no open interval containing x on which F is increasing. Then the complement of P is an open set U and

(1) *F is increasing on the closure ofeach component of U.*

It follows that P is perfect (perhaps without a and b). We show that P is empty. Suppose that $P \neq \emptyset$. For $\sigma_x = \{y : y = x \text{ or } (F(y) - F(x))/(y - x) \geq 0\} \in \mathcal{S}(x)$, let $\delta(x) > 0$, $x \in [a,b]$ be given by condition (I.C.). Let $P_{mn} = \{x \in P : x \in [m/n,(m+1)/n], 1/n < \delta(x) < 1/(n-1)\}$, $n = \overline{2,\infty}$, $m = \pm 1, \pm 2, \ldots$. By Theorem 1.7.1, there exists an open interval (c,d) such that $\emptyset \neq (c,d) \cap P \subset \overline{P}_{mn}$, for some m and n. Let $x_o < y_o$ such that x_o is a right accumulation point of $(c,d) \cap P$, and y_o is a left accumulation point of $(c,d) \cap P$. Let $x_1, y_1 \in P_{mn}$ such that $x_o < x_1 < y_1 < y_o$, $x_1 \in (x_o, x_o + \delta(x_o))$, $y_1 \in (y_o - \delta(y_o), y_o)$. Then $\sigma_{x_o} \cap \sigma_{x_1} \neq \emptyset$, $\sigma_{x_1} \cap \sigma_{y_1} \neq \emptyset$, $\sigma_{y_1} \cap \sigma_{y_o} \neq \emptyset$, hence $F(x_o) \leq F(y_o)$. By (1), F is increasing on $[c,d]$, a contradiction. It follows that $P = \emptyset$.

Remark 4.4.1 *For further results on local monotonicity, and for another proof of Lemma 4.4.1 see [T5] (pp. 115 - 119).*

Corollary 4.4.1 *Let $\mathcal{S}$ be a local system satisfying intersection condition (I.C.), and let $F : [a,b] \mapsto \mathbb{R}$.*

(i) If $\mathcal{S}\,\underline{D}F(x) \geq 0$ a.e. and $\mathcal{S}\,\underline{D}F(x) > -\infty$ everywhere then F is increasing on $[a,b]$.

(ii) If $\mathcal{S}\,DF(x)$ exists, with $\mathcal{S}\,DF(X) \neq -\infty$ everywhere, and $\mathcal{S}\,DF(x) \geq 0$ a.e. then F is increasing on $[a,b]$.

Proof. (i) Let $\epsilon > 0$ and let $Z = \{x \in \mathcal{S}\,DF(x) < 0\}$. Then $|Z| = 0$. By Theorem 2.14.6, there exists an increasing function $g : [a,b] \mapsto [0,\epsilon)$ such that $g'(x) = +\infty$, $x \in Z$, $g(a) = 0$ and $g'(x) > 0$, $x \in [a,b] \setminus Z$. It follows that $F + g$ is strictly $\mathcal{S}$-increasing on $[a,b]$. By Lemma 4.4.1, $F + g$ is increasing on $[a,b]$. Since ϵ is arbitrary, F is increasing on $[a,b]$.

(ii) The proof is similar to that of (i). Let $H = F + g$. Then there exists $\mathcal{S}\,DH(x)$ and $\mathcal{S}\,DH(x) > 0$, $x \in [a,b]$. Hence $F + g$ is strictly $\mathcal{S}$-increasing on $[a,b]$.

4.5 $\mathcal{S}$-derivatives and the Mean Value Theorem

Theorem 4.5.1 *Let $\mathcal{S}$ be a local system satisfying intersection condition (I.C.), and let $F : [a,b] \mapsto \mathbb{R}$ be a function satisfying the following conditions:*

(1) $F \in sCM$ on $[a,b]$;

(2) An exact $\mathcal{S}$-derivative $\mathcal{S}\,DF(x)$ exists, finite or infinite, at each $x \in [a,b]$;

(3) $\mathcal{S}\,DF(x)$ is $\mathcal{B}_1$ on $[a,b]$.

Then we have:

(i) $\{x : \mathcal{S}\,DF(x) < v\}$ is simultaneously $\mathcal{S}_2^+$ and $\mathcal{S}_2^-$ open, and $\{x : \mathcal{S}\,DF(x) > u\}$ is simultaneously $\mathcal{S}_2^+$ and $\mathcal{S}_2^-$ open , whenever $u, v \in \mathbb{R}$;

(ii) $\mathcal{S}\,DF(x)$ is $\mathcal{D}$ and $\{x : u < \mathcal{S}\,DF(x) < v\}$ is simultaneously $\mathcal{S}_2^+$ and $\mathcal{S}_2^-$ open, whenever $u, v \in \mathbb{R}$;

(iii) F fulfills the Mean Value Theorem.

Proof. Let $g(x) = \mathcal{S}\,DF(x)$.

(i) Suppose for example that $E^\alpha = \{x : g(x) < \alpha\}$ is not $\mathcal{S}_2^-$ open. Then there exist $x_o \in E^\alpha$ and $\delta_o > 0$, such that $x_o - \delta_o > 0$ and, for example, $g(x) > \alpha$ on $(x_o - \delta_o, x_o)$. Let $A = \{x \in (x_o = \delta_o, x_o) : g \in \mathcal{C}$ at $x\}$. If $x \in A$ then $g(x) \geq \alpha$ (because, if $g(x) < \alpha$ then there exists $\delta > 0$, with $(x - \delta, x + \delta) \subset (x_o - \delta_o, x_o)$, such that $g(y) < \alpha$ for each $y \in (x - \delta, x + \delta)$, a contradiction). Let $x \in A$. Then there exists $[c,d] \subset (x_o - \delta_o, x_o)$, such that $x \in (c,d)$ and $g(y) > -\infty$ for each $y \in [c,d]$. By Corollary 4.4.1, $F(x) - \alpha x$ is increasing on $[c,d]$. It follows that there exist maximal open intervals (a_n, b_n), $n \geq 1$ such that $F(x) - \alpha x$ is increasing on each (a_n, b_n). By (1), $F(x) - \alpha x$ is increasing on $[a_n, b_n]$, hence $G = \cup_{n\geq 1}(a_n, b_n)$ is dense in $(x_o - \delta_o, x_o)$ and $P = (x_o - \delta_o, x_o) \setminus G$ is a perfect set. Suppose on the contrary that $P \neq \emptyset$. Let $x_1 \in (x_o - \delta_o, x_o) \setminus G$ be a point of continuity for $g_{/P}$ (this is possible by (3)). Then $g(x_1) \geq \alpha$ (because, if $g(x_1) < \alpha$ then by (3), there exists $\delta_1 > 0$ such that $(x_1 - \delta_1, x_1 + \delta_1) \subset (x_o - \delta_o, x_o)$ and $g(y) < \alpha$, for each $y \in (x_1 - \delta_1, x_1 + \delta_1) \cap P$; it follows that there exists $n \geq 1$ such that $(a_n, b_n) \subset (x_1 - \delta_1, x_1 + \delta_1)$, hence $g(a_n) \geq \alpha$ and $g(b_n) \geq \alpha$, a contradiction). Since $g_{/P}$ is $\mathcal{C}$ at x_1, it follows that there exists a closed interval $[c_1, d_1]$ containing x_1, such that $g(y) > \alpha - 1$ on $[c_1, d_1] \cap P$. By Corollary 4.4.1, $F(x) - \alpha x$ is increasing on $[c_1, d_1]$, a contradiction. Hence $P = \emptyset$. Therefore $G = (x_o - \delta_o, x_o)$. By (1), $F(x) - \alpha x$ is increasing on $[x_o - \delta_o, x_o]$, hence $g(x_o) \geq \alpha$, a contradiction.

(ii) This follows by (i), (3) and Corollary 2.6.1.

(iii) Let $c, d \in [a,b]$, $c < d$ and let $\alpha = (F(d) - F(c))/(d - c)$. Suppose on the contrary that there is no $x \in (c,d)$ such that $g(x) = \alpha$. Since $g \in \mathcal{D}$ (see (ii)), it follows that either $g(x) > \alpha$ or $g(x) < \alpha$. In the first situation for example, we obtain that $F(x) - \alpha x$ is increasing on $[c,d]$ (see Corollary 4.4.1). Since $g(x) > \alpha$ on (c,d) it follows that $F(d) - \alpha > F(c)$, a contradiction.

Remark 4.5.1 *(i) In fact (i) $\Leftrightarrow$ (ii) in Theorem 4.5.1.*

(ii) In [Pr1] (see Theorem 6), Preiss proved Theorem 4.5.1 for approximate derivatives and for internal functions (instead of sCM functions). There exists a function which satisfies the hypothesis of Theorem 4.5.1, but not those of Preiss' theorem (see Section 6.16).*

4.6 Relative monotonicity

For more information on this topic see [T5] (pp. 120 - 130). We shall need the following lemma.

Lemma 4.6.1 (Thomson) *([T5]). Let $F : [a,b] \mapsto \mathbb{R}$ and let r be a real number. Let $\mathcal{S}$ be a local system satisfying intersection condition (I.C.). If the inequality $\mathcal{S}DF(x) > r$ holds everywhere on a set X, then there is a denumerable partition $\{X_n\}$ of the set X such that the function $F_r(x) = F(x) - rx$ is strictly increasing on each set X_n.*

Proof. Define for each $x \in X$ the sets $\sigma_x = \{y : y = x \text{ or } (F(y) - F(x))/(y-x) > r\}$ and observe that, because of the hypotheses, each $\sigma_x \in \mathcal{S}(x)$. As we are assuming intersection condition (I.C.), this choice from $\mathcal{S}$ induces a denumerable partition $\{X_n\}$ of the set X in such a way that, whenever $x < y$ belong to the same member of the partition, the intersection $\sigma_x \cap \sigma_y \cap [x,y]$ contains some point z. We consider separately the situations $z = x$, $z = y$ and $x < z < y$. In each case it is easy to verify that this requires $F_r(x) < F_r(y)$, which concludes our lemma.

4.7 An application of Corollary 4.4.1

Theorem 4.7.1 *Let $\mathcal{S}$ be a local system satisfying intersection condition (I.C.), and let $F : [a,b] \mapsto \mathbb{R}$, $F \in \mathcal{C}_i$ (resp. $F \in [\mathcal{C}_i G] \cap uCM$). If an $\mathcal{S}DF(x)$ exists (finite or infinite) n.e. and $\mathcal{S}DF(x) \geq 0$ a.e. then F is increasing on $[a,b]$.*

Proof. We prove the first part. Let $X = \{x : \mathcal{S}DF(x) > -2\}$ and $Y = \{x : \mathcal{S}DF(x) < -1\}$. Let K be the denumerable set of points x, such that $\mathcal{S}DF(x)$ does not exist. By Lemma 4.6.1, there exists a denumerable partition $\{X_n\}$ of the set X, such that the function $F_{-2}(x) = F(x) + 2x$ is strictly increasing on each X_n. By the same lemma, there exists a denumerable partition $\{Y_n\}$ of the set Y such that $F_{-1}(x) = F(x) + x$ is strictly decreasing on each set Y_n. Let E be the collection of all x for which there exists no interval containing x, on which F is nondecreasing. Let U be the complement of E. Then U is an open set and F is nondecreasing on each component of U. Since $F \in \mathcal{C}_i \subset uCM$, F is nondecreasing on the closure of each component of U. Thus E contains no isolated points and is therefore a perfect set. If E is shown to be empty we will have that F is nondecreasing. Suppose that E is a nonempty perfect set. Then we have $(\cup_{n\geq 1}(E \cap X_n)) \cup (\cup_{n\geq 1}(E \cap Y_n)) \cup (E \cap K) = E$. Therefore Theorem 1.7.1 yields the existence of an open interval (c,d) with either (I) $\emptyset \neq (c,d) \cap E \subset \overline{X_n}$ or (II) $\emptyset \neq (c,d) \cap E \subset \overline{Y_n}$. We will show that no such (c,d) can exist. Suppose for example (I). Let $x, y \in (c,d)$, $x < y$. We show that $F(y) - F(x) \geq -2(y-x)$. We have three cases for E:
1) x is right isolated and y is a left accumulation point;
2) x is a right accumulation point and y is left isolated;
3) x is right isolated and y is left isolated.
We prove for example the third case. Let $[c_1, d_1]$ respectively $[c_2, d_2]$ be the closures of the components which contain x respectively y. If $[c_1, d_1] = [c_2, d_2]$ then $F(x) < F(y)$. If $[c_1, d_1] \neq [c_2, d_2]$ then $d_1 < c_2$. Let $r_k \searrow d_1$, $s_k \nearrow c_2$, $r_k < s_k$, $r_k, s_k \in X_n$, $k = \overline{1,\infty}$. It follows that $F_{-2}(s_k) > F_{-2}(r_k)$, hence $F_{-2}(c_2) \geq$

$\limsup_{z\nearrow c_2} F_{-2}(z) \geq \limsup_{k\to\infty} \geq \limsup_{k\to\infty} F_{-2}(r_k) \geq \liminf_{k\to\infty} F_{-2}(r_k) \geq \liminf_{z\searrow d_1} F_{-2}(z) \geq F_{-2}(d_1)$. Hence $F(c_2) - F(d_1) \geq -2(c_2 - d_1)$. Since $F(y) - F(c_2) \geq 0 > -2(y-c_2)$ and $F(d_1) - F(x) \geq -2(d_1 - x)$, it follows that $F(y) - F(x) \geq -2(y-x)$. Hence $F \in \underline{AC}$ on (c,d). Clearly $F'(x) \geq 0$ *a.e.* on (c,d), hence F is increasing on (c,d), a contradiction (see Theorem 2.14.5).

We show the second part. Using the notations from above, we may suppose that $F \in \mathcal{C}_i$ on $[c,d] \cap E$, $c, d \in E$. By Proposition 2.3.2, it follows that $F \in \mathcal{C}_i$ on $[c,d]$.Now the proof continues as in the first part.

Corollary 4.7.1 *Let $F : [a,b] \mapsto \mathbb{R}$, $F \in \mathcal{D}_-$. If $F'(x)$ exists finite or infinite n.e., and $F'(x) \geq 0$ a.e. then F is increasing on $[a,b]$.*

Proof. By Corollary 2.13.2, $F \in VB^*G$ on $[a,b]$; by Theorem 2.10.3, (v), $F \in [\mathcal{C}_i G]$ on $[a,b]$, and by Theorem 2.4.2, (vi), $F \in uCM$ on $[a,b]$. Now Theorem 4.7.1 completes the proof.

Remark 4.7.1 *Particularly, Corollary 4.7.1 is true for $\mathcal{D}$ functions, and this is in fact Theorem I of Zahorski (see [Z1]).*

Theorem 4.7.2 *Let $\mathcal{S}$ be a local system satisfying intersection condition (I.C.) and let $F : [a,b] \mapsto \mathbb{R}$, such that:*

(i) $F \in \mathcal{D}_-\mathcal{B}_1$ on $[a,b]$;

(ii) An $\mathcal{S}\,DF(x)$ exists (finite or infinite) n.e. on $[a,b]$;

(iii) $\mathcal{S}\,DF(x) \geq 0$ a.e. on $[a,b]$.

Then F is increasing on $[a,b]$.

Proof. By (ii) and Lemma 4.6.1, $F \in VBG$ on $[a,b]$. Let $A = int\{x : F \in \mathcal{C}_i \text{ at } x\}$, and let $[c,d] \subset A$. By Theorem 4.7.1, F is increasing on $[c,d]$, hence $F \in B_i$ on $[c,d]$. By Corollary 4.1.1, F is increasing on $[a,b]$.

4.8 A general monotonicity theorem

Lemma 4.8.1 *Let $F : [a,b] \mapsto \mathbb{R}$ such that $F \in uCM$ on $[a,b]$ and $F'(x) \geq 0$ a.e. where F is derivable. Let $P = \{x$: there exists no open interval containing x on which F is increasing$\}$. If $P \cap (a,b) \neq \emptyset$ then P is a perfect set, and whenever Q is a dense subset of a portion of P then:*

(i) $F \notin \underline{AC}(Q \wedge (Q \cup Q_-)) \cap \underline{AC}((Q \cup Q_+) \wedge Q)$;

(ii) there is no $\lambda \in (-\infty, 0)$ such that $F \in \overline{\underline{L}}((Q \cup Q_-) \wedge Q) \cap \overline{\underline{L}}(Q \wedge (Q \cup Q_+))$ with the constant λ.

Proof. Let $U = (a,b) \setminus P$. Then U is an open set and F is increasing on each of its components. Since $F \in uCM$ on $[a,b]$, F is increasing on the closure of each component of U, hence P is perfect. Let $(c,d) \cap P \neq \emptyset$.

(i) Suppose on the contrary that there exists Q, a dense subset of $(c,d) \cap P$, such that $F \in \underline{AC}(Q \wedge (Q \cup Q_-)) \cap \underline{AC}((Q \cup Q_+) \wedge Q)$. We may suppose without loss of generality that $c, d \in P$. By Theorem 2.11.1, (xvi), $F \in \underline{AC}$ on $[c, d]$. Since $F'(x) \geq 0$ *a.e.* where F is derivable on $[a, b]$, by Theorem 2.14.5, it follows that F is increasing on $[c, d]$, a contradiction.

(ii) Suppose on the contrary that there exists Q a dense subset of $(c, d) \cap P$ and $\lambda \in (-\infty, 0)$ such that $F \in \overline{L}((Q \cup Q_-) \wedge Q) \cap \overline{L}(Q \wedge (Q \cup Q_+))$ with the constant λ. Then Q is a nowhere dense set (because, if there exists $(c_1, d_1) \subset Q$ then F is $\overline{L}$ with the constant λ, on $[c_1, d_1]$, and $F'(x) \leq \lambda < 0$ *a.e.* on (c_1, d_1), a contradiction). Let $(r, s) \subset (c, d)$ be a component of U. Then $F(s) - F(r) < \lambda(s - r) < 0$. This contradicts the fact that F is increasing on $[r, s]$.

Lemma 4.8.2 *Let $\mathcal{S}$ be a bilateral local system having intersection condition (I.C.). Let $F : [a, b] \mapsto \mathbb{R}$ and $A = \{x : \mathcal{S}\,\underline{D}F(x) > -\infty\}$. Suppose that:*

(i) $E = [a, b] \setminus A$ is at most countable;

(ii) for each $x \in E$ there exists a set $E_x \in \mathcal{S}(x)$ such that $\limsup_{y \nearrow x, y \in E_x} F(y) \leq F(x) \leq \liminf_{y \searrow x, y \in E_x} F(y)$;

(iii) $\mathcal{S}\,\underline{D}F(x) \geq 0$ a.e. on $[a, b]$.

Then F is increasing on $[a, b]$.

Proof. Clearly $F \in uCM$ on $[a, b]$. Let P and $U = \cup_{n \geq 1}(a_n, b_n)$ be defined as in Lemma 4.8.1. Suppose that $P \neq \emptyset$. Since $F \in uCM$ on $[a, b]$, F is increasing on each $[a_n, b_n]$, hence P is a perfect subset of $[a, b]$. Let $f : A \mapsto \mathbb{R}$ such that $-\infty < f(x) < \mathcal{S}\,\underline{D}F(x)$ for each $x \in A$. Let $\sigma_x = \{y : y = x \text{ or } (F(y) - F(x))/(y - x) > f(x)\} \in \mathcal{S}(x)$ for $x \in a$, and $\sigma_x = E_x$ for $x \in E$. Let $\delta(x)$, $x \in [a, b]$ be a positive function such that, whenever $0 < y - x < \min\{\delta(x), \delta(y)\}$ then $\sigma_x \cap \sigma_y \cap [x, y] \neq \emptyset$. Let $A_n = \{x \in A : f(x) > -n\}$, $n = \overline{1, \infty}$. Let A_{nj}, $j = \overline{1, \infty}$ be a δ-decomposition of A_n. Since $P \subset E \cap (\cup_{n,j} A_{nj})$, by Theorem 1.7.1, it follows that there exists an open interval (c, d), such that $Q = A_{nj} \cap (c, d)$ is a dense subset of $(c, d) \cap P \neq \emptyset$, and $Q \subset \overline{A}_{nj}$ for some n and j. In what follows we prove that $F \in \underline{L}(Q \wedge (Q \cup Q_-)) \cap \underline{L}((Q \cup Q_+) \wedge Q)$ with the constant $-n$, hence $F \in \underline{AC}(Q \wedge (Q \cup Q_-)) \cap \underline{AC}((Q \cup Q_+) \wedge Q)$ (see Proposition 2.13.1). We show for example that $F \in \underline{L}((Q \cup Q_+) \wedge Q)$. We have three cases:

1) Let $x < y$, $x, y \in Q$. Then for $t \in \sigma_x \cap \sigma_y \cap [x, y] \neq \emptyset$ we have $F(t) - F(x) \geq -n(t - x)$ and $F(y) - F(t) \geq -n(y - t)$. Hence $F(y) - F(x) \geq -n(y - x)$.

2) Let $x < y$, $x \in A \cap P_+ \cap (c, d)$, $y \in Q$. By Remark 1.2.1, (ii), let $z_k \in \sigma_x \cap \sigma_y \cap [c, x_k] \neq \emptyset$, hence $F(y) - F(x) = F(y) - F(x_k) + F(x_k) - F(z_k) + F(z_k) - F(x) \geq -n(y - x) - n(x_k - z_k) + f(x)(z_k - x) = -n(y - z_k) + f(x)(z_k - x)$. If $k \to \infty$ then $F(y) - F(x) \geq -n(y - x)$.

3) Let $x < y$, $x \in E \cap P_+ \cap (c, d)$, $y \in Q$. Let $x_k \searrow x$, $x_k \in A_{nj}$, $x_k \in (x, x + \delta(x))$, $x_k < y$. Let $z_k \in \sigma_{x_k} \cap [x, x_k] \neq \emptyset$. Then $F(y) - F(x) > -n(y - x_k)$, $F(x_k) - F(z_k) > -n(x_k - z_k)$ and $\liminf_{k \to \infty} F(z_k) \geq F(x)$. Hence $F(y) - F(x) \geq F(y) - \liminf_{k \to \infty} F(z_k) \geq -n(y - x)$, a contradiction (see Lemma 4.8.1).

It follows that $P = \emptyset$, hence $U = (a, b)$. By the Heine-Borel Covering Theorem, F is increasing on (a, b). Since $F \in uCM$ on $[a, b]$, F is increasing on $[a, b]$.

Lemma 4.8.3 *Let $\mathcal{S}$ be a local system having intersection conditions (I.C.) and E.I.C.[m]. Let $F : [a, b] \mapsto \mathbb{R}$ and $A = \{x : \mathcal{S}\,DF(x) > -\infty\}$. Suppose that:*

(i) $E = [a,b] \setminus A$ *is at most countable;*

(ii) for each $x \in E$ *and for every* $\epsilon > 0$, $\{x \in (z \in (x-\epsilon, x) : F(x) < F(x)+\epsilon\}$ *and* $\{z \in (x, x+\epsilon) : F(z) > F(x) - \epsilon\}$ *are uncountable sets;*

(iii) $S\,\underline{D}F(x) \geq 0$ *a.e. on* $[a,b]$.

Then F is increasing on $[a,b]$.

Proof. We observe that $F \in uCM$ on $[a,b]$ and $F'(x) \geq 0$ *a.e.* where F is derivable. Let P and $U = \cup_{n\geq 1}(a_n, b_n)$ be defined as in Lemma 4.8.2. Suppose on the contrary that $P \neq \emptyset$. Since $F \in uCM$ on $[a,b]$, it follows that F is increasing on each $[a_n, b_n]$, hence P is a perfect subset of $[a,b]$. Let $f : A \mapsto \mathbb{R}$ be a finite function such that $-\infty < f(x) < S\,\underline{D}F(x)$. Let $\sigma_x = \{y : y = x \text{ or } (F(y)-F(x))/(y-x) > f(x)\} \in \mathcal{S}(x)$, for $x \in A$. For each $x \in E$, let σ_x be a fixed set of $\mathcal{S}(x)$. Let $A_n = \{x \in A : f(x) > -n\}$, $n = \overline{1,\infty}$. Let $\delta(x)$, $x \in [a,b]$ be a positive function such that $\sigma_x \cap \sigma_y \cap [x,y] \neq \emptyset$, $\sigma_x \cap \sigma_y \cap (y, y+m(y-x)) \neq \emptyset$ and $\sigma_x \cap \sigma_y \cap (x-m(y-x), x) \neq \emptyset$, whenever $0 < y-x < \min\{\delta(x), \delta(y)\}$, $x < y$. Let $\{A_{nj}\}_j$ be a δ-decomposition of A_n. By Theorem 1.7.1, there exists an open interval (c,d) such that $\emptyset \neq (c,d) \cap P \subset \overline{A}_{nj}$, for some n and j. We prove that $F \in \underline{L}(Q \wedge (Q \cup Q_-)) \cap \underline{L}((Q \cup Q_+) \wedge Q)$, with the constant $-n$, where $Q = (c,d) \cap A_{nj}$. We show for example that $F \in \underline{L}((Q \cup Q_+) \wedge Q)$. We have three cases:

1) If $x, y \in Q$, $x < y$ then $F(y) - F(x) > -n(y-x)$ (see Remark 1.2.1, (i), condition (I.C.) and 1) of the proof of Lemma 4.8.2).

2) If $x \in A \cap P_+ \cap (c,d)$, $y \in Q$, $x < y$ then $F(y) - F(x) \geq -n(y-x)$ (see 1), Remark 1.2.1, (ii) and 2) of the proof of Lemma 4.8.2).

3) If $x \in E \cap P_+ \cap (c,d)$, $y \in Q$, $x < y$ then $F(y) - F(x) \geq -n(y-x)$. Indeed, let $G(x) = F(x) + nx$. Suppose on the contrary that $G(x) > G(y)$. Let $\epsilon < \min\{(y-x)/2, (G(x)-G(y))/2\}$. Since $x \in E$, it follows that $\{z \in (x, x+\epsilon) : G(z) > G(x)-\epsilon\}$ is uncountable. We have two possibilities:

a) There exists $z \in (x, x+\epsilon) \cap P \cap P_o$ such that $G(z) > G(x) - \epsilon$, where $P_o = \{x \in P : x$ is a bilateral accumulation point of $P\}$. Then by 2), $G(x) - \epsilon < G(z) \leq G(y)$, hence $0 < G(x) - G(y) < \epsilon < (G(x) - G(y))/2$. This is a contradiction.

b) There exists $z \in (a_i, b_i) \subset (x, x+\epsilon)$ for some i, such that $G(z) > G(x) - \epsilon$. Since F is increasing on $[a_i, b_i]$, it follows that G is strictly increasing on $[a_i, b_i]$, hence $G(x) - \epsilon < G(z) < G(u)$ for each $u \in [z, b_i]$. Let $t \in A_{nj}$ such that $t > b_i$, $m(t-b_i) < b_i - z$, $t - b_i < \min\{\delta(b_i), \delta(t)\}$ and let $v \in \sigma_t \cap \sigma_{b_i} \cap (b_i - m(t-b_i), b_i) \subset (z, b_i)$ (see condition E.I.C.[m]). Then by 2), $G(x) - \epsilon < G(v) < G(y)$, a contradiction.

It follows that $G(x) \leq G(y)$, a contradiction (see Lemma 4.8.1). Thus $P = \emptyset$ and $U = (a,b)$. By the Heine-Borel Covering Theorem, F is increasing on (a,b). Since $F \in uCM$ on $[a,b]$, it follows that F is increasing on $[a,b]$.

Theorem 4.8.1 *Let* $\mathcal{S}$ *be a local system satisfying intersection conditions (I.C.) and E.I.C.[m], and let* $F : [a,b] \mapsto \mathbb{R}$. *Let* $A = \{x : S\,\underline{D}F(x) > -\infty\}$ *and* $B = \{x : S\,\underline{D}F(x) = -\infty$ *and* $S\,\overline{D}F(x) < 0\}$. *Suppose that:*

(i) $F \in uCM$ *on* $[a,b]$;

(ii) $E = [a,b] \setminus (A \cup B)$ *is at most countable;*

(iii) for each $x \in E$ there exists a set $E_x \in \mathcal{S}(x)$ such that $\limsup_{y \nearrow x, y \in E_x} F(y) \leq F(x) \leq \liminf_{y \searrow x, y \in E_x} F(y)$;

(iv) $\mathcal{S}\underline{D}F(x) \geq 0$ *a.e. on* $[a,b]$.

Then F is increasing on $[a,b]$.

Proof. Let P and $U = \cup_{n\geq 1}(a_n, b_n)$ be defined as in Lemma 4.8.2. Suppose on the contrary that $P \neq \emptyset$. Let $f : A \cup B \mapsto \mathbb{R}$ be a finite function such that $-\infty < f(x) < \mathcal{S}\underline{D}F(x)$ for $x \in A$, and $\mathcal{S}\overline{D}F(x) < f(x) < 0$ for $x \in B$. Let $\sigma_x = \{y : y = x$ or $(F(y) - F(x))/(y - x) > f(x)\} \in \mathcal{S}(x)$ for $x \in A$, $\sigma_x = E_x$ for $x \in E$, and $\sigma_x = \{y : y = x$ or $(F(y) - F(x))/(y - x) < f(x)\} \in \mathcal{S}(x)$ for $x \in B$. Let $\delta(x)$, $x \in [a,b]$ be a positive function such that $\sigma_x \cap \sigma_y \cap [x,y] \neq \emptyset$, $\sigma_x \cap \sigma_y \cap (y, y + m(y - x)) \neq \emptyset$ and $\sigma_x \cap \sigma_y \cap (x - m(y - x), x) \neq \emptyset$, whenever $0 < y - x < \min\{\delta(x), \delta(y)\}$. Let $A_n = \{x \in A : f(x) > -n\}$ and $B_n = \{x \in B : f(x) < -1/n\}$, $n = \overline{1,\infty}$. Let $\{A_{nj}\}_j$ be a δ-decomposition of A_n, and let $\{B_{nj}\}_j$ be a δ-decomposition of B_n. Since $P \subset \cup_{n,j}(A_{nj} \cup B_{nj} \cup E)$, by Theorem 1.7.1, there exists an open interval (c,d) such that $(c,d) \cap P \neq \emptyset$ and either
a) $Q = (c,d) \cap A_{nj}$ is dense in $(c,d) \cap P$ for some n and j, and then it will follow that $F \in \underline{L}(Q \wedge (Q \cup Q_-)) \cap \underline{L}((Q \cup Q_+) \wedge Q)$ with the constant $-n$, or
b) $Q = (c,d) \cap B_{nj}$ is dense in $(c,d) \cap P$ for some n and j, and then it will follow that $F \in \overline{L}((Q \cup Q_-) \wedge Q) \cap \overline{L}(Q \wedge (Q \cup Q_+))$ with the constant $-1/n$.
a) We show for example that $F \in \underline{L}((Q \cup Q_+) \wedge Q)$. We have four situations:
1) If $x < y$, $x, y \in Q$ then $F(y) - F(x) \geq -n(y - x)$ (see Remark 1.2.1, (i) and condition (I.C.)).
2) If $x < y$, $x \in A \cap P_+ \cap (c,d)$, $y \in Q$ then $F(y) - F(x) > -n(y - x)$. Indeed, let $x_k \in (x, x+\delta(x)) \cap A_{nj}$, $x_k \searrow x$, and let $z_k \in \sigma_x \cap \sigma_{x_k} \cap [x, x_k] \neq \emptyset$ (see Remark 1.2.1,(ii) and condition (I.C.)). Then $F(z_k) - F(x) > f(x)(z_k - x)$, $F(x_k) - F(z_k) > -n(x_k - z_k)$ and by 1), $F(y) - F(x_k) > -n(y - x_k)$. Hence $F(y) - F(x) > -n(y - z_k) + f(x)(z_k - x)$. If $k \to \infty$ then $F(y) - F(x) \geq -n(y - x)$.
3) If $x < y$, $x \in B \cap P_+ \cap (c,d)$, $y \in Q$ then $F(y) - F(x) \geq -n(y - x)$. Indeed, let $x_k \in (x, x + \delta(x)) \cap A_{nj} \cap (c,d)$, $x_k \searrow x$, and let $z_k \in \sigma_x \cap \sigma_{x_k} \cap (x - m(x_k - x), x)$ (see Remark 1.2.1, (ii) and condition E.I.C.[m]). Then $F(z_k) - F(x) \geq -f(x)(x - z_k)$, $F(x_k) - F(z_k) \geq -n(x_k - z_k)$ and by 1), $F(y) - F(x_k) > -n(y - x_k)$. Hence $F(y) - F(x) \geq -n(y - z_k) - f(x)(x - z_k)$. If $k \to \infty$ then $z_k \searrow x$, hence $F(y) - F(x) \geq -n(y - x)$.
4) If $x < y$, $x \in E \cap P_+ \cap (c,d)$, $y \in Q$ then $F(y) - F(x) \geq -n(y - x)$. Indeed, let $x_k \in (x, x + \delta(x)) \cap A_{nj} \cap (c,d)$, $x_k \searrow x$, and let $z_k \in E_x \cap \sigma_{x_k} \cap [x, x_k] \neq \emptyset$ (see Remark 1.2.1, (ii) and condition (I.C.)). Then $F(x_k) - F(z_k) \geq -n(x_k - z_k)$ and by 1) $F(y) - F(x_k) \geq -n(y - x_k)$. Since $F(x) \leq \liminf_{k\to\infty} F(z_k)$, it follows that $F(y) - F(x) \geq -n(y - x)$.
b) We show for example that $F \in \overline{L}((Q \cup Q_-) \wedge Q)$. Let $K_o = \{x \in Q : x$ is a bilateral accumulation point for $Q\}$. We have four situations:
1) If $x < y$, $x, y \in Q$ then $F(y) - F(x) < (-1/n)(y - x)$ (see Remark 1.2.1, (i) and condition (I.C.)).
2) If $x < y$, $x \in A \cap P_- \cap (c,d)$, $y \in Q$ then $F(y) - F(x) \leq (-1/n)(y - x)$. Indeed, let $x_k \in B_{nj}$, $x_k \nearrow x$, $x \in (x - \delta(x), x)$ and let $z_k \in \sigma_{x_k} \cap \sigma_x \cap (x_k - m(x - x_k), x_k)$. If $k \to \infty$ then $z_k \nearrow x$, $z_k \in \sigma_x$. It follows that $F(x_k) - F(z_k) \leq (-1/n)(x_k - z_k)$, $F(y) - F(x_k) < (-1/n)(y - x_k)$and $F(x) - F(z_k) > f(x)(x - z_k)$. Therefore we have

$F(y) - F(z_k) + F(z_k) - F(x) < (-1/n)(y - z_k) + f(x)(z_k - x)$. If $k \to \infty$ then $F(y) - F(x) \leq (-1/n)(y - x)$.
3) Since F is increasing on each $[a_n, b_n]$ and $\mathcal{S}$ is bilateral, it follows that $B \cap P \cap (c,d) \subset K_o$. If $x < y$, $x \in B \cap (c,d)$, $y \in Q$ then $F(y) - F(x) \leq (-1/n)(y - x)$. Indeed, let $x_k \in (x, x + \delta(x)) \cap B_{nj}$, $x_k \searrow x$, and let $z_k \in \sigma_x \cap \sigma_{x_k} \cap [c, x_k] \neq \emptyset$ (see Remark 1.2.1, (ii) and condition (I.C.)). Then $F(z_k) - F(x_k) < f(x)(z_k - x)$, $F(x_k) - F(z_k) < (-1/n)(x_k - z_k)$, and by 1), $F(y) - F(x_k) < (-1/n)(y - x_k)$. It follows that $F(y)-F(x) < (-1/n)(y-x)+f(x)(z_k-x)$. If $k \to \infty$ then $F(y)-F(x) \leq (-1/n)(y - x)$.
4) If $x < y$, $x \in E \cap P_- \cap (c,d)$ and $y \in Q$ then $F(y) - F(x) < (-1/n)(y - x)$. Indeed, let $x_k \in (x - \delta(x), x) \cap B_{nj}$ such that $x_k \nearrow x$, and suppose that $z_k \in \sigma_{x_k} \cap E_k \cap (x_k - m(x - x_k), x)$. Then $z_k \nearrow x$, $z_k \in E_x$, $F(x_k) - F(z_k) < (-1/n)(x_k - z_k)$, $F(y) - F(x_k) < (-1/n)(y - x_k)$ and $\limsup_{k\to\infty} F(z_k) \leq F(x)$. Hence $F(y) - F(x) \leq \limsup_{k\to\infty}(F(y) - F(z_k)) \leq \limsup_{k\to\infty}(-1/n)(y - z_k) \leq (-1/n)(y - x)$, a contradiction (see Lemma 4.8.1).
It follows that $P = \emptyset$, hence $U = (a,b)$. By the Heine- Borel Covering Theorem, F is increasing on (a,b). Since $F \in uCM$ on $[a,b]$, F is increasing on $[a,b]$.

Theorem 4.8.2 *Let $\mathcal{S}$ be a bilateral c-dense local system satisfying intersection conditions (I.C.) and E.I.C.[m]. Let $F : [a,b] \mapsto \mathbb{R}$, $F \in uCM$, and let $E \subset [a,b]$, such that $\mathcal{S}\overline{D}F(x) < 0$ whenever $\mathcal{S}\underline{D}F(x) = -\infty$ and $x \in [a,b] \setminus E$. Suppose that:*

(i) $\mathcal{S}\underline{D}F(x) \geq 0$ a.e. on $[a,b]$;

(ii) E is countable;

(iii) $F \in \overline{\mathcal{B}}_1$ on $\overline{E}$;

(iv) For each $x \in E$ and $\epsilon > 0$, the sets $\{z \in (x - \epsilon, x) : F(z) < F(x) + \epsilon\}$ and $\{z \in (x, x + \epsilon) : F(z) > F(x) - \epsilon\}$ are uncountable.

Then F is increasing on $[a,b]$.

Proof. Let $A = \{x : \mathcal{S}\underline{D}F(x) > -\infty\}$ and $B = \{x : \mathcal{S}\underline{D}F(x) = -\infty$ and $\mathcal{S}\overline{D}F(x) < 0\}$. Then $[a,b] = A \cup B \cup E$ and

(1) *for each $x \in A$ and $\epsilon > 0$, the sets $\{z \in (x, x + \epsilon) : F(z) > F(x) - \epsilon\}$ and*

$\{z \in (x - \epsilon, x) : F(z) < F(x) + \epsilon\}$ are uncountable.

We prove (1) Let $x \in A$, $\epsilon > 0$ and let p be a positive integer such that $\mathcal{S}\underline{D}F(x) > -p$. Then $\sigma_x = \{y : y = x$ or $(F(y) - F(x))/(y - x) > -p\}$ is bilaterally c-dense in itself. If $z \in [x, x + \epsilon/p) \cap \sigma_x$ then $F(z) > F(x) - p(z - x) > F(x) - p\epsilon/p + F(x) - \epsilon$. Similarly, if $z \in (x - \epsilon/p, x] \cap \sigma_x$ then $F(z) < F(x) + p(x - z) < F(x) + \epsilon$. Therefore we have (1).
Let P and $U = \cup_{n\geq 1}(a_n, b_n)$ be the sets defined in Lemma 4.8.2. Suppose on the contrary that $P \neq \emptyset$. By Theorem 4.8.1, F is increasing on each component interval of $(a,b) \setminus \overline{E}$, hence $\overline{E} \supset P$. But $E \subset P$, hence $P = \overline{E}$. By (iii), $F \in \overline{\mathcal{B}}_1$ on P. Let $P_o = P \setminus (\cup_{n\geq 1}\{a_n, b_n\} \cup E)$. Since $F \in uCM$ on $[a,b]$, F is increasing on each $[a_n, b_n]$, hence P is a perfect subset of $[a,b]$. We show that:

(2) *if $x \in P_+ \cap E$ (resp. $x \in P_- \cap E$) and $\epsilon > 0$ then $\{z \in (x, x + \epsilon) \cap P_o : F(z)$*

$> F(x) - \epsilon\}$ *(resp.* $\{z \in (x-\epsilon, x) \cap P_o : F(z) < F(x) + \epsilon\}$*) is nonempty.*

Suppose on the contrary that there exists $x_o \in P_+ \cap E$ and $\epsilon_o > 0$, such that $A_o = \{z \in (x_o, x_o + \epsilon_o) \cap P_o : F(z) > F(x_o) - \epsilon_o\}$ is empty. Let $B_o = \{b_k \in (x_o, x_o + \epsilon_o) : F(b_k) > F(x_o) - \epsilon_o/2\}$. For $x \in (x_o, x_o + \epsilon_o)$ let $A_x = \{n : (a_n, b_n) \subset (x_o, x)\}$. Then A_x is infinite (see (iv) and the fact that A_o is empty)). We show that B_o is nonempty and contains no isolated points. Let $\epsilon < \epsilon_o/2$. By (iv), since A_o is empty, it follows that there exists $z \in (a_k, b_k) \subset (x_o, x_o + \epsilon) \subset (x_o, x_o + \epsilon_o)$ for some positive integer $k \in A_{\epsilon + x_o}$ such that $F(z) > F(x_o) - \epsilon$. Since F is increasing on $[a_k, b_k]$ it follows that $F(b_k) \geq F(z) > F(x_o) - \epsilon > F(x_o) - \epsilon_o/2$. Hence $b_k \in B_o$, so B_o is nonempty. Suppose on the contrary that B_o contains an isolated point b_n. Then there exists $0 < \delta < \min\{x_o + \epsilon_o - b_n, F(b_n) - F(x_o) + \epsilon_o/2\}$ such that $(b_n, b_n + \delta) \cap \{z : F(z) > F(x_o) - \epsilon_o/2\} \cap (\cup_{i \geq 1}[a_i, b_i]) \neq \emptyset$. Since $A_o = \emptyset$ we obtain that $(b_n, b_n + \delta) \cap \{z : F(z) > F(x_o) - \epsilon_o/2\} \cap P_o = \emptyset$. Therefore the set $(b_n, b_n + \delta) \cap \{z : F(z) > F(b_n) - \delta\}$ is at most countable (because $F(b_n) - \delta > F(x_o) - \epsilon_o/2$). But this contradicts (1). Since $\mathcal{S}$ is bilateral and F is increasing on $[a_n, b_n]$, it follows that $b_n \in A \cup E$. Hence $\overline{B}_o$ is a nonempty perfect subset of P. Since $F \in \overline{\mathcal{B}}_1$ on P, there exists a sequence $\{Q_n\}_n$ of closed subsets of P, such that $\{x \in \overline{B}_o : F(x) < F(x_o) - \epsilon_o/2\} = \cup_{n \geq 1} Q_n$. Since $A_o = \emptyset$, the set $C = \{x \in \overline{B}_o : F(x) > F(x_o) - \epsilon_o/2\} \subset E \cup (\cup_{n \geq 1}\{a_n, b_n\})$ is countable. Since $\overline{B}_o = C \cup (\cup_{n \geq 1} Q_n)$, by Theorem 1.7.1, there exists an open interval (c, d) such that $\emptyset \neq (c, d) \cap \overline{B}_o \subset Q_n$ for some positive integer n. Let $b_j \in (a, b) \cap B_o$.Then $F(b_j) > F(x_o) - \epsilon_o/2$, a contradiction. It follows that A_o is nonempty, and we have (2).

Let $f : A \cup B \mapsto \mathbb{R}$ such that $-\infty < f(x) < \mathcal{S}\,\underline{D}F(x)$ for $x \in A$, and $\mathcal{S}\,\overline{D}F(x) < f(x) < 0$ for $x \in B$. Let $\sigma_x = \{y : y = x \text{ or } (F(y) - F(x))/(y - x) > f(x)\} \in \mathcal{S}(x)$ for $x \in A$, $\sigma_x = \{y : y = x \text{ or } (F(y) - F(x))/(y - x) < f(x)\} \in \mathcal{S}(x)$ for $x \in B$, and let $\sigma_x \in \mathcal{S}(x)$ be a fixed set for $x \in E$. Let $\delta(x)$, $x \in [a, b]$ be a positive function such that $\sigma_x \cap \sigma_y \cap [x, y] \neq \emptyset$, $\sigma_x \cap \sigma_y \cap (y, y + m(y - x)) \neq \emptyset$ and $\sigma_x \cap \sigma_y \cap (x - m(y - x), x) \neq \emptyset$, whenever $y < x$, $0 < y - x < \min\{\delta(x), \delta(y)\}$. Let $A_n = \{x \in A : f(x) > -n\}$ and $B_n = \{x \in B : f(x) < -1/n\}$, $n = \overline{1, \infty}$. Let $\{A_{nj}\}_j$ be a δ-decomposition of A_n and let $\{B_{nj}\}_j$ be a δ-decomposition of B_n. By Theorem 1.7.1, there exists $(c, d) \cap P \neq \emptyset$ such that either a) $Q = A_{nj} \cap (c, d)$ is dense in $(c, d) \cap P$ for some n and j, or b) $Q = B_{nj} \cap (c, d)$ is dense in $(c, d) \cap P$ for some n and j.

a) Then $F \in \underline{L}(Q \wedge (Q \cup Q_-)) \cap \underline{L}((Q \cup Q_+) \wedge Q)$ with the constant $-n$. We show for example that $F \in \underline{L}((Q \cup Q_+) \wedge Q)$.

1) If $x < y$, $x, y \in Q$ then $F(y) - F(x) \geq -n(y - x)$.

2) If $x < y$, $x \in A \cap P_+ \cap (c, d)$, $y \in Q$ then $F(y) - F(x) \geq -n(y - x)$.

3) If $x < y$, $x \in B \cap P_+ \cap (c, d)$, $y \in Q$ then $F(y) - F(x) \geq -n(y - x)$.

(The proofs of 1), 2), 3) follow similarly to a) 1), 2), 3) of the proof of Theorem 4.8.1.)

4) If $x < y$, $x \in E \cap P_+ \cap (c, d)$, $y \in Q$ then $F(y) - F(x) \geq -n(y - x)$. Indeed, let $\epsilon > 0$, $x + \epsilon < b$. By (2), it follows that there exists $z \in (x, x + \epsilon)$ such that $F(z) > F(x) - \epsilon$. By 2) and 3), $F(y) - F(z) \geq -n(y - z)$, hence $F(y) - F(x) + \epsilon > F(y) - F(z) \geq -n(y - x) - n(x - z)$. Since $|x - z| < \epsilon$ and ϵ is arbitrary, it follows that $F(y) - F(x) \geq -n(y - x)$.

b) Then $F \in \overline{L}((Q \cup Q_-) \wedge Q) \cap \overline{L}(Q \wedge (Q \cup Q_+))$ with the constant $-1/n$. We show for example that $F \in \overline{L}((Q \cup Q_-) \wedge Q)$.

1) If $x < y$, $x, y \in Q$ then $F(y) - F(x) \leq (-1/n)(y - x)$.

2) If $x < y$, $x \in A \cap P_- \cap (c,d)$, $y \in Q$ then $F(y) - F(x) \leq (-1/n)(y-x)$.
3) If $x < y$, $x \in B \cap P_- \cap (c,d)$, $y \in Q$ then $F(y) - F(x) \leq (-1/n)(y-x)$.
(The proofs of 1), 2), 3) follow similarly to b) 1), 2), 3) of the proof of Theorem 4.8.1.)
4) If $x < y$, $x \in E \cap P_- \cap (c,d)$, $y \in Q$ then $F(y) - F(x) \leq (-1/n)(y-x)$. Indeed, let $\epsilon > 0$, $x - \epsilon > a$. By (2), it follows that there exists $z \in (x-\epsilon, x)$ such that $F(z) < F(x) + \epsilon$. By 2) and 3), $F(y) - F(z) \leq (-1/n)(y-z)$, hence $F(y) - F(x) - \epsilon < F(y) - F(z) < (-1/n)(y - x + x - z)$. Since $| x - z | < \epsilon$ and ϵ is arbitrary, it follows that $F(y) - F(x) \leq (-1/n)(y-x)$. By Lemma 4.8.1, we obtain a contradiction.
Hence $P = \emptyset$, so $U = (a,b)$. By the Heine-Borel Covering Theorem, F is increasing on (a,b). Since $F \in uCM$ on $[a,b]$, F is increasing on $[a,b]$.

Remark 4.8.1 *(i) In [Pr1], Preiss proved the following theorem:*
Let $F : [a,b] \mapsto \mathbb{R}$, $F \in$ lower internal. Let $E \subset [a,b]$ such that, if $x \in [a,b] \setminus E$ and $\underline{F}'_{ap}(x) = -\infty$ then $F'_{ap}(x) = -\infty$. If $\underline{F}'_{ap}(x) \geq 0$ a.e. on $[a,b]$, E is countable, $F \in \mathcal{B}_1$ on $\overline{E}$, and if the sets $\{z \in (x-\epsilon, x) : F(z) < F(x) + \epsilon\}$ and $\{z \in (x, x+\epsilon) : F(z) > F(x) - \epsilon\}$ are uncountable whenever $x \in E$ and $\epsilon > 0$ then F is increasing on $[a,b]$.*

Since lower internal $\subset uCM$ and by Remark 2.2.1, (ii), it follows that Theorem 4.8.2 improves Preiss' theorem. However, we do not know if Preiss' theorem or Theorem 4.8.2 follow by Corollary 4.1.1.*

(ii) In Theorem 4.8.2 we cannot give up condition "$F \in \overline{\mathcal{B}}_1$ on $\overline{E}$" (see Section 6.19).

4.9 Monotonicity in terms of extreme derivatives

Theorem 4.9.1 *([S1], p. 203 and [Br2], p. 189). Let $F : [a,b] \mapsto \mathbb{R}$ and let $E = \{x : \overline{D}^+ F(x) \leq 0\}$. If $F \in \underline{Z}_i$ on $[a,b]$ and $int(F(E)) = \emptyset$ then F is increasing on $[a,b]$.*

Corollary 4.9.1 *([S1], p. 204 and [Br2], p. 189). Let $F : [a,b] \mapsto \mathbb{R}$. If $F \in \underline{Z}_i$ on $[a,b]$ and $\overline{D}^+ F(x) \geq 0$ n.e. then F is increasing on $[a,b]$.*

Corollary 4.9.2 *([Br2], p. 189). Let $F : [a,b] \mapsto \mathbb{R}$. If $F \in \underline{Z}_i$ on $[a,b]$ and $\overline{D}^+ F(x) \geq 0$ a.e. and $\overline{D}^+ F(x) \neq -\infty$, n.e. then F is increasing on $[a,b]$.*

Corollary 4.9.3 *Let $F : [a,b] \mapsto \mathbb{R}$, $F \in \mathcal{C}_i$. If $\overline{D}^+ F(x) \geq 0$ a.e. and $\overline{D}^+ F(x) \neq -\infty$ n.e. then F is increasing on $[a,b]$.*

Remark 4.9.1 *(i) In [O1], O'Malley proved the following result (using O'Malley's Theorem 1.14.1):*
Let $f : [a,b] \mapsto \mathbb{R}$ be a function satisfying the following conditions:

(a) $F \in \mathcal{B}_1$ on $[a,b]$;

(b) $ap \limsup_{x \nearrow x_o} f(x) \leq f(x_o) \leq ap \limsup_{x \searrow x_o} f(x)$, for every $x_o \in [a,b]$;

(c) $int(f(\{x : D^+_{ap} f(x) \leq 0\})) = \emptyset$.

Then F is increasing on $[a,b]$.

(ii) *In [MPZ], Maly, Preiss and Zajícek obtained the following main result: Suppose that $f : [a,b] \mapsto \mathbb{R}$ and that $\alpha \in (0,1)$. Suppose further that there is a set $S \subset (a,b)$ such that the following conditions hold:*

(A) *For every $x \in (a,b) \setminus S$,*

(A_1) *there is $y \in (x,b]$ such that $|\{z \in (x,y) : f(z) > f(x)\}| \geq \alpha(y-x)$, or*

(A_2) *there is $y \in [a,x)$ such that $|\{z \in (y,x) : f(z) > f(x)\}| < \alpha(x-y)$.*

(B) *For every $x \in [a,b)$ and every $c < f(x)$, there is $y \in (x,b]$ such that $|\{z \in (x,y) : f(z) > c\}| \geq \alpha(y-x)$.*

(C) *For every $x \in (a,b]$ and every $d > f(x)$, there is $y \in [a,x)$ such that $|\{z \in (y,x) : f(z) > d\}| < \alpha(x-y)$.*

(D) *$f(S)$ contains no interval.*

Then $f(b) \geq f(a)$.

This result has many applications (see [MPZ]) to monotonicity theorems involving preponderant or symmetrical derivatives, and to one-side densities of linear sets. From the later we mention the follow proposition which is similar to O'Malley's Theorem 1.14.1:
Let $\beta \in (0,1)$ and let $\emptyset \neq E \subset (a,b)$ be a measurable set which is not connected. Let $d^+(E,x) > \beta$ for any $x \in E$. Then there exists a point $z \in (a,b) \setminus E$ for which $d_-(E,z) \geq \beta$.

Chapter 5

Integrals

5.1 Descriptive and Perron type definitions for the Lebesgue integral

Definition 5.1.1 *Let $f : [a, b] \mapsto \overline{\mathbb{R}}$. We define the following classes of majorants:*

- $\overline{\mathcal{M}_1^\#}(f) = \{M : [a,b] \mapsto \mathbb{R} : M(a) = 0;\ M \in AC$ *and* $M'(x)$ *exists in* $\overline{\mathbb{R}}$ *everywhere;* $-\infty \neq \underline{D}^\# M(x) \geq f(x),\ x \in [a,b]\}$;
- *(Saks).* $\overline{\mathcal{M}_2^\#}(f) = \{M : [a,b] \mapsto \mathbb{R} : M(a) = 0;\ M \in AC;\ \underline{D}M(x) \geq f(x)$ *everywhere*$\}$;
- *(Saks).* $\overline{\mathcal{M}_3^\#}(f) = \{M : [a,b] \mapsto \mathbb{R} : M(a) = 0;\ M \in VB;\ -\infty \neq \underline{D}M(x) \geq f(x)$ *everywhere*$\}$;
- $\overline{\mathcal{M}_4^\#}(f) = \{M : [a,b] \mapsto \mathbb{R} : M(a) = 0;\ M \in AC;\ -\infty \neq \underline{D}^\# M(x) \geq f(x)$ *everywhere*$\}$;
- $\overline{\mathcal{M}_5^\#}(f) = \{M : [a,b] \mapsto \mathbb{R} : M(a) = 0;\ -\infty \neq \underline{D}^\# M(x) \geq f(x)$ *everywhere*$\}$;
- $\overline{\mathcal{M}_6^\#}(f) = \{M : [a,b] \mapsto \mathbb{R} : M(a) = 0;\ \underline{D}M(x)$ *is lower bounded;* $\underline{D}M(x) \geq f(x)$ *everywhere*$\}$;
- $\overline{\mathcal{M}_7^\#}(f) = \{M : [a,b] \mapsto \mathbb{R} : M(a) = 0;\ \underline{D}M(x) > -\infty$ *everywhere;* $\underline{D}M(x)$ *is lower bounded a.e.;* $\underline{D}M(x) \geq f(x)$ *a.e.*$\}$;
- $\overline{\mathcal{M}_8^\#}(f) = \{M : [a,b] \mapsto \mathbb{R} : M(a) = 0;\ M \in \underline{L};\ M'(x) \geq f(x)$ *a.e.*$\}$;
- $\overline{\mathcal{M}_9^\#}(f) = \{M : [a,b] \mapsto \mathbb{R} : M(a) = 0;\ M \in \underline{AC} \cap C;\ M'(x) \geq f(x)$ *a.e.*$\}$;
- $\overline{\mathcal{M}_{10}^\#}(f) = \{M : [a,b] \mapsto \mathbb{R} : M(a) = 0;\ M \in \underline{AC};\ M'(x) \geq f(x)$ *a.e.*$\}$;
- $\overline{\mathcal{M}_{11}^\#}(f) = \{M : [a,b] \mapsto \mathbb{R} : M(a) = 0;\ M \in \mathcal{D}_{-}\mathcal{B}_1 T_2 \cap N^{-\infty};\ M'(x) \geq f(x)$ *a.e. where* $M'(x)$ *exists*$\}$;
- $\overline{\mathcal{M}_{12}^\#}(f) = \{M : [a,b] \mapsto \mathbb{R} : M(a) = 0;\ M \in (-) \cap N^{-\infty};\ M'(x) \geq f(x)$ *a.e. where* $M'(x)$ *exists*$\}$;

- $\overline{\mathcal{M}}^{\#}_{13}(f) = \{M : [a,b] \mapsto \mathbb{R} : M(a) = 0;\ M \in [\mathcal{C}_i G] \cap \mathcal{D}_-$ *on* $[a,b]$; $M \in \underline{M}$ *on each closed interval contained in* $int\{x : M \in \mathcal{C}_i$ *at* $x\}$; $M'(x) \geq f(x)$ *a.e. where* $M'(x)$ *exists*$\}$;

We define the following classes of minorants: $\underline{\mathcal{M}}^{\#}_k(f) = \{m : [a,b] \mapsto \mathbb{R} : -m \in \overline{\mathcal{M}}^{\#}_k(-f)\}$, $k = \overline{1,13}$.
Let $k \in \{1,2,\ldots,13\}$. *If* $\overline{\mathcal{M}}^{\#}_k(f) \neq \emptyset$ *then we denote by* $\overline{I}_k(b)$ *the lower bound of all* $M(b)$, $M \in \overline{\mathcal{M}}^{\#}_k(f)$. *If* $\underline{\mathcal{M}}^{\#}_k(f) \neq \emptyset$ *then we denote by* $\underline{I}_k(b)$ *the upper bound of all* $m(b)$, $m \in \underline{\mathcal{M}}^{\#}_k(f)$.
Let $(j,k) \in \{1,2,\ldots,10\} \times \{1,2,\ldots,13\}$. *We say that f has a* $(\mathcal{P}^{\#}_{j,k})$*-integral on* $[a,b]$ *if* $\overline{\mathcal{M}}^{\#}_j \times \underline{\mathcal{M}}^{\#}_k(f) \neq \emptyset$ *and* $\overline{I}_j(b) = \underline{I}_k(b) = (\mathcal{P}^{\#}_{j,k}) \int_a^b f(t)dt$.

Lemma 5.1.1 *Let* $f : [a,b] \mapsto \overline{\mathbb{R}}$.

(i) If $\overline{\mathcal{M}}^{\#}_1(f) \neq \emptyset$ *then* $\overline{\mathcal{M}}^{\#}_1(f) \subseteq \overline{\mathcal{M}}^{\#}_k(f) \subseteq \overline{\mathcal{M}}^{\#}_{10}(f)$, $k = \overline{2,9}$.

(ii) If $\overline{\mathcal{M}}^{\#}_{10}(f) \neq \emptyset$ *then* $\overline{\mathcal{M}}^{\#}_{10}(f) \subseteq \overline{\mathcal{M}}^{\#}_{13}(f)$.

(iii) If $\overline{\mathcal{M}}^{\#}_{10}(f) \neq \emptyset$ *then* $\overline{\mathcal{M}}^{\#}_{10}(f) \subseteq \overline{\mathcal{M}}^{\#}_{11}(f) \subseteq \overline{\mathcal{M}}^{\#}_{12}(f)$.

Proof. (i) We show that $\overline{\mathcal{M}}^{\#}_3(f) \subseteq \overline{\mathcal{M}}^{\#}_{10}(f)$. Let $M \in \overline{\mathcal{M}}^{\#}_3(f)$. Since $-\infty \neq \underline{D}M(x) \geq f(x)$, it follows that $M \in \mathcal{C}_i \cap N^{-\infty}$. Since $F \in VB$, by Corollary 2.21.1, (i), $M \in N^{-\infty} \cap \mathcal{C}_i \cap VB = \underline{AC}$ on $[a,b]$. By Theorem 2.14.5, M is derivable *a.e.* on $[a,b]$, hence $M \in \overline{\mathcal{M}}^{\#}_{10}(f)$.
We show that $\overline{\mathcal{M}}^{\#}_5(f) \subset \overline{\mathcal{M}}^{\#}_{10}(f)$. Let $M \in \overline{\mathcal{M}}^{\#}_5(f)$. Since $-\infty \neq \underline{D}^{\#}M(x) \geq f(x)$, by Theorem 2.13.1, $M \in \underline{L}$ on $[a,b]$. By Proposition 2.13.1, (ii), $M \in \underline{AC}$ on $[a,b]$, hence $M \in \overline{\mathcal{M}}^{\#}_{10}(f)$.
We show that $\overline{\mathcal{M}}^{\#}_1(f) \subseteq \overline{\mathcal{M}}^{\#}_6(f)$. Let $M \in \overline{\mathcal{M}}^{\#}_6(f)$. Since $-\infty \neq \underline{D}^{\#}M(x) \geq f(x)$, by Theorem 2.13.1, $M \in \underline{L}$ on $[a,b]$ with a constant λ. Then $\underline{D}M(x) \geq \underline{D}^{\#}M(x) \geq \lambda$, hence $M \in \overline{\mathcal{M}}^{\#}_6(f)$.
We show that $\overline{\mathcal{M}}^{\#}_7(f) \subseteq \overline{\mathcal{M}}^{\#}_8(f)$. Let $M \in \overline{\mathcal{M}}^{\#}_7(f)$. Since $\underline{D}M(x)$ is lower bounded *a.e.* on $[a,b]$, it follows that there exists a real constant λ such that $\underline{D}M(x) \geq \lambda$ *a.e.* on $[a,b]$. Since $\underline{D}M(x) > -\infty$ on $[a,b]$, by Corollary 4.4.1, (i), $M(x) - \lambda x$ is increasing on $[a,b]$, hence $M(y) - M(x) \geq \lambda(y-x)$, whenever $a \leq x < y \leq b$. This implies that $M \in \underline{L}$ on $[a,b]$. Thus $M \in \overline{\mathcal{M}}^{\#}_8(f)$.
The other parts follow by definitions.
(ii) and (iii) are evident.

Lemma 5.1.2 *Let* $f : [a,b] \mapsto \overline{\mathbb{R}}$. *If* $(M,m) \in \overline{\mathcal{M}}^{\#}_j(f) \times \underline{\mathcal{M}}^{\#}_k(f) \neq \emptyset$ *for some* $(j,k) \in \{1,2,\ldots,10\} \times \{1,2,\ldots,13\}$ *then* $M - m$ *is increasing on* $[a,b]$ *and* $m \in \overline{AC}$ *on* $[a,b]$, *hence* $M(b) \geq m(b)$.

Proof. Let $(j,k) \in \{1,2,\ldots,10\} \times \{1,2,\ldots,12\}$ such that $\overline{\mathcal{M}}^{\#}_j(f) \times \underline{\mathcal{M}}^{\#}_k(f) \neq \emptyset$. Let $(M,m) \in \overline{\mathcal{M}}^{\#}_j(f) \times \underline{\mathcal{M}}^{\#}_k(f) \subseteq \overline{\mathcal{M}}^{\#}_{10}(f) \times \underline{\mathcal{M}}^{\#}_{12}(f)$ (see Lemma 5.1.1). Hence $M - m \in \underline{AC} + (N^{-\infty} \cap (-))$ and $(M-m)'(x) \geq 0$ *a.e.* where $M - m$ is derivable. By Theorem 4.3.2, (i), $M - m$ is increasing on $[a,b]$. By Theorem 2.11.1, (vi), $M \in VB$ on $[a,b]$, hence $-m \in VB$ on $[a,b]$. By Corollary 2.21.1, (i), $-m \in \underline{AC}$ on $[a,b]$, hence

$m \in \overline{AC}$ on $[a,b]$.
Let $(j,k) \in \{1,2,\ldots,10\} \times \{13\}$ such that $\overline{\mathcal{M}}_j^{\#}(f) \times \underline{\mathcal{M}}_{13}^{\#}(f) \neq \emptyset$. Let $(M,m) \in \overline{\mathcal{M}}_j^{\#}(f) \times \underline{\mathcal{M}}_{13}^{\#}(f) \subseteq \overline{\mathcal{M}}_{10}^{\#}(f) \times \underline{\mathcal{M}}_{13}^{\#}(f)$ (see Lemma 5.1.1). Hence $M - m \in \underline{AC} + (\mathcal{D}_- \cap [\mathcal{C}_i G]) \subset \mathcal{D}_- \cap [\mathcal{C}_i G]$ (see Theorem 2.11.1, (xx)) and Corollary 2.5.1, (iii)). But $M - m \in \underline{M}$ on each closed interval contained in $A = int\{x : m \in \mathcal{C}_i$ at $x\} = int\{x : M - m \in \mathcal{C}_i$ at $x\}$ (see Theorem 2.23.2, (vi)). Clearly $(M-m)'(x)$ exists at the same points x for which $m'(x)$ exists. Hence $(M-m)'(x) \geq 0$ *a.e.* where $m'(x)$ exists on A. By Corollary 4.2.3, $M - m$ is increasing on $[a,b]$. By Theorem 2.11.1, (vi), $M \in VB$ on $[a,b]$, hence $-m \in VB$ on $[a,b]$. By Theorem 2.10.3, $-m \in \mathcal{C}_i$ on $[a,b]$. Since $-m \in \overline{\mathcal{M}}_{13}^{\#}(f)$, it follows that $-m \in \underline{M}$ on each closed subinterval of (a,b), hence $-m \in [\underline{ACG}]$ on $[a,b]$. By Theorem 2.23.2, (vi), $-m \in \underline{M}$ on $[a,b]$, hence $-m \in \underline{AC}$ on $[a,b]$. It follows that $m \in \overline{AC}$ on $[a,b]$.

Lemma 5.1.3 *Let $f : [a,b] \mapsto \overline{\mathbb{R}}$. The following assertions are equivalent:*

(i) f is $(\mathcal{P}_{j,k}^{\#})$-integrable on $[a,b]$ for some $(j,k) \in \{1,2,\ldots,10\} \times \{1,2,\ldots,13\}$;

(ii) For every $\epsilon > 0$ there exist M and m (depending on ϵ), such that $(M,m) \in \overline{\mathcal{M}}_j^{\#}(f) \times \underline{\mathcal{M}}_k^{\#}(f) \neq \emptyset$ and $M(b) - m(b) < \epsilon$.

Proof. The proof follows by definitions and Lemma 5.1.2.

Lemma 5.1.4 *Let $f : [a,b] \mapsto \overline{\mathbb{R}}$. If $(M,m) \in \overline{\mathcal{M}}_j^{\#}(f) \times \underline{\mathcal{M}}_k^{\#}(f) \neq \emptyset$ for some $(j,k) \in \{1,2,\ldots,10\} \times \{1,2,\ldots,13\}$ then: (i) $M - m$, (ii) $M - \underline{I}_k$, (iii) $\overline{I}_j - m$ and (iv) $\overline{I}_j - \underline{I}_k$ are positive increasing functions on $[a,b]$, hence $M(b) \geq m(b)$, $M(b) \geq \underline{I}_k(b)$, $\overline{I}_j(b) \geq m(b)$ and $\overline{I}_j(b) \geq \underline{I}_k(b)$.*

Proof. (i) See Lemma 5.1.2.

(ii) By (i), $M(y) - M(x) \geq m(y) - \underline{I}_k(x)$, hence $M(y) - M(x) \geq \underline{I}_k(y) - \underline{I}_k(x)$ whenever $a \leq x < y \leq b$. It follows that $M(y) - \underline{I}_k(y) \geq M(x) - \underline{I}_k(x)$.

(iii) $M(y) - \overline{I}_j(x) \geq M(y) - M(x) \geq m(y) - m(x)$, hence $\overline{I}_j(y) - \underline{I}_k(x) \geq m(y) - m(x)$, whenever $a \leq x < y \leq b$.

(iv) $M(y) - \overline{I}_j(x) \geq M(y) - M(x) \geq \underline{I}_k(y) - \underline{I}_k(x)$, hence $\overline{I}_j(y) - \overline{I}_j(x) \geq \underline{I}_k(y) - \underline{I}_k(x)$, whenever $a \leq x < y \leq b$.

Theorem 5.1.1 *Let $f : [a,b] \mapsto \overline{\mathbb{R}}$. The following assertions are equivalent:*

(i) f is summable on $[a,b]$;

(ii) There exists $F : [a,b] \mapsto \mathbb{R}$ such that: a) $F \in AC$ on $[a,b]$; b) $F'(x) = f(x)$ a.e. on $[a,b]$;

(iii) There exists $F : [a,b] \mapsto \mathbb{R}$ such that: a) $F \in \mathcal{DB}_1 \cap (N)$ on $[a,b]$; b) $F'(x) = f(x)$ a.e. where F is derivable on $[a,b]$; c) f is summable on the set $N = \{x : F'(x) \leq 0\}$;

(iv) There exists $F : [a,b] \mapsto \mathbb{R}$ such that: a) $F \in \mathcal{DB}_1 T_2 \cap N^\infty$ on $[a,b]$; b) $F'(x) = f(x)$ a.e. where F is derivable on $[a,b]$; c) f is summable on the set $N = \{x : F'(x) \leq 0\}$;

(v) *There exists* $F : [a,b] \mapsto \mathbb{R}$ *such that: a)* $F \in (+) \cap (-) \cap N^\infty$ *on* $[a,b]$*; b)* $F'(x) = f(x)$ *a.e. where* F *is derivable on* $[a,b]$*; c)* f *is summable on the set* $N = \{x : F'(x) \leq 0\}$*;*

(vi) *There exists* $F : [a,b] \mapsto \mathbb{R}$ *such that: a)* $F \in \mathcal{D} \cap [CG]$ *on* $[a,b]$*; b)* $F \in (M)$ *on each closed interval contained in* $A = int\{x : F \in C \text{ at } x\}$*; c)* $F'(x) = f(x)$ *a.e. where* F *is derivable on* A*;* f *is summable on the set* $A \cap \{x : F'(x) < 0\}$*;*

(vii) *There exists* $F : [a,b] \mapsto \mathbb{R}$ *such that: a)* $F \in \mathcal{DB}_1 \cap (N)$ *on* $[a,b]$*; b)* $\overline{\mathcal{M}}^{\#}_{10}(f) \neq \emptyset$*; c)* $F'(x) = f(x)$ *a.e. where* F *is derivable on* $[a,b]$*;*

(viii) *There exists* $F : [a,b] \mapsto \mathbb{R}$ *such that: a)* $F \in \mathcal{DB}_1T_2 \cap N^\infty$ *on* $[a,b]$*; b)* $\overline{\mathcal{M}}^{\#}_{10}(f) \neq \emptyset$*; c)* $F'(x) = f(x)$ *a.e. where* F *is derivable on* $[a,b]$*;*

(ix) *There exists* $F : [a,b] \mapsto \mathbb{R}$ *such that: a)* $F \in (+) \cap (-) \cap N^\infty$ *on* $[a,b]$*; b)* $\overline{\mathcal{M}}^{\#}_{10}(f) \neq \emptyset$*; c)* $F'(x) = f(x)$ *a.e. where* F *is derivable on* $[a,b]$*;*

(x) *There exists* $F : [a,b] \mapsto \mathbb{R}$ *such that: a)* $F \in \mathcal{D} \cap [CG]$ *on* $[a,b]$*; b)* $F \in (M)$ *on each closed interval contained in* $A = int\{x : F \in C \text{ at } x\}$*; c)* $\overline{\mathcal{M}}^{\#}_{10}(f) \neq \emptyset$*; d)* $F'(x) = f(x)$ *a.e. where* F *is derivable on* A*.*

In all situations we have $(\mathcal{L}) \int_a^x f(t)dt = F(x) - F(a)$*.*

Proof. *(i)* ⇔ *(ii)* This follows by Theorem 2.14.2 and Corollary 2.14.2.

(ii) ⇒ *(iii)* (iii), c) follows by (ii) and Corollary 2.14.2. By Corollary 1.10.1, N is measurable. The other assertions follow easily.

(iii) ⇒ *(iv)* ⇒ *(v)* These follow by Theorem 2.18.9 and Lemma 2.21.1.

(v) ⇒ *(ii)* By Corollary 2.21.2, we have (ii), a). Since an AC function on $[a,b]$ is derivable *a.e.* on $[a,b]$, we have (ii), b).

(ii) ⇒ *(vi)* This is evident.

(vi) ⇒ *(ii)* See Theorem 4.2.3.

(ii) ⇒ *(vii)* Clearly (ii), a) implies (vii), a). Let F be given by (ii). Then $F \in \overline{\mathcal{M}}^{\#}_{10}(f) \neq \emptyset$.

(vii) ⇒ *(viii)* ⇒ *(ix)* See Theorem 2.18.9 and Lemma 2.21.1.

(ix) ⇒ *(ii)* Let $G \in \overline{\mathcal{M}}^{\#}_{10}(f) \neq \emptyset$ and let F be given by (ix). Let $H = G - F$. By (ix), a), $-F \in (+) \cap (-) \cap N^\infty$ on $[a,b]$. Since $G \in \underline{AC}$ and $G'(x) \geq f(x)$ *a.e.* on $[a,b]$, by (ix), c) it follows that $H'(x) \geq 0$ *a.e.* where H is derivable on $[a,b]$. By Theorem 4.3.2, (ii), $-F \in AC$ on $[a,b]$. Since any AC function is derivable *a.e.* on $[a,b]$, we have (ii), b).

(ii) ⇒ *(x)* This is evident.

(x) ⇒ *(ii)* Let $G \in \overline{\mathcal{M}}^{\#}_{10}(f) \neq \emptyset$, and let F be given by (x).Let $H = G - F$. By (x), a) and b), it follows that $-F \in \mathcal{D} \cap [CG]$ on $[a,b]$ and $-F \in (M)$ on each closed interval contained in A. Since $G \in \underline{AC}$ on $[a,b]$, by (x), d), it follows that $H'(x) \geq 0$ *a.e.* where H is derivable. By Corollary 4.2.6, $F \in AC$ on $[a,b]$. Since any AC function is derivable *a.e.*, we have (ii), b).

Lemma 5.1.5 *Let* $f : [a,b] \mapsto \overline{\mathbb{R}}$*. If* f *is summable on* $[a,b]$ *then* f *is* $(\mathcal{P}^{\#}_{4,4})$*-integrable on* $[a,b]$ *and* $(\mathcal{P}^{\#}_{4,4}) \int_a^b f(t)dt = (\mathcal{L}) \int_a^b f(t)dt$*.*

Proof. Let $\epsilon > 0$. By Theorem 2.14.2, there exists $u : [a,b] \mapsto (-\infty,+\infty]$ such that u is lower semicontinuous, u is summable, $u(x) \geq f(x)$ and $(\mathcal{L})\int_a^b u(t)dt \leq \epsilon + (\mathcal{L})\int_a^b f(t)dt$. Let $M(x) = (\mathcal{L})\int_a^x u(t)dt$. Clearly $M(a) = 0$. By Lemma 2.14.1, $\underline{D}^{\#}M(x) \geq u(x)$, hence $f(x) \leq \underline{D}^{\#}M(x) \neq -\infty$. Since u is summable on $[a,b]$, it follows that $M \in AC$ on $[a,b]$, hence $M \in \overline{\mathcal{M}_4^{\#}}(f)$. We have $M(b) = (\mathcal{L})\int_a^b u(t)dt < \epsilon + (\mathcal{L})\int_a^b f(t)dt$. Then $\overline{I}_4(b) \leq (\mathcal{L})\int_a^b f(t)dt$. Similarly $\underline{I}_4(b) \geq (\mathcal{L})\int_a^b f(t)dt$. By Lemma 5.1.4, $\underline{I}_4(b) \leq \overline{I}_4(b)$, hence $\underline{I}_4(b) = \overline{I}_4(b) = (\mathcal{P}_{4,4}^{\#})\int_a^b f(t)dt = (\mathcal{L})\int f(y)dt$.

Lemma 5.1.6 *Let $f : [a,b] \mapsto \overline{\mathbb{R}}$. If f is summable on $[a,b]$ then f is $(\mathcal{P}_{1,1}^{\#})$-integrable on $[a,b]$ and $(\mathcal{P}_{1,1}^{\#})\int_a^b f(t)dt = (\mathcal{L})\int_a^b f(t)dt$.*

Proof. By Lemma 5.1.5, f is $(\mathcal{P}_{4,4}^{\#})$-integrable. It follows that for $\epsilon > 0$ there exists $M \in \overline{\mathcal{M}_4^{\#}}(f)$ such that $M(b) < \epsilon/4 + (\mathcal{L})\int_a^b f(t)dt$. Let $E = \{x : M'(x)$ does not exist in $\overline{\mathbb{R}}\}$. Then $|E| = 0$. Let Z be a G_δ-set, $|Z| = 0$, $Z \supset E$. By Theorem 2.14.6, there exists $G : [a,b] \mapsto \mathbb{R}$, $G(a) = 0$, $G(b) < \epsilon/4$, G is increasing and AC on $[a,b]$, $G'(x) = +\infty$ for $x \in Z$, $G'(x) > 0$ for $x \in [a,b] \setminus Z$. Since G is increasing it follows that $\underline{D}^{\#}G(x) \geq 0$ on $[a,b]$. Let $H = M + G$. By Remark 1.12.1,(ii), we have $-\infty \neq \underline{D}^{\#}H(x) \geq \underline{D}^{\#}M(x) + \underline{D}^{\#}G(x) \geq f(x)$ on $[a,b]$. Clearly $H \in AC$ on $[a,b]$, $H'(x) = +\infty$ for $x \in Z$, and $H'(x)$ exists on $[a,b] \setminus Z$. It follows that $H \in \overline{\mathcal{M}_1^{\#}}(f)$ and $H(b) < \epsilon/2 + (\mathcal{L})\int_a^b f(t)dt$, hence $\overline{I}_1(b) \leq (\mathcal{L})\int_a^b f(t)dt$. Similarly we obtain that $\underline{I}_1(b) \geq (\mathcal{L})\int_a^b f(t)dt$. By Lemma 5.1.4, (iv), $\underline{I}_1(b) \leq \overline{I}_1(b)$, hence $\underline{I}_1(b) = \overline{I}_1(b) = (\mathcal{P}_{1,1}^{\#})\int_a^b f(t)dt$.

Theorem 5.1.2 *Let $f : [a,b] \mapsto \overline{\mathbb{R}}$ and let $(j,k) \in \{1,2,\ldots,10\} \times \{1,2,\ldots,13\}$. The following assertions are equivalent:*

(i) f is summable on $[a,b]$;

(ii) f is $(\mathcal{P}_{j,k}^{\#})$-integrable on $[a,b]$

and $(\mathcal{P}_{j,k}^{\#})\int_a^b f(t)dt = (\mathcal{L})\int_a^b f(t)dt$.

Proof. $(i) \Rightarrow (ii)$ By Lemma 5.1.6 and Lemma 5.1.1, we have $\overline{I}_j(b) \leq \overline{I}_1(b) = (\mathcal{L})\int_a^b f(t)dt$, for $j \in \{1,2,\ldots,10\}$ and $\underline{I}_k(b) \geq \underline{I}_1(b) = (\mathcal{L})\int_a^b f(t)dt$, for $k \in \{1,2,\ldots,13\}$. By Lemma 5.1.4, (iv), $\overline{I}_j(b) = \underline{I}_k(b) = (\mathcal{P}_{j,k}^{\#})\int_a^b f(t)dt = (\mathcal{L})\int_a^b f(t)dt$.

$(ii) \Rightarrow (i)$ Let $F(x) = (\mathcal{P}_{j,k}^{\#})\int_a^x f(t)dt$. For $\epsilon > 0$ let $(M,m) \in \overline{\mathcal{M}_j^{\#}}(f) \times \underline{\mathcal{M}_k^{\#}}(f) \neq \emptyset$ such that $M(b) - F(b) < \epsilon/2$ and $F(b) - m(b) < \epsilon/2$. By Lemma 5.1.1, $M \in \overline{\mathcal{M}_{10}^{\#}}(f)$, and by Lemma 5.1.2, $M \in \overline{AC}$ on $[a,b]$. Since any $\overline{AC}$ function on $[a,b]$ is derivable a.e., it follows that $m'(x) \leq f(x) \leq M'(x)$ a.e. on $[a,b]$, and $m \in \underline{\mathcal{M}_{10}^{\#}}(f)$. By Lemma 5.1.2, $g = M - F$, $h = F - m$ and $M - m$ are increasing functions on $[a,b]$. For ϵ let $\delta > 0$ be given by the fact that $M \in \underline{AC}$ and $m \in \overline{AC}$ on $[a,b]$. Let $\{[a_i,b_i]\}$, $i = \overline{1,n}$ be a finite set of nonoverlapping closed subintervals of $[a,b]$, such that $\sum_{i=1}^n (b_i - a_i) < \delta$. Then we have $\sum_{i=1}^n (F(b_i) - F(a_i)) = \sum_{i=1}^n (M(b_i) - M(a_i)) - \sum_{i=1}^n (g(b_i) - g(a_i)) > -\epsilon/2 - (g(b) - g(a)) > -\epsilon$. It follows that $F \in \underline{AC}$ on $[a,b]$. We also obtain that $\sum_{i=1}^n (F(b_i) - F(a_i)) = \sum_{i=1}^n (m(b_i) - m(a_i)) + \sum_{i=1}^n (h(b_i) - h(a_i)) < \epsilon/2 + (h(b) - h(a)) < \epsilon$, hence $F \in \overline{AC}$ on $[a,b]$. It follows that $F \in AC$ on $[a,b]$. We show that $F'(x) = f(x)$ a.e. on $[a,b]$. Let $\epsilon > 0$ such that $M(b) - m(b) < \epsilon^2$. By Theorem 2.14.5, $M'(x)$ and $m'(x)$ are summable on $[a,b]$. Since $m'(x) \leq f(x) \leq$

$M'(x)$, it follows that f is finite *a.e.* on $[a,b]$. Let $E = \{x : f(x),\ F'(x),\ M'(x)$ and $m'(x)$ are finite$\}$. Then E is measurable and $|E| = b - a$. Let $A_\epsilon = \{x \in E : |F'(x) - f(x)| > \epsilon\}$ and $B_\epsilon = \{x \in E : M'(x) - m'(x) > \epsilon\}$. Then B_ϵ is measurable and $A_\epsilon \subset B_\epsilon$. We have $\epsilon \cdot |B_\epsilon| \le (\mathcal{L}) \int_{B_\epsilon} (M-m)'(t)dt \le (\mathcal{L}) \int_a^b (M-m)'(t)dt \le M(b) - m(b) < \epsilon^2$ (see Theorem 2.14.5). Hence $|B_\epsilon| < \epsilon$ and $|A_\epsilon| < \epsilon$. Let $A = \{x \in E : |F'(x) - f(x)| > 0\}$. Then $A = \cup_{k=1}^{\infty} A_{\epsilon/2^k}$, hence $|A| < \epsilon$. Since ϵ is arbitrary it follows that $|A| = 0$, hence $F'(x) = f(x)$ *a.e.* on $[a,b]$.
By Theorem 5.1.1, (i), (ii), f is summable on $[a,b]$ and $(\mathcal{L}) \int_a^b f(t)dt = F(b) - F(a) = F(b)$.

5.2 Ward type definitions for the Lebesgue integral

Definition 5.2.1 *Let $f : [a,b] \mapsto \mathbb{R}$. We define the following classes of majorants:*

- $\overline{\mathcal{W}_1^\#}(f) = \{M : [a,b] \mapsto \mathbb{R} : M(a) = 0;$ *there exists* $\delta : [a,b] \mapsto (0,\infty)$ *such that* $M(z) - M(y) > f(x)(z-y)$, *whenever* $[y,z] \subset (x - \delta(x), x + \delta(x))\}$;
- $\overline{\mathcal{W}_2^\#}(f) = \{M : [a,b] \mapsto \mathbb{R} : M \in \overline{\mathcal{W}_1^\#}(f) \cap AC\}$;
- $\overline{\mathcal{W}_3^\#}(f) = \{M : [a,b] \mapsto \mathbb{R} : M \in \overline{\mathcal{W}_1^\#}(f) \cap AC;\ M'(x)$ *exists finite or infinite on* $[a,b]$.

We define the following classes of minorants: $\underline{\mathcal{W}_j^\#}(f) = \{m : [a,b] \mapsto \mathbb{R} : -m \in \overline{\mathcal{W}_j^\#}(-f)\}$, $j = 1,2,3$.
If $\overline{\mathcal{W}_j^\#}(f) \ne \emptyset$ *then we denote by* $\overline{J}_j(b)$ *the lower bound of all* $M(b)$, $M \in \overline{\mathcal{W}_j^\#}(f)$. *If* $\underline{\mathcal{W}_j^\#}(f) \ne \emptyset$ *then we denote by* $\underline{J}_j(b)$ *the upper bound of all* $m(b)$, $m \in \underline{\mathcal{W}_j^\#}(f)$.
Let $(j,k) \in \{1,2,3\} \times \{1,2,3\}$. *We say that f has a* $(\mathcal{W}_{j,k}^\#)$*-integral on* $[a,b]$, *if* $\overline{\mathcal{W}_j^\#}(f) \times \underline{\mathcal{W}_k^\#}(f) \ne \emptyset$ *and* $\overline{J}_j(b) = \underline{J}_k(b) = (\mathcal{W}_{j,k}^\#) \int_a^b f(t)dt$.

Lemma 5.2.1 *Let* $f : [a,b] \mapsto \mathbb{R}$ *and let* $(j,k) \in \{1,2,3\} \times \{1,2,3\}$ *such that* $\overline{\mathcal{W}_j^\#}(f) \times \underline{\mathcal{W}_k^\#}(f) \ne \emptyset$. *If* $(M,m) \in \overline{\mathcal{W}_j^\#}(f) \times \underline{\mathcal{W}_k^\#}(f)$ *then: (i)* $M - m$; *(ii)* $M - \underline{J}_k$; *(iii)* $\overline{J}_j - m$ *and (iv)* $\overline{J}_j - \underline{J}_k$ *are positive increasing functions. Hence* $M(b) \ge m(b)$; $M(b) \ge \underline{J}_k(b)$; $\overline{J}_j(b) \ge m(b)$ *and* $\overline{J}_j(b) \ge \underline{J}_k(b)$.

Proof. (i) Clearly $(M,m) \in \overline{\mathcal{W}_1^\#}(f) \times \underline{\mathcal{W}_1^\#}(f)$. Then there exists $\delta : [a,b] \mapsto (0,+\infty)$ such that $M(z) - M(y) \ge f(x)(z-y) \ge m(z) - m(y)$, whenever $[y,z] \subset (x - \delta(x), x + \delta(x))$. It follows that $M - m$ is $\mathcal{S}_o$ increasing on $[a,b]$. By Lemma 4.4.1, $M - m$ is increasing on $[a,b]$.

(ii), (iii) and (iv) follow similarly to Lemma 5.1.4, (ii), (iii), (iv).

Theorem 5.2.1 *Let* $f : [a,b] \mapsto \mathbb{R}$ *and let* $(j,k) \in \{1,2,3\} \times \{1,2,3\}$. *The following assertions are equivalent:*

(i) f is summable on $[a,b]$;

(ii) f is $(\mathcal{W}_{j,k}^\#)$*-integrable on* $[a,b]$

and $(\mathcal{W}^{\#}_{j,k})\int_a^b f(t)DT = (\mathcal{L})\int_a^b f(t)dt.$

Proof. $(i) \Rightarrow (ii)$ By Theorem 5.1.2, f is $(\mathcal{P}^{\#}_{1,1})$-integrable on $[a,b]$ and $(\mathcal{L})\int_a^b f(t)dt = (\mathcal{P}^{\#}_{1,1})\int_a^b f(t)dt$. Let $\epsilon > 0$. Then there exists $G \in \overline{\mathcal{M}^{\#}_1}(f)$ such that $G(b) < \epsilon/2 + (\mathcal{L})\int_a^b f(t)dt$. It follows that $G \in AC$, $G'(x)$ exists on $[a,b]$ and $-\infty \neq \underline{D}^{\#}G(x) \geq f(x)$ on $[a,b]$. Hence there exists $\delta : [a,b] \mapsto (0,+\infty)$, such that $(G(z)-G(y))/(z-y) \geq f(x) - \epsilon/(2(b-a))$, whenever $[y,z] \subset (x-\delta(x), x+\delta(x))$, $x \in [a,b]$. Let $M : [a,b] \mapsto \mathbb{R}$, $M(x) = G(x) + (\epsilon(x-a))/(2(b-a))$. Then $M(z) - M(y) \geq f(x)(z-y)$, hence $M \in \overline{\mathcal{W}^{\#}_3}(f) \neq \emptyset$. Clearly we have $M(b) < \epsilon + (\mathcal{L})\int_a^b f(t)dt$, so $\overline{J}_3(b) \leq (\mathcal{L})\int_a^b f(t)dt$. We obtain that $\overline{\mathcal{M}^{\#}_5}(f) \supset \overline{\mathcal{W}^{\#}_1}(f) \supset \overline{\mathcal{W}^{\#}_2}(f) \supset \overline{\mathcal{W}^{\#}_3}(f)$. By Theorem 5.1.2, it follows that f is $(\mathcal{P}^{\#}_{5,5})$-integrable on $[a,b]$ and $(\mathcal{L})\int_a^b f(t)dt = \overline{I}_5(b) \leq \overline{J}_1(b) \leq \overline{J}_2(b) \leq \overline{J}_3(b) \leq (\mathcal{L})\int_a^b f(t)dt$, hence we obtain that $\overline{J}_1(b) = \overline{J}_2(b) = \overline{J}_3(b) = (\mathcal{L})\int_a^b f(t)dt$. Similarly $\underline{J}_1(b) = \underline{J}_2(b) = \underline{J}_3(b) = (\mathcal{L})\int_a^b f(t)dt$. Therefore f is $(\mathcal{W}^{\#}_{j,k})$- integrable on $[a,b]$ and $(\mathcal{W}^{\#}_{j,k})\int_a^b f(t)dt = (\mathcal{L})\int_a^b f(t)dt$.

$(ii) \Rightarrow (i)$ Since f is $(\mathcal{W}^{\#}_{j,k})$- integrable, it follows that $\overline{\mathcal{W}^{\#}_j}(f) \times \underline{\mathcal{W}^{\#}_k}(f) \neq \emptyset$, and $\overline{J}_j(b) = \underline{J}_k(b) = (\mathcal{W}^{\#}_{j,k})\int_a^b f(t)dt$. Clearly $\overline{\mathcal{W}^{\#}_j}(f) \times \underline{\mathcal{W}^{\#}_k}(f) \subset \overline{\mathcal{M}^{\#}_5}(f) \times \underline{\mathcal{M}^{\#}_5}(f)$, hence $\overline{J}_5(b) \geq \overline{I}_5(b)$ and $\underline{J}_k(b) \leq \underline{I}_5(b)$. By Lemma 5.1.4, $\underline{I}_5(b) \leq \overline{I}_5(b)$, hence $\underline{I}_5(b) = \overline{I}_5(b) = (\mathcal{P}^{\#}_{5,5})\int_a^b f(t)dt = (\mathcal{L})\int_a^b f(t)dt = (\mathcal{W}^{\#}_{j,k})\int_a^b f(t)dt$ (see Theorem 5.1.2).

5.3 Henstock variational definitions for the Lebesgue integral

Definition 5.3.1 *Let* $f : [a,b] \mapsto \mathbb{R}$.

- *f is said to be* $(\mathcal{V}^{\#}_1)$*-integrable on* $[a,b]$*, if there exists* $H : [a,b] \mapsto \mathbb{R}$ *such that for every* $\epsilon > 0$ *there exist* $\delta : [a,b] \mapsto (0,+\infty)$ *and* $G : [a,b] \mapsto \mathbb{R}$ *with the following properties:* $G(a) = 0$, $G(b) < \epsilon$, *G is increasing on* $[a,b]$ *and* $\mid H(z) - H(y) - F(x)(z-y) \mid < G(z) - G(y)$, *whenever* $[y,z] \subset (x-\delta(x), x+\delta(x))$.
- *f is said to be* $(\mathcal{V}^{\#}_2)$*- integrable on* $[a,b]$*, if f is* $(\mathcal{V}^{\#}_1)$*-integrable and* $G \in AC$ *on* $[a,b]$.
- *f is said to be* $\mathcal{V}^{\#}_3)$*-integrable on* $[a,b]$*, if f is* $(\mathcal{V}^{\#}_1)$*-integrable,* $G \in AC$ *and* $G'(x)$ *(finite or infinite) on* $[a,b]$.

H is called the $(\mathcal{V}^{\#}_i)$*-indefinite integral of f on* $[a,b]$*, and* $(\mathcal{V}^{\#}_i)\int_a^b f(t)dt = H(b) - H(a)$, $i = 1,2,3$.

Lemma 5.3.1 *Let* $f : [a,b] \mapsto \mathbb{R}$. *If* $H_1, H_2 : [a,b] \mapsto \mathbb{R}$ *are* $(\mathcal{V}^{\#}_1)$*-indefinite integrals of f on* $[a,b]$ *then* $H_1(b) - H_1(a) = H_2(b) - H_2(a)$.

Proof. For $\epsilon > 0$ there exist $\delta : [a,b] \mapsto (0,+\infty)$ and $G_1, G_2 : [a,b] \mapsto \mathbb{R}$, with $G_1(a) = G_2(a) = 0$, $G_1(b) < \epsilon$, $G_2(b) < \epsilon$, G_1, G_2 are increasing on $[a,b]$, $\mid H_1(z) - H_1(y) - f(x)(z-y) \mid < G_1(z) - G_1(y)$, and $\mid H_2(z) - H_2(y) - f(x)(z-y) \mid < G_2(z) - G_2(y)$, whenever $[y,z] \subset (x-\delta(x), x+\delta(x))$, $x \in [a,b]$. It follows that $\mid (H_1 - H_2)(z) - (H_1 - H_2)(y) \mid < (G_1+G_2)(z) - (G_1+G_2)(y)$. By Lemma 2.24.1, there exists a partition $a = x_o < x_1 < \ldots < x_n = b$ and $t_i \in [x_{i-1}, x_i]$, $i = \overline{1,n}$ such that $[x_{i-1}, x_i] \subset (t_i - \delta(t_i), t_i + \delta(t_i))$. It follows that $| \sum_{i=1}^{n}((H_1 - H_2)(x_i) - (H_1 - H_2)(x_{i-1})) | \leq \sum_{i=1}^{n} | (H_1 - H_2)(x_i) - (H_1 - H_2)(x_{i-1}) | \leq \sum_{i=1}^{n}((G_1 + G_2)(x_i) - (G_1 + G_2)(x_{i-1})) < 2\epsilon$, so $| (H_1 - H_2)(b) - (H_1 - H_2)(a) | < 2\epsilon$. Since ϵ is arbitrary we have $H_1(b) - H_1(a) = H_2(b) - H_2(a)$.

Remark 5.3.1 *Lemma 5.3.1 shows that the $(\mathcal{V}_i^{\#})$-integrals, $i = 1,2,3$ are well defined.*

Lemma 5.3.2 *Let $f : [a,b] \mapsto \mathbb{R}$. If f is $(\mathcal{W}_{3,3}^{\#})$-integrable on $[a,b]$ then f is $(\mathcal{V}_3^{\#})$-integrable on $[a,b]$ and $(\mathcal{W}_{3,3}^{\#}) \int_a^b f(t)dt = (\mathcal{V}_3^{\#}) \int_a^b f(t)dt$.*

Proof. Let $H : [a,b] \mapsto \mathbb{R}$, $H(x) = (\mathcal{W}_{3,3}^{\#}) \int_a^b f(t)dt$. For $\epsilon > 0$ let $(M,m) \in (\overline{\mathcal{W}_3^{\#}}(f) \times \underline{\mathcal{W}_3^{\#}}(f)$ such that $H(b) - \epsilon/2 < m(b)$ and $M(b) < H(b) + \epsilon/2$. Let $G = M - m$ and $\delta : [a,b] \mapsto (0, +\infty)$, such that $M(z) - M(y) \geq f(x)(z-y) \geq m(z) - m(y)$, whenever $[y,z] \subset (x - \delta(x), x + \delta(x))$. Then $H(z) - H(y) = (M(z) - M(y)) - ((M - H)(z) - (M - H)(y)) \geq f(x)(z-y) - (G(z) - G(y))$ (see Lemma 5.2.1, (ii)), and $H(z) - H(y) = (m(z) - m(y)) - ((m - H)(z) - (m - H)(y)) \leq f(x)(z-y) + (G(z) - G(y))$ (see Lemma 5.2.1, (iii)). It follows that f is $(\mathcal{V}_3^{\#})$-integrable on $[a,b]$ and $(\mathcal{W}_{3,3}^{\#}) \int_a^b f(t)dt = (\mathcal{V}_3^{\#}) \int_a^b f(t)dt$.

Lemma 5.3.3 *Let $f : [a,b] \mapsto \mathbb{R}$. If f is $(\mathcal{V}_1^{\#})$-integrable on $[a,b]$ then f is $(\mathcal{W}_{1,1}^{\#})$-integrable on $[a,b]$ and $(\mathcal{V}_1^{\#}) \int_a^b f(t)dt = (\mathcal{W}_{1,1}^{\#}) \int_a^b f(t)dt$.*

Proof. Let $H : [a,b] \mapsto \mathbb{R}$, $H(x) = (\mathcal{V}_1^{\#}) \int_a^b f(t)dt$. For $\epsilon > 0$ there exist $\delta : [a,b] \mapsto (0, +\infty)$ and $G : [a,b] \mapsto \mathbb{R}$ such that $G(a) = 0$, $G(b) < \epsilon$, G is increasing on $[a,b]$, and $| H(z) - H(y) - f(x)(z-y) | < G(z) - G(y)$, whenever $[y,z] \subset (x - \delta(x), x + \delta(x))$. Let $M = H + G$, $m = H - G$. Then $(M,m) \in \overline{\mathcal{W}_1^{\#}}(f) \times \underline{\mathcal{W}_1^{\#}}(f)$. It follows that $\overline{J}_1(b) \leq H(b)$ and $\underline{J}_1(b) \geq H(b)$. By Lemma 5.2.1, (iv), $\overline{J}_1(b) \geq \underline{J}_1(b)$, hence $\overline{J}_1(b) = \underline{J}_1(b) = (\mathcal{W}_{1,1}^{\#}) \int_a^b f(t)dt = H(b) - H(a)$.

Theorem 5.3.1 *Let $f : [a,b] \mapsto \mathbb{R}$ and let $i \in \{1,2,3\}$. The following assertions are equivalent:*

(i) f is summable on $[a,b]$;

(ii) F is $(\mathcal{V}_i^{\#})$-integrable on $[a,b]$,

and $(\mathcal{V}_i^{\#}) \int_a^b f(t)dt = (\mathcal{L}) \int_a^b f(t)dt$.

Proof. $(i) \Rightarrow (ii)$ By Theorem 5.2.1, it follows that f is $(\mathcal{W}_{3,3}^{\#})$-integrable on $[a,b]$ and $(\mathcal{W}_{3,3}^{\#}) \int_a^b f(t)dt = (\mathcal{L}) \int_a^b f(t)dt$. By Lemma 5.3.2, f is $(\mathcal{V}_3^{\#})$-integrable on $[a,b]$ and $(\mathcal{V}_3^{\#}) \int_a^b f(t)dt = (\mathcal{L}) \int_a^b f(t)dt$. But the $(\mathcal{V}_3^{\#})$- integrability implies the $(\mathcal{V}_2^{\#})$-integrability, which implies the $(\mathcal{V}_1^{\#})$-integrability. By Lemma 5.3.1, it follows that $(\mathcal{V}_i^{\#}) \int_a^b f(t)dt = (\mathcal{L}) \int_a^b f(t)dt$.

$(ii) \Rightarrow (i)$ Clearly f is $(\mathcal{V}_1^{\#})$- integrable. By Lemma 5.3.1, $(\mathcal{V}_i^{\#}) \int_a^b f(t)dt = (\mathcal{V}_1^{\#}) \int_a^b f(t)dt$. By Lemma 5.3.3, f is $(\mathcal{W}_{1,1}^{\#})$- integrable on $[a,b]$ and $(\mathcal{V}_1^{\#}) \int_a^b f(t)dt = (\mathcal{W}_{1,1}^{\#}) \int_a^b f(t)dt$. By Theorem 5.2.1, f is summable on $[a,b]$ and $(\mathcal{W}_{1,1}^{\#}) \int_a^b f(t)dt = (\mathcal{L}) \int_a^b f(t)dt$.

5.4 Riemann type definitions for the Lebesgue integral (The McShane integral)

The relations between the Riemann integral and the Lebesgue integral were investigated in many respects. There exist many approaches to the Lebesgue integral (for example Denjoy [D], Davies and Schuss [DaS], McShane [McS3]), and there exists a Lebesgue approach to the Riemann integral (Marcus [Ma2], [Ma3]). In what follows we expose the McShane approach for the Lebesgue integral.

Definition 5.4.1 *Let $\delta : [a,b] \mapsto (0,+\infty)$ and let $E \subseteq [a,b]$. Let $\beta^{\#}_{\delta}[E] = \{([y,z];x) : [y,z] \subset (x-\delta(x), x+\delta(x)),\ x \in E\}$. Let P be a finite set of pairs $([c_i,d_i];t_i) \in \beta^{\#}_{\delta}[E]$ such that $\{[c_i,d_i]\}_i$ are nonoverlapping, nondegenerate closed intervals, and let $\sigma(P) = \cup_i[c_i,d_i]$. We denote by $\mathcal{P}^{\#}(E;\delta)$ the collection of all P defined as above. Let $f : [a,b] \mapsto \mathbb{R}$. Let $\sigma(f;P) = \sum_i f(t_i)(d_i - c_i)$ and $S(f;P) = \sum_i (f(d_i) - f(c_i))$, $P \in \mathcal{P}^{\#}(E;\delta)$. If $E = [a,b]$ and $\sigma(P) = [a,b]$ then we denote the collection of these P by $\mathcal{P}^{\#}_1([a,b];\delta)$.*

Remark 5.4.1 *Recall that $D^{\#}[E] = \{\beta^{\#}_{\delta}[E] : \delta : [a,b] \mapsto (0,+\infty)\}$ is called the sharp derivation basis on the set E (see Section 2.24).*

Definition 5.4.2 (McShane) *Let $f : [a,b] \mapsto \mathbb{R}$. f is said to be $(RD^{\#})$-integrable on $[a,b]$, if there exists a real number I with the following property: for $\epsilon > 0$ there exists $\delta : [a,b] \mapsto (0,+\infty)$ such that $\mid \sigma(f;P) - I \mid < \epsilon$, whenever $P \in \mathcal{P}^{\#}_1([a,b];\delta)$. Then $(RD^{\#}) \int_a^b f(t)dt = I$. Let $E \subset [a,b]$. f is said to be $(RD^{\#})$-integrable on E, if $f \cdot K_E$ is $(RD^{\#})$-integrable on $[a,b]$.*

Remark 5.4.2 *The number I in the above definition is unique. Indeed, suppose that there exist two such numbers I_1 and I_2. Then for $\epsilon > 0$ there exist the corresponding δ_1 and δ_2. Let $\delta : [a,b] \mapsto (0,+\infty)$, $\delta(x) = \min\{\delta_1(x), \delta_2(x)\}$. Let $P \in \mathcal{P}^{\#}_1([a,b];\delta)$. Then $\mid \sigma(f;P) - I_1 \mid < \epsilon$ and $\mid \sigma(f;P) - I_2 \mid < \epsilon$. Hence $\mid I_1 - I_2 \mid < 2\epsilon$, so $I_1 = I_2$.*

Lemma 5.4.1 *Let $f : [a,b] \mapsto \mathbb{R}$. The following assertions are equivalent:*

- *(i) f is $(RD^{\#})$-integrable on $[a,b]$;*
- *(ii) For every $\epsilon > 0$ there exists $\delta : [a,b] \mapsto (0,+\infty)$ such that $\mid \sigma(f;P) - \sigma(f;P') \mid < \epsilon$, whenever $P, P' \in \mathcal{P}^{\#}_1([a,b];\delta)$.*

Proof. $(i) \Rightarrow (ii)$ This is evident.

$(ii) \Rightarrow (i)$ For $1/n$, $n = \overline{2,\infty}$, let δ_n be given by (ii). We may suppose that $\delta_n < \delta_{n-1}$, $n > 3$. For $\epsilon > 0$ there exists n_ϵ such that $1/n_\epsilon < \epsilon$. Let $\delta = \delta_{n_\epsilon}$. Let $m,n > n_\epsilon$ and $P_n \in \mathcal{P}^{\#}_1([a,b];\delta_n)$, $P_m \in \mathcal{P}^{\#}_1([a,b];\delta_m)$, $P_{n_\epsilon} \in \mathcal{P}^{\#}_1([a,b];\delta)$. We have $\mid \sigma(f;P_n) - \sigma(f;P_{n_\epsilon}) \mid < 1/n_\epsilon < \epsilon$ and $\mid \sigma(f;P_m) - \sigma(f;P_{n_\epsilon}) \mid < 1/n_\epsilon < \epsilon$. It follows that $\mid \sigma(f;P_n) - \sigma(f;P_m) \mid < 2\epsilon$. Hence $\{\sigma(f;P_n)\}_n$ is a Cauchy sequence with the limit I. If $m \to \infty$ then $\mid \sigma(f;P_n) - I \mid < 2\epsilon$. Let $P \in \mathcal{P}^{\#}_1([a,b];\delta)$. Then $\mid \sigma(f;P) - I \mid < \mid \sigma(f;P) - \sigma(f;P_n) \mid + \mid \sigma(f;P_n) - I \mid < 3\epsilon$.

Lemma 5.4.2 *Let $f,g : [a,b] \mapsto \mathbb{R}$, $\alpha, \beta \in \mathbb{R}$, $\alpha \neq 0$, $\beta \neq 0$.*

(i) *If f and g are $(RD^{\#})$-integrable then $\alpha f+\beta g$ is $(RD^{\#})$-integrable and* $(RD^{\#})\int_a^b(\alpha f+\beta g)(t)dt = \alpha\cdot(RD^{\#})\int_a^b f(t)dt + \beta\cdot(RD^{\#})\int_a^b g(t)dt$.

(ii) *If $a<c<b$ and f is $(RD^{\#})$-integrable on $[a,c]$ and $[c,b]$ then f is $(RD^{\#})$-integrable on $[a,b]$ and* $(RD^{\#})\int_a^c f(t)dt + (RD^{\#})\int_c^b f(t)dt = (RD^{\#})\int_a^b f(t)dt$.

(iii) *If $a\le c<d\le b$ and f is $(RD^{\#})$-integrable on $[a,b]$ then f is $(RD^{\#})$-integrable on $[c,d]$.*

Proof. (i) Let $A=(RD^{\#})\int_a^b f(t)dt$ and $B=(RD^{\#})\int_a^b g(t)dt$. For $\epsilon>0$ there exist $\delta_1,\delta_2:[a,b]\mapsto(0,+\infty)$ such that $\mid\sigma(f;P)-A\mid<\epsilon/(2\mid\alpha\mid)$, $P\in\mathcal{P}_1^{\#}([a,b];\delta_1)$ and $\mid\sigma(g;P)-B\mid<\epsilon/(2\mid\beta\mid)$, $P\in\mathcal{P}_1^{\#}([a,b];\delta_2)$. Let $\delta:[a,b]\mapsto(0,+\infty)$, $\delta(x)=\min\{\delta_1(x),\delta_2(x)\}$ and let $P\in\mathcal{P}_1^{\#}([a,b];\delta)$. We have $\mid\sigma(\alpha f+\beta g;P)-(\alpha A+\beta B)\mid\le\mid\alpha\mid\cdot\mid\sigma(f;P)-A\mid+\mid\beta\mid\cdot\mid\sigma(g;P)-B\mid<\epsilon$. So we have (i).

(ii) Let $A=(RD^{\#})\int_a^c f(t)dt$ and $B=(RD^{\#})\int_c^b f(t)dt$. For $\epsilon>0$ there exist $\delta_1:[a,c]\mapsto(0,+\infty)$ and $\delta_2:[c,b]\mapsto(0,+\infty)$ such that $\mid\sigma(f;P)-A\mid<\epsilon/2$, $P\in\mathcal{P}_1^{\#}([a,b];\delta_1)$ and $\mid\sigma(f;P)-B\mid<\epsilon/2$, $P\in\mathcal{P}_1^{\#}([c,b];\delta_2)$. Let $\delta:[a,b]\mapsto(0,+\infty)$, $\delta(x)=\min\{\delta_1(x),c-x\}$ for $x\in[a,c)$, $\delta(x)=\min\{\delta_1(c),\delta_2(c)\}$ for $x=c$, $\delta(x)=\min\{\delta_2(x),x-c\}$ for $x\in(c,b]$. Let $P\in\mathcal{P}_1([a,b];\delta)$. Then there exists $([u,v];c)\in P$ such that one of the following happens: $c\in(u,v)$, $c=u$ or $c=v$. If $c\in(u,v)$ then $f(c)(v-u)=f(c)(v-c)+f(c)(c-u)$. It follows that $\mid\sigma(f;P)-(A+B)\mid=\mid\sum_{\substack{([u,v];t)\in P\\ v<a}}(f(t)(v-u)+f(c)(v-c)-A\mid+\mid\sum_{\substack{([u,v];t)\in P\\ u>b}}(f(t)(v-u)+f(c)(c-u))-B\mid<\epsilon/2+\epsilon/2=\epsilon$. If $c=u$ or $c=v$ then the proofs follow similarly.

(iii) Suppose that $a<c<d<b$ and let $\epsilon>0$. By Lemma 5.4.1 there exists $\delta:[a,b]\mapsto(0,+\infty)$ such that $\mid\sigma(f;P)-\sigma(f;P')\mid<\epsilon$, whenever $P,P'\in\mathcal{P}_1^{\#}([a,b];\delta)$. Let $P_1,P_2\in\mathcal{P}_1^{\#}([c,d];\delta_{/[c,d]})$. Let $P_3\in\mathcal{P}_1^{\#}([a,c];\delta_{/[a,c]})$ and $P_4\in\mathcal{P}_1^{\#}([d,b];\delta_{/[d,b]})$. Then $P_1\cup P_3\cup P_4$ and $P_2\cup P_3\cup P_4$ belong to $\mathcal{P}_1^{\#}([a,b];\delta)$. We have $\mid\sigma(f;P_1)-\sigma(f;P_2)\mid=\sigma(f;P_1\cup P_3\cup P_4)-\sigma(f;P_2\cup P_3\cup P_4)\mid<\epsilon$. By Lemma 5.4.1, f is $(RD^{\#})$-integrable on $[a,b]$.

Lemma 5.4.3 (Henstock) *Let $f:[a,b]\mapsto\mathbb{R}$ such that f is $(RD^{\#})$-integrable on $[a,b]$. Let $F:[a,b]\mapsto\mathbb{R}$, $F(x)=(RD^{\#})\int_a^x f(t)dt$. For $\epsilon>0$ let $\delta:[a,b]\mapsto(0,+\infty)$ such that $\mid\sigma(f;P)-F(b)\mid<\epsilon$, whenever $P\in\mathcal{P}_1([a,b];\delta)$. Then we have:*

(i) *$\mid\sigma(f;P)-S(F;P)\mid<3\epsilon/2$ whenever $P\in\mathcal{P}^{\#}([a,b];\delta)$;*

(ii) *$\sum_{([y,z];x)\in P}\mid f(x)(z-y)-(F(z)-F(y))\mid<3\epsilon$, whenever $P\in\mathcal{P}^{\#}([a,b];\delta)$.*

Proof. (i) Let $P\in\mathcal{P}^{\#}([a,b];\delta)$. Let $\{(c_i,d_i)\}$, $i=\overline{1,n}$ be the components of the open set $(a,b)\setminus\sigma(P)$. By Lemma 5.4.2, (iii), (ii), it follows that $(RD^{\#})\int_{c_i}^{d_i}f(t)dt=F(d_i)-F(c_i)$. Then there exist $\delta_i:[c_i,d_i]\mapsto(0,+\infty)$, $i=\overline{1,n}$ such that $\delta_i(x)\le\delta(x)$, $x\in[a,b]$ and $\mid F(d_i)-F(c_i)-\sigma(f;P_i)\mid<\epsilon/(2n)$, whenever $P_i\in\mathcal{P}_1^{\#}([c_i,d_i];\delta_i)$, $i=\overline{1,n}$. Let $P'=P\cup(\cup_{i=1}^n P_i)$. Then $P'\in\mathcal{P}_1^{\#}([a,b];\delta)$. Since $F(b)=S(F;P')$, it follows that $\mid\sigma(f;P')-S(F;P')\mid<\epsilon$. We have $\mid\sigma(f;P)-S(F;P)+\sum_{i=1}^n(F(d_i)-F(c_i)-\sigma(f;P_i))\mid<\epsilon$, hence $\mid\sigma(f;P)-S(F;P)\mid<3\epsilon/2$.

(ii) Let $P\in\mathcal{P}^{\#}([a,b];\delta)$, $P'=\{([y,z];x)\in P: f(x)(z-y)>F(z)-F(y)\}$ and $P"=P\setminus P'$. Then P' and $P"$ belong to $\mathcal{P}^{\#}([a,b];\delta)$. By (i), $\mid\sigma(f;P')-S(F;P')\mid<3\epsilon/2$ and $\mid\sigma(f;P")-S(F;P")\mid<3\epsilon/2$. Thus $\sum_{([y,z];x)\in P}\mid f(x)(z-y)-F(z)-F(y))\mid=\mid\sigma(f;P')-S(F;P')\mid+\mid\sigma(f;P")-S(F;P")\mid<3\epsilon$.

Lemma 5.4.4 *Let $f : [a,b] \mapsto \mathbb{R}$. If f is $(RD^{\#})$-integrable on $[a,b]$ and $F(x) = (RD^{\#})\int_a^x f(t)dt$ then for every $\epsilon > 0$ there exist $\delta : [a,b] \mapsto (0,+\infty)$, such that $G(a) = 0$, $G(b) < \epsilon$, G is increasing on $[a,b]$, and $G(z)-G(y) > \mid f(x)(z-y)-(F(z)-F(y)) \mid$ whenever $[y,z] \subset (x-\delta(x), x+\delta(x))$, $x \in [a,b]$.*

Proof. For $\epsilon > 0$ there exists $\delta : [a,b] \mapsto (0,+\infty)$ such that $\mid \sigma(f;P) - F(b) \mid < \epsilon/3$ whenever $P \in \mathcal{P}_1^{\#}([a,b];\delta)$. By Lemma 5.4.3, (ii) we have $\sum_{([y,z];x)\in P} \mid f(x)(z-y) - (F(y)-F(z)) \mid < \epsilon$, whenever $P \in \mathcal{P}^{\#}([a,b];\delta)$. Let $G(x) = \sup\{\sum_{([y,z];x)\in P} \mid f(x)(z-y)-(F(z)-F(y)) \mid : P \in \mathcal{P}^{\#}([a,b];\delta)$ and $\sigma(P) = [a,x]\}$. Then G has the required properties.

Lemma 5.4.5 (Gordon) *([G5], p. 160). Let $f : [a,b] \mapsto \mathbb{R}$ and let $Z \subset [a,b]$, $\mid Z \mid = 0$. Then for every $\epsilon > 0$ there exists $\delta : [a,b] \mapsto (0,+\infty)$ such that $\mid \sigma(f;P) \mid < \epsilon$ whenever $P \in \mathcal{P}^{\#}(Z;\delta_{/Z})$.*

Proof. For each positive integer n, let $Z_n = \{t \in Z : n-1 \leq \mid f(t) \mid < n\}$. Then $Z = \cup_{n=1}^{\infty} Z_n$. For $\epsilon > 0$ and for each n there exists an open set E_n such that $Z_n \subset E_n$ and $\mid E_n \mid < \epsilon/(n \cdot 2^n)$. Let $\delta(t) = d(t;[a,b] \setminus E_n)$, $t \in Z$. Let $P \in \mathcal{P}^{\#}(Z;\delta_{/Z})$ and let $P_n = \{([y,z];x) \in P : x \in Z_n\}$, $n = \overline{1,\infty}$. Then $\mid \sigma(f;P) \mid \leq \sum_{n=1}^{\infty} \mid \sigma(f;P_n) \mid < n \cdot \mid E_n \mid < \sum_{n=1}^{\infty} \epsilon/2^n = \epsilon$.

Theorem 5.4.1 *Let $f : [a,b] \mapsto \mathbb{R}$. The following assertions are equivalent:*

(i) f is summable on $[a,b]$;

(ii) f is $(\mathcal{V}_1^{\#})$-integrable on $[a,b]$;

(iii) f is $(RD^{\#})$-integrable on $[a,b]$;

(iv) there exists $F : [a,b] \mapsto \mathbb{R}$, $F \in Y_{D\#}$ such that $F'(x) = f(x)$ a.e. on $[a,b]$,

and $(\mathcal{L})\int_a^b f(t)dt = (\mathcal{V}_1^{\#})\int_a^b f(t)dt = (RD^{\#})\int_a^b f(t)dt = F(b)-F(a)$.

Proof. (i) $\Leftrightarrow$ (ii) and $(\mathcal{L})\int_a^b f(t)dt = (\mathcal{V}_1^{\#})\int_a^b f(t)dt$ follow by Theorem 5.3.1.

(ii) $\Rightarrow$ (iii) Let $H : [a,b] \mapsto \mathbb{R}$, $H(x) = (\mathcal{L})\int_a^x f(t)dt$. For $\epsilon > 0$ there exist $\delta : [a,b] \mapsto (0,+\infty)$ and $G : [a,b] \mapsto \mathbb{R}$ such that $G(a) = 0$, $G(b) < \epsilon$, G is increasing and $\mid H(z)-H(y)-f(x)(z-y) \mid < G(z)-G(y)$, whenever $[y,z] \subset (x-\delta(x), x+\delta(x))$, $x \in [a,b]$. Let $P \in \mathcal{P}_1^{\#}([a,b];\delta)$. It follows that $\mid \sigma(f;P) - H(b) \mid \leq \sum_{([y,z];x)\in P} \mid f(x)(z-y) - (H(z)-H(y)) \mid < \sum_{([y,z];x)\in P}(G(z)-G(y)) = G(b)-G(a) < \epsilon$, hence f is $(RD^{\#})$-integrable and $(RD^{\#})\int_a^b f(t)dt = (\mathcal{V}_1^{\#})\int_a^b f(t)dt$.

(iii) $\Rightarrow$ (ii) Let $F : [a,b] \mapsto \mathbb{R}$, $F(x) = (RD^{\#})\int_a^x f(t)dt$. By Lemma 5.4.4, for $\epsilon > 0$ there exist $\delta : [a,b] \mapsto (0,+\infty)$ and $G : [a,b] \mapsto \mathbb{R}$ such that $G(a) = 0$, $G(b) < \epsilon$, G is increasing on $[a,b]$ and $G(z)-G(y) > \mid f(x)(z-y) - (F(z)-F(x)) \mid$, whenever $[y,z] \subset (x-\delta(x), x+\delta(x))$, $x \in [a,b]$. Hence f is $(\mathcal{V}_1^{\#})$-integrable on $[a,b]$ and $(\mathcal{V}_1^{\#})\int_a^b f(t)dt = F(b)$.

(iii) $\Rightarrow$ (iv) Let $F : [a,b] \mapsto \mathbb{R}$, $F(x) = (RD^{\#})\int_a^x f(t)dt$. We show that $F \in Y_{D\#}$ on $[a,b]$. For $\epsilon > 0$ there exists $\delta : [a,b] \mapsto (0,+\infty)$ such that $\mid \sigma(f;P) - F(b) \mid < \epsilon$ whenever $P \in \mathcal{P}_1^{\#}([a,b];\delta)$. Let $Z \subset [a,b]$, $\mid Z \mid = 0$. Let $P \in \mathcal{P}^{\#}(Z;\delta_{/Z})$. By Lemma 5.4.3,(i) and Lemma 5.4.4, it follows that $\mid S(F;P) \mid \leq \mid S(F;P) - \sigma(f;P) \mid + \mid \sigma(f;P) \mid < 3\epsilon/2 + \epsilon = 5\epsilon/2$. Hence $F \in Y_{D\#}$ on $[a,b]$.
We show that $F'(x) = f(x)$ a.e. on $[a,b]$ (this part is due to Gordon, see [G1], pp. 561

- 563). Let $A^+ = \{t \in [a,b) : DF^+(t)$ does not exist or $DF_+(t) \neq f(t)\}$. For each $t \in A^+$ there exist $\mu_t > 0$ with the following property: for each $\beta > 0$ there exists $v_{\beta,t} \in (a,b] \cap (t, y+\beta)$ such that $| F(v_{\beta,t}) - F(t) - f(t)(v_{\beta,t} - t) | \geq \mu_t(v_{\beta,t} - t)$. Let $A_n^+ = \{t \in A_+ : \mu_t \geq 1/n\}$, $n = \overline{1,\infty}$. Then $A^+ = \cup_{n=1}^{\infty} A_n^+$. We show that $| A_n^+ |= 0$, $n = \overline{1,\infty}$. Fix a positive integer n and let $\epsilon > 0$. Choose $\delta : [a,b] \mapsto (0,+\infty)$ such that $| \sigma(f;P) - F(b) | < \epsilon/(6n)$, whenever $P \in \mathcal{P}_1^{\#}([a,b];\delta)$. The intervals $[t, v_{\beta,y}]$, $t \in A_n^+$, $\beta < \delta(t)$ form a Vitali covering of A_n^+. By Theorem 1.8.1, there exists a finite set $\{[c_i, d_i]\}$, $i = \overline{1,N}$ of disjoint intervals, such that $| A_n^+ | < \sum_{i=1}^{N}(d_i - c_i) + \epsilon/2$. Then $[c_i, d_i] \subset (c_i - \delta(c_i), c_i + \delta(c_i))$ and $(d_i - x_i) \cdot \mu_{c_i} \leq | F(d_i) - F(c_i) - f(c_i)(d_i - c_i) |$. By Lemma 5.4.3, (ii), it follows that $\sum_{i=1}^{N}(d_i - c_i) \leq \sum_{i=1}^{N}(1/\mu_{c_i}) \cdot | F(d_i) - F(c_i) - f(c_i)(d_i - c_i) | \leq n \cdot \sum_{i=1}^{N} | f(c_i)(d_i - c_i) - (F(d_i) - F(c_i)) | \leq n \cdot (3\epsilon)/(6\epsilon) = \epsilon/2$. Therefore $| A_n^+ | \leq \epsilon$. Since ϵ was arbitrary we conclude that $| A_n^+ |= 0$, hence $| A^+ |= 0$. Similarly $| A^- |= 0$, where $A^- = \{t \in (a,b] : DF^-(t)$ does not exist or $DF^-(t) \neq f(t)\}$. Hence $F'(x) = f(x)$ *a.e.* on $[a,b]$.

$(iv) \Rightarrow (i)$ By Corollary 2.27.2, $F \in AC$ on $[a,b]$, hence f is summable on $[a,b]$ and $(\mathcal{L}) \int_a^b F'(t)dt = (\mathcal{L}) \int_a^b f(t)dt = F(b) - F(a)$ (see Corollary 2.14.2).

Remark 5.4.3 *In [P2], Pfeffer showed that in the definition of the $(RD^{\#})$-integral the function δ may be supposed to be summable on $[a,b]$, and upper semicontinuous, when restricted to a suitable subset whose complement has measure 0.*

5.5 Theorems of Marcinkiewicz type for the Lebesgue integral

Theorem 5.5.1 *Let $f : [a,b] \mapsto \overline{\mathbb{R}}$. The following assertions are equivalent:*

(i) f is summable on $[a,b]$;

(ii) f is measurable and there exists $(j,k) \in \{1,2,\ldots,10\} \times \{1,2,\ldots,13\}$ such that $\overline{\mathcal{M}}_j^{\#}(f) \times \underline{\mathcal{M}}_k^{\#}(f) \neq \emptyset$.

Proof. $(i) \Rightarrow (ii)$ See the definition of summability and Theorem 5.1.2.

$(ii) \Rightarrow (i)$ Let $(M,m) \in \overline{\mathcal{M}}_j^{\#}(f) \times \underline{\mathcal{M}}_k^{\#}(f)$. By Lemma 5.1.1 and Lemma 5.1.2, $M \in \underline{AC}$ and $m \in \overline{AC}$ on $[a,b]$. Also $m'(x) \leq f(x) \leq M'(x)$ *a.e.* on $[a,b]$. By Theorem 2.14.5, m' and M' are summable on $[a,b]$. Since f is measurable, it follows that f is summable on $[a,b]$.

Theorem 5.5.2 *Let $f : [a,b] \mapsto \mathbb{R}$. The following assertions ar equivalent:*

(i) f is summable on $[a,b]$;

(ii) f is measurable and there exists $(j,k) \in \{1,2,3\} \times \{1,2,3\}$ such that $\overline{\mathcal{W}}_j^{\#}(f) \times \underline{\mathcal{W}}_k^{\#}(f) \neq \emptyset$.

Proof. $(i) \Rightarrow (ii)$ See Theorem 5.2.1.

$(ii) \Rightarrow (i)$ Let $(M,m) \in \overline{\mathcal{W}}_j^{\#}(f) \times \underline{\mathcal{W}}_k^{\#}(f) \subset \overline{\mathcal{M}}_5^{\#}(f) \times \underline{\mathcal{M}}_5^{\#}(f)$. Now the proof continues as in Theorem 5.5.1, using Lemma 5.1.2.

5.6 Bounded Riemann# sums and locally small Riemann# sums

Definition 5.6.1 (Bullen and Vyborny) *Let $f : [a, b] \mapsto \mathbb{R}$. f is said to satisfy the $(BRD^{\#}S)$ (bounded $RD^{\#}$ sum) property on $[a, b]$, if there exist $\delta : [a, b] \mapsto (0, +\infty)$ and a positive constant k (depending on δ), such that $\mid \sigma(f; P) \mid < k$, whenever $P \in \mathcal{P}_1^{\#}([a, b]; \delta)$ (see Definition 5.4.1).*

Definition 5.6.2 (Schurle) *Let $f : [a, b] \mapsto \mathbb{R}$. f is said to satisfy the $(LSRD^{\#}S)$ (locally small $RD^{\#}$ sum) property on $[a, b]$, if for every $\epsilon > 0$ there exists $\delta : [a, b] \mapsto (0, +\infty)$ with the following property: if $x \in [y, z] \subset (x - \delta(x), x + \delta(x))$ for each $x \in [a, b]$ then $\mid \sigma(f; P) \mid < \epsilon$, whenever $P \in \mathcal{P}_1^{\#}([y, z]; \delta)$.*

Lemma 5.6.1 *Let $f : [a, b] \mapsto \mathbb{R}$.*

(i) If f is $(RD^{\#})$-integrable on $[a, b]$ then $f \in (LSRD^{\#}S)$ on $[a, b]$.

(ii) $(LSRD^{\#}S) \subseteq (BRD^{\#}S)$ on $[a, b]$.

Proof. (i) Let $F : [a, b] \mapsto \mathbb{R}$, $F(x) = (RD^{\#}) \int_a^x f(t)dt$ and let $\epsilon > 0$. Then there exists $\delta : [a, b] \mapsto \mathbb{R}$ such that $\mid \sigma(f; P) - F(b) \mid < \epsilon/4$, whenever $P \in \mathcal{P}_1^{\#}([a, b]; \delta)$. But $F \in \mathcal{C}$ on $[a, b]$, so F is uniformly continuous on $[a, b]$. Let $\mu > 0$ such that $\mid F(y) - F(x) \mid < \epsilon/4$, whenever $\mid x - y \mid < \mu$. Let $\delta_1(x) = \min\{\mu, \delta(x)\}$. Let $[u, v] \subset [t - \delta(t), t + \delta(t)), t \in (u, v)$, and let $P \in \mathcal{P}_1^{\#}([u, v]; \delta_1)$. By Lemma 5.4.3, (i), it follows that $\mid \sigma(f; P) \mid \leq \mid \sigma(f; P) - S(F; P) \mid + \mid S(F; P) \mid \leq 3\epsilon/4 + \mid F(v) - F(u) \mid < 3\epsilon/4 + \epsilon/4 = \epsilon$.

(ii) Since $f \in (LSRD^{\#}S)$ on $[a, b]$, it follows that for $\epsilon = 1$ there exists $\delta : [a, b] \mapsto (0, +\infty)$ with the following property: if $x \in [y, z] \subset (x - \delta(x), x + \delta(x))$ for each $x \in [a, b]$ then $\mid \sigma(f; P) \mid < 1$, whenever $P \in \mathcal{P}_1^{\#}([y, z]; \delta)$. By Lemma 2.24.1, there exist a partition $a = x_o < x_1 < \ldots < x_n = b$ and $t_i \in [x_{i-1}, x_i]$ such that $[x_{i-1}, x_i] \subset (t_i - \delta(t_i), t_i + \delta(t_i))$, $i = \overline{1, n}$. Let $\delta_1 : [a, b] \mapsto (0, +\infty)$ be defined as follows: $\delta_1(x) = \min\{(x_1 - a)/4, \delta(a)\}$ for $x = a$; $\delta_1(x) = \min\{(x_i - x_{i-1})/4, (x_{i+1} - x_i)/4, \delta(x_i)\}$, for $x = x_i$, $i = \overline{1, n - 1}$; $\delta_1(x) = \min\{(b - x_{n-1})/4, \delta(b)\}$ for $x = b$; $\delta_1(x) = \min\{\delta(x), x - x_{i-1}, x_i - x\}$ for $x \in (x_{i-1}, x_i)$, $i = \overline{1, n - 1}$. Let $P \in \mathcal{P}_1^{\#}([a, b]; \delta_1)$. Then for each $i \in \{1, 2, \ldots, n - 1\}$, P contains either a pair $([y_i, z_i]; x_i)$, or two pairs $([y_i, x_i]; x_i)$ and $([x_i, z_i]; x_i)$ with $x_i < \delta(x_i) < y_i < x_i < z_i < x_i + \delta(x_i)$. Since $f(x_i)(z_i - y_i) = f(x_i)(z_i - x_i) + f(x_i)(x_i - y_i)$, we may suppose without loss of generality only the second case. But P contains also a pair of the form $([a, z_o]; a)$ with $a < z_o < a + (x_1 - a)/4$ and a pair of the form $([z_n, b]; b)$ with $b - (b - x_{n-1})/4 < z_n < b$. Let $P_i = \{([y, z]; x) \in P : x \in [x_{i-1}, x_i] \text{ and } [y, z] \subset [x_{i-1}, x_i]\}$. Then $P = \cup_{i=1}^n P_i$ and $P_i \in \mathcal{P}_1^{\#}([x_{i-1}, x_i]; \delta_1)$, hence $\mid \sigma(f; P) \mid \leq \sum_{i=1}^n \mid \sigma(f; P_i) \mid < n$.

Lemma 5.6.2 *Let $f : [a, b] \mapsto \mathbb{R}$. The following assertions are equivalent:*

(i) $\overline{W}_1^{\#}(f) \times \underline{W}_1^{\#}(f) \neq \emptyset$;

(ii) $f \in (BRD^{\#}S)$ on $[a, b]$.

Proof. *(i)* $\Rightarrow$ *(ii)* Let $(M, m) \in \overline{W}_1^{\#}(f) \times \underline{W}_1^{\#}(f)$. Then $M(a) = m(a) = 0$, and there exists $\delta : [a, b] \mapsto (0, +\infty)$ such that $M(z) - M(y) \geq f(x)(z - y) \geq m(z) - m(y)$, whenever $[y, z] \subset (x - \delta(x), x + \delta(x))$, $x \in [a, b]$. Let $P \in \mathcal{P}_1^{\#}([a, b]; \delta)$. Then $m(b) \leq \sigma(f; P) \leq M(b)$, hence $| \sigma(f; P) | \leq | m(b) | + | M(b) |$.

(ii) $\Rightarrow$ *(i)* By (ii) there exist $\delta : [a, b] \mapsto (0, +\infty)$ and a positive constant k such that $| \sigma(f; P) | < k$, whenever $P \in \mathcal{P}_1^{\#}([a, b]; \delta)$. Let $a < c < b$. Let $P_o, P \in \mathcal{P}([a, b]; \delta)$ with $\sigma(P) = \sigma(P_o) = [a, c]$, and let $P' \in \mathcal{P}_1^{\#}([c, b]; \delta)$. Then $P_o \cup P'$ and $P \cup P'$ belong to $\mathcal{P}_1^{\#}([a, b]; \delta)$. It follows that $| \sigma(f; P) - \sigma(f; P_o) | = | \sigma(f; P \cup P') - \sigma(f; P \cup P') | \leq | \sigma(f; P_o \cup P') | + | \sigma(f; P \cup P') | < 2k$, hence $| \sigma(f; P) | \leq | \sigma(f; P_o) | + 2k$. Let $M(x) = \sup\{\sigma(f; P) : P \in \mathcal{P}([a, b]; \delta) \text{ and } \sigma(P) = [a, x]\}$ and $m(x) = \inf\{\sigma(f; P) : P \in \mathcal{P}^{\#}([a, b]; \delta) \text{ and } \sigma(P) = [a, x]\}$. Then $(M, m) \in \overline{W}_1^{\#}(f) \times \underline{W}_1^{\#}(f) \neq \emptyset$.

Theorem 5.6.1 *Let $f : [a, b] \mapsto \mathbb{R}$. The following assertions are equivalent:*

(i) f is summable on $[a, b]$;

(ii) f is measurable and $f \in (LSRD^{\#}S)$ on $[a, b]$;

(iii) f is measurable and $f \in (BRD^{\#}S)$ on $[a, b]$.

Proof. *(i)* $\Rightarrow$ *(ii)* Clearly f is measurable on $[a, b]$. By Theorem 5.4.1, f is $(RD^{\#})$-integrable on $[a, b]$. By Lemma 5.6.1, (i), $f \in (LSRD^{\#}S)$ on $[a, b]$.

(ii) $\Rightarrow$ *(iii)* See Lemma 5.6.1, (ii).

(iii) $\Rightarrow$ *(i)* By Lemma 5.6.2, it follows that $\overline{W}_1^{\#}(f) \times \underline{W}_1^{\#}(f) \neq \emptyset$, and by Theorem 5.5.1, f is summable on $[a, b]$.

Remark 5.6.1 *Theorem 5.6.1, (i) $\Leftrightarrow$ (ii) was shown by Schurle in [Sc3], and Theorem 5.6.1, (i) $\Leftrightarrow$ (iii) by Bullen and Vyborny in [BV] (our proof is different).*

5.7 Descriptive and Perron type definitions for the $\mathcal{D}^*$- integral

Definition 5.7.1 *Let $f : [a, b] \mapsto \overline{\mathbb{R}}$. We define the following classes of majorants:*

- $\overline{\mathcal{M}}_1(f) = \{M : [a, b] \mapsto \mathbb{R} : M(a) = 0,\ M \in AC^*G \cap C;\ M'(x)$ *exists (finite or infinite); $f(x) \leq M'(x) \neq -\infty\}$;*

- *(<u>Skljarenko</u>).* $\overline{\mathcal{M}}_2(f) = \{M : [a, b] \mapsto \mathbb{R} : M(a) = 0,\ M \in AC^*G \cap C;\ f(x) \leq \underline{D}M(x) \neq -\infty\}$;

- *(<u>Bauer</u>).* $\overline{\mathcal{M}}_3(f) = \{M : [a, b] \mapsto \mathbb{R} : M(a) = 0,\ M \in C;\ f(x) \leq \underline{D}M(x) \neq -\infty\}$;

- *(<u>Tolstoff</u>).* $\overline{\mathcal{M}}_4(f) = \{M : [a, b] \mapsto \mathbb{R} : M(a) = 0,\ M \in C;\ M'(x)$ *exists (finite or infinite); $f(x) \leq M'(x) \neq -\infty\}$;*

- *(<u>Saks</u>).* $\overline{\mathcal{M}}_5(f) = \{M : [a, b] \mapsto \mathbb{R} : M(a) = 0,\ M \in C \cap VB^*G \cap N^{-\infty}$ *on* $[a, b];\ f(x) \leq M'(x)$ *a.e. on* $[a, b]\}$;

- $\overline{\mathcal{M}}_6(f) = \{M : [a,b] \mapsto \mathbb{R} : M(a) = 0,\ M \in C_i^* \cap VB^*G \cap N^{-\infty}$ *on* $[a,b]$; $f(x) \leq M'(x)$ *a.e. on* $[a,b]\}$;
- *(Saks).* $\overline{\mathcal{M}}_7(f) = \{M : [a,b] \mapsto \mathbb{R} : M(a) = 0,\ f(x) \leq \underline{D}M(x) \neq -\infty\}$;
- $\overline{\mathcal{M}}_8(f) = \{M : [a,b] \mapsto \mathbb{R} : M(a) = 0,\ M \in C_i \cap \underline{AC^*}G$ *on* $[a,b]$; $f(x) \leq M'(x)$ *a.e. on* $[a,b]\}$;
- $\overline{\mathcal{M}}_9(f) = \{M : [a,b] \mapsto \mathbb{R} : M(a) = 0,\ M \in C_i \cap T_2 \cap N^{-\infty}$ *on* $[a,b]$; $f(x) \leq M'(x)$ *a.e. where* $M'(x)$ *exists on* $[a,b]\}$;
- $\overline{\mathcal{M}}_{10}(f) = \{M : [a,b] \mapsto \mathbb{R} : M(a) = 0,\ M \in C_i \cap \underline{\mathcal{M}}$ *on* $[a,b]$; $f(x) \leq M'(x)$ *a.e. where* $M'(x)$ *exists on* $[a,b]\}$;
- $\overline{\mathcal{M}}_{11}(f) = \{M : [a,b] \mapsto \mathbb{R} : M(a) = 0,\ M \in \mathcal{D}_-\mathcal{B}_1T_2 \cap N^{-\infty}$ *on* $[a,b]$; $f(x) \leq M'(x)$ *a.e. where* $M'(x)$ *exists on* $[a,b]\}$;
- $\overline{\mathcal{M}}_{12}(f) = \{M : [a,b] \mapsto \mathbb{R} : M(a) = 0,\ M \in \mathcal{D}_- \cap [CG]$ *on* $[a,b]$; $M \in \underline{\mathcal{M}}$ *on each closed interval contained in* $A = int\{x : M \in C_i \text{ at } x\}$; $f(x) \leq M'(x)$ *a.e. where* $M'(x)$ *exists on* $A\}$;
- *(Hake).* $\overline{\mathcal{M}}_{13}(f) = \{M : [a,b] \mapsto \mathbb{R} : M(a) = 0,\ M \in C;\ f(x) \leq \underline{D}^+M(x) \neq -\infty\}$;
- *(McShane).* $\overline{\mathcal{M}}_{14}(f) = \{M : [a,b] \mapsto \mathbb{R} : M(a) = 0,\ M \in C$ *on* $[a,b]$; $f(x) \leq \underline{D}^+M(x)$ *a.e.*; $\underline{D}^+M(x) \neq -\infty$ *n.e.*$\}$;
- $\overline{\mathcal{M}}_{15}(f) = \{M : [a,b] \mapsto \mathbb{R} : M(a) = 0,\ M \in C_i^*;\ f(x) \leq \max\{\underline{D}^+M(x), \underline{D}^-M(x)\}$ *a.e.*; $\max\{\underline{D}^+M(x), \underline{D}^-M(x)\} \neq -\infty$ *n.e.*$\}$;
- $\overline{\mathcal{M}}_{16}(f) = \{M : [a,b] \mapsto \mathbb{R} : M(a) = 0,\ M \in C_i;\ f(x) \leq \max\{\underline{D}^+M(x), \underline{D}^-M(x)\}$ *a.e.*; $\max\{\underline{D}^+M(x), \underline{D}^-M(x)\} \neq -\infty$ *n.e.*$\}$;

We define the following classes of minorants: $\underline{\mathcal{M}}_j(f) = \{m : [a,b] \mapsto \mathbb{R} : -m \in \overline{\mathcal{M}}_j(-f)\}$, $j = \overline{1,16}$.
Let $j \in \{1,2,\ldots,16\}$. *If* $\overline{\mathcal{M}}_j(f) \neq \emptyset$ *then we denote by* $\overline{I}_j(b)$ *the lower bound of all* $M(b)$, $M \in \overline{\mathcal{M}}_j(f)$. *If* $\underline{\mathcal{M}}_j(f) \neq \emptyset$ *then we denote by* $\underline{I}_j(b)$ *the upper bound of all* $m(b)$, $m \in \underline{\mathcal{M}}_j(f)$.
Let $(j,k) \in (\{1,2,\ldots,6\}\times\{1,2,\ldots,12\}) \cup (\{7,8\}\times\{1,2,\ldots,10\}) \cup (\{13,14,15,16\}\times\{13,14,15,16\})$. f *is said to have a* $(\mathcal{P}_{j,k})$*-integral on* $[a,b]$, *if* $\overline{\mathcal{M}}_j \times \underline{\mathcal{M}}_k(f) \neq \emptyset$ *and* $\overline{I}_j(b) = \underline{I}_k(b) = (\mathcal{P}_{j,k}) \int_a^b f(t)dt$.

Definition 5.7.2 *Let* $f : [a,b] \mapsto \overline{\mathbb{R}}$ *and let* $P \subseteq [a,b]$. f *is said to have a* $\mathcal{D}^*$*-integral (Denjoy* - integral) on* $[a,b]$, *if there exists* $F : [a,b \mapsto \mathbb{R}$ *such that* $F \in AC^*G \cap C$ *on* $[a,b]$ *and* $F'(x) = f(x)$ *a.e. on* $[a,b]$. *Then* $(\mathcal{D}^*)\int_a^b f(t)dt = F(b) - F(a)$.
f *is said to be* $\mathcal{D}^*$*-integrable on* P, *if* $f \cdot K_P$ *is* $\mathcal{D}^*$*-integrable on* $[a,b]$.

Remark 5.7.1 *The above defined* $\mathcal{D}^*$*-integral is well-defined (see Corollary 4.3.3).*

Definition 5.7.3 *Let* $F : [a,b] \mapsto \mathbb{R}$. *Let* $F^* : [a,b] \mapsto \mathbb{R}$ *be defined as follows:* $F^*(x) = F'(x)$ *where* F *is derivable, and* $F^*(x) = 0$ *elsewhere.*

Theorem 5.7.1 *Let $f : [a,b] \mapsto \overline{\mathbb{R}}$. The following assertions are equivalent:*

(i) f is $\mathcal{D}^*$-integrable on $[a,b]$;

(ii) There exists a function $F : [a,b] \mapsto \mathbb{R}$ such that : a) $F \in \mathcal{DB}_1 \cap (N)$ on $[a,b]$; b) $F'(x) = f(x)$ a.e. where F is derivable; c) F^* is $\mathcal{D}^*$-integrable on $[a,b]$;

(iii) There exists a function $F : [a,b] \mapsto \mathbb{R}$ such that : a) $F \in \mathcal{DB}_1 T_2 \cap N^\infty$ on $[a,b]$; b) $F'(x) = f(x)$ a.e. where F is derivable; c) F^* is $\mathcal{D}^*$- integrable on $[a,b]$;

(iv) There exists a function $F : [a,b] \mapsto \mathbb{R}$ such that : a) $F \in \mathcal{C} \cap (N)$ on $[a,b]$; b) $F'(x) = f(x)$ a.e. where $F'(x)$ exists (finite or infinite); c) $\overline{\mathcal{M}}_8(f) \neq \emptyset$;

(v) There exists a function $F : [a,b] \mapsto \mathbb{R}$ such that : a) $F \in \mathcal{C} \cap T_2 \cap N^\infty$ on $[a,b]$; b) $F'(x) = f(x)$ a.e. where $F'(x)$ exists (finite or infinite); c) $\overline{\mathcal{M}}_8(f) \neq \emptyset$;

(vi) There exists a function $F : [a,b] \mapsto \mathbb{R}$ such that : a) $F \in \mathcal{DB}_1 T_2 \cap (N)$ on $[a,b]$; b) $F'(x) = f(x)$ a.e. where $F'(x)$ exists (finite or infinite); c) $\overline{\mathcal{M}}_6(f) \neq \emptyset$;

(vii) There exists a function $F : [a,b] \mapsto \mathbb{R}$ such that : a) $F \in \mathcal{DB}_1 T_2 \cap N^\infty$ on $[a,b]$; b) $F'(x) = f(x)$ a.e. where $F'(x)$ exists (finite or infinite); c) $\overline{\mathcal{M}}_6(f) \neq \emptyset$;

(viii) There exists a function $F : [a,b] \mapsto \mathbb{R}$ such that : a) $F \in \mathcal{D} \cap [\mathcal{C}G]$ on $[a,b]$; b) $F \in (M)$ on each closed interval contained in $A = int\{x : F \in \mathcal{C}$ at $x\}$; c) $F'(x) = f(x)$ a.e. where F is derivable on A; d) F^* is $\mathcal{D}^*$-integrable on $[a,b]$;

(ix) There exists a function $F : [a,b] \mapsto \mathbb{R}$ such that : a) $F \in (M) \cap \mathcal{C}$ on $[a,b]$; b) $F'(x) = f(x)$ a.e. where $F'(x)$ exists (finite or infinite); c) $\overline{\mathcal{M}}_8(f) \neq \emptyset$;

(x) There exists a function $F : [a,b] \mapsto \mathbb{R}$ such that : a) $F \in \mathcal{D} \cap [\mathcal{C}G]$ on $[a,b]$; b) $F \in (M)$ on each closed interval contained in $A = int\{x : F \in \mathcal{C}$ at $x\}$; c) $F'(x) = f(x)$ a.e. where $F'(x)$ exists (finite or infinite) on A; d) $\overline{\mathcal{M}}_5(f) \neq \emptyset$.

In all situations we have $(\mathcal{D}^*) \int_a^b f(t)dt = F(b) - F(a)$.

Proof. $(i) \Rightarrow (ii)$ Let $F(x) = (\mathcal{D}^*) \int_a^x f(t)dt$. Since $\mathcal{C} \cap AC^*G \subset \mathcal{DB}_1 \cap (N)$ we have (ii).

$(ii) \Rightarrow (iii)$ This is evident.

$(iii) \Rightarrow (i)$ Let $G(x) = (\mathcal{D}^*) \int_a^x F^*(t)dt$, and let $H = G - F$ on $[a,b]$. Then $G \in \mathcal{C} \cap AC^*G$ on $[a,b]$ and $G'(x) = F^*(x)$ *a.e.* on $[a,b]$. It follows that $\{x : F$ is derivable at $x\} = \{x : H$ is derivable at $x\}$ *a.e.*. Also, $H'(x) = 0$ *a.e.* where H is derivable. By Corollary 4.3.6, (v), $-F \in \mathcal{C} \cap AC^*G$ on $[a,b]$, hence $F \in \mathcal{C} \cap AC^*G$ on $[a,b]$. Then F is derivable *a.e.* on $[a,b]$, hence by (iii), b), c), $F'(x) = f(x) = F^*(x)$ *a.e.* on $[a,b]$. Also, f is $\mathcal{D}^*$-integrable and $(\mathcal{D}^*) \int_a^x f(t)dt = F(x) - F(a)$, $x \in [a,b]$.

$(i) \Rightarrow (iv)$ Let $F(x) = (\mathcal{D}^*) \int_a^x f(t)dt$. Then $F \in AC^*G \cap \mathcal{C}$ on $[a,b]$ and $F'(x) = f(x)$ *a.e.* on $[a,b]$. It follows that $F \in \mathcal{C} \cap (N)$ on $[a,b]$ and $F \in \overline{\mathcal{M}}_8(f) \neq \emptyset$.

$(iv) \Rightarrow (v)$ This is evident.

$(v) \Rightarrow (i)$ Let $G : [a,b] \mapsto \mathbb{R}$, $G \in \overline{\mathcal{M}}_8(f)$. Then $G \in \underline{AC}^*G \cap \mathcal{C}_i$ and $G'(x) \geq f(x)$ *a.e.* on $[a,b]$. Let $H = G - F$ on $[a,b]$. Then $\{x : H$ is derivable at $x\} = \{x : F$ is derivable at $x\}$ *a.e.*. Since $F'(x) = f(x)$ *a.e.* where $F'(x)$ exists (finite or infinite), it follows that $H'(x) \geq 0$ *a.e.* where $H'(x)$ exists (finite or infinite). By Corollary 4.3.6, (vi), $F \in AC^*G \cap \mathcal{C}$ on $[a,b]$, hence F is derivable *a.e.* on $[a,b]$. Then $F'(x) = f(x)$ *a.e.* on $[a,b]$, hence f is $\mathcal{D}^*$-integrable on $[a,b]$ and $(\mathcal{D}^*)\int_a^x f(t)dt = F(x) - F(a)$, $x \in [a,b]$.

(i) ⇒ *(vi)* The proof is similar to that of *(i)* ⇒ *(iv)*.

(vi) ⇒ *(vii)* This is evident.

(vii) ⇒ *(i)* The proof is similar to that of *(v)* ⇒ *(i)*, using Corollary 4.3.6, (v).

(i) ⇒ *(viii)* This is evident.

(viii) ⇒ *(i)* Let $G(x) = (\mathcal{D}^*)\int_a^x F^*(t)dt$, and let $H = G - F$ on $[a,b]$. Then $G \in \mathcal{C} \cap AC^*G$ on $[a,b]$ and $G'(x) = F^*(x)$ *a.e.* on $[a,b]$. It follows that $\{x : F$ is derivable at $x\} = \{x : H$ is derivable at $x\}$ *a.e.*. By (viii), $H'(x) = 0$ *a.e.* where $H'(x)$ exists (finite or infinite). By Corollary 2.5.1,(iv), $H \in \mathcal{D} \cap [\mathcal{C}G]$ on $[a,b]$. Clearly $int\{x : F \in \mathcal{C}$ at $x\} = int\{x : H \in \mathcal{C}$ at $x\} = A$. Let $[c,d] \subset A$. Then $F \in (M)$ on $[c,d]$. By Theorem 2.23.2, (vi), $H \in (M)$ on $[c,d]$. By Corollary 4.2.4, H is AC and increasing on $[a,b]$. Since H is derivable *a.e.* on $[a,b]$, and $H'(x) = 0$ *a.e.* where $H'(x)$ exists, it follows that H is constant on $[a,b]$, hence $F(x) = G(x)$ on $[a,b]$. So $F \in AC^*G \cap \mathcal{C}$ on $[a,b]$ and $F'(x) = F^*(x) = f(x)$ *a.e.* on $[a,b]$. It follows that f is $\mathcal{D}^*$-integrable on $[a,b]$ and $(\mathcal{D}^*)\int_a^x f(t)dt = F(x) - F(a)$, $x \in [a,b]$.

(i) ⇒ *(ix)* Let $F(x) = (\mathcal{D}^*)\int_a^x f(t)dt$. Then $F \in AC^*G \cap \mathcal{C}$ on $[a,b]$ and $F'(x) = f(x)$ *a.e.* on $[a,b]$. It follows that $F \in (M) \cap \mathcal{C}$ on $[a,b]$, and $F \in \overline{\mathcal{M}}_8(f) \neq \emptyset$.

(ix) ⇒ *(i)* Let $G : [a,b] \mapsto \mathbb{R}$, $G \in \overline{\mathcal{M}}_8(f)$ and let $H = G - F$ on $[a,b]$. Then $G \in \underline{AC}^*G \cap \mathcal{C}_i$ on $[a,b]$, $G'(x) \geq f(x)$ *a.e.* on $[a,b]$, and $\{x : H$ is derivable at $x\} = \{x : F$ is derivable at $x\}$ *a.e.*. Since $F'(x) = f(x)$ *a.e.* where $F'x)$ exists (finite or infinite), it follows that $H'(x) \geq 0$ *a.e.* where $H'(x)$ exists (finite or infinite). By Proposition 2.3.1, (iv), $H \in \mathcal{C}_i$ on $[a,b]$, and by Theorem 2.23.2,(vi), $H \in \underline{M}$ on $[a,b]$. By Theorem 4.2.1, H is increasing on $[a,b]$. By Proposition 2.12.1, (iv), $G \in VB^*G$ on $[a,b]$. Since H is increasing on $[a,b]$, $F \in VB^*G$ on $[a,b]$. Because $F \in (M) \cap \mathcal{C}$ on $[a,b]$, it follows that $F \in ACG \cap \mathcal{C} \subset (N)$. By Corollary 2.22.1, (v), $F \in AC^*G \cap \mathcal{C}$ on $[a,b]$, hence F is derivable *a.e.* on $[a,b]$. By (ix), b), $F'(x) = f(x)$ *a.e.* on $[a,b]$. Hence F is $\mathcal{D}^*$-integrable on $[a,b]$ and $(\mathcal{D}^*)\int_a^x f(t)dt = F(x) - F(a)$, $x \in [a,b]$.

(i) ⇒ *(x)* This is evident.

(x) ⇒ *(i)* Let $G : [a,b] \mapsto \mathbb{R}$, $G \in \overline{\mathcal{M}}_5(f)$. Then $G \in \mathcal{C} \cap VB^*G \cap N^{-\infty} \subset [\underline{AC}^*G] \cap \mathcal{C} \subset [\underline{ACG}] \cap \mathcal{C}$ (the first inclusion follows by Corollary 2.22.1, (ii)). Let $H = G - F$ on $[a,b]$. By Theorem 2.23.2,(vi), $H \in \underline{M}$ on $[a,b]$. Since $G \in \mathcal{C}$ on $[a,b]$, by Corollary 2.5.1, (iv), it follows that $H \in [\mathcal{C}G] \cap \mathcal{D}$ on $[a,b]$. But $A = int\{x : F \in \mathcal{C}$ at $x\} = int\{x : H \in \mathcal{C}$ at $x\}$. Since $G \in VB^*G$ on $[a,b]$, G is derivable *a.e.* on $[a,b]$, hence $\{x \in A : H$ is derivable at $x\} = \{x \in A : F$ is derivable at $x\}$ *a.e.*. Because $G'(x) \geq f(x)$ *a.e.* on $[a,b]$, it follows that $H'(x) \geq 0$ *a.e.* where $H'(x)$ exists (finite or infinite) on A. By Corollary 4.2.4, H is increasing and $\mathcal{C}$ on $[a,b]$. It follows that $F = G - H \in VB^*G \cap \mathcal{C} \cap (M) = AC^*G \cap \mathcal{C}$. Hence F is derivable *a.e.* on $[a,b]$, and $F'(x) = f(x)$ *a.e.*on $[a,b]$. Thus f is $\mathcal{D}^*$- integrable on $[a,b]$ and $(\mathcal{D}^*)\int_a^x f(t)dt = F(x) - F(a)$, $x \in [a,b]$.

Lemma 5.7.1 *Let* $f : [a,b] \mapsto \overline{\mathbb{R}}$.

(i) If $(M,m) \in \overline{\mathcal{M}}_{16}(f) \times \underline{\mathcal{M}}_{16}(f)$ *then* $(M,m) \in \overline{\mathcal{M}}_8(f) \times \underline{\mathcal{M}}_8(f)$ *and* $M - m$ *is increasing on* $[a,b]$.

(ii) (Sarkhel). *If* $(M,m) \in \overline{\mathcal{M}}_{15}(f) \times \underline{\mathcal{M}}_{15}(f)$ *then* $(M,m) \in \overline{\mathcal{M}}_6(f) \times \underline{\mathcal{M}}_6(f)$ *and* $M - m$ *is increasing on* $[a,b]$.

Proof. (i) We have $\max\{\underline{D}^+M(x), \underline{D}^-M(x)\} \geq f(x) \geq \min\{\overline{D}^+m(x), \overline{D}^-m(x)\}$ *a.e.* on $[a,b]$, $\max\{\underline{D}^+M(x), \underline{D}^-M(x)\} \neq -\infty$ *n.e.* and $\min\{\overline{D}^+m(x), \overline{D}^-m(x)\} \neq +\infty$ *n.e.*. There are four situations:

(I) We have

$$(1) \quad \max\{\underline{D}^+M(x), \underline{D}^-M(x)\} = \underline{D}^+M(x) \text{ and}$$

$$(2) \quad \min\{\overline{D}^+m(x), \overline{D}^-m(x)\} = \overline{D}^+m(x).$$

Let $E_{+,+} = \{x : \underline{D}^+M(x) > -\infty, \overline{D}^+m(x) < +\infty\}$. We have $0 \leq \underline{D}^+M(x) - \overline{D}^+m(x) \leq \underline{D}^+(M-m)(x) \leq \overline{D}^+(M-m)(x)$ *a.e.* on $A_{+,+}$, where $A_{+,+} = \{x : x$ satisfies (1) and (2)$\}$ (see Lemma 1.10.1, (vii)).

(II) We have

$$(3) \quad \max\{\underline{D}^+M(x), \underline{D}^-M(x)\} = \underline{D}^-M(x) \text{ and}$$

$$(4) \quad \min\{\overline{D}^+m(x), \overline{D}^-m(x)\} = \overline{D}^-m(x).$$

Let $E_{-,-} = \{x : \underline{D}^-M(x) > -\infty, \overline{D}^-m(x) < +\infty\}$. We have $0 \leq \underline{D}^-M(x) - \overline{D}^-m(x) \leq \underline{D}^-(M-m)(x) \leq \overline{D}^+(M-m)(x)$ *a.e.* on $A_{-,-}$, where $A_{-,-} = \{x : x$ satisfies (3) and (4)$\}$ (see Lemma 1.10.1, (vii) and Theorem 1.10.1, (iii)).

(III) We have

$$(5) \quad \max\{\underline{D}^-M(x), \underline{D}^+M(x)\} = \underline{D}^+M(x) \text{ and}$$

$$(6) \quad \min\{\overline{D}^-m(x), \overline{D}^+m(x)\} = \overline{D}^-m(x).$$

Let $E_{+,-} = \{x : \underline{D}^+M(x) > -\infty, \overline{D}^-m(x) < +\infty\}$. Then $\underline{D}^+M(x) \leq \overline{D}^-M(x)$ *n.e.* (see Theorem 1.10.1, (iii)), hence $E_{+,-} \subset E_{-,-}$ *n.e.*. We have $0 \leq \underline{D}^+M(x) - \overline{D}^-m(x) \leq \underline{D}^+M(x) - \underline{D}^+m(x) \leq \overline{D}^+(M-m)(x)$ *a.e.* on $A_{+,-}$, where $A_{+,-} = \{x : x$ satisfies (5) and (6)$\}$ (see Lemma 1.10.1, (viii) and Theorem 1.10.1, (iii)).

(IV) We have

$$(7) \quad \max\{\underline{D}^-M(x), \underline{D}^+M(x)\} = \underline{D}^-M(x) \text{ and}$$

$$(8) \quad \min\{\overline{D}^-m(x), \overline{D}^+m(x)\} = \overline{D}^+m(x).$$

Let $E_{-,+} = \{x : \underline{D}^-M(x) > -\infty, \overline{D}^+m(x) < +\infty\}$. Then $\underline{D}^-M(x) \leq \overline{D}^+M(x)$ *n.e.* (see Theorem 1.10.1, (iii)), hence $E_{-,+} \subset E_{+,+}$ *n.e.*. We have $0 \leq \underline{D}^-M(x) - \overline{D}^+m(x) \leq \overline{D}^+M(x) - \overline{D}^+m(x) \leq \overline{D}^+(M-m)(x)$ *a.e.* on $A_{-,+}$, where $A_{-,+} = \{x : x$ satisfies (7) and (8)$\}$ (see Lemma 1.10.1, (vi) and Theorem 1.10.1, (iii)).

Clearly $A_{+,+} \cup A_{-,-} \cup A_{+,-} \cup A_{-,+} = [a,b]$ *a.e.*, hence $\overline{D}^+(M-m)(x) \geq 0$ *a.e.*. Also $E_{+,+} \cup E_{-,-} \cup E_{+,-} \cup E_{-,+} = [a,b]$ *n.e.*. Since $E_{+,-} \subset E_{-,-}$ *n.e.* and $E_{-,+} \subset E_{+,+}$ *n.e.*, it follows that $E_{+,+} \cup E_{-,-} = [a,b]$ *n.e.*. Since $\max\{\underline{D}^+M(x), \underline{D}^-M(x)\} \neq -\infty$ *n.e.* on $[a,b]$, and $\overline{D}^+M(x) \geq \underline{D}^-M(x)$ *n.e.* (see Theorem 1.10.1, (iii)), it follows that $\overline{D}^+M(x) > -\infty$ *n.e.* on $[a,b]$. Similarly $\underline{D}^+m(x) < +\infty$ *n.e.* on $[a,b]$. By Lemma 1.10.1, (v), it follows that $-\infty < \overline{D}^+M(x) - \underline{D}^+m(x) \leq \overline{D}^+(M-m)(x)$ *n.e.* on $[a,b]$. Since $M - m \in \mathcal{C}_i$ on $[a,b]$, by Corollary 4.9.3, $M - m$ is increasing on $[a,b]$. We show that $M \in VB^*G$ on $E_{-,-}$. Let $E_n = \{x \in E_{-,-} : M(x) - M(y) > -n(x-y)$

and $m(x) - m(y) < n(x - y)$, whenever $y \in (x - 1/n, x)\}$, $n = \overline{1,\infty}$. Let $x \in E_n$ and $y \in (x - 1/n, x)$. Since $h = M - m$ is increasing on $[a,b]$, it follows that $-n(x-y) - h(x) + h(y) < M(x) - M(y) = n(x-y) + h(x) - h(y)$, hence

$$(9) \qquad | M(x) - M(y) | < n(x-y) + h(x) - h(y).$$

Let $E_{n,i} = E_n \cap [i/n, (i+1)/n)$, $i = 0, \pm 1, \pm 2, \ldots$. Clearly $E_{-,-} = \cup_{n=1}^{\infty} E_n = \cup_{n,i} E_{n,i}$. Let $x_1, x_2 \in E_{n,i}$, $x_1 < x_2$. Let $\alpha_1, \alpha_2 \in [x_1, x_2]$. By (9) we have $| M(\alpha_2) - M(\alpha_1) | \leq | M(\alpha_2) - M(x_2) | + | M(x_2) - M(\alpha_1) | < n(x_2 - \alpha_2) + h(x_2) - h(\alpha_2) - n(x_2 - \alpha_1) + h(x_2) - h(\alpha_1) \leq 2(n(x_2 - x_1) + h(x_2) - h(x_1))$. It follows that

$$(10) \qquad \mathcal{O}(M; [x_1, x_2]) \leq 2(n(x_2 - x_1) + h(x_2) - h(x_1)).$$

Let $\{[a_k, b_k]\}$, $k = \overline{1,p}$ be a finite set of nonoverlapping closed intervals with endpoints in $E_{n,i}$. By (10) we have $\sum_{k=1}^{p} \mathcal{O}(M; [a_k, b_k]) < 2n \cdot (1/n) + 2(h((i+1)/n) - h(i/n)) < 2(1 + h(b) - h(a))$, hence $M \in VB^*$ on $E_{n,i}$. Thus $M \in VB^*G$ on $E_{-,-}$. Similarly $M \in VB^*G$ on $E_{+,+}$, hence $M \in VB^*G$ on $[a,b]$. It follows that $m = M - h \in VB^*G$ on $[a,b]$. Clearly $M \in N^{-\infty}$ on $[a,b]$. It follows that $M \in \mathcal{C}_i \cap VB^*G \cap N^{-\infty} = \mathcal{C}_i \cap \underline{AC}^*G$ on $[a,b]$ (see Corollary 2.22.1, (iv)). Since $M \in VB^*G$ on $[a,b]$, M is derivable a.e. on $[a,b]$. Hence $M'(x) \geq f(x)$ a.e. on $[a,b]$. Therefore $M \in \overline{\mathcal{M}}_8(f)$. Similarly $m \in \underline{\mathcal{M}}_8(f)$.

(ii) The proof is similar to that of (i).

Lemma 5.7.2 *Let $f : [a,b] \mapsto \overline{\mathbb{R}}$.*

(i) If $\overline{\mathcal{M}}_1(f) \neq \emptyset$ then $\overline{\mathcal{M}}_1(f) \subseteq \overline{\mathcal{M}}_2(f) \subseteq \overline{\mathcal{M}}_3(f)$; $\overline{\mathcal{M}}_1(f) \subseteq \overline{\mathcal{M}}_2(f) \subseteq \overline{\mathcal{M}}_4(f) \subseteq \overline{\mathcal{M}}_7(f)$; $\overline{\mathcal{M}}_1(f) \subseteq \overline{\mathcal{M}}_i(f) \subseteq \overline{\mathcal{M}}_8(f)$, $i = \overline{2,7}$; $\overline{\mathcal{M}}_1(f) \subseteq \overline{\mathcal{M}}_i(f) \subseteq \overline{\mathcal{M}}_6(f)$, $i = 2,3,4,5$.

(ii) If $\overline{\mathcal{M}}_8(f) \neq \emptyset$ the $\overline{\mathcal{M}}_8(f) \subseteq \overline{\mathcal{M}}_9(f) \subseteq \overline{\mathcal{M}}_{11}(f)$ and $\overline{\mathcal{M}}_8(f) \subseteq \overline{\mathcal{M}}_{10}(f) \subseteq \overline{\mathcal{M}}_{12}(f)$.

(iii) If $\overline{\mathcal{M}}_1(f) \neq \emptyset$ then $\overline{\mathcal{M}}_1(f) \subseteq \overline{\mathcal{M}}_{13}(f) \subseteq \overline{\mathcal{M}}_{14}(f) \subseteq \overline{\mathcal{M}}_{15}(f) \subseteq \overline{\mathcal{M}}_{16}(f)$.

Proof. We show that $\overline{\mathcal{M}}_7(f) \subseteq \overline{\mathcal{M}}_8(f)$. Let $M \in \overline{\mathcal{M}}_7(f)$. Since $\underline{D}M(x) > -\infty$, it follows that $M \in \mathcal{C}_i \cap N^{-\infty} \cap VB^*G$ on $[a,b]$ (see Corollary 2.13.2). The other parts follow similarly.

Lemma 5.7.3 *Let $f : [a,b] \mapsto \overline{\mathbb{R}}$.*

(i) Let $(j,k) \in (\{1,2,\ldots,8\} \times \{1,2,\ldots,10\}) \cup (\{13,14,15,16\} \times \{13,14,15,16\})$. If $(M,m) \in \overline{\mathcal{M}}_j(f) \times \underline{\mathcal{M}}_k(f) \neq \emptyset$ then $M - m$ is increasing on $[a,b]$ and $(M,m) \in \overline{\mathcal{M}}_8(f) \times \underline{\mathcal{M}}_8(f)$.

(ii) Let $(j,k) \in (\{1,2,\ldots,6\} \times \{1,2,\ldots,12\}) \cup (\{13,14,15\} \times \{13,14,15\})$. If $(M,m) \in \overline{\mathcal{M}}_j(f) \times \underline{\mathcal{M}}_k(f) \neq \emptyset$ then $M - m$ is increasing on $[a,b]$ and $(M,m) \in \overline{\mathcal{M}}_6(f) \times \underline{\mathcal{M}}_6(f)$.

Proof. (i) Let $(j,k) \in \{1,2,\ldots,8\} \times \{1,2,\ldots,10\}$. By Lemma 5.7.2, (i), (ii), we have two cases: a) $(M,m) \in \overline{\mathcal{M}}_8(f) \times \underline{\mathcal{M}}_9(f)$. By Corollary 4.3.6, (i), $M - m$ is increasing on $[a,b]$ and $-m \in [\underline{AC}^*G]$ on $[a,b]$. Since $-m \in \mathcal{C}_i$ on $[a,b]$, it follows that $m \in \underline{\mathcal{M}}_8(f)$.

b) $(M,m) \in \overline{\mathcal{M}}_8(f)\times\underline{\mathcal{M}}_{10}(f)$. By Theorem 4.2.1, $M-m$ is increasing on $[a,b]$. Since $M \in \overline{\mathcal{M}}_8(f)$, $M \in VB^*G$ on $[a,b]$. Hence $m = M-(M-m)$ is VB^*G on $[a,b]$. Since $-m \in \mathcal{C}_i \cap VB^*G \cap \underline{M}$ on $[a,b]$, it follows that $-m \in \underline{AC}^*G$ on $[a,b]$, so $m \in \underline{\mathcal{M}}_8(f)$. Let $(j,k) \in \{13,14,15,16\} \times \{13,14,15,16\}$. By Lemma 5.7.2, (iii), $(M,m) \in \overline{\mathcal{M}}_{16}(f)\times\underline{\mathcal{M}}_{16}(f)$. By Lemma 5.7.1, (i),$M-m$ is increasing on $[a,b]$ and $(M,m) \in \overline{\mathcal{M}}_8(f)\times\underline{\mathcal{M}}_8(f)$.

(ii) Let $(j,k) \in \{1,2,\dots,6\} \times \{1,2,\dots,12\}$. By Lemma 5.7.2, (i), (ii), we have two cases: a) $(M,m) \in \overline{\mathcal{M}}_6(f)\times\underline{\mathcal{M}}_{11}(f)$. By Corollary 4.3.6, (ii), $M-m$ is increasing on $[a,b]$ and $-m \in \underline{AC}^*G \cap \mathcal{C}_i$ on $[a,b]$, hence $m \in \underline{\mathcal{M}}_6(f)$.
b) $(M,m) \in \overline{\mathcal{M}}_6(f)\times\underline{\mathcal{M}}_{12}(f)$. By Corollary 4.2.1, $H = M-m$ is increasing on $[a,b]$. Since $M \in VB^*G$ on $[a,b]$, it follows that $m \in VB^*G$ on $[a,b]$. Since $M \in \mathcal{C}_i^*$ on $[a,b]$, $M(x+)$ exists and is finite for each $x \in [a,b)$. Since H is increasing on $[a,b]$, $H(x+)$ exists and is finite. It follows that $m(x+) = M(x+) - H(x+)$ exists and is finite for each $x \in [a,b)$. Similarly $m(x-)$ exists and is finite for each $x \in (a,b]$. Since $-m \in \mathcal{D}_- \subset$ *lower internal* on $[a,b]$, by Remark 2.4.1, $-m \in \mathcal{C}_i^*$ on $[a,b]$. It follows that $-m \in \mathcal{C}_i^* \cap VB^*G \cap \underline{M}$ on $[a,b]$, hence $-m \in \mathcal{C}_i^* \cap [\underline{AC}^*G]$ on $[a,b]$. Thus $m \in \underline{\mathcal{M}}_6(f)$.
Let $(j,k) \in \{13,14,15\}\times\{13,14,15\}$. By Lemma 5.7.2, (iii), it follows that $(M,m) \in \overline{\mathcal{M}}_{15}(f)\times\underline{\mathcal{M}}_{15}(f)$. By Lemma 5.7.1, (ii), $M-m$ is increasing on $[a,b]$ and $(M,m) \in \overline{\mathcal{M}}_6(f)\times\underline{\mathcal{M}}_6(f)$.

Lemma 5.7.4 *Let $f : [a,b] \mapsto \overline{\mathbb{R}}$. Let $(j,k) \in (\{1,2,\dots,6\} \times \{1,2,\dots,12\}) \cup (\{7,8\}\times\{1,2,\dots,10\}) \cup (\{13,14,15,16\}\times\{13,14,15,16\})$. Then: (i) $M-m$, (ii) $M-\underline{I}_k$, (iii) $\overline{I}_j - m$, and (iv) $\overline{I}_j - \underline{I}_k$ are positive increasing functions on $[a,b]$. Hence $M(b) \geq m(b)$, $M(b) \geq \underline{I}_k(b)$, $\overline{I}_j(b) \geq m(b)$ and $\overline{I}_j(b) \geq \underline{I}_k(b)$.*

Proof. For (i) see Lemma 5.7.3. The other parts follow similarly to Lemma 5.1.4.

Lemma 5.7.5 *Let $f : [a,b] \mapsto \overline{\mathbb{R}}$. Let $(j,k) \in (\{1,2,\dots,6\} \times \{1,2,\dots,12\}) \cup (\{7,8\}\times\{1,2,\dots,10\}) \cup (\{13,14,15,16\}\times\{13,14,15,16\})$. The following assertions are equivalent:*

(i) f is $(\mathcal{P}_{j,k})$-integrable on $[a,b]$;

(ii) For $\epsilon > 0$ there exists $(M,m) \in \overline{\mathcal{M}}_j(f)\times\underline{\mathcal{M}}_k(f) \neq \emptyset$ such that $M(b)-m(b) < \epsilon$.

Proof. This follows by definitions and Lemma 5.7.4.

5.8 An improvement of the Hake Theorem

Lemma 5.8.1 *Let $F, H : [a,b] \mapsto \mathbb{R}$, $H(x) = \mathcal{O}(F;[a,x])$. If $F \in AC^*G\cap\mathcal{C}$ then H is AC and increasing on $[a,b]$.*

Proof. Clearly H is increasing on $[a,b]$. For $x_1, x_2 \in [a,b]$, $x_1 < x_2$ we have

(1) $\quad H(x_2) - H(x_1) \leq \mathcal{O}(F;[x_1,x_2])$.

But $F \in \mathcal{C}$, so $H \in \mathcal{C}$ on $[a,b]$. We show that $H \in AC^*G$ on $[a,b]$. Since $F \in AC^*G\cap\mathcal{C}$ on $[a,b]$, there exists a sequence $\{P_n\}_n$ of closed sets such that $P = \cup_{n=1}^{\infty} P_n$ and $F \in AC^*$ on each P_n. For $\epsilon > 0$ let δ_n be given by the fact that $F \in AC^*$ on P_n. Let

$\{[a_i, b_i]\}$, $i = \overline{1,p}$ be a finite set of nonoverlapping closed intervals with endpoints in P_n, such that $\sum_{i=1}^{n}(b_i - a_i) < \delta_n$. Since H is increasing, by (1) we have $\mathcal{O}(H; [a_i, b_i]) = H(b_i) - F(a_i) \le \mathcal{O}(F; [a_i, b_i])$. Then $\sum_{i=1}^{p} \mathcal{O}(H; [a_i, b_i]) \le \sum_{k=1}^{p} \mathcal{O}(F; [a_i, b_i]) < \epsilon$, hence $H \in AC^*$ on P_n. It follows that $H \in AC^*G$ on $[a, b]$. By Theorem 2.18.9, (vii), $H \in AC$ on $[a, b]$.

Lemma 5.8.2 *Let $F, H, G : [a, b] \mapsto \mathbb{R}$, and let $H(x) = \mathcal{O}(F; [a, b]) - \mathcal{O}(F; [x, b]) + \mathcal{O}(G; [a, x])$, $G = F + H$. If $F \in AC^*G \cap C$ on $[a, b]$ then we have:*

(i) $H(a) = 0$ and $H(b) = 2 \cdot \mathcal{O}(F; [a, b])$;

(ii) H is increasing and AC on $[a, b]$;

(iii) $G \in AC^*G \cap C$ and $G(a) \le G(b)$ on $[a, b]$.

Proof. (i) This is evident.

(ii) By Lemma 5.8.1, H is AC on $[a, b]$. Let $x_1, x_2 \in [a, b]$, $x_1 < x_2$. It follows that $H(x_2) - H(x_1) = \mathcal{O}(F; [a, x_2]) - \mathcal{O}(F; [a, x_1]) + \mathcal{O}(F; [x_1, b]) - \mathcal{O}(F; [x_2, b]) \ge 0$, hence H is increasing on $[a, b]$.

(iii) Clearly $G \in AC^*G \cap C$ on $[a, b]$. Let $x \in [a, b]$. We have $G(x) - G(a) = F(x) - F(a) + \mathcal{O}(F; [a, x]) + \mathcal{O}(F; [a, b]) - \mathcal{O}(F; [x, b]) \ge 0$ and $G(b) - G(x) = F(b) - F(x) + 2 \cdot \mathcal{O}(F; [a, b]) - \mathcal{O}(F; [a, b]) + \mathcal{O}(F; [x, b]) - \mathcal{O}(F; [a, x]) = F(b) - F(x) + \mathcal{O}(F; [x, b]) + \mathcal{O}(F; [a, b]) - \mathcal{O}(F; [a, x]) \ge 0$. It follows that $G(a) \le G(x) \le G(b)$.

Lemma 5.8.3 *Let $\{r_k\}_k$ be a sequence of positive numbers such that $\sum_{k=1}^{\infty} r_k = r < +\infty$. Let $F_k : [a, b] \mapsto [0, r_k]$ such that F_k is increasing and AC on $[a, b]$. Let $F : [a, b] \mapsto \mathbb{R}$, $F(x) = \sum_{k=1}^{\infty} F_k(x)$. Then we have:*

(i) $F(a) = 0$ and $F(b) < r$;

(ii) F is increasing and AC on $[a, b]$.

Proof. (i) This is evident.

(ii) Clearly F is increasing and C on $[a, b]$. Let $\epsilon > 0$, and let N be a positive integer such that $\sum_{k=N+1}^{\infty} r_k < \epsilon/2$. For each $j \in \{1, 2, \ldots, N\}$ and for $\epsilon/(2N)$, let δ_j be given by the fact that $F_j \in AC$ on $[a, b]$. Let $\delta = \min\{\delta_j : j = \overline{1, N}\}$. Let $\{[a_i, b_o]\}$, $i = \overline{1,p}$ be a finite set of nonoverlapping closed intervals such that $\sum_{i=1}^{p}(b_i - a_i) < \delta$. Then $\sum_{i=1}^{p}(F(b_i) - F(a_i)) = \sum_{j=1}^{N} \sum_{i=1}^{p}(F_j(b_i) - F_j(a_i)) + \sum_{j=N+1}^{\infty} \sum_{i=1}^{p}(F_j(b_i) - F_j(a_i)) \le N\epsilon/(2N) + \sum_{j=N+1}^{\infty}(F_j(b) - F_j(a)) \le \epsilon/2 + \epsilon/2 = \epsilon$.

Lemma 5.8.4 *Let $F : [a, b] \mapsto \mathbb{R}$, $F \in AC^*G \cap C$. Let E be a countable subset of $[a, b]$ and $r > 0$. Then there exists $H : [a, b] \mapsto \mathbb{R}$ such that:*

(i) $H(a) = 0$ and $H(b) < r$;

(ii) H is increasing and AC on $[a, b]$;

(iii) $G = F + H$ is $AC^*G \cap C$ on $[a, b]$ and $\underline{D}G(x) \ge 0$ on E.

Proof. We may suppose without loss of generality that $E = \{x_1, x_2, \ldots\} \subset (a, b)$. Since $F \in C$, for each positive integer n there exist $a \le a_n < x_n < b_n \le b$ such that $\mathcal{O}(F; [a_n, x_n]) + \mathcal{O}(F; [x_n, b_n]) < r/2^n$. Let $H_n : [a, b] \mapsto [0, r/2^n]$ be defined as follows: $H_n(x) = 0$ for $x \in [a, a_n]$; $H_n(x) = \mathcal{O}(F; [a_n, x_n]) - \mathcal{O}(F; [x, x_n])$ for $x \in [a_n, x_n])$; $H_n(x) = \mathcal{O}(F; [a_n, x_n]) + \mathcal{O}(F; [x_n, x])$ for $x \in [x_n, b_n]$; $H_n(x) = \mathcal{O}(F; [a_n, x_n]) + \mathcal{O}(F; [x_n, b_n])$ for $x \in [b_n, b]$. Let $H(x) = \sum_{n=1}^{\infty} H_n(x)$, $x \in [a, b]$. By Lemma 5.8.1, H_n is increasing and AC on $[a, b]$. Now by Lemma 5.8.3 we obtain (i) and (ii).

(iii) Since $F \in AC^*G \cap C$ on $[a, b]$ it follows that $G \in AC^*G$ on $[a, b]$. Fix a positive integer n. For each $x \in [a_n, x_n]$ we have $G(x_n) - G(x) \ge F(x_n) - F(x) + H(x_n) - H(x_n) = F(x_n) - F(x) + \mathcal{O}(F; [x, x_n]) \ge 0$, and for each $x \in [x_n, b_n]$ we have $G(x) - G(x_n) \ge F(x) - F(x_n) + H(x_n) - H(x_n) = F(x) - F(x_n) + \mathcal{O}(F; [x_n, x]) \ge 0$. It follows that $\underline{D}G(x_n) \ge 0$.

Lemma 5.8.5 *Let $F : [a, b] \mapsto \mathbb{R}$, $F \in AC^*G \cap C$. Let P be a perfect subset of $[a, b]$ and $r > 0$. If $F \in AC^*$ on P then there exists $H : [a, b] \mapsto \mathbb{R}$ such that:*

(i) $H(a) = 0$ and $H(b) < r$;

(ii) H is increasing and AC on $[a, b]$;

(iii) $G = F + H$ is $AC^*G \cap C$ and $\underline{D}G$ is bounded below on P.

Proof. Let $c = \inf(P)$, $d = \sup(P)$, and let $\{(a_n, b_n)\}_n$ be the intervals contiguous to P. Since $F \in AC^*$ on P, it follows that there exists a positive integer N such that $\sum_{n=N+1}^{\infty} \mathcal{O}(F; [a_n, b_n]) < r/6$. Let $f_k : [a, b] \mapsto \mathbb{R}$, $f_k(x) = 0$ for $x \in [a, a_k]$; $f_k(x) = \mathcal{O}(F; [a_k, x]) + \mathcal{O}(F; [a_k, b_k]) - \mathcal{O}(F; [x, b_k])$ for $x \in [a_k, b_k]$; $f_k(x) = 2 \cdot \mathcal{O}(F; [a_k, b_k])$ for $x \in [b_k, x]$. By Lemma 5.8.2, f_k is increasing and AC on $[a, b]$. Let $H_1 : [a, b] \mapsto \mathbb{R}$, $H_1(x) = \sum_{k=N+1}^{\infty} f_k(x)$. By Lemma 5.8.3, H_1 is increasing and AC on $[a, b]$, $H_1(a) = 0$, $H_1(b) < r/3$. Let $G_1 : [a, b] \mapsto \mathbb{R}$, $G_1 = F + H_1$. By Lemma 5.8.2 it follows that

(1) $\quad G_1(a_k) \le G_1(x) \le G_1(b_k)$, $x \in [a_k, b_k]$, $k \ge N + 1$ *and* $G_1 \in AC^*G \cap C$

on $[a, b]$.

Let $K == \{x \in P : x$ is a bilateral accumulation point of $P\}$ and let $E = P \setminus K$. Since E is countable, by Lemma 5.8.4, there exists $H_2 : [a, b] \mapsto \mathbb{R}$ such that $H_2(b) = 0$, $H_2(b) < r/3$, H_2 is increasing and AC on $[a, b]$. Also $G_2 = F + H_2$ is $AC^*G \cap C$ and

(2) $\quad \underline{D}G_2(x) \ge 0$ *on* E.

By Theorem 2.11.1, (xviii), $F_P \in AC$ on $[c, d]$, and by Corollary 2.14.2, $\underline{D}F_P$ is summable on $[c, d]$. By Theorem 2.14.2, it follows that there exists $u : [c, d] \mapsto (-\infty, +\infty]$ such that u is lower semicontinuous, $u(x) \ge \underline{D}F_P(x)$, u is summable and $(\mathcal{L}) \int_c^d u(t)dt \le r/3 + (\mathcal{L}) \int_c^d \underline{D}F_P(t)dt$. By Proposition 1.15.5, u is bounded below on $[c, d]$. Let $U : [c, d] \mapsto \mathbb{R}$, $U(x) = (\mathcal{L}) \int_c^x u(t)dt$. By Lemma 2.14.1, $\underline{D}^{\#}U(x) \ge u(x)$ on $[c, d]$. Since $\underline{D}U(x) \ge \underline{D}^{\#}U(x)$, it follows that $\underline{D}U(x)$ is bounded below on $[c, d]$. Let $H_3 : [a, b] \mapsto \mathbb{R}$ be defined as follows: $H_3(x) = 0$ for $x \in [a, c]$; $H_2(x) = F(c) + U(x) - F_P(x)$ for $x \in [c, d]$; $H_3(x) = F(c) + H(d) - F(d)$ for $x \in [d, b]$. It follows that $H_3(a) = 0$, $H_3(b) < r/3$, H_3 is increasing and AC on $[a, b]$. Then $H = H_1 + H_2 + H_3$ satisfies (i) and (ii).

(iii) Clearly $G \in AC^*G \cap C$ on $[a, b]$. By (2), $\underline{D}G(x) \ge 0$ on E. Let $q = \inf\{\underline{D}U(x) :$

$x \in [c,d]\}$. Then $q \neq -\infty$. We have two situations:
a) Suppose that $q > 0$. Let $x_o \in K$. Then there exists $\delta > 0$ such that $(x_o-\delta, x_o+\delta) \subset (c,d)$ and $(U(x)-U(x_o))/(x-x_o) > 0$, whenever $x \neq x_o$, $x \in (x_o-\delta, x_o+\delta)$. Suppose for example that $x \in (x_o, x_o+\delta)$. If $x \in P$ then

$$(3) \quad G(x)-G(x_o) = (H_1(x)-H_1(x_o)) + (H_2(x)-H_2(x_o)) + (U(x)-U(x_o)) > 0.$$

If $x \notin P$ then there exists $k \geq N+1$ such that $x \in (a_k, b_k) \subset (x_o, x_o+\delta)$ and

$$(4) \quad G(x)-G(x_o) = (G(x)-G(a_k)) + (G(a_k)-G(x_o)) > 0 \ (\textit{see } (1) \textit{ and } (3)).$$

b) Suppose that $q \leq 0$. Let $x_o \in K$. Then there exists $\delta > 0$ such that $(x_o-\delta, x_o+\delta) \subset (c,d)$ and $(U(x)-U(x_o))/(x-x_o) > q-1$, whenever $x \neq x_o$, $x \in (x_o-\delta, x_o+\delta)$. Suppose for example that $x \in (x_o, x_o+\delta)$. If $x \in P$ then

$$(5) \quad G(x)-G(x_o) = (H_1(x)-H_1(x_o)) + (H_2(x)-H_2(x_o)) + (U(x)-U(x_o)) > (q-1)(x-x_o).$$

If $x \notin P$ then there exists $k \geq N+1$ such that $x_o \in (a_k, b_k] \subset (x_o, x_o+\delta)$. We have

$$(6) \quad G(x)-G(x_o) = (G(x)-G(a_k)) + (G(a_k)-G(x_o)) > (q-1)(a_k-x_o) > (q-1)(x-x_o).$$

By (3), (4), (5) and (6), it follows that $\underline{D}G(x_o) \geq q-1$. Hence $\underline{D}G$ is bounded below on P.

Corollary 5.8.1 *Let $F : [a,b] \mapsto \mathbb{R}$, $F \in AC^*G \cap C$ on $[a,b]$ and let $r > 0$. Then there exists $H : [a,b] \mapsto \mathbb{R}$ such that:*

(i) $H(a) = 0$ and $H(b) < r$;

(ii) H is increasing and AC on $[a,b]$;

(iii) $G = F + H$ is $AC^*G \cap C$ and $\underline{D}G(x) \neq -\infty$ on $[a,b]$.

Proof. Since $F \in AC^*G \cap C$ on $[a,b]$, there exists a sequence $\{P_n\}_n$ of perfect subsets of $[a,b]$, and a countable subset E of $[a,b]$ such that $[a,b] = E \cup (\cup_{n=1}^{\infty} P_n)$. By Lemma 5.8.4, there exists $H_1 : [a,b] \mapsto \mathbb{R}$ such that $H_1(a) = 0$, $H_1(b) < r/3$, H_1 is increasing and AC on $[a,b]$, $G_1 = F + H_1$ is $AC^*G \cap C$ on $[a,b]$, and $\underline{D}G_1(x) \geq 0$ on E. By Lemma 5.8.5, for each positive integer n there exists $h_n : [a,b] \mapsto [0, r/2^{n+2})$ such that $\underline{D}(F+h_n)(x) \neq -\infty$, $x \in P_n$, and h_n is AC and increasing on $[a,b]$. Let $H_2 : [a,b] \mapsto \mathbb{R}$, $H_2(x) = \sum_{n=1}^{\infty} h_n(x)$. Then $H_2(a) = 0$ and $H_2(b) < r/2$. By Lemma 5.8.3, H_2 is increasing and AC on $[a,b]$. Then $H = H_1 + H_2$ has the required properties.

Theorem 5.8.1 *Let $f : [a,b] \mapsto \overline{\mathbb{R}}$. If f is $\mathcal{D}^*$- integrable on $[a,b]$ then f is $(\mathcal{P}_{1,1})$-integrable on $[a,b]$ and $(\mathcal{P}_{1,1}) \int_a^b f(t)dt = (\mathcal{D}^*) \int_a^b f(t)dt$.*

Proof. Let $\epsilon > 0$ and $F : [a,b] \mapsto \mathbb{R}$, $F(x) = (\mathcal{D}^*) \int_a^x f(t)dt$. Then $F \in AC^*G \cap C$ on $[a,b]$ and $F'(x) = f(x)$ *a.e.* on $[a,b]$. By Corollary 5.8.1, there exists $H_1 : [a,b] \mapsto [0,\epsilon/2]$ such that $H_1(a) = 0$, H_1 is AC and increasing on $[a,b]$ and $\underline{D}(F + H_1)(x) \neq -\infty$ on $[a,b]$. Let $A_1 = \{x : F$ is derivable at x and $F'(x) = f(x)\}$. Then A_1 is measurable and $| A_1 |= b - a$. Let $A_2 = \{x : H_1$ is derivable at $x\}$. Since $H_1 \in AC$ on $[a,b]$, it follows that A_2 is measurable and $| A_2 |= b - a$. Let $Z = [a,b] \setminus (A_1 \cup A_2)$. Then $| Z |= 0$. Let Y be a G_δ-set such that $Z \subset Y$, $| Y |= 0$. By Theorem 2.14.6, there exists $H_2 : [a,b] \mapsto [0,\epsilon/2)$ such that $H_2(a) = 0$, H_2 is AC and increasing on $[a,b]$, $H_2'(x) = +\infty$ for $x \in Y$ and $H_2'(x) > 0$ for $x \in [a,b] \setminus Y$. Let $H, M : [a,b] \mapsto \mathbb{R}$, $H = H_1 + H_2$, $M = F + H$. Then $H(a) = 0$, $H(b) < \epsilon$, H is increasing and AC on $[a,b]$, $M(a) = 0$, $M \in AC^*G \cap C$ on $[a,b]$, $M'(x)$ exists on $[a,b]$, and $-\infty \neq M'(x) \geq f(x)$ on $[a,b]$. hence $M \in \overline{\mathcal{M}}_1(f)$. Since ϵ is arbitrary, $\overline{I}_1(b) \leq F(b)$. Similarly $\underline{I}_1(b) \geq F(b)$. By Lemma 5.7.4, $\overline{I}_1(b) \geq \underline{I}_1(b)$, hence $F(b) = \overline{I}_1(b) = \underline{I}_1(b) = (\mathcal{D}^*) \int_a^b f(t)dt$.

Corollary 5.8.2 *Let $f : [a,b] \mapsto \overline{\mathbb{R}}$ and let $(j,k) \in (\{1,2,\ldots,6\} \times \{1,2,\ldots,12\}) \cup (\{7,8\} \times \{1,2,\ldots,10\}) \cup (\{13,14,15,16\} \times \{13,14,15,16\})$. If f is $\mathcal{D}^*$-integrable on $[a,b]$ the f is $(\mathcal{P}_{j,k})$-integrable on $[a,b]$ and we have $(\mathcal{P}_{j,k}) \int_a^b f(t)dt = (\mathcal{D}^*) \int_a^b f(t)dt$.*

Proof. By Theorem 5.8.1, f is $(\mathcal{P}_{1,1})$- integrable and $(\mathcal{P}_{1,1}) \int_a^b f(t)dt = (\mathcal{D}^*) \int_a^b f(t)dt$. Thus we obtain that $\overline{calM}_1(f) \times \underline{\mathcal{M}}_1(f) \neq \emptyset$. By Lemma 5.7.2, $\overline{\mathcal{M}}_1(f) \subseteq \overline{\mathcal{M}}_j(f)$. It follows that $\overline{I}_j(b) \leq \underline{I}_1(b) = (\mathcal{D}^*) \int_a^b f(t)dt$ and $\underline{I}_k(b) \geq \underline{I}_1(b) = (\mathcal{D}^*) \int_a^b f(t)dt$. By Lemma 5.7.4, $\overline{I}_j(b) \geq \underline{I}_k(b)$, so $\overline{I}_j(b) = \underline{I}_k(b) = (\mathcal{P}_{j,k}) \int_a^b f(t)dt = (\mathcal{D}^*) \int_a^b f(t)dt$.

Remark 5.8.1 *In 1921 Hake proved that the Perron integral (in our terms the $(\mathcal{P}_{7,7})$-integral) generalizes the $\mathcal{D}^*$-integral. This result was derived from the constructive definition of the $\mathcal{D}^*$-integral. The same result was also obtained by Saks (see [S3]) from the descriptive definition of the $\mathcal{D}^*$-integral. However, Saks asked the $\overline{\mathcal{M}}_7(f)$ majorants to be continuous. In 1939 Tolstoff improved the above result, obtaining that each $\mathcal{D}^*$-integrable function is $(\mathcal{P}_{4,4})$-integrable. In 1982 Skljarenko (see [Skl]) obtained that any $\mathcal{D}^*$ integrable function is also $(\mathcal{P}_{2,2})$-integrable.*
In Theorem 5.8.1 we improve all these results, showing that any $\mathcal{D}^*$-integrable function is $(\mathcal{P}_{1,1})$- integrable, and even $(\mathcal{P}_{j,k})$-integrable. In fact, in the following section we show that the $\mathcal{D}^*$-integral is equivalent with each $(\mathcal{P}_{j,k})$-integral. For more information on this topic, see [S3] and [B1].

5.9 An improvement of the Looman-Alexandroff Theorem. The equivalence of the $\mathcal{D}^*$-integral and the $(\mathcal{P}_{j,k})$-integral

Theorem 5.9.1 *Let $f : [a,b] \mapsto \overline{\mathbb{R}}$ and let $(j,k) \in (\{1,2,\ldots,6\} \times \{1,2,\ldots,12\}) \cup (\{7,8\} \times \{1,2,\ldots,10\}) \cup (\{13,14,15,16\} \times \{13,14,15,16\})$. If f is $(\mathcal{P}_{j,k})$-integrable on $[a,b]$ then f is $\mathcal{D}^*$-integrable on $[a,b]$ and we have $(\mathcal{P}_{j,k}) \int_a^b f(t)dt = (\mathcal{D}^*) \int_a^b f(t)dt$.*

Proof. Let $F : [a,b] \mapsto \mathbb{R}$, $F(x) = (\mathcal{P}_{j,k}) \int_a^x f(t)dt$. If $(M,m) \in \overline{\mathcal{M}}_j(f) \times \underline{\mathcal{M}}_k(f) \neq \emptyset$ then by Lemma 5.7.3, it follows that $(M,m) \in \overline{\mathcal{M}}_8(f) \times \underline{\mathcal{M}}_8(f)$. By Lemma 5.7.4, the functions $M - m$, $M - F$ and $F - m$ are positive and increasing on $[a,b]$, and $M(a) = m(a) = F(a) = 0$. We show that $F \in C$ on $[a,b]$. Let $x_o \in [a,b)$.

For $\epsilon > 0$ let $(M,m) \in \overline{\mathcal{M}}_8(f) \times \underline{\mathcal{M}}_8(f)$ such that $M(b) - m(b) < \epsilon/2$. Since $M \in \mathcal{C}_i$ and $m \in \mathcal{C}_d$ on $[a,b]$, there exists $\delta > 0$ (depending on ϵ and x_o) such that $M(x)-M(x_o) > -\epsilon/2$ and $m(x)-m(x_o) < \epsilon/2$, $x \in [x_o, x_o+\delta)$. Hence $F(x)-F(x_o) = (M(x) - M(x_o)) + (M(x_o) - F(x_o)) + (F(x) - M(x)) > \epsilon/2 + 0 - \epsilon/2 = -\epsilon$ and $F(x)-F(x_o) = (m(x)-m(x_o))+(m(x_o)-F(x_o))+(F(x)-m(x)) < \epsilon/2+0+\epsilon/2 = \epsilon$. It follows that F is continuous at the right side of $x_o \in [a,b)$. Similarly F is continuous at the left side of $x_o \in (a,b]$. Hence $F \in \mathcal{C}$ on $[a,b]$.
We show that $F \in AC^*G$ on $[a,b]$. Let $(M_o, m_o) \in \overline{\mathcal{M}}_8(f) \times \underline{\mathcal{M}}_8(f)$. Then there exists $\{Q_i\}_i$ a sequence of closed sets such that $M_o \in \underline{AC} \cap VB^*$ on Q_i and $m_o \in \overline{AC} \cap VB^*$ on Q_i. Let $(M,m) \in \overline{\mathcal{M}}_8(f) \times \underline{\mathcal{M}}_8(f)$. Then $M - m$, $M - m_o$ and $M_o - m$ are positive increasing functions on $[a,b]$. Since $M_o \in VB^*$ on Q_i, it follows that $m = M_o - (M_o - m) \in VB^*$ on Q_i, and $M = m + (M - m) \in VB^*$ on Q_i. But $-m, M \in \mathcal{C}_i \cap \underline{M}^*$ on $[a,b]$. By Theorem 2.22.2, it follows that $M \in \underline{AC}$ and $m \in \overline{AC}$ on Q_i. We show that $F \in AC$ on Q_i. For $\epsilon > 0$ let $(M,m) \in \overline{\mathcal{M}}_8(f) \times \underline{\mathcal{M}}_8(f)$ such that $M(b)-m(b) < \epsilon/2$, and let $\delta > 0$ be given by the fact that $M \in \underline{AC}$ on Q_i and $m \in \overline{AC}$ on Q_i. Let $\{[a_n, b_n]\}$, $n = \overline{1,p}$ be a finite set of nonoverlapping closed intervals, $a_n, b_n \in Q_i$, such that $\sum_{n=1}^p (b_n - a_n) < \delta$. Then $\sum_{n=1}^p (M(b_n) - M(a_n)) > -\epsilon/2$ and $\sum_{n=1}^p (m(b_n) - m(a_n)) < \epsilon/2$. It follows that $\sum_{n=1}^p (F(b_n) - F(a_n)) = \sum_{n=1}^p (M(b_n) - M(a_n)) - \sum_{n=1}^p ((M-F)(b_n) - (M-F)(a_n)) > -\epsilon/2 - (M-F)(b) > -\epsilon/2 - \epsilon/2 = -\epsilon$, hence $F \in \underline{AC}$ on Q_i. Also we have $\sum_{n=1}^p (F(b_n) - F(a_n)) = \sum_{n=1}^p (m(b_n) - m(a_n)) + \sum_{n=1}^p ((F-m)(b_n) - (F-m)(a_n)) < \epsilon/2 + (F-m)(b) < \epsilon/2 + \epsilon/2 = \epsilon$. Hence $F \in \overline{AC}$ on Q_i. Thus $F \in AC$ on Q_i. By Theorem 2.12.1, (i), (ii), $F \in AC^*$ on Q_i, hence $F \in AC^*G$ on $[a,b]$.
We show that $F'(x) = f(x)$ a.e. on $[a,b]$. For $\epsilon > 0$ let $(M,m) \in \overline{\mathcal{M}}_8(f) \times \underline{\mathcal{M}}_8(f)$ such that $M(b) - m(b) < \epsilon^2$. Since $M, m \in VB^*G$ on $[a,b]$, it follows that M and m are derivable a.e. on $[a,b]$, and $m'(x) \le f(x) \le M'(x)$ a.e. on $[a,b]$. So f is finite a.e.. Let $E = \{x : f(x), F'(x), M'(x), m'(x) \text{ are finite}\}$. Then E is measurable and $|E| = b-a$. Let $A_\epsilon = \{x \in E : |F'(x) - f(x)| > \epsilon\}$ and $B_\epsilon = \{x \in E : M'(x) - m'(x) > \epsilon\}$. Then B_ϵ is measurable and $A_\epsilon \subset B_\epsilon$. We have $\epsilon \cdot |B_\epsilon| \le (\mathcal{L}) \int_{B_\epsilon} (M-m)'(t)dt \le (\mathcal{L}) \int_a^b (M-m)'(t)dt \le M(b) - m(b) < \epsilon^2$ (this follows by Theorem 2.14.5). Hence $|B_\epsilon| < \epsilon$ and $|A_\epsilon| < \epsilon$. Let $A = \{x \in E : |F'(x) - f(x)| > 0\}$. Then $A = \cup_{n=1}^\infty A_{\epsilon/2^n}$, hence $|A| < \epsilon$. Since ϵ is arbitrary, it follows that $|A| = 0$, hence $F'(x) = f(x)$ a.e. on $[a,b]$.

Corollary 5.9.1 *Let $f : [a,b] \mapsto \overline{\mathbb{R}}$ and let $(j,k) \in (\{1,2,\ldots,6\} \times \{1,2,\ldots,12\}) \cup (\{7,8\} \times \{1,2,\ldots,10\}) \cup (\{13,14,15,16\} \times \{13,14,15,16\})$. The following assertions are equivalent:*

(i) f is $\mathcal{D}^*$-integrable on $[a,b]$;

(ii) f is $(\mathcal{P}_{j,k})$-integrable on $[a,b]$

and $(\mathcal{D}^*) \int_a^b f(t)dt = (\mathcal{P}_{j,k}) \int_a^b f(t)dt$.

Proof. See Corollary 5.8.2 and Theorem 5.9.1.

Remark 5.9.1 *In 1924-1925, P. Alexandroff and H. Looman obtained independently that, if a function f is Perron integrable (in our terms $(\mathcal{P}_{7,7})$-integrable) then f is $\mathcal{D}^*$-integrable (see [S3]). This result was obtained from the constructive definition of the $\mathcal{D}^*$-integral. In [S1] Saks showed that any $(\mathcal{P}_{5,5})$-integrable function is $\mathcal{D}^*$- integrable. For more information on this topic see [S3], [N], [B1], [R1].*

5.10 Ward type definitions for the $\mathcal{D}^*$-integral

Definition 5.10.1 *Let $f : [a,b] \mapsto \mathbb{R}$. We define the following classes of majorants:*

- $\overline{\mathcal{W}}_1(f) = \{M : [a,b] \mapsto \mathbb{R} : M(a) = 0;$ *there exists* $\delta : [a,b] \mapsto (0,+\infty)$ *such that* $M(z) - M(y) \geq f(x)(z-y)$, *whenever* $x \in [y,z] \subset (x-\delta(x), x+\delta(x))\}$;
- $\overline{\mathcal{W}}_2(f) = \{M : [a,b] \mapsto \mathbb{R} : M \in \overline{\mathcal{M}}_1(f);\ M \in AC^*G \cap C$ *on* $[a,b]\}$;
- $\overline{\mathcal{W}}_3(f) = \{M : [a,b] \mapsto \mathbb{R} : M \in \overline{\mathcal{M}}_1(f);\ M \in AC^*G \cap C$ *on* $[a,b]$; $M'(x)$ *exists finite or infinite on* $[a,b]\}$.

We define the following classes of minorants: $\underline{\mathcal{W}}_j(f) = \{m : [a,b] \mapsto \mathbb{R} : -m \in \overline{\mathcal{W}}_j(-f)\}$, $j = 1,2,3$. *If* $\overline{\mathcal{W}}_j(f) \neq \emptyset$ *then we denote by* $\overline{J}_j(b)$ *the lower bound of all* $M(b)$, $M \in \overline{\mathcal{W}}_j(f)$. *If* $\underline{\mathcal{W}}_k(f) \neq \emptyset$ *then we denote by* $\underline{J}_k(b)$ *the upper bound of all* $m(b)$, $m \in \underline{\mathcal{W}}_k(f)$. *Let* $(j,k) \in \{1,2,3\} \times \{1,2,3\}$. *We say that f has a* $(\mathcal{W}_{j,k})$*-integral on* $[a,b]$ *if* $\overline{\mathcal{W}}_j(f) \times \underline{\mathcal{W}}_k(f) \neq \emptyset$ *and* $\overline{J}_j(b) = \underline{J}_k(b) = (\mathcal{W}_{j,k}) \int_a^b f(t)dt$.

Lemma 5.10.1 *Let $f : [a,b] \mapsto \mathbb{R}$ and let $(j,k) \in \{1,2,3\} \times \{1,2,3\}$ such that $\overline{\mathcal{W}}_j(f) \times \underline{\mathcal{W}}_k(f) \neq \emptyset$. If $(M,m) \in \overline{\mathcal{W}}_j(f) \times \underline{\mathcal{W}}_k(f)$ then $M - m$, $M - \underline{J}_k$, $\overline{J}_j - m$ and $\overline{J}_j - \underline{J}_k$ are positive increasing functions. Hence $M(b) \geq m(b)$, $M(b) \geq \underline{J}_k(b)$, $\overline{J}_j(b) \geq m(b)$ and $\overline{J}_j(b) \geq \underline{J}_k(b)$.*

Proof. Clearly $(M,m) \in \overline{\mathcal{W}}_1(f) \times \underline{\mathcal{W}}_1(f)$. It follows that there exists $\delta : [a,b] \mapsto (0,+\infty)$ such that $M(z) - M(y) \geq f(x)(z-y) \geq m(z) - m(y)$, whenever $x \in [y,z] \subset (x-\delta(x), x+\delta(x))$, $x \in [a,b]$. So $M - m$ is $\mathcal{S}_o$-increasing on $[a,b]$. By Lemma 4.4.1, $M - m$ is increasing on $[a,b]$. The other parts follow similarly to the proof of Lemma 5.1.4.

Theorem 5.10.1 *Let $f : [a,b] \mapsto \mathbb{R}$ and let $(j,k) \in \{1,2,3\} \times \{1,2,3\}$. The following assertions are equivalent:*

(i) f is $\mathcal{D}^*$-integrable on $[a,b]$;

(ii) f is $(\mathcal{W}_{j,k})$-integrable on $[a,b]$,

and $(\mathcal{W}_{j,k}) \int_a^b f(t)dt = (\mathcal{D}^*) \int_a^b f(t)dt$.

Proof. $(i) \Rightarrow (ii)$ By Corollary 5.9.1, f is $(\mathcal{P}_{1,1})$-integrable on $[a,b]$ and $(\mathcal{D}^*) \int_a^b f(t)dt = (\mathcal{P}_{1,1}) \int_a^b f(t)dt$. Then for $\epsilon > 0$ there exists $G \in \overline{\mathcal{M}}_1(f)$ such that $G(b) < \epsilon/2 + (\mathcal{D}^*) \int_a^b f(t)dt$. It follows that $G \in AC^*G \cap C$ on $[a,b]$, $M'(x)$ exists (finite or infinite), and $f(x) \leq M'(x) \neq -\infty$ on $[a,b]$. Hence there exists $\delta : [a,b] \mapsto (0,+\infty)$ such that $(G(z)-G(y))/(z-y) \geq f(x) - \epsilon/(2(b-a))$, whenever $x \in [y,z] \subset (x-\delta(x), x+\delta(x))$, $x \in [a,b]$. Let $M : [a,b] \mapsto \mathbb{R}$, $M(x) = G(x) + (\epsilon(x-a))/(2(b-a))$. Then $M(z) - M(y) \geq f(x)(z-y)$, hence $M \in \overline{\mathcal{W}}_3(f) \neq \emptyset$. Clearly $M(b) < \epsilon + (\mathcal{D}^*) \int_a^b f(t)dt$, hence $\overline{J}_3(b) \leq (\mathcal{D}^*) \int_a^b f(t)dt$. It follows that $\overline{\mathcal{M}}_1(f) \supset \overline{\mathcal{W}}_1(f) \supset \overline{\mathcal{W}}_2(f) \supset \overline{\mathcal{W}}_3(f)$. By Corollary 5.9.1, f is $(\mathcal{P}_{7,7})$-integrable. Then $(\mathcal{D}^*) \int_a^b f(t)dt = \overline{I}_7(b) \leq \overline{J}_1(b) \leq \overline{J}_2(b) \leq \overline{J}_3(b) \leq (\mathcal{D}^*) \int_a^b f(t)dt$, hence $\overline{J}_1(b) = \overline{J}_2(b) = \overline{J}_3(b) = (\mathcal{D}^*) \int_a^b f(t)dt$. Similarly $\underline{J}_1(b) = \underline{J}_2(b) = \underline{J}_3(b) = (\mathcal{D}^*) \int_a^b f(t)dt$. Hence f is $(\mathcal{W}_{j,k})$- integrable on $[a,b]$ and $(\mathcal{W}_{j,k}) \int_a^b f(t)dt = (\mathcal{D}^*) \int_a^b f(t)dt$.

$(ii) \Rightarrow (i)$ Since f is $(\mathcal{W}_{j,k})$- integrable, it follows that $\overline{\mathcal{W}}_j(f) \times \underline{\mathcal{W}}_k(f) \neq \emptyset$

and $\overline{J}_j(b) = \underline{J}_k(b) = (\mathcal{W}_{j,k}) \int_a^b f(t)dt$. Clearly $\overline{\mathcal{W}}_j(f) \times \underline{\mathcal{W}}_k(f) \subset \overline{\mathcal{M}}_7(f) \times \underline{\mathcal{M}}_7(f)$, hence $\overline{J}_j(b) \geq \overline{I}_7(b)$ and $\underline{J}_k(b) \leq \underline{I}_k(b)$. By Lemma 5.7.4, $\underline{I}_7(b) \leq \overline{I}_7(b)$, hence $\underline{I}_7(b) = \overline{I}_7(b) = (\mathcal{P}_{7,7}) \int_a^b f(t)dt = (\mathcal{D}^*) \int_a^b f(t)dt = (\mathcal{W}_{j,k}) \int_a^b f(t)dt$ (see Corollary 5.9.1.

5.11 Henstock variational definitions for the $\mathcal{D}^*$-integral

Definition 5.11.1 *Let $f : [a,b] \mapsto \mathbb{R}$.*

- *(<u>Kubota</u>). f is said to be $(\mathcal{V}_1)$- integrable on $[a,b]$ if there exists $H : [a,b] \mapsto \mathbb{R}$ with the following property: for every $\epsilon > 0$ there exist $\delta : [a,b] \mapsto (0,+\infty)$ and $G : [a,b] \mapsto \mathbb{R}$ such that $G(a) = 0$, $G(b) < \epsilon$, G is increasing on $[a,b]$, and $\mid H(z) - H(y) - f(x)(z-y) \mid < G(z) - G(y)$, whenever $x \in [y,z] \subset (x-\delta(x), x+\delta(x))$.*

- *(<u>Skljarenko</u>). f is said to be $(\mathcal{V}_2)$- integrable on $[a,b]$, if f is $(\mathcal{V}_1)$-integrable and $G \in AC$ on $[a,b]$.*

- *f is said to be $(\mathcal{V}_3)$-integrable on $[a,b]$, if f is $(\mathcal{V}_1)$-integrable, $G \in AC$ and $G'(x)$ exists (finite or infinite) on $[a,b]$.*

H is called the indefinite $(\mathcal{V}_i)$-integral on $[a,b]$, and $(\mathcal{V}_i) \int_a^b f(t)dt = H(b) - H(a)$, $i = 1,2,3$.

Lemma 5.11.1 *Let $f : [a,b] \mapsto \mathbb{R}$. If $H_1, H_2 : [a,b] \mapsto \mathbb{R}$ are indefinite $(\mathcal{V}_1)$-integrals of f on $[a,b]$ then $H_1(b) - H_1(a) = H_2(b) - H_2(a)$.*

Proof. The proof is similar to that of Lemma 5.3.1.

Remark 5.11.1 *(i) Lemma 5.11.1 shows that the $(\mathcal{V}_i)$- integrals, $i = 1,2,3$ are well-defined.*

(ii) The properties of the $(\mathcal{V}_1)$-integral are studied by Kubota in [Kub3].

Lemma 5.11.2 *Let $f : [a,b] \mapsto \mathbb{R}$. If f is $(\mathcal{W}_{3,3})$-integrable on $[a,b]$ then f is $(\mathcal{V}_3)$-integrable and $(\mathcal{W}_{3,3}) \int_a^b f(t)dt = (\mathcal{V}_3) \int_a^b f(t)dt$.*

Proof. The proof is similar to that of Lemma 5.3.2.

Lemma 5.11.3 *Let $f : [a,b] \mapsto \mathbb{R}$. If f is $(\mathcal{V}_1)$-integrable on $[a,b]$ then f is $(\mathcal{W}_{1,1})$-integrable and $(\mathcal{V}_1) \int_a^b f(t)dt = (\mathcal{W}_{1,1}) \int_a^b f(t)dt$.*

Proof. The proof is similar to that of Lemma 5.3.3.

Theorem 5.11.1 *Let $f : [a,b] \mapsto \mathbb{R}$ and let $i \in \{1,2,3\}$. The following assertions are equivalent:*

(i) f is $\mathcal{D}^*$-integrable on $[a,b]$;

(ii) f is $(\mathcal{V}_i)$-integrable on $[a,b]$,

and $(\mathcal{V}_i) \int_a^b f(t)dt = (\mathcal{D}^*) \int_a^b f(t)dt$.

Proof. The proof is similar to that of Theorem 5.3.1.

5.12 Riemann type definitions for the $\mathcal{D}^*$-integral (The Kurzweil-Henstock integral)

Definition 5.12.1 *Let $\delta : [a,b] \mapsto (0,+\infty)$ and $E \subset [a,b]$. Let $\beta^o_\delta[E] = \{[y,z];x) : x \in E$ and $x \in [y,z] \subset (x-\delta(x), x+\delta(x))\}$. Let P be a finite set of pairs $\{[c_i,d_i];t_i) \in \beta^o_\delta[E]$, such that $\{[c_i,d_i]\}_i$ is a set of nonoverlapping nondegenerate closed intervals, and let $\sigma(P) = \cup_i[c_i,d_i]$. We denote by $\mathcal{P}(E;\delta)$ the collection of all P defined as above. Let $f : [a,b] \mapsto \mathbb{R}$ and let $\sigma(f;P) = \sum_i f(t_i)(d_i - c_i)$, $S(f;P) = \sum_i(f(d_i) - f(c_i))$, $P \in \mathcal{P}(E;\delta)$. If $E = [a,b]$ and $\sigma(P) = [a,b]$ then we denote the collection of all these P by $\mathcal{P}_1([a,b];\delta)$. Clearly $\mathcal{P}(E;\delta) \subset \mathcal{P}^\#(E;\delta)$.*

Remark 5.12.1 *Recall that $D^o[E] = \{\beta^o_\delta[E] : \delta : [a,b] \mapsto (0,+\infty)\}$ is called the ordinary derivation basis on the set E (see Section 2.24).*

Definition 5.12.2 (Kurzweil-Henstock) *Let $f : [a,b] \mapsto \mathbb{R}$. f is said to be (RD^o)-integrable on $[a,b]$ if there exists a real number I with the following property: for $\epsilon > 0$ there exists $\delta : [a,b] \mapsto (0,+\infty)$ such that $| \sigma(f;P) - I | < \epsilon$, whenever $P \in \mathcal{P}_1([a,b];\delta)$. Then $(RD^o)\int_a^b f(t)dt = I$.*
Let $E \subset [a,b]$. f is said to be (RD^o)-integrable on E, if $f \cdot K_E$ is (RD^o)-integrable on $[a,b]$.

Remark 5.12.2 *In the above definition, the real number I is unique (the proof is similar to that in Remark 5.4.2).*

Lemma 5.12.1 *([H4]). Let $f : [a,b] \mapsto \mathbb{R}$. The following assertions are equivalent:*

(i) f is (RD^o)-integrable on $[a,b]$;

(ii) for every $\epsilon > 0$ there exists $\delta : [a,b] \mapsto (0,+\infty)$ such that $| \sigma(f;P) - \sigma(f,P') | < \epsilon$, whenever $P, P' \in \mathcal{P}_1([a,b];\delta)$.

Proof. The proof is similar to that of Lemma 5.4.1.

Lemma 5.12.2 *([H4]). Let $f, g : [a,b] \mapsto \mathbb{R}$, $\alpha, \beta \in \mathbb{R}$, $\alpha \neq 0$, $\beta \neq 0$.*

(i) If f and g are (RD^o)-integrable then $\alpha f + \beta g$ is (RD^o)-integrable on $[a,b]$ and $(RD^o)\in_a^b *\alpha f + \beta g)(t)dt = \alpha \cdot (RD^o)\int_a^b f(t)dt + \beta \cdot (RD^o)\in_a^b g(t)dt$.

(ii) If $a < c < b$ and f is (RD^o)-integrable on $[a,c]$ and $[c,b]$ then f is (RD^o)-integrable on $[a,b]$ and $(RD^o)\int_a^c f(t)dt + (RD^o)\int_c^b f(t)dt = (RD^o)\int_a^b f(t)dt$.

(iii) If $a \le c < d \le b$ and f is (RD^o)-integrable on $[a,b]$ then f is (RD^o)-integrable on $[c,d]$.

Proof. The proof is similar to that of Lemma 5.4.2.

Lemma 5.12.3 (Henstock) *Let $f : [a,b] \mapsto \mathbb{R}$ such that f is (RD^o)-integrable. Let $F : [a,b] \mapsto \mathbb{R}$, $F(x) = (RD^o)\int_a^x f(t)dt$. For $\epsilon > 0$ let $\delta : [a,b] \mapsto (0,+\infty)$ such that $| \sigma(f;P) - F(b) | < \epsilon$, whenever $P \in \mathcal{P}_1([a,b];\delta)$. Then we have:*

(i) $| \sigma(d;P) - S(F;P) | < 3\epsilon/2$, whenever $P \in \mathcal{P}([a,b];\delta)$;

(ii) $\sum_{([y,z];x)\in P} | f(x)(z-y) - (F(z)-F(y)) | < 3\epsilon$, *whenever* $P \in \mathcal{P}([a,b];\delta)$.

Proof. The proof is similar to that of Lemma 5.4.3.

Lemma 5.12.4 *Let* $f : [a,b] \mapsto \mathbb{R}$ *such that* f *is* (RD^o)*-integrable on* $[a,b]$*. Let* $F : [a,b] \mapsto \mathbb{R}$, $F(x) = (RD^o)\int_a^x f(t)dt$*. Then for every* $\epsilon > 0$ *there exist* $\delta : [a,b] \mapsto (0,+\infty)$ *and* $G : [a,b] \mapsto \mathbb{R}$ *such that* $G(a) = 0$, $G(b) < \epsilon$, G *is increasing on* $[a,b]$*, and* $G(z) - G(y) > | f(x)(z-y) - (F(z)-F(y)) |$*, whenever* $x \in [y,z] \subset (x-\delta(x), x+\delta(x))$, $x \in [a,b]$.

Proof. The proof is similar to that of Lemma 5.4.4.

Theorem 5.12.1 *Let* $f : [a,b] \mapsto \mathbb{R}$*. The following assertions are equivalent:*

(i) f *is* $\mathcal{D}^*$*-integrable on* $[a,b]$*;*

(ii) f *is* $(\mathcal{V}_1)$*-integrable on* $[a,b]$*;*

(iii) f *is* (RD^o)*-integrable on* $[a,b]$*;*

(iv) There exists $F : [a,b] \mapsto \mathbb{R}$, $F \in Y_{D^o}$ *such that* $F'(x) = f(x)$ *a.e. on* $[a,b]$*,*

and $(\mathcal{D}^*)\int_a^b f(t)dt = (\mathcal{V}_1)\int_a^b f(t)dt = (RD^o)\int_a^b f(t)dt$.

Proof. The proof is similar to that of Theorem 5.4.1 (using Corollary 2.27.1 instead of Corollary 2.27.2). However it is possible to prove directly that $(iv) \Rightarrow (iii)$. Indeed, let $F : [a,b] \mapsto \mathbb{R}$, $F \in Y_{D^o}$ such that $F'(x) = f(x)$ a.e. on $[a,b]$. Let $Z = \{z : F'(x) \neq f(x)\}$. Then $| Z | = 0$. Let $\epsilon > 0$. It follows that for each $x \in [a,b] \setminus Z$ there exists $\delta : [a,b] \setminus Z \mapsto (0,+\infty$, such that $| F(y) - F(x) - f(x)(y-x) | < \epsilon | y - x |$, whenever $| y - x | < \delta(x)$. Since $F \in Y_{D^o}$, by Lemma 5.4.5, there exists $\delta : Z \mapsto (0,+\infty)$ such that $| S(F;P) | < \epsilon$ and $| \sigma(f;P) | < \epsilon$, whenever $P \in \mathcal{P}(Z;\delta)$. Hence $\delta : [a,b] \mapsto (0,+\infty)$. Let $P \in \mathcal{P}_1([a,b]\delta)$ and $P_Z = \{([y,z];x) \in P : x \in Z\}$. It follows that $| \sigma(f;P) - (F(b)-F(a)) | = | \sigma(f;P) - S(F;P) | \le | \sigma(f;P \setminus P_Z) - S(F;P \setminus P_Z) | + | \sigma(f;P_Z) | + | S(F;P_Z) | < \epsilon(b-a) + \epsilon + \epsilon$. Therefore f is (RD^o)-integrable on $[a,b]$ and $(RD^o)\int_a^b f(t)dt = F(b) - F(a)$.

Remark 5.12.3 *In [B5], Bullen asked if the function* δ *in the definition of the* (RD^o)*-integral may be supposed to be measurable. A positive answer was given independently by Foran and Meinershagen in [FM], Genquian in [Ge], Gordon in [G3], and Pfeffer in [P2] (the proofs are not easy). Moreover Foran and Meinershagen showed that* δ *may be supposed to be measurable on* $[a,b]$ *and* $\mathcal{B}_2$ *on the complement of a null set of* G_δ*-type (but* δ *cannot be* $\mathcal{B}_2$ *on* $[a,b]$*, see [FM]). Also Pfeffer showed that* δ *may be supposed to be measurable on* $[a,b]$ *and upper semicontinuous when restricted to a suitable subset whose complement has measure zero.*

5.13 Cauchy and Harnak extensions of the $\mathcal{D}^*$ - integral

Theorem 5.13.1 (Cauchy extension) *Let* $f : [a,b] \mapsto \overline{\mathbb{R}}$ *such that* f *is* $\mathcal{D}^*$*- integrable on each closed interval* $[c,d] \subset (a,b)$*. If* $\lim_{\substack{c\to a\\ d\to b}} (\mathcal{D}^*)\int_c^d f(t)dt = \ell$ *exists then* f *is* $\mathcal{D}^*$*-integrable and* $(\mathcal{D}^*)\int_a^b f(t)dt = \ell$.

Proof. ([S1], p. 252). Let $F : [a,b] \mapsto \mathbb{R}$ be defined as follows: $F(a) = 0$, $F(b) = \ell$, and $D(x) = \lim_{\epsilon\to 0}^{x} f(t)dt$, $x \in (a,b)$. Then $F \in AC^*G \cap \mathcal{C}$ on $[a+1/n, b-1/n]$ and $F'(x) = f(x)$ a.e. on $[a+1/n, b-1/n]$, $n = \overline{1,\infty}$. It follows that $F \in AC^*G \cap \mathcal{C}$ on $[a,b]$, $F'(x) = f(x)$ a.e. on $[a,b]$ and $(\mathcal{D}^*)\int_a^b f(t)dt = F(b) - F(a) = \ell$.

Theorem 5.13.2 (Harnak extension) *Let P be a closed subset of $[a,b]$, $a,b \in P$, and let $\{(a_k,b_k)\}_k$ be the intervals contiguous to P. Let $f : [a,b] \mapsto \overline{\mathcal{R}}$ such that f is $\mathcal{D}^*$-integrable on P and on each interval $[a_k,b_k]$. Let $G_k(x) = (\mathcal{D}^*)\int_{a_k}^{x} f(t)dt$. If $\sum_{k=1}^{\infty} \mathcal{O}(G_k;[a_k,b_k]) < +\infty$ then f is $\mathcal{D}^*$- integrable on $[a,b]$ and $(\mathcal{D}^*)\int_a^b f(t)dt = (\mathcal{D}^*)\int a^b(f\cdot K_P)(t)dt + \sum_{k=1}^{\infty}(\mathcal{D}^*)\int_{a_k}^{b_k} f(t)dt$.*

Proof. See [S3], p. 257.

Remark 5.13.1 *In the above theorems, the $\mathcal{D}^*$-integral may be replaced by the (RD°)-, the $(\mathcal{V}_1)$-, or the $(\mathcal{P}_{7,7})$- integral (since these integrals are equivalent). Direct proofs of these facts were given by Gordon in [G1], Kubota in [Kub3], and Saks in [S3] (p. 249), respectively.*

5.14 A theorem of Marcinkiewicz type for the $\mathcal{D}^*$-integral

Lemma 5.14.1 (Bullen and Vyborny) *([BV]). Let $f : [a,b] \mapsto \overline{\mathbb{R}}$. The following assertions are equivalent:*

(i) f is $\mathcal{D}^*$-integrable;

(ii) f is measurable on $[a,b]$, and for every $\epsilon > 0$ there exist $M, m : [a,b] \mapsto \mathbb{R}$ (depending on ϵ) with the following properties:

(a) $M(a) = m(a) = 0$;

(b) $M, m \in VB^*G$ on $[a,b]$;

(c) $m'(x) \le f'(x) \le M'(x)$ a.e. on $[a,b]$;

(d) there exists $\delta : [a,b] \mapsto (0,+\infty)$ such that $M(z) - M(y) < \epsilon$ and $m(z) - m(y) \ge -\epsilon$, whenever $[y,z] \subset (x-\delta(x), x+\delta(x))$, $x \in [a,b]$;

(e) for any interval $[c,d] \subset [a,b]$ on which f is $\mathcal{D}^*$-integrable we have $M(d) - M(c) \ge F(d) - F(c) \ge m(d) - m(c)$, where $F(x) = (\mathcal{D}^*)\int_a^x f(t)dt$, $x \in [c,d]$.

Proof. $(i) \Rightarrow (ii)$ This is evident.

$(ii) \Rightarrow (i)$ Let G be a maximal open subset of (a,b) such that f is $\mathcal{D}^*$-integrable on the closure of each component of G. Let $Q = \overline{(a,b)\setminus G}$. Suppose on the contrary that Q contains an interval $[a_1,b_1]$. By 2) and Theorem 1.7.1, there exists a closed interval $[a_2,b_2] \subset [a_1,b_1]$ such that $M, m \in VB^*$ on $[a_2,b_2]$. By Theorem 2.14.3, M' and m' are summable on $[a_2,b_2]$. Since f is measurable on $[a,b]$, by 3), f is summable on $[a_2,b_2]$, hence f is $\mathcal{D}^*$-integrable on $[a_2,b_2]$, a contradiction. So Q is nowhere dense. We show that Q is perfect. Let (u,v) be a component of G, and let $[u',v'] \subset (u,v)$. By 4), for $\epsilon > 0$ there exists $\delta : [a,b] \mapsto (0,+\infty)$ such that $M(z) - M(y) < \epsilon$ and $m(z) - m(y) > -\epsilon$, whenever $v - \delta(v) < y < z < v$. By 5), $-\epsilon < m(z) - m(y) \le$

$(\mathcal{D}^*)\int_a^x f(t)dt \leq M(z) - M(y) < \epsilon$. It follows that $|(\mathcal{D}^*)\int_y^z f(t)dt| < \epsilon$, hence $\lim_{v'\to v}(\mathcal{D}^*)\int_{u'}^{v'} f(t)dt$ exists and is finite. By Theorem 5.13.1, f is $\mathcal{D}^*$-integrable on $[u,v]$. It follows that Q has no isolated points, so Q is perfect. By Theorem 1.7.1, there exist $c,d \in Q$, $c<d$ such that $(c,d)\cap Q \neq \emptyset$ and $M,m \in VB^*$ on $P=[c,d]\cap Q$. Then M_P and m_p (see Definition 1.1.3) are VB on $[c,d]$, and $M_P'(x) \geq f(x) \geq m_P'(x)$ *a.e.* on P. But M_P' and m_P' are summable on $[c,d]$. Since f is measurable it follows that f is summable on P. Let $\{(c_j,d_j)\}_j$ be the components of G contained in (c,d). Then f is $\mathcal{D}^*$-integrable on each $[c_j,d_j]$. Let $F_j : [c_j,d_j] \mapsto \mathbb{R}$, $F_j(x) = (\mathcal{D}^*)\int_{c_j}^x f(t)dt$. By 5), $m(y)-m(x) \leq F_j(y)-F_j(x) \leq M(y)-M(x)$, whenever $c_j \leq x < y \leq b_j$, hence $|F_j(y)-F_j(x)| < |M(y)-M(x)| + |m(y)-m(x)|$. It follows that $\mathcal{O}(F_j;[c_j,d_j]) < \mathcal{O}(M;[c_j,d_j]) + \mathcal{O}(m;[c_j,d_j])$. Since $M,m \in VB^*$ on Q, $\sum_{j=1}^{\infty}\mathcal{O}(F_j;[c_j,d_j]) < +\infty$. By Theorem 5.13.2, f is $\mathcal{D}^*$-integrable on $[c,d]$, a contradiction, hence $G=(a,b)$. By Theorem 5.13.1, f is $\mathcal{D}^*$-integrable on $[a,b]$.

Theorem 5.14.1 *Let $f:[a,b]\mapsto \overline{\mathbb{R}}$. The following assertions are equivalent:*

(i) f is $\mathcal{D}^*$-integrable on $[a,b]$;

(ii) f is measurable; there exists $(j,k) \in (\{1,2,\ldots,6\}\times\{1,2,\ldots,12\})\cup(\{13,14,15\}\times\{13,14,15\})$ such that $\overline{\mathcal{M}}_j(f)\times\underline{\mathcal{M}}_k(f)\neq\emptyset$;

(iii) f is measurable; there exists $(j,k) \in \{7,8,16\}\times\{7,8,16\}$ such that $\overline{\mathcal{M}}_j(f)\times\underline{\mathcal{M}}_k(f)\neq\emptyset$; for $\epsilon>0$ there exists $\delta:[a,b]\mapsto(0,+\infty)$ and $(M,m)\in\overline{\mathcal{M}}_j(f)\times\underline{\mathcal{M}}_k(f)$ such that $-\epsilon < m(z)-m(y) < M(z)-M(y) < \epsilon$, whenever $[y,z]\subset(x-\delta(x),x+\delta(x))\setminus\{x\}$, $x\in[a,b]$.

Proof. $(i)\Rightarrow(ii)$ There exists $F:[a,b]\mapsto\mathbb{R}$, $F\in AC^*G\cap C$ such that $F'(x)=f(x)$ *a.e.* on $[a,b]$. But F' is measurable on $[a,b]$, so f is measurable on $[a,b]$. by Corollary 5.8.2, $\overline{\mathcal{M}}_j(f)\times\underline{\mathcal{M}}_k(f)\neq\emptyset$.

$(ii)\Rightarrow(i)$ Let $(M,m)\in\overline{\mathcal{M}}_j(f)\times\underline{\mathcal{M}}_k(f)$. By Lemma 5.7.3, (ii), $(M,m)\in\overline{\mathcal{M}}_6(f)\times\underline{\mathcal{M}}_6(f)$. It follows that: (a) $M(a)=m(a)=0$; (b) $M,m\in VB^*G$ on $[a,b]$; (c) $m'(x)\leq f(x)\leq M'(x)$ *a.e.* on $[a,b]$;
(d) $M\in C_i^*$ and $m\in C_d^*$ on $[a,b]$. By Proposition 2.3.1, (vii), there exists $\delta:[a,b]\mapsto(0,+\infty)$ such that $|M(z)-M(y)|<\epsilon$ and $|m(z)-m(y)|<\epsilon$, whenever $a\leq y<z\leq b$ and $[y,z]\subset(x-\delta(x),x+\delta(x))\setminus\{x\}$, $x\in[a,b]$;
(e) Let $[c,d]\subset[a,b]$ such that f is $\mathcal{D}^*$-integrable on $[c,d]$. Let $F:[c,d]\mapsto\mathbb{R}$, $F(x)=(\mathcal{D}^*)\int_a^x f(t)dt$. Since $M\in AC^*G\cap C_i^*$ on $[a,b]$ and $M'(x)\geq f(x)$ *a.e.* on $[c,d]$, it follows that $M-F\in AC^*G\cap C_i^*\subset\underline{M}\cap C_i$ on $[c,d]$ and $(M-F)'(x)\geq 0$ *a.e.* on $[c,d]$. By Theorem 4.2.1, $M-F$ is increasing on $[c,d]$, hence $M(d)-M(c)\geq F(d)-F(c)$. Similarly $F(d)-F(c)\geq m(d)-m(c)$.
By Lemma 5.14.1, f is $\mathcal{D}^*$-integrable on $[a,b]$.

$(i)\Rightarrow(iii)$ The proof is similar to that of $(i)\Rightarrow(ii)$.

$(iii)\Rightarrow(i)$ Let $(M,m)\in\overline{\mathcal{M}}_j(f)\times\underline{\mathcal{M}}_k(f)\neq\emptyset$. By Lemma 5.7.3, $(M,m)\in\overline{\mathcal{M}}_8(f)\times\underline{\mathcal{M}}_8(f)$. Similarly to $(ii)\Rightarrow(i)$, we can verify (a) - (e) of Lemma 5.14.1, hence f is $\mathcal{D}^*$-integrable on $[a,b]$.

Remark 5.14.1 *In 1937, Marcinkiewicz showed that, if f is measurable and $\overline{\mathcal{M}}_3(f)\times\underline{\mathcal{M}}_3(f)\neq\emptyset$ then f is $(\mathcal{P}_{3,3})$-integrable, and so f is $\mathcal{D}^*$-integrable (see [S3], p. 253). This result was also proved independently by Tolstoff in 1939.*

In 1978, Sarkhel (see [Sa]) generalizes Marcinkiewicz' theorem, but this generalization is also contained in Theorem 5.14.1 (see (i) $\Rightarrow$ (iii) for $(j,k) = (16,16)$). For more information on this topic see [B6] and [BLMMP].

5.15 Bounded Riemann sums and locally small Riemann sums

Definition 5.15.1 (Bullen and Vyborny) *Let $f : [a,b] \mapsto \mathbb{R}$. f is said to satisfy the BRS (bounded Riemann sum) property on $[a,b]$, if there exists a function $\delta : [a,b] \mapsto (0,+\infty)$ and a positive constant k (which depends on δ) such that $\mid f(P) \mid < k$, whenever $P \in \mathcal{P}_1([a,b];\delta)$.*

Definition 5.15.2 (Schurle) *([Sc3]). Let $f : [a,b] \mapsto \mathbb{R}$. F is said to satisfy the $(LSRS)$ (locally small Riemann sum) property on $[a,b]$, if for every $\epsilon > 0$ there exists $\delta : [a,b] \mapsto (0,+\infty)$ with the following property: if $x \in [y,z] \subset (x-\delta(x), x+\delta(x))$ for each $x \in [a,b]$ then $\mid \sigma(f;P) \mid < \epsilon$, whenever $P \in \mathcal{P}_1([y,z];\delta)$.*

Lemma 5.15.1 (Bullen and Vyborny) *Let $f : [a,b] \mapsto \mathbb{R}$.*

(i) $(LSRS) \subseteq (BRS)$ on $[a,b]$.

(ii) If $\overline{\mathcal{M}}_7(f) \times \underline{\mathcal{M}}_7(f) \neq \emptyset$ then $f \in (BRS)$ on $[a,b]$.

(iii) If $f \in (BRS)$ on $[a,b]$ then for the δ given by this fact there exist M_δ, $m_\delta : [a,b] \mapsto \mathbb{R}$, defined as follows: $M_\delta(a) = m_\delta(a) = 0$, $M_\delta(x) = \sup\{f(P) : P \in \mathcal{P}_1([a,x];\delta)$ and $m_\delta(x) = \inf\{f(P) : P \in \mathcal{P}_1([a,x];\delta)$, $x \in (a,b]$. Moreover, $(M_\delta, m_\delta) \in \overline{\mathcal{M}}_7(f) \times \underline{\mathcal{M}}_7(f)$.

(iv) Let $f \in (LSRS)$ on $[a,b]$, $\epsilon > 0$ and consider the δ given by this fact for $\epsilon/2$. Then there exist M_δ and m_δ defined as above and $(M_\delta, m_\delta) \in \overline{\mathcal{M}}_7(f) \times \underline{\mathcal{M}}_7(f)$. Moreover $-\epsilon < m_\delta(z) - m_\delta(y) \leq M_\delta(z) - M_\delta(y) < \epsilon$, whenever $[y,z] \subset (x-\delta(x), x+\delta(x) \setminus \{x\}$, $x \in [a,b]$.

Proof. (i) The proof is similar to that of lemma 5.6.1, (ii).

(ii) ([BV], pp. 167 -168). Let $(M,m) \in \overline{\mathcal{M}}_7(f) \times \underline{\mathcal{M}}_7(f)$. For $\epsilon = 1$ let $\delta : [a,b] \mapsto (0,+\infty)$ such that $M(v) - M(u) \geq (v-u)(f(x)-1)$ and $m(v) - m(u) \leq (v-u)(f(x)+1)$, whenever $x - \delta(x) < u \leq x \leq v < x + \delta(x)$. For $P = \{([y_i,z_i];x_i) : i = \overline{1,n}\} \in \mathcal{P}_1([a,b];\delta)$ we have $M(b) \geq \sigma(f;P) - (b-a)$ and $m(b) \leq \sigma(f;P) + (b-a)$, hence $m(b) - (b-a) \leq \sigma(f;P) \leq M(b) + (b-a)$. Let $k = \max\{\mid M(b)+(b-a) \mid, \mid m(b) - (b-a) \mid\}$. It follows that $\mid \sigma(f;P) \mid < k$, hence $f \in (BRS)$ on $[a,b]$.

(iii) ([BV], pp. 167 - 168). Let $\delta : [a,b] \mapsto (0,+\infty)$ and let k (depending on δ) such that $\mid \sigma(f;P) \mid < k$, whenever $P \in \mathcal{P}_1([a,b];\delta)$. Let $x \in (a,b]$, $P_x \in \mathcal{P}_1([a,x];\delta)$, $Q_x \in \mathcal{P}_1([x,b];\delta)$ and $P \in \mathcal{P}_1([a,x];\delta)$. Then $P_x \cup Q_x$ and $P \cup Q_x$ belong to $\mathcal{P}_1([a,b];\delta)$. It follows that $\mid \sigma(f;P_x) + \sigma(f;Q_x) \mid < k$ and $\mid \sigma(f;P) + \sigma(f;Q_x) \mid < k$, hence $\mid \sigma(f;P) - \sigma(f;P_x) \mid < 2k$. Therefore m_δ and M_δ exist and are well defined. If $x - \delta(x) < u \leq x \leq v < x + \delta(x)$ then $M_\delta(v) - M_\delta(u) \geq f(x)(v-u)$, so $M_\delta \in \overline{\mathcal{M}}_7(f)$. Similarly $m_\delta \in \underline{\mathcal{M}}_7(f)$.

(iv) The first part follows by (i) and (iii). Let $x \in [a,b]$ and $[y,z] \subset (x-$

$\delta(x), x+\delta(x)) \setminus \{x\}$. By Lemma 5.7.3, (i), the function $M_\delta - m_\delta$ is increasing, hence $M_\delta(z) - M_\delta(y) \geq m_\delta(z) - m_\delta(y)$. Suppose for example that $[y,z] \subset (x-\delta(x), x)$. Let $\epsilon > 0$. Since $f \in (LSRS)$, for $\epsilon/6$ there exists a positive function δ such that $\mid \sigma(f;P_1) \mid < \epsilon/6$, whenever $P_1 \in \mathcal{P}_1([y,x];\delta)$, and $\mid \sigma(f;P_2) \mid < \epsilon/6$, whenever $P_2 \in \mathcal{P}_1([z,x];\delta)$. It follows that $\mid \sigma(f;P) \mid < \epsilon/3$, whenever $P \in \mathcal{P}_1([y,z];\delta)$. Let $P_z = \{([x_{i-1},x_i];t_i) : i = \overline{1,n}\} \in \mathcal{P}_1([a,z];\delta)$, where $a = x + o < x_1 < \ldots < x_n = z$. We have cases:

1) Suppose that there exists a $j \in \{1,2,\ldots,n-1\}$ such that $x_j = y$. It follows that $\sigma(f;P_z) = \sum_{i=1}^{j} f(t_i)x_i - x_{i-1}) + \sum_{i=j+1}^{n} f(t_i)(x_i - x_{i-1}) < M_\delta(y) + \epsilon/3$.

2) Suppose that there exists a $j \in \{1,2,\ldots,n-2\}$ such that $x_j < y < x_{j+1}$. Let $P_j \in \mathcal{P}_1([x_j,y];\delta)$. It follows that $\sigma(f;P_z) = \sum_{i=1}^{j} f(t_i)(x_i - x_{i-1}) + \sigma(f;P_j) - \sigma(f;P_j) + f(t_{j+1})(x_{j+1} - x_j) + \sum_{i=j+2}^{n} d(t_i)(x_i - x_{i-1}) < M_\delta(y) + \epsilon/3 + \epsilon/3 + \epsilon/3 = M_\delta(y) + \epsilon$. Therefore $M_\delta(z) \leq M_\delta(y) + \epsilon$.

Theorem 5.15.1 (Schurle) *([Sc2] and [BV]). Let $f : [a,b] \mapsto \mathbb{R}$. The following assertions are equivalent:*

(i) f is $\mathcal{D}^*$-integrable on $[a,b]$;

(ii) f is measurable and has the $(LSRS)$ property on $[a,b]$.

Proof. $(i) \Rightarrow (ii)$ By Theorem 5.14.1, (i), f is measurable, and there exists a pair $(M,m) \in \overline{\mathcal{M}}_2(f) \times \underline{\mathcal{M}}_2(f) \neq \emptyset$. Let $\epsilon > 0$. Since $M, m \in \mathcal{C}$ on $[a,b]$, there exists $\delta_1 : [a,b] \mapsto (0,+\infty)$ such that $\mid M(z) - M(y) \mid < \epsilon/2$ and $\mid m(z) - m(y) \mid < \epsilon/2$, whenever $x \in [y,z] \subset (x - \delta_1(x), x + \delta_1(x))$, $x \in [a,b]$. Also there exists $\delta_2 : [a,b] \mapsto (0,+\infty)$ such that $M(z) - M(y) \geq (z-y)(f(x) - \epsilon/(2(b-a)))$ and $m(z) - m(y) \leq (z-y)(f(x) + \epsilon/(2(b-a)))$, whenever $x \in [y,z] \subset (x - \delta_2(x), x + \delta_2(x))$, $x \in [a,b]$. Let $\delta(x) = \inf\{\delta_1(x), \delta_2(x)\}$, $x \in [a,b]$, and let $x_o \in [y_o, z_o] \subset (x_o - \delta(x_o), x_o + \delta(x_o))$. Let $P \in \mathcal{P}_1([y_o,z_o];\delta)$. Then $m(z_o) - m(y_o) - \epsilon/(2(b-a))(z_o - y_o) < \sigma(f;P) < M(z_o) - M(y_o) + \epsilon/(2(b-a))(z_o - y_o)$. It follows that $\mid \sigma(f;P) \mid < \epsilon$, hence $f \in (LSRS)$ on $[a,b]$.

$(ii) \Rightarrow (i)$ See Lemma 5.15.1, (iv) and Theorem 5.14.1, (i), (iii).

5.16 Riemann type integrals and local systems

Definition 5.16.1 (Wang and Ding) *([WD], p. 248). Let $\mathcal{S}$ be a bilateral, filtering local system, satisfying intersection condition (I.C.). Let $f : [a,b] \mapsto \mathbb{R}$. F is said to be $\mathcal{S}$-integrable on $[a,b]$ if there exists a real number A with the following property: for every $\epsilon > 0$ there is a set value function $\Delta : [a,b] \mapsto \mathcal{S}[[a,b]]$ such that, if $\Pi = \{([c_i,d_i];x_i) : i = \overline{1,n}\} \in \Pi([a,b];\Delta)$ (see Definition 2.38.1) then $\mid \sum_{i=1}^{n} f(x_i)(d_i - c_i) - A \mid < \epsilon$. The number A is unique because $\mathcal{S}$ is filtering. We write $(\mathcal{S})\int_a^b f(t)dt = A$.*

Theorem 5.16.1 (Wang and Ding) *([WD], pp. 250 - 251). Let $\mathcal{S}$ be a bilateral, filtering local system, satisfying intersection condition (I.C.). Let $f, g : [a,b] \mapsto \mathbb{R}$.*

(i) If f and g are $\mathcal{S}$-integrable on $[a,b]$ then so is $f+g$, and we have $(\mathcal{S})\int_a^b (f(t) + g(t))dt = (\mathcal{S})\int_a^b f(t)dt + (\mathcal{S})\int_a^b g(t)dt$.

(ii) *f is $\mathcal{S}$-integrable on $[a,b]$ if and only if for every $\epsilon > 0$ there exists a set value function $\Delta : [a,b] \mapsto \mathcal{S}[[a,b]]$ such that $\mid \sum_{i=1}^{n} f(x_i)(d_i - c_i) - \sum_{i=1}^{n'} f(x'_i)(d'_i - c'_i) \mid < \epsilon$, whenever $\{([c_i,d_i];x_i) : i = \overline{1,n}\} \in \Pi([a,b];\Delta)$ and $\{([c'_i,d'_i];x'_i) : i = \overline{1,n'}\} \in \Pi([a,b];\Delta)$.*

(iii) *(Saks-Henstock). If f is $\mathcal{S}$- integrable on $[a,b]$ and $F(x) = (\mathcal{S})\int_a^x f(t)dt$ then for every $\epsilon > 0$, there exists a set value function $\Delta : [a,b] \mapsto \mathcal{S}[[a,b]]$ such that, for any partial partition $\Pi = \{([c_i,d_i];x_i) : i = \overline{1,n}\} \in \Pi_p([a,b];\Delta)$ we have $\sum_{i=1}^{n} \mid F(d_i) - F(c_i) - f(x_i)(d_i - c_i) \mid < 2\epsilon$.*

Theorem 5.16.2 *Let $\mathcal{S}$ be a bilateral filtering local system satisfying intersection condition (I.C.). Let $f : [a,b] \mapsto \mathbb{R}$ be $\mathcal{S}$-integrable. Then $F(x) = (\mathcal{S})\int_a^x f(t)dt$ is $\mathcal{S}$-continuous on $[a,b]$.*

Proof. Let $x_o \in (a,b)$ (if $x_o = a$ or $x_o = b$ then the proof is similar). Let $\epsilon > 0$. Choose $\Delta : [a,b] \mapsto \mathcal{S}[[a,b]]$ such that $\mid \sum_{i=1}^{n} f(x_i)(d_i - c_i) - F(b) \mid < \epsilon/4$, whenever $\{([c_i,d_o];x_i) : i = \overline{1,n}\} \in \Pi([a,b];\Delta)$. Let $\eta < (\epsilon/2)/(1+ \mid f(x_o) \mid)$ and $\sigma_{x_o} = \Delta(x_o) \cap (x_o - \eta, x_o + \eta) \in \mathcal{S}(x)$. Suppose for example that $t \in (x_o, x_o + \eta) \cap \sigma_{x_o}$. By Theorem 2.38.1, there exist $\Pi_1 \in \Pi([a,x_o];\Delta)$ and $\Pi_2 \in \Pi([t,b];\Delta)$. Hence $\Pi = \Pi_1 \cup \Pi_2 \cup \{([x_o,t];x_o)\} \in \Pi([a,b];\Delta)$. By Theorem 5.16.1, (iii), it follows that $\mid F(t) - F(x_o) \mid \leq \mid F(t) - F(x_o) - f(x_o)(t - x_o) \mid + \mid f(x_o)(t - x_o) \mid < \epsilon/2 + \mid f(x_o) \mid \cdot\eta < \epsilon$. This completes the proof.

Theorem 5.16.3 *Let $\mathcal{S}$ be a bilateral filtering local system satisfying intersection condition (I.C.). Let $f : [a,b] \mapsto \mathbb{R}$ be $\mathcal{S}$-integrable. Then $F(x) = (\mathcal{S})\int_a^x f(t)dt$ is $\mathcal{S}ACG$ on $[a,b]$ and $\mathcal{S}DF(x) = f(x)$ a.e. on $[a,b]$.*

Proof. We show that F is $\mathcal{S}ACG$ on $[a,b]$ (using Gordon's technique of [G5], p. 160). For each positive integer k let $E_k = \{x \in [a,b] : k - 1 \leq\mid f(x) \mid< k\}$. Fix k and let $\epsilon > 0$. Then there exists a set value function $\Delta : [a,b] \mapsto \mathcal{S}[[a,b]]$ such that $\mid \sum_{i=1}^{n} f(x_i)(d_i - c_i) - F(b) \mid < \epsilon$, whenever $\{([c_i,d_i];x_i) :< i = \overline{1,n}\} \in \Pi([a,b];\Delta)$. Let $\eta = \epsilon/k$, $\Delta_k = \Delta_{/E_k}$, and let $\{[(c_i,d_i];x_i) : i = \overline{1,m}\} \in \Pi_p(E_k;\Delta_k)$ such that $\sum_{i=1}^{m}(d_i - c_i) < \eta$. By Theorem 5.16.1, it follows that $\mid \sum_{i=1}^{m}(F(d_i) - F(c_i)) \mid \leq \mid \sum_{i=1}^{m}(F(d_i) - F(c_i) - f(x_i)(d_i - c_i)) \mid + \mid \sum_{i=1}^{m} f(x_i)(d_i - c_i) \mid < \epsilon + k\cdot\eta = 2\epsilon$, hence $F \in \mathcal{S}AC$ on E_k.
We show the second part (using the technique of Wang and Ding of [WD],p. 251). Let $Z = \{x : \mathcal{S}DF(x)$ does not exist or if it does, it is not equal to $f(x)\}$. For each $x \in Z$ there exists $\eta(x) > 0$ with the following property: for every $\sigma \in \mathcal{S}(x)$ there exists $y \in \sigma$, $y \neq x$ such that

(1) $\quad \mid F(y) - F(x) - f(x)(y - x) \mid > \eta(x)\cdot \mid y - x \mid .$

For each positive integer k let $Z_k = \{x \in Z : \eta(x) \geq 1/k\}$. Then $Z = \cup_{k=1}^{\infty} Z_k$. It suffices to show that each Z_k is of measure zero. The family of closed intervals with endpoints x and y, satisfying (1), forms a Vitali cover of Z_k. Let $\epsilon > 0$. By Theorem 5.16.1, (iii), there exists a set-value function $\Delta : [a,b] \mapsto \mathcal{S}[[a,b]]$ such that $\mid \sum_{i=1}^{m}(F(d_i) - F(c_i) - f(x_i)(d_i - c_i)) \mid< \epsilon$, whenever $\{([c_i,d_i];x_i) : i = \overline{1,m}\} \in \Pi_p([a,b];\Delta)$. By Theorem 1.8.1, there exist $[u_i,v_i]$, $i = \overline{1,q}$ such that for example, $v_i \in \Delta(u_i)$, $u_i \in Z_k$ and $\mid Z_k \mid< \epsilon + \sum_{i=1}^{q}(v_i - u_i)$. By (1), it follows that $\sum_{i=1}^{q}(v_i - u_i) < \sum_{i=1}^{q} \mid F(v_i) - F(u_i) - f(u_i)(v_i - u_i) \mid /(\eta(u_i)) \leq k\cdot\sum_{i=1}^{q} \mid F(v_i) - F(u_i) - f(u_i)(v_i - u_i) \mid < k\cdot\epsilon$, hence $\mid E_k \mid< n\cdot\epsilon + \epsilon$. Since ϵ is arbitrary, we obtain that $\mid E_k \mid= 0$.

Corollary 5.16.1 *Let $\mathcal{S}$ be a bilateral, filtering local system, satisfying intersection condition (I.C.). Let $f : [a,b] \mapsto \mathbb{R}$ be $\mathcal{S}$-integrable. Then f is a measurable function.*

Proof. Let $F(x) = (\mathcal{S})\int_a^x f(t)dt$. By Theorem 5.16.2, F is $\mathcal{S}$-continuous on $[a,b]$. By Theorem 2.2.5, $F \in \mathcal{B}_1$. It follows that F is measurable on $[a,b]$. By Theorem 2.15.6, F is approximately derivable *a.e.* on $[a,b]$. It follows that f is also measurable on $[a,b]$ (see Corollary 1.11.1).

Theorem 5.16.4 *Let $\mathcal{S}$ be a bilateral, filtering local system, satisfying intersection condition (I.C.). Let $F, f : [a,b] \mapsto \mathbb{R}$ such that F is $\mathcal{S}$-continuous on $[a,b]$ and $\mathcal{S}DF(x) = f(x)$ n.e. on $[a,b]$. Then f is $\mathcal{S}$-integrable on $[a,b]$ and $(\mathcal{S})\int_a^x f(t)dt = F(x) - F(a)$, $x \in [a,b]$.*

Proof. We are using Gordon's technique of [G5] (p. 159). Let $E = \{x \in [a,b] : \mathcal{S}DF(x) \neq f(x)\} = \{x_1, x_2, \ldots\}$. For $\epsilon > 0$ let $\Delta : [a,b] \mapsto \mathcal{S}[[a,b]]$ be a set-value function, defined as follows: let $\sigma_{x_k} = \{y : y = 0 \text{ or } \mid F(y) - F(x_k) \mid < \epsilon/2^k\}$; since F is $\mathcal{S}$- continuous on $[a,b]$,$\sigma_{x_k} \in \mathcal{S}(x_k)$; let $\delta(x_k) > 0$ such that $\mid f(x_k) \mid \cdot \delta(x_k) < \epsilon/2^k$; now we define $\Delta(x_k) = \sigma_{x_k} \cap (x_k - \delta(x_k), x_k + \delta(x_k)) \in \mathcal{S}(x_k)$ (see Definition 1.13.1, (iii)); for $x \in [a,b] \setminus E$, let $\Delta(x) \in \mathcal{S}(x)$ such that $\mid f(x)(d-c) - (F(d) - F(c)) \mid < \epsilon \cdot (d-c)$, whenever $x \in [c,d]$, $c, d \in \Delta(x)$ (this is possible because $\mathcal{S}DF(x) = f(x)$). Let $\Pi = \{([c_i,d_i]; y_i) : i = \overline{1,n}\} \in \Pi([a,b];\Delta)$, and let Π_E be the set formed of all members of Δ with tags in E. Let $\Pi' = \Pi \setminus \Pi_E$. Then we have $\mid \sum_{i\in\Pi_E} f(y_i)(d_i - c_i) \mid \leq \sum_{k=1}^{\infty} \mid f(x_k) \mid \cdot 2 \cdot \delta(x_k) < \sum_{k=1}^{\infty} 2\epsilon/2^k = 2\epsilon$ and $\sum_{i\in\Pi_E} \mid F(d_i) - F(c_i) \mid < \sum_{k=1}^{\infty} \epsilon/2^k = \epsilon$. Then $\mid \sum_{i=1}^{n} f(y_i)(d_i - c_i) - (F(b) - F(a)) \mid$ $\leq \mid \sum_{i\in\Pi_E} f(y_i)(d_i - c_i) - (F(d_i) - F(c_i)) \mid + \mid \sum_{i\in\Pi'} f(y_i)(d_i - c_i) - F(d_i) - F(c_i)) \mid$ $< 2\epsilon + \epsilon + (b-a)\cdot\epsilon$, hence $(\mathcal{S})\int_a^b f(t)dt = F(b) - F(a)$.

Theorem 5.16.5 *Let $\mathcal{S}$ be a bilateral, filtering local system, satisfying intersection condition (I.C.), and let $f : [a,b] \mapsto \mathbb{R}$. The following assertions are equivalent:*

(i) f is $\mathcal{S}$-integrable on $[a,b]$;

(ii) There exists $F : [a,b] \mapsto \mathbb{R}$, $F \in \mathcal{S}ACG$ such that $\mathcal{S}DF(x) = f(x)$ a.e. on $[a,b]$;

(iii) There exists $G : [a,b] \mapsto \mathbb{R}$, $G \in \mathcal{S}Y$ such that $\mathcal{S}DG(x) = f(x)$ a.e. on $[a,b]$;

(iv) For every $\epsilon > 0$ there exist $H : [a,b] \mapsto \mathbb{R}$, and an increasing function $h : [a,b] \mapsto [0,\epsilon]$, and a set-value function $\Delta : [a,b] \mapsto \mathcal{S}[[a,b]]$ such that $\mid H(z) - H(y) - f(x)(z-y) \mid < h(z) - h(y)$, whenever $x \in [y,z]$, $y, z \in \Delta(x)$;

(v) For every $\epsilon > 0$ there exist $M, m : [a,b] \mapsto \mathbb{R}$ and a set-value function $\Delta : [a,b] \mapsto \mathcal{S}[[a,b]]$ such that $0 \leq M(b) - m(b) \leq \epsilon$, $M(a) = m(a) = 0$ and $M(z) - M(y) \geq f(x)(z-y) \geq m(z) - m(y)$, whenever $x \in [y,z]$ and $y, z \in \Delta(x)$. We denote by $I_5(b) = \inf_{\epsilon\to 0}\{M(b)\} = \sup_{\epsilon\to 0}\{m(b)\}$;

(vi) For $\epsilon > 0$ there exist $M, m : [a,b] \mapsto \mathbb{R}$ such that $M(a) = m(a) = 0$, $M(b) - m(b) < \epsilon$, and $-\infty \geq \mathcal{S}\underline{D}M(x) \geq f(x) \geq \mathcal{S}\overline{D}m(x) \neq +\infty$ for each $x \in [a,b]$. We denote by $I_6(b) = \inf_{\epsilon\to 0}\{M(b)\} = \sup_{\epsilon\to 0}\{m(b)\}$;

(vii) *For $\epsilon > 0$ there exist $M, m : [a,b] \mapsto \mathbb{R}$, $M \in \mathcal{S}\underline{ACG}$, $m \in \mathcal{S}\overline{ACG}$ such that $M(a) = m(a) = 0$, $M(b) - m(b) < \epsilon$ and $\mathcal{S}\,\underline{D}M(x) \geq f(x) \geq \mathcal{S}\,\overline{D}m(x)$ a.e. on $[a,b]$. We denote by $I_7(b) = \inf_{\epsilon\to 0}\{M(b)\} = \sup_{\epsilon\to 0}\{m(b)\}$;*

(viii) *For $\epsilon > 0$ there exist $M, m : [a,b] \mapsto \mathbb{R}$, $M \in \mathcal{S}\underline{Y}$, $m \in \mathcal{S}\overline{Y}$, such that $M(a) = m(a) = 0$, $M(b) - m(b) < \epsilon$ and $\mathcal{S}\,\underline{D}M(x) \geq f(x) \geq \mathcal{S}\,\overline{D}m(x)$ a.e. on $[a,b]$. We denote by $I_8(b) = \inf_{\epsilon\to 0}\{M(b)\} = \sup_{\epsilon\to 0}\{m(b)\}$;*

Moreover we have $(\mathcal{S})\int_a^b f(t)dt = F(b) - F(a) = G(b) - G(a) = H(b) - H(a) = I_5(b) = I_6(b) = I_7(b) = I_8(b)$.

Proof. $(i) \Rightarrow (ii)$ Let $F(x) = (\mathcal{S})\int_a^b f(t)dt$ and see Theorem 5.16.3.

$(ii) \Rightarrow (iii)$ Let $G = F$ and see Lemma 2.38.1.

$(iii) \Rightarrow (i)$ Let G be given by (iii) and let $Z = \{x : \mathcal{S}\,DG(x) \neq f(x)\}$. Then $\mid Z \mid = 0$. For $\epsilon > 0$ let $\Delta : [a,b] \mapsto \mathcal{S}[[a,b]]$ be a set-value function defined as follows: for $x \in [a,b] \setminus Z$ let $\Delta(x) = \{y :$ either $y = x$ or $\mid F(y) - F(x) - f(x)(y - x) \mid < \epsilon(y - x)\} \in \mathcal{S}(x)$; by Lemma 5.4.5, there exists $\delta : Z \mapsto (0, +\infty)$ such that $\mid \sum_{i=1}^m f(x_i)(d_i - c_i) \mid < \epsilon$, whenever $\{([c_i, d_i]; x_i) : i = \overline{1,m}\} \in \Pi_1(Z;\delta)$; since $G \in \mathcal{S}Y$, there exists $\Delta_1 : Z \mapsto \mathcal{S}[Z]$ such that $\mid \sum_{i=1}^m (G(d_i) - G(c_i)) \mid < \epsilon$, whenever $\{([c_i, d_i]; x_i) : i = \overline{1,m}\} \in \Pi(Z;\Delta_1)$; Let $\Delta(x) = \Delta_1(x) \cap (x - \delta(x), x + \delta(x))$ for each $x \in Z$. Let $\Pi = \{([a_i, b_i]; y_i) : i = \overline{1,n}\} \in \Pi([a,b];\Delta)$. Let P_Z be the set formed of all members of Π with tags in Z, and let $\Pi' = \Pi \setminus \Pi_Z$. It follows that $\mid \sum_{i=1}^n f(y_i)(b_i - a_i) - (G(b) - G(a)) \mid \leq \mid \sum_{i\in\Pi'} (f(y_i)(b_i - a_i) - (G(b_i) - G(a_i)) \mid + \sum_{i\in\Pi_Z} f(y_i)(b_i - a_i) + \sum_{i\in\Pi_Z} \mid G(b_i) - G(a_i) \mid < \epsilon(b - a) + \epsilon + \epsilon$, hence f is $\mathcal{S}$-integrable on $[a,b]$ and $(\mathcal{S})\int_a^b f(t)dt = G(b) - G(a)$.

$(i) \Rightarrow (iv)$ We shall use Pu's technique of [Pu] (pp. 107 - 108). Let $H(x) = (\mathcal{S})\int_a^x f(t)dt$ For $\epsilon > 0$ let $\Delta : [a,b] \mapsto \mathcal{S}[[a,b]]$ be a set-value function such that $\mid \sum_{i=1}^m f(x_i)(d_i - c_i) - H(b) \mid < \epsilon/2$, whenever $\{([c_i, d_i]; x_i) : i = \overline{1,m}\} \in \Pi([a,b];\Delta)$. Let $a < x \leq b$ and $\{[a_j, b_j]; y_j) : j = \overline{1,n}\} \in \Pi([a,x];\Delta)$. Then $\sum_{j=1}^n \mid f(y_j)(b_j - a_j) - (H(b_j) - H(a_j)) \mid < \epsilon$ (see Theorem 5.16.1, (iii)). Let $h : [a,b] \mapsto [0,\epsilon]$ such that $h(a) = 0$, and for each $x \in (a,b]$ let's define $h(x) = \sup\{\sum_{i=1}^n \mid f(y_j)(b_j - a_j) - (H(b_j) - H(a_j)) \mid : \{([a_j, b_j]; y_j) : j = \overline{1,n}\} \in \Pi([a,x];\Delta)\}$. Then h is increasing and $h(y) + \mid H(z) - H(y) - f(x)(z - y) \mid \leq h(z)$, hence $\mid H(z) - H(y) - f(x)(z - y) \mid \leq h(z) - h(y)$, whenever $x \in [y,z]$, $y, z \in \Delta(x)$.

$(iv) \Rightarrow (v)$ For $\epsilon/2$ let H and h be given by (iv). Then $M = H + h$ and $m = H - h$ have the required properties.

$(v) \Rightarrow (vi)$ This is evident.

$(vi) \Rightarrow (vii)$ See Lemma 2.38.1, (iii).

$(vii) \Rightarrow (viii)$ See Lemma 2.38.1.

$(viii) \Rightarrow (i)$ For $\epsilon > 0$ let $M \in \mathcal{S}\underline{Y}$ such that $M(b) - I_8(b) < \epsilon/2$ and $\mathcal{S}\,\underline{D}M(x) \geq f(x)$ a.e. on $[a,b]$. Let $A = \{x : \mathcal{S}\underline{D}M(x) \geq f(x)\}$ and $Z = [a,b] \setminus A$. Then $\mid Z \mid = 0$. Let $\Delta : [a,b] \mapsto \mathcal{S}[[a,b]]$ be a set-value function, defined as follows: for each $x \in A$ let $\Delta(x) \in \mathcal{S}(x)$ such that $(M(z) - M(y))/(z - y) > f(x) - \epsilon/(2(b - a))$, $x \in [y,z]$ and $y, z \in \Delta(x)$; by Lemma 5.4.5, for each $x \in Z$ there exists $\delta : Z \mapsto (0, +\infty)$ such that $\mid \sum_{i=1}^m f(x_i)(d_i - c_i) \mid < \epsilon$, whenever $\{([c_i, d_i]; x_i) : i = \overline{1,m}\} \in \Pi_1(Z;\delta)$; since $M \in \mathcal{S}\underline{Y}$, there exists a set-value function $\Delta_1 : Z \mapsto \mathcal{S}[Z]$ such that $\sum_{i=1}^n (M(d_i) - M(c_i)) > -\epsilon$, whenever $\{([c_i, d_i]; x_i) : i = \overline{1,n}\} \in \Pi(Z;\Delta_1)$; for each $x \in Z$ let $\Delta(x) = \Delta_1(x) \cap (x - \delta(x), x + \delta(x))$. Let $\Pi = \{([a_j, b_j]; y_j) : j = \overline{1,n}\} \in \Pi([a,b];\Delta)$. Let Π_Z be the set formed of all members of Π with tags in Z, and let $\Pi' = \Pi \setminus \Pi$. It follows that $M(b) = \sum_{j=1}^{n}(M(b_j)-M(a_j)) = \sum_{j\in\Pi'}(M(b_j)-M(a_j)) + \sum_{j\in\Pi_Z}(M(b_j)-M(a_j)) \geq \sum_{j\in\Pi'}(f(x_j) - \epsilon/(b-a))(b_j - a_j) - \epsilon > \sum_{j\in\Pi'} f(x_j)(b_j - a_j) - 2\epsilon = \sum_{j=1}^{n} f(x_j)(b_j - a_j) - \sum_{j\in\Pi_Z} f(x_j)(b_j - a_j) > \sum_{j=1}^{n} f(x_j)(b_j - a_j) - 3\epsilon$. It follows that $I_8(b) + (7/2)\epsilon \geq \sum_{j=1}^{n} f(x_j)(b_j - a_j)$. Similarly we obtain that $I_8(b) - (7/2)\epsilon \leq \sum_{j=1}^{n} f(x_j)(b_j - a_j)$. Therefore $(\mathcal{S})\int_a^b f(t)dt = I_8(b)$. The last part is obvious.

Remark 5.16.1 *In Theorem 5.16.5, $I_j(b)$, $j = 5,6,7,8$ are well defined, because for each case $M - m$ is increasing on $[a,b]$ (see Lemma 2.38.1 and Lemma 2.38.2).*

Remark 5.16.2 *In Theorem 5.16.5, (i) is a Riemann (Kurzweil-Henstock) type definition, (ii) and (iii) are descriptive type definitions, (iv) is a variational Henstock type definition, (v) is a Ward type definition, and (vi), (vii) and (viii) are Perron type definitions for the $\mathcal{S}$-integral.*

Remark 5.16.3 *(i) If in Theorem 5.16.5, (vi), $\mathcal{S}$ is replaced by $\mathcal{S}_{ap}$, and M, m are supposed to be $\mathcal{S}_{ap}$- continuous (i.e. approximately continuous) then we obtain the Burkill integral, denoted by (AP) (see [Bu]).*

(ii) If on Definition 5.16.1, $\mathcal{S}$ is replaced by $\mathcal{S}_{ap}$, we obtain Bullen's R_{ap}-integral (see [B3]). This integral was also studied by Gordon (see [G5]).

(iii) By Theorem 5.16.5, (i), (vi), it follows that the AP integral is contained in the R_{ap} integral.

5.17 The $< LPG >$ and $< LDG >$ integrals

Let $< uL >$ be an upper semilinear space of real functions defined on $[a,b]$ (i.e., if $F, G : [a,b] \mapsto \mathbb{R}$, $F, G \in < uL >$ and $\alpha, \beta \geq 0$ then $\alpha F + \beta G \in < uL >$), and let $< L > = \{F : F$ and $-F$ belong to $< uL >\}$. In what follows we consider upper semilinear spaces $< uL >$ with the following properties:

(a) $\mathcal{C} \subseteq < uL > \subset uCM$ on $[a,b]$;

(b) each function $F \in < uL >$ is Lebesgue measurable on $[a,b]$;

(c) $< uL >$ is closed under uniform convergence (i.e., $F_n \to F$ $[unif]$ and $F_n \in < uL >$ on $[a,b]$ implies that $F \in < uL >$).

Definition 5.17.1 (C.M. Lee) *([Le1]). Let $f : [a,b] \mapsto \overline{\mathbb{R}}$. A function $M : [a,b] \mapsto \mathbb{R}$ is said to be an $< LPG >$ major function for f if:*

(1) $M(a) = 0$;

(2) $M \in < uL >$;

(3) $M \in [\underline{ACG}]$ on $[a,b]$;

(4) $M'_{ap}(x) \geq f(x)$ a.e. on $[a,b]$ ($M'_{ap}(x)$ exists a.e., see Remark 2.15.4).

A function $m : [a,b] \mapsto \mathbb{R}$ is said to be an $< LPG >$ minor function for f if $-m$ is and $< LPG >$ major function for $-f$ on $[a,b]$.

Lemma 5.17.1 *Let $f : [a,b] \mapsto \overline{\mathbb{R}}$. If M is an $< LPG >$ major function and m is an $< LPG >$ minor function for f then $M - m$ is increasing on $[a,b]$, hence $M(b) \geq m(b)$.*

Proof. Clearly $(M-m)'_{ap}(x)$ a.e. on $[a,b]$ and $M - m \in [\underline{AC}G] \cap uCM$. By Remark 4.3.4, it follows that $M - m$ is increasing on $[a,b]$.

Definition 5.17.2 (C.M. Lee) *Let $f : [a,b] \mapsto \overline{\mathbb{R}}$. Suppose that f has $< LPG >$ major and minor functions. We denote by $\overline{I}(b) = \overline{< LPG >} \int_a^b f(t)dt = \inf\{M(b) :$ M is an $< LPG >$ major function for $f\}$ and $\underline{I}(b) = \underline{< LPG >} \int_a^b f(t)dt = \sup\{m(b) : m$ is an $< LPG >$ minor function for $f\}$. Then $\overline{I}(b)$ and $\underline{I}(b)$ are finite and $\overline{I}(b) \geq \underline{I}(b)$ (see Lemma 5.17.1). F is said to have an $< LPG >$ integral on $[a,b]$ if $\overline{I}(b) = \underline{I}(b) = < LPG > \int_a^b f(t)dt$.*

Lemma 5.17.2 *Let $f : [a,b] \mapsto \overline{\mathbb{R}}$. The following assertions are equivalent:*

(i) *f is $< LPG >$-integrable on $[a,b]$;*

(ii) *For every $\epsilon > 0$ there exist an $< LPG >$ major function M and an $< LPG >$ minor function m such that $M(b) - m(b) < \epsilon$.*

Definition 5.17.3 (C.M. Lee) *Let $f : [a,b] \mapsto \overline{\mathbb{R}}$. f is said to be $< LDG >$-integrable on $[a,b]$ if there exists $F : [a,b] \mapsto \mathbb{R}$ such that $F \in < L > \cap [ACG]$ and $F'_{ap}(x) = f(x)$ a.e. on $[a,b]$ ($F'_{ap}(x)$ exists a.e., see Remark 2.15.4, (i)). We denote by $< LDG > \int_a^b f(t)dt = F(b) - F(a)$.*

Remark 5.17.1 *(i)* *By Remark 4.3.4, it follows that the $< LDG >$ integral is well defined.*

(ii) *If in Definition 5.17.3, $< L > = \mathcal{C}$ then we obtain the $(\mathcal{D})$ integral (the wide Denjoy integral). If $< L > =$ the class of all approximately continuous functions, then we obtain Ridder's β-integral (see Definition 7 of [R3], p. 148, and Theorem 2.11.1, (xviii)). For other particular cases see [Le1].*

(iii) *If in Definition 5.17.3, we choose $< L >$ such that $(\mathcal{C}) \subset < L > \subset \mathcal{DB}_1$, and $[ACG]$ is replaced by ACG then we obtain another well-defined integral (see Corollary 4.3.4), that we denote by LDG.*

Theorem 5.17.1 *Let $f, g : [a,b] \mapsto \overline{\mathbb{R}}$ and $c \in (a,b)$.*

(i) *If f is $< LPG >$ integrable on $[a,b]$ then f is finite a.e. on $[a,b]$;*

(ii) *f is $< LPG >$ integrable on $[a,b]$ if and only if f is $< LPG >$ integrable on both $[a,c]$ and $[c,b]$. Moreover $< LPG > \int_a^b f(t)dt = < LPG > \int_a^c f(t)dt + < LPG > \int_c^b f(t)dt$;*

(iii) *If f and g are $< LPG >$ integrable on $[a,b]$, and $\alpha, \beta \in \mathbb{R}$ then $\alpha f + \beta g$ is $< LPG >$ integrable and $< LPG > \int_a^b (\alpha f + \beta g)(t)dt = \alpha \cdot < LPG > \int_a^b f(t)dt + \beta \cdot < LPG > \int_a^b f(t)dt$;*

(iv) *If f is $< LPG >$ integrable on $[a,b]$, $F(x) = < LPG > \int_a^x f(t)dt$, M is an $< LPG >$ major function and m is an $< LPG >$ minor function for f then $M - F$ and $F - m$ are increasing on $[a,b]$;*

(v) *If f is* $< LPG >$ *integrable on* $[a,b]$ *and* $F(x) = < LPG > \int_a^x f(t)dt$ *then* $F'_{ap}(x) = f(x)$ *a.e. on* $[a,b]$*, hence f is Lebesgue measurable on* $[a,b]$.

Proof. (i), (ii) and (iii) follow by definitions.

For $\epsilon > 0$ there exist an $< LPG >$ major function M and an $< LPG >$ minor function m, such that $M(b) - m(b) < \epsilon$. But $M - F$, $F - m$ and $M - m$ are increasing on $[a,b]$, so they are also derivable *a.e.* on $[a,b]$. Since M and m are approximately derivable *a.e.*, it follows that F is approximately derivable *a.e.* and $M'_{ap}(x) \geq F'_{ap}(x) \geq m'_{ap}(x)$ *a.e.* on $[a,b]$. But $M'_{ap}(x) \geq f(x) \geq m'_{ap}(x)$ *a.e.* on $[a,b]$, hence $(\mathcal{L}) \int_a^b | F'_{ap}(t) - f(t) | dt \leq (\mathcal{L}) \int_a^b | M'_{ap}(t) - m'_{ap}(t) | dt \leq M(b) - m(b) < \epsilon$. Since ϵ is arbitrary, it follows that $(\mathcal{L}) \int_a^b | F'_{ap}(t) - f(t) | dt = 0$, hence $F'_{ap}(x) = f(x)$ *a.e.* on $[a,b]$. By Theorem 1.11.4, f is Lebesgue measurable on $[a,b]$.

Lemma 5.17.3 *Let* $f : [a,b] \mapsto \overline{\mathbb{R}}$. *If f is* $< LDG >$ *integrable on* $[a,b]$ *then f is* $< LPG >$ *integrable on* $[a,b]$ *and* $< LDG > \int_a^b f(t)dt = < LPG > \int_a^b f(t)dt$.

Proof. Let $F(x) = < LDG > \int_a^x f(t)dt$. Then F is simultaneously an $< LPG >$ major and an $< LPG >$ minor function for f. Now the proof follows easily.

Theorem 5.17.2 *Let* $f : [a,b] \mapsto \overline{\mathbb{R}}$ *be an* $< LPG >$ *integrable function. Suppose that there exist an* $< LPG >$ *major function* M_o *and an* $< LPG >$ *minor function* m_o *for f, such that* $M_o, m_o \in [CG]$. *The following assertions are equivalent:*

(i) *f is* $< LDG >$ *integrable on* $[a,b]$;

(ii) *f is* $< LPG >$ *integrable on* $[a,b]$,

and $< LPG > \int_a^b f(t)dt = < LDG > \int_a^b f(t)dt$.

Proof. (i) $\Rightarrow$ (ii) See Lemma 5.17.3.

(ii) $\Rightarrow$ (i) Let $F(x) = < LPG > \int_a^b f(t)dt$. For each positive integer n there exist an $< LPG >$ major function M_n and an $< LPG >$ minor function m_n such that $M_n(b) - m_n(b) < 1/n$ (see Lemma 5.17.2). By Theorem 5.17.1, (iv), we obtain that $0 \leq M_n(x) - F(x) < 1/n$ and $0 \leq F(x) - m_n(x) < 1/n$. It follows that $M_n, m_n \to F$ $[unif]$, hence F and $-F$ belong to $< uL >$ (see (c)). Therefore $F \in < L >$. Since $-m_o, M_o \in [CG] \cap [\underline{AC}G]$ on $[a,b]$, it follows that there exists a sequence $\{Q_j\}_j$ of closed sets, covering $[a,b]$, such that $M_o \in C \cap \underline{AC}$ and $m_o \in C \cap \overline{AC}$ on each Q_j. Fix some Q_j and denote it by Q. Let M be any $< LPG >$ major function and let m be any $< LPG >$ minor function for f. Then $M = (M - m_o) + m_o$. By Theorem 2.11.1, it follows that $(m_o)_Q \in C \cap VB$ (see Definition 1.1.3). Since $M - m_o$ is increasing (see Lemma 5.17.1), $M_Q \in C_i \cap VB$ on $[c,d]$, where $c = \inf(Q)$ and $d = \sup(Q)$. Clearly $M_Q \in [\underline{AC}G]$ on $[c,d]$. It follows that $M_Q \in \underline{AC}$ on $[c,d]$ (see Corollary 2.21.1), hence $M \in \underline{AC}$ on Q. Similarly $m_Q = M_Q - (M_Q - m_Q) \in C_d \cap VB \cap [\overline{AC}G] = \overline{AC}$ on $[c,d]$. Therefore $m \in \overline{AC}$ on Q. Particularly, each $M_n \in \underline{AC}$ and each $m_n \in \overline{AC}$ on Q. By Theorem 2.11.1, (xxii), $F \in \underline{AC} \cap \overline{AC}$ on Q. It follows that $F \in [ACG]$ on $[a,b]$. By Theorem 5.17.1, (v), we have that f is $< LDG >$ integrable on $[a,b]$ and $F(b) = < LDG > \int_a^b f(t)dt$.

Theorem 5.17.3 *Let* $f : [a,b] \mapsto \overline{\mathbb{R}}$ *and let* $< uL > = C_i$ *on* $[a,b]$. *Then* $< L > = C$, *and the following assertions are equivalent:*

(i) f is $(\mathcal{D})$*-integrable on* $[a,b]$*;*

(ii) f is $< LPG >$ *integrable on* $[a,b]$,

and $(\mathcal{D}) \int_a^b f(t)dt = < LPG > \int_a^b f(t)dt$.

Proof. That C_i satisfies conditions (a), (b) and (c) follows by Proposition 2.3.4, Proposition 2.3.1, (iv) and Theorem 2.4.2, (vi). That $< L >= \mathcal{C}$ follows by Proposition 2.3.1, (i).

$(i) \Rightarrow (ii)$ See Theorem 5.17.2.

$(ii) \Rightarrow (i)$ As in the proof of Theorem 5.17.2, since $M \in C_i$ on $[a,b]$ and by Proposition 2.3.2, i), it follows that $M_Q \in C_i$. Now the proof continues as that of Theorem 5.17.2.

Remark 5.17.2 *In [Le1] C.M. Lee stated the following theorem:*
A function $f : [a,b] \mapsto \overline{\mathbb{R}}$ *is* $< LDG >$ *integrable on* $[a,b]$ *if and only if f is* $< LPG >$ *integrable on* $[a,b]$.
However C.M. Lee's proof is not correct (because the assertion "$M \in uCM$ *on* $[a,b]$ *implies* $M_P \in uCM$ *for a closed set* $P \subset [a,b]$*" is not true). So the question is if C.M. Lee's theorem is true? Partial answers to this question are Theorem 5.17.2 and Theorem 5.17.3.*
For $< uL >=< L >=$ *the class of all approximately continuous functions on* $[a,b]$*, C.M. Lee's theorem was stated before by Ridder in [R3] (pp. 148 - 149) and [R2] (p. 7). The proof is based on a different idea, but it isn't clear either (because his proof is based on the following fact, about we do not know if it is true: for an* $< LPG >$ *integrable function f on* $[a,b]$*, there exists a sequence of majorants* $\{M_k\}_k$*, a sequence of minorants* $\{m_k\}_k$*, a countable set of perfect sets* $\{E_j\}_j$*, and a countable set H such that* $H \cup (\cup_{j=1}^{\infty}) = [a,b]$ *and* $M_k \in \underline{AC}$, $m_k \in \overline{AC}$ *on each* E_j *and for each k).*

Remark 5.17.3 *By the proof of Lemma 2.38.5 and by Corollary 2.38.1,* $S_{ap}Y \subseteq [ACG] \cap (S_{ap}\text{-continuous}) = [ACG] \cap (\text{approximately continuous})$*. By Theorem 5.16.5,(i), (iii), we have: (AP)-integral* $\subsetneq$ *(*R_{ap}*-integral)* $\subset$ *(Ridder's* β*-integral). The inclusions follow by Remark 5.16.3 and Remark 5.17.1. That the last inclusion is strict follows by the following argument: the* $(\mathcal{D})$*-integral* $\subsetneq$ β*-integral, but there exists a function which is* $(\mathcal{D})$*-integrable without being* R_{ap}*-integrable (see Skvortsov [Skv]).*

5.18 The chain rule for the derivative of a composite function

Definition 5.18.1 *Let* $g : [a,b] \mapsto \mathbb{R}$*. We denote the following property of g by*

(*) For each $A \subseteq [a,b]$, $|A| > 0$ *there exists a point* $x_o \in A \cap (\inf(A), \sup(A))$ *such that* $g(x_o - \delta, x_o + \delta)) \subset [\inf(A), \sup(A)]$*, for some* $\delta > 0$ *with* $(x_o - \delta, x_o + \delta) \subset (\inf(A), \sup(A))$.

Remark 5.18.1 *Let* $g : [a,b] \mapsto \mathbb{R}$*. If* $g \in \mathcal{D}$ *(resp. g is continuous a.e.) then g satisfies property (*).*

Lemma 5.18.1 *Let $g : [a,b] \mapsto [c,d]$, $A \subseteq [a,b]$, $|A| > 0$ and $g(A) = B$. Let $f : [c,d] \mapsto \mathbb{R}$ such that f is derivable on P, $f'(y) > 0$ for each $y \in B$, $f \circ g$ is derivable on A, and $(f \circ g)'(x) > 0$ for each $x \in A$. If g has the property (*) then there exists a point $x_o \in A$ such that g is derivable at x_o.*

Proof. Since $(f \circ g)(x) \in (0, +\infty)$ for each $x \in B$, it follows that there exist a sequence of sets $\{A_n\}_n$ and two sequences of positive numbers $\{m_n\}_n$ and $\{M_n\}_n$ such that $A = \cup_{n=1}^{\infty} A_n$, $m_n < M_n$ for each n, and

(1) $\quad m_n < ((f \circ g)(y) - (f \circ g)(x))/(y - x) < M_n,$

whenever $\inf(A_n) \le x < y \le \sup(A_n)$ and $\{x, y\} \cap A_n \ne \emptyset$. Since $g'(x) \in (0, +\infty)$ for each $y \in B$, there exist a sequence of sets $\{B_k\}_k$ and two sequences of positive numbers $\{s_k\}_k$ and $\{S_k\}_k$ such that $B = \cup_{k=1}^{\infty} B_k$, $s_k < S_k$ for each k, and

(2) $\quad s_k < (f(z) - f(t))/(z - t) < S_k,$

whenever $\inf(B_k) \le t < z \le \sup(B_k)$ and $\{t, z\} \cap B_k \ne \emptyset$. Let $A_{nk} = A_n \cap g^{-1}(B_k)$ and $B_{nk} = g(A_{nk}) \subset B_k$. Because $|A| > 0$, it follows that $|A_{nk}| > 0$ for some n and k. Let $a_{nk} = \inf(A_{nk})$ and $b_{nk} = \sup(A_{nk})$. But g satisfies (*), so there exist $x_o \in A_{nk} \cap (a_{nk}, b_{nk})$ and $\delta > 0$ such that $(x_o - \delta, x_o + \delta) \subset (a_{nk}, b_{nk})$ and $g((x_o - \delta, x_o + \delta)) \subset [g(a_{nk}), g(b_{nk})] = [\inf(B_{nk}), \sup(B_{nk})]$. Let $x \in (x_o, x_o + \delta)$. It follows that $g(x) \in [\inf(B_{nk}), \sup(B_{nk})]$. By (1) and (2) we have $m_n < ((f \circ g)(x) - (f \circ g)(x_o))/(x - x_o) < M_n$ and $s_k < (f(g(x)) - f(g(x_o)))/(g(x) - g(x_o)) < S_k$. Then $m_n/S_k < (g(x) - g(x_o))/(x - x_o) < M_n/s_k$, hence $g(x) - g(x_o) > 0$ and g is continuous to the right at x_o. It follows that $\lim_{x \searrow x_o}(g(x) - g(x_o))/(x - x_o) = (f \circ g)'(x_o) \cdot (1/(f'(g(x_o))))$. Similarly we obtain the left version, so g is derivable at x_o.

Theorem 5.18.1 *Let $g : [a,b] \mapsto [c,d]$ satisfy (*), $A = \{x : g'(x) = 0\}$ and $B = \{x : g'(x)$ does not exist$\}$. Let $F, f : [c,d] \mapsto \mathbb{R}$ such that $F \in (N)$ on $g(A \cup B)$, F is derivable a.e. on $g(B)$, and $f(x) = F^*(x)$ a.e. on $[c,d]$ (see Definition 5.7.3. Then we have*

(1) $\quad (F \circ g)^*(x) = (f \circ g)(x) \cdot g^*(x)$ *a.e. on* $[a,b]$.

Proof. Clearly $g^*(x) = 0$ for each $x \in A$. Let $A_1 = \{x \in A : (F \circ g)^*(x) \ne 0\}$. Then $(F \circ g)^*(x) = 0$ for each $x \in A \setminus A_1$, so (1) is valid on $A \setminus A_1$. By Theorem 1.10.4, it follows that $|g(A)| = 0$. Since $F \in (N)$ on $g(A \cup B)$, $|(F \circ g)(A)| = 0$. By Lemma 2.17.1, it follows that $|A_1| = 0$, so (1) is valid *a.e.* on A.
Clearly $g^*(x) = 0$ for each $x \in B$. Let $B_o = \{x \in B : (F \circ g)'(x)$ does not exist or if it does equals $0\}$. Then $(F \circ g)^*(x) = 0$ for each $x \in B_o$, hence (1) is valid on B_o. Suppose on the contrary that (1) does not hold *a.e.* on $B \setminus B_o$. Then there exists $B_1 \subset B \setminus B_o$, $|B_1| > 0$ such that $F \circ g$ is derivable on B_1 and $(F \circ g)'(x) \ne 0$, $x \in B_1$. Suppose for example that $(F \circ g)'(x) > 0$, $x \in B_1$. Let $B_2 = \{x \in B_1 : F'(g(x)) = f(g(x)) \ne 0\}$. We show that $|B_2| > 0$. Suppose that $|B_2| = 0$. By Corollary 2.13.3, $F \circ g \in AC^*G \subset (N)$ on B, hence $|F(g(B_2))| = 0$. Let $B_3 = \{x \in B_1 : F'(x)) = f(g(x)) = 0\}$. By Theorem 1.10.4, $|F(g(B_3))| = 0$. Let $B_4 = \{x \in B_1 : F'(g(x)) \ne f(g(x))\}$. Since F is derivable *a.e.* on $g(B)$ and $f = F^*$ *a.e.* on $g(B)$, it follows that $|g(B_4)| = 0$. Since $F \in (N)$ on $g(B)$, $|F(g(B_4))| = 0$. But $B_1 = B_2 \cup B_3 \cup B_4$, hence $|(F \circ g)(B_1)| = 0$.

By Lemma 2.17.1, it follows that $| B_1 |= 0$, a contradiction. Therefore $| B_2 |> 0$. We may suppose without loss of generality that $F'(g(t)) = f(g(t)) > 0$ on B_2. By Lemma 5.18.1, there exists $x_o \in B_2$ such that g is derivable at x_o, a contradiction. Hence (1) holds *a.e.* on $B \setminus B_o$. Let $C = \{x \in [a,b] : g \text{ is derivable at x and } g'(x) \neq 0\}$. We may suppose without loss of generality that $g'(x) > 0$ for each $x \in C$. Let $Q = g(C)$ and $Q_1 = \{y \in Q : F^*(y) \neq f(y)\}$. By hypothesis, $| Q_1 |= 0$. Let $C_1 = g^{-1}(Q_1)$. Then $| g(C_1) |= 0$. By Lemma 2.17.1, it follows that $g'(x) = 0$ *a.e.* on C_1, hence $| C_1 |= 0$. Let $Q_2 = \{y \in Q : F'(y) = +\infty\}$ and $C_2 = g^{-1}(Q_2)$. By Corollary 2.15.1, $| Q_2 |= 0$, and by Lemma 2.17.1, $| C_2 |= 0$. Let $Q_3 = \{y \in Q : F'(y) = f(y)\}$ and $C_3 = g^{-1}(Q_3)$. Let $x_o \in C_3$. Since $(F(g(x)) - F(g(x_o)))/(x - x_o) = ((F(g(x)) - F(g(x_o))/(g(x) - g(x_o))) \cdot (g(x) - g(x_o))/(x - x_o)$, $x \neq x_o$, $x \in [a,b]$, it follows that (1) is valid *a.e.* on C. Let $C_\infty = \{x \in [a,b] : g'(x) = \pm\infty\}$. By Corollary 2.15.1, $| C_\infty |= 0$, and this completes the proof.

Corollary 5.18.1 *Let $g : [a,b] \mapsto [c,d]$ satisfy property (*) and let $F : [c,d] \mapsto \mathbb{R}$ such that $F \in (N)$ on $[c,d]$ and let F be derivable a.e. on $[c,d]$. Let $f : [c,d] \mapsto \mathbb{R}$, $f(x) = F'(x)$ a.e.. Then $(F \circ g)^* = f(g(x)) \cdot g^*(x)$ a.e. on $[a,b]$.*

Remark 5.18.2 *By Remark 5.18.1, it follows that Corollary 5.18.1 holds if $g \in \mathcal{C}$ a.e. on $[a,b]$ (this result is due to Goodman, [Go1], [Go2], [F9]), and also if $g \in \mathcal{D}$ on $[a,b]$.*

Corollary 5.18.2 (Serrin and Varberg) *([SerV]). Let $g : [a,b] \mapsto [c,d]$ and let $F : [c,d] \mapsto \mathbb{R}$. If F, g and $F \circ g$ are derivable a.e. on their domains, and $F \in (N)$ on $[c,d]$ then the chain rule $(F \circ g)' = (f \circ g) \cdot g'$ holds a.e. on $[a,b]$, whenever $f : [a,b] \mapsto \mathbb{R}$ and $F' = f$ a.e. on $[a,b]$.*

5.19 The chain rule for the approximate derivative of a composite function

Lemma 5.19.1 (Foran) *([F9]). Let $g : [a,b] \mapsto \mathbb{R}$, $A \subset [a,b]$ and $0 < m < M < +\infty$ such that $m(y - x) < g(y) - g(x) < M(y - x)$ for $x, y \in A$, $x < y$. Then there exists $G : [a,b] \mapsto \mathbb{R}$ such that $G = g$ on A and $m(y - x) \leq G(y) - G(x) \leq M(y - x)$ whenever $x, y \in A$, $x < y$. Moreover, if A is a measurable set and $x_o \in A$ is a point of density for A then $g(x_o)$ is a point of density for $g(A)$.*

Proof. Let $c = \inf(A)$ and $d = \sup(A)$. For $x \in A$ we define $G(x) = g(x)$. Let $x \in \overline{A} \setminus A$. If x is a left accumulation point for A, put $G(x) = \sup\{g(t) : t \in A \cap [a,x)\}$. If x is a right accumulation point for A, put $G(x) = \inf\{g(t) : t \in A \cap (x,b]\}$. Extending G linearly on each interval contiguous to $\overline{A}$ we have G defined and continuous on $[c,d]$. If $a \neq c$ or $b \neq d$ then we extend G linearly on both $[a,c]$ and $[c,d]$ with the slopes in (m,M), such that G is continuous on $[a,b]$. Then G has the required properties.
Suppose that A is measurable and $x_o \in A$ is a point of density for A. For $\epsilon > 0$ let $\delta > 0$ such that $| [x_o, x_o - h] \setminus A | < (\epsilon m/M) \cdot h$, whenever $0 < h < \delta$. Let $\delta_1 = m\delta$ and let $x_1 > x_o$ such that $G(x_1) - g(x_o) < \delta_1$. It follows that $x_1 - x_o \leq (1/m)(G(x_1) - G(x_o)) < \delta$. Clearly G is strictly increasing and $G \in L \subset (N)$ on $[a,b]$. Since A and $[x_o, x_1] \setminus A$ are measurable sets, by Theorem 2.18.2, we obtain that $G(A)$

and $G([x_o, x_1] \setminus A)$ are also measurable and disjoint sets Now by Theorem 1.10.4, (ii), it follows that $| [G(x_o), G(x_1)] \setminus G(A) | = | G([x_o, x_1] \setminus A) | \leq M \cdot | [x_o, x_1] \setminus A | < \epsilon m(x_1 - x_o) \leq \epsilon(G(x_1) - G(x_o))$. Therefore $G(x_o)$ is a right point of density for $G(A)$. The left version follows similarly.

Definition 5.19.1 *Let $F : [a, b] \mapsto \mathbb{R}$. Let $F^*_{ap} : [a, b] \mapsto \mathbb{R}$ be defined as follows: $F^*_{ap}(x) = F'_{ap}(x)$ where F is approximately derivable, and 0 elsewhere.*

Theorem 5.19.1 (Foran) *([F9]). Let $g : [a, b] \mapsto [c, d]$ be a measurable function, $A = \{x \in [a, b] : g'_{ap}(x) = 0\}$ and $B = \{x \in [a, b] : g'_{ap}(x)$ does not exist$\}$. Let $F : [c, d] \mapsto \mathbb{R}$ be a measurable function and let $f(x) = F^*_{ap}(x)$ a.e. on $[c, d]$. If $F \in (N)$ on $g(A \cup B)$, F is approximately derivable a.e. on $g(B)$ and $F \circ g$ is measurable on $[a, b]$ then*

(1) $\quad (F \circ g)^*_{ap}(x) = (f \circ g)(x) \cdot g^*_{ap}(x)$ *a.e. on $[a, b]$.*

Proof. Clearly $g^*_{ap}(x) = 0$ on A. Suppose that (1) does not hold *a.e.* on A. Then there exists $A_1 \subset A$ such that $| A_1 | > 0$ and $(F \circ g)'_{ap}(x) \neq 0$ on A_1. By Theorem 1.11.3, $| g(A) | = 0$, hence $| (F \circ g)(A_1) | = 0$ (because $F \in (N)$ on $g(A)$). Now by Lemma 2.17.2, $(F \circ g)'_{ap}(x) = 0$ *a.e.* on A_1, a contradiction. Therefore (1) holds *a.e.* on A. Clearly $g^*_{ap}(x) = 0$ on B. By [S3] (p. 299), it follows that $[a, b] \setminus B$ is measurable, hence B is also a measurable set. Let $B_o = \{x \in B : (F \circ g)'_{ap}(x)$ does not exist or if it does equals $0\}$. Then $(F \circ g)'_{ap}(x) = 0$ for each $x \in B_o$. Hence (1) holds on B_o. Suppose on the contrary that (1) does not hold on $B \setminus B_o$. Then there exists $B_1 \subset B_o$, $| B_1 | > 0$ such that $F \circ g$ is approximately derivable on B_1 and $(F \circ g)'_{ap}(x) \neq 0$ for each $x \in B_1$. Suppose for example that $(F \circ g)'_{ap}(x) > 0$ for each $x \in B_1$. Let $B_2 = \{x \in B_1 : F'_{ap}(g(x)) = f(g(x)) \neq 0\}$. We show that $| B_2 | > 0$. Suppose that $| B_2 | = 0$. By Theorem 2.15.4, $F \circ g \in ACG \subset (N)$ on B, hence $| F(g(B_2)) | = 0$. Let $B_3 = \{x \in B_1 : F'_{ap}(g(x)) = f(g(x)) = 0\}$. By Theorem 1.11.3, $| F(g(B_3)) | = 0$. Let $B_4 = \{x \in B_1 : F'_{ap}(g(x)) \neq f(g(x))\}$. Since F is approximately derivable *a.e.* on $g(B)$ and $f = F^*_{ap}$ *a.e.* on $g(B)$, it follows that $| g(B_4) | = 0$. Because $F \in (N)$ on $g(B)$, $| F(g(B_4)) | = 0$. But $B_1 = B_2 \cup B_3 \cup B_4$, hence $| (F \circ g)(B_1) | = 0$. By Lemma 2.17.2, it follows that $| B_1 | = 0$, a contradiction. Thus $| B_2 | > 0$.
We may suppose without loss of generality that $F'_{ap}(g(x)) = f(g(x)) > 0$ on B_2. By Corollary 2.15.2, g is approximately derivable *a.e.* on B_2, a contradiction. Hence (1) holds *a.e.* on B. Let $C = \{x \in [a, b] : g$ is approximately derivable at x and $g'_{ap}(x) \neq 0\}$. We may suppose without loss of generality that $g'_{ap}(x) > 0$ for each $x \in C$. Let $Q = g(C)$ and $Q_1 = \{y \in Q : F^*_{ap}(y) \neq f(y)\}$. By hypothesis, $| Q_1 | = 0$. Let $C_1 = g^{-1}(Q_1)$. Then $g(C_1) = 0$. By Lemma 2.17.2, it follows that $g'_{ap}(x) = 0$ *a.e.* on C_1, hence $| C_1 | = 0$. Let $Q_2 = \{y \in Q : F'_{ap}(y) = +\infty\}$ and $C_2 = g^{-1}(Q_2)$. By Corollary 2.15.3, $| Q_2 | = 0$, and by Lemma 2.17.2, $| C_2 | = 0$. Let $Q_3 = \{y \in Q : F'_{ap}(y) = f(y)\}$ and $C_3 = g^{-1}(Q_3)$. Since g is approximately derivable with $g'_{ap}(x) > 0$ on C_3, it follows that there exist a sequence of sets $\{P_n\}_n$ with $C_3 = \cup_{n=1}^{\infty} P_n$, and two sequences of numbers $\{m_n\}_n$ and $\{M_n\}_n$ with $0 < m_n < M_n < +\infty$ such that $m_n < (g(y) - g(x))/(y - x) < M_n$, whenever $x, y \in P_n$, $x < y$. By Lemma 5.19.1, for each n there exist $G_n : [a, b] \mapsto \mathbb{R}$, $G_n \in \mathcal{C}$, such that $m_n < (G_n(y) - G_n(x))/(y - x) < M_n$, $x < y$ and $G_n = g$ on P_n. Let $E_n = \{x \in [a, b] : G_n(x) = g(x)\}$. Since g is measurable and G_n is continuous on $[a, b]$, it follows that E_n is measurable. Let $x_o \in E_n$ be a point of density for E_n. Then $g(x_o)$ is a point

of density for $g(E_n)$ (see Lemma 5.19.1). But $(F(G_n(x)) - F(G_n(x_o)))/(x - x_o) = ((F(G_n(x) - F(G_n(x_o)))/(G_n(x) - G_n(x_o))) \cdot ((G_n(x) - G_n(x_o))/(x - x_o))$. It follows that $(F \circ g)^*_{ap}(x_o) = F^*_{ap}(g(x_o)) \cdot g'_{ap}(x_o) = f(g(x_o)) \cdot g'_{ap}(x_o)$. Therefore (1) holds a.e. on C. Let $C_\infty = \{x \in [a,b] : g'_{ap}(x) = \pm\infty\}$. By Corollary 2.15.3, $| C_\infty |= 0$), and this completes the proof.

Corollary 5.19.1 (Foran) *([F9]). Let $g : [a,b] \mapsto \mathbb{R}$, $F : [c,d] \mapsto \mathbb{R}$ and $F \circ g$ be measurable functions, such that $F \in (N)$ on $[c,d]$ and F is approximately derivable a.e. on $g([a,b])$. Let $f : [c,d] \mapsto \mathbb{R}$ such that $f = F^*_{ap}$ a.e.. Then we have*

(1) $\quad (F \circ g)^*_{ap}(x) = (f \circ g)(x) \cdot g^*_{ap}(x)$ *a.e. on* $[a,b]$.

Remark 5.19.1 *Foran asserts in fact that Theorem 5.19.1 and Corollary 5.19.1 are true even if one drops the measurability condition for functions, but the proof isn't quite clear, because of Lemma 2.17.2 (see Remark 2.17.2). None of the hypotheses on the sets $g(A)$ and $g(B)$ can be dropped from Theorem 5.19.1 (see Section 6.66).*

5.20 Change of variable formula for the Lebesgue integral

Theorem 5.20.1 *Let $g : [a,b] \mapsto [c,d]$ be a function having the property (*) and let $F : [c,d] \mapsto \mathbb{R}$ such that $F, F \circ g \in \mathcal{DB}_1 \cap T_2 \cap N^\infty$. Let $f : [c,d] \mapsto \mathbb{R}$ such that $f = F^*$ a.e.. If $\overline{\mathcal{M}}^{\#}_{10}(f) \neq \emptyset$ and $\overline{\mathcal{M}}^{\#}_{10}((F \circ g)^*) \neq \emptyset$ then $(\mathcal{L}) \int_a^b (f \circ g)(t) \cdot g^*(t)dt = (\mathcal{L}) \int_{g(a)}^{g(b)} f(t)dt$ (see Definitions 5.18.1, 5.7.3 and 5.1.1).*

Proof. By Theorem 5.1.1, (viii), (i), f is summable on $[g(a), g(b)]$ and $(\mathcal{L}) \int_{g(a)}^{g(b)} f(t)dt = F(g(b)) - F(g(a))$. It also follows that $(F \circ g)^*$ is summable on $[a,b]$ and $(\mathcal{L}) \int_a^b (F \circ g)^*(t)dt = F(g(b)) - F(g(a))$. By Theorem 5.1.1, (ii), $F \in AC$ on $[g(a), g(b)]$ and $F'(t) = f(t)$ a.e.. By Corollary 5.18.1, we obtain $(F \circ g)^*(x) = (f \circ g)(x) \cdot g^*(x)$ a.e. on $[a,b]$. Now the theorem follows easily.

Corollary 5.20.1 *Let $g : [a,b] \mapsto [c,d]$ and $F : [c,d] \mapsto \mathbb{R}$, $F, g \in \mathcal{DB}_1 \cap T_2 \cap N^\infty$ and let $f : [c,d] \mapsto \mathbb{R}$ such that $f = F^*$ a.e.. If $\overline{\mathcal{M}}^{\#}_{10}(f) \neq \emptyset$ and $\overline{\mathcal{M}}^{\#}_{10}((F \circ g)^*) \neq \emptyset$ then $(\mathcal{L}) \int_a^b (f \circ g)(t) \cdot g^*(t)dt = (\mathcal{L}) \int_{g(a)}^{g(b)} f(t)dt$*

Proof. By Theorem 2.22.3, $F \circ g \in \mathcal{D} \cap N^\infty$, and by the proof of Theorem 5.20.1, $F \in AC \subset (N)$. By Corollary 2.5.1, (vii), $F \circ g \in \mathcal{DB}_1$. If $A = \{z : F^{-1}(z)$ is uncountable$\}$ and $B = \{y : g^{-1}(y)$ is uncountable$\}$ then $| A |=| B |= 0$. But $F \in (N)$, hence $| F(B) |= 0$. Since $C = \{z : (F \circ g)^{-1}(z)$ is uncountable$\} \subset A \cup B$, it follows that $| C |= 0$. Therefore $F \circ g \in T_2$. Now the proof follows by Theorem 5.20.1 and Remark 5.18.1.

Corollary 5.20.2 *Let $g : [a,b] \mapsto [c,d]$ and $f : [c,d] \mapsto \overline{\mathbb{R}}$ such that g is derivable a.e. and f is Lebesgue integrable. If $F(x) = (\mathcal{L}) \int_c^x f(t)dt$ then the following assertions are equivalent:*

(i) $F \circ g \in AC$ on $[a,b]$;

(ii) $F \circ g \in \mathcal{DB}_1 \cap T_2 \cap N^\infty$, *$(f \circ g) \cdot g'$ is Lebesgue integrable on $[a, b]$ and* $(\mathcal{L}) \int_{g(a)}^{g(b)} f(t)dt = (\mathcal{L}) \int_a^x f(g(t)) \cdot g'(t)dt$;

(iii) $F \circ g \in \mathcal{DB}_1 \cap T_2 \cap N^\infty$ *and $(f \circ g) \cdot g'$ is Lebesgue integrable on $[a, b]$.*

Proof. $(i) \Rightarrow (ii)$ See Theorem 5.20.1 and Remark 5.18.1.

$(ii) \Rightarrow (iii)$ This is evident.

$(iii) \Rightarrow (i)$ By Corollary 5.18.2, $(F \circ g) \cdot g' = (F \circ g)'$. Hence $(F \circ g)'$ is summable on $[a, b]$, and by Theorem 5.1.1, (iv), (i), (ii), it follows that $F \circ g \in AC$ on $[a, b]$.

Corollary 5.20.3 (Serrin and Varberg) *([SerV]). Let $g : [a, b] \mapsto [c, d]$ and $f : [c, d] \mapsto \overline{\mathbb{R}}$ such that f and $(f \circ g) \cdot g'$ are Lebesgue integrable functions. Then $(\mathcal{L}) \int_{g(a)}^{g(b)} f(t)dt = (\mathcal{L}) \int_a^b f(g(t)) \cdot g'(t)dt$.*

Proof. Let $F(x) = (\mathcal{L}) \int_a^x f(t)dt$. Then $F \in AC$, and the proof follows by Corollary 5.20.1.

5.21 Change of variable formula for the Denjoy* integral

Theorem 5.21.1 *Let $g : [a, b] \mapsto [c, d]$ be a function having the property (*), and let $F : [c, d] \mapsto \mathbb{R}$ such that $F, F \circ g \in \mathcal{DB}_1 \cap T_2 \cap N^\infty$. Let $f : [c, d] \mapsto \mathbb{R}$ such that $f = F^*$ a.e.. If $\overline{\mathcal{M}}_6(f) \neq \emptyset$ and $\overline{\mathcal{M}}_6((F \circ g)^*) \neq \emptyset$ then $(\mathcal{D}^*) \int_a^b (f \circ g)(t) \cdot g^*(t)dt = (\mathcal{D}^*) \int_{g(a)}^{g(b)} f(t)dt$ (see Definitions 5.18.1, 5.7.3 and 5.7.1).*

Proof. By Theorem 5.7.1, f is $\mathcal{D}^*$-integrable on $[g(a), g(b)]$, $F \in AC^*G$ on $[c, d]$, $F'(x) = f(x)$ a.e. and $(\mathcal{D}^*) \int_{g(a)}^{g(b)} f(t)dt = F(g(b)) - F(g(a))$. We also obtain that $(F \circ g)^*$ is $\mathcal{D}^*$-integrable on $[a, b]$ and $(\mathcal{D}^*) \int_a^b (f \circ g)(t) \cdot g^*(t)dt = F(g(b)) - F(g(a))$. By Corollary 5.18.1, $(F \circ g)^*(x) = (f \circ g)(x) \cdot g^*(x)$ a.e. on $[a, b]$. Now the theorem follows easily.

Corollary 5.21.1 *Let $g : [a, b] \mapsto [c, d]$ and $F : [c, d] \mapsto \mathbb{R}$, $F, g \in \mathcal{DB}_1 \cap T_2 \cap N^\infty$ and let $f : [c, d] \mapsto \mathbb{R}$ such that $f = F^*$ a.e.. If $\overline{\mathcal{M}}_6(f) \neq \emptyset$ and $\overline{\mathcal{M}}_6((F \circ g)^*) \neq \emptyset$ then $(\mathcal{D}^*) \int_a^b (f \circ g)(t) \cdot g^*(t)dt = (\mathcal{D}^*) \int_{g(a)}^{g(b)} f(t)dt$.*

Proof. The proof is similar to that of Corollary 5.20.1 (because $F \in AC^*G \subset (N)$).

Corollary 5.21.2 (Goodman) *([Go1]). Let $g : [a, b] \mapsto [c, d]$, $F : [c.d] \mapsto \mathbb{R}$ such that $F, g \in \mathcal{C} \cap (N)$. If F^* and $(F \circ g)^*$ are $\mathcal{D}^*$-integrable then $(\mathcal{D}^*) \int_a^b (f \circ g)(t) \cdot g^*(t)dt = (\mathcal{D}^*) \int_{g(a)}^{g(b)} f(t)dt$.*

Corollary 5.21.3 *Let $g : [a, b] \mapsto [c, d]$ and $f : [c, d] \mapsto \overline{\mathbb{R}}$ such that g is derivable a.e. and f is $\mathcal{D}^*$-integrable on $[a, b]$. If $F(x) = (\mathcal{D}^*) \int_c^x f(t)dt$ then the following assertions are equivalent:*

(i) $F \circ g \in AC^*G$ *on $[a, b]$;*

(ii) $F \circ g \in \mathcal{DB}_1 \cap T_2 \cap N^\infty$, *$(f \circ g) \cdot g'$ is $\mathcal{D}^*$-integrable on $[a, b]$ and $(\mathcal{D}^*) \int_a^b (f \circ g)(t) \cdot g^*(t)dt = (\mathcal{D}^*) \int_{g(a)}^{g(b)} f(t)dt$.*

(iii) $F \circ g \in \mathcal{DB}_1 \cap T_2 \cap N^\infty$ *and* $(f \circ g) \cdot g'$ *is* $\mathcal{D}^*$*-integrable on* $[a, b]$.

Proof. (*i*) ⇒ (*ii*) See Theorem 5.21.1 and Remark 5.18.1.

(*ii*) ⇒ (*iii*) This is evident.

(*iii*) ⇒ (*i*) By Corollary 5.18.2, $(f \circ g) \cdot g' = (F \circ g)'$. Hence $(F \circ g)'$ is $\mathcal{D}^*$-integrable on $[a, b]$, and by Theorem 5.7.1, (iii), $F \circ g \in AC^*G$ on $[a, b]$.

Remark 5.21.1 *Corollary 5.21.3 extends some results of Krzyzewski (see Theorem 1 and Theorem 2 of [Kr1], I).*

5.22 Change of variable formula for the $< LDG >$ integral

Lemma 5.22.1 *Let* $F : [a, b] \mapsto \mathbb{R}$, $F \in [CG] \cap \overline{M}$ *(resp.* $F \in [CG] \cap (M)$*). If* F^*_{ap} *has an* $< LPG >$ *major function* G *and* $G - F \in uCM$ *then* $F \in [\overline{ACG}]$ *(resp.* $F \in [ACG]$*) on* $[a, b]$ *(see Definition 5.19.1). If in addition, in both cases* $F \in \mathcal{D}$ *and* $G \in \mathcal{C}$ *then* $F \in \mathcal{C}$ *on* $[a, b]$.

Proof. Let $H = G - F$. Then $H'(x) \geq 0$ *a.e.* where H is derivable. But $H \in uCM$, so by Theorem 4.2.2, it follows that H is increasing on $[a, b]$. Hence $F \in [CG] \cap [VBG] \cap \overline{M} \subset [\overline{ACG}]$ (resp. $F \in [CG] \cap [VBG] \cap (M) \subset [ACG]$) on $[a, b]$. If $F \in \mathcal{D}$ and $G \in \mathcal{C}$ then by Corollary 2.5.1, (iv), $H \in \mathcal{D}$. Since H is increasing it follows that $H \in \mathcal{C}$, hence $F \in \mathcal{C}$ on $[a, b]$.

Corollary 5.22.1 *Let* $F : [a, b] \mapsto \mathbb{R}$, $F \in < L >$ *(resp.* $F \in \mathcal{D}$*) on* $[a, b]$*. The following assertions are equivalent:*

(*i*) *F is an indefinite* $< LDG >$ *integral (resp.* $(\mathcal{D})$*- integral on* $[a, b]$*;*

(*ii*) $F \in [CG] \cap (M)$ *and there exists* $f : [a, b] \mapsto \overline{\mathbb{R}}$ *such that f has an* $< LPG >$ *major function (resp. a continuous* $< LPG >$ *major function) and* $F'_{ap}(x) = f(x)$ *a.e. on* $[a, b]$.

Remark 5.22.1 *Corollary 5.22.1 extends Ridder's Theorem 7 of [R4], p. 178 (because he stated the theorem for the* β*-integral and asked at (ii) the function F to be* T_2*).*

Theorem 5.22.1 *Let* $g : [a, b] \mapsto [c, d]$ *and* $F : [c, d] \mapsto \mathbb{R}$ *such that* $F, F \circ g \in < L > \cap (M) \cap [CG]$*. If* F^*_{ap} *has an* $< LPG >$ *major function on* $[c, d]$ *and* $(F \circ g)^*_{ap}$ *has an* $< LPG >$ *major function on* $[a, b]$ *then F and* $F \circ g$ *are approximately derivable a.e. and*

$$(1) \qquad < LDG > \int_a^b (F^*_{ap} \circ g)(t) \cdot g^*_{ap}(t)dt = < LDG > \int_{g(a)}^{g(b)} F^*_{ap}(t)dt.$$

Proof. By Lemma 5.22.1, F, $F \circ g \in [ACG]$ and by Remark 2.15.4, F, $F \circ g$ are approximately derivable *a.e.*. By Definition 5.17.3, it follows that F^*_{ap} and $(F \circ g)^*_{ap}$ are $< LDG >$ integrable with $< LDG > \int_{g(a)}^{g(b)} F^*_{ap}(t)dt = F(g(b)) - F(g(a))$ and $< LDG > \int_a^b (F \circ g)^*_{ap}(t)dt = F(g(b)) - F(g(a))$. But $(F^*_{ap} \circ g)(x) \cdot g^*_{ap}(x) = (F \circ g)^*_{ap}(x)$ *a.e.* on $[a, b]$ (see Corollary 5.21.1), hence we obtain (1).

Corollary 5.22.2 (Foran) *([F9]). Let $g : [a,b] \mapsto [c,d]$ and $F : [c,d] \mapsto \mathbb{R}$ such that $F, F \circ g \in \mathcal{C} \cap (M)$. If F^*_{ap} and $(F \circ g)^*_{ap}$ are $(\mathcal{D})$-integrable then F and $F \circ g$ are approximately derivable a.e. and $(\mathcal{D}) \int_a^b (F^*_{ap} \circ g)(t) \cdot g^*_{ap}(t)dt = (\mathcal{D}) \int_{g(a)}^{g(b)} F^*_{ap}(t)dt$.*

5.23 Integrals of Foran type

Let $< L >$ be a real linear space of real-valued functions defined on $[a,b]$ such that $\mathcal{C} \subseteq < L > \subset \mathcal{DB}_1$.

Definition 5.23.1 *Let $f : [a,b] \mapsto \overline{\mathbb{R}}$. f is said to be $L\mathcal{F}$- integrable on $[a,b]$ if there exists $F : [a,b] \mapsto \mathbb{R}$ such that $F \in < L > \cap \mathcal{F}$, F is approximately derivable a.e. and $F'_{ap}(x) = f(x)$ a.e. on $[a,b]$. Then we denote by $(L\mathcal{F}) \int_a^b f(t)dt = F(b) - F(a)$. By Corollary 4.3.4, it follows that the $L\mathcal{F}$-integral is well- defined).*
If $< L > = \mathcal{C}$ then we obtain the $\mathcal{F}$-integral, i.e., the Foran integral (see [F2]).

Theorem 5.23.1 *Let $\mathcal{C} \subseteq < L > \subset \mathcal{DB}_1$. Then the class of all LDG integrable functions is strictly contained in the class of all $L\mathcal{F}$-integrable functions. Moreover, if a function is integrable in both senses the integrals are equal.*

Proof. See Corollary 2.28.1, (iii), (x) and Corollary 4.3.4.

Theorem 5.23.2 *Let $f, g : [a,b] \mapsto \overline{\mathbb{R}}$, $c \in (a,b)$ and $\alpha, \beta \in \mathbb{R}$.*

(i) f is $L\mathcal{F}$-integrable on $[a,b]$ if and only if f is $L\mathcal{F}$-integrable on $[a,c]$ and $[c,b]$. Then $(L\mathcal{F}) \int_a^c f(t)dt + (L\mathcal{F}) \int_c^b f(t)dt = (L\mathcal{F}) \int_a^b f(t)dt$.

(ii) If f and g are $L\mathcal{F}$-integrable on $[a,b]$ then $\alpha f + \beta g$ is $L\mathcal{F}$-integrable on $[a,b]$ and $(L\mathcal{F}) \int_a^b (\alpha f + \beta g)(t)dt = \alpha \cdot (L\mathcal{F}) \int_a^b f(t)dt + \beta \cdot (L\mathcal{F}) \int_a^b g(t)dt$.

Proof. See Corollary 2.28.1, (x).

Lemma 5.23.1 *Let $F, G, H : [a,b] \mapsto \mathbb{R}$ such that $F \in VB$, G is continuous and $H(x) = F(x) \cdot G(x) - \int_a^x G(t)dF(t)$. Let $P \subseteq [a,b]$ and let n be a positive integer. If $G \in AC_n$ on P then $H \in \mathcal{C}$ on $[a,b]$ and $H \in AC_{n^2}$ on P. Moreover, if F is monotone then $H \in AC_n$ on P. (Here $\int_a^x G(t)dF(t)$ is the Riemann Stieltjes integral).*

Proof. Suppose that F is increasing on $[a,b]$, $F(a) = 0$ and $F(b) = \beta$. Let $[c,d] \subset [a,b]$. It follows that $H(d) - H(c) = (G(d) - G(c)) \cdot F(d) + (F(d) - F(c)) \cdot G(c) - \int_a^x G(t)dF(t)$. From the definition of the Riemann Stieltjes integral (see for example [N] (pp. 227 - 236), it follows that there exists α between the bounds of G on $[c,d]$ such that $\int_c^d G(t)dF(t) = \alpha(F(d) - F(c))$. We have $H(d) - H(c) = (G(d) - G(c))F(d) + (F(d) - F(c))(G(c) - \alpha)$, hence

$$(1) \qquad | H(d) - H(c) | \leq \beta | G(d) - G(c) | + \mathcal{O}(G; [c,d]) \cdot (F(d) - F(c)).$$

Since $G \in \mathcal{C}$, by (1) it follows that $H \in \mathcal{C}$ on $[a,b]$. Let $\epsilon > 0$. Since $G \in AC_n$ on P, by Proposition 2.28.1, it follows that there exists a $\delta > 0$ with the following property: if $\{I_k\}$, $k = \overline{1,s}$ are nonoverlapping closed intervals, with each $P \cap I_k \neq \emptyset$ and $\sum_{k=1}^s | I_k | < \delta$, then for each k there exist P_{kj}, $j = \overline{1,n}$ such that $P \cap I_k = \cup_{j=1}^n P_{kj}$ and $\sum_{k=1}^s \sum_{j=1}^n \mathcal{O}(F; P_{kj}) < \epsilon/(2\beta)$. Let $\eta > 0$ such that $\mathcal{O}(G; I) \leq \epsilon/(2n\beta)) = \epsilon_1$, for each

closed subinterval I of $[a,b]$ with $|I| < \eta$ (this is possible because $G \in \mathcal{C}$ on $[a,b]$). Let $\delta_1 = \min\{\delta, \eta\}$. Then for each k, $\mathcal{O}(G; I_k) < \epsilon_1$. By (1), $\mathcal{O}(H; P_{kj}) \le \beta \cdot \mathcal{O}(G; P_{kj}) + \mathcal{O}(G; I_k) \cdot |F(I_k)|$. Hence $\sum_{k=1}^{s}\sum_{j=1}^{n} \mathcal{O}(H; P_{kj}) \le \beta \cdot \sum_{k=1}^{s}\sum_{j=1}^{n} \mathcal{O}(G; P_{kj}) + n \cdot \epsilon_1 \cdot \sum_{k=1}^{s} |F(I_k)| \le \beta \cdot \epsilon/(2\beta) + n \cdot \epsilon_1 \cdot \beta < \epsilon$. Thus $H \in AC_n$ on P.
Suppose that $F \in VB$ on $[a,b]$. Then there exist F_1 and F_2 increasing on $[a,b]$ such that $F = F_1 - F_2$. Let $H_i(x) = F_i(x) \cdot G(x) - \int_a^x G(t)dF_i(t)$, $i = 1, 2$. It follows that $H = H_1 - H_2$. Since H_1 and H_2 are AC_n on P, by Theorem 2.28.1, (ix), we obtain that $H \in AC_{n^2}$ on P.

Theorem 5.23.3 *(The integration by parts formula for the Foran integral). Let $F, g : [a,b] \mapsto \mathbb{R}$. If $F \in VB$ and g is $\mathcal{F}$- integrable on $[a,b]$ then $F \cdot g$ is $\mathcal{F}$-integrable and $(\mathcal{F}) \int_a^b F(t)g(t)dt = F(b) \cdot G(b) - \int_a^b G(t)dF(t)$, where $G(x) = (\mathcal{F}) \int_a^x g(t)dt$.*

Proof. We have that $G \in \mathcal{F} \cap \mathcal{C}$ and $G'_{ap}(x) = g(x)$ *a.e.* on $[a,b]$. By Lemma 5.23.1, $H(x) = F(x) \cdot G(x) - \int_a^x G(t)dF(t)$ belongs to $\mathcal{F} \cap \mathcal{C}$ on $[a,b]$. Let $x_o \in (a,b)$ such that $G'_{ap}(x_o) = g(x_o)$ and F is derivable at x_o. Then there exists a measurable subset Q_{x_o} of $[a,b]$ such that $d(Q_{x_o}; x_o) = 1$ and $\lim_{x \to x_o, x \in Q_{x_o}} (G(x) - G(x_o))/(x - x_o) = g(x_o)$. Let $x \in Q_{x_o}$, $x > x_o$ and $\lambda_x \in [x_o, x]$ such that $\int_{x_o}^x G(t)dF(t) = G(\lambda_x) \cdot (F(x) - F(x_o))$ (this is possible because $G \in \mathcal{C}$ on $[a,b]$). If $x \to x_o$ then $\lambda_x \to x_o$, and we have

$$\lim_{\substack{x \to x_o \\ x \in Q_{x_o}}} \frac{H(x) - H(x_o)}{x - x_o} = \lim_{\substack{x \to x_o \\ x \in Q_{x_o}}} \frac{G(x) - G(x_o)}{x - x_o} \cdot F(x) + \lim_{x \to x_o} \frac{F(x) - F(x_o)}{x - x_o} \cdot G(x_o) -$$

$$\lim_{x \to x_o} \frac{G(\lambda_x)(F(x) - F(x_o))}{x - x_o} = G'_{ap}(x_o) \cdot F(x_o) + F'(x_o) \cdot G(x_o) - G(x_o) \cdot F'(x_o) =$$

$$G'_{ap}(x_o) \cdot F(x_o) = g(x_o) \cdot F(x_o).$$

It follows that $H'_{ap}(x) = g(x) \cdot F(x)$ *a.e.* on $[a,b]$. Thus $F \cdot g$ is $\mathcal{F}$-integrable on $[a,b]$ and $(\mathcal{F}) \int_a^b F(t) \cdot g(t)dt = H(b) - H(a) = F(b) \cdot G(b) - \int_a^b G(t)dF(t)$.

Theorem 5.23.4 *Let $F : [a,b] \mapsto \mathbb{R}$ be an increasing function, and let $g : [a,b] \mapsto \overline{\mathbb{R}}$ be an $\mathcal{F}$-integrable function. Then there exists $c \in [a,b]$ such that $(\mathcal{F}) \int_a^b g(t)F(t)dt = F(a) \cdot (\mathcal{F}) \int_a^c g(t)dt + F(b) \cdot (\mathcal{F}) \int_c^b g(t)dt$.*

Proof. We shall use Saks' technique of [S1] (p. 247). Let $G(x) = (\mathcal{F}) \int_a^x g(t)dt$. By Theorem 5.23.3, we have $(\mathcal{F}) \int_a^b g(t)F(t)dt = G(b) \cdot F(b) - \int_a^b G(t)dF(t)$. Since $G \in \mathcal{C}$ on $[a,b]$, it follows that there exists $c \in [a,b]$ such that $\int_a^b G(t)dF(t) = G(c) \cdot (F(b) - F(a))$. Therefore $(\mathcal{F}) \int_a^b F(t)g(t)dt = G(b)F(b) - G(c) \cdot (F(b) - F(a)) = F(a) \cdot (\mathcal{F}) \int_a^c g(t)dt + F(b) \cdot (\mathcal{F}) \int_c^b g(t)dt$.

Lemma 5.23.2 *Let $F : [a,b] \mapsto \mathbb{R}$ and let F^*_{ap} be defined as in Definition 5.19.1.*

(i) If $F \in \mathcal{DB}_1 \cap \mathcal{E}$ and F^*_{ap} is $\mathcal{F}$-integrable on $[a,b]$ then F is approximately derivable a.e. and $(\mathcal{F}) \int_a^b F^*_{ap}(t)dt = F(b) - F(a)$.

(ii) If $F \in \langle L \rangle$ and F^*_{ap} is $L\mathcal{F}$-integrable on $[a,b]$ then F is approximately derivable a.e. and $(L\mathcal{F}) \int_a^b F^*_{ap}(t)dt = F(b) - F(a)$.

Proof. (i) Let $G(x) = (\mathcal{F})\int_a^x F^*_{ap}(t)dt$. Then $G \in \mathcal{F} \cap \mathcal{C}$ on $[a,b]$ and $G'_{ap}(x) = F^*_{ap}(x)$; a.e. on $[a,b]$. By Corollary 2.5.1, (ii) and Corollary 2.37.1, (ix), it follows that $F-G \in \mathcal{DB}_1 \cap \mathcal{E}$. But $(F-G)'_{ap}(x) = 0$ a.e. where F is approximately derivable, hence by Corollary 4.3.4, $F-G$ is constant on $[a,b]$. Thus F is approximately derivable a.e. on $[a,b]$, $F \in \mathcal{F} \cap \mathcal{C}$ on $[a,b]$, and $(\mathcal{F})\int_a^b F^*_{ap}(t)dt = G(b) = F(b) - F(a)$.

(ii) The proof is similar to that of (i).

Theorem 5.23.5 *(the change of variable formula for the $\mathcal{LF}$-integral). Let $g : [a,b] \mapsto [c,d]$ be a measurable function, and let $F : [c,d] \mapsto \mathbb{R}$ such that $F, F \circ g \in\ < L > \cap \mathcal{E}$ and F^*_{ap}, $(F \circ g)^*_{ap}$ are $\mathcal{LF}$-integrable on $[a,b]$. Then F and $F \circ g$ are approximately derivable a.e. and $(\mathcal{LF})\int_a^b (F^*_{ap} \circ g)(t) \cdot g^*_{ap}(t)dt = (\mathcal{LF})\int_{g(a)}^{g(b)} F^*_{ap}(t)dt$.*

Proof. By Lemma 5.23.2, (i), it follows that $(\mathcal{LF})\int_{g(a)}^{g(b)} F^*_{ap}(t)dt = F(g(b)) - F(g(a))$ and $(\mathcal{LF})\int_a^b (F^*_{ap} \circ g)(t) \cdot g^*_{ap}(t)dt = (F \circ g)(b) - (F \circ g)(a)$. It also follows that F and $F \circ g$ are approximately a.e.. But $(F^*_{ap} \circ g)(x) \cdot g^*_{ap}(x) = (F \circ g)^*_{ap}(x)$ a.e. on $[a,b]$. Hence we obtain the asserted equality.

Definition 5.23.2 *Let $f : [a,b] \mapsto \overline{\mathbb{R}}$. A function $M : [a,b] \mapsto \mathbb{R}$ is said to be a $\mathcal{PF}$-major function for f if $M(a) = 0$, $M \in \mathcal{C} \cap \underline{\mathcal{F}}$ on $[a,b]$, M is approximately derivable a.e. on $[a,b]$ and $M'_{ap}(x) \geq f(x)$ a.e. on $[a,b]$.*
A function $m : [a,b] \mapsto \mathbb{R}$ is said to be $\mathcal{PF}$-minor function for f if $-m$ is $\mathcal{PF}$-major function for $-f$.

Lemma 5.23.3 *Let $f : [a,b] \mapsto \overline{\mathbb{R}}$. If M is a $\mathcal{PF}$- major function and m is a $\mathcal{PF}$-minor function for f then $M - m$ is increasing on $[a,b]$.*

Proof. Clearly $(M-m)'_{ap}(x) \geq 0$ a.e. on $[a,b]$ and $M - m \in \mathcal{C} \cap \underline{\mathcal{F}}$ on $[a,b]$. By Corollary 2.37.1, (xvi) and Theorem 4.2.1, $M - m$ is increasing on $[a,b]$.

Definition 5.23.3 *Let $f : [a,b] \mapsto \overline{\mathbb{R}}$ having $\mathcal{PF}$-major and $\mathcal{PF}$ minor functions. We denote by $\overline{I}(b) = \inf\{M(b) : M$ is a $\mathcal{PF}$-major function for $f\}$ and $\underline{I}(b) = \sup\{m(b) : m$ is a $\mathcal{PF}$-minor function for $f\}$. Then $\overline{I}(b)$ and $\underline{I}(b)$ are finite and $\overline{I}(b) \geq \underline{I}(b)$ (see Lemma 5.23.3).*
f is said to be $\mathcal{PF}$-integrable on $[a,b]$ if $\overline{I}(b) = \underline{I}(b)$, and we denote their common value by $(\mathcal{PF})\int_a^b f(t)dt$.

Lemma 5.23.4 *Let $f : [a,b] \mapsto \overline{\mathbb{R}}$. If f is $\mathcal{F}$-integrable on $[a,b]$ then f is $\mathcal{PF}$-integrable on $[a,b]$ and $(\mathcal{F})\int_a^b f(t)dt = (\mathcal{PF})\int_a^b f(t)dt$.*

Proof. Let $F(x) = (\mathcal{F})\int_a^x f(t)dt$. Then F is both, a $\mathcal{PF}$-major and a $\mathcal{PF}$-minor function for f.

Remark 5.23.1 *We do not know if the $\mathcal{PF}$-integral is a strict generalization of the $\mathcal{F}$-integral.*

5.24 Integrals which extend both, Foran's integral and Iseki's integral

Let $< L >$ be a real linear space of real-valued functions on $[a,b]$ such that $\mathcal{C} \subseteq < L > \subset \mathcal{DB}_1$.

Definition 5.24.1 *Let $f : [a,b] \mapsto \overline{\mathbb{R}}$. f is said to be $L\mathcal{SF}$- integrable on $[a,b]$ if there exists $F : [a,b] \mapsto \mathbb{R}$, $F \in < L > \cap \mathcal{F}$ such that F is approximately derivable a.e. and $F'_{ap}(x) = f(x)$ a.e. on $[a,b]$. Then we denote by $(L\mathcal{SF}) \int_a^b f(t)dt = F(b) - F(a)$. By Corollary 4.3.4, it follows that this integral is well defined. If $< L > = \mathcal{C}$ then we denote this integral by $\mathcal{SF}$.*

Remark 5.24.1 *If in the above definition, the class $< L > \cap \mathcal{SF}$ is replaced by $\mathcal{C} \cap SACG$ then we obtain Iseki's sparse Integral (see [I3], [I4]).*

Theorem 5.24.1 *Let $\mathcal{C} \subseteq < L > \subset \mathcal{DB}_1$. Let's denote the following classes of functions by:*

- $\mathcal{H}_1 = \{f : [a,b] \mapsto \overline{\mathbb{R}} : f$ *is $\mathcal{F}$-integrable*$\}$;
- $\mathcal{H}_2 = \{f : [a,b] \mapsto \overline{\mathbb{R}} : f$ *is Sparse integrable in Iseki's sense*$\}$;
- $\mathcal{H}_3 = \{f : [a,b] \mapsto \overline{\mathbb{R}} : f$ *is $\mathcal{SF}$-integrable*$\}$;
- $\mathcal{H}_4 = \{f : [a,b] \mapsto \overline{\mathbb{R}} : f$ *is $L\mathcal{SF}$-integrable*$\}$;

Then we have:

(i) $\mathcal{H}_1 \setminus \mathcal{H}_2 \neq \emptyset$;

(ii) $\mathcal{H}_2 \setminus \mathcal{H}_1 \neq \emptyset$;

(iii) $\mathcal{H}_1 \subset \mathcal{H}_3$ *and* $\mathcal{H}_2 \subset \mathcal{H}_3$;

(iv) $\mathcal{H}_3 \subseteq \mathcal{H}_4$.

Moreover, if a function f is integrable in several senses then it integrals are equal.

Proof. (i) See Section 6.34.
(ii) See Section 6.42.
(iii) This follows by (i), (ii) and Corollary 2.34.1.
(iv) This is evident.
The last part follows by Corollary 2.34.1 and Corollary 4.3.4.

Theorem 5.24.2 *Let $f, g : [a,b] \mapsto \overline{\mathbb{R}}$, $c \in (a,b)$ and $\alpha, \beta \in \mathbb{R}$.*

(i) f is $\mathcal{SF}$-integrable on $[a,b]$ if and only if f is $L\mathcal{SF}$-integrable on $[a,c]$ and $[c,b]$. Then $(L\mathcal{SF}) \int_a^c f(t)dt + (L\mathcal{SF} \int_c^b f(t)dt = (L\mathcal{SF}) \int_a^b f(t)dt$.

(ii) If f and g are $L\mathcal{SF}$-integrable on $[a,b]$ then $\alpha f + \beta g$ is $L\mathcal{SF}$-integrable and $(L\mathcal{SF}) \int_a^b (\alpha f + \beta g)(t)dt = \alpha \cdot (L\mathcal{SF}) \int_a^b f(t)dt + \beta \cdot (L\mathcal{SF}) \int_a^b g(t)dt$.

Proof. See Corollary 2.34.1.

Lemma 5.24.1 *Let $F, H, G : [a,b] \mapsto \mathbb{R}$ such that $F \in VB$, G is continuous and $H(x) = F(x) \cdot G(x) - \int_a^b G(t)dF(t)$. Let $P \subseteq [a,b]$, and let n be a positive integer. If $G \in SAC_n$ on $[a,b]$ then $H \in C$ and $H \in SAC_{n^2}$ on P. Moreover, if F is monotone then $H \in SAC_n$ on P. (Here $\int_a^x G(t)dF(t)$ is the Riemann Stieltjes integral.)*

Proof. The proof is similar to that of Lemma 5.23.1, using Proposition 2.34.1 instead of Proposition 2.28.1, and Corollary 2.34.1, (iv) instead of Theorem 2.28.1, (ix).

Theorem 5.24.3 *(The integration by parts formula for the $\mathcal{SF}$-integral). Let $F, g : [a,b] \mapsto \mathbb{R}$. If $F \in VB$ and g is $\mathcal{SF}$-integrable on $[a,b]$ then $F \cdot g$ is $\mathcal{SF}$-integrable on $[a,b]$ and $(\mathcal{SF})\int_a^b F(t)g(t)dt = F(b) \cdot G(b) - \int_a^b G(t)dF(t)$, where $G(x) = (\mathcal{SF})\int_a^b g(t)dt$.*

Proof. The proof is similar to that of Theorem 5.23.3, using Lemma 5.24.1 instead of Lemma 5.23.1.

Theorem 5.24.4 *Let $F : [a,b] \mapsto \mathbb{R}$ be an increasing function and let $g : [a,b] \mapsto \overline{\mathbb{R}}$ be an $\mathcal{SF}$-integrable function. Then there exists $c \in [a,b]$ such that $(\mathcal{SF})\int_a^b g(t)F(t)dt = F(a) \cdot (\mathcal{SF})\int_a^c g(t)dt + F(b) \cdot (\mathcal{SF})\int_a^b g(t)dt$.*

Proof. The proof is similar to that of Theorem 5.23.4.

Theorem 5.24.5 *(The change of variable formula for the $L\mathcal{SF}$-integral). Let $g : [a,b] \mapsto [c,d]$ be a measurable function and $F : [c,d] \mapsto \mathbb{R}$. If F, $F \circ g$ are approximately derivable a.e., and F^*_{ap}, $(F \circ g)^*_{ap}$ are $L\mathcal{SF}$-integrable then $(L\mathcal{SF})\int_a^b (F^*_{ap} \circ g)(t) \cdot g^*_{ap}(t)dt = (L\mathcal{SF})\int_{g(a)}^{g(b)} F^*_{ap}(t)dt$ (see Definition 5.19.1).*

Proof. See Corollary 5.19.1.

Chapter 6

Examples

6.1 The Cantor ternary set, a perfect nowhere dense set

Let $C = \{x \in [0,1] : x = \sum_{i=1}^{\infty} c_i/3^i$, with $c_i \in \{0,2\}$ for each $i\}$. Each point $x \in C$ is uniquely represented by $\sum_{i=1}^{\infty} c_i(x)/3^i = 0, c_1(x)c_2(x)\ldots c_i(x)\ldots$ in base 3. The set C can also be obtained by the ternary process of Cantor: at the first step we exclude from the interval $[0,1]$ the middle third $(1/2, 2/3)$; then the intervals $[0,1/3]$ and $[2/3,1]$ will be called the remained closed intervals from the first step. At the second step we exclude the middle third of each remained closed interval from the first step. Continuing infinitely many times, C will be the set left after excluding all middle thirds. *Then C is a perfect nowhere dense null set.*
Let $I_{c_1\ldots c_i} = [\sum_{k=1}^{i} c_k/3^k, \sum_{k=1}^{i} c_k 3^k/+\sum_{k=1+i}^{\infty} 2/3^k] = [\sum_{k=1}^{i} c_k/3^k, \sum_{k=1}^{i} c_k/3^k + 1/3^i]$ be the remained closed intervals from the step i (we have 2^i such intervals). Then $I_{c_1\ldots c_i} \subset I_{c_1\ldots c_{i-1}}$. Let $J_{c_1\ldots c_{i-1}} = (\sum_{k=1}^{i-1} c_k/3^k + \sum_{k=i+1}^{\infty} 2/3^k, \sum_{k=1}^{i-1} c_k/3^k + 2/3^i) = (\sum_{k=1}^{i-1} c_k/3^k + 1/3^i, \sum_{k=1}^{i-1} c_k/3^k + 2/3^i)$ be the excluded open intervals from the step i (we have 2^{i-1} such open intervals). Then $J_{c_1\ldots c_i} \subset J_{c_1\ldots c_{i-1}}$.

Remark 6.1.1 *Let $a, b \in C$. Then there exists a positive integer n such that $1/3^{n+1} \leq b - a < 1/3^n$.*

(i) If $b - a = 1/3^{n+1}$ then we have two possibilities:

a) There exist $c_1, \ldots, c_{n+1} \in \{0,2\}$ such that $a = 0, c_1c_2\ldots c_{n+1}000\ldots$ and $b = 0, c_1\ldots c_{n+1}222\ldots$ in base 3. It follows that $[a,b]$ is a remained closed interval from the step $n+1$.

b) There exist $c_1, \ldots, c_n \in \{0,2\}$ such that $a = 0, c_1\ldots c_n0222\ldots$ and $b = 0, c_1\ldots c_n2000\ldots \in \{0,2\}$ in base 3. It follows that (a,b) is an excluded open interval from the step $n+1$.

(ii) If $1/3^{n+1} < b - a < 1/3^n$ then there exist $c_1, \ldots, c_n \in \{0,2\}$ such that $a = 0, c_1\ldots c_n0c_{n+2}(a)c_{n+3}(a)\ldots$ and $b = 0, c_1\ldots c_n2c_{n+2}(b)c_{n+3}(b)\ldots$ in base 3.

6.2 The Cantor ternary function φ

Let $\varphi : [0,1] \mapsto [0,1]$, $\varphi(x) = \sum_{i=1}^{\infty} c_i(x)/2^{i=1}$, $x \in C$ (see Section 6.1). Then φ is continuous on C. Extending φ linearly on the closure of each interval contiguous to C, we have φ defined and continuous on $[0,1]$ (see Figure 6.1). It follows that φ is constant on the closure of each interval contiguous to C, hence $\varphi'(x) = 0$ on $[0,1] \setminus C$. Then we have:

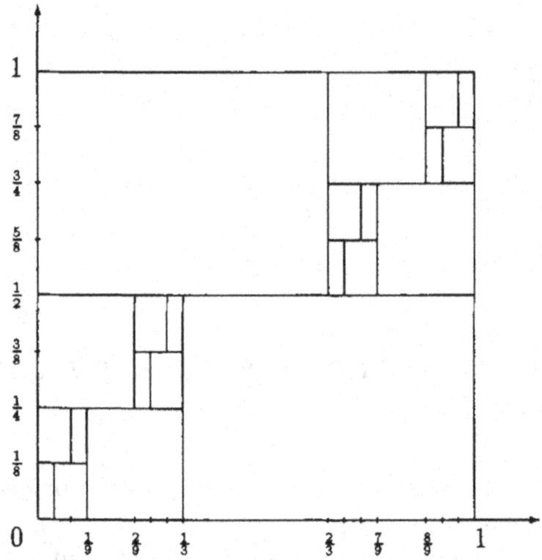

Figure 6.1: The Cantor ternary function φ

(i) $\varphi \in VB \subseteq VBG \subseteq \mathcal{B}$ *on* $[0,1]$*;*

(ii) $\varphi \in T_1 \subset T_2$ *on* $[0,1]$*;*

(iii) $\varphi \notin \overline{M}$*, hence* $\varphi \notin (M)$, $\varphi \notin (N)$, $\varphi \notin \mathcal{E}$, $\varphi \notin \mathcal{F}$, $\varphi \notin ACG$, $\varphi \notin AC$ *on* $[0,1]$.

Remark 6.2.1 *The utility of this example follows by Theorem 2.11.1, (v); Corollary 2.29.1, (xvi); Remark 2.37.3; Remark 3.1.1.*

6.3 A real bounded S_o^+ closed set which is not of F_σ-type

Let C be the Cantor ternary set (see Section 6.1), and let (a_i, b_i), $i = \overline{1,\infty}$ be the intervals contiguous to C. Let $G_1 = \cup_{i=1}^{\infty}[a_i, b_i)$ and $C_1 = C \setminus \{a_1, a_2, \ldots\} = [0,1] \setminus G_1$. Then we have:

(i) C_1 *is a bounded* S_o^+ *closed set;*

(ii) C_1 *is a set of* G_δ*-type;*

(iii) C_1 *is not of* F_σ*-type.*

Proof. (i) Clearly G_1 is S_o^+ open. By Proposition 1.14.2, C_i is S_o^+ closed.

(ii) Clearly G_1 is of F_σ-type, hence C_1 is of G_δ-type.

(iii) Suppose on the contrary that there exists a sequence of closed sets $\{K_j\}_j$ such that $C_1 \subset \cup_{j=1}^\infty K_j$. By Theorem 1.7.1, there exist $a, b \in C_1$ such that $(a,b) \cap C_1 \neq \emptyset$ and $[a,b] \cap C_1 \subset K_j$, for some j. It follows that $[a,b] \cap C_1 \subset K_j \cap [a,b] \subset [a,b] \cap C_1$, hence $[a,b] \cap C_1 = K_j \cap [a,b]$, a contradiction (because $[a,b] \cap C_1$ is not closed, but $K_j \cap [a,b]$ is closed).

Remark 6.3.1 *The utility of this example follows by Remark 1.14.1, (v).*

6.4 An S_o^+ lower semicontinuous function which is not $\overline{\mathcal{B}}_1$

Let P be a bounded S_o^+ closed set which is not of F_σ-type (see Section 6.3), $a = \inf(P)$, $b = \sup(P)$. Let $F : [a,b] \mapsto \{0,1\}$, $F(x) = 0$ for $x \in P$, and $F(x) = 1$ for $x \notin P$. Then we have:

(i) F is S_o^+ lower semicontinuous on $[a,b)$ (see Proposition 1.15.3);

(ii) $F \notin \overline{\mathcal{B}}_1$ on $[a,b]$.

Remark 6.4.1 *The utility of this example follows by Remark 2.2.2.*

6.5 A function $F \in \mathcal{C}_i$, $F \notin \mathcal{C}_i^*$

Let $F : [0,1] \mapsto \mathbb{R}$, $F(0) = 0$ and $F(x) = 1/x$, $x \in (0,1]$. Then we have:

(i) $F \in \mathcal{C}_i$ on $[0,1]$;

(ii) $F \notin \mathcal{C}_i^*$ on $[0,1]$.

Remark 6.5.1 *The utility of this example follows by Proposition 2.3.1, (ii).*

6.6 A function $F \in \mathcal{D}$, $F \in [\mathcal{C}_i^*G]$, $F \notin [\mathcal{C}G]$

Let C be the Cantor ternary set (see Section 6.1), and let (a_i, b_i), $i = \overline{1,\infty}$ be the intervals contiguous to C, $c_i = a_i + (b_i - a_i)/3$, $d_i = a_i + 2(b_i - a_i)/3$. Let $F : [0,1] \mapsto \mathbb{R}$ be a function defined as follows: $F(x) = 0$ if $x \in C \setminus \{a_1, b_1, a_2, b_2, \ldots\}$; $F(x) = 1/2^{n-1}$ if $x = a_i$ and (a_i, b_i) is from the step n; $F(x) = -1/2^{n-1}$ if $x = b_i$ and (a_i, b_i) is from the step n; $F(x) = 1$ if $x \in \{x_1, c_2, \ldots\}$; $F(x) = -1$ if $x \in \{d_1, d_2, \ldots\}$; F is linear on each interval $[a_i, c_i]$, $[c_i, d_i]$, $[d_i, b_i]$, $i = \overline{1,\infty}$. Then we have:

(i) $F \in \mathcal{C}_i^*$ on C, hence $F \in [\mathcal{C}_i^*G]$ on $[0,1]$;

(ii) $F \in \mathcal{D}$ on $[0,1]$;

(iii) $F \notin [\mathcal{C}G]$ on C.

Remark 6.6.1 *The utility of this example follows by Proposition 2.3.5, (v).*

6.7 A function $F \in \mathcal{DB}_1$, $F \notin [C_iG]$

Let C be the Cantor ternary set (see Section 6.1), and let (a_i, b_i), $i = \overline{1,\infty}$ be the intervals contiguous to C, $c_i = a_i + (b_i - a_i)/3$, $d_i = a_i + 2(b_i - a_i)/3$. Let $F : [0,1] \mapsto \mathbb{R}$ be a function defined as follows:
$F(x) = 0$ if $x \in C \setminus \{a_1, b_1, a_2, b_2, \ldots\}$;
$F(x) = 1/2^{n-1}$ if $x = a_i$, where (a_i, b_i) is from the step n and n is odd;
$F(x) = -1/2^{n-1}$ if $x = b_i$, where (a_i, b_i) is from the step n and n is odd;
$F(x) = -1/2^{n-1}$ if $x = a_i$, where (a_i, b_i) is from the step n and n is even;
$F(x) = 1/2^{n-1}$ if $x = b_i$, where (a_i, b_i) is from the step n and n is even;
$F(x) = 1$ if $x \in \{c_1, c_2, \ldots\}$;
$F(x) = -1$ if $x \in \{d_1, d_2, \ldots\}$;
F is linear on each interval $[a_i, c_i]$, $[c_i, d_i]$, $[d_i, b_i]$, $i = \overline{1,\infty}$.
Then we have:

(i) $F \in [0,1]$;

(ii) $F \notin [C_iG]$ *on* C, *hence* $F \notin [C_iG]$ *on* $[0,1]$.

Remark 6.7.1 *The utility of this example follows by Proposition 2.3.5.*

6.8 A function $F \in uCM$; $F \notin \ell CM$

Let C be the Cantor ternary set (see Section 6.1), and let (a_i, b_i), $i = \overline{1,\infty}$ be the intervals contiguous to C, $c_i = (a_i + b_i)/2$. By Theorem 2.14.6, there exists a function $g : [0,1] \mapsto [0,1]$, such that $g(0) = 0$, $g(1) = 1$, g is continuous and strictly increasing on $[0,1]$, and $g'(x) = +\infty$ if $x \in C$. Let $F : [0,1] \mapsto [0,1]$ be defined as follows: $F(x) = g(x)$ if $x \in C$; $F(x) = g(c_i)$ if $x \in (a_i, b_i)$, $i = \overline{1,\infty}$. Then we have:

(i) $F(0) = 0$ *and* $F(1) = 1$;

(ii) F *is increasing on* $[0,1]$;

(iii) $F'(x) = +\infty$ *for each* $x \in C$;

(iv) F *is constant on each* (a_i, b_i);

(v) $F \in uCM$ *on* $[0,1]$;

(vi) $F \notin \ell CM$ *on* $[0,1]$, *hence* $F \notin CM$ *on* $[0,1]$.

Proof. We show (iii) (the other statements are obvious). Let $x_o \in C$. Since $g'(x_o) = +\infty$ it follows that for $\alpha > 0$ there exists $\delta > 0$ such that

(1) $g(x) - g(x_o) > \alpha(x - x_o)$, *for each* $x \in [x_o, x_o + \delta)$.

We have three situations:
(I) If $x \in C \cap [x_o, x_o + \delta)$ then by (1) it follows that $F(x) - F(x_o) > \alpha(x - x_o)$;
(II) If $x \in (a_i, c_i) \cap [x_o, x_o + \delta)$ for some i then by (1) it follows that $F(x) - F(x_o) = g(c_i) - g(x_o) \ge g(x) - g(x_o) > \alpha(x - x_o)$;
(III) If $x \in [c_i, b_i] \cap [x_o, x_o + \delta)$ then by (1) it follows that $F(x) - F(x_o) = g(c_i) - g(x_o) > \alpha(c_i - x_o) \ge (\alpha/2)(x - x_o)$. Thus $F'_+(x_o) = +\infty$. Similarly $F'_-(x_o) = +\infty$.

Remark 6.8.1 *(i) Following Theorem 2.14.6, Preiss constructed in 1971 an example as above, but the proof is difficult (see [Pr1]).*

(ii) The utility of this example follows by Theorem 2.4.2, (vii).

6.9 A function concerning conditions $\mathcal{D}_+$, $\mathcal{D}_-$, CM, sCM, lower internal

Let $\alpha, \lambda \in \mathbb{R}$, and let $F : [0,1] \mapsto \mathbb{R}$ be a function defined as follows: $F(x) = 1 - x$ if $x \in [0,1)$; $F(1) = \alpha$. Then we have:

(i) If $\alpha > 0$ then $F \notin \mathcal{D}_+$, but $F \in \mathcal{D}_-$ on $[0,1]$;

(ii) If $\alpha < 0$ then:

- *a) $F \in \mathcal{D}_+$, but $F \notin \mathcal{D}_-$ on $[0,1]$;*
- *b) $F \notin$ lower internal on $[0,1]$;*
- *c) $F(x) + \lambda x \in CM$ on $[0,1]$, hence $F \in CM$, whenever $\lambda < 1$;*
- *d) $F(x) + \lambda x \notin uCM$ on $[0,1]$, whenever $\lambda \geq 1$;*
- *e) $F \notin sCM$ on $[0,1]$.*

Remark 6.9.1 *The utility of this example follows by Remark 2.4.2 and Theorem 2.4.2, (vii).*

6.10 A function concerning conditions: $\mathcal{D}_-$, $\mathcal{D}$, internal, $\underline{\mathcal{B}}_1$, $\overline{\mathcal{B}}_1$, $\mathcal{B}_1$, $w\mathcal{B}_1$, $[VBG]$, $(-)$, T_1, T_2 (Bruckner)

Let C be the Cantor ternary set (see Section 6.1), and let (a_i, b_i), $i = \overline{1,\infty}$ be the intervals contiguous to C. Let $C_1 = C \setminus \{a_1, b_1, a_2, b_2, \ldots\}$ and let $\alpha \in \mathbb{R}$. Let $F : [0,1] \mapsto \mathbb{R}$ be a function defined as follows: $F(0) = \alpha$; $F(x) = 0$ if $x \in C_1 \setminus \{0,\}$; $F(x) = -1$ if $x \in \{a_1, a_2, \ldots\}$; $F(x) = 1$ if $x \in \{b_1, b_2, \ldots\}$; F is linear on each $[a_i, b_i]$.

Example 6.10.1 *If $\alpha = 0$ then we have:*

(i) $F \in \mathcal{D}$ on $[0,1]$, hence F is internal on $[0,1]$;

(ii) F is discontinuous at each $x \in C$;

(iii) $F \notin \underline{\mathcal{B}}_1$ and $F \notin \overline{\mathcal{B}}_1$ on $[0,1]$ on $[0,1]$;

(iv) F in wB_1 on $[0,1]$;

(v) $F \in \underline{L}(P \wedge (P \cup P_-)) \cap \underline{L}((P \cup P_+) \wedge P) \subseteq \underline{AC}(P \wedge (P \cup P_-)) \cap \underline{AC}((P \cup P_+) \wedge P)$ on C_1;

(vi) $F \in [VBG]$ on $[0,1]$;

(vii) $-F \in \underline{L}((P \cup P_-) \wedge P) \cap \underline{L}(P \wedge (P \cup P_+)) \subseteq \underline{AC}((P \cup P_-) \wedge P) \cap \underline{AC}(P \wedge (P \cup P_+))$ *on* C_1;

(viii) $-F \in (-)$ *and* $F \notin (-)$ *on* $[0,1]$;

(ix) $F \in T_2$ *and* $F \notin T_1$ *on* $[0,1]$.

Remark 6.10.1 *The utility of Example 6.10.1 follows by Proposition 2.3.5, (v); Theorem 2.4.2, (vi), (vii); Remark 2.6.1; Theorem 2.18.9, (iv), (xvii), (xviii); Remark 2.11.2.*

Example 6.10.2 *If* $\alpha = 2$ *then we have:*

(i) $F \notin \mathcal{D}_-$ *on* $[0,1]$;

(ii) F *satisfies Theorem 2.5.1, (v);*

(iii) F *is discontinuous at each* $x \in C$;

(iv) $F \notin \underline{\mathcal{B}}_1$ *and* $F \notin \overline{\mathcal{B}}_1$ *on* $[0,1]$.

Remark 6.10.2 *The utility of Example 6.10.2 follows by Remark 2.5.2 and Theorem 2.18.9, (xvii).*

6.11 A function concerning conditions: $\overline{\mathcal{B}}_1$, $\underline{\mathcal{B}}_1$, $\mathcal{D}_-$, $\mathcal{D}_+$, lower internal, internal, internal* (Dirichlet)

Let Q be the set of all rational numbers in $[0,1]$ and $\alpha \in \mathbb{R}$. Let $F : [0,1] \mapsto \mathbb{R}$ be a function defined as follows: $F(0) = \alpha$; $F(x) = 1$ if $x \in Q \setminus \{0\}$; $F(x) = 0$ if $x \notin Q$. Then we have:

(i) F *is discontinuous at each point* $x \in [0,1]$;

(ii) $F \notin \overline{\mathcal{B}}_1$, *but* $F \in \underline{\mathcal{B}}_1$ *on* $[0,1]$;

(iii) $F \notin \mathcal{D}_-$ *and* $F \notin \mathcal{D}_+$ *on* $[0,1]$;

(iv) *If* $\alpha < 1$ *then* $F \in$ *lower internal on* $[0,1]$;

(v) *If* $\alpha > 1$ *then* $F \notin$ *lower internal on* $[0,1]$;

(vi) *If* $\alpha = 1$ *then* $F \in$ *internal on* $[0,1]$;

(vii) $\lim_{x \to x_o} F(x)$ *exists for no* $x_o \in [0,1]$, *hence* $F \in$ *internal* on* $[0,1]$;

(viii) F *satisfies the second part of Theorem 2.5.1,(v), but* F *does not satisfy the first part of the same theorem.*

Remark 6.11.1 *The utility of this example follows by Theorem 2.4.2, (vii); Remark 2.5.2.*

6.12 A function concerning conditions: $\mathcal{D}$, $\mathcal{D}_-$, $\mathcal{B}_1$, C_i, C_i^*, lower internal, internal*, VB, VB^*G, $N^{-\infty}$

Let $\alpha \in \mathbb{R}$, and let $F : [0,1] \mapsto \mathbb{R}$ be a function defined as follows: $F(0) = \alpha$, $F(x) = \sin(2\pi/x)$ if $x \in [0,1]$. Then we have:

(i) *$F \in C$ on $(0,1]$, hence $F \in \mathcal{B}_1$ on $[0,1]$ and $\lim_{x\searrow 0} F(x)$ does not exist;*

(ii) *$F \in VB^*G$ on $[0,1]$, but $F \notin VB$ and $F \notin T_1$ on $[0,1]$;*

(iii) *$F \in N^{-\infty}$ on $[0,1]$;*

(iv) *If $\alpha \in [-1,1]$ then $F \in \mathcal{D}$ on $[0,1]$;*

(v) *If $\alpha < -1$ then $F \notin \mathcal{D}$ and $F \notin C_i^*$ on $[0,1]$, but $F \in C_i$ on $[0,1]$;*

(vi) *If $\alpha > 1$ then $F \in$ internal* on $[0,1]$, but $F \notin$ lower internal, hence $F \notin \mathcal{D}_-\mathcal{B}_1$ on $[0,1]$;*

(vii) *If $\alpha \in (-1,1)$ then $F \notin C_i$ on $[0,1]$.*

Remark 6.12.1 *The utility of this example follows by Theorem 2.4.2, (vi), (vii), (viii), (ix); Theorem 2.10.3, (i), (ii); Theorem 2.18.9, (iv); Remark 2.18.7, (iv); Corollary 2.22.1, (ii).*

6.13 A function $F \in \mathcal{D}$, $F \in \underline{\mathcal{B}}_1 \setminus \overline{\mathcal{B}}_1$; $-F \in \overline{\mathcal{Z}}_i \setminus C_i$, $-F \in \mathcal{D}_-\overline{\mathcal{B}}_1 \setminus \overline{\mathcal{Z}}_i$

Let C be the Cantor ternary set (see Section 6.1), and let (a_i, b_i), $\overline{1,\infty}$ be the intervals contiguous to C. Let $\alpha \in \mathbb{R}$, and let $F : [0,1] \mapsto \mathbb{R}$ be a function defined as follows: $F(1) = \alpha$; $F(x) = 0$ if $x \in C \setminus \{1, b_1, b_2, \ldots\}$; $F(x) = 1$ if $x \in \{b_1, b_2, \ldots\}$; F is linear on each $[a_i, b_i]$. Then we have:

(i) *If $\alpha = 0$ then:*

- a) *$F \in \mathcal{D}$ on $[0,1]$;*
- b) *$F \in \underline{\mathcal{B}}_1$, but $F \notin \overline{\mathcal{B}}_1$ on $[0,1]$;*
- c) *$-F \in \overline{\mathcal{Z}}_i$, but $F \notin C_i$ on $[0,1]$.*

(ii) *If $\alpha = 1$ then $-F \in \mathcal{D}_-\overline{\mathcal{B}}_1$, but $-F \notin \overline{\mathcal{Z}}_i$ on $[0,1]$.*

Remark 6.13.1 *The utility of this example follows by Theorem 2.4.2, (v).*

6.14 A function $F \in \underline{\mathcal{B}}_1 \setminus \overline{\mathcal{B}}_1$, $F \in$ lower internal, $F \notin \mathcal{D}_-$

Let C be the Cantor ternary set (see Section 6.1), and let (a_i, b_i), $i = \overline{1,\infty}$ be the intervals contiguous to C. Let $F : [0,1] \mapsto \mathbb{R}$, $F(x) = 1$ if $x \in \{b_1, b_2, \ldots\}$, $F(x) = 0$ otherwise. Then we have:

(i) $F'_{ap}(x) = 0$ *at each point* $x \in [0,1] \setminus \{b_1, b_2, \ldots\}$;

(ii) $F \in \underline{\mathcal{B}}_1$ *and* $F \notin \overline{\mathcal{B}}_1$ *on* $[0,1]$;

(iii) $F \in$ *lower internal on* $[0,1]$;

(iv) $F \notin \mathcal{D}_-$ *on* $[0,1]$.

Remark 6.14.1 *The utility of this example follows by Theorem 2.4.2, (vi).*

6.15 A function $F \in sCM$, $F \notin$ internal*

Let $\alpha, \lambda \in \mathbb{R}$, and let Q be the set of all rational numbers in $[0,1]$. Let $F : [0,1] \mapsto \mathbb{R}$ be a function defined as follows: $F(x) = 1 - x$ if $x \in Q \setminus \{1\}$, $F(x) = 0$ if $x \notin Q$, $F(1) = \alpha$. Then we have:

(i) $\lim_{x \to x_o} F(x)$ *exists for no* $x_o \in [0,1)$, *but* $\lim_{x \to 1} F(x) = 0$;

(ii) $F \in \underline{\mathcal{B}}_1$, *but* $F \notin \overline{\mathcal{B}}_1$ *on* $[0,1]$;

(iii) $F \notin \mathcal{D}_-$ *and* $F \notin \mathcal{D}_+$ *on* $[0,1]$;

(iv) If $\alpha \geq 0$ *then* $F \in$ *lower internal on* $[0,1]$;

(v) If $\alpha < 0$ *then* $F \notin$ *lower internal* on* $[0,1]$, *hence* $F \notin$ *internal* on* $[0,1]$;

(vi) $F(x) + \lambda x \in CM$ *on* $[0,1]$, *hence* $F \in sCM$ *on* $[0,1]$.

Remark 6.15.1 *The utility of this example follows by Theorem 2.4.2, (vii).*

6.16 A function $F \in AC^*G \setminus AC$, $F \in C_i^* \setminus \mathcal{D}$, $F \in sCM \setminus$internal*

Let $\alpha, \lambda \in \mathbb{R}$ and let $F : [0,1] \mapsto \mathbb{R}$ be a function defined as follows: $F(0) = \alpha$, $F(x) = x \cdot \sin(2\pi)/x$ if $x \neq 0$. Then we have:

(i) F is continuous and derivable on $(0,1]$, *hence* $F \in \mathcal{B}_1$ *on* $[0,1]$;

(ii) If $\alpha = 0$ *then:*

- *a) F is continuous on* $[0,1]$;
- *b)* $F'(0)$ *does not exist (finite or infinite);*

c) $F \notin AC$ *on* $[0,1]$;

d) $F \in AC^*G$ *on* $[0,1]$;

(ii) If $\alpha < 0$ *then:*

a) $F \notin \mathcal{D}$ *on* $[0,1]$;

b) $F \in C_i^*$ *on* $[0,1]$;

c) $F'(0) = +\infty$;

(iii) If $\alpha > 0$ *then:*

a) $F \notin$ *lower internal* on* $[0,1]$, *hence* $F \notin$ *internal* on* $[0,1]$;

b) $F(x) + \lambda x \in CM$ *on* $[0,1]$, *hence* $F \in sCM$ *on* $[0,1]$.

Remark 6.16.1 *The utility of this example follows by Theorem 2.4.2, (vi), (viii), (ix); Theorem 2.18.9, (vi).*

6.17 A function $F \in (D.C.)$, $F \in \mathcal{B}_1$, $F \notin m_2$, $F \notin \mathcal{D}$

Let $F : [0,1] \mapsto \overline{\mathbb{R}}$, $F(1/2) = 0$, and $F(x) = +\infty$ if $x \neq 1/2$. Then we have:

(i) $F \in (D.C.)$ *and* $F \in \mathcal{B}_1$ *on* $[0,1]$;

(ii) $F \notin m_2$ *and* $F \notin \mathcal{D}$ *on* $[0,1]$.

Remark 6.17.1 *The utility of this example follows by Remark 2.6.2*

6.18 A function $F \in (+) \cap (-)$; $F \notin \mathcal{D}\mathcal{B}_1 T_2$

Let C be the Cantor ternary set (see Section 6.1), and let (a_i, b_i), $i = \overline{1,\infty}$ be the intervals contiguous to C. Let $F : [0,1] \mapsto [-1,1]$ be a function defined as follows:
$F(a_i) = 1$ and $F(b_i) = -1$, whenever (a_i, b_i) is from an even step;
$F(b_i) = 1$ and $F(a_i) = -1$, whenever (a_i, b_i) is from an odd step;
$F(x) = 0$, whenever $x \in C \setminus \{a_1, b_1, a_2, b_2, \ldots\}$;
F is linear on each $[a_i, b_i]$.
Then we have:

(i) $F \in \mathcal{D}$ *on* $[0,1]$;

(ii) $F \notin \mathcal{B}_1$ *on* $[0,1]$;

(iii) $F \in T_2$ *on* $[0,1]$;

(iv) $F \in (+) \cap (-)$ *on* $[0,1]$.

Remark 6.18.1 *The utility of this example follows by Theorem 2.18.9, (xviii).*

6.19 A function $G \in \mathcal{D}$, $G \notin \underline{\mathcal{B}}_1$, $G \notin \overline{\mathcal{B}}_1$, $G'_{ap}(x)$ exists *n.e.*, $G'_{ap}(x) \geq 0$ *a.e.* (Preiss)

Let C be the Cantor ternary set (see Section 6.1), and let (a_i, b_i), $i = \overline{1,\infty}$ be the intervals contiguous to C. Let F be the function defined in Section 6.8, and let $G : [0,1] \mapsto \mathbb{R}$ be a function defined as follows:
$G(x) = 1 - F(x)$ if $x \in C \setminus \{a_1, b_1, a_2, b_2, \ldots\}$;
$G(x) = 1 - F(x)$ if $x \in [a_i + (b_i - a_i)/2^{i+1}, b_i - (b_i - a_i)/2^{i+1}]$, $i \geq 1$;
$G(x) = 0$ if $x \in \{a_1, a_2, \ldots\}$;
$G(x) = 1$ if $x \in \{b_1, b_2, \ldots\}$.
On each interval $(a_i, a_i + (b_i - a_i)/2^{i+1})$ and $(b_i - (b_i - a_i)/2^{i+1}, b_i)$ we define G such that it is continuous and increasing on each $[a_i, b_i]$ and $F'(x)$ exists on each (a_i, b_i). Then we have:

(i) $G \in \mathcal{D}$ *on* $[0,1]$;

(ii) $G \notin \overline{\mathcal{B}}_1$, $G \notin \underline{\mathcal{B}}_1$ *on* $[0,1]$;

(iii) $G'_{ap}(x)$ *exists (finite or infinite) n.e., and* $G'_{ap}(x) \geq 0$ *a.e. on* $[0,1]$.

Proof. (i) and (ii) are evident.

(iii) Let $E_* = \cup_{i=1}^{\infty}[a_i, b_i - (b_i - a_i)/2^{i+1}]$ and $E^* = \cup_{i=1}^{\infty}[a_i + (b_i - a_i)/2^{i+1}, b_i]$. If $x_o \in C$ and $x_o \neq a_i$, $i \geq 1$ then $G'_{ap}(x_o) = \lim_{x \searrow x_o, x \in E_*}(G(x) - G(x_o))/(x - x_o) \leq \lim_{x \searrow x_o, x \in E_*}(F(x) - F(x_o))/(x - x_o) = -\infty$. Similarly, if $x_o \in C$ and $x_o \neq b_i$, $i \geq 1$ then $G'_{ap}(x_o) = -\infty$. Hence $G'_{ap}(x) = -\infty$, for each $x \in C \setminus \{a_1, b_1, a_2, b_2, \ldots\}$. If $x_o \in (a_i, b_i)$ then $G'_{ap}(x_o) = G'(x_o) \geq 0$ (by construction).

Remark 6.19.1 *The utility of this example follows by Remark 4.8.1, (ii).*

6.20 A function $H \in \mathcal{D}$, $H \notin \overline{\mathcal{B}}_1$, $H \notin \underline{\mathcal{B}}_1$, $H'_{ap}(x)$ exists on $(0,1)$ (Preiss)

Let C be a Cantor ternary set (see Section 6.1) and let (a_i, b_i), $i = \overline{1,\infty}$ be the intervals contiguous to C. Let F be the function defined in Section 6.8, and let $H : [0,1] \mapsto \overline{\mathbb{R}}$ be a function defined as follows:
$H(x) = 1 - F(x)$ if $x \in C \setminus \{a_1, b_1, a_2, b_2, \ldots\}$;
$H(x) = 1 - F(x)$ if $x \in [a_i + (b_i - a_i)/2^{i+1}, b_i - (b_i - a_i)/2^{i+1}]$, $i \geq 1$;
$H(x) = 0$ if $x \in \{a_1, a_2, \ldots\}$;
$H(x) = 1$ if $x \in \{b_1, b_2, \ldots\}$;
$H(x) = -1$ if $x = a_i + (b_i - a_i)/2^{i+2}$, $i \geq 1$;
$H(x) = 1$ if $x = b_i - (b_i - a_i)/2^{i+2}$, $i \geq 1$;
on the intervals $(a_i, a_i + (b_i - a_i)/2^{i+2})$, $(a_i + (b_i - a_i)/2^{i+2}, a_i - (b_i - a_i)/2^{i+1})$, $(b_i - (b_i - a_i)/2^{i+1}, b_i - (b_i - a_i)/2^{i+2})$ and $(b_i - (b_i - a_i)/2^{i+2}, b_i)$, $i \geq 1$ we define H such that:

- $H \in \mathcal{C}$ on $[a_i, b_i]$, $i \geq 1$;
- $H'(x)$ exists on (a_i, b_i), $i \geq 1$;

- $H'_+(a_i) = -\infty$ and $H'_-(b_i) = -\infty$, $i \geq 1$;
- H is increasing on $[a_i + (b_i - a_i)/2^{i+2}, b_i - (b_i - a_i)/2^{i+2}]$, $i \geq 1$;
- H is decreasing on $[a_i, a_i + (b_i - a_i)/2^{i+2}]$ and $[b_i - (b_i - a_i)/2^{i+2}, b_i]$, $i \geq 1$.

Then we have:

(i) $H \in \mathcal{D}$ *on* $[0,1]$*;*

(ii) $H \notin \overline{\mathcal{B}}_1$ *and* $H \notin \underline{\mathcal{B}}_1$ *on* $[0,1]$*;*

(iii) $H'_{ap}(x)$ *exists (finite or infinite) for each* $x \in (0,1)$*.*

Remark 6.20.1 *The utility of this example follows by Remark 2.2.3.*

6.21 A function $F \in \mathcal{DB}_1$, $F(x) = 0$ *a.e.*, F is not identically zero (Croft)

For each $x \in (0,1)$ corresponds its binary expansion $\sum_{n\geq 1} a_n/2^n$, $a_n \in \{0,1\}$. For each real z denote by $< z >$ the fractional part of z. Define $f(x) = \inf\{, 2x >, < 2^2 x >, \ldots, < 2^{2^n} x >, \ldots\}$ (thus $f(x)$ is the infimum of those numbers obtained from x by shifting the "binary point" 2^n places and looking of what remains to the right of it). If $x = \sum_{n\geq 1} a_n/2^n$ then $1 - x = \sum_{n\geq 1} b_n/2^n$, where $b_n = 0$ if $a_n = 1$, and $b_n = 1$ if $a_n = 0$. Define $F(x) = \min\{f(x), f(1-x)\}$. The value $F(x)$ does not depend on which expansion one uses for x, if two expansions are possible (see [Br2], p. 12). Then we have:

(i) $F \in \mathcal{DB}_1$ *on* $[0,1]$*;*

(ii) $F(x) = 0$ *a.e. on* $[0,1]$*, but F is not identically zero.*

Remark 6.21.1 *The utility of this example follows by Remark 2.5.3.*

6.22 A function $F \in \mathcal{D}$, $F \in [CG]$, $F \in [VBG]$, $F \notin VB^*G$, $F \notin C$ (Bruckner)

Let C be the Cantor ternary set (see Section 6.1), and let (a_i, b_i), $i = \overline{1,\infty}$ be the intervals contiguous to C, $c_i = (a_i + b_i)/2$. Let $\alpha_i \in \mathbb{R}$, $i = \overline{1,\infty}$.

Example 6.22.1 *Let* $F : [0,1] \mapsto \mathbb{R}$ *be a function defined as follows:* $F(x) = 0$ *if* $x \in C$*;* $F(x) = \alpha$ *if* $x \in \{c_1, c_2, \ldots\}$*;*$F(x)$ *is linear on each* $[a_i, c_i]$ *and* $[c_i, b_i]$ *(see [Br2], p. 12). Then we have:*

(i) $F \in \mathcal{D}$ *on* $[0,1]$*;*

(ii) $F \in [CG]$ *on* $[0,1]$*;*

(iii) $F \in C$ *on* $[0,1]$ *if and only if* $\lim_{i\to\infty} F(c_i) = 0$*.*

Remark 6.22.1 *The utility of this example follows by Proposition 2.3.5, (v).*

Example 6.22.2 *Let $F : [0,1] \mapsto \mathbb{R}$ be a function defined as follows: $F(x) = 0$ if $x \in C \setminus \{1\}$; $F(x) = 1$ if $x \in \{1, c_1, c_2, \ldots\}$; $F(x)$ is linear on each $[a_i, c_i]$ and $[c_i, b_i]$. Then we have:*

(i) $F \in \mathcal{D} \cap [\mathcal{C}G]$ *on* $[0,1]$, *hence* $F \in \mathcal{B}_1$ *on* $[0,1]$;

(ii) $F \in VB$ *on* C;

(iii) $F_{/C} \notin \mathcal{C}$ *at* $x = 1$;

(iv) $F \in [VBG]$ *on* $[0,1]$;

(v) $F \notin VB^*G$ *on* $[0,1]$.

Remark 6.22.2 *The utility of this example follows by Theorem 2.10.3, (i), (vi); Remark 2.10.1.*

6.23 A function $F \in AC^*$, $F \notin VB^*$

Let $P = [-1, -1/2] \cup [1/2, 1]$, and let $F : [-1,1] \mapsto \mathbb{R}$ be a function defined as follows: $F(0) = 0$; $F(x) = 1/x$ if $x \neq 0$. Then we have:

(i) $F \in AC^*$ *on* P;

(ii) $F \notin VB^*$ *on* P.

Remark 6.23.1 *The utility of this example follows by Remark 2.12.2.*

6.24 A function $F \in \mathcal{C}$, $F \in T_1$, $F \in VBG$, $F \notin VB^*G$

Let C be the Cantor ternary set (see Section 6.1), and let (a_i, b_i), $i = \overline{1,\infty}$ be the intervals contiguous to C, $c_i = (a_i + b_i)/2$. Let $F : [0,1] \mapsto \mathbb{R}$ be a function defined as follows: $F(x) = 0$ if $x \in C$; $F(x) = 1/2^{k-1}$ if $x = c_i$ and (a_i, b_i) is from the step k; $F(x)$ is linear on each $[a_i, c_i]$ and $[c_i, b_i]$. Then we have:

(i) $F \in \mathcal{C}$ *on* $[0,1]$;

(ii) $F \in T_1$ *on* $[0,1]$;

(iii) $F \in VBG$ *on* $[0,1]$;

(iv) $F \notin VB^*G$ *on* $[0,1]$.

Remark 6.24.1 *The utility of this example follows by Proposition 2.8.1, (iii); Theorem 2.18.9, (iv).*

6.25 A function $F \in [bAC^*G] \cap VB^*G \cap N^{-\infty}$, $F \notin$ lower internal

Let $F : [0,1] \mapsto \mathbb{R}$ be a function defined as follows: $F(1/2) = -1$, $F(x) = 0$ if $x \neq 1/2$. Then we have:

(i) $F \in [bAC^*G]$ *on* $[0,1]$*;*

(ii) $F \in VB^*G$ *on* $[0,1]$*;*

(iii) $F \in N^{-\infty}$ *on* $[0,1]$*;*

(iv) $F \notin$ *lower internal on* $[0,1]$.

Remark 6.25.1 *The utility of this example follows by Corollary 2.22.1, (ii).*

6.26 A function $F \in \mathcal{C} \cap (S) \cap LG$, $F \notin AC^*G$, $F'(x)$ does not exist on a set of positive measure, $F(x) + x \in LG$, $F(x) + x \notin T_1$

We shall construct a perfect set of Cantor type as follows: at the first step we exclude from $[0,1]$ the open centered interval of length $1/4$; at the second step we exclude from each remained closed interval from the first step, the open centered interval of length $1/4^2$; at the nth step we exclude from each closed interval from the step $(n-1)$, the open centered interval of length $1/4^n$ (there are 2^{n-1} such centered intervals). It follows that there are 2^n remained closed intervals at this step, and each has the length

$$\frac{1 - (\frac{1}{2^2} + \frac{1}{2^3} + \ldots + \frac{1}{2^n})}{2^n} \in \left(\frac{1}{2^{n+1}}, \frac{1}{2^n}\right).$$

Continuing infinitely many times, P will be the set left after all exclusions. Then P is a perfect nowhere dense set with $\mid P \mid = 1/2$. Let (a_i, b_i), $i = \overline{1,\infty}$ be the intervals contiguous to P and let $c_i = (a_i + b_i)/2$. Let $F : [0,1] \mapsto \mathbb{R}$ be a function defined as follows: $F(x) = 0$ if $x \in P$; $F(x) = 1/2^{n-1}$ if $x = c_i$ and (a_i, b_i) is from the step n; F is linear on each $[a_i, c_i]$ and $[c_i, b_i]$. Then we have:

(i) $F \in \mathcal{C}$ *on* $[0,1]$*;*

(ii) $F \in (S) \subset T_1$ *on* $[0,1]$*;*

(iii) $F \in LG \subset ACG$ *on* $[0,1]$*;*

(iv) $F \notin AC^*G$ *on* $[0,1]$*;*

(v) $F'(x)$ *does not exist whenever* $x \in P$*;*

(vi) $F(x) + x \in LG \subset (N) \subset T_2$ *on* $[0,1]$*;*

(vii) $F(x) + x \notin T_1$ *on* $[0,1]$*, hence* $F(x) + x \notin (S)$ *on* $[0,1]$*; but* $F(x) + x \in ACG$ *on* $[0,1]$.

Proof. (v) Let $x_o \in P$. Then x_o belongs to some remained closed interval J from the step n. Let I be an excluded interval from the step n such that $\overline{I}$ joins J. Let c be the middle point of I. Then $F(c) = 1/2^{n-1}$, $F(x_o) = 0$, $| c - x_o | < | I | + | J | < 1/2^n + 1/4^n$. It follows that $(F(x) - F(x_o))/ | c - x_o | > (1/2^{n-1})/(1/2^n + 1/4^n) \to 2$, when $n \to \infty$. But $\lim_{x \to x_o, x \in P}(F(x) - F(x_o))/(x - x_o) = 0$, hence $F'(x_o)$ does not exist.

(vii) Let $G(x) = F(x) + x$. Then $G'(x)$ does not exist whenever $x \in P$, and $G(P) = P$. By Theorem 2.18.1, $G \notin T_1$ on $[0,1]$. That $G \notin (S)$ follows by Theorem 2.18.8.

The other statements follow similarly.

Remark 6.26.1 *(i) In 1930 Nina Bary constructed first an example with the properties (v) and (vi) (see [Ba], pp. 212 - 214; [S3], p. 224).*

(ii) The utility of this example follows by Theorem 2.18.9, (iv), (vi), (xv); Corollary 2.33.1, (xiv).

6.27 A function $F \in (S) \cap C$ such that the sum of F and any linear nonconstant function does not satisfy (N) (Mazurkiewicz)

Let $n_1 \geq 2$ be a positive integer, $n_k = 2^{k-1}(4n_1 + 5) \cdot \ldots \cdot (4n_{k-1} + 5)$ for $k \geq 2$, and $m_k = 2(n_k + 1)(2n_k + 1) + n_k$ for $k \geq 1$. Let $a_k = 2/((2m_1 + 1) \cdot \ldots \cdot (2m_k + 1))$ and $b_k = 2/((2n_1 + 1) \cdot \ldots \cdot (2n_k + 1))$ for $k \geq 1$. We observe that $a_j/2 = \sum_{i=j+1}^{\infty} a_i m_i$ and $b_j/2 = \sum_{i=j+1}^{\infty} b_i n_i$. Let $P = \{x : x = \sum_{i=1}^{\infty} a_i p_i,\ p_i \in \{0, 1, \ldots, m_i\}\}$ and $Q = \{y : y = \sum_{i=1}^{\infty} b_i q_i,\ q_i \in \{0, 1, \ldots, n_i\}\}$. Each $x \in P$ is uniquely represented by $\sum_{i=1}^{\infty} a_i p_i(x)$, and each $y \in Q$ is uniquely represented by $\sum_{i=1}^{\infty} b_i q_i(x)$. Then P and Q are perfect sets, $| P | = \lim_{j \to \infty}(a_j/2) \cdot (m_1 + 1) \cdot \ldots \cdot (m_j + 1) = 0$ and $| Q | = \lim_{j \to \infty}(b_j/2) \cdot (n_1 + 1) \cdot \ldots \cdot (n_j + 1) = 0$. We denote by $r(x; y)$ the remainder of the quotient x/y. Let $F : P \mapsto Q$, $F(x) = F(\sum_{i=1}^{\infty} a_i p_i(x)) = \sum_{i=1}^{\infty} b_i r(p_i(x); n_i + 1)$. Extending F linearly on the closure of each interval contiguous to P, we have F defined and continuous on $[0,1]$ (see Figure 6.2). Clearly $F(P) = Q$. For any real number $t \neq 0$, let $G_t : [0,1] \mapsto \mathbb{R}$, $G_t(x) = F(x) + tx$. Then we have:

(i) $F \in (S) \subset (N) \subset T_2$ *on* $[0,1]$;

(ii) $G_t \notin (N)$ *on* $[0,1]$;

(iii) $F \notin \mathcal{E}$ *on* $[0,1]$.

Proof. (i) Since $F(P) = Q$ it follows that $| F(P) | = | Q | = 0$, hence $F \in (N)$ on $[0,1]$. Since F is linear on each interval contiguous to P, it follows that the set $F(\{x : F'(x)$ does not exist (finite or infinite)$\})$ has measure zero. By Theorem 2.18.1, $F \in T_1$ on $[0,1]$, and by Theorem 2.18.8, $F \in (S) \subset (N) \subset T_2$ on $[0,1]$.

(ii) Let's define $I_{i_1 \ldots i_k} = [c_{i_1 \ldots i_k}, d_{i_1 \ldots i_k}]$, where $c_{i_1 \ldots i_k} = \sum_{p=1}^{k} i_p a_p$ and $d_{i_1 \ldots i_k} = c_{i_1 \ldots i_k} + \sum_{p=k+1}^{\infty} m_p a_p$, $i_p \in \{0, 1, \ldots, m_j\}$, $j = \overline{1, k}$. If $t > 0$ then there exists a positive integer k such that $t \in [1/2^k, 2^k]$. Since $2^k(2n_k + 1) < b_k/a_k < (1/2^k)(m_k - n_k - 1)$, we have $G_t(I_{i_1 \ldots i_{k-1}, (n_k+1)j+i}) \cap (I_{i_1 \ldots i_{k-1}, (n_k+1)(j+1)+i}) \neq \emptyset$, $i = \overline{0, n_k}$, $j = \overline{0, 4n_k + 1}$ and $G_t(c_{i_1 \ldots i_{k-1}, i+1}) < G_t(c_{i_1 \ldots i_{k-1}, m_k - n_k + i})$, $i = \overline{0, n_k - 1}$. Then $G_t(I_{i_1 \ldots i_{k-1}} \cap P) =$

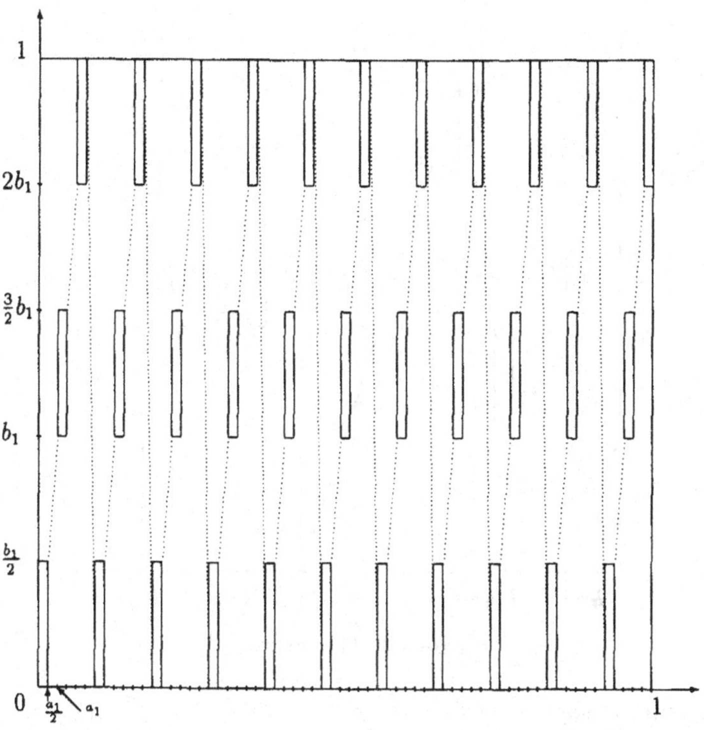

Figure 6.2: The function F

$G_t(I_{i_1\dots i_{k-1}})$. Since $\mid G_t(I_{i_1\dots i_{k-1}}) \mid > 0$, it follows that $G_t \notin (N)$ on $[0,1]$.
If $t < 0$ then there exists a positive integer k such that $-t \in [1/2^k, 2^k]$. Since $2^k(2n_k+3) < b_k/a_k < (1/2^k)(m_k - n_k + 1)$, it follows that $G_t(I_{i_1\dots i_{k-1},(n_k+1)j+i}) \cap G_t(I_{i_1\dots i_{k-1},(n_k+1)(j+1)+i}) \neq \emptyset$, $i = \overline{0,n_k}$, $j = \overline{0,4n_k+1}$ and we have $G_t(c_{i_1\dots i_{k-1},i}) > G_t(c_{i_1\dots i_{k-1},m_k-n_k+i+1})$, $i = \overline{0,n_k-1}$. Then $G_t(I_{i_1\dots i_{k-1}} \cap P) = G_t(I_{i_1\dots i_{k-1}})$, hence $G_t \notin (N)$ on $[0,1]$.

(iii) Suppose that $F \in \mathcal{E}$ on $[0,1]$. By Corollary 2.37.1, (ix), (xvii), $G_t \in \mathcal{E} \subset (N)$, a contradiction.

Remark 6.27.1 *The utility of this example follows by Remark 2.18.7, (ii), (iii); Corollary 2.37.1, (xviii).*

6.28 A function $F \in (M)$, $F \notin T_2$

Let C be the Cantor ternary set (see Section 6.1), and let (a_i, b_i), $i = \overline{1,\infty}$ be the intervals contiguous to C. Let $F : C \mapsto \mathbb{R}$ be a function defined as follows: $F(x) = F(\sum_{i=1}^{\infty} c_i(x)/3^i) = \sum_{i=1}^{\infty}(c - 3i(x)/2^{2i} + c_{3i-1}(x)/2^{2i+1})$. Extending F linearly on each $[a_i, b_i]$ we have F defined and continuous on $[0,1]$ (see Figure 6.3). Then we have:

(i) $F \notin T_2$ on $[0,1]$, hence $F \notin (N)$ on $[0,1]$;

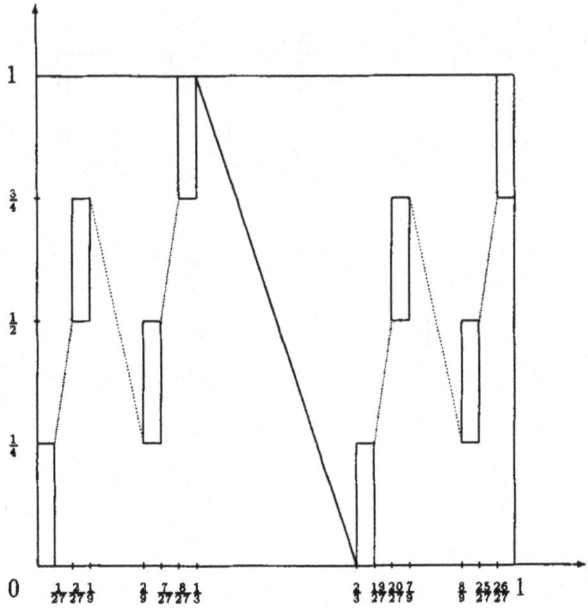

Figure 6.3: The function F

(ii) $F \in (M)$ *on* $[0,1]$.

Proof. (i) Let $y \in [0,1]$. We consider the infinite expansion in base 2 for y, $y = \sum_{i=1}^{\infty} y_i/2^i$. Then $F^{-1}(y) \cap C = \{x \in C : c_{3i}(x) = 2y_{2i}$ an $c_{3i-1}(x) = 2y_{2i-1}\}$ is a perfect set. It follows that $F^{-1}(y) \cap C$ is uncountable, hence $F \notin T_2$ on $[0,1]$. By Theorem 2.18.7, $G \notin (N)$ on $[0,1]$.

(ii) On each remained closed interval I from the step $3i$, $B(F;I)$ is contained in a rectangle of length $|I| = 1/3^{3i}$ and height $1/4^i$. In fact $B(F; I \cap C)$ is contained in eight rectangles, each of length $|I| = 1/3^{3i+3}$ and height $1/4^{i+1}$ (the projection of each such rectangle is a remained closed interval from the step $3i+3$ (see Figure 6.3). Applying Remark 2.23.2, (i), we show that $F \in \overline{M}$ on C. Let E be a closed subset of C such that F is $\underline{L}$ with a positive constant on E. Then $B(F;E)$ is contained in at most three of the eight rectangles from the step $3i+3$. It follows that $|F(E)| < (3/4)^{i+1} \to 0$. Hence $F \in (N)$ on E and $F \in \overline{M}$ on C. Similarly $F \in \underline{M}$ on C, hence $F \in (M)$ on C. It follows that $F \in (M)$ on $[0,1]$.

Remark 6.28.1 *(i) In 1979 Foran constructed first an example with the above properties (see [F7]).*

(ii) The utility of this example follows by Theorem 2.23.2, (ii).

6.29 Functions concerning conditions (M), AC, T_1, T_2, (S), (N), L, L_2G, VBG, $S\mathcal{F}$, quasi- derivable

Let C be the Cantor ternary set (see Section 6.1). Let $p_k = 1/2^{k+2} + 3/4^{k+1}$, $k = \overline{1,\infty}$. Then $\{p_k\}_k$ is a strictly decreasing sequence and $2p_{k+1} < p_k < 4p_{k+1}$. Let $P = \{x \in [0,1] : x = \sum_{i=1}^{\infty} x_k p_k$, where $c_k \in \{0,2\}\}$. Then P is a perfect nowhere dense set, and each $x \in P$ is uniquely represented by $\sum_{k=1}^{\infty} c_k(x)p_k$. The set P can also be obtained by a process similarly to the Cantor ternary process: at the first step we exclude the open centered interval $(3/8, 5/8)$ of length $1/4$; at the second step we exclude of each remained closed interval from the first step the centered interval of length $1/4^2$, etc. It follows that for each remained closed interval from the step i (we have 2^i such intervals), there exist $c_1, c_2, \ldots, c_i \in \{0,2\}$ such that this interval can be written as $I'_{c_1\ldots c_i} = [\sum_{k=1}^{i} c_k p_k, \sum_{k=1}^{i} c_k p_k + \sum_{k=i+1}^{\infty} 2p_k] = [\sum_{k=1}^{i} c_k p_k, \sum_{k=1}^{i} c_k p_k + 1/2^{i+1} + 1/(2\cdot 4^i)]$. We denote by $J'_{c_1\ldots c_{i-1}} = (\sum_{k=1}^{i-1} c_k p_k + \sum_{k=i+1}^{\infty} 2p_k, \sum_{k=1}^{i-1} c_k p_k + 2p_i) = (\sum_{k=1}^{i-1} c_k p_k + 1/2^{i+1} + 1/(2\cdot 4^i), \sum_{k=1}^{i-1} c_k p_k + 1/2^{i+1} + 3/(2\cdot 4^i))$ the excluded open intervals from the step i, $i \geq 2$. It follows that $P \subset \bigcup_{(c_1,\ldots,c_i)} I'_{1\ldots c_i}$, hence $|P| = \lim_{i\to\infty} 2^i \cdot (1/2^{i+1} + 1/(2\cdot 4^i)) = 1/2$.
We define the following functions:

$I : [0,1] \mapsto [0,1],\quad I(x) = x;$

$$F_{CP} : C \mapsto P,\quad F_{CP}(x) = F_{CP}\left(\sum_{k=1}^{\infty} \frac{c_k(x)}{3^k}\right) = \sum_{k=1}^{\infty} c_k(x)p_k;$$

$$F_{PC} : P \mapsto C,\quad F_{PC}(x) = F_{PC}\left(\sum_{k=1}^{\infty} c_k(x)p_k\right) = \sum_{k=1}^{\infty} \frac{c_k(x)}{3^k}$$

(see Figure 6.4);

$$G_{CC} : C \mapsto C,\quad G_{CC}(x) = G_{CC}\left(\sum_{k=1}^{\infty} \frac{c_k(x)}{3^k}\right) = \sum_{k=1}^{\infty} \left(\frac{c_{2k-1}(x)}{3^{2k}} + \frac{c_{2k}(x)}{3^{2k-1}}\right)$$

(see Figure 6.5);

$$G_{CP} : C \mapsto P,\quad G_{CP}(x) = G_{CP}\left(\sum_{k=1}^{\infty} \frac{c_k(x)}{3^k}\right) = \sum_{k=1}^{\infty} (c_{2k-1}(x)p_{2k} + c_{2k}(x)p_{2k-1});$$

$$G_{PC} : P \mapsto C,\quad G_{PC}(x) = G_{PC}(\sum_{k=1}^{\infty} c_k(x)p_k) = \sum_{k=1}^{\infty} \left(\frac{c_{2k-1}(x)}{3^{2k}} + \frac{c_{2k}(x)}{3^{2k-1}}\right);$$

$$H_{CC} : C \mapsto C,\quad H_{CC}(x) = H_{CC}\left(\sum_{k=1}^{\infty} \frac{c_k(x)}{3^k}\right) = \sum_{k=1}^{\infty} \frac{c_{2k-1}(x)}{3^{2k-1}}$$

(see Figure 6.6, where $\eta = \sum_{k=1}^{\infty} 2/3^{2k-1}$ and $\gamma = \sum_{k=2}^{\infty} 2/3^{2k-1}$);

$$H_{CP} : C \mapsto P,\quad H_{CP}(x) = H_{CP}\left(\sum_{k=1}^{\infty} \frac{c_k(x)}{3^k}\right) = \sum_{k=1}^{\infty} c_{2k-1}(x)p_{2k-1};$$

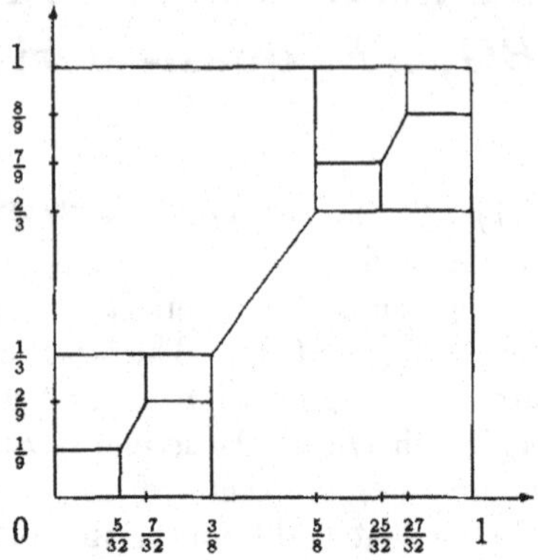

Figure 6.4: The function F_{PC}

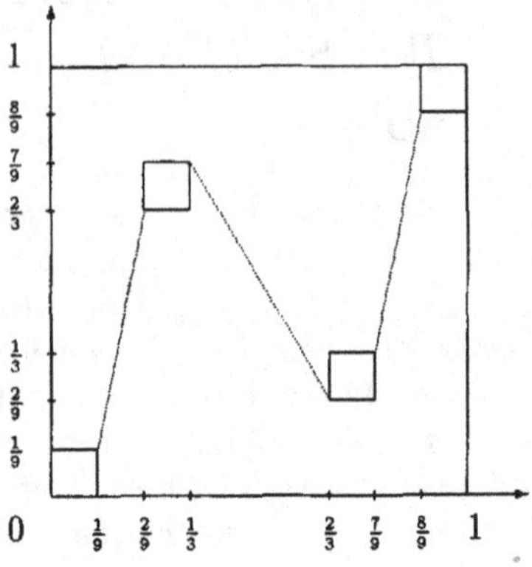

Figure 6.5: The function G_{CC}

$$H_{PC} : P \mapsto C, \quad H_{PC}(x) = H_{PC}(\sum_{k=1}^{\infty} c_k(x)p_k) = \sum_{k=1}^{\infty} \frac{c_{2k-1}(x)}{3^{2k-1}};$$

$$H_{PP} : P \mapsto P, \quad H_{PP}(x) = H_{PP}(\sum_{k=1}^{\infty} c_k(x)p_k) = \sum_{k=1}^{\infty} c_{2k-1}(x)p_{2k-1};$$

$$K_{CC} : C \mapsto C, \quad K_{CC}(x) = K_{CC}\left(\sum_{k=1}^{\infty} \frac{c_k(x)}{3^k}\right) = \sum_{k=1}^{\infty} \frac{c_{2k}(x)}{3^{2k}}$$

(see Figure 6.7, where $\alpha = \sum_{k=1}^{\infty} 2/9^k$ and $\beta = \sum_{k=2}^{\infty} 2/9^k$);

$$K_{CP} : C \mapsto P, \quad K_{CP}(x) = K_{CP}\left(\sum_{k=1}^{\infty} \frac{c_k(x)}{3^k}\right) = \sum_{k=1}^{\infty} c_{2k}(x)p_{2k};$$

$$K_{PC} : P \mapsto C, \quad K_{PC}(x) = K_{PC}(\sum_{k=1}^{\infty} c_k(x)p_k) = \sum_{k=1}^{\infty} \frac{c_{2k}(x)}{3^{2k}};$$

$$K_{PP} : P \mapsto P, \quad K_{PP}(x) = K_{PP}(\sum_{k=1}^{\infty} c_k(x)p_k) = \sum_{k=1}^{\infty} c_{2k}(x)p_{2k}.$$

Extending F_{CP}, G_{CC}, G_{CP}, H_{CC}, H_{CP}, K_{CC}, K_{CP} (resp. F_{PC}, G_{PP}, G_{PC}, H_{PP}, H_{PC}, K_{PP}, K_{PC}) linearly on the closure of each interval contiguous to C (resp. P), we have these functions defined and continuous on $[0, 1]$. It follows that:

(1) $F_{CP} = H_{CP} + K_{CP}$;

(2) $F_{PC} = H_{PC} + K_{PC}$;

(3) $K_{CC} + H_{CC} = I$;

(4) $K_{PP} + H_{PP} = I$;

(5) $G_{CC}(I_{c_1 c_2 \ldots c_{2i}}) = I_{c_2 c_1 c_4 c_3 \ldots c_{2i} c_{2i-1}}$;

(6) $G_{CP}(I_{c_1c_2\ldots c_{2i}}) = I'_{c_2c_1c_4c_3\ldots c_{2i}c_{2i-1}}$;

(7) $G_{PC}(I'_{c_1c_2\ldots c_{2i}}) = I_{c_2c_1c_4c_3\ldots c_{2i}c_{2i-1}}$;

(8) $G_{PP}(I'_{c_1c_2\ldots c_{2i}}) = I'_{c_2c_1c_4c_3\ldots c_{2i}c_{2i-1}}$;

(9) $F_{PC} \circ F_{CP} = I$;

(10) $F_{CP} \circ F_{PC} = I$;

(11) $G_{PP} \circ G_{PP} = I$;

(12) $G_{CC} \circ G_{CC} = I$;

(13) $G_{CP} \circ G_{PC} = I$;

(14) $G_{PC} \circ G_{CP} = I$;

(15) $G_{PP} \circ G_{CP} = F_{CP}$;

(16) $G_{CP} \circ G_{CC} = F_{CP}$;

(17) $G_{CC} \circ G_{PC} = F_{PC}$;

(18) $G_{CP} \circ G_{PP} = F_{PC}$

(19) $G_{CC} \circ F_{PC} = G_{PC}$;

(20) $G_{PP} \circ F_{CP} = G_{CP}$;

(21) $H_{CC} \circ F_{PC} = H_{PC}$;

(22) $K_{CC} \circ F_{PC} = K_{PC}$.

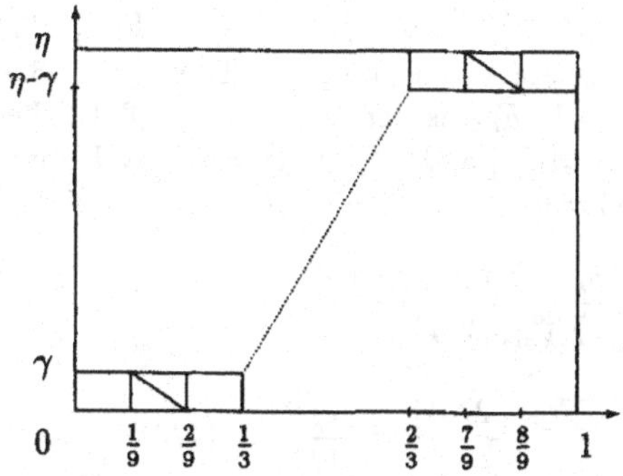

Figure 6.6: The function H_{CC}

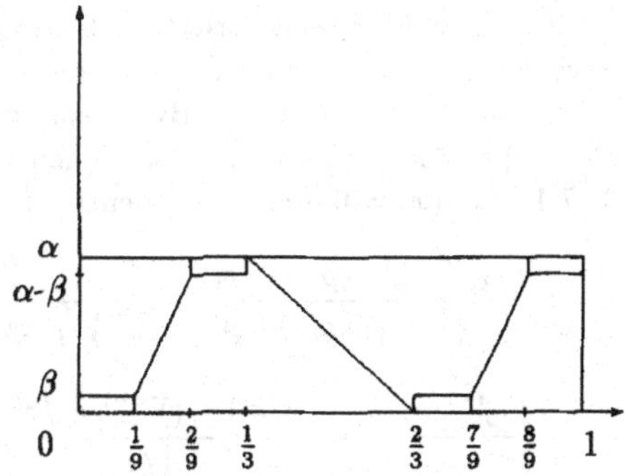

Figure 6.7: The function K_{CC}

Then we have:

(i) F_{CP} *is strictly increasing on* $[0,1]$; $F_{CP} \notin (M)$ *on* $[0,1]$;

(ii) F_{PC} *is strictly increasing on* $[0,1]$; $F_{PC} \in AC$ *on* $[0,1]$; $F'_{PC}(x) = 0$ *a.e.on* P; $F_{PC} \notin L$ *on* $[0,1]$;

(iii) $G_{CC} \in L$ *on* $[0,1]$*;* $(G_{CC})_{/C}$ *is a bijection;* G_{CC} *is monotone on no portion of* C*;*

(iv) $(G_{CP})_{/C}$ *is aa bijection;* $G_{CP} \in T_2$ *on* $[0,1]$*;* $G_{CP} \notin (N)$ *on* $[0,1]$*;* G_{CP} *is monotone on no portion of* C*;* $G_{CP} \in (M)$ *on* $[0,1]$*;* $G_{CP} \notin T_1$ *on* $[0,1]$*;*

(v) $G_{PC} \notin L$ *on* $[0,1]$*;* $G_{PC} \in AC$ *on* $[0,1]$*;* $(G_{PC})_{/P}$ *is a bijection;* G_{PC} *is monotone on no portion of* P*;* $G'_{PC}(x) = 0$ *a.e. on* P*;*

(vi) $G_{PP} \in L_2G$ *on* $[0,1]$*;* $G_{PP} \in (N)$ *on* $[0,1]$*;* $(G_{PP})_{/P}$ *is a bijection;* G_{PP} *is monotone on no portion of* P*;* $G'_{PP}(x)$ *does not exist (finite or infinite) at each* $x \in P$*;* $(G_{PP})'_{ap}(x)$ *does not exist at almost all points* $x \in P$*;* $G_{PP} \notin VBG$ *on* $[0,1]$*;* $G_{PP} \notin T_1$ *on* $[0,1]$*;* $G_{PP} \notin (S)$ *on* $[0,1]$*;* G_{PP} *is quasi-derivable on* $[0,1]$*;* $G_{PP} \in T_1G$ *on* $[0,1]$*;* $G_{PP} \in SG$ *on* $[0,1]$*;*

(vii) K_{CC}, $H_{CC} \in L$ *on* $[0,1]$*;* K_{CC}, H_{CC} *are monotone on no portion of* C*;*

(viii) K_{PC}, $H_{PC} \notin L$ *on* $[0,1]$*;* K_{PC}, $H_{PC} \in AC$ *on* $[0,1]$*;* K_{PC}, H_{PC} *are monotone on no portion of* P*;* $K'_{PC}(x) = H'_{PC}(x) = 0$ *a.e. on* P*;*

(ix) K_{CP}, $H_{CP} \in (S) = (N) \cap T_1$ *on* $[0,1]$*; at least one of the functions* K_{CP} *or* H_{CP} *does not belong to* $S\mathcal{F}$ *on* $[0,1]$*;* K_{CP}, H_{CP} *are monotone on no portion of* C*;*

(x) K_{PP}, $H_{PP} \in L_1G$ *on* $[0,1]$*;* K_{PP}, $H_{PP} \in (S)$ *on* $[0,1]$*; the sets* $\{x \in P : K'_{ap}(x)$ *exists*$\}$ *and* $\{x \in P : H'_{ap}(x)$ *exists*$\}$ *have measure zero;* K_{PP}, $H_{PP} \notin VBG$ *on* $[0,1]$*;* K_{PP}, H_{PP} *are monotone on no portion of* P.

Proof. (i) Clearly F_{CP} is increasing on $[0,1]$, hence $F_{CP} \in VB$ on $[0,1]$. We have $|C| = 0$, $|P| = 1/2$, $F_{CP}(C) = P$, hence $F_{CP} \notin (N)$ on $[0,1]$. By Theorem 2.18.9, (vii), $F_{CP} \notin AC$ on $[0,1]$. By the definition of (M) it follows that $F_{CP} \notin (M)$ on $[0,1]$.

(ii) Clearly F_{PC} is strictly increasing on $[0,1]$, hence $F_{PC} \in VB$ on $[0,1]$. We have $F_{PC}(P) = C$ and $|C| = 0$, hence $F_{PC} \in (N)$ on $[0,1]$. By Theorem 2.18.9, (vii), $F_{PC} \in AC$ on $[0,1]$. By Theorem 2.14.1, F_{PC} is derivable *a.e.* on $[0,1]$. Let $P_o = \{x \in P : F'_{PC}(x) \text{ exists}\}$. Then $|P_o| = |P|$ and $|F_{PC}(P_o)| = 0$. By Lemma 2.17.1. $F'_{PC}(x) = 0$ *a.e.* on P_o, hence $F'_{PC}(x) = 0$ *a.e.* on P. We have

$$\frac{F_{PC}(\sum_{k=1}^{i-1} c_k p_k + 2p_i) - F_{PC}(\sum_{k=1}^{i-1} c_k p_k + \sum_{k=i+1}^{\infty} 2p_k)}{(\sum_{k=1}^{i-1} c_k p_k + 2p_i) - (\sum_{k=1}^{i-1} c_k p_k + \sum_{k=i+1}^{\infty} 2p_k)} =$$

$$\frac{(\sum_{k=1}^{i-1} c_k/3^k + 2/3^i) - (\sum_{k=1}^{i-1} c_k/3^k + \sum_{k=i+1}^{\infty} 2/3^k)}{1/4^i} = \frac{1/3^i}{1/4^i} = (4/3)^i \to \infty,$$

hence $F_{PC} \notin L$ on $[0,1]$.

(iii) Clearly $G_{CC}(C) = C$. Let $x \in [0,1] \setminus C$. Then either $x \in J = (1/3, 2/3)$, or there exist $c_1, c_2, \ldots, c_{i-1} \in \{0,2\}$, $i \geq 2$, such that $x \in J_{c_1 c_2 \ldots c_{i-1}}$. If $x \in J$ then $(G_{CC}(2/3) - G_{CC}(1/3))/(2/3 - 1/3) = -5/3$, hence

(23) $\quad G'_{CC}(x) = -5/3.$

If $x \in J_{c_1c_2\ldots c_{i-1}}$, $i \geq 2$ then we have two situations: 1) $i = 2j+1$, $j = \overline{1,\infty}$. Then

$$\frac{G_{CC}(\sum_{k=1}^{2j} c_k/3^k + 2/3^{2j+1}) - G_{CC}(\sum_{k=1}^{2j} c_k/3^k + \sum_{k=2j+2}^{\infty} 2/3^k)}{(\sum_{k=1}^{2j} c_k/3^k + 2/3^{2j+1}) - (\sum_{k=1}^{2j} c_k/3^k + \sum_{k=2j+2}^{\infty} 2/3^k)} =$$

$$\frac{2/3^{2j+2} - (2/3^{2j+1} + \sum_{k=2j+3}^{\infty} 2/3^k)}{1/3^{2j+1}} = -5/3,$$

hence

(24) $\quad G'_{CC}(x) = -5/3;$

2) $i = 2j$, $j = \overline{1,\infty}$. Then

$$\frac{G_{CC}(\sum_{k=1}^{2j-1} c_k/3^k + 2/3^{2j}) - G_{CC}(\sum_{k=1}^{2j-1} c_k/3^k + \sum_{k=2j+1}^{\infty} 2/3^k)}{(\sum_{k=1}^{2j-1} c_k/3^k + 2/3^{2j}) - (\sum_{k=1}^{2j-1} c_k/3^k + \sum_{k=2j+1}^{\infty} 2/3^k)} =$$

$$\frac{(2/3^{2j-1} + c_{2j-1}/3^{2j}) - (c_{2j-1}/3^{2j} + \sum_{k=2j+1}^{\infty} 2/3^k)}{1/3^{2j}} = 5,$$

hence

(25) $\quad G'_{CC}(x) = 5.$

By (23), (24) and (25), $| G'_{CC}(x) | \leq 5$ on $[0,1] \setminus C$. By Lemma 2.13.1, $G_{CC} \in L$ with the constant 5 on $[0,1]$. Since $G_{CC}(C) = C$ it follows that $(G_{CC})_{/C}$ is a surjection. Let $y \in C$, $y = \sum_{k=1}^{\infty} c_k/3^k$. Then $(G_{CC})^{-1}(y) = \{\sum_{k=1}^{\infty}(c_{2k-1}/3^{2k} + c_{2k}/3^{2k-1})\}$, hence $(G_{CC})_{/C}$ is a bijection. Let $(u,v) \cap C$ be a portion of C. Then there exist $c_1, c_2, \ldots, c_{2i} \in \{0,2\}$ for some i, such that $I_{c_1c_2\ldots c_{2i}} \subset (u,v)$. Let $x_1 = \sum_{k=1}^{2i} c_k/3^k$, $x_2 = x_1 + 2/3^{2i+2}$, $x_3 = x_1 + 2/3^{2i+1}$ Then $x_1, x_2, x_3 \in I_{c_1c_2\ldots c_{2i}}$, $x_1 < x_2 < x_3$ and $G_{CC}(x_1) < G_{CC}(x_2) > G_{CC}(x_3)$, hence G_{CC} is not monotone on $(u,v) \cap C$.

(iv) Clearly $G_{CP}(C) = P$, hence $(G_{CP})_{/C}$ is a surjection. Let $y \in P$. Then $y = \sum_{k=1}^{\infty} c_k p_k$ and $(G_{CP})^{-1}(y) = \{\sum_{k=1}^{\infty}(c_{2k-1}/3^{2k} + c_{2k}/3^{2k-1})\}$, hence $(G_{CP})_{/C}$ is a bijection. It follows that $G_{CP} \in T_2$ on $[0,1]$. Since $| G_{CP}(C) |=| P |= 1/2$, $G_{CP} \notin (N)$ on $[0,1]$. Similarly to (iii), we obtain that G_{CP} is monotone on no portion of C.

Applying Theorem 2.23.1, (ii), (vi) we show that $G_{CP} \in \overline{M}$ on C. Let $E \subset C$ such that G_{CP} is $\underline{L}$ with a positive constant on $\overline{E}$. Let $A_i = \{(c_1, c_2, \ldots, c_{2i}) : \overline{E} \cap I_{c_1c_2\ldots c_{2i}} \neq \emptyset\}$. By (6) it follows that A_i contains at most 3^i elements and $| I'_{c_2c_1\ldots c_{2i}c_{2i-1}} |= 1/2^{2i+1} + 1/(2 \cdot 4^{2i}) < 1/4^i$. Then $G_{CP}(E) |< (3/4)^i \to 0$, hence $G_{CP} \in (N)$ on $\overline{E}$, so $G_{CP} \in \overline{M}$ on C. Similarly $G_{CP} \in \underline{M}$ on C, hence $G_{CP} \in (M)$ on C. It follows that $G_{CP} \in (M)$ on $[0,1]$.

Suppose on the contrary that $G_{CP} \in T_1$ on $[0,1]$. By Theorem 2.23.2, (iii), $G_{CP} \in (N)$ on $[0,1]$, a contradiction.

(v) Let $i = 2j+1$. Then for $J_{c_1c_2\ldots c_{i-1}}$ we have

$$\frac{G_{PC}(2p_{2j+1} + \sum_{k=1}^{2j} c_k p_k) - G_{PC}(\sum_{k=1}^{2j} c_k p_k + \sum_{k=2j+2}^{\infty} 2p_k)}{(2p_{2j+1} + \sum_{k=1}^{2j} c_k p_k) - (\sum_{k=1}^{2j} c_k p_k + \sum_{k=2j+2}^{\infty} 2p_k)} =$$

$$\frac{2/3^{2j+1} - (2/3^{2j+1} + \sum_{k=2j+3}^{\infty} 2/3^k)}{1/4^{2j+1}} = \frac{1/3^{2j+2} - 2/3^{2j+1}}{1/4^{2j+1}} = \frac{-5/3^{2j+2}}{1/4^{2j+1}} \to -\infty,$$

hence $G_{PC} \notin L$ on $[0,1]$. By (iii), $G_{CC} \in L$ and by (ii), $F_{PC} \in AC$ on $[0,1]$. By (19) and Theorem 2.32.1, (xvii), $G_{PC} \in AC$ on $[0,1]$. Since F_{PC} is strictly increasing on $[0,1]$ and $(G_{CC})_{/C}$ is a bijection, by (19) it follows that $(G_{PC})_{/P}$ is a bijection too. Similarly G_{PC} is monotone on no portion of P. That $G'_{PC}(x) = 0$ *a.e.* on P follows similarly to (ii).

(vi) We show that $G_{PP} \in L_2$ on P. Let $a, b \in P$, $a < b$. We have two situations: 1) $(3/8, 5/8) \subset [a,b]$. Then $\mathcal{O}(G_{PP}; P \cap [a,b]) \leq 1$, hence

$$(26) \quad \mathcal{O}^2(G_{PP}; P \cap [a,b]) \leq (5/8 - 3/8) \cdot 4 < 4(b-a).$$

2) $(3/8, 5/8) \not\subset [a,b]$. Let s be the largest positive integer such that $[a,b]$ is contained in some interval $I'_{c_1c_2\ldots c_s} = [\sum_{k=1}^{s} c_kp_k, \sum_{k=1}^{s} c_kp_k + \sum_{k=s+1}^{\infty} 2p_k]$. Let $(a_1, b_1) = (\sum_{k=1}^{s} c_kp_k + \sum_{k=s+2}^{\infty} 2p_k, \sum_{k=1}^{s} c_kp_k + 2p_{s+1})$. Let $x \in [a, a_1] \cap P$, $y \in [b_1, b] \cap P$. Then $x = \sum_{k=1}^{s} c_kp_k + \sum_{k=s+2}^{\infty} c_k(x)p_k$ and $y = \sum_{k=1}^{s} c_kp_k + 2p_{s+1} + \sum_{k=s+2}^{\infty} c_k(y)p_k$. Let $A_1 = \{k : k \text{ is odd and } k \geq s+2\}$ and $A_2 = \{k : k \text{ is even and } k \geq s+2\}$. It follows that $0 \leq G_{PP}(a_1) - G_{PP}(x) = \sum_{k\in A_1}(2 - c_k(x))p_{k+1} + \sum_{k\in A_2}(2 - c_k(x))p_{k-1} < \sum_{k\in A_1}(2 - c_k(x))p_k + \sum_{k\in A_2}(2 - c_k(x)) \cdot 4p_k < 4 \cdot \sum_{k=s+2}^{\infty}(2 - c_k(x))p_k = 4(a_1 - x) < 4(a_1 - a)$ and $0 \leq G_{PP}(y) - G_{PP}(b_1) = \sum_{k\in A_1} c_k(y)p_{k+1} + \sum_{k\in A_2} c_k(y)p_{k-1} < \sum_{k\in A_1} c_k(y)p_k + \sum_{k\in A_2} c_k(y) \cdot 4p_k < 4 \cdot \sum_{k=s+2}^{\infty} c_k(y)p_k = 4(y - b_1) < 4(b - b_1)$. We have

$$(27) \quad \mathcal{O}^2(G_{PP}; P \cap [a,b]) \leq 4(a_1 - a) + 4(b - b_1) < 4(b-a).$$

By (26) and (27) it follows that $G_{PP} \in L_2$ on P, hence $G_{PP} \in L_2G$ on $[0,1]$. By Corollary 2.32.1, (iv), $G_{PP} \in (N)$ on $[0,1]$. Similarly to (iii) it follows that $(G_{PP})_{/P}$ is a bijection and G_{PP} is monotone on no portion of P.
We show that $G'_{PP}(x)$ does not exist (finite or infinite), whenever $x \in P$. Let $x_o \in P$, $x_o = \sum_{k=1}^{\infty} c_kp_k$. Let n be a positive integer, $n > 1$. Then we have four cases:
(I) $c_{2n-1} = c_{2n} = 0$. Let $x \in I'_{c_1\ldots c_{2n-2}02}$ and $y \in I_{c_1\ldots c_{2n-2}2000}$. Then $x_o < x < y$, $G_{PP}(x) > G_{PP}(x_o)$, $G_{PP}(y) > G_{PP}(x_o)$ and

$$\frac{G_{PP}(x) - G_{PP}(x_o)}{x - x_o} - \frac{G_{PP}(y) - G_{PP}(x_o)}{y - x_o} > \frac{G_{PP}(x) - G_{PP}(y)}{y - x_o} >$$

$$\frac{G_{PP}(\sum_{k=1}^{2n-2} c_kp_k + 2p_{2n}) - G_{PP}(\sum_{k=1}^{2n-2} c_kp_k + 2p_{2n-1} + \sum_{k=2n+3}^{\infty} 2p_k)}{(\sum_{k=1}^{2n-2} c_kp_k + 2p_{2n-1} + \sum_{k=2n+3}^{\infty} 2p_k) - \sum_{k=1}^{2n-2} c_kp_k} =$$

$$\frac{2p_{2n-1} - (2p_{2n} + \sum_{k=2n+3}^{\infty} 2p_k)}{2p_{2n-1} + \sum_{k=2n+3}^{\infty} 2p_k} > \frac{(2p_{2n-1} - \sum_{2n}^{\infty} 2p_k) + 2p_{2n+1} + 2p_{2n+2}}{2p_{2n-1} + 2p_{2n}} >$$

$$\frac{p_{2n+1} + p_{2n+2}}{p_{2n-1} + p_{2n}} > 1/16.$$

But if $x_n = x_o + 2p_{2n-1}$ then

$$\frac{G_{PP}(x_n) - G_{PP}(x_o)}{x_n - x_o} = \frac{2p_{2n}}{2p_{2n-1}} \to 1/2, \quad n \to \infty.$$

(II) $c_{2n-1} = 0$ and $c_{2n} = 2$. Let $x \in I'_{c_1\ldots c_{2n-2}00}$ and $y \in I'_{c_1\ldots c_{2n-2}20}$. Then $x < x_o < y$, $G_{PP}(x) < G_{PP}(x_o)$, $G_{PP}(y) < G_{PP}(x_o)$ and

$$\frac{G_{PP}(x_o) - G_{PP}(x)}{x_o - x} >$$

$$\frac{G_{PP}(\sum_{k=1}^{2n-2} c_k p_k + 2p_{2n}) - G_{PP}(\sum_{k=1}^{2n-2} c_k p_k + \sum_{k=2n+1}^{\infty} 2p_k)}{2p_{2n}} >$$

$$\frac{2p_{2n-1} - \sum_{k=2n+1}^{\infty} 2p_k}{2p_{2n}} > \frac{p_{2n-1} - p_{2n}}{p_{2n}} > 1, \text{ but } \frac{G_{PP}(y) - G_{PP}(x_o)}{y - x_o} < 0.$$

(III) $c_{2n-1} = 2$ and $c_{2n} = 0$. Let $x \in I'_{x_1\ldots c_{2n-2}02}$ and $y \in I'_{c_1\ldots c_{2n-2}22}$. Then $x < x_o < y$, $G_{PP}(x) > G_{PP}(x_o)$, $G_{PP}(y) > G_{PP}(x_o)$ and $(G_{PP}(y) - G_{PP}(x_o))/(y - x_o) > 1$ (this follows similarly to (II)). But $(G_{PP}(x) - G_{PP}(x_o))/(x - x_o) < 0$.
(IV) $c_{2n-1} = c_{2n} = 0$. Let $x \in I'_{c_1\ldots c_{2n-2}20}$ and $y \in I'_{c_1\ldots c_{2n-2}0222}$. Then $y < x < x_o$, $G_{PP}(x_o) > G_{PP}(x)$, $G_{PP}(x_o) > G_{PP}(y)$ and $(G_{PP}(x_o) - G_{PP}(x))/(x - x_o) - (G_{PP}(x_o) - G_{PP}(y))/(x_o - y) > (G_{PP}(y) - G_{PP}(x))/(x_o - y) > 1/16$ (this follows similarly to (I)). But if $x_n = \sum_{k=1}^{2n-2} c_k p_k + 2p_{2k+2} + \sum_{k2n+3}^{\infty} 2p_k$ then $(G_{PP}(x_o) - G_{PP}(x_n))/(x_o - x_n) \to 1/2$, $n \to \infty$.
By (I), (II), (III) and (IV) it follows that $G'_{PP}(x_o)$ does not exist (finite or infinite). We show that $(G_{PP})'_{ap}(x)$ doe not exist (finite) *a.e.* on P. Let $P_o = \{x \in P : d(P; x) = 1\}$. Then $| P_o |=| P |$ (see Theorem 1.6.1). Let $x_o \in P_o$. Then (I), (II), (III) and (IV) are valid for x_o. It follows that $x_o \in I'_{c_1\ldots c_{2n-2}}$. Clearly $| P \cap I'_{c_1\ldots c_{2n-2}} | / | I'_{c_1\ldots c_{2n-2}} |\to 1$. We consider again the four cases:

$$(I) \qquad \frac{| P \cap I'_{c_1\ldots c_{2n-2}02} |}{| I'_{c_1\ldots c_{2n-2}} |} \to \frac{1}{4} \; and \; \frac{| P \cap I'_{c_1\ldots c_{2n-2}20} |}{I'_{c_1\ldots c_{2n-2}} |} \to \frac{1}{16};$$

$$(II) \qquad \frac{| P \cap I'_{c_1\ldots c_{2n-2}00} |}{| I'_{c_1\ldots c_{2n-2}} |} \to \frac{1}{4} \; and \; \frac{| P \cap I'_{c_1\ldots c_{2n-2}20} |}{I'_{c_1\ldots c_{2n-2}} |} \to \frac{1}{4};$$

$$(III) \qquad \frac{| P \cap I'_{c_1\ldots c_{2n-2}02} |}{| I'_{c_1\ldots c_{2n-2}} |} \to \frac{1}{4} \; and \; \frac{| P \cap I'_{c_1\ldots c_{2n-2}20} |}{I'_{c_1\ldots c_{2n-2}} |} \to \frac{1}{4};$$

$$(IV) \qquad \frac{| P \cap I'_{c_1\ldots c_{2n-2}20} |}{| I'_{c_1\ldots c_{2n-2}} |} \to \frac{1}{4} \; and \; \frac{| P \cap I'_{c_1\ldots c_{2n-2}0222} |}{I'_{c_1\ldots c_{2n-2}} |} \to \frac{1}{16};$$

By (I), (II), (III) and (IV) it follows that $(G_{PP})'_{ap}(x_o)$ does not exist (finite).
By Theorem 2.15.3, $G_{PP} \notin VBG$ on $[0,1]$, and by Theorem 2.18.1, $G_{PP} \notin T_1$ on $[0,1]$ (because $G_{PP}(P) = P$). By Theorem 2.18.8, $G_{PP} \notin (S)$ on $[0,1]$. The fact that $G_{PP} \in T_1G$ on $[0,1]$ is evident. By Theorem 2.18.8, $G_{PP} \in (S)$ on P, hence $G_{PP} \in SG$ on $[0,1]$.

(vii) Clearly $K_{CC} = C$. We show that $K_{CC} \in L$ on $[0,1]$. Let $x \in [0,1] \setminus C$. Then either $x \in J = (1/3, 2/3)$, or there exist $c_1, c_2, \ldots, c_{i-1} \in \{0,2\}$, $i \geq 2$, such that $x \in J_{c_1\ldots c_{i-1}}$. If $x \in J$ then $(K_{CC}(2/3) - K_{CC}(1/3))/(2/3 - 1/3) = -3/4$, hence

$$(28) \qquad K'_{CC}(x) = -3/4.$$

If $x \in J_{c_1 \ldots c_{i-1}}$, $i \geq 2$ then we have two situations:
1) $i = 2j+1$, $j = \overline{1, \infty}$. Then

$$\frac{K_{CC}(\sum_{k=1}^{2j} c_k/3^k + \sum_{k=2j+2}^{\infty} 2/3^k) - K_{CC}(\sum_{k=1}^{2j} c_k/3^k + 2/3^{2j+1})}{(\sum_{k=1}^{2j} c_k/3^k + \sum_{k=2j+2}^{\infty} 2/3^k) - (\sum_{k=1}^{2j} c_k/3^k + 2/3^{2j+1})} =$$

$$-\frac{\sum_{k=j+1}^{\infty} 2/3^{2k}}{1/3^{2j+1}} = -\frac{3}{4},$$

hence

(29) $\quad K'_{CC}(x) = -3/4.$

2) $i = 2j$, $j = \overline{1, \infty}$. Then

$$\frac{K_{CC}(\sum_{k=1}^{2j-1} c_k/3^k + 2/3^{2j}) - K_{CC}(\sum_{k=1}^{2j-1} c_k/3^k + \sum_{k=2j+1}^{\infty} 2/3^k)}{1/3^{2j}} =$$

$$\frac{2/3^{2j} - \sum_{k=j+1}^{\infty} 2/3^{2k}}{1/3^{2j}} = \frac{3}{2},$$

hence

(30) $\quad K'_{CC}(x) = 3/2.$

By (28), (29) and (30), $| K'_{CC}(x) | \leq 3/4$ on $[0,1] \setminus C$. By Lemma 2.13.1, $K_{CC} \in L$ with the constant 3/2 on $[0,1]$. By (3) and Theorem 2.32.1, (ix), $H_{CC} \in L$ on $[0,1]$. Similarly to (iii) it follows that K_{CC} and H_{CC} are monotone on no portion of C.

(viii) Similarly to (v) it follows that K_{PC} and H_{PC} does not satisfy L on $[0,1]$. By (21), (22) and Theorem 2.32.1, (xvii), it follows that $K_{PC}, H_{PC} \in AC$ on $[0,1]$. Similarly to (iii), H_{PC} and K_{PC} are monotone on no portion of P. The fact that $K'_{PC}(x) = H'_{PC}(x) = 0$ *a.e.* on P follows similarly to (ii).

(ix) Let $P_1 = H_{CP}(C)$ and $P_2 = K_{CP}(C)$. Then $P_1 = \{y \in P : y = \sum_{k=1}^{\infty} d_k(y) p_{2k-1},\ d_k(y) \in \{0,2\}\}$ and $P_2 = \{y \in P : y = \sum_{k=1}^{\infty} d_k(y) p_{2k},\ d_k(y) \in \{0,2\}\}$. It follows that P_1 and P_2 are symmetric perfect sets and $| P_1 |=| P_2 |= 0$. Hence $K_{CP}, H_{CP} \in (N)$ on $[0,1]$. By Theorem 2.18.1, it follows that $K_{CP}, H_{CP} \in T_1$ on $[0,1]$, and by Theorem 2.18.8, $K_{CP}, H_{CP} \in (S)$ on $[0,1]$. By Corollary 2.34.1, (ii), (vi), we have $\mathcal{SF} + \mathcal{SF} = \mathcal{SF} \subseteq (N)$. But by (1), $K_{CP} + H_{CP} = F_{CP} \notin (N)$. It follows that at least one of the functions K_{CP} or H_{CP} does not belong to $\mathcal{SF}$ on $[0,1]$. Similarly to (iii), K_{CP} and H_{CP} are monotone on no portion of C.

(x) Let $a, b \in P$, $a < b$. Then we have two situations:
1) $(3/8, 5/8) \subset [a,b]$. Then $\mathcal{O}^2(K_{PP}; P \cap [a,b]) \leq \mathcal{O}^2(G_{PP}; P) \leq G_{PP}(1) - G_{PP}(0) = G_{PP}(1)$. It follows that $(\mathcal{O}^2(K_{PP}; P \cap [a,b]))/(b-a) \leq 4 \cdot G_{PP}(1)$.
2) $(3/8, 5/8) \not\subset [a,b]$. Let s, (a_1, b_1), x, y, A_1 and A_2 be defined as in the proof of (vi), 2). Then we have $0 \leq K_{PP}(a_1) - K_{PP}(x) = \sum_{k \in A_2} (2 - c_k(x)) p_k \leq \sum_{k=s+2}^{\infty} (2 - c_k(x)) p_k = a_1 - x \leq a_1 - a$ and $0 \leq K_{PP}(y) - K_{PP}(b_1) = \sum_{k \in A_2} c_k(y) p_k \leq \sum_{k=s+2}^{\infty} c_k(y) p_k = y - b_1 \leq b - b_1$. Hence $\mathcal{O}^2(K_{PP}; P \cap [a,b]) \leq a_1 - a + b - b_1 < b - a$. By 1) and 2) it follows that $K_{PP} \in L_2$ on P, hence $K_{PP} \in L_2G$ on $[0,1]$. Similarly (using A_1 instead of A_2) we obtain that $H_{PP} \in L_2G$ on $[0,1]$.
Let P_1 and P_2 be defined as in the proof of (ix). Then $K_{PP}(P) = P_2$, $H_{PP}(P) = P_1$

and $|P_1|=|P_2|=0$. Hence $K_{PP}, H_{PP} \in (N)$ on $[0,1]$. By Theorem 2.18.1, $K_{PP}, H_{PP} \in T_1$ on $[0,1]$, and by Theorem 2.18.8, $K_{PP}, H_{PP} \in (S)$ on $[0,1]$.
Let $B = \{x \in P : (K_{PP})'_{ab}(x)$ exists (finite)$\}$. By (4) it follows that $\{x \in P : (H_{PP})'_{ap}(x)$ exists (finite)$\} = B$. Suppose on the contrary that $|B| > 0$. By Lemma 2.17.2, since $|K_{PP}(P)| = |H_{PP}(P)| = 0$, it follows that $(K_{PP})'_{ap}(x) = (H_{PP})'_{ap}(x) = 0$ *a.e.* on B. But $(K_{PP})'_{ap}(x) + (H_{PP})'_{ap}(x) = 1$ *a.e.* on B, a contradiction. Hence $|B| = 0$. By Theorem 2.15.3, $K_{PP}, H_{PP} \notin VBG$ on $[0,1]$. Similarly to (iii), K_{PP} and H_{PP} are monotone on no portion of P.

Corollary 6.29.1 *(i)* $G_{PP} \circ G_{CP} \in L_2 G \circ (M) \subseteq AC_2 G \circ (M)$, *but* $G_{PP} \circ G_{CP} \notin (M)$ *on* $[0,1]$;

(ii) $G_{CP} \circ G_{CC} \in (M) \circ L$, *but* $G_{CP} \circ G_{CC} \notin (M)$ *on* $[0,1]$.

Remark 6.29.1 *(i) The functions* F_{PC} *and* G_{PC} *are used in Proposition 2.13.1, (iii).*

(ii) The functions G_{PP}, H_{PP} *and* K_{PP} *are used in Remark 2.15.4.*

(iii) The function G_{PP} *is used in Theorem 2.18.9, (i), (x); Remark 2.34.2, (ii); Remark 3.1.1, (ii), (iii), (v).*

(iv) The function G_{CP} *is used in Theorem 2.23.2, (ii).*

(v) Corollary 6.29.1, (i) is used in Theorem 2.28.1;Corollary 2.33.1, (xiii).

(vi) Corollary 6.29.1, (ii) is used in Theorem 2.23.2, (viii).

6.30 A function $G \in N^\infty$, $F \notin (M)$, $F \notin (+)$

Let C be the Cantor ternary set (see Section 6.1), and let $F : C \mapsto [0,2]$ be a function defined as follows: $F(x) = F(\sum_{i=1}^{\infty} c_i(x)/3^i) = \sum_{i=1}^{\infty} c_{2i}(x)/2^i$. Extending F linearly on the closure of each interval contiguous to C we have F defined and continuous on $[0,1]$ (see Figure 6.8). Then we have:

(i) $F'(x) \leq 0$ *on* $[0,1] \setminus C$, *and* $F'(x)$ *exists for no* $x \in C$;

(ii) $F \in N^\infty$ *on* $[0,1]$;

(iii) $F \notin (M)$ *on* $[0,1]$;

(iv) $F \notin (+)$ *on* $[0,1]$.

Proof. (i) Note that if I is an interval contiguous to C from the step $2k$ then F is constant on I, and if I is an interval contiguous to C from the step $2k+1$ then F is strictly decreasing on I (see Figure 6.8). It follows that F is derivable on $[0,1] \setminus C$ and $F'(x) \leq 0$ on $[0,1] \setminus C$. Let $x_o \in C$ and let $c_i \in \{0,2\}$, $i = \overline{1,\infty}$ such that $x_o = \sum_{i=1}^{\infty} c_i/3^i$. Let $x_n = \sum_{i \neq 2n+1} c_i/3^i + (2 - c_{2n+1})/3^{2n+1}$ and $y_n = \sum_{i \neq 2n} c_i/3^i + (2 - c_{2n})/3^{2n}$, $n = \overline{1,\infty}$. Then $x_n, y_n \in C$, $x_n \to x_o$, $y_n \to x_o$, $|x_n - x_o| = 2/3^{2n+1}$, $y_n - x_o| = 2/9^n$, $F(x_n) = F(x_o)$ and $|F(y_n) - F(x_o)| = 2/2^n$. Hence $|(F(x_n) - F(x_o))/(x_n - x_o)| = 0$ and $\lim_{n \to \infty} |(F(y_n) - F(x_o))/(y_n - y_o)| = +\infty$.

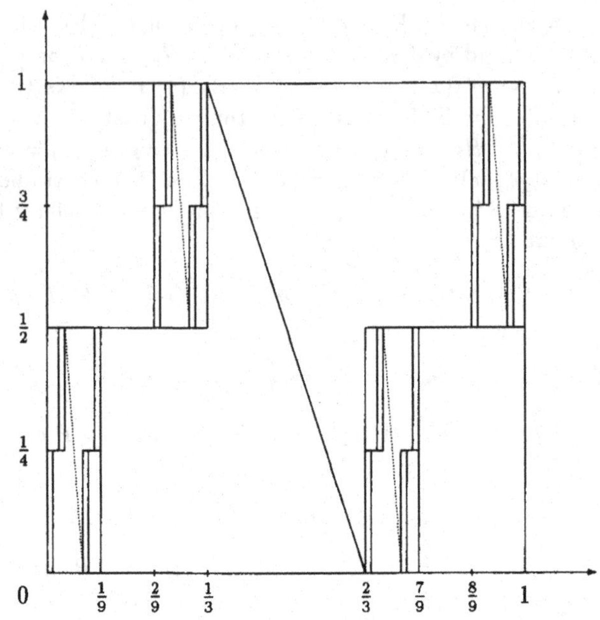

Figure 6.8: The function F

Therefore $F'(x_o)$ does not exist.

(ii) By (i), $E^\infty = \{x : F'(x) = \pm\infty\} = \emptyset$, hence $F \in N^\infty$ on $[0,1]$.

(iii) Let $Q = \{x \in C : c_{2n+1}(x) = 0,\ n = \overline{0,\infty}\}$. Then Q is a perfect subset of C and $F(Q) = [0,2]$. We show that F is increasing on Q, hence $F \in VB$ on Q. Let $x, y \in Q$, $x < y$, and let m be the first positive integer such that $c_{2m}(x) < c_{2m}(y)$. Then $c_i(x) = c_i(y)$, $i = \overline{1, 2m-1}$ and $F(y) - F(x) \geq 2/2^m - \sum_{i=1}^\infty 2/2^{m+i} = 0$. Hence $F(y) \geq F(x)$. By Theorem 2.18.9, (vii), it follows that $F \notin AC$ on Q, hence $F \notin (M)$ on $[0,1]$.

(iv) Since $[F(0), F(1/3)] = [0,2]$ and $| F(P \cap [0,1/3]) | = 0$, where $P = \{x \in P : F'(x) \geq 0\}$, it follows that $F \notin (+)$ on $[0,1]$.

Remark 6.30.1 *The utility of this example follows by Remark 4.3.2.*

6.31 Functions concerning conditions (S), (N), (M), T_1, T_2, ACG, AC_n, SAC_n, VB_2, VBG, SVB, $\mathcal{F}$, $S\mathcal{F}$

Let C be the Cantor ternary set and let φ be the Cantor ternary function (see Sections 6.1 and 6.2). Let $F_1, F_2, G, G_n, H_n : C \mapsto [0,1]$, $n \geq 1$ be functions defined as follows:

- $F_1(x) = F_1(\sum_{i=1}^\infty c_i(x)/3^i) = \sum_{i=1}^\infty c_{2i-1}(x)/4^i$ (see Figure 6.9);
- $F_2(x) = F_2(\sum_{i=1}^\infty c_i(x)/3^i) = (1/2) \cdot \sum_{i=1}^\infty c_{2i}(x)/4^i$ (see Figure 6.10);
- $G(x) = G(\sum_{i=1}^\infty c_i(x)/3^i) = \sum_{i=1}^\infty (c_{2i-1}(x)/2^{2i+1} + c_{2i}(x)/2^{2i})$ (see Figure 6.11);

- $G_n(x) = F_1(x) + ((4^n - 1)/3)\varphi(x)$ (see Figure 6.12 for $n = 1$);
- $H_n(x) = F_2(x) + ((4^n - 1)/3)\varphi(x)$ (see Figure 6.13 for $n = 1$).

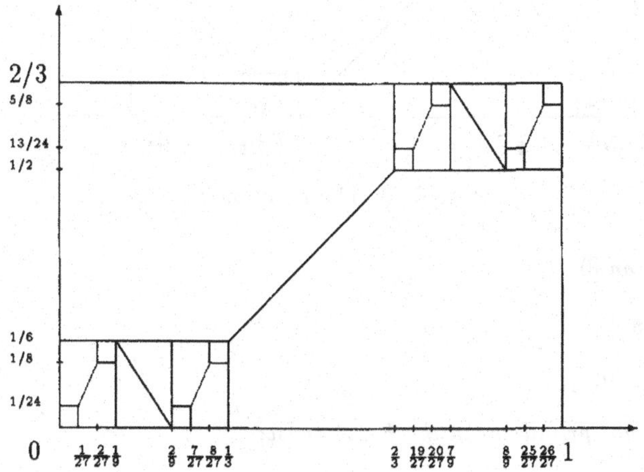

Figure 6.9: The function F_1

Extending F_1, F_2, and G linearly on the closure of each interval contiguous to C, we have F_1, F_2, G, G_n and H_n defined and continuous on $[0,1]$.

Then we have

(i) $\varphi = F_1 + F_2$ *on* $[0,1]$;

(ii) *We have*

- $F_1(x) = (1/2)F_2(3x)$ *if* $x \in [0,1/3]$;
- $F_1(x) = x - 1/6$ *if* $x \in (1/3,2/3)$;
- $F_1(x) = 1/2 + 1/2 \cdot F_2(3x-2)$ *if* $x \in [2/3,1]$;

(iii) $2G = 3F_2 + \varphi = 4\varphi - 3F_1 = 4F_2 + F_1$ *on* $[0,1]$;

(iv) $F_1, F_2 \in (N) \cap T_1 = (S)$ *on* $[0,1]$;

(v) $F_1, F_2 \notin S\mathcal{F}$ *on* $[0,1]$, *hence* $F_1, F_2 \notin \mathcal{F}$ *and* $F_1, F_2 \notin ACG$ *on* $[0,1]$;

(vi) F_1, F_2 *are* SAC_n *on no portion of* C *and for no positive integer* n, *hence* F_1, F_2 *are* AC_n *on no portion of* C *and for no positive integer* n;

(vii) $F_1, F_2 \in VB_2$ *on* C;

(viii) $F_1, F_2 \in SVB$ *on* C:

(ix) $F_1, F_2, G \notin VBG$ *on* $[0,1]$;

(x) $G(C) = [0,1]$, *hence* $G \notin (N)$ *on* $[0,1]$;

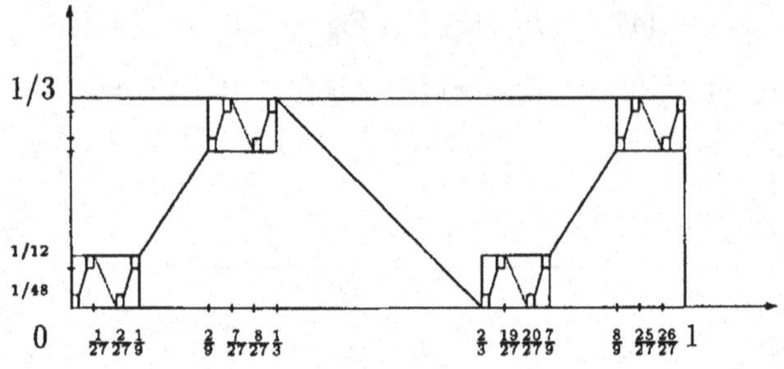

Figure 6.10: The function F_2

(xi) $G \in (M)$ *on* $[0,1]$;

(xii) $G \notin T_1$ *on* $[0,]$;

(xiii) $G \in T_2$ *on* $[0,1]$;

(xiv) F_1, F_2 *are not primitives in the Foran sense;*

(xv) $G_n, H_n \in (N)$ *on* $[0,1]$; $H_n'(x) = -G_n'(x) = F_2'(x)$ *a.e. on* $[0,1]$; $G_n(0) = H_n(0) = F_2(0) = 0$; $G_n(1) \to +\infty$ *and* $H_n(1) \to +\infty$.

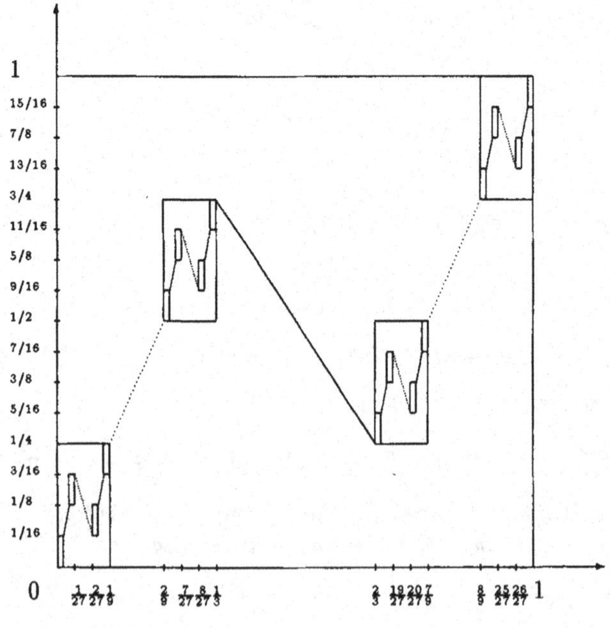

Figure 6.11: The function G

Proof. (i), (ii), (iii) and (x) are evident.

(iv) $F_1(C) = \{y : y = \sum_{i=1}^{\infty} d_i/4^i,\ d_i \in \{0,2\},\ i = \overline{1,\infty}\}$ and $F_2(C) = \{y : y = \sum_{i=1}^{\infty} d_i/4^i,\ d_i \in \{0,1\},\ i = \overline{1,\infty}\}$ are perfect sets of measure 0 (It can be proved

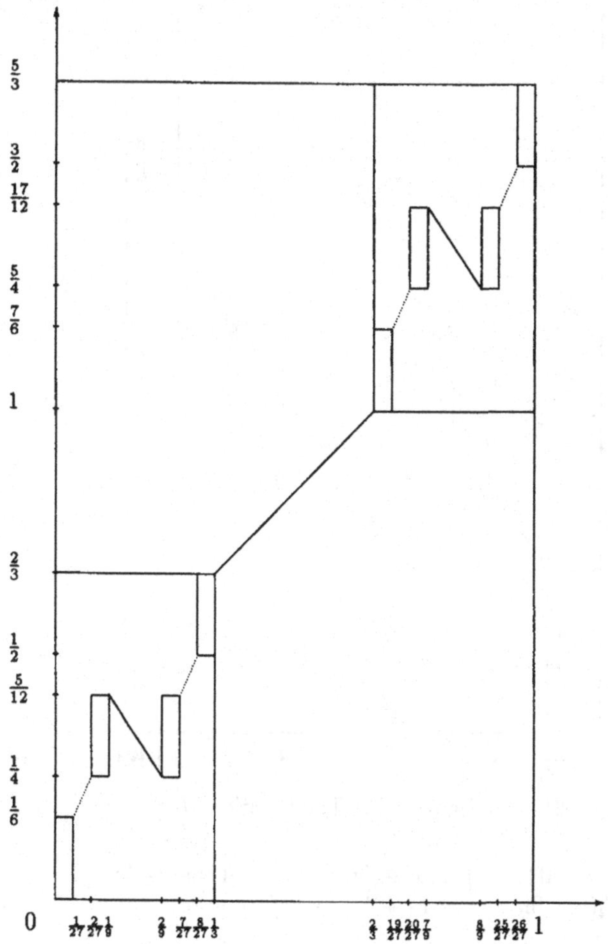

Figure 6.12: The function G_1

that for each positive integer $b \geq 2$ a set of the form $\{y : y = \sum_{i=1}^{\infty} d_i/b_i,\ d_i \in E \subset \{0, 1, \ldots, b-1\},\ i = \overline{1, \infty}$, with E having at most $b-1$ elements$\}$ is perfect and has the measure 0).

(v) Suppose on the contrary that $F_2 \in \mathcal{SF}$ on $[0,1]$. By (ii) it follows that $F_1 \in \mathcal{SF}$ on $[0,1]$. By Corollary 2.34.1, (vi), (ii), $\varphi = F_1 + F_2 \in \mathcal{SF} \subseteq (N)$ on $[0,1]$, a contradiction. Hence $F_1, F_2 \notin \mathcal{SF}$ on $[0,1]$. By Corollary 2.34.1, (iii), $F_1, F_2 \notin \mathcal{F}$ on $[0,1]$. By Corollary 2.28.1, (iii), $F_1, F_2 \notin ACG$ on $[0,1]$.

(vi) Let $a, b \in C$, $a < b$. Then there exists a maximal open interval (c,d) (and only one) excluded in the Cantor ternary process, such that $[c,d] \subset [a,b]$. Suppose that (c,d) is from the step n. Then $c = \sum_{i=1}^{n} c_i/3^i + \sum_{i=1}^{\infty} 2/3^{n+i}$, with $c_i \in \{0,2\}$, $i = \overline{1, n-1}$, $c_n = 0$. We have $c - a \leq 1/3^n$ and $b - d \leq 1/3^n$. If $x \in [a,c] \cap C$ then $c_i = c_i(x)$ for each $i = \overline{1, n}$. Hence

(1) $\quad F_2(c) - F_2(x) \geq 0$ *and* $F_1(c) - F_1(x) \geq 0.$

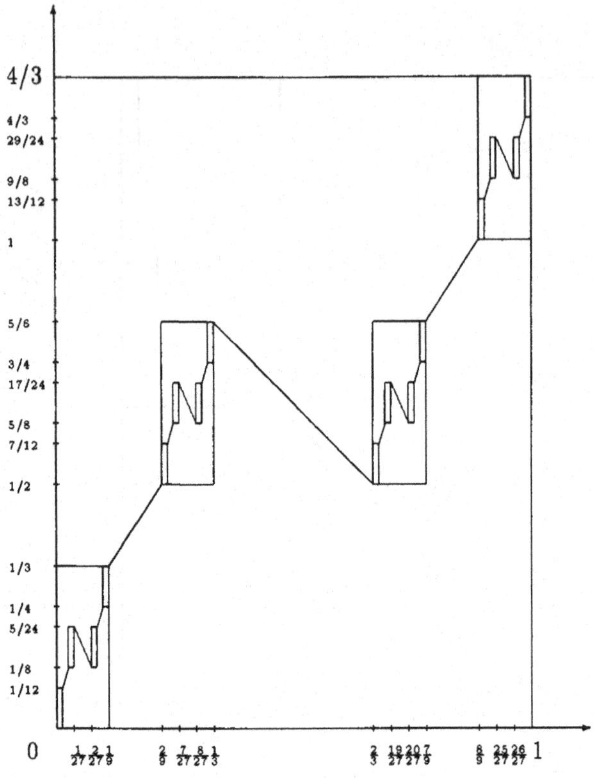

Figure 6.13: The function H_1

Let $F_2(x_o) = \inf\{F_2(x) : x \in [a,c] \cap C\}$. Then $F_2([a,c] \cap C) \subset J_1$, where $J_1 = [F_2(x_o), F_2(c)]$. Since $\varphi = F_1 + F_2$, by (1) we have

(2) $\quad | J_1 | \leq \varphi(c) - \varphi(a)$.

If $x \in [d,b] \cap C$ then $c_i = c_i(x)$ for each $i = \overline{1, n-1}$, and $c_n(x) = 2$. Hence

(3) $\quad F_2(x) - F_2(d) \geq 0$ *and* $F_1(x) - F_1(d) \geq 0$.

Let $F_2(x_1) = \sup\{F_2(x) : x \in [d,b] \cap C\}$. Then $F_2([d,b] \cap C) \subset J_2$, where $J_2 = [F_2(d), F_2(x_1)]$. Since $\varphi = F_1 + F_2$, by (3) we have

(4) $\quad | J_2 | \leq \varphi(b) - \varphi(d)$.

By (2) and (4), $F_2([a,b] \cap C) \subset J_1 \cup J_2$ and $| J_1 | + | J_2 | \leq \varphi(b) - \varphi(a)$. It follows that $\mathcal{O}^2(F_2[a,b] \cap C) \leq \varphi(b) - \varphi(a)$. Let (a_i, b_i), $i = \overline{1,p}$ be a finite set of nonoverlapping closed intervals with $a_i, b_i \in C$. Then $\sum_{i=1}^{p} \mathcal{O}^2(F_2; [a_i, b_i] \cap C) \leq \sum_{i=1}^{p} (\varphi(b_i) - \varphi(a_i)) \leq \varphi(1) - \varphi(0) = 1$. Hence $F_2 \in VB_2$ on C. Similarly $F_1 \in VB_2$ on C.

(viii) Let $W = \cup_{i=1}^{p} [a_k, b_k]$ be a sparse figure for C. Then there exists a positive integer n such that $\max\{b_k - a_k : k = \overline{1,p}\} \in [1/9^{n+1}, 1/9^n)$. Let E_{2n} be the union of the 2^{2n} remained closed intervals from the step $2n$ (each interval has the length $1/3^{2n} = 1/9^n$). Each of the $2^{2n} - 1$ intervals contiguous to E_{2n} has the length at least

$1/9^n$. Since $b_k - a_k < 1/9^n$, $k = \overline{1,p}$, it follows that each $[a_k, b_k]$ is contained in one of the closed intervals which form the set E_{2n}. Let's denote this interval J_k. Then $|J_k| = 1/9^n$. Since W is a sparse figure, it follows that each of the closed intervals which form the set E_{2n} contains at most eight components of W, hence $p \le 8 \cdot 2^{2n} = 8 \cdot 4^n$. We have $\mathcal{O}(F_2; [a_k, b_k] \cap C) \le \mathcal{O}(F_2; J_k \cap C) \le \mathcal{O}(F_2; J_k) = (1/3) \cdot (1/4^n)$. It follows that $\sum_{k=1}^{p} \mathcal{O}(F_2; [a_k, b_k] \cap C) \le 8 \cdot 4^n \cdot (1/3) \cdot (1/4^n) = 8/3$, hence $F_2 \in SVB$ on C. Similarly $F_1 \in SVB$ on C.

(ix) By (iv), (v) and Theorem 2.18.9, (vii), it follows that $F_1, F_2 \notin VBG$ on $[0,1]$. By (iii), $G \notin VBG$ on $[0,1]$.

(xi) On each remained closed interval I from the step $2i$, $B(G; I)$ is contained in a rectangle of length $|I| = 1/3^{2i}$ and height $1/4^i$ (see Figure 6.11). In fact $B(G; I \cap C)$ is contained in four rectangles of length $|I|/3^2 = 1/3^{2i+2}$ and height $1/4^{i+1}$ (the projection of each such rectangle on the x-line is a remained closed interval from the step $2i+2$). Applying Theorem 2.23.1, (ii), (vi) we show that $G \in \overline{M}$ on C. Let E be a closed subset of C such that G is $\underline{L}$ with a positive constant on E. Then $B(G; E)$ is contained in at most three of the four rectangles from the step $2i+2$. It follows that $|G(E)| < (3/4)^{i+1} \to 0$, hence $G \in \overline{M}$ on C. Similarly $G \in \underline{M}$ on C, hence $G \in (M)$ on C. It follows that $G \in (M)$ on $[0,1]$.

(xii) Suppose on the contrary that $G \in T_1$ on $[0,1]$. Then by Theorem 2.23.2, (iii), $G \in (N)$ on $[0,1]$, a contradiction.

(xiii) Let $y \in [0,1]$ be an irrational number. Then y is uniquely represented in base two by $\sum_{i=1}^{\infty} y_i/2^i$, with $y_i \in \{0,1\}$, and $C \cap G^{-1}(y) = \{x\}$, where $x = \sum_{i=1}^{\infty} c_i(x)/3^i$, with $c_{2i-1}(x) = 2y_{2i}$ and $c_{2i}(x) = 2y_{2i-1}$. It follows that $G \in T_2$ on $[0,1]$.

(xiv) Suppose on the contrary that there exists a continuous function F on $[0,1]$, $F \in \mathcal{F}$, such that $F'_{ap}(x) = F'_2(x)$ *a.e.* on $[0,1]$.Then there exists a continuous function h on $[0,1]$ with $h(0) = 0$, which is constant on each interval contiguous to C such that

$$(5) \quad F = F_2 + h.$$

By (ii) and (5) it follows that

$$(6) \quad \varphi(x) + (1/2)\varphi(3x) + h(x) = 0, \ x \in [0,1/3].$$

Indeed, let $H(x) = F(x) + (1/2)F(3x) = \varphi(x) + (1/2)h(3x) + h(x)$, $x \in [0,1/3]$. Then $H'_{ap}(x) = 0$ *a.e.* on $[0,1/3]$. $F \in \mathcal{F}$ implies that $H \in \mathcal{F}$, hence H is constant on $[0,1/3]$. But $H(0) = 0$, so we have (6).
Since $\varphi(x) = (1/2)\varphi(3x)$, (6) becomes

$$(7) \quad \varphi(x)/2 + h(x) + (1/2)(\varphi(3x)/2 + h(3x)) = 0, \ x \in [0,1/3].$$

By the fact that

$$(8) \quad F_2(x) = F_2(x + 2/3), \ x \in [0.1/3],$$

it follows that $h(x + 2/3) - h(x) = a$, $x \in [0,1/3]$, where a is a constant (indeed, let $R(x) = F(x+2/3) - F(x) = h(x+2/3) - h(x)$, $x \in [0,1/3]$; since $F'_{ap}(x) = F'_2(x)$ *a.e.* on $[0,1/3]$, it follows that $R'_{ap}(x) = 0$ *a.e.* on $[0,1/3]$; but $R \in \mathcal{F}$ implies that R is a constant). Since $h(1/3) = h(2/3)$ and $h(0+2/3) - h(0) = a$, we have that $h(2/3) = a$. It follows that $h(1/3 + 2/3) - h(1/3) = a$, so $h(1) = 2a$. Thus $h(x) = -(1/2)\varphi(x)$ on $[1/3, 3/3]$. Continuing, by (7) and (8) it follows that $h(x) = -(1/2)\varphi(x)$ on the closure of each interval contiguous to C. By the continuity of h we have $h(x) = -(1/2)\varphi(x)$ on $[0,1]$. Thus $F(x) = F_2(x) - (1/2)\varphi(x)$ on $[0,1]$. Moreover, for each $x \in C$ we have $F(x) = 1/6 - (1/6) \cdot \sum_{i=1}^{\infty}(c_{2k-1}(x) + c_{2k}(x)/2 + 1)/4^k$. Hence $F(C) = [-1/3, 1/6]$. It follows that $F \notin (N)$ on $[0,1]$, a contradiction.

(xv) Let $I_{c_1c_2\ldots c_{2nq}} = [\sum_{i=1}^{2nq} c_i/3^i, \sum_{i=1}^{2nq} c_i/3^i + \sum_{i=2nq+1}^{\infty} 2/3^i]$, $q = \overline{1,\infty}$. We have

$$(9) \qquad H_n(I_{c_1\ldots c_{2n(q-1)}20\ldots0}) = H_n(I_{c_1\ldots c_{2n(q-1)}02\ldots2}),$$

where 0 and2 appear $2n-1$ times, and

$$(10) \qquad |H_n(I_{c_1\ldots c_{2nq}})| = \frac{1}{2} \cdot \sum_{k=nq+1}^{\infty} \frac{2}{4^k} + \frac{4^n - 1}{3} \cdot \sum_{k=2nq+1}^{\infty} \frac{2}{2^{k+1}} = \frac{1}{3} \cdot \frac{1}{4^{n(q-1)}}.$$

By (9) and (10) it follows that $H_n(C)$ can be covered with $(4^n-1)^q$ closed intervals of length $(1/3) \cdot (1/4^{n(q-1)})$. Hence $|H_n(C)| = \lim_{q\to\infty}(1/3) \cdot (1/4^{n(q-1)}) \cdot (4^n - 1)^q = ((4^n - 1)/3) \cdot \lim_{q\to\infty}((4^n - 1)/4^n)^{q-1} = 0$, so $H_n \in (N)$ on $[0,1]$.
Since $\varphi(x) = (1/2)\varphi(3x)$, $x \in [0,1/3]$ and $\varphi(x) = 1/2 + (1/2)\cdot\varphi(3x-2)$, $x \in [2/3,1]$, by (ii) it follows that

- $G_n(x) = (1/2) \cdot H_n(3x)$ if $x \in [0,1/3]$;
- $G_n(x) = x + (4^n - 2)/6$ if $x \in [1/3,2/3]$;
- $G_n(x) = (4^n + 2)/6 + (1/2) \cdot H_n(3x - 2)$ if $x \in [2/3,1]$.

Therefore $G_n \in (N)$ on $[0,1]$.
By (i), $H_n'(x) = F_2'(x) = -F_1'(x) = -G_n'(x)$ *a.e.* on $[0,1]$.

Remark 6.31.1 *(i) Let $F(x) = F_2(x) - (1/2)\varphi(x)$. Then $F \in (M)$ and $F \notin (N)$ on $[0,1]$ (see (xiv)).*

(ii) The utility of this example follows by Theorem 2.18.9, (v), (vi); Remark 2.18.7, (i); Theorem 2.23.2, (ii); Remark 2.29.2, (i); Corollary 2.34.1,(ii); Remark 2.35.3, (i).

6.32 A function $F \in$ lower semicontinuous, $F \in AC_2$, $F \notin \underline{AC}$

Let C be the Cantor ternary set (see Section 6.1). Let $F : [0,1] \mapsto \mathbb{R}$, $F(x) = 0$ if $x \in C$, and $F(x) = 1$ if $x \notin C$. Then we have:

(i) $F \in$ lower semicontinuous on $[0,1]$;

(ii) $F \in AC_2 \subseteq \underline{AC}_2$ on $[0,1]$;

(iii) $F \notin \underline{AC}$ on $[0,1]$.

Remark 6.32.1 *The utility of this example follows by Remark 2.28.2.*

6.33 A function $F_n \in L_{n+1}$ on a perfect set, $F_n \in VB_n$ on no portion of this set, $F_n \in L_{n+1}G$, $F_n \notin AC_nG$ on $[0,1]$

Let C be the Cantor ternary set (see Section 6.1), and let n be a positive integer. Let $C_{2n+1} = \{x : x = \sum_{i=1}^{\infty} c_i/(2n+1)^i,\ c_i \in \{0,2,4,\ldots,2n\}\}$. Then C_{2n+1} is a perfect set and each $x \in C_{2n+1}$ is uniquely represented by $\sum_{i=1}^{\infty} c_i(x)/(2n+1)^i$, $c_i(x) \in \{0,2,4,\ldots,2n\}$. Clearly $C_3 = C$ and $| C_{2n+1} |= 0$, $n = \overline{1,\infty}$. Let $\{j_k\}$, $k = \overline{0,\infty}$ be an increasing sequence of natural numbers, $j_o = 0$. Let $F_n : C_{2n+1} \mapsto \mathbb{R}$, $F_n(x) = F_n(\sum_{i=1}^{\infty} c_i(x)/(2n+1)^i) = \sum_{k=1}^{\infty} c_{j_k}(x)/(2n+1)^{j_{k-1}+1}$. Extending F_n linearly on the closure of each interval contiguous to C_{2n+1}, we have F_n defined and continuous on $[0,1]$ (for $n = 2$, $j_k = 3k$, $\alpha = (1/5)\cdot\sum_{k=1}^{\infty} 4/5^{3(k-1)}$ and $\beta = (1/5)\cdot\sum_{k=2}^{\infty} 4/5^{3(k-1)}$, see Figure 6.14).
Then we have

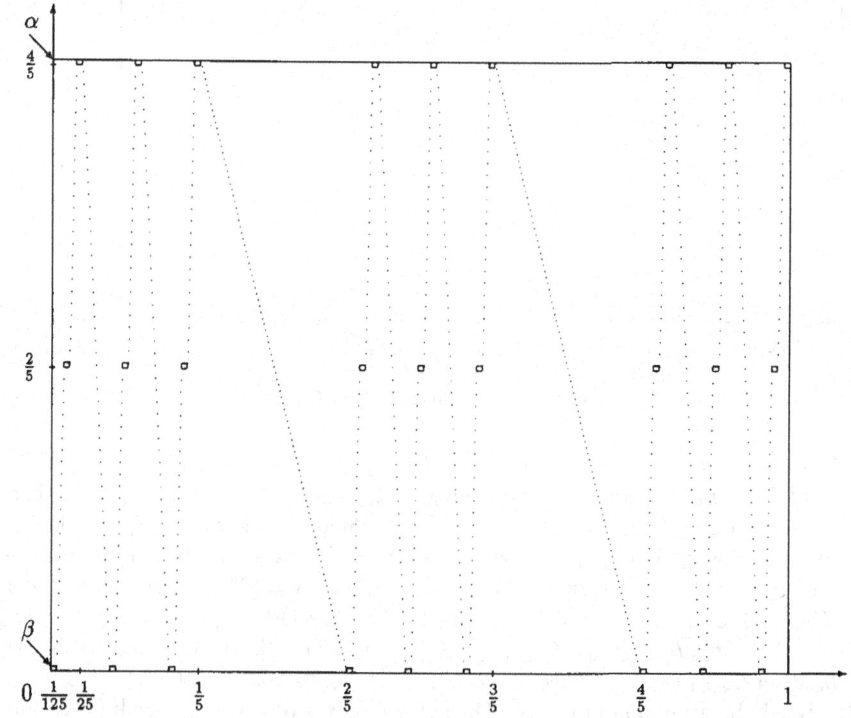

Figure 6.14: The function F_2

(i) $F_n \in L_{n+1} \subseteq AC_{n+1}$ *on* C_{2n+1}*;*

(ii) If $j_k = 3^k$, $k = \overline{1,\infty}$ *then* $F_n \in VB_n$ *on no portion of* C_{2n+1}*, hence* $F_n \in AC_n$ *on no portion of* C_{2n+1}*;*

(iii) $F_n \in L_{n+1}G \subseteq AC_{n+1}G$ *on* $[0,1]$ *and* $F_n \notin AC_nG$ *on* $[0,1]$.

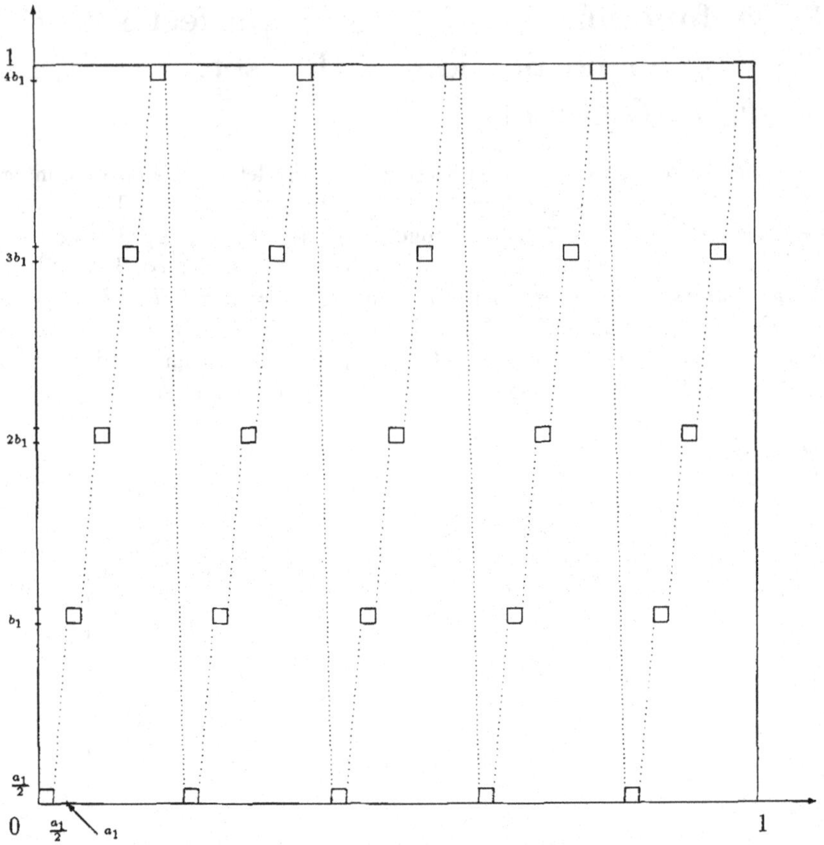

Figure 6.15: The function G_4

Proof. (i) Let I be an interval such that $I \cap C_{2n+1} \neq \emptyset$. Then $1/(2n+1)^{m+1} \leq |\, I \,| < 1/(2n+1)^m$ for some positive integer m. Since $|\, I \,| < 1/(2n+1)^m$, there exist $c_1, c_2, \ldots, c_m$ such that $c_i(x) = c_i$, $i = \overline{1,m}$, whenever $x \in I \cap C_{2n+1}$. Let k be a positive integer such that $j_{k-1} \leq m < j_k$. Let $J_i = [F_n(A + 2i/(2n+1)^{j_k}), F_n(A + 2i/(2n+1)^{j_k} + \sum_{p=j_k+1}^{\infty} 2n/(2n+1)^p)]$, $i = \overline{0,n}$, where $A = \sum_{i=1}^{m} c_i/(2n+1)^{j_k}$. Then $F_n(I \cap C_{2n+1}) \subset \cup_{i=0}^{n} J_i$ and $|\, J_i \,| = (2n/(2n+1)) \cdot (1/(2n+1)^{j_k} + 1/(2n+1)^{j_k+1} + \ldots) \leq |I|$. Then $\mathcal{O}^{n+1}(F_n; I \cap C_{2n+1}) \leq \sum_{i=0}^{n} |\, J_i \,| \leq (n+1) \cdot |I|$. By Remark 2.32.1, (i), $F_n \in L_{n+1}$ on C_{2n+1}, and by Theorem 2.32.1, (xv), $F_n \in AC_{n+1}$ on C_{2n+1}.

(ii) Let K be a portion of C_{2n+1}. Then there exist a positive integer k (fixed and depending on K) and $c_1, c_2, \ldots, c_k \in \{0, 2, 4, \ldots, 2n\}$ such that $K \supset I' \cap C_{2n+1}$, where $I' = [\sum_{i=1}^{j_k} c_i/(2n+1)^i, \sum_{i=1}^{j_k} c_i/(2n+1)^i + \sum_{i=j_k+1}^{\infty} 2n/(2n+1)^i]$. Let $n_k = j_{k-1} - j_k - 1$. Let $\{I_j\}$, $j = \overline{1,(n+1)^{n_k}}$ be the remained closed intervals from the step $j_{k+1} - 1$ in a $(2n+1)$-ary process (similarly to the Cantor ternary process), which are contained in I'. Let $I_j = [a_j, b_j]$. It follows that for each $j = \overline{1,(n+1)^{n_k}}$ there exist $1 \leq i \leq n_k$ and $c_{j_k+1}, c_{j_k+2}, \ldots, c_{j_k+i} \in \{0, 2, \ldots, 2n\}$ such that $a_j = \sum_{m=1}^{j_k+1} c_m/(2n+1)^m$ and $b_j = \sum_{m=j_k+1}^{\infty} 2n/(2n+1)^m$. If $F_n(C_{2n+1} \cap I_j) \subset J_{j,1} \cup J_{j,2} \cup \ldots \cup J_{j,n}$ then at least one of the intervals $J_{j,m}$, $m = \overline{1,n}$ has the length greater than $(2/(2n+1)) \cdot (1/(2n+1)^{j_k} - 2n \cdot$

$(1/(2n+1)^{j_{k-1}} + 1/(2n+1)^{j_{k+2}}) > 1/(2n+1)^{j_k+1}$. It follows that $\mathcal{O}^n(F_n; C_{2n+1} \cap I_j) > 1/(2n+1)^{j_k+1}$. We have $\sum_{j=1}^{(n+1)^{n_k}} \mathcal{O}^n(F_n; C_{2n+1} \cap I_j) > (n+1)^{n_k}/(2n+1)^{j_k+1} = (n+1)^{2j_k-1}/(2n+1)^{j_k+1} = (1/(n+1)(2n+1)) \cdot ((n^2+2n+1)/(2n+1))^{j_k} \to \infty$, $k \to \infty$ (because $j_{k+1} = 3j_k$). Hence $F_n \notin VB_n$ on K. By Theorem 2.29.1, (xv), $F_n \notin AC_n$ on K.

(iii) See (i) and Lemma 2.28.1.

Remark 6.33.1 *We can define another function G_n on a perfect set P_n, which has the same properties as F_n, using the following construction: let n be a positive integer and let*

- $m_1 = (n+1)^2 - 1$;
- $m_k = (n+1)^{k+1}(2m_1+1) \cdot \ldots \cdot (2m_{k-1}+1) - 1$, $k = \overline{2,\infty}$;
- $a_k = 2/((2m_1+1) \cdot \ldots \cdot (2m_k+1))$, $k = \overline{1,\infty}$;
- $b_k = (m_k/n) \cdot a_k$, $k = \overline{1,\infty}$;
- $p_k = (n+1)^k((2m_1+1) \cdot \ldots \cdot (2m_{k-1}+1) - 1)$, $k = \overline{2,\infty}$ *and* $p_1 = n$.

Then we have

- $m_k = (n+1)p_k + n$, $k = \overline{1,\infty}$;
- $\sum_{i=k+1}^{\infty} nb_i = \sum_{i=k+1}^{\infty} m_i a_i = a_k/2$, $k = \overline{1,\infty}$.

Let $P_n = \{x : x = \sum_{i=1}^{\infty} a_i t_i,\ t_i \in \{0,1,\ldots,m_i\}\}$. Then P_n is a perfect set of measure 0 and each point $x \in P_n$ is uniquely represented by $\sum_{i=1}^{\infty} a_i t_i(x)$. For each $x \in P_n$, let $G_n(x) = \sum_{i=1}^{\infty} b_i \cdot r(t_i(x); n+1)$, where $r(\alpha;\beta)$ is the remainder of the quotient α/β. Extending G_n linearly on the closure of each interval contiguous to P_n, we have G_n defined and continuous on $[0,1]$ (see Figure 6.15 for $n = 4$). Then we have:

(i) $G_n \in L_{n+1} \subsetneq AC_{n+1}$ *on* P_n;

(ii) $G_n \in VB_n$ *on no portion of* P_n, *hence* $G_n \in AC_n$ *on no portion of* P_n;

(iii) $G_n \in L_{n+1}G \subsetneq AC_{n+1}G$ *on* $[0,1]$ *and* $G_n \notin AC_nG$ *on* $[0,1]$.

Remark 6.33.2 *The utility of this example follows by Theorem 2.28.1, (vi) and Corollary 2.28.1, (iii).*

6.34 Functions $F \in L_2G$, $G_s \in (N)$, $G'_s = F'$ a.e., $G_s - F$ is not identically zero, $F \notin SACG$

Let C be the Cantor ternary set (see Section 6.1), and let φ be the Cantor ternary function (see Section 6.2). Let $\{j_k\}_k$ be a strictly increasing sequence of positive integers, such that $j_o = 0$ and

(1) $3^{j_k} < 2^{j_{k+1}}$, *and set* $n_k = j_{k+1} - j_k - 1$.

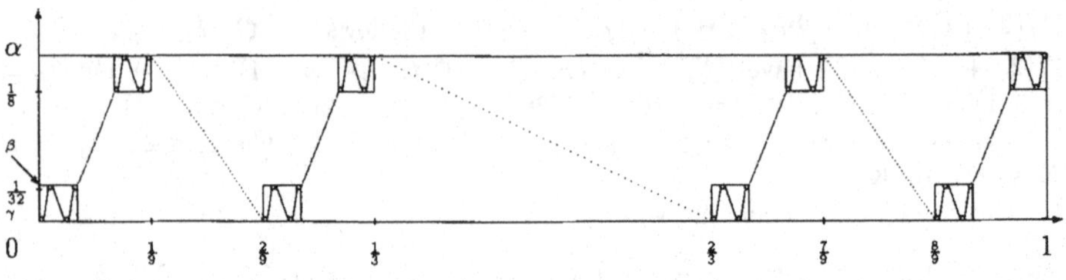

Figure 6.16: The function F

Let $F, G : C \mapsto [0,1]$ be defined as follows: $F(x) = \sum_{i=1}^{\infty} c_{j_i}(x)/2^{j_i+1}$ and $G(x) = \sum_{i=0}^{\infty}\sum_{k=1}^{n_i} c_{j_i+k}(x)/2^{j_i+k+1}$. Extending F and G linearly on the closure of each interval contiguous to C, we have F and G defined and continuous on $[0,1]$ (see Figure 6.16 and Figure 6.17, where $j_1 = 3$, $j_2 = 5$, $j_3 = 8$, $j_4 > 13$ etc.; $\alpha = 1/8 + 1/32 + \gamma$; $\beta = 1/32 + \gamma$; $\gamma = 1/256 + \sum_{i=4}^{\infty} 1/2^{j_i}$; $\alpha_o = 1-\alpha$; $\alpha_1 = 1/4 - \alpha$; $\alpha_2 = 1/2 - \alpha$; $\alpha_3 = 3/4 - \alpha$; $\beta_1 = 1/16 - -beta$; note that $1/4 - \alpha = 1/8 - \beta$ and $1/32 - \gamma = 1/16 - \beta$). Clearly

(2) $\quad F + G = \varphi$ *on* $[0,1]$.

Let $G_s : [0,1] \mapsto \mathbb{R}$, $G_s(x) = -G(x) + (1 - 2^{n_s})\varphi(x)$, $s = \overline{1,\infty}$. By (2) we have

(3) $\quad G_s = F - 2^{n_s}\varphi$ *on* $[0,1]$.

Then we have

(i) $F \in L_2$ *on* C, *hence* $F \in L_2G \subset AC_2G \subset \mathcal{F} \subset (N)$ *on* $[0,1]$;

(ii) G_s *is differentiable a.e. on* $[0,1]$;

(iii) $G, G_s \in (N)$ *on* $[0,1]$;

(iv) $G_s' = F'$ *a.e. on* $[0,1]$;

(v) $G_s(0) = F(0) = 0$ *and* $G_s - F$ *is not identically zero;*

(vi) $F \notin SACG$ *on* C, *hence* $F \notin SACG$ *on* $[0,1]$.

Proof. (i) Let $[a,b]$ be a closed interval $a, b \in C$, and choose n such that $1/3^{n+1} \leq \mid I \mid < 1/3^n$. Then there exist $c_1, c_2, \dots, c_n \in \{0,2\}$ such that $c_i(x) = c_i$, $i = \overline{1,n}$, for each $x \in [a,b] \cap C$. Let k be a positive integer such that $j_{k-1} \leq n < j_k$ and let $A = \sum_{i=1}^{n} c_i/3^i$. Let's denote by $J_1 = [F(A), F(A + 1/3^{j_k})]$ and $J_2 = [F(A + 2/3^{j_k}), F(A + 3/3^{j_k})]$. Then $F([a,b] \cap C) \subset J_1 \cup J_2$, and by (1), $\mid J_1 \mid = \mid J_2 \mid = \sum_{i=k+1}^{\infty} 2/2^{j_i+1} < 2/2^{j_{k+1}} < 2/3^{j_k} < 2/3^n < 6(b-a)$. Hence $\mathcal{O}^2(F; [a,b] \cap C) < 6(b-a)$, and $F \in L_2$ on C. It follows that $F \in L_2G$ on $[0,1]$. By Corollary 2.32.1, (iv) and Corollary 2.28.1, (iii), $L_2G \subset AC_2G \subset \mathcal{F} \subset (N)$ on $[0,1]$.

(ii), (iv) and (v) are evident.

(iii) Given a positive integer p, $G(C)$ can be covered by $2^{n_0} \cdot 2^{n_1} \cdot \ldots 2^{n_p}$ intervals each of length at most $2/2^{j_{p+1}}$. Hence $\mid G(C) \mid = 0$ and $G \in (N)$ on $[0,1]$. Note that for each $x \in C$ and $s \geq 1$ we have

$$-G_s(x) = 2^{n_s} \cdot \sum_{i=0}^{s-1}\sum_{k=1}^{n_i} \frac{c_{j_i+k}(x)}{2^{j_i+k+1}} + (2^{n_s} - 1) \cdot \sum_{i=0}^{s-1} \frac{c_{j_i}(x)}{2^{j_i+1}} +$$

$$\sum_{i=0}^{\infty} \frac{(2^{n_s}-1)\cdot c_{j_{s+i}}(x) + \sum_{k=1}^{n_s} 2^{n_s-k}\cdot c_{j_{s+i}+k}(x)}{2^{j_{s+i}+1}} +$$

$$2^{n_s}\cdot \sum_{i=0}^{\infty}\sum_{k=1}^{n_{s+i}-n_s} \frac{c_{j_{s+i}+n_s+k}(x)}{2^{j_{s+i}+n_s+k+1}}.$$

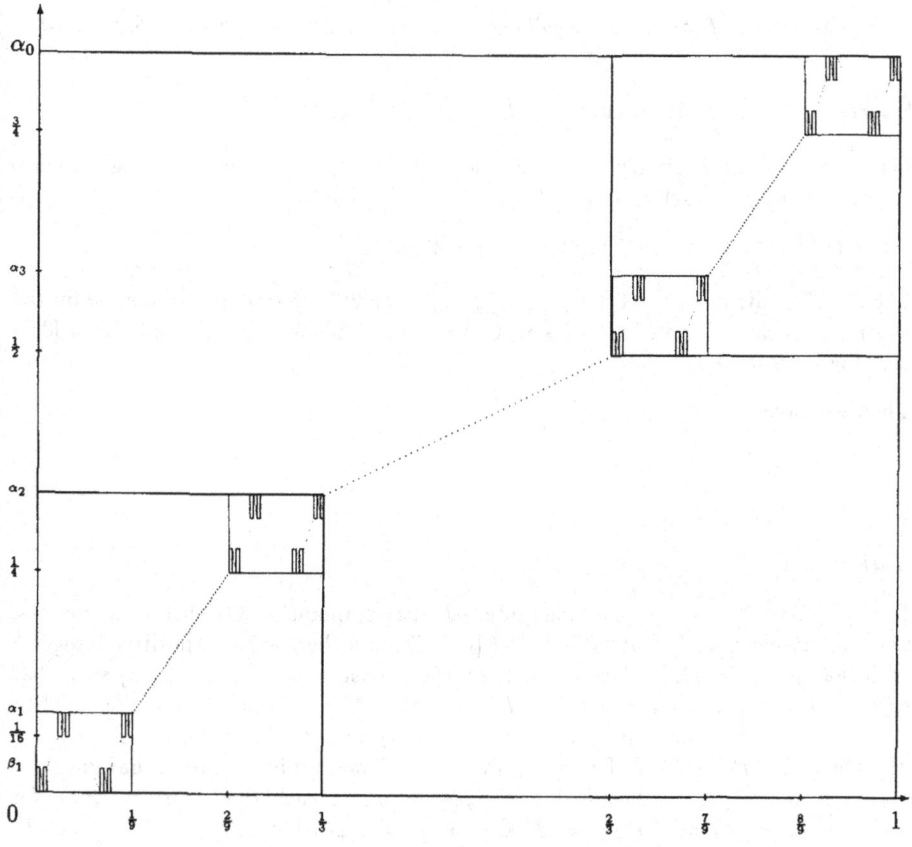

Figure 6.17: The function G

Since $1+2+\ldots+2^{n_s-1} = 2^{n_s}-1$, it follows that $-G_s(C)$ can be covered by $2^{j_s-1}\cdot(2^{n_s+1}-1)\cdot((2^{n_s+1}-1)\cdot 2^{n_{s+1}-n_s})\cdot\ldots\cdot((2^{n_s+1}-1)\cdot 2^{n_{s+p}-n_s}) = ((2^{n_s+1}-1)/2^{n_s+1})^{p+1}.2^{j_{s+p+1}-1}$ intervals, each of length at most $2^{n_s+1}/2^{j_{s+p+1}}$. Hence $|-G_s(C)| = |G_s(C)| = 0$, and $G_s \in (N)$ on $[0,1]$.

(vi) This assertion is true even with a weaker hypothesis than (1), namely: $j_{k+1} - j_k \geq 2$ for each k. Let $(a,b)\cap C$ be a portion of C. Then there exists $[c,d]$ be a remained closed interval from the step k_o such that $[c,d] \subset (a,b)$. Let $j_k > k_o$ and let E_k be the sparse figure for $[c,d]\cap C$ formed by the closures of all the intervals contiguous to C from the step j_k which are contained in $[c,d]$. Then E_k has $2^{j_k-k_o-1}$ components. Let's denote them by $[\alpha_j,\beta_j]$, $j = \overline{1, 2^{j_k-k_o-1}}$. It follows that $F(\beta_j) - F(\alpha_j) = F(2/3^{j_k}) - F(1/3^{j_k}) = 2/2^{j_k+1} - \sum_{i=1}^{\infty} 2/2^{j_{k+i}+1} > 1/2^{j_k} - 2/2^{j_{k+1}}$ for each j. Then

$\sum_{i=1}^{2^{j_k-k_o-1}}(F(\beta_j)-F(\alpha_j)) > 1/2^{k_o+1} - 1/2^{j_{k+1}-j_k+k_o+1} > 1/2^{k_o+2}$. For $\epsilon_o = 1/2^{k_o+3}$, since $|E_k| \to 0$, $k \to \infty$, it follows that $F \notin SAC$ on $[c,d] \cap C$. Hence $F \in SAC$ on no portion of C. By Lemma 2.34.3, $F \notin SACG$ on C.

Remark 6.34.1 *(i) Answering to a problem mentioned by D.W.Solomon, Foran constructed in [F6] two functions $f_1, f_2 \in (N)\cap(C)$ on $[0,1]$, differentiable a.e., $f_1' = f_2'$ a.e. such that $f_1 - f_2$ is not identically constant.*

(ii) The utility of this example follows by Remark 4.2.2, (i); Theorem 5.24.1, (i).

6.35 A function $F \in \underline{L}_2$, $F \notin T_2$, $F \notin \mathcal{B}$

Let C be the Cantor ternary set (see Section 6.1). Let $\{j_k\}_k$ be an increasing sequence of positive integers such that $j_o = 0$, $j_1 = 2$, $j_2 = 6$ and

(1) $(1/2)\cdot(1/3^{j_k}) \geq 1/2^{j_{k+1}-k}$, *for* $k = \overline{3,\infty}$.

Let $F : C \mapsto \mathbb{R}$, $F(x) = (1/2)\cdot\sum_{k=0}^{\infty}\sum_{i=j_k+1}^{j_{k+1}-1} c_i(x)/2^{i-k}$. Extending F linearly on the closure of each interval contiguous to C, we have F defined and continuous on $[0,1]$ (see Figure 6.18).

Then we have:

(i) $F \in \underline{L}_2 \subseteq \underline{AC}_2$ on C;

(ii) $F \notin T_2$ on C;

(iii) $F \notin \mathcal{B}$.

Proof. (i) Let $I \subset [0,1]$ be a closed interval with endpoints in C. Then there exists a positive integer n such that $1/3^{n+1} \leq |I| < 1/3^n$, and there exists a positive integer k such that $j_k \leq n < j_{k+1}$. Since $|I| < 1/3^n$, there exist $c_1, c_2, \dots, c_n \in \{0,2\}$ such that $c_i(x) = c_i$ for each $i = \overline{1,n}$, $x \in C \cap I$. Let $a = \sum_{i=1}^{j_k} c_i/3^i$ and $b = a + 1/3^{j_k}$. Then $I \subset [a,b]$. Let $P_1 = \{x \in [a,b] \cap C : c_{j_{k+1}}(x) = 0\}$ and $P_2 = \{x \in [a,b]\cap C : c_{j_{k+1}} = 2\}$. Then $P_1 \cup P_2 = C \cap I$. Let $x, y \in P_1$, $x < y$. Then we have three situations:
1) $y - x > 1/3^{j_{k+2}-1}$. Let $j_k + 1 \leq i_o \leq j_{k+2} - 1$ such that $c_i(y) = c_i(x)$, for each $i < i_o$, $c_{i_o}(x) = 0$, $c_{i_o}(y) = 2$. Clearly $i_o \neq j_{k+1}$. Let $a_1 = a + \sum_{i=j_k+1}^{i_o-1} c_i/3^i$. Then $x = a + \sum_{i=i_o+1}^{\infty} c_i(x)/3^i$ and $y = a_1 + 2/3^{i_o} + \sum_{i=i_o+1}^{\infty} c_i(y)/3^i$. We have two possibilities:
a) $j_{k+1} \leq i_o \leq j_{k+1} - 1$.Then $F(y) - F(x) \geq F(a_1 + 2/3^{i_o}) - F(a_1 + \sum_{i=i_o+1}^{\infty} 2/3^i) = F(a_1) + 1/2^{i_o-k} - F(a_1) - 1/3^{i_o-k} = 0$.
b) $j_{k+1}+1 \leq i_o \leq j_{k+2}-2$. Then $F(y)-F(x) \geq F(a_1+2/3^{i_o}) - F(a_1+\sum_{i=i_o+1}^{\infty} 2/3^i) = F(a_1) + 1/2^{i_o-k-1} - F(a_1) - 1/2^{i_o-k-1} = 0$.
By a) and b) it follows that $F(y) - F(x) \geq 0$.
2) $y - x = 1/3^{j_{k+2}-1}$. We have two possibilities:
a) $x = \sum_{i=1}^{j_{k+2}-2} c_i/3^i + \sum_{i=j_{k+2}}^{\infty} 2/3^i$ and $y = \sum_{i=1}^{j_{k+2}-2} c_i/3^i + 2/3^{j_{k+2}-1}$. Then $F(y) - F(x) = 0$.
b) $x = \sum_{i=1}^{j_{k+2}-2} c_i/3^i$ and $y = \sum_{i=1}^{j_{k+2}-2} c_i/3^i + \sum_{i=j_{k+2}}^{\infty} 2/3^i$. Then $F(y) - F(x) \geq 1/2^{j_{k+2}-k-1}$.
By a) and b) it follows that $F(y) - F(x) \geq 0$.

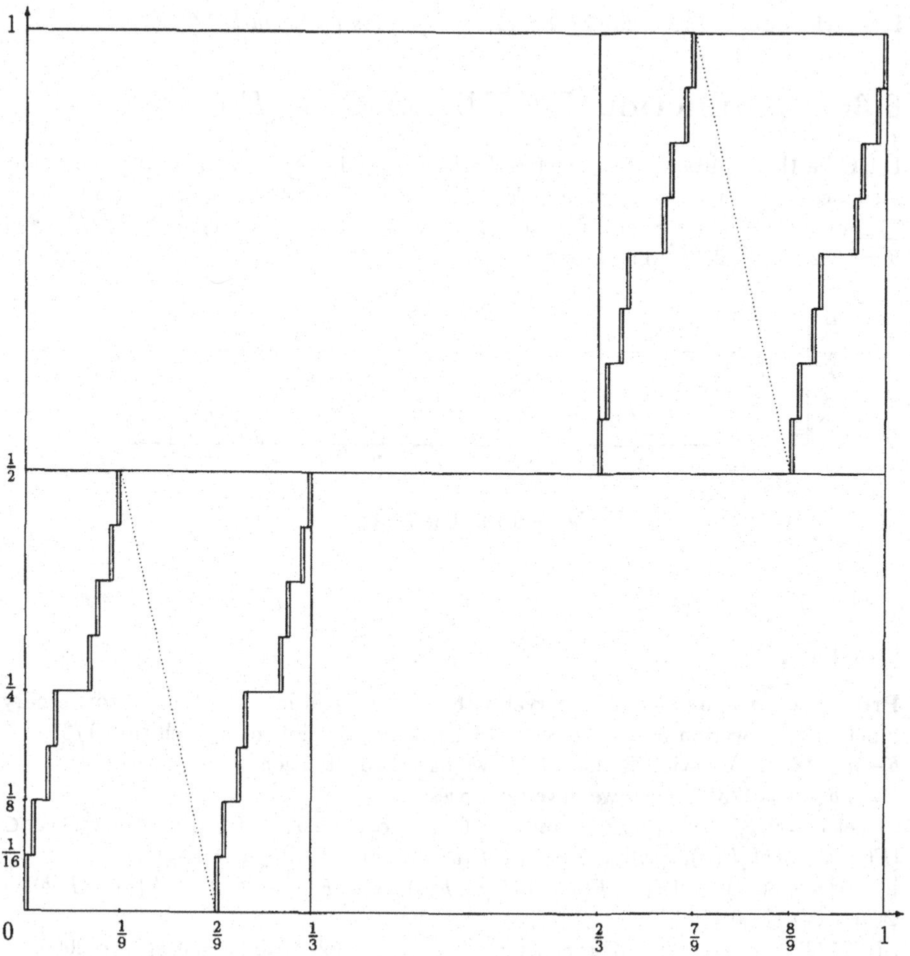

Figure 6.18: The function F

3) $y - x < 1/3^{j_{k+2}-1}$. Let $a_2 = a + \sum_{i=j_k+1}^{j_{k+32}-1} c_i/3^i$. Then $F(a_2) \leq F(y) \leq F(a_2) + 1/2^{j_{k+2}-k-1}$. Hence $| F(y) - F(x) | \leq 1/2^{j_{k+2}-k-1}$ and $F(y_1) - F(x_1) = -1/2^{j_{k+2}-k-1}$, where $x_1 = a_2 + \sum_{i=j_{k+2}+1}^{\infty} 2/3^i$ and $y_1 = a_2 + 2/3^{j_{k+2}}$.

By 1), 2) and 3) we obtain that $| \mathcal{O}_-(F; P_1) | \leq (1/2)\cdot(1/3^{j_{k+2}}) \leq (1/2)\cdot(1/3^{n+1}) \leq | I | /2$. Similarly $| \mathcal{O}_-(F; P_2) | \leq | I | /2$. It follows that $\mathcal{O}_-^2(F; C \cap I) = \sup\{\mathcal{O}_-(F; P_1) + \mathcal{O}_-(F; P_2) : P_1 \cup P_2 = C \cap I\} \geq -| I |$. By Remark 2.32.1, (i), $F \in \underline{L}_2$ on C, and by Theorem 2.32.1, (xiv), $\underline{L}_2 \subseteq \underline{AC}_2$ on C.

(ii) Let $y \in [0,1]$. Then y is uniquely represented in base two by $\sum_{i=1}^{\infty} y_i/2^i$, $y_i \in \{0,1\}$ (we always take the infinite representation). Let $c_i = 2y_{i-k}$ for $j_k \leq i \leq j_{k+2} - 2$. Then $j_k - k \leq i - k \leq j_{k+1} - k - 2$. Let $C_y = \{x \in C : c_i(x) = c_i, j_k \leq i \leq j_{k+1} - 2\}$. Then C_y is a perfect set and $F^{-1}(y) = C_y$. Hence $F \notin T_2$ on C.

(iii) By Corollary 2.29.1, (xvi) it follows that $\mathcal{B} \subset \sigma.f.l. \subset T_2$. Since $F \notin T_2$ it follows that $F \notin \mathcal{B}$.

Remark 6.35.1 *The utility of this example follows by Remark 2.32.2.*

6.36 A function $F \in VB_2$ on C, $V_2(F;C) \leq 1$

Let C be the Cantor ternary set (see Section 6.1). Let $\{j_i\}_i$ be a strictly increasing sequence of positive integers, and let $F : C \mapsto \mathbb{R}$, $F(x) = F(\sum_{i=1}^{\infty} c_i(x)/3^i) = \sum_{i=1}^{\infty} c_{j_i}(x)/2^{j_i+1}$ (see Figure 6.19 for $j_k = 3k$, $k = \overline{1,\infty}$, $\alpha = (1/2)\cdot\sum_{i=1}^{\infty} 2/2^{3i}$ and $\beta = (1/2)\cdot\sum_{i=2}^{\infty} 2/2^{3i}$).Then we have:

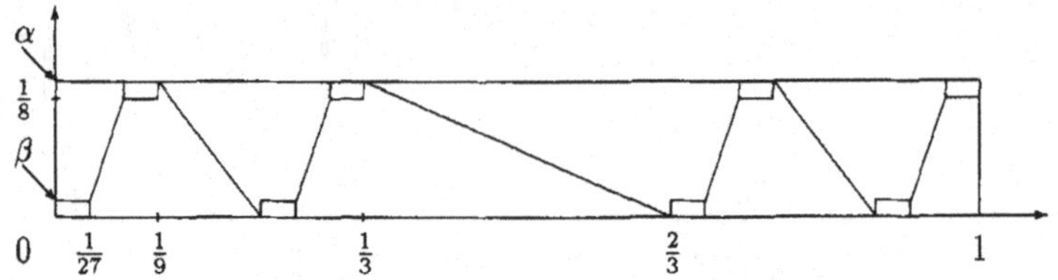

Figure 6.19: The function F

(i) $F \in VB_2$ *on C;*

(ii) $V_2(F;C) \leq 1$.

Proof. Let $[a,b]$ be a closed interval with $a,b \in C$, and let φ be the Cantor ternary function (see Section 6.2). Then there exists a positive integer n such that $1/3^{n+1} \leq b-a < 1/3^n$. Following Remark 6.1.1 we have two situations:
(I) If $b-a = 1/3^{n+1}$ then we have two cases:
a) Let $c = 0, c_1 \ldots c_{n+1}0222\ldots$ and $d = 0, c_1 \ldots c_{n+1}2000\ldots$. Clearly $F([a,b]\cap C) \subset [F(a),F(c)] \cup [F(d),F(b)]$ and $F(c)-F(a)+F(b)-F(d) \leq \varphi(b)-\varphi(a)$.
b) $F([a,b]\cap C) \subset [F(a),F(a)] \cup [F(b),F(b)]$ and $F(a)-F(a)+F(b)-F(b) = \varphi(b)-\varphi(a) = 0$.
(ii) If $1/3^{n+1} < b-a) < 1/3^n$ then let $c = 0, c_1 \ldots c_n 0222\ldots$ and $d = 0, c_1 \ldots c_n 2000\ldots$. Let $\alpha \in [a,c]\cap C$ such that $F(\alpha) = \inf(F([a,c]\cap C))$, and let $\beta \in [d,b]$ such that $F(\beta) = \sup(F([d,b]\cap C))$. Clearly $F([a,b]\cap C) \subset [F(\alpha),F(c)] \cup [F(d),F(\beta)]$ and $F(c)-F(\alpha+F(\beta)-F(d) \leq \varphi(b)-\varphi(a)$.
Let $\{[a_k,b_k]\}$, $k = \overline{1,p}$ be a finite set of nonoverlapping closed intervals such that $a_k, b_k \in C$. Then $\sum_{k=1}^{p} \mathcal{O}^2(F;[a_k,b_k]\cap C) \leq \sum_{k=1}^{p}(\varphi(b_k)-\varphi(a_k)) \leq \varphi(1)-\varphi(0) = 1$. Hence $F \in VB_2$ on C, and $V_2(F;C) \leq 1$.

Remark 6.36.1 *The utility of this example follows by Sections 6.37 and 6.39.*

6.37 A function $F_p \in L_{2^p}$, $F_p \notin AC_{2^p-1}$, $F_p \in VB_2$ on C; $V_2(F_p;C) \leq 1$

Let C be the Cantor ternary set (see Section 6.1), and let p be a positive integer. Let $\{j_i\}_i$ be a sequence of positive integer such that

(1) $3^{j_i+p} < 2^{j_{i+1}-p}$ *and* $j_1 = p$.

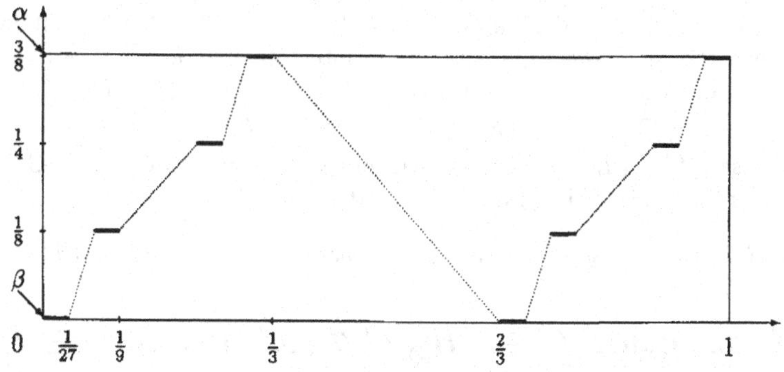

Figure 6.20: The function F_2

Let $F_p : C \mapsto \mathbb{R}$, $F_p(x) = F_p(\sum_{i=1}^{\infty} c_i(x)/3^i) = \sum_{i=1}^{\infty}\sum_{k=0}^{p-1} c_{j_i+k}(x)/2^{j_i+k+1}$. Extending F_p linearly on the closure of each interval contiguous to C, we have F_p defined and continuous on $[0,1]$ (see Figure 6.20 for $p=2$, $j_1 = 2$, $j_2 = 9$, $\alpha = \sum_{i=1}^{\infty}\sum_{k=0}^{1} 2/2^{j_i+k+1}$ and $\beta = \sum_{i=2}^{\infty}\sum_{k=0}^{1} 2/2^{j_i+k+1}$). Moreover $F_p : [0,1] \mapsto [0, 1/2^{p-1})$. Then we have:

(i) $F_p \in L_{2^p} \subseteq AC_{2^p}$ *on C, hence* $F_p \in \mathcal{F}$ *on* $[0,1]$*;*

(ii) $F_p \notin AC_{2^p-1}$ *on C;*

(iii) $F_p \in VB_2$ *on C and* $V_2(F_p; C) \leq 1$.

Proof. (i) Let $[a,b]$ be a closed interval with $a, b \in C$, and choose a positive integer n such that $1/3^{n+1} \leq b-a < 1/3^n$. Then there exist $_1, c_2, \ldots, c_n \in \{0,2\}$ such that $c_i(x) = c_i$, $i = \overline{1,n}$, for each $x \in [a,b] \cap C$. Let k be a positive integer such that $j_{k-1} \leq n < j_k$. Suppose that $k > 1$ (for $k=1$ the proof is similar). Then we have two possibilities:

a) $j_{k+1} \leq n < j_{k-1} + p - 1$. Let $j = n - j_{k-1}$. Clearly $j < p-1$. By (1) we obtain

$$(2) \quad \frac{1}{3^{n+1}} = \frac{1}{3^{j_{k-1}+j+1}} = \frac{3^{p-j-1}}{3^{j_{k-1}+p}} > \frac{3^{p-j-1}}{2^{j_k-p}} = \frac{2^p \cdot 2^{j-1} \cdot 3^{p-j-1}}{2^{j_k+j+1}} > \frac{2^p}{2^{j_k+j-1}}.$$

Let $A = \{x \in C : c_i(x) = c_i$ if $i \in \{1,2,\ldots,n\}\}$, and $c_i(x) = 0$ if $i \in \{j_{k+1} + p, \ldots, j_k - 1\} \cup \{j_k + j, j_k + j + 1, \ldots\}\}$. Let $x_1 < x_2 < \ldots < x_{2^p}$ be the elements of the set A and let $y_i = x_i + \sum_{t=j_k+j+1}^{\infty} 2/2^t$, $i = \overline{1, 2^p}$. Clearly $F_p([a,b] \cap C) \subset \cup_{i=1}^{2^p}[F_p(x_i), F_p(y_i)]$, and by (2), $\sum_{i=1}^{2^p}(F_p(y_i) - F_p(x_i)) \leq 2^p/2^{j_k+j-1} < 1/3^{n+1} < b-a$.

b) $j_{k+1} + p - 1 \leq n < j_k$. Let $A = \{x \in C : c_i(x) = c_i$ if $i \in \{1,2,\ldots,n\}$ and $c_i(x) = 0$ if $i \in \{n+1, n+2, \ldots, j_k - 1\} \cup \{j_k + p, j_k + p + 1, \ldots\}\}$. Let $x_1 < x_2 < \ldots < x_{2^p}$ be the elements of the set A, and let $y_i = x_i + \sum_{t=j_k+p}^{\infty} 2/2^t$. Then $F_p([a,b] \cap C) \subset \cup_{i=1}^{2^p}[F_p(x_i), F_p(y_i)]$, and by (1), $\sum_{i=1}^{2^p}(F_p(y_i) - F_p(x_i)) \leq 2^p/2^{j_{k+1}-1} = 2/2^{j_{k+1}-p} < 2/3^{j_k+p} \leq (2/3^p) \cdot (1/3^{n+1}) \leq (2/3^p) \cdot (b-a) < b-a$.

By a) and b), $\mathcal{O}^{2^p}(F_p; [a,b] \cap C) \leq b-a$, and by Remark 2.32.1, (i), $F_p \in L_{2^p}$ on C. By Theorem 2.32.1, (xv), $F_p \in AC_{2^p}$ on C.

(ii) Let $I_m = [a_m, b_m]$, $m = \overline{1, 2^{j_k-1}}$ be the remained closed intervals from the step $j_k - 1$, and let $I_m^i = [a_m^i, b_m^i]$ be the remained closed intervals from the step $j_k - 1 + p$ which are contained in I_m. Clearly (b_n^i, a_m^{i+1}), $i = \overline{1, 2^p - 1}$ are the excluded open intervals from the steps $j_k, j_k+1, \ldots, j_k+p-1$. By Remark 6.1.1 we have $F_p(C \cap I_m^i) \subset$

$[F_p(a^i_m), F_p(b^i_m)]$, $i = \overline{1, 2^p}$ and $F_p(a^{i+1}_m) - F_p(b^i_m) \geq 2/2^{j_k+p} - 2/2^{j_k+1}$. It follows that, if we cover the set $F_p(C \cap I_m)$ with $2^p - 1$ intervals J_{mj}, $j = \overline{1, 2^p - 1}$ then for some j, J_{mj} contains the closed interval $[F(b^i_m), F(a^{i+1}_m)]$. Hence $\mathcal{O}^{2^p-1}(F_p; C \cap I_m) \geq 2/2^{j_k+p} - 2/2^{j_k+1}$ and $\sum_{m=1}^{2^{j_k-1}} \mathcal{O}^{2^p-1}(F_p; C \cap I_m) \geq 2^{j_k-1} \cdot (2/2^{j_k+1}) - 2/2^{j_k+p}) \to 1/2^p$, $k \to \infty$. Because $\sum_{m=1}^{2^{j_k-1}} \mid I_m \mid \to 0$, $k \to \infty$, it follows that $F_p \notin AC_{2^p-1}$ on C.

(iii) See Section 6.36 (F_p is a function of that type).

Remark 6.37.1 *The utility of this example follows by Remark 2.29.2, (ii).*

6.38 A function $G \in VB_2$, $G \notin AC_n$ on C, $G \in \mathcal{F}$ on $[0, 1]$

Let C be the Cantor ternary set (see Section 6.1). Let $x_p, y_p \in C$, $p = \overline{1, \infty}$, such that $c_i(x_p) = c_i(y_p) = 2$ if $i = \overline{1, p}$; $c_i(x_p) = 0$ if $i = \overline{p+1, \infty}$; $c_{p+1}(y_p) = 0$; $c_i(y_p) = 2$ if $i = \overline{p+2, \infty}$. For each $p = \overline{1, \infty}$ let F_p be the function defined in Section 6.37. Let G be a function defined as follows: $G(1) = 0$; $G(x) = (1/2^p) \cdot F_p(3^p(x - x_o))$, $x \in [x_p, y_p]$. Extending G linearly on each interval $[y_p, x_{p+1}]$, we have G defined and continuous on $[0, 1]$. Then we have:

(i) $G \in VB_2$ on C;

(ii) $G \in \mathcal{F}$ on $[0, 1]$;

(iii) $G \in AC_n$ on C, for no positive integer n.

Proof. (i) Clearly $G(x) \in [0, 1/2^p)$, $x \in [x_n, y_n]$. Let $\{I_k\}_k$ be a finite set of nonoverlapping closed intervals with endpoints in C. Let $A_p = \{k : I_k \subset [x_p, y_p]\}$ and let $A'_p = \{k :$ there exists no p such that $I_k \subset [x_p, y_p]\}$. By Section 6.37, (iii) it follows that $\sum_{k=1}^{\infty} \mathcal{O}^2(G; I_k \cap C) = \sum_{k \in A_p} \mathcal{O}^2(G; I_k \cap C) + \sum_{k \in A'_p} \mathcal{O}^2(G; I_k \cap C) \leq V_2(G; [x_p, y_p] \cap C) + \sum_{k=1}^{\infty} 1/2^k \leq \sum_{p=1}^{\infty} (1/2^p) \cdot V_2(F_p; [x_p, y_p] \cap C) + 2 \leq 2 + 2 = 4$, hence $G \in VB_2$ on C.

(ii) Since $F_p \in \mathcal{F}$ on $[0, 1]$ (see Section 6.37, (i)), it follows that $G \in \mathcal{F}$ on $[0, 1]$.

(iii) See Section 6.37, (ii).

Remark 6.38.1 *The utility of this example follows by Remark 2.29.2, (ii).*

6.39 A function $F_1 \in VB_2$ on C, $V_2(F_1; [0, x] \cap C) = \varphi(x)$ (G. Ene)

Let C be the Cantor ternary set (see Section 6.1) and let φ be the Cantor ternary function (see Section 6.2). Let F_1 and F_2 be the functions defined in Section 6.31. Then we have:

(i) $F_1, F_2 \in VB_2$ on C;

(ii) $V_2(F_1; C) = 1$ and $V_2(F_2; C) = 1$;

(iii) $V_2(F_1; [0, x]) = \varphi(x)$ and $V_2(F_2; [0, x]) = \varphi(x)$, whenever $x \in C$.

Proof. (i) See Section 6.31 (the proof of (vii)) or Section 6.36.

(ii) Let $[a,b]$ be a closed interval such that $a, b \in C$. From the proof of Section 6.31, (viii) it follows that there exist two intervals J_1 and J_2 such that

(1) $B(F_2; C \cap [a,b]) \subset [a,b] \times (J_1 \cup J_2)$ *and* $|J_1| + |J_2| \leq \varphi(b) - \varphi(a)$.

It follows that

(2) $V_2(F_2; C) \leq \varphi(1) - \varphi(0) = 1$.

Let $[a_i, b_i]$, $i = \overline{1,16}$ be the remained closed intervals from the step 4. Let $V = V_2(F_2; C)$. Since $F_2(x) = (1/16) \cdot F_2(3^4(x - a_i)) + F_2(a_i)$, $x \in [a_i, b_i]$, we have

(3) $V_2(F_2; [a_i, b_i] \cap C) = (1/16)V$.

Consider the closed interval $[b_7, a_{10}]$. Then $F_2([b_7, a_{10}] \cap C) \subset J_1 \cup J_2$, where $J_1 = [F(b_7), F_2(b_8)]$ and $J_2 = [F_2(a_9), F_2(a_{10})]$. It follows that

(4) $|J_1| + |J_2| = 2 \cdot (1/16)$,

and this sum is minimal. By Lemma 2.30.2 (applied to the intervals $[b_7, a_{10}]$, $[a_i, b_i]$, $i \in \{1, 2, \ldots, 7\} \cup \{10, 11, \ldots, 16\}$, and by (3) and (4) it follows that $(14/16)V + (2/16)V \leq V$, hence $V \geq 1$. By (2), $V = 1$. By (1), Section 6.31, (ii) and Lemma 2.30.2, it follows that $V_2(F_1; C) = 1$.

(iii) Let $I_{k_1 \ldots k_n} = [a_{k_1 \ldots k_n}, b_{k_1, \ldots k_n}]$, $k_i = 1, 2, 3, 4$, $i = \overline{1,n}$ be the remained closed intervals from the step $2n$ (numbered from the left to the right). Let $x \in C$. Then $x = I_{k_1(x)} \cap I_{k_1(x)k_2(x)} \cap \ldots$ and

(5) $F_2(x) = (1/4^n) \cdot F_2(9^n(x - a_{k_1(x) \ldots k_n(x)})) + F_2(a_{k_1(x) \ldots k_n(x)})$.

By (1), $V_2(F_2; [0,x] \cap C) \leq \varphi(x)$. By (ii) and (5), $V_2(F_2; I_{k_1(x) \ldots k_n(x)} \cap C) = 1/4^n$. By Lemma 2.30.2, $V_2(F_2; [0,x] \cap C) \geq k_1(x)/4 + k_2(x)/4^2 + \ldots = \varphi(x)$. By (1), (5) and Lemma 2.30.2, $V_2(F_1; [a,x] \cap C) = \varphi(x)$.

Remark 6.39.1 *The utility of this example follows by Remark 2.30.2.*

6.40 A function $F_q \in (N)$ on $[0,1]$, $F_q \notin VB_n$ on C, $F_q \in VB_\omega$ on C (G. Ene)

Let C be the Cantor ternary set (see Section 6.1), and let $q \in (2,4)$. Let $F_q : C \mapsto \mathbb{R}$, $F_q(x) = F_q(\sum_{i=1}^\infty c_i(x)/3^i) = \sum_{i=1}^\infty c_{2i}(x)/q^i$. Extending F_q linearly on the closure of each interval contiguous to C, we have F_q defined and continuous on $[0,1]$ (see Figure 6.21 for $q = 3$).

Then we have:

(i) $F_q \in (N)$ *on* $[0,1]$*;*

(ii) $F_q \in VB_n$ *on* C *for no positive integer* n*;*

(iii) $F \in VB_\omega$ *on* C.

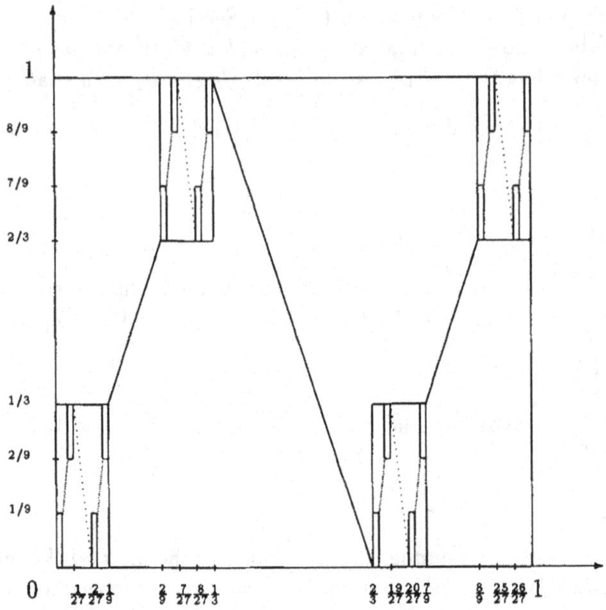

Figure 6.21: The function F_3

Proof. (i) $F_q(C)$ can be covered by 2^n closed intervals, each having the length at most $2/q^n$. It follows that $| F_q(C) | \le 2^n \cdot (2/q^n)$, hence $| F_q(C) | = 0$. Thus $F_q \in (N)$ on $[0,1]$.

(ii) Suppose on the contrary that $F_q \in VB_n$ on C for some positive integer n, and let $V = V_n(F_q; C)$. Let $[a_i, b_i]$, $i = 1,2,3,4$ be the remained closed intervals from the second step. Then $F_q(x) = (1/q) \cdot F_q(9(x - a_i)) + F_q(a_i)$, $x \in [a_i, b_i]$. It follows that $V_i = V_n([a_i, b_i] \cap C) = V/q$. By Lemma 2.30.2, $V_1 + V_2 + V_3 + V_4 \le V$. Hence $4 \cdot (V/q) \le V$ and $q \ge 4$, a contradiction.

(iii) This is evident.

Remark 6.40.1 *(i) For $q = 3$, $F_q \in \Lambda Z$ and $F_q \notin [\mathcal{E}]$ (see Section 6.57). The function F_3 was constructed by Fleissner and Foran in [FlF1] (they showed that $F_3 \in (N)$).*

(ii) The utility of this example follows by Theorem 2.29.1, (vi).

6.41 A function $G_1 \in L_2G$, $G_1 \in \mathcal{F}$, $G_1 \notin SVBG$, $G_1 \notin SACG$, $(G_1)'_{ap}$ does not exist on a set of positive measure

Let C be the Cantor ternary set (see Section 6.1), and let φ be the Cantor ternary function (see Section 6.2). Let $F_1, F_2 : [0,1] \mapsto [0,1]$ be the functions defined in Section 6.31. Let $a \in (0,1)$ and $P = \{y : y = \sum_{i=1}^{\infty} d_i \cdot ((1-a)/2^i + 3a/4^i),\ d_i \in \{0,1\}\}$. Each point $y \in P$ is uniquely represented by $\sum_{i=1}^{\infty} d_i(y) \cdot ((1-a)/2^i + 3a/4^i)$. Then

P is a perfect set and $|P| = 1 - a$. (The set P can also be obtained by a process similarly to the Cantor ternary process.) Let $h : P \mapsto C$, $h(y) = \sum_{i=1}^{\infty} 2d_i(y)/3^i$. Clearly $h(P) = C$. Extending h linearly on the closure of each interval contiguous to P, we have h defined and continuous on $[0,1]$. In fact h is a homeomorphism (because h is strictly increasing on $[0,1]$). Let $g, G_1, G_2 : [0,1] \mapsto [0,1]$, $g = \varphi \circ h$, $G_1 = F_1 \circ h$, $G_2 = F_2 \circ h$ (see Figure 6.22 and Figure 6.23 for $a = 1/2$). Clearly $g = G_1 + G_2$ on $[0,1]$, $g(y) = \sum_{i=1}^{\infty} d_i(y)/2^i$ for $y \in P$, and g is constant on each interval contiguous to P.
Then we have:

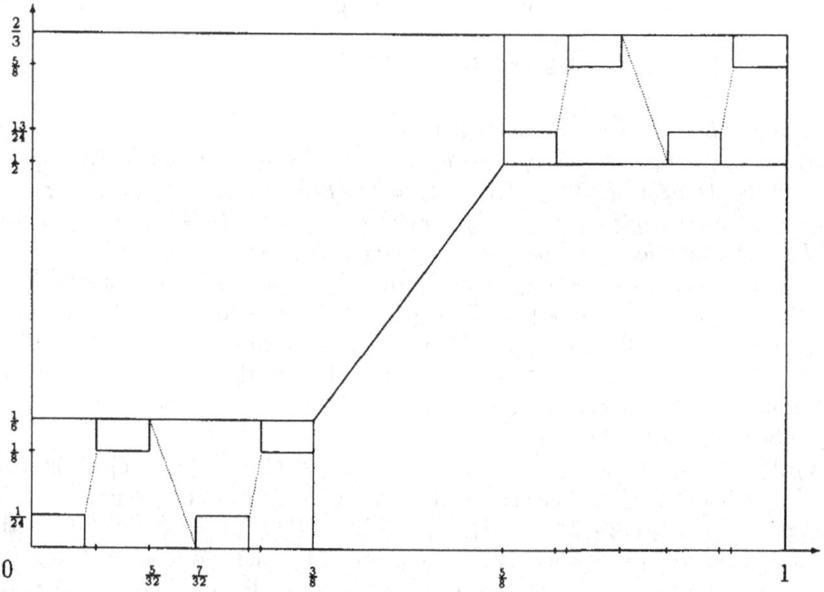

Figure 6.22: The function G_1

(i) $g \in L$ *on* P *with the constant* $1/(1-a)$, *hence* $g \in AC$ *on* $[0,1]$;

(ii) $G_1, G_2 \in L_2$ *on* P, *hence* $G_1, G_2 \in L_2G \subseteq AC_2G \subseteq \mathcal{F}$ *on* $[0,1]$;

(iii) G_1, G_2 *are not approximately derivable a.e. on* P;

(iv) g *is derivable a.e. on* $[0,1]$;

(v) $G_1, G_2 \notin SVBG$ *on* P, *hence* $G_1, G_2 \notin SACG$, $G_1, G_2 \notin VBG$ *and* $G_1, G_2 \notin ACG$ *on* P.

Proof. (i) Let $x, y \in P$, $x < y$. Then there exists a positive integer k such that $d_i(x) = d_i(y)$, $i = \overline{1,k}$. Let $S = \sum_{i=k+2}^{\infty}(d_i(y) - d_i(x))/4^{i-k-1}$. Then $S \in [-1/3, 1/3]$ and $y - x = \sum_{i=1}^{\infty}(d_i(y) - d_i(x)) \cdot ((1-a)/2^i + 3a/4^i) = (1-a)(g(y) - g(x)) + 3a/4^{k+1} + 3a/4^{k+1} \cdot S > (1-a)(g(y) - g(x))$. It follows that

(1) $\quad g(y) - g(x) < (y-x)/(1-a), \quad x, y \in P, \ x < y.$

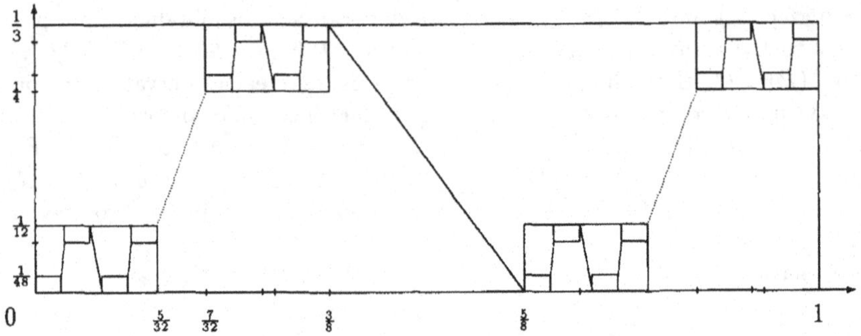

Figure 6.23: The function G_2

Hence $g \in L$ on P with the constant $1/(1-a)$.

(ii) Let $x, y \in P$, $x < y$, $h(x) = x_1$, $h(y) = y_1$. Then by Section 6.31 (the proof of (vii)) and by (1) we have $\mathcal{O}^2(G_2; P \cap [x,y]) = \mathcal{O}^2(F_2; C \cap [x_1, y_1]) \le \varphi(y_1) - \varphi(x_1) = \varphi(h(y)) - \varphi(h(x)) = g(y) - g(x) < (y-x)/(1-a)$. Hence $G_2 \in L_2$ on P. Similarly $G_1 \in L_2$ on P. For the second part see Corollary 2.32.1, (iv).

(iii) Suppose on the contrary that G_1 and G_2 are approximately derivable a.e. on P. By Lemma 2.17.2, since $| G_1(P) | = | G_2(P) | = 0$, it follows that $(G_1)'_{ap}(x) = (G_2)'_{ap}(x) = 0$ a.e. on P. But $G_1' = -G_2'$ on $[0,1] \setminus P$, hence $(G_1 + G_2)'_{ap}(x) = 0$ a.e. on $[0,1]$. By (i), $G_1 + G_2 = g \in AC$ on $[0,1]$. It follows that g is constant on $[0,1]$ (see Corollary 2.14.1), a contradiction.

(iv) See Corollary 2.14.2, (i).

(v) At the step $2n + 2k - 1$, $k = \overline{1,n}$ we have 4^{n+k-1} excluded open intervals, each of length $1/4^{2n+2k-1}$. The remained closed intervals from the step $4n-1$ have the length $(1 - \sum_{i=2}^{4n} 1/2^i)/2^{4n} > (1/2)/2^{4n} \ge 1/4^{2n+2k-1}$, $k = \overline{1,n}$. It follows that W_n is a sparse figure on P, where W_n is the union of the closures of all excluded open intervals from the steps $2n + 2k - 1$, $k = \overline{1,n}$. The oscillation of G_2 on the closure of an open interval from the step $2n + 2k - 1$, $k = \overline{1,n}$ is $(1/3) \cdot (1/4^{n+k})$. It follows that $\sum_{I \in W_n} \mathcal{O}(G_n; P \cap I) = (1/3) \cdot n$, hence $G_2 \notin SVBG$ on P (see Lemma 2.35.2 and Theorem 2.35.1, (v). By Lemma 2.34.3 and Theorem 2.34.1, (v), $G_2 \notin SACG$ on P. For G_1 the proof is similar.

Remark 6.41.1 *The utility of this example follows by Remark 2.28.3; Remark 2.29.2, (iii); Remark 2.34.2, (i); Remark 2.35.3, (ii).*

6.42 A function $F \in SACG$, $F \notin \mathcal{F}$, $F \notin ACG$

Let $\alpha \in (0,1]$ and let $S = [a,b] \times [c,d] \subseteq S_o = [0,1] \times [0,1]$ be a square with side length α. Let m, n and p be positive integers such that $\alpha/2 \ge 1/m$, $n \ge 2$, $p = mn$, and let $\beta = \alpha/(p \cdot 2^{p-1})$.

For S, α, m, n, p and β we define: $A_k = [a_k, b_k]$ for each $k = \overline{1,p}$, and $B_k = (b_k, a_{k+1})$ for each $k = \overline{1,p-1}$, where $a_k = a + \beta(k-1) + \sum_{i=1}^{k-1} \alpha/2^i$ and $b_k = a_k + \beta$, $k = \overline{1,p}$. Let $C_j = [c_j, d_j]$, $j = \overline{1,m}$, where $c_j = c + (j-1)(\alpha - \beta)/(m-1)$ and $d_j = c_j + \beta$. Let's denote by $S_{ij} = A_{mi+j} \times C_j$, $i = \overline{0,n-1}$, $j = \overline{1,m}$. Let $E = \cup_{k=1}^{p} A_k = \cup_{i=0}^{n-1} \cup_{j=1}^{m} A_{mi+j}$. Let $g : [a,b] \mapsto [c,d]$ be a function such that

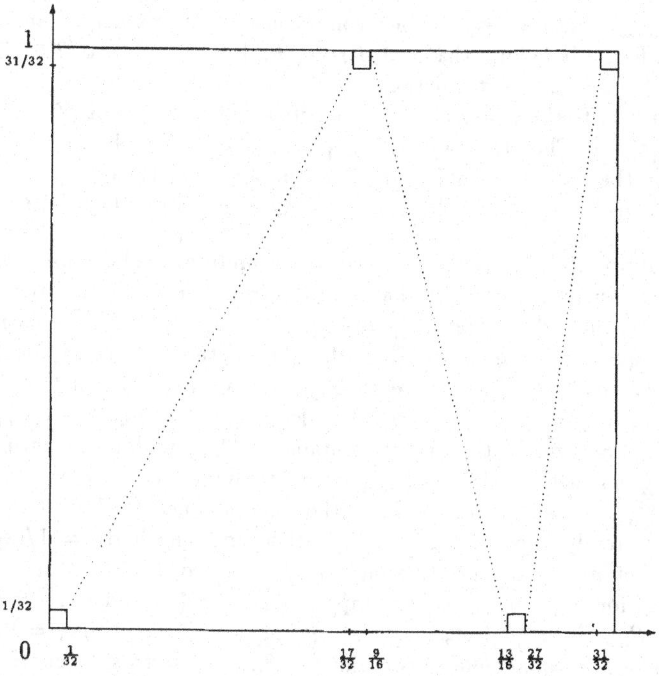

Figure 6.24: The function g

$B(g;E) \subset \cup_{i=1}^{n-1} \cup_{j=1}^{m} S_{ij}$, and g is linear on each $\overline{B}_k$, $k = \overline{1,p-1}$. Clearly $\mid A_k \mid = \beta$, $k = \overline{1,p}$; $\mid B_k \mid = \alpha/2^k$, $k = \overline{1,p-1}$; $\mid C_j \mid = \beta$, $j = \overline{1,m}$; $[c,d] \setminus \cup_{j=1}^{m} C_j$ is an open set with $m-1$ components of equal length $(\alpha - m\beta)/(m-1)$; S_{ij} are equal squares contained in S with side length β (see Figure 6.24 for $S = [0,1] \times [0,1]$, $\alpha = 1$, $m = n = 2$, $p = 4$, $\beta = 1/32$).
In what follows we take $n = m^3$. Then $p = m^4$.

Lemma 6.42.1 *Let W be a sparse figure for E. Suppose that $W \not\subset E$. Then:*

(i) there exist $k_o \in \{1, 2, \ldots, p-1\}$ and a unique component K_o of W such that $K_o \subset [a_{k_o}, b]$ and each A_i, $i = \overline{1, k_o - 1}$ contains at most one component of W;

(ii) $W \subset \cup_{i=1}^{k_o-1}[a_i, b_i] \cup [a_{k_o}, b]$;

(iii) $\sum_{K \in W} \mathcal{O}(g;K) < 2\alpha$.

Proof. (i) Since $W \not\subset E$, there exists at least one positive integer $k \in \{1, \ldots, p-1\}$ such that $M_k \subset W$. Let $k_o = \inf\{k : B_k \subset W\}$. Let K_o be the component of W such that $B_{k_o} \subset K_o$. Then $\mid K_o \mid \geq \mid B_{k_o} \mid = \alpha/2^{k_o}$. Since K_o has the endpoints in E it follows that $K_o \subset [a_{k_o}, b]$. Suppose that there exists another component K of W such that $K \subset [a_{k_o}, b]$. Since W is a sparse figure it follows that $d(K_o, K) = \inf\{\mid x - y \mid : x \in K_o,\ y \in K\} > \mid K_o \mid \geq \alpha/2^{k_o}$. Then $2\alpha/2^{k_o} < \mid K_o \mid + d(K_o;K) \leq b - a_{k_o} = \alpha/2^{k_o-1} - (k_o - 1)\cdot\beta < \alpha/2^{k_o-1}$, a contradiction. If $k_o = 1$ then W consists of only one component. If $k_o \geq 2$ then we show that each A_i, $i = \overline{1, k_o - 1}$ contains at most one component of W. Suppose on the contrary that there exists $i_o \in \{1, 2, \ldots, k_o - 1\}$

such that $[a_{i_o}, b_{o_o}]$ contains two distinct components of W, namely J and J'. Since W is a sparse figure it follows that $d(J, J') > | K_o | \geq \alpha/2^{k_o} > \alpha/2^{p-1} > \beta$. But $\beta = b_{i_o} - a_{i_o} \geq d(J, J')$, a contradiction.

(ii) Suppose on the contrary that there exists a component J of W such that $J \not\subset \cup_{i=1}^{k_o-1}[a_i, b_i] \cup [a_{k_o}, b]$. Then there exists $i_o \in \{1, 2, \dots, k_o - 1\}$ such that $B_{i_o} \subset J \subset W$, a contradiction (because $k_o = \inf\{k : B_k$ is a component of $W\}$).

(iii) $\sum_{k \in W} \mathcal{O}(g; K) \leq \sum_{i=1}^{k_o-1} \mathcal{O}(g; [a_i, b_i]) + \mathcal{O}(g; [a_{k_o}, b]) < (k_o - 1)\beta + \alpha < 2\alpha$.

We shall repeat infinitely many times the construction from the beginning. Let $\alpha_o = 1$ and let m_1 be a positive integer such that $\alpha_o/2 \geq 1/m_1$. For $S, \alpha_0,\ m_1,\ n_1 = m_1^3,\ p_1 = m_1^4$ and $\alpha_1 = \alpha_o/(p_1 \cdot 2^{p_1-1})$, let $S_{i_1 j_1} = A_{m_1 i_1 + j_1} \times C_{j_1}$, $i_1 = \overline{0, n_1 - 1}$, $j_1 = \overline{1, m_1}$ be the equal squares contained in S_o, with side length α_1. Let m_2 be a positive integer such that $\alpha_1/2 \geq 1/m_2$. For each $S_{i_1 j_1}$,, for $\alpha_1,\ m_2,\ n_2 = m_2^3,\ p_2 = m_2^4$ and $\alpha_2 = \alpha_1/(p_2 \cdot 2^{p_2-1}) = 1/(p_1 \cdot p_2 \cdot 2^{p_1+p_2-2})$, let $S_{i_1 j_1 i_2 j_2} = A_{m_1 i_1 + j_1,, m_2 i_2 + j_2} \times C_{j_1 j_2}$, $i_2 = \overline{0, n_2 - 1}$, $j_2 = \overline{1, m_2}$ be the squares contained in $S_{i_1 j_1}$ with side length α_2, where $A_{m_1 i_1 + j_1, m_2 i_2 + j_2}$ is constructed in $A_{m_1 i_1 + j_1}$ in the same manner as A_k was constructed in $[a, b]$, and $C_{j_1 j_2}$ is constructed in C_{j_1} in the same manner as C_j in $[c, d]$. Suppose that we have already constructed $S_{i_1 j_1 \dots i_k j_k}$ with side length $\alpha_k = 1/(p_1 p_2 \cdots p_k \cdot 2^{p_1 + \dots + p_k - k})$. Let m_{k+1} be a positive integer such that $\alpha_k/2 \geq 1/m_{k+1}$. For every $i_1, j_1, \dots, i_k, j_k$, for $\alpha_k,\ m_{k+1},\ n_{k+1} = m_{k+1}^3,\ p_{k+1} = m_{k+1}^4$ and $\alpha_{k+1} = \alpha_k/(p_{k+1} \cdot 2^{p_{k+1}-1})$. Let $S_{i_1 j_1 \dots i_{k+1} j_{k+1}} = A_{m_1 i_1 + j_1, \dots, m_{k+1} i_{k+1} + j_{k+1}} \times C_{j_1 \dots j_{k+1}}$, $i_{k+1} = \overline{0, n_{k+1} - 1}$, $j_{k+1} = \overline{1, n_{k+1}}$ be the equal squares contained in $S_{i_1 j_1 \dots i_k j_k}$ with side length α_{k+1}. Let $P_k = \cup_{(i_1, j_1, \dots, i_k, j_k)} A_{m_1 i_1 + j_1, \dots, m_k i_k + j_k}$, $k = \overline{1, \infty}$ and $Q_k = \cup_{(j_1, \dots, j_k)} C_{j_1 \dots j_k}$, $k = \overline{1, \infty}$. Let $P = \cap_{k=1}^{\infty} P_k$ and $Q = \cap_{k=1}^{\infty} Q_k$. Then P and Q are perfect sets of measure zero. For each $x \in P$ there exist two unique sequences $\{i_k(x)\}_k$ and $\{j_k(x)\}_k$ such that $x = \cap_{k=1}^{\infty} A_{m_1 i_1(x) + j_1(x), \dots, m_k i_k(x) + j_k(x)}$. Let $y_x = \cap_{k=1}^{\infty} C_{j_1(x) \cdot j_k(x)}$. Then $y_x \in Q$ and y_x is unique. Let $F(x) = y_x$ if $x \in P$. Then F is continuous on P. Extending F linearly on the closure of each interval contiguous to P, we have $F : [0, 1] \mapsto [0, 1]$ and $F \in \mathcal{C}$ on $[0, 1]$. Then we have:

(i) $F \in SAC$ on P, hence $F \in SACG$ on $[0, 1]$;

(ii) $F \notin \mathcal{B}$ on P, hence $F \notin \mathcal{F}$ and $F \notin ACG$ on $[0, 1]$.

Proof. (i) By the construction (see Figure 6.24) it follows that

(1) $\quad F_{/A_{m_1 i_1 + j_1, \dots, m_k i_k + j_k}} \subset \cup_{i_{k+1}=0}^{n_{k+1}-1} \cup_{j_{k+1}=1}^{m_{k+1}} S_{i_1 j_1 \dots i_k j_k i_{k+1} j_{k+1}}$.

Let $\epsilon > 0$ and let k_o be a positive integer such that

(2) $\quad 1/2^{p_1 + \dots p_{k_o} - k_o} < \epsilon/2$.

Let $0 < \delta_\epsilon < \alpha_{k_o}$ and let W be a sparse figure for P such that $| W | < \delta_\epsilon$. Then $W \subset P_{k_o}$. Let k_1 be the positive integer for which $W \subset P_{k_1}$ and $W \not\subset P_{k_1+1}$. Then

(3) $\quad k_1 < k_o$.

Let $W_{i_1 j_1 \dots i_{k_1} j_{k_1}}$ be the collection of all components of W which are contained in the set $A_{m_1 i_1 + j_1, \dots, m_{k_1} o_{k_1} + j_{k_1}}$. So $W_{i_1 j_1 \dots i_{k_1} j_{k_1}}$ is a sparse figure for $\cup_{j=1}^{p_{k_1}+1} A_{m_1 i_1 + j_1, \dots, m_{k_1} i_{k_1} + j_{k_1} + j}$. By Lemma 6.42.1 and (1), it follows that $\sum_{K \in W} \mathcal{O}(F; K \cap P) \leq \sum_{K \in W} \mathcal{O}(F; K) \leq \sum_{(i_1, j_1, \dots, i_{k_1}, j_{k_1})} \sum_{K \in W_{i_1 j_1 \dots i_{k_1} j_{k_1}}} \mathcal{O}(F; K) < n_1 \cdot m_1 \cdot \dots \cdot n_{k_1} \cdot m_{k_1} \cdot 2\alpha_{k_1} = p_1 \cdot p_{k_1} \cdot 2\alpha_{k_1} =$

$2/2^{p_1+\dots p_{k_1}-k_1} < \epsilon$ (see (2) and (3)), hence $F \in SAC$ on P.

(ii) Suppose that $F \in \mathcal{B}$ on P. By Lemma 2.29.2, there exist a portion P_o of P and a positive integer n_o such that $F \in VB_{n_o}$ on P_o. Let $M_o > 0$ be given by this fact. Since $\{m_k\}_k$ be a strictly increasing sequence it follows that there exists a positive integer i_o such that $m_{i_o} - 1 > \max\{n_o, M_o\}$. Since P_o is a portion of P, there exist a positive integer $s_o \geq i_o$ and $i_1, j_1, \dots, i_{s_o}, j_{s_o}$ such that $P' = P \cap A_{m_1 i+1+j_1,\dots,m_{s_o} i_{s_o}+j_{s_o}} \subset P_o$. Let J_1 be the smallest closed interval containing $A_{m_1 i_1+j_1,\dots,m_{s_o} i_{s_o}+j_{s_o}, m_{s_o+1} i+j}$, $i = \overline{0, n_{s_o+1}-1}$, $j = \overline{1, m_{s_o+1}}$. Then $F(P \cap J_i) \subset \cup_{j=1}^{m_{s_o+1}} C_{j_1\dots j_{s_o} j}$. But $C_{j_1\dots j_{s_o}} \setminus \cup_{j=1}^{m_{s_o+1}} C_{j_1\dots j_{s_o} j}$ is an open set with $m_{s_o+1}-1$ components of equal length $(\alpha_{s_o} - m_{s_o+1} \cdot \alpha_{s_o+1})/(m_{s_o+1}-1)$. It follows that $\mathcal{O}(F; P \cap J_i) \geq (\alpha_{s_o} - m_{s_o+1} \cdot \alpha_{s_o+1})/(m_{s_o+1}-1)$. Since there are n_{s_o+1} nonoverlapping intervals J_i contained in $A_{m_1 i_1+j_1,\dots,m_{s_o} i_{s_o}+j_{s_o}}$, it follows that $V_{n_o}(F; P') \geq V_{m_{s_o}-1}(F; P') \geq n_{s_o+1} \cdot (\alpha_{s_o} - m_{s_o+1} \cdot \alpha_{s_o+1})/(m_{s_o+1}-1) > (\alpha_{s_o}/(2 \cdot m_{s_o+1})) \cdot m^3_{s_o+1} = m^2_{s_o+1} \cdot \alpha_{s_o}/2 > m_{s_o+1} > M$, a contradiction.

Remark 6.42.1 *The utility of this example follows by Corollary 2.34.1, (ii), (iii).*

6.43 A function $F \in DW_1$, $F \notin DW^*$

Let $F : [0,1] \mapsto \mathbb{R}$ be a nonconstant continuous function on $[0,1]$. Let P be a countable dense subset of $[0,1]$.Then we have:

(i) $F \in DW_1$ *on* P;

(ii) $F \notin DW^*$ *on* P.

Remark 6.43.1 *The utility of this example follows by Theorem 2.36.1, (viii).*

6.44 A function $F \in AC^*; DW_1G$, $F \notin AC^*; DW^*G$

Let C be the Cantor ternary set (see Section 6.1), and let φ be the Cantor ternary function (see Section 6.2). Let $I^k_n = [a^k_n, b^k_n]$, $n = \overline{1, 2^{k-1}}$ be the open intervals excluded at the step k. Let c^k_n be the middle point of I^k_n. Let $F : [0,1] \mapsto \mathbb{R}$ be a function defined as follows: $F(x) = 0$ if $x \in C$, $F(x) = 1/2^k$ if $x = c^k_n$, F is linear on each $[a^k_n, c^k_n]$ and $[c^k_n, b^k_n]$. It follows that F is continuous on $[0,1]$. Then we have:

(i) $F \in AC^*; DW_1G$ *on* $[0,1]$;

(ii) $F \notin AC^*; DW^*G$ *on* $[0,1]$.

Proof. (i) This is evident.

(ii) Clearly $F \in AC^*G$ on $[0,1] \setminus C$. Suppose on the contrary that $F \in AC^*; DW^*G$ on C. Then there exists a sequence of sets $\{E_n\}_n$ such that $C = \cup_n E_n$, and either $F \in AC^*$ on E_n or $F \in DW^*$ on E_n. Let p be a positive integer such that $F \in AC^*$ on E_p. Since $F \in \mathcal{C}$, it follows that $F \in AC^*$ on $\overline{E}_p$. We show that $F \in DW^*$ on $\overline{E}_p$. Let $\epsilon > 0$ and let $\delta > 0$ be given by the fact that $F \in AC^*$ on $\overline{E}_p$. Since $F \in \mathcal{C}$ and $| \overline{E}_p | = 0$, we can cover the set E_p with a sequence of nonoverlapping intervals $\{I_k\}_k$, such that $\sum_{n=1}^{\infty} | I_n | < \delta$ and $\sum_{n=1}^{\infty} \mathcal{O}(F; I_n) < \epsilon$. Hence $F \in DW^*$ on $\overline{E}_p$. It follows that $F \in DW^*G = DW^*$ on C. We shall obtain a contradiction by showing that $F \notin DW^*$ on C. Let $C \subset \cup_{i=1}^{\infty} (a_i, b_i)$. For each i let J_i be the greatest excluded open interval contained in $[\alpha, \beta]$, where $\alpha = \inf((a_i, b_i)) \cap C)$ and $\beta = \sup((a_i, b_i) \cap C)$. Suppose that J_i is excluded at the step k. Then $J_i = (\sum_{i=1}^{k-1} c_i/3^i + \sum_{i=k}^{\infty} 2/3^i, \sum_{i=1}^{k-1} c_i/3^i + 2/3^k)$. Let $J_i' = [\sum_{i=1}^{k-1} c_i/3^i, \sum_{i=1}^{k-1} c_i/3^i + \sum_{i=k}^{\infty}]$. Then $[\alpha, \beta] \subset J_i'$, hence $C \subset_{i=1}^{\infty} J_i'$. We have $\mathcal{O}(F; J_i) = \mathcal{O}(F; J_i') = \mid \varphi(J_i') \mid = 1/2^k$, and $[0,1] = \varphi(C) \subset \sum_{i=1}^{\infty} \varphi(J_i')$, hence $\sum_{i=1}^{\infty} \mathcal{O}(F; (a_i, b_i)) \geq \sum_{i=1}^{\infty} \mathcal{O}(F; J_i) = \sum_{i=1}^{\infty} \mid \varphi(J_i') \mid \geq 1$. It follows that $F \notin DW^*$ on C.

Remark 6.44.1 *The utility of this example follows by Theorem 2.36.1, (ix).*

6.45 Functions $F_1, F_2 \in C \cap AC^*; DW^*G$, F_1, F_2 are derivable *a.e.*, $F_1' = F_2'$ *a.e.*, F_1 and F_2 do not differ by a constant

Let C be the Cantor ternary set (see Section 6.1), and let φ be the Cantor ternary function (see Section 6.2). By Theorem 3.1.1, (i), there exist $F_1, F_2 : [0,1] \mapsto \mathbb{R}$ such that:

(i) $F_1, F_2 \in C \cap AC^*; DW^*G$ *on* $[0,1]$;

(ii) F_1, F_2 *are derivable a.e. on* $[0,1]$;

(iii) $F_1 - F_2 = \varphi$ *on* $[0,1]$, *hence* $F_1' = F_2'$ *a.e. on* $[0,1]$.

6.46 Functions $F_1, F_2 \in C \cap AC^*; DW_1G$, F_1, F_2 are approximately derivable *a.e.*, $F_1 + F_2 \notin$ quasi-derivable

Let P be the symmetric, perfect, nowhere dense subset of $[0,1]$ constructed in Section 6.26. Then $0, 1 \in P$ and $\mid P \mid = 1/2$. Let $P_1 = P$ and let $I_{n_1} = (a_{n_1}, b_{n_1})$, $n = \overline{1, \infty}$ be the intervals contiguous to P_1. Let $P_2 = P_1 \cup S_2$, where $S_2 = \cup_{i=1}^{\infty} P^{n_1}$. Let $I_{n_1 n_2} = (a_{n_1 n_2}, b_{n_1 n_2})$, $n_1, n_2 = \overline{1, \infty}$ be the intervals contiguous to P_2, $I_{n_1 n_2} \subset I_{n_1}$. Let $f : [0,1] \mapsto [0,1]$, $f(x) = 0$, $x \in P_1$; $f(x) = (1/2^{n-1})(1/ \mid P^{n_1} \mid) \cdot (\mathcal{L}) \int_{a_{n_1}}^{x} (K_{[a_{n_1}, c_{n_1}] \cap P^{n_1}} - K_{[c_{n_1}, b_{n_1}] \cap P^{n_1}})(t) dt$, $x \in \overline{I}_{n_1}$. where I_{n_1} is an excluded interval from the step n. Then we have:

(1) $f \in C$ on $[0,1]$;

(2) $f(x) = 0$ on P;

(3) $f'(x)$ does not exist whenever $x \in P$ (this follows similarly to (v) of Section 6.26);

(4) $f \in L$ on each I_{n_1}, $n_1 = \overline{1, \infty}$;

(5) f is constant on each $I_{n_1 n_2}$, $n_1, n_2 = \overline{1, \infty}$;

(6) $f \in ACG$ on $[0,1]$.

Suppose that $P_1, P_2, \ldots, P_{i-1}, i \geq 3$ have already been defined and let's define P_i. Let $P^{n_1 \ldots n_{i-1}} = a_{n_1 \ldots n_{i-1}} + (b_{n_1 \ldots n_{i-1}} - a_{n_1 \ldots n_{i-1}}) \cdot P$, $S_i = \cup_{(n_1,\ldots,n_{i-1})} P^{n_1 \ldots n_{i-1}}$ and $P_i = P_{i-1} \cup S_i$. Let $f_i : [0,1] \mapsto [0, 1/2^i]$, $f_i(x) = 0$ if $x \in P_i$ and $f_i(x) = (1/2^{n_1+\ldots+n_i}) \cdot f((x - a_{n_1 \ldots n_i})/(b_{n_1 \ldots n_i} - a_{n_1 \ldots n_i}))$ if $x \in I_{n_1 \ldots n_i}$, where $I_{n_1 \ldots n_i} = (a_{n_1 \ldots n_i}, b_{n_1 \ldots n_i})$, $n_1, \ldots, n_i = \overline{1, \infty}$ are the intervals contiguous to P_i. Then we have:

(7) $f_i \in \mathcal{C}$ on $[0,1]$;

(8) $f_i(x) = 0$ on P_i;

(9) $f_i(x) = 0$ on S_{i+1};

(10) f_i is not derivable *a.e.* on S_{i+1};

(11) $f_i \in L \subset AC^*$ on each $I_{n_1 \ldots n_{i+1}}$;

(12) f_i is constant on each $I_{n_1 \ldots n_{i+2}}$;

(13) $f_i \in ACG$ on $[0,1]$;

(14) $| f_i(x) | < 1/2^{n_1+\ldots+n_i}$, $x \in I_{n_1 \ldots n_i}$.

Let $A = P \cup (\cup_{i=1}^{\infty} S_{2i+1})$, $B = \cup_{i=1}^{\infty} S_{2i}$, $E = [0,1] \setminus (A \cup B) = [0,1] \setminus (\cup_{i=1}^{\infty} P_i)$. Then $| A \cup B | = 1$. Let $F_1, F_2 : [0,1] \mapsto \mathbb{R}$ be functions defined as follows: $F_1(x) = \sum_{i=1}^{\infty} f_{2i-1}(x)$ and $F_2(x) = \sum_{i=1}^{\infty} f_{2i}(x)$. Let $R_{2i+1}(x) = \sum_{k=i}^{\infty} f_{2k+1}(x)$ and $R_{2i}(x) = \sum_{k=i+1}^{\infty} f_{2k}(x)$. It follows that $F_1(x) = \sum_{k=1}^{i} f_{2k-1}(x) + R_{2i+1}(x)$ and $F_2(x) = \sum_{k=1}^{i-1} f_{2k}(x) + R_{2i}(x)$. Then we have

(15) $R_{2i+1}(x) = 0$ on P_{2i+1}, and $R_{2i}(x) = 0$ on P_{2i};

(16) $\sum_{(n_1,\ldots,n_{2i+1})} \mathcal{O}(R_{2i+1}; I_{n_1 \ldots n_{2i+1}}) < 1/2^{2i-1}$ and $\sum_{(n_1,\ldots,n_{2i})} \mathcal{O}(R_{2i}; I_{n_1 \ldots n_{2i}}) < 1/2^{2i-2}$;

(17) $R_{2i+1} \in AC^*$ on P_{2i+1} and $R_{2i} \in AC^*$ on P_{2i};

(18) $F(x) = \sum_{k=1}^{i} f_{2k-1}(x)$ on P_{2i+1}, and $F_2(x) = \sum_{k=1}^{i-1} f_{2k}(x)$ on P_{2i};

(19) $\mathcal{O}(F_1; I_{n_1 \ldots n_{2i+1}}) = \mathcal{O}(R_{2i+1}; I_{n_1 \ldots n_{2i+1}})$ and $\mathcal{O}(F_2; I_{n_1 \ldots n_{2i}}) = \mathcal{O}(R_{2i}; I_{n_1 \ldots n_{2i}})$.

Proof. (15), (16), (17) follow by (8), (14) and Lemma 3.1.1; (18) follows by (15); (19) follows by (12).

Then we have:

(i) F_1, F_2, $F_1 + F_2 \in \mathcal{C}$ *on* $[0,1]$;

(ii) $F_1, F_2 \in ACG$ *on* $A \cup B$, *hence* F_1, F_2 *are approximately derivable a.e. on* $[0,1]$;

(iii) $F_1 \in AC^*G$ *on* A, *and* F_1 *is not derivable a.e. on* B;

(iv) $F_2 \in AC^*G$ *on* B, *and* F_2 *is not derivable a.e. on* A;

(v) $F_1 + F_2$ *is not derivable a.e. on* $[0,1]$, *hence* $F_1 + F_2 \notin$ *quasi-derivable on* $[0,1]$;

(vi) $F_1, F_2 \in DW^*$ *on* E;

(vii) $F_1 \in DW_1G$ *on* S_{2i}, *and* $F_2 \in DW_1G$ *on* S_{2i-1};

(viii) $F_1, F_2 \in AC^*; DW_1G$ *on* $[0,1]$.

Proof. (i) See (14).

(ii) By (13) and (18) it follows that $F_1 \in ACG$ on P_{2i+1} and $F_2 \in ACG$ on P_{2i}. Hence $F_1, F_2 \in ACG$ on $A \cup B$. By Theorem 2.15.3, F_1, F_2 are approximately derivable *a.e.* on $[0,1]$.

(iii) Since $F_1 = R_1$ on $[0,1]$, by (17) it follows that $F_1 \in AC^*$ on P_1. Since $f_1, f_3, \ldots, f_{2i-1} \in L \subset AC^*$ on each $I_{n_1\ldots n_{2i}}$ (see (11)), it follows that $\sum_{k=1}^{i} f_{2k-1}(x)$ is AC^* on each $P^{n_1\ldots n_{2i}}$, hence it is AC^*G on S_{2i+1}. By (17), $R_{2i+1} \in AC^*$ on P_{2i+1}, hence $R_{2i+1} \in AC^*$ on S_{2i+1}. It follows that $F_1 \in AC^*G$ on S_{2i+1}, hence $F_1 \in AC^*G$ on A. By (12) and (11) we obtain that $f'_{2i-1}(x)$ does not exist whenever $x \in S_{2i}$. Since R_{2i+1} is derivable *a.e.* on $P_{2i+1} \supset S_{2i+1} \supset S_{2i}$ (see (17) and Theorem 2.15.1), it follows that F_1 is not derivable *a.e.* on S_{2i}. Hence F_1 is not derivable *a.e.* on B.

(iv) The proof is similar to that of (iii).

(v) See (iii) and (iv).

(vi) Let $\epsilon > 0$, and let i be a positive integer such that $1/2^{2i-1} < \epsilon$. Since $E \subset \bigcup_{(n_1,\ldots,n_{2i+1})} I_{n_1\ldots n_{2i+1}}$, by (16) and (19), $\sum_{(n_1,\ldots,n_{2i+1})} \mathcal{O}(F_1; I_{n_1\ldots n_{2i+1}}) < 1/2^{2i-1} < \epsilon$, hence $F_1 \in DW^*$ on E. Similarly $F_2 \in DW^*$ on E.

(vii) By (12), since $f_{2i-1}(x) = 0$ on $S_{2i} \subset \bigcup_{(n_1,\ldots,n_{2i-1})}$, it follows that $F_1 \in DW_1G$ on S_{2i}. Similarly $F_2 \in DW_1G$ on S_{2i-1}.

(viii) See (iii), (iv) and (vii).

Remark 6.46.1 *The utility of this example follows by Remark 3.1.1, (i); Remark 3.1.3, (i).*

6.47 Functions $F_1, F_2 \in \mathcal{E} \cap \mathcal{B}$, $F_1, F_2 \notin \mathcal{F}$, $F_1 + F_2 \notin \mathcal{E}$

Let C be the Cantor ternary set (see Section 6.1) and let φ be the Cantor ternary function (see Section 6.2). Let $\{j_i\}_i$ be a strictly increasing sequence of natural number such that $j_o = 0$,

(1) $\quad 3^{j_i} < 2^{j_{i+1}}$ *and* $n_i = j_{i+1} - j_i$.

Let $F_1, F_2 : C \mapsto [0,1]$, $F_1(x) = \sum_{i=0}^{\infty}\sum_{k=1}^{n_{2i+1}} c_{j_{2i+1}+k}(x)/2^{j_{2i+1}+k+1}$ and $F_2(x) = \sum_{i=0}^{\infty}\sum_{k=1}^{n_{2i}} c_{j_{2i}+k}(x)/2^{j_{2i}+k+1}$. Extending F_1 and F_2 linearly on the closure of each interval contiguous to C, we have F_1 and F_2 defined and continuous on $[0,1]$. Clearly

(2) $\quad F_1 + F_2 = \varphi$ *on* $[0,1]$.

Then we have:

(i) F_1, F_2 *are derivable a.e. on* $[0,1]$ *and* $F_1' = -F_2'$ *a.e. on* $[0,1]$, *but* $F_1 + F_2$ *is not constant on* $[0,1]$;

(ii) $F_1, F_2 \in E_1$ *on C, hence* $F_1, F_2 \in E_1G \subseteq \mathcal{E} \subseteq (N)$ *on* $[0,1]$;

(iii) $F_1, F_2 \in VB_2$ *on C, hence* $F_1, F_2 \in VB_2G \subseteq \mathcal{B}$ *on* $[0,1]$;

(iv) $F_1 + F_2 \notin \mathcal{E}$ *on* $[0,1]$;

(v) $F_1, F_2 \notin \mathcal{F}$ *on* $[0,1]$.

Proof. (i) This is evident.

(ii) It is sufficient to show that $F_2 \in DW_1$ on C. Let $\epsilon > 0$ and let p be a positive integer such that

(3) $(2/3)^{j_{2p+1}} < \epsilon$.

Let $I_m = [a_m, b_m]$, $m = \overline{1, 2^{j_{2p+1}}}$ be the remained closed intervals from the step j_{2p+1}. Then $| I_m | = 1/3^{j_{2p+1}}$. If $x \in I_m \cap C$ then $c_i(x) = c_i(b_m) = c_i(a_m)$ for $i = \overline{1, j_{2p+1}}$; $c_i(a_m) = 0$ and $c_i(b_m) = 2$ for $i > j_{2p+1}$. Let $J_m = [F_2(a_m), F_2(b_m)]$. By (1), $| J_m | = \sum_{i>p+1} \sum_{k=1}^{n_{2i}} 2/2^{j_{2i}+k+1} < 1/2^{j_{2p-2}} < 1/3^{j_{2p+1}} = | I_m |$ and $\mathcal{O}(F_2; I_m \cap C) = | J_m |$. By (3), $\sum_{m=1}^{2^{j_{2p+1}}} \mathcal{O}(F_2; I_m \cap C) < \sum_{m=1}^{2^{j_{2p+1}}} 1/3^{j_{2p+1}} = (2/3)^{j_{2p+1}} < \epsilon$, hence $F_2 \in DW_1$ on C. It follows that $F_2 \in E_1$ on C, and $F_2 \in E_1 G$ on $[0,1]$. By Corollary 2.37.1, (ii), $F_2 \in \mathcal{E} \subseteq (N)$ on $[0,1]$. For F_1 the proof is similar.

(iii) In fact F_1 and F_2 are functions as in Section 6.36, hence $F_1, F_2 \in VB_2$ on C. It follows that $F_1, F_2 \in VB_2 G \subseteq \mathcal{B}$ on $[0,1]$.

(iv) Since $\mathcal{E} \subseteq (N)$ and $\varphi \notin (N)$, it follows that $F_1 + F_2 \notin \mathcal{E}$.

(v) By Corollary 2.37.1, (ix), $\mathcal{F} + \mathcal{E} = \mathcal{E}$. By (ii) and (iv), $F_1, F_2 \notin \mathcal{F}$.

Remark 6.47.1 *The utility of this example follows by Corollary 2.37.1, (xvii); Remark 2.37.3, (i); Remark 4.2.2, (ii), (iii).*

6.48 A function $G_n \in E_{n+1}$, $G_n \notin E_n$, $G_n \in L_{n^2+2n+1}$, $G_n \notin VB_{n^2+2n}$

Let C be the Cantor ternary set (see Section 6.1), and let φ be the Cantor ternary function (see Section 6.2). Let $n \geq 1$ be a positive integer and let $C_n = \{x : x = \sum_{i=1}^{\infty} c_i/(2n+1)^i,\ c_i \in \{0, 2, \dots, 2n\},\ i = \overline{1, \infty}\}$. Each $x \in C_n$ is uniquely represented by $\sum_{i=1}^{\infty} c_i(x)/(2n+1)^i$. Then C_n is a symmetric, perfect, nowhere dense subset of $[0,1]$ and $| C_n | = 0$. Clearly $C_1 = C$. Let $\varphi_n : C_n \mapsto [0,1]$ be a function defined as follows: $\varphi_n(x) = (1/2) \sum_{i=1}^{\infty} c_i(x)/(n+1)^i$. Extending φ_n linearly on the closure of each interval contiguous to C_n, we have φ_n defined and continuous on $[0,1]$. Clearly $\varphi_1 = \varphi$. For each positive integer k let $R_k = \sum_{i=k}^{\infty} 2n/(2n+1)^i$. Then we have:

(i) φ_n *is increasing on* $[0,1]$;

(ii) φ_n *is constant on each interval contiguous to* C_n.

Proof. (i) Let $x, y \in C_n$, $x < y$ and let k be a positive integer such that $c_k(x) + 2 \leq c_k(y)$. Then $c_i(x) = c_i(y)$ for each $i = \overline{1, k-1}$. We have $\varphi_n(y) - \varphi_n(x) \geq (1/2) \cdot (2/(n+1)^k + \sum_{i=k+1}^{\infty} (c_i(y) - c_i(x))/(n+1)^i) \geq 0$. It follows that φ_n is increasing on $[0,1]$.

(ii) Let $J = (a, b)$ be an interval contiguous to C_n. Then there exist a positive integer m and $c_1, c_2, \dots, c_m \in \{0, 2, \dots, 2n\}$, $c_m \geq 2$ such that $b = \sum_{i=1}^{m} c_i/(2n+1)^i$ and $a = \sum_{i=1}^{m-1} c_i/(2n+1)^m + R_{m+1}$. It follows that $\varphi_n(a) = \varphi_n(b)$, hence φ_n is constant on (a, b).

Let $\{j_k\}$, $k=\overline{1,\infty}$ be an increasing sequence of positive integers, $j_o=0$. Let $\{a_k\}_k$, $k=\overline{0,\infty}$ be a strictly increasing sequence of positive real numbers such that $a_0=0$ and $\lim_{k\to\infty} a_k=0$. Let $G_n : C_n \mapsto \mathbb{R}$, $G_n(x) = (1/2^n)\cdot\sum_{k=0}^{\infty} c_{j_k+2}(x)\cdot(a_k-a_{k+1})$. Extending G_n linearly on the closure of each interval contiguous to C_n, we have G_n defined and continuous on $[0,1]$. Then we have:

(iii) If $a_k \le 1/(2n+1)^{j_k}$, $k=\overline{1,\infty}$ *then* $G_n \in E_{n+1}$ *on* C_n*;*

(iv) If $a_k = 1/(2n+1)^{j_k}$ *and* $j_k-2 \ge 2(j_{k-1}+1)$ *then* $G_n \in E_n$ *on no portion of* C_n*;*

(v) If $a_k \le 1/(2n+1)^{j_k}$ *then* $G_n \in L_{n^2+2n+1} \subseteq AC_{n^2+2n+1}$ *on* C_n*;*

(vi) If $a_k = 1/(2n+1)^{j_k}$ *and* $j_{k+2}-j_k \ge 2j_{k+1}+2$ *then* $G_n \in VB_{n^2+2n}$ *on no portion of* C_n*.*

Proof. (iii) Let $\epsilon>0$ and let p be a positive integer such that $(n+1)^{j_p+1}/(2n+1)^{j_p} < \epsilon$. Let $A_p = \{x\in C_n : x=\sum_{i=1}^{j_p} c_i(x)/(2n+1)^i\}$. For each $x\in A_p$ let $I_{x,p}=[x, x+R_{j_p+1}]$. Then A_p has $(n+1)^{j_p}$ elements and $|I_{x,p}| = 1/(2n+1)^{j_p} = R_{j_p+1}$. For each $x\in A_p$ let $J^j_{x,p} = [G_n(x+2j/(2n+1)^{j_p+1}), G_n(x+2j/(2n+1)^{j_p+1}+R_{j_{p+1}+1})]$, $j=\overline{0,n}$. Then $|J^j_{x,p}| = (1/2n)\cdot\sum_{k=p}^{\infty} 2n(a_k-a_{k+1}) = a_p \le 1/(2n+1)^{j_p} = |I_{x,p}|$ and $B(G_n;C_n)\subset \cup_{x\in A_p}\cup_{j=0}^{n}(I_{x,p}\times J^j_{x,p})$. Therefore $B(G_n;C_n)$ is contained in $(n+1)(n_1)^{j_p}$ squares, each of side length $1/(2n+1)^{j_p}$. It follows that $\sum_{x\in A_p}\mathcal{O}^{n+1}(G_n;C_n\cap I_{x,p}) \le (n+1)^{j_p}\cdot|I_{x,p}|\cdot(n+1) = (n+1)^{j_p+1}/(2n+1)^{j_p} < \epsilon$. Hence $G_n\in DW_{n+1}$ on C_n. Since $|C_n|=0$ it follows that $G_n\in E_{n+1}$ on C_n.

(iv) Let K be a portion of C_n and let $s\ge 2$ be a natural number such that $K\supset K' = I'\cap C_n$, where $I' = [\sum_{i=1}^{j_s} c_i/(2n+1)^i, \sum_{i=1}^{j_s} c_i/(2n+1)^i + R_{j_n+1}]$, $c_i\in\{0,2,\dots,2n\}$. Then $|\varphi_n(K')| = 1/(2n+1)^{j_s}$. We show that $G_n\notin E_n$ on K'. Let $\epsilon>0$, $\epsilon<1/(2n+1)^{j_s}$, and let $I=[a,b]$ be a closed interval such that $a,b\in K'$. Then $I\cap C_n = I\cap K'$. We claim that, if $G_n(I\cap K')\subset\cup_{i=1}^{n} J_i$, where J_i are intervals then

$$(1)\qquad |\varphi_n(I)| \le \sum_{i=1}^{n} |J_i|.$$

Let $\{I_k\}_k$ be a sequence of nonoverlapping closed intervals such that $K'\subset\cup_{k=1}^{\infty} I_k$. Then for $D_{k,i} = I_k\times J_{kn}$, $i=\overline{1,n}$, with $B(G_n;K')\subset\cup_{k=1}^{\infty}\cup_{i=1}^{n} D_{ki}$, we have $\sum_{k=1}^{\infty}\mathcal{O}^n(G_n;I_k\cap K') \ge \sum_{k=1}^{\infty}\sum_{i=1}^{n}|J_{ki}| \ge \sum_{k=1}^{\infty}|\varphi_n(K')| = 1/(2n+1)^{j_s} > \epsilon$ (see (1)). Hence $G_n\notin DW_n$ on K'. It follows that $G_n\notin E_n$ on K'. It remains to show (1). Let $I=[a,b]$, $a,b\in K'$. Then there exists a positive integer m such that $1/(2n+1)^{m+1} \le |I| < 1/(2n+1)^m$. Since $|I| < 1/(2n+1)^m$, there exist $c_1,c_2,\dots,c_m\in\{0,2,\dots,2n\}$ such that for each $x\in I\cap K'$ we have $c_i(x)=c_i$, $i=\overline{1,m}$. Since $|I|\ge 1/(2n1)^{m+1}$, we have four possibilities: 1) $c_{m+1}(a)=c_{m+1}(b)=c_{m+1}$; 2) $c_{m+1}(b)-c_{m+1}(a)\ge 4$; 3) $c_{m+1}(a)+2=c_{m+1}(b)$ and $b-a=1/(2n+1)^{m+1}$; 4) $c_{m+1}(a)+2=c_{m+1}(b)$ and $b-a>1/(2n+1)^{m+1}$.
1) We have $a=\sum_{i=1}^{m+1} c_i/(2n+1)^i$ and $b=a+R_{m+2}$. Now the proof follows as in 2).
2) Let $c_{m+1}=c_{m+1}(a)+2$, $A=\sum_{i=1}^{m+1} c_i/(2n+1)^i$ and $B=A+R_{m+2}$. Then $[a,b]\supset[A,B]$ and $B(G_n;I\cap I')\supset B(G_n;[A,B]\cap K')$. Let k be a positive integer such that $j_k < m+2 \le j_{k+1}$. Let $D_j = G_n(A+2j/(2n+1)^{j_{k+1}})$ and $E_j = G_n(A+2j/(2n+1)^{j_{k+1}}+R_{j_{k+1}+1})$, $j=\overline{0,n}$. Then $G_n([A,B]\cap K')\subset\cup_{j=0}^{n}[D_j,E_j]$

and $D_{j+1} > E_j$, $j = \overline{0, n-1}$. Indeed, if we cover the set $G_n([A,B] \cap K')$ with n intervals $J_1, J_2, \ldots, J_n$, then at least one of them contains an interval $[E_j, D_{j+1}]$ for some $j \in \{0,1,\ldots,n-1\}$. Hence $\sum_{i=1}^{n} | J_i | \geq D_{j+1} - E_j = (a_{k+1} - a_k)/n - a_k = (a_{k-1}/n)\cdot(1-(n+1)/(2n+1)^{j_k - j_{k-1}}) > (a_{k-1}/n)\cdot(1-(n+1)/(2n+1)) = a_{k-1}/(2n+1)$. Therefore we have

$$(2) \quad \sum_{i=1}^{n} | J_i | > \frac{a_{k-1}}{2n+1}.$$

Let $I_1 = [\sum_{i=1}^{m} c_i/(2n+1)^i, \sum_{i=1}^{m} c_i/(2n+1)^i + R_{m+1}]$. Then $I \subset I_1$ and $| \varphi_n(I) | \leq | \varphi_n(I_1) | = 1/(2n+1)^m \leq 1/(n+1)^{j_k - 2} \leq 1/(n^2+2n+1)^{j_{k-1}+1} \leq a_{k-1}/(2n+1)$. Now by (2)we obtain (1).

3) Let $c_{m+1} = c_{m+1}(a)$. Then $a = \sum_{i=1}^{m+1} c_i/(2n+1)^i + R_{m+2}$ and $b = \sum_{i=1}^{m} c_i/(2n+1)^i + (c_{m+1}+2)/(2n+1)^{m+1}$, hence $\varphi_n(a) = \varphi_n(b)$. Now (1) follows easily.

4) Let $c_{m+1} = c_{m+1}(a)$, $A = \sum_{i=1}^{m+1} c_i/(2n+1)^i + R_{m+2}$ and $B = \sum_{i=1}^{m} c_i/(2n+1)^i + (c_{m+1}+2)/(2n+1)^{m+1}$. Then $a = \sum_{i=1}^{m+1} c_i/(2n+1)^i + \sum_{i=m+2}^{\infty} c_i(a)/(2n+1)^i$ and $b = \sum_{i=1}^{m} c_i/(2n+1)^i + (c_{m+1}+2)/(2n+1)^{m+1} + \sum_{i=m+2}^{\infty} c_i(b)/(2n+1)^i$. We have two situations:

a) $A - a \leq b - B$ and $b \neq B$. Since $b \neq B$ it follows that there exists a positive integer $p = \inf\{i : i \geq m+2, c_i(b) \geq 2\}$. Then $p \geq m+2$ and $B(G_n; I \cap K') \supset B(G_n; [B, B+R_{p+1}] \cap K')$. Let k be a positive integer such that $j_k < p+1 \leq j_{k+1}$. Similarly to 2), if we cover the set $G_n([B, B+R_{p+1}] \cap K')$ with n intervals $J_1, J_2, \ldots, J_n$ then $\sum_{i=1}^{n} | J_i | > a_{k-1}/(2n+1)$. We have $b - B \leq R_p = 1/(2n+1)^{p-1}$. Let $A_2 = A - R_p$, $B_2 = B + R_p$ and $I_2 = [A_2, B_2]$.Then $I \subset I_2$ and $| \varphi_n(I_2) | = 2/(n+1)^{p-1} \leq 1/(n+1)^{j_k - 2} < a_{k-1}/(2n+1)$. Hence we obtain (1).

b) $A - a \geq b - B$ and $a \neq A$. Since $a \neq A$ it follows that there exists a positive integer p such that $p = \inf\{i : i \geq m+2, c_i(a) \leq 2n-2\}$. Then $p \geq m+2$. Let $A_3 = A - R_{p+1} = \sum_{i=1}^{m} c_i/(2n+1)^i + \sum_{i=m+2}^{p} 2n/(2n+1)^i$. Then $[a, A] \supset [A_3, A]$ and $A = A_3 + R_{p+1}$. It follows that $B(G_n; I \cap K') \supset B(G_n; [A_3, A_3 + R_{p+1}] \cap K')$. Similarly to a) we obtain (1).

(v) Let $I = [a,b]$, $a, b \in C_n$, $I \cap C_n \neq \emptyset$. Then there exists a positive integer m such that $1/(2n+1)^{m+1} \leq | I | < 1/(2n+1)m$. Since $| I | < 1/(2n+1)^m$, there exist $c_1, c_2, \ldots, c_m \in \{0, 2, \ldots, 2n\}$ such that $c_i(x) = c_i$, $i = \overline{1, m}$, whenever $x \in I \cap C_n$. We may suppose without loss of generality that $m \geq j_2$. Let $A_1 = \sum_{i=1}^{m} c_i/(2n+1)^i$ and $B_1 = A_1 + R_{m+1}$. Let k be the first positive integer such that $j_k \geq m+1$, and let $P = \{x \in C : x = c_{j_k}(x)/(2n+1)^{j_k} + c_{j_{k+1}}(x)/(2n+1)^{j_{k+1}}\}$. Then P has n^2+2n+1 elements. For each $x \in P$ let $J_x = [G_n(A_1 + x), G_n(A_1 + x + R_{j_{k+1}+1})]$. It follows that $G_n(I \cap C_n) \subset G_n([A_1, B_1] \cap C_n) \subset \cup_{x \in P} J_x$ and $| J_x | = (1/2n) \cdot \sum_{i=k}^{\infty} 2n(a_i - a_{i+1}) = a_k \leq 1/(2n+1)^{j_k} \leq 1/(2n+1)^{m+1} \leq | I |$. Hence $\mathcal{O}^{n^2+2n+1}(G_n; I \cap C_n) \leq \sum_{x \in P} | J_x | \leq (n^2+2n+1) \cdot | I |$ and $G_n \in L_{n^2+2n+1}$ on C_n. By Theorem 2.32.1, (xv), $G_n \in AC_{n^2+2n+1}$ on C_n.

(vi) Let K be a portion of C_n. Then there exist $c_i \in \{0, 2, \ldots, 2n\}$, $i = \overline{1, j_p - 1}$ such that $K \supset K_1$, where $K_1 = C_n \cap [S_p, S_p + R_{j_p}]$, $S_p = \sum_{i=1}^{j_p - 1} c_i/(2n+1)^i$. We show that $G_n \notin VB_{n^2+2n}$ on K_1. Let $p \geq 2$ be a natural number and let $A_p = \{x \in C_n : x = \sum_{i=j_p}^{j_{p+2}-1} c_i(x)/(2n+1)^i\}$. For each $x \in A_p$ let $I_{p,x} = [S_p + x, S_p + x + R_{j_{p+2}}]$. Clearly $I_{p,x} \neq \emptyset$ and A_p has $(n+1)^{j_{p+2} - j_p}$ elements. Let $B_p = \{y \in C_n : y = c_{j_{p+2}}(y)/(2n+1)^{j_{p+1}} + c_{j_{p+3}}(y)/(2n+1)^{j_{p+3}}\}$. Then B_p has n^2+2n+1 elements, namely $y_1 < y_2 < \ldots < y_{n^2+2n+1}$. For each $x \in A_p$ and $y \in B_p$, let $A_{xy} = G_n(S_p + x + y)$ and

$B_{xy} = G_n(S_p + x + y + R_{j_{p+3}+1})$. We have

(3) $\quad G_n(I_{p,x} \cap C_n) \subset \cup_{y \in B_p}[A_{xy}, B_{xy}]$.

Let $y, z \in B_p$, $y < z$. Then we have two situations:
a) $c_{j_{p+2}}(y) < c_{j_{p+2}}(z)$. Then $A_{xz} - B_{xy} \geq (2/2^n)\cdot(a_p - a_{p+1}) - 1/2^n$ and $\sum_{i=p+1}^{\infty} 2n(a_i - a_{i+1}) \geq (1/n)(a_p - a_{p+1}) - a_{p+1} = T_p$, where $T_p = a_p - (n+1)\cdot a_{p+1}/n$.
b) $c_{j_{p+2}}(y) = c_{j_{p+2}}(z)$ and $c_{j_{p+3}}(y) < c_{j_{p+3}}(z)$. Similarly to a) we obtain that $A_{xz} - B_{xy} \geq T_{p+1}$.
Since $T_p > T_{p+1}$ it follows in both cases that

(4) $\quad A_{xz} - B_{xy} \geq T_{p+1}$.

By (3) and (4) it follows that, if $G_n(I_{p,x} \cap C_n)$ is covered by $n^2 + 2n$ intervals J_{ni}, $i = \overline{1, n^2 + 2n}$ then there exists at least one $y_i \in B_p$ such that at least one of the intervals J_{ni} contains the interval $[B_{xy_i}, A_{xy_{i+1}}]$. Hence $\sum_{x \in A_p} \mathcal{O}^{n^2+2n}(G_n; I_{p,x} \cap C_n) \geq \sum_{x \in A_p} \sum_{i=1}^{n^2+2n} \mid J_{x,i} \mid \geq (n+1)^{j_{p+2}-j_p} \cdot T_{p+1} \geq (n^2 + 2n + 1)^{j_{p+1}+1} \cdot (1/n) \cdot (1/(2n+1)^{j_{p-1}} - (n+1)/(2n+1)^{j_{p+1}+1}) = ((n^2 + 2n + 1)/(2n+1))^{j_{p+1}+1} \to +\infty$, $p \to \infty$.

Remark 6.48.1 *The utility of this example follows by Theorem 2.37.1, (v).*

6.49 Functions concerning conditions $\underline{\mathcal{L}}$, $\underline{\mathcal{F}}$, $\overline{\mathcal{F}}$, VB_2G, $\mathcal{B}$, E_1G

Let C be the Cantor ternary set (see Section 6.1), and let φ be the Cantor ternary function (see Section 6.2). Let $\{j_k\}_k$ be an increasing sequence of positive integers, $j_o = 0$. Let p be a positive integer such that $2^{j_{2k+4}-p} \geq 2^{j_{2k+4}-k} \geq 3^{j_{2k+2}-1}$, $k \geq p$. Let $F, G_p : C \mapsto \mathbb{R}$, $F(x) = \sum_{i=0}^{\infty} \sum_{k=j_{2i}+1}^{j_{2i+1}} c_k(x)/2^{k+2}$ and $G_p(x) = F(x) + \varphi(x)/(2^p - 1)$. Extending F and G_p linearly on the closure of each interval contiguous to C, we have F and G_p defined and continuous on $[0,1]$. Then we have:

(i) $G_p' = F'$ *a.e. on* $[0,1]$, *and* $G_p \to F$ $[unif]$, $p \to \infty$;

(ii) $G_p \in \underline{L}_{2^p} \subsetneq \underline{AC}_{2^p}$ *on C, hence* $G_p \in \underline{\mathcal{L}} \subseteq \underline{\mathcal{F}}$ *on* $[0,1]$;

(iii) $F, G_p \in VB_2G \subseteq \mathcal{B}$ *on* $[0,1]$;

(iv) If $3^{j_i} < 2^{j_{i+1}}$ *then* $F \in E_1G \subseteq \mathcal{E}$ *and* $F \notin \mathcal{F}$ *on* $[0,1]$;

(v) If $3^{j_i} < 2^{j_{i+1}}$ *then* $F \notin \underline{\mathcal{F}}$ *and* $F \notin \overline{\mathcal{F}}$ *on* $[0,1]$.

Proof. (i) This is evident.

(ii) Let $\epsilon > 0$ and let $\delta_p = \min\{\epsilon; 1/3^{j_{2p}+1}\}$. Let $I \subset [0,1]$ be an interval with endpoints in C such that $\mid I \mid < 1/3^{j_{2p}}$. Let n be a positive integer such that $1/3^{n+1} \leq \mid I \mid < 1/3^n$, and let k be a positive integer such that $j_{2k} \leq n < j_{2k+2}$. Clearly $k \geq p$. Since $\mid I \mid < 1/3^n$, there exist $c_1, c_2, \ldots, c_n \in \{0,2\}$ such that $c_i(x) = c_i$, $i = \overline{1, \infty}$, whenever $x \in I \cap C$. Let $a = \sum_{i=1}^{j_{2k}} c_i/3^i$ and $b = a + 1/3^{j_{2k}}$. Let $d_1, d_2, \ldots, d_p \in \{0,2\}$ and let $E_{d_1 \ldots d_p} = \{x \in [a,b] \cap C : c_{j_{2k+2}}(x) = d_1, c_{j_{2k+2}-1}(x) = d_2, \ldots, c_{j_{2k+2}-p+1}(x) = d_p\}$. Let $x < y$, $x, y \in E_{d_1 \ldots d_p}$. Then we have three situations: 1) $y - x > 1/3^{j_{2k+3}}$; 2) $y - x = 1/3^{j_{2k+3}}$; 3) $y - x < 1/3^{j_{2k+3}}$.

1) Let $j_{2k}+1 \le i_o \le j_{2k+3}$ such that $c_i(x) = c_i(y) = c_i$, $i \le i_o - 1$, $c_{i_o}(x) = 0$, $c_{i_o}(y) = 2$. Clearly $i_o \in \{j_{2k+2}, j_{2k+2}-1, \dots, j_{2k+2}-p+1\}$. Let $a_1 = \sum_{i=1}^{i_o-1} c_i/3^i$. Then $x = a_1 + \sum_{i=i_o+1}^{\infty} c_i(x)/3^i$ and $y = a_1 + 2/3^{i_o} + \sum_{i=i_o+1}^{\infty} c_i(y)/3^i$. We have three possibilities:
a) $j_{2k}+1 \le i_o \le j_{2k+2}$. Then $G_p(y) - G_p(x) = F(y) - F(x) + (\varphi(y) - \varphi(x))/(2^p - 1) \ge F(y) - F(x) \ge F(a_1 + 2/3^{i_o}) - F(a_1 + \sum_{i=i_o+1}^{\infty} 2/3^i) > 0$.
b) $j_{2k+1}+1 \le i_o \le j_{2k+2}-p$. Then $G_p(y) - G_p(x) \ge G_p(a_1 + 2/3^{i_o} + \sum_{i=j_{2k+2}-p+1}^{j_{2k+2}} (d_i - d_{j_{2k+2}+p})/3^i) - G_p(a_1 + \sum_{i=i_o+1}^{j_{2k+2}-p} 2/3^i + \sum_{i=j_{2k+2}-p+1}^{j_{2k+2}} (d_i - d_{j_{2k+2}-p})/3^i + \sum_{i=j_{2k+2}}^{\infty})2/3^i = G_p(2/3^{i_o}) - G_p(\sum_{i=i_o+1}^{j_{2k+2}-p} 2/3^i + \sum_{i=j_{2k+2}+1}^{\infty} 2/3^i) = F(2/3^{i_o}) - F(\sum_{i=i_o+1}^{j_{2k+2}-p} 2/3^i) - F(\sum_{i=j_{2k+2}+1}^{\infty} 2/3^i) + (1/(2^p-1)) \cdot (\varphi(2/3^{i_o}) - \varphi(\sum_{i=i_o+1}^{j_{2k+2}-p} 2/3^i) - \varphi(\sum_{i=j_{2k+2}+1}^{\infty} 2/3^i)) = -F(\sum_{i=j_{2k+2}+1}^{\infty} 2/3^i) + (1/(2^p-1)) \cdot (1/2^{j_{2k+2}-p} - 1/2^{j_{2k+2}}) \cdot ((2^p-1)/2^{j_{2k+2}}) = 0$.
c) $j_{2k+2}+1 \le i_o \le j_{2k+3}$. Then $G_p(y) - G_p(x) > 0$ (the proof is similar to a)).
2) Let $a_2 = \sum_{i=1}^{j_{2k+3}-1} c_i/3^i$. We have two possibilities:
a) $x = a_2 + 1/3^{j_{2k+3}}$ and $y = a_2 + 2/3^{j_{2k+3}}$. Then $G_p(y) - G_p(x) = F(y) - F(x) > 0$.
b) $x = a_2$ and $y = a_2 + 1/3^{j_{2k+3}}$. Then $G_p(y) - G_p(x) > 0$.
3) Let $i_o \ge j_{2k+3}+1$ such that $c_{i_o}(x) = 0$, $c_{i_o}(y) = 2$, $c_i(x) = c_i(y) = c_i$, $i = \overline{1, i_o - 1}$. Let $a_3 = \sum_{i=1}^{i_o-1} c_i/3^i$. Then $G_p(y) - G_p(x) \ge G_p(a_3 + 2/3^{i_o}) - G_p(a_3 + 1/3^{i_o}) = F(2/3^{i_o}) - F(1/3^{i_o})$. Let m be a positive integer such that $j_{2k+m+2}+1 \le i_o \le j_{2k+m+4}$. We have two possibilities:
a) m is even. Then $F(2/3^{i_o}) - F(1/3^{i_o}) = -F(\sum_{i=j_{2k+m+4}}^{\infty} 2/3^i) > -\varphi(1/3^{j_{2k+m+4}}) = -1/2^{j_{2k+m+4}}$.
b) m is odd. Then $F(2/3^{i_o}) - F(1/3^{i_o}) > 0$.
By 1), 2) and 3) it follows that $| \mathcal{O}_-(G_p; E_{d_1 \dots d_p}) | \ge 1/2^{j_{2k+4}}$. Since $| I | > 1/3^{n+1} \ge 1/3^{j_{2k+2}-1} \ge 2^p/2^{j_{2k+4}}$ and $k \ge p$, it follows that $\sum_{(d_1,\dots,d_p)} | \mathcal{O}_-(G_p; E_{d_1 \dots d_p}) | < | I |$, hence $\mathcal{O}_-^{2^p}(G_p; I \cap C) > -1 \cdot | I |$. Therefore $G_p \in \underline{L}_{2^p} \subseteq \underline{AC}_{2^p}$ on C.

(iii) Clearly F is an example as in Section 6.36. It follows that $F \in VB_2$ on C, hence $F \in VB_2G$ on $[0,1]$. Since $\varphi \in VB$, by Corollary 2.29.1, (vi), we obtain that $G_p \in VB_2G$ on $[0,1]$.

(iv) Let F_1 and F_2 be the functions defined in Section 6.47. Then $F = F_2$ and $\varphi - F = F_1$. If $3^{j_i} < 2^{j_i+1}$ then $F, \varphi - F \in E_1G \subset \mathcal{E}$, and $F, \varphi - F \notin \mathcal{F}$ on $[0,1]$.

(v) Suppose on the contrary that $F \in \overline{\mathcal{F}}$ on $[0,1]$. By Corollary 2.37.1, (viii), $\overline{\mathcal{F}} + \overline{\mathcal{E}} = \overline{\mathcal{E}}$ on $[0,1]$. It follows that $F + (\varphi - F) = \varphi \in \overline{\mathcal{E}} \subseteq \overline{\mathcal{M}}$ (see Corollary 2.37.1, (xvi)), a contradiction (see Section 6.2). Therefore $F \notin \overline{\mathcal{F}}$ on $[0,1]$.
We show that $F \notin \underline{\mathcal{F}}$ on $[0,1]$. By Remark 2.28.4, (ii), it is sufficient to show that $F \in \underline{AC}_{2^n-1}$ on no portion of C and for no positive integer n. Let P be a portion of C. Suppose on the contrary that there exists a positive integer n such that $F \in \underline{AC}_{2^n-1}$ on P. Let $[a_o, b_o]$ be a remained closed interval from the step q, such that $[a_o, b_o] \cap C \subset P$ (we take the first q with this property). Then $F \in \underline{AC}_{2^n-1}$ on $[a_o, b_o] \cap C$. We may suppose without loss of generality that $j_{2k+1} < j_{2k+2} - n$ and $j_{2k+2} - n > q$. Let $n_i = j_{i+1} - j_i$. Then $n - n_{2k+1} < 0$. Let $I = [a, b]$ be a remained closed interval from the step $j_{2k+2} - n$ such that $I \subset [a_o, b_o]$ (we have $2^{j_{2k+2}-n-q}$ such intervals). Then $a = \sum_{i=1}^{j_{2k+2}-n} c_i/3^i$ and $b = a + 1/3^{j_{2k+2}-n}$. Let $\{E_i\}$, $i = \overline{1, 2^n - 1}$ be sets such that $E_i = \overline{E}_i \subset I \cup C$ and $\cup_{i=1}^{2^n-1} E_i = I \cap C$. Then

$$(1) \quad \sum_{i=1}^{2^n-1} | \mathcal{O}_-(F; E_i) | > \frac{2}{2^{j_{2k+2}+2}}.$$

It follows that $\sum_I \sum_{i=1}^{2^n-1} | \mathcal{O}_-(F;E_i) | > 2^{j_{2k+2}-n-q} \cdot 2/2^{j_{2k+2}+2} = 1/2^{n+q+1}$. Since $| I | \cdot 2^{j_{2k+2}-n-q} \to 0$, $k \to \infty$, we obtain that $F \notin \underline{AC}_{2^n-1}$ on $[a_o, b_o] \cap C$, a contradiction. It remains to show (1). Let $I_t = [a_t, b_t]$, $t = \overline{1, 2^n}$, $a_1 = a$, be the remained closed intervals from the step j_{2k+2} which are contained in I (numbered from the left to the right). Then $| I_t | = 1/3^{j_{2k+2}}$ and $F(a_1 + x) = F(a + x)$, for each $x \in [0, 1/3^{j_{2k+2}}] \cap C$, $t = \overline{1, 2^n}$. Let $R_i = E_i \setminus (\cup_t I_t)$, $i = \overline{1, 2^n - 1}$.Then there exists $i \in \{1, 2, \ldots, 2^n - 1\}$ such that $b_i \in E_i$ and $R_i \neq \emptyset$. Let i_1 be the first i with this property. Let $x_i = b_i$, $i = \overline{1, i_1}$. Let $m_{i_1} = \inf\{F(x) : x \in R_{i_1}\}$. Then $| \mathcal{O}_-(F; E_{i_1}) | \geq M_{i_1} - m_{i_1}$, where $M_{i_1} = F(x_{i_1})$.
a_1) If $M_{i_1} = F(a)$ then $\sum_{i=1}^{2^n-1} | \mathcal{O}_-(F;E_i) | > | \mathcal{O}_-(F;E_1) | \geq M_{i_1} - m_{i_1} = F(b) - F(a)$.
b_1) If $m_{i_1} > F(a)$ then $p_i^{(1)} = \sup\{x \in I_i : F(x) \leq m_{i_1}\}$, $i = \overline{i_1 + 1, 2^n}$. It follows that $F(p_{i_1+1}^{(1)}) = \ldots = F(p_{2^m}^{(1)})$. Let $i_2 \in \{i_1 + 1, \ldots, 2^n - 1\}$ be the first index such that $R_{i_2} \neq \emptyset$ and $p_{i_2}^{(1)} \in E_{i_2}$. Let $x_i \in P_i^{(1)}$, $i = \overline{i_1 + 1, i_2}$ and let $m_{i_2} = \inf\{F(x) : x \in R_{i_2}\}$. Then $| \mathcal{O}_-(F; E_{i_2}) | \geq M_{i_2} - m_{i_2}$, where $M_{i_o} = F(x_{i_o})$.
a_2) If $m_{i_2} = F(a)$ then $\sum_{i=1}^{2^n-1} | \mathcal{O}_-(F;E_i) | \geq | \mathcal{O}_-(F;E_{i_1}) | + | \mathcal{O}_-(F;E_{i_2}) | \geq F(b) - F(a) - (m_{i_1} - M_{i_2})$.
b_2) If $m_{i_2} > F(a)$ then let $p_i^{(2)} = \sup\{x \in I_i : F(x) < m_{i_2}\}$, $i = \overline{i_2 + 1, 2^n}$. Then $F(p_{i_2+1}^{(2)} = \ldots = F(p_{2^m}^{(2)})$. Let $i_3 \in \{i_2 + 1, \ldots, 2^n - 1\}$ be the first subscript such that $R_{i_3} \neq \emptyset$ and $p_{i_3}^{(2)} \in E_{i_3}$. Let $x_i \in p_i^{(2)}$, $i = \overline{i_2, i_3}$. Let $m_{i_3} = \inf\{F(x) : x \in R_{i_3}\}$. Then $| \mathcal{O}_-(F; E_{i_3}) | M_{i_3} - m_{i_3}$, where $M_{i_3} = F(x_{i_3})$.
Continuing, it follows that there exists some $j_o \in \{1, 2, \ldots, 2^{n-1}\}$ such that $R_{i_{j_o}} \neq \emptyset$ and $R_i = \emptyset$ for each $i > i_{j_o}$. Then $a_{2^n} \in R_{i_{j_o}}$. (indeed, since $a_{2^n} \notin E_i$ for $R_i = \emptyset$ it follows that $a_{2^n} \in E_{i_o}$, with $R_{i_o} \neq \emptyset$ for some i_o; hence $i_o = i_{j_o}$.)
$a_{i_{j_o}}$) Then $m_{i_{j_o}} = F(a_{2^n}) = F(a)$ and

$$(2) \quad \sum_{i=1}^{2^n-1} | \mathcal{O}_-(F;E_i) | \geq \sum_{t=1}^{j_o} | \mathcal{O}_-(F;E_{i_t}) | \geq F(b) - F(a) = -\sum_{t=1}^{j_o-1} (m_{i_t} - M_{i_{t+1}})$$

$$> \frac{F(b) - F(a)}{2} \geq \frac{2}{2^{j_{2k+2}+2}}.$$

So we have (1). It remains to show (2). Let $Q = F(I \cap C) = F(I_1 \cap C) = \ldots = F(I_n \cap C)$. If $m_{i_t} \neq M_{i_{t+1}}$, $t = \overline{1, j_o - 1}$ then $(M_{i_{t+1}}, m_{i_o})$ are intervals contiguous to $Q \subset [F(a), F(b)]$. Let $I'_m = [a'_m, b'_m]$, $m = \overline{1, 2^{n_{2k+2}}}$ be the remained closed intervals from the step j_{2k+3} which are contained in I. Then $Q = \cup_{m=1}^{2^{n_{2k+2}}} F(I'_m \cap C)$. It follows that $F(b'_m) - F(a'_m) = B$, where $B = \sum_{t=k+1}^{\infty} \sum_{m=j_{2t+2}+1}^{j_{2t+3}} 2/3^{m+1}$, and $F(a'_{m+1}) - F(b'_m) = A$, where $A = 2/2^{j_{2k+3}+1} - B$. Clearly $A > B$ and $(F(b'_m), F(a'_{m+1}))$, $m = \overline{1, 2^{n_{2k+2}} - 1}$ are intervals contiguous to $Q \subset [F(a), F(b)]$ of length A. Since $n < n_{2k+1}$ it follows that $j_o - 1 < 2^n - 2 < 2^{n_{2k+2}} - 2$. Hence $2(2^n - 2) < 2^{n_{2k+1}+1} - 4 < 2^{n_{2k+2}} - 1$ (since $\{n_k\}_k$ is strictly increasing). Then $F(b) - F(a) = 2^{n_{2k+2}} \cdot B + (2^{n_{2k+2}} - 1) \cdot A > 2(2^n - 2) \cdot A$. It follows that $(2^n - 2) \cdot A < (F(b) - F(a))/2$. We also have that $\sum_{i=t}^{j_o-1} (m_{i_t} - M_{i_{t+1}}) \leq (2^n - 2) \cdot A$ and $\sum_{i=1}^{2^n-1} | \mathcal{O}_-(F;E_i) | \geq F(b) - F(a) - (2^n - 2) \cdot A > (F(b) - F(a))/2$. Thus we obtain (2).

Remark 6.49.1 *The utility of this example follows by Corollary 2.37.1, (xvi).*

6.50 A function $F \in \mathcal{E} \cap VB_\omega G$, $F \notin \mathcal{B}$

Let $P = \{x \in [0,1] : x = \sum_{k=1}^{\infty} d_k/(2k + 1)^k,\ d_k \in \{0, 2, \ldots, 2k\}\}$. Each point $x \in P$ is uniquely represented by $\sum_{k=1}^{\infty} d_k(y)/(2k + 1)^k$. Then P is a perfect nowhere dense subset of $[0,1]$ and $|P| = 0$. Let $a = \inf(P)$ and $b = \sup(P)$. Let $\{j_k\}_k$ be a strictly increasing sequence of positive integers, $j_o = 0$. Let $F : P \mapsto \mathbb{R}$, $F(x) = \sum_{k=1}^{\infty} d_{j_k}(x)/(2j_k+1)^{j_{k-1}}$. Extending F linearly on the closure of each interval contiguous to $P \subset [a,b]$, we have F defined and continuous on $[a,b]$. Then we have:

(i) $F \in E_1$ *on* P, *hence* $F \in \mathcal{E}$ *on* $[a,b]$;

(ii) If $j_{k+1} = 3j_k + 1$ *then* $F \in VB_n$ *on no portion of* P *and for no positive integer* n, *hence* $F \notin \mathcal{B}$ *on* $[a,b]$;

(iii) $F \in VB_\omega$ *on* P, *hence* $F \in VB_\omega G$ *on* $[a,b]$.

Proof. (i) Let $I_{d_1\dots d_{j_k}} = [\sum_{i=1}^{j_k} d_i/(2i+1)^i, \sum_{i=1}^{j_k} d_i/(2i+1)^i + \sum_{i=j_k+1}^{\infty} 2i/(2i+1)^i]$, $d_i \in \{0,2,\dots,2i\}$, $i = \overline{1,j_k}$. Then $B(F;P)$ can be covered by $(j_k+1)!$ rectangles $I_{d_1\dots d_{j_k}} \times [\sum_{i=1}^{k} d_{j_i}/(2j_i+1)^{j_i-1}, \sum_{i=1}^{k} d_{i_j}/(2j_i+1)^{j_i-1} + \sum_{i=k+1}^{\infty} d_{j_i}/(2j_i+1)^{j_i-1}]$ and $\mathcal{O}(F; P \cap I_{d_1\dots d_{j_k}}) < \sum_{i=j_k+1}^{\infty} 2i/(2i+1)^i < (3/2)/(2j_k+3)^{j_k}$. Hence $\sum_{(d_1,\dots,d_{j_k})} \mathcal{O}(F; P \cap I_{d_1\dots d_{j_k}}) < (1+j_k)! \cdot (3/2)/(2j_k+3)^{j_k} < (3/2) \cdot (1/2) j_k \to 0$. It follows that $F \in DW_1$ on P, hence $F \in E_1$ on P. Thus $F \in \mathcal{E}$ on $[0,1]$.

(ii) Let K be a portion of P. Then there exist a positive integer k and $d_i \in \{0,2,\dots,2i\}$, $i = \overline{1,j_k}$ such that $K \supset E = P \cap I_{d_1\dots d_{j_k}}$. Then the intervals $I_{d_1\dots d_{j_{k+1}-1}}$, $d_i \in \{0,2,\dots,2i\}$, $i = \overline{j_k+1, j_{k+1}-1}$ are nonoverlapping and contain points of E. If $j_{k+1} = 3j_k + 1$ then we have $(j_k+1)(j_k+2)\dots(j_{k+1}-1) > (j_k+1)(j_{k+1}-1)^{j_k}$ such intervals. If the set $F(P \cap I_{d_1\dots d_{j_{k+1}-1}})$ is covered by j_{k+1} intervals then at least one of them has the length greater than $1/(2j_{k+1}+1)^{j_k}$. Therefore $\mathcal{O}^{j_{k+1}}(F; P \cap I_{d_1\dots d_{j_k} d_{j_k+1}\dots d_{j_{k+1}-1}}) \geq 1/(2j_{k+1}+1)^{j_k}$. Hence $\sum_{(d_1,\dots,d_{j_{k+1}-1})} \mathcal{O}(F; P \cap I_{d_1\dots d_{j_{k+1}-1}}) \geq ((j_k+1)(j_{k+1}-1)/(2j_{k+1}+1))^{j_k} \to \infty$, $k \to \infty$. It follows that $F \notin VB_{j_k}$ on P, hence $F \in VB_n$ for no positive integer n. Thus $F \notin \mathcal{B}$ on $[a,b]$.

(iii) Let $\{I_k\}$, $k = \overline{1,p}$ be a finite set of nonoverlapping closed intervals with endpoints in P. Since $F \in \mathcal{E} \subset (N)$ on $[a,b]$, it follows that $|F(P)| = 0$. Then $|F(P \cap I_k)| = 0$. By Remark 2.29.1, (v), $F \in VB_\omega$ on P.

Remark 6.50.1 *The utility of this example follows by Remark 2.37.3, (ii).*

6.51 A function $F \in (N)$, $F \notin \Lambda Z$ (Foran)

Let F and G_t be the functions defined in Section 6.27. Then we have:

(i) $F \in (S) \subset (N)$ *on* $[0,1]$;

(ii) $F \notin \Lambda Z$ *on* $[0,1]$.

Proof. (i) See Section 6.27.

(ii) Suppose on the contrary that $F \in \Lambda Z$ on $[0,1]$. Then by Theorem 2.33.2, (iv), (vi), $G_t \in \Lambda Z \subset (N)$, a contradiction.

Remark 6.51.1 *The utility of this example follows by Theorem 2.33.2, (vi).*

6.52 A function $F \in AC \circ \Lambda Z$, $F \notin \Lambda Z$ (Foran)

Let F be the function defined in Section 6.27. Then we have:

(i) $F \in (S) \subset AC \circ \Lambda Z$ *on* $[0,1]$;

(ii) $F \notin \Lambda Z$ *on* $[0,1]$.

Proof. See Section 6.27, Theorem 2.18.9, (xii), Theorem 2.33.2,(ii) and Section 6.51, (ii).

Remark 6.52.1 *The utility of this example follows by Theorem 2.33.2.*

6.53 A function $H \in AC + \Lambda Z$, $H \notin \Lambda Z$ (Foran)

Let F, P and Q be defined as in Section 6.27. By Lemma 2.18.3, it follows that there exists a strictly increasing function $g : [0,1] \mapsto \mathbb{R}$ such that g and g^{-1} are AC and

(1) $\Lambda(B(F \circ g^{-1}; g(P)) = 0.$

Let $F_1 = g^{-1}$, $F_2 = F \circ g^{-1}$ and $H = F_1 + F_2$. Then we have:

(i) F_1 *is* AC *and strictly increasing;*

(ii) $F_2 \in \Lambda Z$;

(iii) $H \in AC + \Lambda Z$, *but* $H \notin (N)$, *hence* $H \notin \Lambda Z$.

Proof. (i) This is evident.

(ii) Since F_2 is linear on each interval contiguous to $g(P)$, by (1) it follows that $F_2 \in \Lambda Z$.

Clearly $H \in AC + \Lambda Z$. By Section 6.27, $F + I \notin (N)$ on P, where $I : [0,1] \mapsto [0,1]$, $I(x) = x$. Then $(F+I)(P) = H(g(P))$. Since $|P| = 0$ and $g \in AC \subset (N)$, it follows that $|g(P)| = 0$. Since $F + I \notin (N)$, $|H(g(P))| \neq 0$. Hence $H \notin (N)$ and by Theorem 2.33.2, vi), $H \notin \Lambda Z$.

Remark 6.53.1 *The utility of this example follows by Theorem 2.33.2, (xi); Remark 2.33.1.*

6.54 A function $G \in AC \cdot \Lambda Z$, $G \notin \Lambda Z$ (Foran)

Let F_1 and F_2 be the functions defined in Section 6.53. Let $G_1(x) = e^{F_1(x)}$, $G_2(x) = e^{F_2(x)}$ and $G = G_1 \cdot G_2$. Then we have:

(i) $G_1 \in AC$;

(ii) $G_2 \in \Lambda Z$;

(iii) $G \in AC \cdot \Lambda Z$, *but* $G \notin (N)$, *hence* $G \notin \Lambda Z$.

Proof. (i) Since $e^x \in L$, by Theorem 2.33.2, (xvii) it follows that $G_1 \in AC$.

(ii) Since $e^x \in L$, by Theorem 2.32.1, (ix) it follows that $G_2 \in \Lambda Z$.

(iii) Clearly $G \in AC \cdot \Lambda Z$. We have $G(x) = e^{F_1(x)+F_2(x)}$. Since $F_1 + F_2 \notin (N)$ (see Section 6.53) and $e^x \in L \subset (N)$, by Theorem 2.18.9, (xi), it follows that $G \notin (N)$. By Theorem 2.33.2, (vi), $G \notin \Lambda Z$.

Remark 6.54.1 *The utility of this example follows by Theorem 2.33.2, (xii).*

6.55 A function $F_1 \in AC$, $F_1 \notin L_n$, $F_1 \notin \mathcal{L}$

Let F_1, F_2, g and P be defined as in Section 6.53. Let $c = \inf(g(P))$ and $d = \sup(g(P))$. Then we have:

(i) $F_1 \in AC \subseteq AC_n \subseteq AC_nG \subseteq \mathcal{F}$ *on* $g(P)$;

(ii) $F_1 \in L_n$ *on* $g(P)$ *for no positive integer* n.

(iii) $F_1 \notin \mathcal{L}$ *on* $[c,d]$, *hence* $F_1 \in L_nG$ *on* $[c,d]$ *for no positive integer* n.

Proof. (i) See Section 6.53.

(ii) Suppose on the contrary that $F_1 \in L_n$ on $g(P)$ for some n. Since $F_2 \in \Lambda Z$, by Theorem 2.33.2, (vi), it follows that $F_1 + F_2 \in \Lambda Z$, a contradiction (see Section 6.53, (iii)). By Corollary 2.32.1, (iii), $F_1 \notin L_nG$ on $[c,d]$.

(iii) Suppose on the contrary that $F_1 \in \mathcal{L}$ on $[c,d]$. Since $F_2 \in \Lambda Z$, by Corollary 2.33.1, (iii), it follows that $F_1 + F_2 \in \Lambda Z$, a contradiction (see Section 6.53, (iii)).

Remark 6.55.1 *The utility of this example follows by Theorem 2.32.1, (xv); Corollary 2.32.1, (iv); Corollary 2.37.1, (xvii).*

6.56 Functions $F_1 \in AC_2G$, $F_2 \in \Lambda Z$, $F_1 + F_2 \notin (M)$, $F_1' = -F_2'$ *a.e.*

Let C be the Cantor ternary set (see Section 6.1), and let φ be the Cantor ternary function (see Section 6.2). Let F and G be the functions defined in Section 6.34. By Lemma 2.18.3 there exists a function $g : [0,1] \mapsto \mathbb{R}$ such that

(1) g and g^{-1} are AC;

(2) $\Lambda(B(G \circ g^{-1}; g(C)) = 0$;

(3) g is linear on each interval contiguous to C;

(4) g^{-1} is linear on each interval contiguous to $g(C)$.

Let $a = \inf(g(C))$, $b = \sup(g(C))$, $F_1 = F \circ g^{-1}$, $F_2 = G \circ g^{-1}$ and $\varphi_1 = \varphi \circ g^{-1}$. Then we have:

(i) $F_1 \in AC_2$ *on* $g(C)$, *hence* $F_1 \in AC_2G \subseteq \mathcal{F}$ *on* $[a,b]$;

(ii) $F_2 \in \Lambda Z$ *on* $[a,b]$;

(iii) $F_1 + F_2 = \varphi_1 \notin (M)$ *on* $[a,b]$*;*

(iv) $F_1' = -F_2'$ *a.e. on* $[a,b]$.

Proof. (i) By Theorem 2.32.1, (xvii), $F_1 \in AC_2$ on $g(C)$.

(ii) See (2).

(iii) We have $\varphi_1(g(C)) = \varphi(C) = [0,1]$, hence $\varphi \notin (N)$. But $|g(C)| = 0$ (because $g \in AC \subset (N)$), φ_1 is constant on each interval contiguous to $g(C)$, and φ is increasing on $[a,b]$. It follows that $F_1 + F_2 = \varphi \notin (M)$.

(iv) We have $(F_1 + F_2)'(x) = \varphi_1'(x) = 0$ on $[a,b] \setminus g(C)$, hence $F_1' = -F_2'$ a.e. on $[a,b]$.

Remark 6.56.1 *The utility of this example follows by Remark 2.33.2, (iii).*

6.57 A function $F \in \Lambda Z$, $F \notin [\mathcal{E}]$

Let C be the Cantor ternary set (see Section 6.1), and let φ be the Cantor ternary function (see Section 6.2). Let $F : C \mapsto [0,1]$, $F(x) = \sum_{i=1}^{\infty} c_{2i}(x)/3^i$. Extending F linearly on the closure of each interval contiguous to C, we have F defined and continuous on $[0,1]$ (see Figure 6.21). Let $R_k = \sum_{i=k}^{\infty} 2/3^i$, $k = \overline{1,\infty}$. Then we have:

(i) $F \in \Lambda Z$ *on* C*;*

(ii) $F \in E_n$ *on no portion of* C *and for no positive integer* n*. Hence* $F \notin [\mathcal{E}]$ *on* $[0,1]$.

Proof. (i) Let $\epsilon > 0$ and let p be a positive integer such that $(\sqrt{2}/9^p)\cdot 8^p < \epsilon$. Let $A_p = \{x \in C : x = \sum_{i=1}^{2p} c_i(x)/3^i\}$. It follows that A_p has 2^{2p} elements. For each $x \in A_p$ let $I_{p,x} = [x, x + R_{2p+1}]$. Let $B_p = \{y \in C : y = \sum_{i=p+1}^{2p} c_{2i}(y)/3^{2i}\}$. It follows that B_p has 2^p elements. For each $x \in A_p$ and $y \in B_p$ let $J_{p,x,y} = [F(x+y), F(x+y+R_{4p+1})]$. Then $|I_{p,x}| = |J_{p,x,y}| = 1/3^{2p} = 1/9^p$. Hence $B(F;C) \subset \cup_{x\in A_p} \cup_{y \in B_p} (I_{p,x} \times J_{p,x,y})$. Therefore $B(F;C)$ is contained in $2^{2p} \cdot 2^p = 8^p$ squares, each of side length $1/9^p$. Now it follows easily that $F \in \Lambda Z$ on C.

(ii) Let K be a portion of C and let $I' = [\sum_{i=1}^{p} c_i/3^i, \sum_{i=1}^{p} c_i/3^i + R_{p+1}]$, $p = \overline{1,\infty}$. Then $K \supset K' = I' \cap C$ for some p. We show that $F \notin E_{2^q-1}$ on K', where $q \geq 1$ is a positive integer. We may suppose without loss of generality that $p \geq 8q + 13$. Let $I = [a,b]$, $a, b \in K'$.Then $I \cap C = I \cap K'$. We claim that if $F(O \cap K') \subset \cup_{i=1}^{2^q-1} J_i$ then

$$(1) \quad |\varphi(I)| \leq \sum_{i=1}^{2^q-1} |J_i|, \ \textit{hence} \ \mathcal{O}^{2^q-1}(F; C \cap I) \geq |\varphi(I)|.$$

Let $\{I_k\}_k$ be a sequence of nonoverlapping closed intervals such that $K' \subset \cup_{k=1}^{\infty} I_k$. By (1), $\sum_{k=1}^{\infty} \mathcal{O}^{2^q-1}(F; C \cap I_k) \geq \sum_{k=1}^{\infty} |\varphi(I_k)| \geq |\varphi(K')| = 1/2^p$. Hence $F \notin DW_{2^q-1}$ on K'. It follows that $F \notin E_{2^q-1}$ on K'. It remains to prove (1). Let $I = [a,b]$, $a, b \in K'$. Then there exists a positive integer m such that $1/3^{m+1} \leq |I| < 1/3^m$. Since $|I| < 1/3^m$, it follows that there exist $c_1, c_2, \ldots, c_m \in \{0,2\}$ such that $c_i(x) = c_i$, $i = \overline{1,m}$, whenever $x \in I \cap C$. Since $|I| \geq 1/3^{m+1}$, we have three situations: a) $c_{m+1}(a) = c_{m+1}(b) = c_{m+1}$; b) $c_{m+1}(a) = 0$, $c_{m+1}(b) = 2$ and $b = 1/3^{m+1}$; c)

$c_{m+1}(a) = 0$, $c_{m+1}(b) = 2$ and $b - a > 1/3^{m+1}$.
a) We have $a = \sum_{i=1}^{m+1} c_i/3^i$, $b = a + R_{m+2}$ and

(2) $\quad \varphi(b) - \varphi(a) = 1/2^{m+2}$.

Let s be the first positive integer such that $m + 2 \leq 2s$ and let $A_{sq} = \{x \in C : x = \sum_{i=1}^{q} c_{2s+2i}/3^{2s+2i}\}$. Then A_{sq} has 2^q elements, namely $x_1 < x_2 < \ldots < x_{2^q}$. Let $A_x = F(a + x)$and $B_x = F(a + x + R_{2s+2q+1})$, $x \in A_{sq}$. Then we have

(3) $\quad F([a, b] \cap C) \subset \cup_{x \in A_{sq}}[A_x.B_x]$ *and*

(4) $\quad A_y - B_x \geq 1/3^{s+q}, \quad x, y \in A_{sq}, \quad x < y$.

Indeed, let $k \in \{1, 2, \ldots, q\}$ such that $c_{2s+2j}(x) = c_{2s+2j}(y)$, $j = \overline{1, k-1}$, $c_{2s+2k}(x) = 0$ and $c_{2s+2k}(y) = 2$. Then $A_y - B_x \geq 2/3^{s+k} - R_{k+1} = 1/3^{s+k} \geq 1/3^{s+q}$ and we have (4). If we cover the set $F([a, b] \cap C)$ by $2^q - 1$ intervals J_i,$i = \overline{1, 2^p - 1}$ then at least one of them contains an interval $[B_{x_i}, A_{x_{i+1}}]$ for some $i \in \{1, 2, \ldots, 2^q - 1\}$ (see (3) and (4)). Hence

(5) $$\sum_{i=1}^{2^q-1} | J_i | \geq 1/3^{s+q}.$$

Clearly $2s - 2 \leq m + 1 \leq 2s - 1$. Since $m + 1 \geq p \geq 8q + 13$, it follows that $2s \geq m + 1 \geq p + 1 \geq 8q + 14$. Hence $s \geq 4q + 7$. We have

(6) $$\frac{2^{m+1}}{3^{s+q}} \geq \frac{2^{2s-2}}{3^{s+q}} = \frac{1}{4} \cdot \left(\frac{4}{3}\right)^s \cdot \left(\frac{1}{3}\right)^q \geq \frac{1}{4} \cdot \left(\left(\frac{4}{3}\right)^s \cdot \left(\frac{1}{3}\right)\right)^q \cdot \left(\frac{4}{3}\right)^7 > 1.$$

By (2), (5) and (6) we obtain (1).
b) We have $a = \sum_{i=1}^{m} c_i/3^i + R_{m+2}$, $b = \sum_{i=1}^{m} c_i/3^i + 2/3^{m+1}$ and $\varphi(a) = \varphi(b)$, so it follows (1).
c) Let $A = \sum_{i=1}^{m} c_i/3^i + R_{m+2}$ and $B = \sum_{i=1}^{m} c_i/3^i + 2/3^{m+1}$. Then $a = \sum_{i=1}^{m} c_i/3^i + \sum_{i=m+2}^{\infty} c_i(a)/3^i$ and $b = \sum_{i=1}^{m} c_i/3^i + 2/3^{m+1} + \sum_{i=m+2}^{\infty} c_i(b)/3^i$. We have two possibilities:
(I) $A - a \leq b - B$ and $b \neq B$. Since $b \neq B$, there exists a positive integer s such that $s = \inf\{i : i \geq m + 2,\ c_i(b) = 2\}$. Then $s \geq m + 2$ and $B(F; I \cap K') \supset B(F; [B, B + R_{s+1}] \cap K')$. Let t be a positive integer such that $2t - 1 \leq s + 1 \leq 2t$. Then $\varphi(I) \subset \varphi([A - R_s, B + R_s])$ and

(7) $\quad | \varphi(I) | \leq 2/2^{s-1}$.

If we cover the set $F([B, B + R_{s+1}] \cap K')$ by $2^q - 1$ intervals J_i, $i = \overline{1, 2^q - 1}$ then

(8) $$\sum_{i=1}^{2^q-1} | J_i | \geq 1/3^{t+q}.$$

Clearly $2t \geq s + 1 \geq m + 3 \geq p + 2 \geq 8q + 15$. Hence $t > 4q + 8$ and

(9) $$\frac{2^{s-2}}{3^{t+q}} \geq \frac{2^{2t-4}}{3^{t+q}} = \frac{1}{16} \cdot \left(\frac{4}{3}\right)^t \cdot \left(\frac{1}{3}\right)^q \geq \frac{1}{16} \cdot \left(\left(\frac{4}{3}\right)^4 \cdot \frac{1}{3}\right)^q \cdot \left(\frac{4}{3}\right)^8 \geq 1.$$

By (7), (8) and (9) we obtain (1).
(II) $A - a \geq b - B$ and $a \neq A$. Since $a \neq A$, there exists a positive integer s such that $s = \inf\{i : i \geq m + 2,\ c_i(a) = 0\}$. Clearly $s \geq m + 2$. Let $A_1 = A - R_{s+1} = \sum_{i=1}^{m} c_i/3^i + \sum_{i=m+2}^{s} 2/3^i$. Then $[a, A] \supset [A_1, A]$ and $A = A_1 + R_{s+1}$. Therefore $B(F; I \cap K') \supset B(F; [A_1, A_1 + R_{s+1}] \cap K')$. Similarly to (I) we obtain again (1). By Lemma 2.37.3 it follows that $F \in [\mathcal{E}]$ on $[0, 1]$.

Remark 6.57.1 *The utility of this example follows by Remark 6.40.1.*

6.58 Functions $F_1 \in (S)$, $F_1 \in AC \circ \sigma.f.l.$, $F_1 \notin \sigma.f.l.$, $F_2 \in L$, $F_1 + F_2 \notin T_2$

Let n be a positive integer and let $\{n_k\}$, $k = \overline{1,\infty}$ be a sequence of positive integers, $n_o = 0$. Let $P = \{x \in [0,1] : x = \sum_{k=1}^{\infty}(c_{2k-1}/((2n+1)^k \cdot (2\cdot n_1+1)(2\cdot n_2+1)\dots(2\cdot n_{k-1}+1)) + c_{2k}/((2n+1)^k\cdot(2\cdot n_1+1)(2\cdot n_2+1)\dots(2\cdot n_k+1))),\ c_{2k} \in \{0,2,4,\dots,2\cdot n_k\}$, $c_{2k-1} \in \{0,2,\dots,2n\}$, $k = \overline{1,\infty}\}$. Each point $x \in P$ is uniquely represented by $\sum_{k=1}^{\infty}(c_{2k-1}(x)/((2n+1)^k\cdot(2\cdot n_1+1)(2\cdot n_2+1)\dots(2\cdot n_{k-1}+1)) + c_{2k}(x)/((2n+1)^k\cdot(2\cdot n_1+1)(2\cdot n_2+1)\dots(2\cdot n_k+1)))$, and we denote this by $0.c_1(x)c_2(x)\dots c_{2k-1}(x)c_{2k}(x)\dots$. We observe that

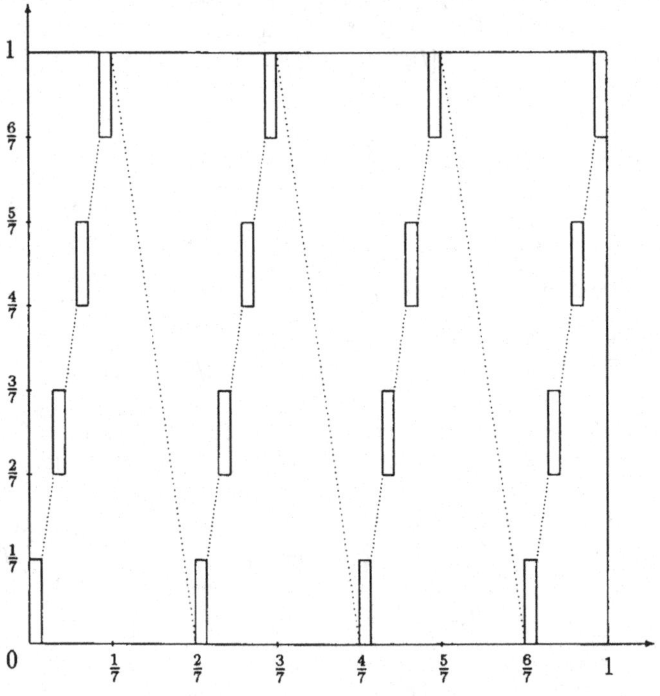

Figure 6.25: The function F_1

- $0.c_1c_2\dots c_{2k-1}1 = 0.c_1c_2\dots c_{2k-1}0(2n)(2n_{k+1})(2n)(2n_{k+2})\dots$ and
- $0.c_1c_2\dots c_{2k-2}1 = 0.c_1c_2\dots c_{2k-2}0(2n_k)(2n)(2n_{k+1})\dots$.

Let

- $I_{c_1\dots c_{2k-1}} = [0.c_1\dots c_{2k-1}, 0.c_1\dots c_{2k-1}(2n_k)(2n)(2n_{k+1})(2n)\dots]$ and
- $I_{c_1\dots c_{2k}} = [0.c_1\dots c_{2k}, 0.c_1\dots c_{2k}(2n)(2n_{k+1})(2n)(2n_{k+2})\dots]$.

Each interval contiguous to P has one of the following forms:

- $J_{c_1\dots c_{2k-1}} = (0.c_1\dots c_{2k-1}(c_{2k}-1), 0.c_1\dots c_{2k-1}c_{2k})$, $c_{2k} \neq 0$;

- $J_{c_1 \ldots c_{2k-2}} = (0.c_1 \ldots c_{2k-2}(c_{2k-1} - 1), c_1 \ldots c_{2k-2}c_{2k-1})$, $c_{2k-1} \neq 0$.

We obtain that $P = \cap_{k=1}^{\infty}(\cup_{(c_1,\ldots,c_{2k-1})} I_{c_1 \ldots c_{2k-1}}) = \cap_{k=1}^{\infty}(\cup_{(c_1 \ldots c_{2k})} I_{c_1 \ldots c_{2k}})$ and $|I_{c_1 \ldots c_{2k-1}}| = 1/((2n+1)^k(2n_1+1)\ldots(2n_k+1))$. Then P is a perfect set and $|P| = \lim_{k\to\infty}(n+1)^k(n_1+1)\ldots(n_{k-1}+1)/((2n+1)^k(2n_1+1)\ldots(2n_k+1)) = 0$. Let $P_1 = \{y \in [0,1] : y = \sum_{k=1}^{\infty} c_k/((2\cdot n_1+1)(2\cdot n_2+1)\ldots(2\cdot n_k+1)),\ c_k \in \{0,2,\ldots,2\cdot n_k\},\ k = \overline{1,\infty}\}$. Each point $y \in P_1$ is uniquely represented by $\sum_{k=1}^{\infty} c_k(y)/((2\cdot n_1+1)(2\cdot n_2+1)\ldots(2\cdot n_k+1))$. Then P_1 is a perfect set and $|P_1| = 0$ (the proof is similar to that for P). For $n_k \geq n$, $k = \overline{1,\infty}$, let $P_2 = \{y \in [0,1] : y = \sum_{k=1}^{\infty} c_k/((2\cdot n_1+1)(2\cdot n_2+1)\ldots(2\cdot n_k+1)),\ c_k \in \{0,2,\ldots,2n\},\ k = \overline{1,\infty}\}$. Each point $y \in P_2$ is uniquely represented by $\sum_{k=1}^{\infty} c_k(y)/((2\cdot n_1+1)(2\cdot n_2+1)\ldots(2\cdot n_k+1))$. As above it follows that P_2 is a perfect set and $|P_2| = 0$. Let $F_1 : P \mapsto P_1$, $F_1(x) = \sum_{k=1}^{\infty} c_{2k}(x)/((2\cdot n_1+1)(2\cdot n_2+1)\ldots(2\cdot n_k+1))$, and let $F_2 : P \mapsto P_2$, $F_2(x) = (1/2)\cdot\sum_{k=1}^{\infty} c_{2k-1}(x)/((2\cdot n_1+1)(2\cdot n_2+1)\ldots(2\cdot n_k+1))$. Then $F_1(P) = P_1$ and $F_2(P) = P_2$. Extending F_1 and F_2 linearly on the closure of each interval contiguous to P, we have F_1 and F_2 defined and continuous on $[0,1]$. (See Figures 6.25, 6.26 and 6.27 for $n = 3$ and $2n_k + 1 = 7^k$). Then we have:

(i) $F_1 \in (S) \subset (N) \subset T_2$, *hence* $F_1 \in AC \circ \sigma.f.l.$ *on* $[0,1]$;

(ii) $F_2 \in L \subset (N)$ *on* $[0,1]$, *whenever* $(2n+1)^k = 2n_k + 1$, $k = \overline{1,\infty}$;

(iii) $F_1 + F_2 \notin T_2$ *on* $[0,1]$ *for* $n = 3$ *and* $2n_k + 1 = 7^k$, $k = \overline{1,\infty}$;

(iv) $F_1 \notin \sigma.f.l.$ *on* $[0,1]$, *for* $n = 3$ *and* $2n_k + 1 = 7^k$, $k = \overline{1,\infty}$.

Proof. (i) Since $F_1(P) = P_1$, $| F_1(P) | = | P_1 | = 0$, hence $F_1 \in (N)$ on $[0,1]$. Because F is linear on each interval contiguous to P it follows that the set $F_1(\{x : F_1'(x)$ does not exist finite or infinite$\})$ has measure zero. By Theorem 2.18.1, $F_1 \in T_1$ on $[0,1]$, and by Theorem 2.18.8, $F_1 \in (S) \subset T_2$ on $[0,1]$. That $(S) = AC \circ AC \subseteq AC \circ \sigma.f.l.$ follows by Theorem 2.18.9, (xii), Corollary 2.28.1, (iii) and Corollary 2.29.1, (xvi).

(ii) Since $F_2(P) = P_2$ it follows that $| F_2(P) | = | P_2 | = 0$, hence $F_2 \in (N)$ on $[0,1]$.

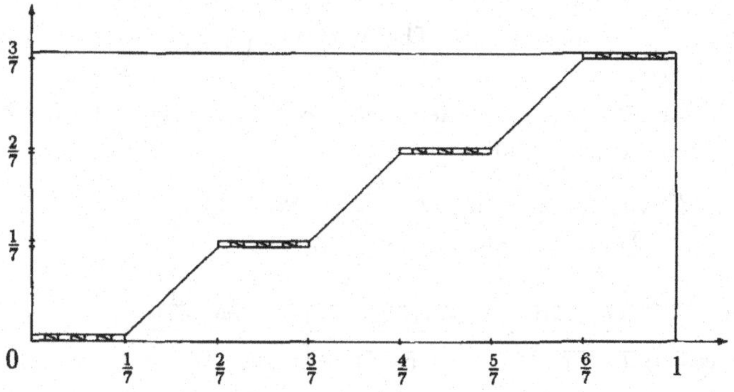

Figure 6.26: The function F_2

Let $E = [0,1] \setminus P$ and $x \in E$. Then we have two situations:

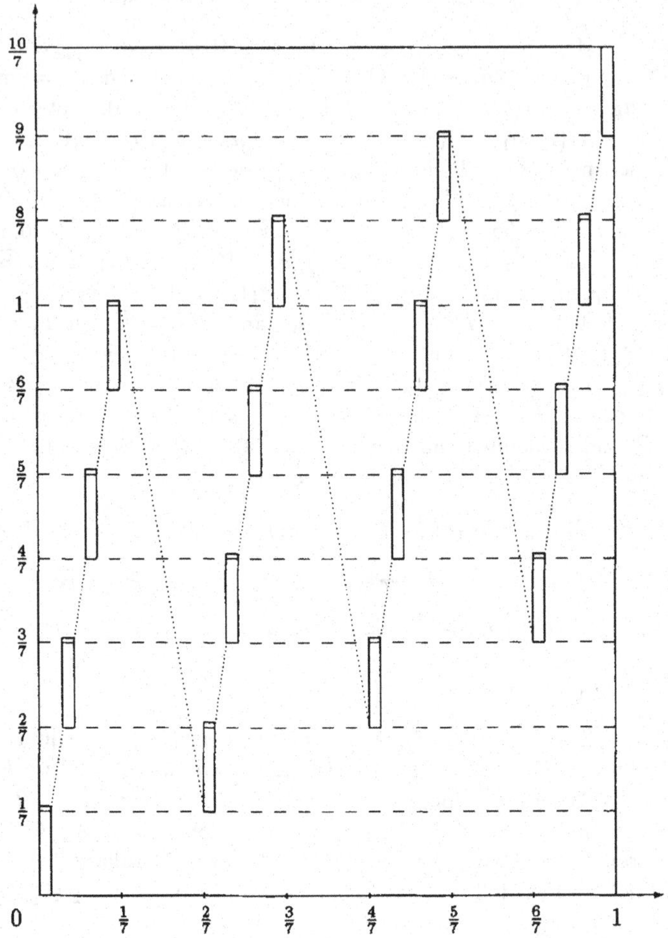

Figure 6.27: The function $F_1 + F_2$

1) Suppose that $x \in J_{c_1 \ldots c_{2k-1}}$. Then $0.c_1 \ldots c_{2k-1}(c_{2k} - 1) = 0.c_1 \ldots c_{2k-1}(c_{2k} - 2)(2n)(2n_{k+1})(2n)(2n_{k+2}) \ldots$ and

$$|F_2'(x)| = \left| \frac{F_2(0.c_1 \ldots c_{2k}) - F_2(0.c_1 \ldots c_{2k-1}(c_{2k} - 1)}{0.c_1 \ldots c_{2k} - 0.c_1 \ldots c_{2k-1}(c_{2k} - 1)} \right| =$$

$$\frac{2n \sum_{i=1}^{\infty} \frac{1}{(2n_1+1)\ldots(2n_{k+i}+1)}}{\frac{1}{(2n+1)^k(2n_1+1)\ldots(2n_k+1)}} < \frac{\frac{2n}{(2n_1+1)\ldots(2n_{k+1}+1)} \cdot \sum_{i=0}^{\infty} \frac{1}{(2n+1)^i}}{\frac{1}{(2n+1)^k(2n+1)\ldots(2n_k+1)}} = 1$$

2) Suppose that $x \in J_{c_1 \ldots c_{2k-2}}$. Then $0.c_1 \ldots c_{2k-2}(c_{2k-1} - 1) = 0.c_1 \ldots c_{2k-2}(c_{2k-1} - 2)(2n_k)(2n)(2n_{k+1}) \ldots$ and

$$F_2'(x) = \frac{\frac{2}{(2n_1)\ldots(2n_k+1)} - \sum_{i=1}^{\infty} \frac{2n}{(2n_1+1)\ldots(2n_{k+i}+1)}}{\frac{1}{(2n+1)^k(2n_1+1)\ldots(2n_{k-1}+1)}} < 2.$$

Therefore $|F_2'(x)| < 2$ for each $x \in E$. By Lemma 2.13.1, $F \in L$ with the constant 2 on $[0,1]$. (iii) Let $F = F_1 + F_2$. Then

$$F(x) = \sum_{k=1}^{\infty} \frac{\frac{1}{2} \cdot c_{2k-1}(x) + c_{2k}}{(2 \cdot n_1 + 1) \dots (2 \cdot n_k + 1)} \quad \textit{for each} \quad x \in P.$$

Let $Q = \{y \in [0,1] : y = \sum_{k=1}^{\infty} d_k/((2 \cdot n_1 + 1) \dots (2 \cdot n_k + 1)),\ d_k \in \{2,3,4,\dots,2 \cdot n_k\}$, $k = \overline{1,\infty}$ (some $y \in Q$ may have two representations). Then Q is a perfect set and

$$|Q| = \frac{2 \cdot n_1 - 1}{2 \cdot n_1 + 1} \cdot \frac{2 \cdot n_2 - 1}{2 \cdot n_2 + 1} \dots > \frac{5}{7} \cdot \left(1 - \frac{1}{2^2}\right) \cdot \left(1 - \frac{1}{3^2}\right) \dots > \frac{5}{7} \cdot \frac{1}{2} = \frac{5}{14}.$$

Let $y_o \in Q$. Then $y_o = \sum_{k=1}^{\infty} d_k/(2 \cdot n_1 + 1) \dots (2 \cdot n_k + 1)$ for some $d_k \in \{2,3,\dots,2 \cdot n_k\}$, $k = \overline{1,\infty}$ and $(F_{/P})^{-1}(y_o) = \{x \in P : x = \sum_{k=1}^{\infty} (c_{2k-1}/(7^k \cdot (2 \cdot n_1 + 1) \dots (2 \cdot n_{k-1} + 1)) + c_{2k}/(7^k \cdot (2 \cdot n_1 + 1) \dots (2 \cdot n_k + 1)))$, where for d_k even, either $c_{2k-1} = 0$, $c_{2k} = d_k$ or $c_{2k-1} = 4$, $c_{2k} = d_k - 2$, $k = \overline{1,\infty}$; and for d_k odd, $d_k \geq 3$, either $c_{2k-1} = 2$, $c_{2k} = d_k - 1$ or $c_{2k-1} = 6$, $c_{2k} = d_k - 3$, $k = \overline{1,\infty}\}$. It follows that $(F_{/P})^{-1}(y_o)$ is a perfect set, therefore uncountable. Hence $F \notin T_2$ on $[0,1]$.

(iv) Suppose on the contrary that $F_1 \in \sigma.f.l.$ on $[0,1]$. By Corollary 2.33.1, (vii), $F_1 + F_2 = F \in \sigma.f.l.$. By Theorem 2.18.9, (ii), $F \in T_2$, a contradiction.

Remark 6.58.1 *The utility of this example follows by Corollary 2.29.1, (xvi); Corollary 2.33.1, (x).*

6.59 Functions $G_1 \in \sigma.f.l.$, $G_2 \in AC$, $G_1 + G_2 \notin \sigma.f.l.$

Let F_1 and F_2 be the functions defined in Section 6.58. By Lemma 2.18.3 it follows that there exists a strictly increasing function $g : [0,1] \mapsto \mathbb{R}$ such that $g, g^{-1} \in AC$ and

(1) $\Lambda(B(F_1 \circ g^{-1}; g(P)) = 0.$

It also follows that g is linear on each interval contiguous to P and g^{-1} is linear on each interval contiguous to $g(P)$. Let $G_1 = F_1 \circ g^{-1}$ and $G_2 = F_2 \circ g^{-1}$. Then we have:

(i) $G_1 \in \Lambda Z \subset \sigma.f.l.$ *on* $[0,1]$;

(ii) $G_2 \in AC$ *on* $[0,1]$;

(iii) $G_1 + G_2 \notin T_1$, *hence* $G_1 + G_2 \notin \sigma.f.l.$ *on* $[0,1]$.

Proof. (i) Since G_1 is linear on each interval contiguous to $g(P)$ and by (1), it follows that $G_1 \in \Lambda Z$ on $[0,1]$. By Theorem 2.33.1, $\Lambda Z \subseteq \sigma.f.l.$.

(ii) By Theorem 2.32.1, (xvii), $G_2 \in AC$ on $[0,1]$.

(iii) $(F_1 + F_2)(P) = (G_1 + G_2)(g(P))$. Since $F_1 + F_2 \notin T_2$, it follows that $G_1 + G_2 \notin T_2$ on $[0,1]$. By Theorem 2.18.9, (ii), $G_1 + g_2 \notin \sigma.f.l.$ on $[0,1]$.

Remark 6.59.1 *The utility of this example follows by Corollary 2.33.1, (xi).*

6.60 Functions $H_1 \in \sigma.f.l.$, $H_2 \in AC$, $H_1 \cdot H_2 \notin \sigma.f.l.$

Let G_1 and G_2 be the functions defined in Section 6.59. Let $H_1(x) = e^{G_1(x)}$ and $H_2(x) = e^{G_2(x)}$. Then we have:

(i) $H_1 \in \sigma.f.l.$ *on* $[0,1]$;

(ii) $H_2 \in AC$ *on* $[0,1]$;

(iii) $H_1 \cdot H_2 \notin T_2$, *hence* $H_1 \cdot H_2 \notin \sigma.f.l.$ *on* $[0,1]$.

Proof. Clearly $e^x \in L$ on a closed interval.

(i) By Corollary 2.33.1, (ix), $H_1 \in \sigma.f.l.$ on $[0,1]$.

(ii) By Theorem 2.32.1, (xvii), $H_2 \in AC$ on $[0,1]$.

(iii) Clearly $(H_1 \cdot H_2)(x) = e^{(G_1+G_2)(x)}$. Since $G_1 + G_2 \notin T_2$, it follows that $H_1 \cdot H_2 \notin T_2$. By Theorem 2.18.9, (ii), $H_1 \cdot H_2 \notin \sigma.f.l.$ on $[0,1]$.

Remark 6.60.1 *The utility of this example follows by Corollary 2.33.1, (xii).*

6.61 A function $F \in \sigma.f.l.$, $F \in T_1$, $F \notin \mathcal{B}$, F is nowhere approximately derivable, (Foran)

Construction 6.61.1 (Foran) *([F1]). Let* $S = [0,1] \times [0,1]$. *Consider a collection* A *of closed rectangles with sides parallel to the coordinate axes, each of which is denoted by a finite sequence consisting of the letter* s *which appears at least once in a sequence, and of the subscripted letters* r_i, $i = 0,1,2$, *any or all of which may be missing in the designated sequence for a rectangle.(In the construction that follows, e.g. we shall take* s *by itself to represent* S.*) Each sequence beginning with an* s *represents a square, and each beginning with an* r_i *a non-square. We associate two new collections* RA *and* SA *with* A. *We associate with each* $q = [a, a+p] \times [b, b+p]$ *in* A *whose sequence begins with an* s, *the three rectangles:* $r_o q = [a, a+p/3] \times [b, b+p]$, $r_1 q = [a+p/3, a+2p/3] \times [b, b+p]$ *and* $r_2 q = [a+2p/3, a+p] \times [b, b+p]$. *These associated together with the rectangles in* A *whose sequence begins with some* r_i *constitute the collection* RA. *To form* SA, *proceed as follows: Let* $q = [a,b] \times [c,d]$ *be any member of* A. *We associate with* q *the three rectangles:*

- $r_o q = [a, a+\frac{1}{3}(b-a)] \times [c, \frac{c+d}{2} - \frac{b-a}{6}]$,
- $s_o q = [a - \frac{1}{2}(b-a), a+\frac{2}{3}(b-a)] \times [\frac{c+d}{2} - \frac{b-a}{6}, \frac{c+d}{2} + \frac{b-a}{6}]$,
- $r_2 q = [a+\frac{2}{3}(b-a), b] \times [\frac{c+d}{2} + \frac{b-a}{6}, d]$,

if the sum of the subscripts of the r_i *in the sequence for* q *is even, and we choose the three rectangles:*

- $r_o q = [a, a+\frac{1}{3}(b-a)] \times [\frac{c+d}{2} + \frac{b-a}{6}, d]$,
- $sq = [a+\frac{1}{3}(b-a), a+\frac{2}{3}b] \times [\frac{c+d}{2} - \frac{b-a}{6}, \frac{c+d}{2} + \frac{b-a}{6}]$,
- $r_2 q = [a+\frac{2}{3}(b-a), b] \times [c, \frac{c+d}{2} - \frac{b-a}{6}]$,

if the sum of the subscripts of the r_i in the sequence of q is odd. The collection of associated rectangles constitutes SA. Note that sq is a square of side length $(b-a)/2$ and has the same center as q. If A is a collection of sets let σA be the union of the sets in A. Let $A = \{S\}$ and associate with S the sequence $\{s\}$. Let

$$E = \sigma RA \cap \sigma SRA \cap \sigma RSRA \cap \sigma SRSRA \cap \dots$$

(see Figure 6.28 of the first steps in the construction of E). Then we have:

(i) E is the graph of a continuous function F on $[0,1]$;

(ii) $F \in \sigma.f.l.$ on $[0,1]$;

(iii) $F \in T_1$ on $[0,1]$;

(iv) F is nowhere approximately derivable on $[0,1]$, hence F is not approximately quasi-derivable on $[0,1]$;

(v) $F \notin \mathcal{B}$ on $[0,1]$.

Proof. (i) Since each $\sigma RS\dots SRA$ or $\sigma SR\dots SRA$ is the union of finitely many closed rectangles, E is a compact set. Thus in order to show that E is the graph of a continuous function, it suffices to show that E is the graph of a function. But for any $x \in [0,1]$ one readily observes that the set of points above x in $RS\dots SRA$ consists of at most the points of two adjoining rectangles and that the points above x in $RS\dots SRA$ form a line segment of length less than $(2/3)^n$, where n is the number of times S appears in the sequence $RS\dots SRA$. Since $(2/3)^n$ approaches 0 there is exactly one point above x in E, and E is the graph of a function.

(ii) Let E_1 be the set of points of E which are in infinitely many squares from $S \cup SRA \cup SRSRA \cup \dots$. Given two squares from $S \cup SRA \cup SRSRA \cup \dots$, note that if one square is not contained in the other, then neither lie above the same $x \in [0,1]$. Given $\epsilon > 0$, let U'_ϵ be the collection of squares of diameter less than ϵ. Let U_ϵ be the collection of squares of U'_ϵ which are contained in no larger squares of U'_ϵ. Then U_ϵ covers E_1 and no two squares of U_ϵ lie above the same $x \in [0,1]$. Thus $\sum diam(S) \le \sum side\ length(S) \cdot \sqrt{2} \le \sqrt{2}$. Consequently, E_1 has length $\le \sqrt{2}$. Given a non-square rectangle q in $RA \cup RSRA \cup RSRSRA \cup \dots$, let E_q be the sets of points in q which are not in any squares of $SRA \cup SRSRA \cup SRSRSRA \cup \dots$ contained in q. Each point in E_q is contained in $r_o q \cup r_2 q$ and $diam(r_o q) + diam(r_2 q) \le 2((w/3)^2 + (w/2)^2)^{1/2} < (w^2 + h^2)1/2 < diam(q)$, where $q = [a,b] \times [c,d]$, $w = b - a$ and $h = d - c$. By the same argument E_q is contained in $\cup r_{i_1} r_{i_2} \dots t_{i_n} q$, where the union is over the 2^n rectangles with $i_j = 0, 2$ and the sum of the diameters of these rectangles is less than $diam(q)$. Since E_q can then be covered by arbitrarily small rectangles so that the sum of their diameters is less than $diam(q)$, it follows that each E_q has Hausdorff length less than $diam(q)$. Since each point of E belongs to E_1 or to one of the countable many sets E_q, it follows that E has σ-finite length.

(iii) First consider $(x, F(x)) \notin E_1$. For such a point $(x, F(x)) \in E_{\bar{q}} = E_{r, s\bar{q}}$, where $s\bar{q}$ is the smallest square from the collection $A \cup SRA \cup SRSRA \cup \dots$ which contains $(x, F(x))$. If $(a, F(a))$ and $(b, F(b))$ (with $a < b$) belong to E_q, either $(a, F(a)) \in r_o q$ and $(b, F(b)) \in r_2 q$, so $F(a) \ne F(b)$, or there are $r_{i_1}, r_{i_2}, \dots, r_{i_n}$, $i_j = 0, 2$, such that

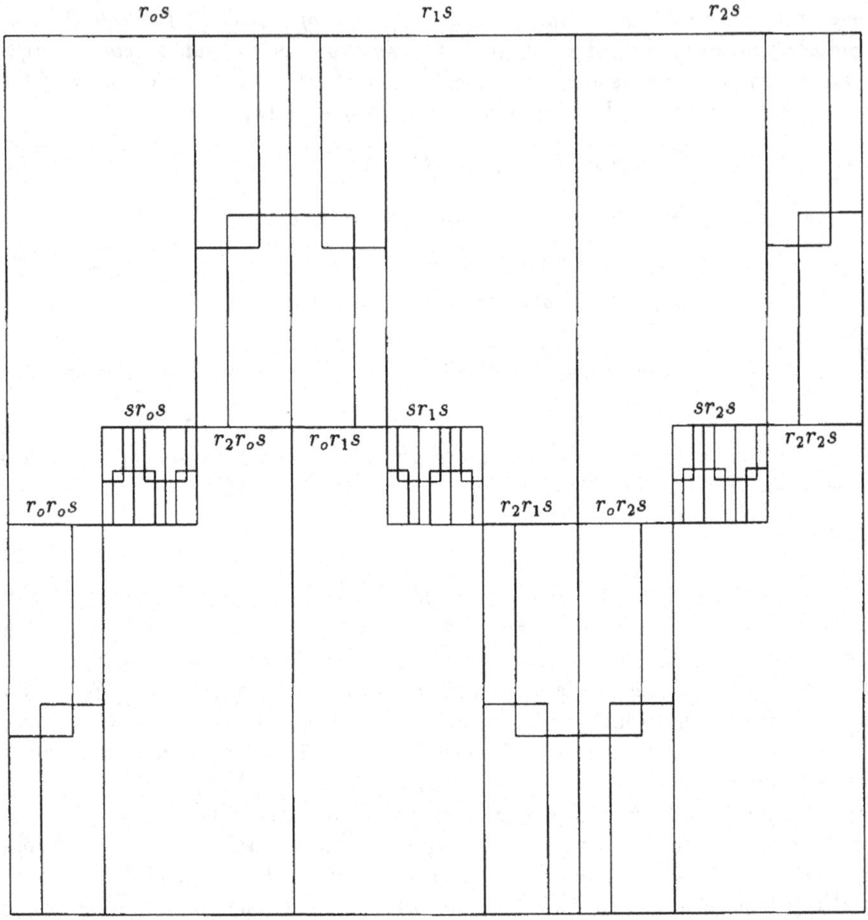

Figure 6.28: The rectangles $SRSRA$

$(a, F(a)) \in r_o r_{i_1} r_{i_2} \ldots r_{i_n} q$ and $(b, F(b)) \in r_o r_{i_1} r_{i_2} \ldots r_{i_n} q$ and again $F(a) \neq F(b)$,. Thus F is one-to-one on $\{x : (x, F(x) \in E_q\}$. (In fact it may be observed that if the sum of the subscripts in q is even (resp. odd) then F is increasing (resp. decreasing) on this set.) Any point $(a, F(a))$ such that $F(a) = F(x)$ belongs to some q' in $B = RSRS \ldots SRA$, where this is the collection of rectangles which contains q. Moreover, $\{y : (x, y) \in q\} = \{y : (x, y) \in q'\}$ and since each successive rectangle in Q' containing $(a, F(a))$ corresponds to a similar one in q containing $(x, F(x))$, it follows that $(a, F(a))$ must belong to E_q. There are only finitely many such rectangles in $B = RSRS \ldots SRA$ (in fact at most 3^n, n = the number of times S occurs in $RSRS \ldots SRA$). Each point $(a, F(a))$ with $F(a) = F(x)$ belongs to some E_q with q' in B. Since F is one-to-one on $\{x : (x, y) \in E_q\}$, it follows that $F^{-1}(F(x))$ is finite whenever $(x, F(x)) \in E \setminus E_1$. Thus F satisfies T_1 on $\{x : (x, F(x)) \in E \setminus E_1\}$. By Theorem 2.18.9, (i) and the fact that E_1 has finite length, $F \in T_1$ on $\{x : (x, F(x)) \in E_1\}$. Consequently, $F \in T_1$ on $[0, 1]$. (One can in fact observe that if $(x, F(x)) \in E_1$

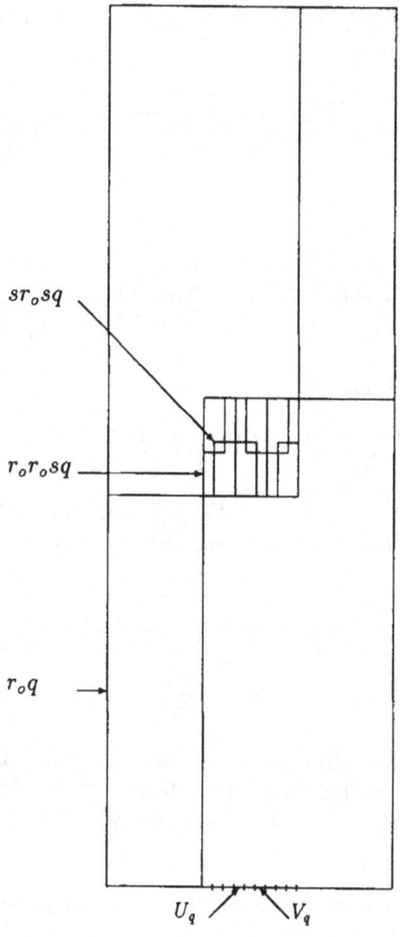

Figure 6.29: case (i)

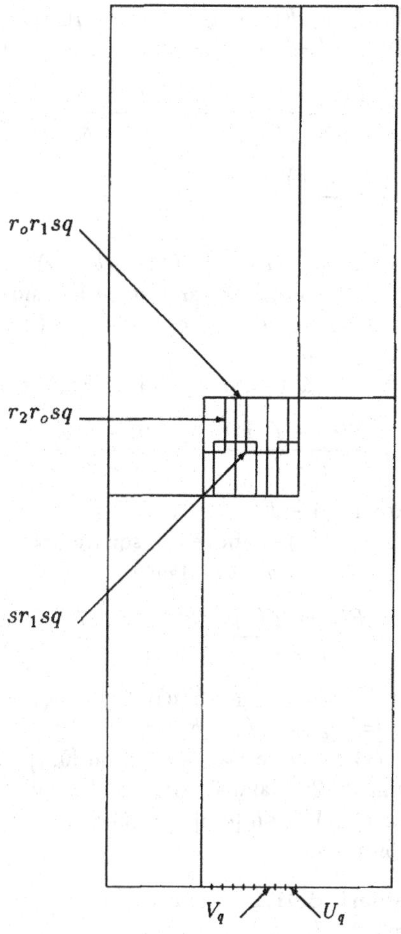

Figure 6.30: case (ii)

then $F^{-1}(F(x))$ is a perfect set, by Theorem 1.3.2, $|(E_1)_n| \leq \Lambda(E_1) \cdot (1/n)$ and thus $\{F(x) : (x, F(x)) \in E_1\}$ has Lebesgue measure 0, where $(E_1)_n = \{c :$ the line $y = c$ meets E_1 in n or more points$\}$.)

(iv) Let $x_o \in [0,1]$ and let $q = [a,b] \times [c,d]$ be a rectangle containing $(x_o, F(x_o))$ with $q \in RA \cup RSRA \cup RSRSRA \cup \ldots$. It will be shown that there are sets U_q and V_q contained in $[a,b]$ such that $|U_q| = |V_q| = (1/27)(b-a)$ and

$$\left| \frac{F(u) - F(x_o)}{u - u_o} - \frac{F(v) - F(x_o)}{v - x_o} \right| \geq \frac{1}{27},$$

whenever $u \in U_q$ and $v \in V_q$. This implies that F is not approximately derivable at x_o, since any set which has x_o as a point of density will have to contain points of U_q and points of V_q, if q is chosen small enough. By symmetry it will be sufficient to assume that q has an even number for the sum of its subscripts. Also by symmetry of the successive subdivisions of q it will be sufficient to consider $(x_o, F(x_o))$ belonging to

the left side of q. Two cases occurs and they are illustrated in Figure 6.29 and Figure 6.30: 1) $(x_o, F(x_o)) \in r_o q \cup r_o r_o sq \cup sr_o sq$ and 2) $(x_o, F(x_o)) \in r_2 r_o sq \cup r_o r_1 sq \cup sr_1 sq$.
1) If $(u, F(u)) \in r_o r_1 sq$ and $(v, F(v)) \in r_2 r_1 sq$ then

$$\frac{F(u)-F(x_o)}{u-x_o} - \frac{F(u)-F(x_o)}{v-x_o} \geq \frac{F(u)-F(x_o)}{v-x_o} - \frac{F(v)-F(x_o)}{v-x_o}$$

$$= \frac{F(u)-F(v)}{v-x_o} \geq \frac{1}{27},$$

since $F(u) - F(v) \geq (1/27)(b-a)$ (because $F(u)$ is above the square $sr_2 sq$ and $F(v)$ is below $sr_2 sq$, and $sr_2 sq$ has side length $(1/27)(b-a)$). Note that for $U_q = \{u : (u, F(u)) \in r_o r_1 sq\}$ and $V_q = \{v : (V, F(v)) \in r_2 r_1 sq\}$ we have $| U_q |=| V_q |= (1/27)(b-a)$.
2) If $(u, F(u)) \in r_2 r_2 sq$ and $(v, F(v)) \in r_o r_2 sq$ then

$$\frac{F(u)-F(x_o)}{u-x_o} - \frac{F(v)-F(x_o)}{v-x_o} \geq \frac{F(u)-F(x_o)}{u-x_o} - \frac{F(v)-F(x_o)}{u-x_o},$$

since $F(v) - F(x_o) \leq 0$ and $u - x_o > v - x_o$. But $F(u) - F(v) \geq (1/27)(b-a)$, because $F(u)$ is above the square $sr_2 sq$ and $F(v)$ is below $sr_2 sq$, and $sr_2 sq$ has side length $(1/27)(b-a)$. Hence

$$\frac{F(u)-F(v)}{u-x_o} \geq \frac{(b-a)/27}{b-a} = \frac{1}{27}.$$

For $U_q = \{u : (u, F(u)) \in r_2 r_2 sq\}$ and $V_q = \{v : (v, F(v)) \in r_o r_2 sq\}$ we have $| U_q |=| V_q |= (1/27)(b-a)$.

(v) Suppose that $F \in \mathcal{B}$ on $[0,1]$. Then $F \in [\mathcal{B}]$ on $[0,1]$ (because $F \in \mathcal{C}$). By Lemma 2.29.2 and Remark 2.29.3, (ii), it follows that there exists $[c,d] \subset [0,1]$ such that $F \in VB$ on $[c,d]$ (see Theorem 2.29.1, (ii)). Hence F is derivable $a.e.$ on $[c,d]$, a contradiction.

Remark 6.61.1 *The utility of this example follows by Corollary 2.29.1, (xvi); Remark 3.1.1, (vi).*

6.62 A function $G \in \sigma.f.l.$, $G \in T_1$, G is nowhere derivable, $G'_{ap}(x) = 0$ a.e., $G \notin W$, $G \in W^*$ (Foran)

Construction 6.62.1 (Foran) *([F1]). The same construction as in Section 6.61 is employed except that each square is constructed with side length $w/(1+2w)$, where w is the width of the rectangle in which the square is placed. Thus, given a collection A of rectangles, we modify the construction of SA as follows to obtain $S'A$. Let $q = [a,b] \times [c,d]$ be in A. Let $w = b - a$. Referring to the construction of SA in the even case associate*

- $r_o q = [a, \frac{a+b}{2} - \frac{w}{2+4w})] \times [c, \frac{c+d}{2} - \frac{w}{2+4w}]$,
- $sq = [\frac{a+b}{2} - \frac{w}{2+4w}, \frac{a+b}{2} + \frac{w}{2+4w}] \times [\frac{c+d}{2} - \frac{w}{2+4w}, \frac{c+d}{2} + \frac{w}{2+4w}]$,

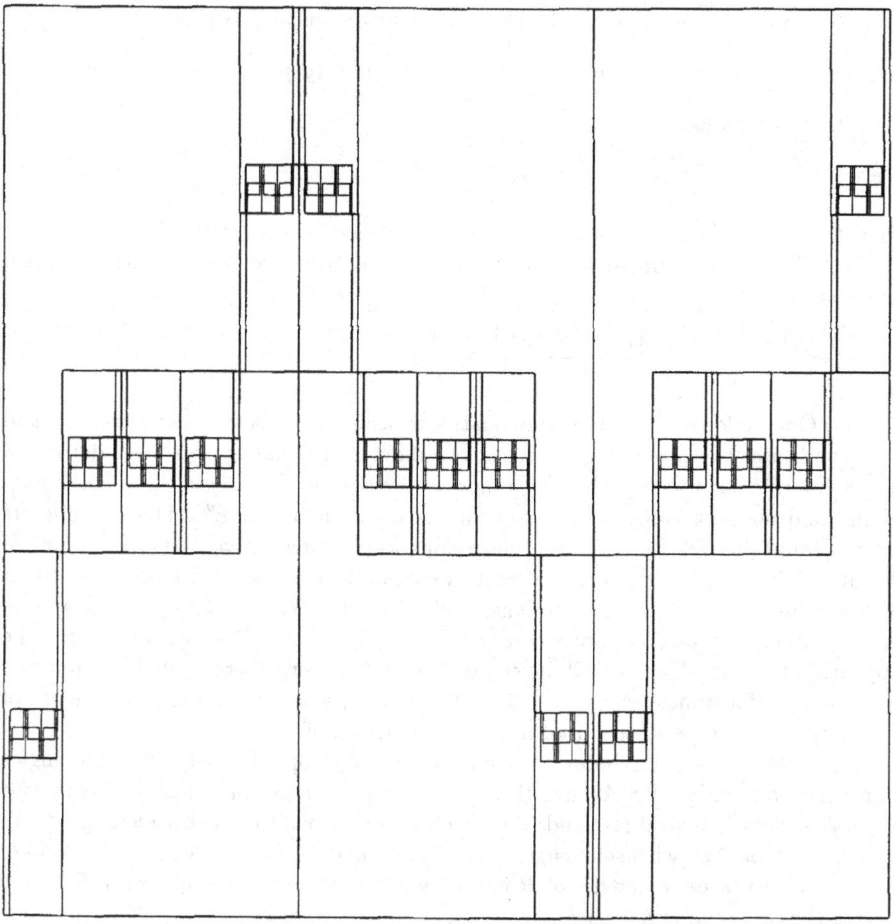

Figure 6.31: Location of the graph of G

- $r_2 q = [\frac{a+b}{2} + \frac{w}{2+4w}, b] \times [\frac{c+d}{2} + \frac{w}{2+4w}, d]$,

and in the odd case

- $r_o q = [a, \frac{a+b}{2} - \frac{w}{2+4w}] \times [\frac{c+d}{2} + \frac{w}{2+4w}, d]$,
- $sq = [\frac{a+b}{2} - \frac{w}{2+4w}, \frac{a+b}{2} + \frac{w}{2+4w}] \times [\frac{c+d}{2} - \frac{w}{2+4w}, \frac{c+d}{2} + \frac{w}{2+4w}]$,
- $r_2 q = [\frac{a+b}{2} + \frac{w}{2+4w}, b] \times [c, \frac{c+d}{2} - \frac{w}{2+4w}]$.

Let $E' = \sigma RA \cap \sigma S'RA \cap \sigma RS'RA \cap \sigma S'RS'RA \cap \ldots$ *The same proof employed before for* E *shows that* E' *is the graph of a function* G *and* G *is continuous on* $[0,1]$ *(see Figure 6.31). Then we have:*

(i) $G \in \sigma.f.l.$ *on* $[0,1]$*;*

(ii) $G \in T_1$ *on* $[0,1]$*;*

(iii) G is nowhere derivable on $[0,1]$*, hence G is not quasi-derivable;*

(iv) $G'_{ap}(x) = 0$ *a.e. on* $[0,1]$*, hence G is approximately quasi-derivable on* $[0,1]$*;*

(v) $G \notin W$ *on* $[0,1]$*;*

(vi) $G \in W^*$ *on* $[0,1]$.

Proof. (i), (ii) The proofs are similar to those of Section 6.61, (ii), (iii).

(iii) The proof is similar to that of Section 6.61, (iv), that is, there exist points u and v such that

$$\left|\frac{G(u)-G(x_o)}{u-x_o} - \frac{G(v)-G(x_o)}{v-x_o}\right| \geq \frac{1}{27}.$$

(iv) One no longer have the large relative measure of the sets of these points u and v in small intervals containing x_o. In fact to show that G has approximate derivative a.e. it is sufficient to show that for almost every $x \in [0,1]$, $G(x)$ is the y-coordinate of the mid-point of some square used in the construction of E'. These midpoints form a countable set $\{y_n\}$ and will have approximate derivative 0 at every point of density of $B_n = G^{-1}(y_n)$, i.e., at almost every point of B_n (since almost every point of a measurable set is a point of density of that set). Since $| \cup B_n |= 1$, G will be approximately derivable at almost every point in $[0,1]$. Since G is constant on B_n, the approximate derivative of G will be equal to 0 almost everywhere on $[0,1]$. It remains to show that for almost every $x \in [0,1]$, $G(x) = y_n$, where y_n is the y-coordinate of the midpoint of some square used in the construction of E'.

Given ϵ with $1 > \epsilon > 0$, choose the pairwise nonoverlapping collection of rectangles from $B_\epsilon = RS'RS' \ldots RS'A$, such that each rectangle in B_ϵ has width less than $\epsilon/3$. Let q represent a typical rectangle and w represent its width. In this rectangle there is a square p of $S'B_\epsilon$ with side length $w/(1+2w)$. Let $1/k = w/(1+2w)$. This square is divided into three rectangles of $RS'B_\epsilon$ of width $1/3^k$ and three squares of $S'RS'B_\epsilon$ having the same center as p and side length $(1/3^k)/(2/3^k+1) = 1/(2+3^k)$. Similarly there are 9 squares in $S'RS'RS'B_\epsilon$ with the same center as p having side length

$$\frac{\frac{1}{3(2+3^k)}}{\frac{2}{3(2+3^k)}+1} = \frac{1}{2+2\cdot 3+3^2\cdot k}.$$

In general there are 3^n squares of side length

$$\frac{1}{2+2\cdot 3+2\cdot 3^2+\ldots 2\cdot 3^{n-1}+3^n\cdot k} = \frac{1}{3^n-1+3^n\cdot k}.$$

Consequently the measure of the points x with $(x, G(x))$ in q at the level of the center of p is

$$\lim_{n\to\infty}\frac{3^n}{3^n-1+r^n\cdot k} = \lim_{n\to\infty}\frac{3^n}{3^n(1+k)-1} = \frac{1}{1+k}.$$

This is true because the set of points in q at the level of the center of p is the intersection of the sets of points in these 3^n squares. Corresponding to these squares are sets E_n on the line, with $\{E_n\}$ decreasing so that $|\cap E_n| = \lim_{n\to\infty}|E_n|$. Thus the points x having $(x, G(x))$ in q at the level of the center of p form a set of measure

$1/(1+k) = w/(3w+1) > w/(\epsilon+1) > w(1-\epsilon)$. Since this holds for all rectangles in the cover of E', the set of x whose image is a y-coordinate of the midpoint of some square has measure larger that $1-\epsilon$. Since ϵ was arbitrary, it follows that this set has measure 1. This establishes the fact that the approximate derivative of G exists and equals 0 at almost every point in $[0,1]$.

(v) See (iv) and Corollary 3.3.1.

(vi) See Corollary 2.16.1.

Remark 6.62.1 *The utility of this example follows by Remark 3.1.1, (i); Remark 3.3.3.*

6.63 A wrinkled function F on a perfect nowhere dense set of positive measure with each level set perfect, and F is nowhere approximately derivable

Construction 6.63.1 *Let* $p_i = 1/2^{i+2} + 3/4^{i+1}$ *and let* $P = \{x \in [0,1] : x = \sum_{k=1}^{\infty} c_k p_k,\ c_k \in \{0,2\}\}$. *Each point* $x \in P$ *is uniquely represented by* $\sum_{k=1}^{\infty} c_k(x) p_k$. *Then P is a perfect nowhere dense subset of* $[0,1]$. *Let* $V_n = \cup_{(c_1,\ldots,c_n)} [\sum_{k=1}^{n} c_k p_k, \sum_{k=1}^{n} c_k p_k + \sum_{k=n+1}^{\infty} 2p_k]$, $n = \overline{1,\infty}$. *Then* $P = \cap_{n=1}^{\infty} V_n$ *and* $|V_n| = 2^n \cdot \sum_{k=n+1}^{\infty} 2p_k \to 1/2$. *Hence* $|P| = 1/2$. *Let* $F_1, F_2 : P \mapsto \mathbb{R}$ *be defined as follows:* $F_1(x) = \sum_{k=1}^{\infty} c_{2k}(x)/2^{k+1}$, $F_2(x) = \sum_{k=1}^{\infty} c_{2k-1}(x)/2^{k+1}$. *Extending F_1 and F_2 linearly on the closure of each interval contiguous to P, we have F_1 and F_2 defined and continuous on $[0,1]$ (see Figures 6.32 and 6.33). Let* $F : [0,1] \mapsto [0,1] \times [0,1]$, $F(x) = (F_1(x), F_2(x))$. *Clearly* $F(P) = [0,1] \times [0,1]$. *For each irrational number $z \in (0,1)$ represented uniquely in base two by* $\sum_{i=1}^{\infty} d_i(z)/2^i$, *let* $A_z = \{x \in P : F_1(x) = z\} = \{x \in P : c_{2i}(x) = 2d_i(z)\} = a_z + A$, *where* $a_z = \sum_{i=1}^{\infty} d_i(z) p_{2i}$ *and* $A = \{x \in P : c_{2i}(x) = 0,\ i = \overline{1,\infty}\}$. *Let* $B_z = \{x \in P : F_2(x) = z\} = \{x \in P : c_{2i-1}(x) = 2d_i(z)\} = b_z + B$, *where* $b_z = \sum_{i=1}^{\infty} 2d_i(z) p_{2i-1}$ *and* $B = \{x \in P : c_{2i-1}(x) = 0,\ i = \overline{1,\infty}\}$. *It follows that A_z and B_z are perfect null sets. Then we have:*

(i) $F_2(A_z) = [0,1]$ *and* $F_1(B_z) = [0,1]$;

(ii) F_2 *is increasing on* A_z *and* F_1 *is increasing on* B_z;

(iii) For each measurable subset M of P the set $F(M)$ is measurable and $|F(M)| = 2 \cdot |M|$;

(iv) $F_1, F_2 \in W$ *on* P;

(v) $(F_1)_{/P}$ *and* $(F_2)_{/P}$ *have a finite or infinite derivative at no point of P;*

(vi) $(F_1)_{/P}$*and* $(F_2)_{/P}$ *have a finite approximate derivative at no point of P.*

Proof. (i) This is evident.

(ii) Let $x, y \in A_z$, $x < y$. It follows that $x = a_z + \sum_{k=1}^{\infty} x_{2k-1}(x) p_{2k-1}$ and $y = a_z + \sum_{k=1}^{\infty} c_{2k-1}(y) p_{2k-1}$. Let n be a positive integer such that $c_{2n-1}(x) = 0$, $c_{2n-1}(y) = 2$ and $c_{2k-1}(x) = c_{2k-1}(y)$ for each $k \le n-1$. We have $F_2(x) \le \sum_{k=1}^{n-1} c_{2k-1}(x)/2^{k+1} +$

$\sum_{k=n+2}^{\infty} 1/2^{k+1} = \sum_{k=1}^{n-1} c_{2k-1}(x)/2^{k+1} + 1/2^{n+1} \le F_2(y)$, hence F_2 is increasing on A_z. Similarly F_1 is increasing on B_z.

(iii) Let $I_{c_1 \ldots c_{2n}} = [\sum_{k=1}^{2n} c_k p_k, \sum_{k=1}^{2n} c_k p_k + \sum_{k=2n+1}^{\infty} 2p_{2k}]$. Then $F(P \cap I_{c_1 \ldots c_{2n}}) =$

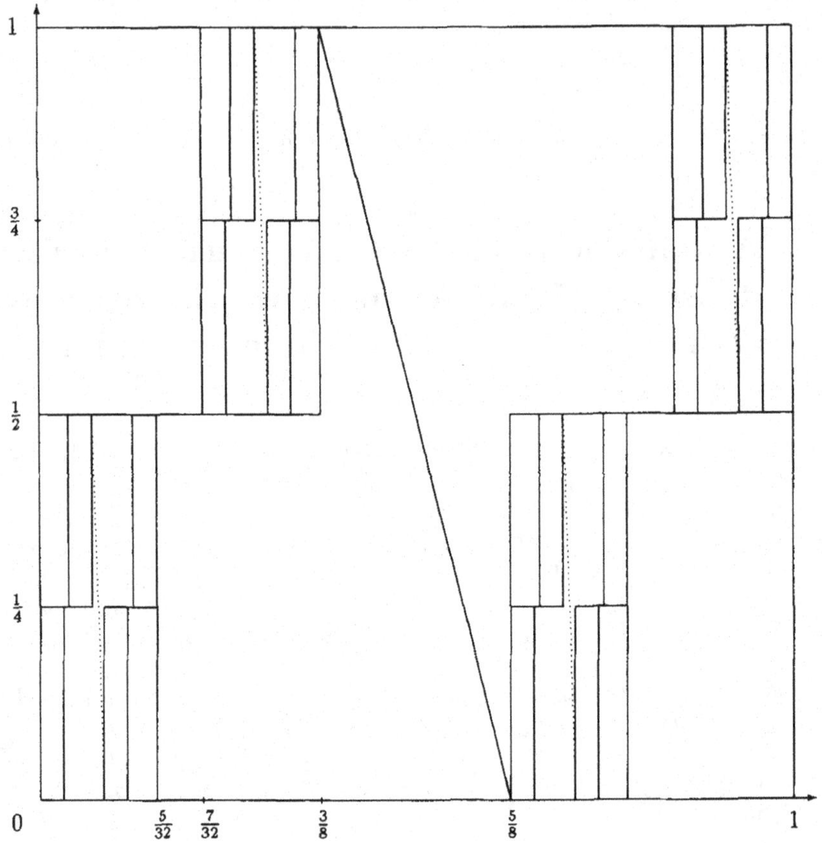

Figure 6.32: The function F_1

$[\sum_{k=1}^{n} c_{2k}/2^k, \sum_{k=1}^{n} c_{2k}/2^k + 1/2^n] \times [\sum_{k=1}^{n} c_{2k-1}/2^k, \sum_{k=1}^{n} c_{2k-1}/2^k + 1/2^n] = J_{c_1 \ldots c_{2n}}$ and $int(J_{c_1 \ldots c_{2n}}) \cap int(J_{c'_1 \ldots c'_{2n}}) = \emptyset$ for $(c_1, \ldots, c_{2n}) \neq (c'_1, \ldots, c'_{2n})$. We have

(1) $\quad |J_{c_1 \ldots c_{2n}}| = 2 \cdot |P \cap I_{c_1 \ldots c_{2n}}| = 1/4^n$.

Let $\epsilon > 0$ and let M be a measurable subset of P. Let E_1 and E_2 be open sets such that $M \subset E_1$ and $F(M) \subset E_2$, $|F(M)| > |E_2| - \epsilon$, $|M| > |E_1| - \epsilon$. Then $E_1 \cap F^{-1}(E_2)$ is an open neighborhood of M. Since $P = \cap_{n=1}^{\infty}(\cup_{(c_1,\ldots,c_{2n})} P \cap I_{c_1 \ldots c_{2n}})$ and $|I_{c_1 \ldots c_{2n}}| \to 0$, $n \to \infty$, it follows that $E_1 \cap F^{-1}(E_2) \cap P$ can be represented as a countable union of disjoint sets $P \cap I_{c'_1 \ldots c'_{2n}}$, $n = \overline{1, \infty}$. By (1) it follows that $2 \cdot |E_1 \cap F^{-1}(E_2) \cap P)| = |F(E_1 \cap F^{-1}(E_2) \cap P)|$. Since $M \subset E_1 \cap F^{-1}(E_2) \cap P \subset E_1$, $F(M) \subset F(E_1 \cap F^{-1}(E_2) \cap P) \subset E_2$, and ϵ is arbitrary, we obtain that $|F(M)| = 2 \cdot |M|$. Thus $F \in (N)$ on P. By the continuity of F and the measurability of M, it follows that $F(M) = F(Q) \cup F(Z)$, where Q and $F(Q)$ are F_σ sets and Z and $F(Z)$ are null sets. Hence $F(M)$ is measurable.

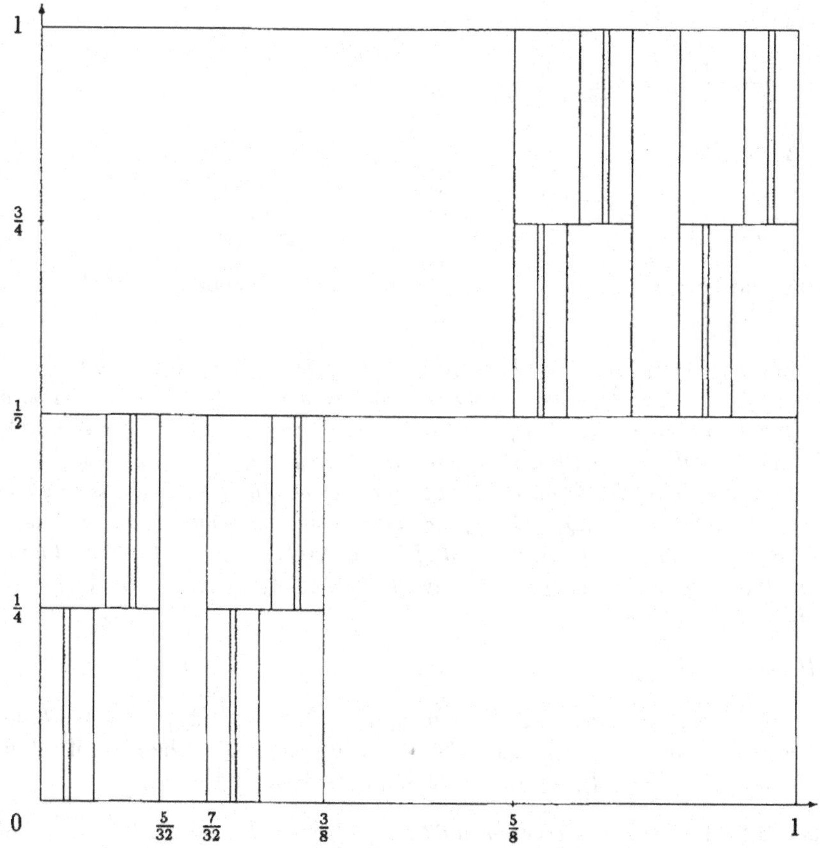

Figure 6.33: The function F_2

(iv) It is known that if a set $E \subset R \times R$ is of positive measure then by Fubini's theorem there exist uncountably many lines, parallel to a fixed line d, such that the intersection of E with each of these parallel lines is a set of positive measure. Let M be a measurable set of positive measure. By (iii), $|F(M)| > 0$. But $|A_z| = 0$ for all irrational numbers z. Let $x \in (0,1)$ be an irrational number such that $|F_2(E_z)| > 0$, where $E_z = \{x \in M : F_1(x) = z\}$. Since $E_z \subset A_z$ and $|E_z| = 0$, by (ii), F_2 is increasing on A_z and on E_z. For F_1 the proof is similar.

(v) By (i) it follows that 0 is a derived number for $(F_1)_{/P}$ at each $x \in P$. Let $x_o \in P$, $x_o = \sum_{i=1}^{\infty} c_i p_i$, $c_i \in \{0,2\}$. Let $x_n = (2 - c_{2n}) + \sum_{i=1, i\neq 2n} c_i p_i$. Then

$$\lim_{n\to\infty} \frac{|F(x_n) - F(x_o)|}{x_n - x_o} = \lim_{n\to\infty} \frac{(2 - c_{2n} - c_{2n}) \cdot \frac{1}{2^n}}{(2 - c_{2n} - c_{2n}) \cdot p_{2n}} = +\infty,$$

hence $(F_1)_{/P}$ has finite or infinite derivative at no point $x \in P$. For $(F_2)_{/P}$ the proof is similar.

(vi) Let $x_o \in P$, $x_o = \sum_{i=1}^{\infty} c_i p_i$, $c_i \in \{0,2\}$. Let $P_{2n}(x_o) = \{x \in P : x = (2 - c_{2n}) + \sum_{i=1, i\neq 2n}^{2n+2} c_i p_i + \sum_{i=2n+3}^{\infty} c_i(x) p_i\}$. Let $A_{2n}(x_o) = [\sum_{i=1}^{2n-1} c_i p_i, \sum_{i=1}^{2n-1} c_i p_i + \sum_{i=2n}^{\infty} 2p_i]$.

It follows that

$$\lim_{n\to\infty}\frac{|P_{2n}(x_o)|}{|A_{2n}(x_o)|} = \lim_{n\to\infty}\frac{\frac{1}{2^{2n+4}}}{\sum_{i=1}^{2n} 2p_i} = \frac{1}{16}.$$

If $x \in P_{2n}(x_o)$ then

$$\frac{|F_1(x) - F_1(x_o)|}{x - x_o} > \frac{\frac{2-c_{2n}-c_{2n}}{2^n} - 2\sum_{i=n+2}^{\infty}\frac{1}{2^{n+2}}}{\frac{1}{2^{2n}} + \frac{2}{4^{2n}}} = \frac{\frac{1}{2^n}}{\frac{1}{2^{2n}} + \frac{2}{4^{2n}}} \to \infty.$$

It follows that $(F_1)_{/P}$ has finite approximate derivative at no point $x \in P$. For $(F_2)_{/P}$ the proof is similar.

Construction 6.63.2 *We can glean the following from Construction 6.63.1. For each interval $I \subset [0,1]$ there exist a perfect nowhere dense subset Q of I and a continuous function g on I such that $|Q| = |I|/2$, $g \in W$ on Q, $g(x) = 0$ if $x \in [0,1] \setminus I$. Let Q_1 and g_1 be the set and function associated with I_1 such that $\sup_{x\in I_1} |g_1 x)| = 1$. Let I_2 be an interval of $I_1 \setminus Q$ of maximal length. Associate Q_2 and g_2 this way to I_2. Let I_3 be an interval of $I_1 \setminus (Q_1 \cup Q_2)$ of maximal length. Associate P_3 and g_3 this way with I_3. In general let I_{k+1} be an interval of $I_1 \setminus (Q_1 \cup Q_2 \cup \ldots Q_k)$ of maximal length. Associate Q_{k+1} and g_{k+1} with I_{k+1} this way. Then $|\cup_{n=1}^{\infty} Q_n| = 1$. Let $H : [0,1] \mapsto \mathbb{R}$, $H(x) = \sum_{n=1}^{\infty} g_n(x)/2^n$. Then we have:*

(vii) $H \in W$ on $[0,1]$.

Proof. Let E be a measurable subset of $[0,1]$, $|E| > 0$. Since $|\cup_{n=1}^{\infty} Q_n| = 1$ it follows that there exists some Q_n such that $|E \cap Q_n| > 0$. Since H is the sum of a linear function and $g_n(x)/2^n$ on P_n, by Lemma 3.3.2, $H \in W$ on $[0,1]$.

Remark 6.63.1 *Using Construction 6.63.1 it follows that there exists a continuous function $F : [0,1] \mapsto [0,1]$ with the following properties:*

(i) $F^{-1}(y)$ is a perfect set whenever $y \in [0,1]$;

(ii) $F \in W$ on $[0,1]$;

(iii) F has a finite or infinite derivative at no point of $[0,1]$;

(iv) F is nowhere approximately derivable on $[0,1]$ (see [E12]).

Nina Bary constructed first (see [Ba]) an example with the properties (i) and (ii) (it probably has also the properties (iii) and (iv)). In [Ca] Cater also constructed a function with the properties (ii) and (iii), bases on the same Construction 2.5.1.

Remark 6.63.2 *The utility of this example follows by Remark 3.3.1, (ii); Remark 3.3.2, (ii).*

6.64 A function $G_1 \in DW_1 \cap \mathcal{C}$, G_1 is not approximately derivable *a.e.* on a set of positive measure

Let F_1 and P as in Section 6.63. Then $F_1 \in W \cap \mathcal{C}$ on P and P is a nowhere dense subset of $[0,1]$, $|P| > 0$. By Remark 3.3.2, there exist $G_1, G_2 \in DW_1 \cap \mathcal{C}$ on P such that $F_1 = G_1 + G_2$ on P. By Theorem 3.3.1, G_1 and G_2 are not approximately derivable *a.e.* on P.

Remark 6.64.1 *The utility of this example follows by Remark 3.3.2, (iii).*

6.65 A function $F \in \mathcal{C}$, F is quasi-derivable, $F \notin AC \circ AC + AC$

Let $F : [a,b] \mapsto \mathbb{R}$, $F \in \mathcal{C}$, $F \in VB \setminus AC$ (for example we can take the Cantor ternary function, see Section 6.2). Then we have:

(i) F is quasi-derivable on $[a,b]$;

(ii) $F \notin AC \circ AC + AC$ on $[a,b]$.

Proof. (i) See Remark 3.1.1,(iv).

(ii) Suppose on the contrary that $F \in AC \circ AC + AC = (S) + AC \subset (N) + AC \subset (M) + AC = (M)$. It follows that $F \in AC$ (because $F \in VB$), a contradiction.

Remark 6.65.1 *The utility of this example follows by Remark 3.1.3,(i).*

6.66 Examples concerning the chain rule for the approximate derivative of a composite function

Example 6.66.1 (Foran) *([F9]). There exist F and g continuous, F approximately derivable a.e. so that $F \in (N)$ on $g(B)$ but not on $g(A)$, and the formula (1) of Theorem 5.19.1 does not hold a.e. (A and B are defined in Theorem 5.19.1).*

Proof. Let $F = F_{CP}$ and $g = F_{PC}$ (these functions are defined in Section 6.29). By (9) of Section 6.29, it follows that $F = g^{-1}$. Then $|A| = |P| = 1/2$ and $B \subset P \setminus A$. Hence $|B| = 0$. But $F(g(A)) = A$ and $|g(A)| = 0$. Therefore $F \notin (N)$ on A. Since $F(g(B)) = B$, $F \in (N)$ on $g(B)$. Because $(F \circ g)(x) = x$ it follows that $(F \circ g)'(x) = 1$ on A. Thus formula (1) of Theorem 5.19.1 does not hold on A.

Example 6.66.2 (Foran) *([F9]). There are $F, g : [0,1] \mapsto [0,1]$, $F, g \in \mathcal{C}$ so that $F \in (N)$ on $[0,1]$, F is not approximately derivable a.e. on $g(B)$, and the formula (1) of Theorem 5.19.1 does not hold a.e..*

Proof. Let $F = g = G_{PP}$ (see Section 6.29). By (11) of Section 6.29, it follows that $|B| = |P| = 1/2$, $g(B) \subset P$, hence F is not approximately derivable *a.e.* on $g(B)$. Since $(F \circ g)'(x) = 1$ on $[0,1]$, and $g'_{ap}(x) = 0$ *a.e.* on P, it follows that the formula (1) of Theorem 5.19.1 does not hold *a.e.* on P.

Example 6.66.3 (Foran) *([F9]). There exist F, g such that $F \notin (N)$ on $g(B)$, $F \in (N)$ on $g(A)$, F is approximately derivable on $g(B)$, and the formula (1) of Theorem 5.19.1 does not hold a.e. on B.*

Proof. Let P be a closed set of positive measure. It is then easy to define g on P so that g is one-to-one on P, $|g(P)| = 0$, and the set B (see Theorem 5.19.1) has positive measure. Let $F(y) = g^{-1}(y)$ if $y \in g(B)$ and $F(y) = 0$ otherwise. Then $(F \circ g)(x) = x$ on B. Therefore formula (1) of Theorem 5.19.1 does not hold *a.e.* on B.

Remark 6.66.1 *(i) For other proofs of Example 6.66.1 and Example 6.66.2 see also [F9].*

(ii) The utility of this example follows by Remark 5.19.1.

Bibliography

[A] Agronski,S.J.: *Associated sets and continuity roads.* Real Analysis Exchange, (1983 - 1984), 195-205.

[Ba] Bary,N.: *Mémoire sur la représentation finie des fonctions continues.* Math. Ann., 103 (1930), 185-248 and 598-653.

[BoPr] Bongiorno,B. and Preiss,D.: *An unusual descriptive definition of integral.* Contemporary Mathematics. Am. Math. Soc., 42 (1985), 13-23.

[BoDi] Bongiorno,B. and Di Piazza,L.: *Convergence theorems for generalized Riemann Stieltjes integrals.* Real Analysis Exchange, 17 (1991-1992), 339-362.

[Br1] Bruckner,A.M.: *An affirmative answer to a problem of Zahorski and some consequences.* Mich. Math. J., 13 (1966), 15-26.

[Br2] Bruckner,A.M.: *Differentiation of real functions.* Lect.Notes in Math., 659, Springer-Verlag (1978).

[Br3] Bruckner,A.M.: *Approximate differentiation.* Real Analysis Exchange, 6 (1980-1981), 9-65.

[Br4] Bruckner,A.M.: *Some new simple proofs of old difficult theorems.* Real Analysis Exchange,9 (1983-1984), 67-78.

[BrOT] Bruckner,A.M., O'Malley,R.J. and Thomson,B.S.: *Path derivatives: a unified view of certain generalized derivatives.* Trans. Amer. Math. Soc., 283 (1984), 97-125.

[BrLPT] Bruckner,A.M., Laczkovitch,M., Petruska,G. and Thomson,B.S.: *Porosity and approximate derivatives.* Can. J. Math., vol. XXXVIII, No.5 (1986), 1149-1180.

[B1] Bullen,P.S.: *Non Absolute Integrals:A Survey.* Real Analysis Exchange. 5 (1979-1980), 195-259.

[B2] Bullen,P.S.: *The Burkill approximately continuous integral.* J. Austral. Math. Soc. (Ser.A), 35 (1983), 236-253.

[B3] Bullen,P.S.: *The Burkill approximately continuous integral II.* Math. Chron., 12 (1983), 93-98.

[B4] Bullen,P.S.: *A survey of integration by parts for Perron integrals.* J.Austral. Math. Soc. (Ser.A), 40 (1986), 343-63.

[B5] Bullen,P.S.: *Queries 178.* Real Analysis Exchange, 12 (1986-1987), 393.

[B6] Bullen,P.S.: *Some applications of a theorem of Marcinkiewicz.* Real Analysis Exchange, 14 (1988-1989), 12-13.

[BLMMP] Bullen,P.S., Lee,P.E., Mawhin,J.L., Muldowney,P. and Pfeffer,W.F.: *New integrals.* Lect. Notes in Math., Springer-Verlag LN 1419 (1990).

[BV] Bullen,P.S. and Vyborny,R.: *Some applications of a theorem of Marcinkiewicz.* Canad. Math. Bull., vol. 34 (2), (1991), 165-174.

[Bu] Burkill,J.C.: *The approximately continues Perron integral.* Math. Zeit. 34 (1931), 270-278.

[Ca] Cater,F.S.: *On nondifferntiable wrinkled functions.* Real Analysis Exchange, 14 (1988-1989), 175-188.

[CeDz] Čelidze,V.G. and Džvaršeišvili,A.G.: *The theory of the Denjoy integral and some applications.* World Scientific. Singapore, New Jersey, London, Hongkong (1978) (in russian - translated by Bullen, P.S. in 1989).

[Cr1] Cross,G.E.: *Generalized integrals as limits of Riemann like sums.* Real Analysis Exchange, 13 (1987-1988), 390-403.

[Cr2] Cross,G.E.: *Integration by parts for the Perron integral.* Real Analysis Exchange, 16 (1990-1991), 546-549.

[CL] Császár,A. and Laczkovich,M.: *Discrete and Equal Convergence.* Studia Sci. Math. Hungar., 10 (1975), 463-472.

[D] Denjoy,A. : *Sur la définition Riemannienne de l'intégrale de Lebesgue.* C.R. Acad. Sci Paris, 193 (1931), 695-698.

[Da] Davies,R.O.: *Subsets of finite measure in analytic sets.* Indag Math., 14 (1952), 488-489.

[DaS] Davies,R.O. and Schuss,Z.: *A proof that Henstock's integral includes Lebesgue's.* J.London Math. Soc., (2) 2 (1970), 561-562.

[DoS] Dongfu,X. and Shipan,L.: *Henstock integrals and Lusin's condition (N).* Real Analysis Exchange,13 (1987-1988), 451-454.

[El] Ellis,H.W.: *Darboux Properties and Applications to non-absolute convergent integrals.* Can. J. Math., 3 (1951), 471-485.

[En1] Ene,G.: *An extension of ordinary variation.* Real Analysis Exchange, 10 (1984-1985), 149-154.

[En2] Ene,G.: *Monotonicity and Darboux properties.* Rev. Roumaine Math. Pures Appl. Tome XXX,9 (1985), 737-743.

[EnE] Ene,G. and Ene,V.: *Nonabsolute Convergent Integrals.* Real Analysis Exchange, 11 (1986), 121-134.

[E1] Ene,V.: *On Lusin's Condition (N)*. Studii si Cercetari Matematice, Tom.36 (1984) 22, 83-85.

[E2] Ene,V.: *On Foran's Conditions A(N),B(N) and (M)*. Real Analysis Exchange, 9 (1984), 495-502.

[E3] Ene,V.: *On Foran's property (M) and its relations to Lusin's property (N)*. Real Analysis Exchange, 9 (1984), 558-563.

[E4] Ene,V.: *A study of Foran's conditions A(N) and B(N) and his class $\mathcal{F}$*. Real Analysis Exchange, 10 (1985), 194-212.

[E5] Ene,V.: *On some classes of continuous functions*. Real Analysis Exchange, 11 (1986), 452-463.

[E6] Ene,V.: *Monotonicity theorems*. Real Analysis Exchange, 12(1987),420-454.

[E7] Ene,V.: *A theorem which implies both Tolstoff's theorem and Zahorski's theorem on monotonicity*. Rev. Roumaine Math. Pures Appl. Tome xxx, 10 (1987), 121-129.

[E8] Ene,V.: *Construction of a wrinkled function*. Real Analysis Exchange, 14 (1989), 224-228.

[E9] Ene,V.: *Answers to three questions of Foran*. Real Analysis Exchange, 14 (1989), 243-248.

[E10] Ene,V.: *Integrals of Lusin and Perron type*. Real Analysis Exchange, 14 (1989), 115-140.

[E11] Ene,V.: *Semicontinuity conditions and monotonicity*. Real Analysis Exchange,14 (1989), 393-413

[E12] Ene,V.: *Finite representation of continuous functions, Nina Bary's wrinkled functions and Foran's condition (M)*. Real Analysis Exchange, 15 (1990), 445-469.

[E13] Ene,V.: *On some questions raised by J.Foran*. Real Analysis Exchange, 15 (1990), 559-581.

[E14] Ene,V.: *Monotonicity and local systems*. Real Analysis Exchange, 17(1991-1992), 291-313.

[E15] Ene,V.: *Integration by parts for the Foran integral*. Real Analysis Exchange, 17 (1991-1992), 402-404.

[E16] Ene,V.: *A fundamental lemma for monotonicity*. Real Analysis Exchange, 19, 2 (1993-94), 579-589.

[E17] Ene,V.: *A generalization of the Banach-Zarecki theorem*. Real Analysis Exchange 20, (1994) (to appear).

[E18] Ene,V.: *Characterization of $AC^*G \cap C$, underline$AC^* \cap C_i$, AC and $\underline{AC}$ functions*. Real Analysis Exchange, 19, 2 (1993- 94), 491-510.

[Fi] Filipczak,F.: *Sur les fonctions continues relativement monotones*. Fund. Math., LVIII (1966), 75-87.

[FlF1] Fleissner,R. and Foran,J.: *Transformations of differentiable functions*. Coll. Math., 39 (1978), 277-284.

[FlF2] Fleissner,R.: and Foran,J.: *A note on differentiable functions*. Proc. Amer. Math. Soc., 69 (1978), 56.

[FlO] Fleissner,R. and O'Malley,R.J.: *Conditions implying summability of approximate derivatives*. Coll. Math., 41 (1979), 257-263.

[F1] Foran,J.: *On continuous functions with graphs of σ–finite linear measure*. Proc.Cambridge Philos. Soc., 76 (1974), 33-43.

[F2] Foran,J.: *An extension of the Denjoy integral*. Proc. Amer. Math. Soc., 49 (1975), 359-365.

[F3] Foran,J.: *A note on Lusin's condition (N)*. Fund. Math., 90 (1976), 181-186.

[F4] Foran,J.: *On functions whose graph is of linear measure* 0 *on sets of measure* 0. Fund Math. XCVI (1977), 31-36.

[F5] Foran,J.: *Continuous functions*. Real Analysis Exchange 2 (1977), 85-103.

[F6] Foran,J.: *Differentiation and Lusin's condition (N)*. Real Analysis Exchange, 3 (1977-1978), 34-37.

[F7] Foran,J.: *A generalization of absolute continuity*. Real Analysis Exchange, 5 (1979-1980), 82-91.

[F8] Foran,J.: *On extending the Lebesgue integral*. Amer. Math. Soc., 81, No.1, (1981), 85-88.

[F9] Foran,J.: *A chain rule for the approximate derivatives and change of variables for the (D)-integral*. Real Analysis Exchange, 8 (1982-1983), 443-454.

[F10] Foran,J.: *The structure of continuous functions which satisfy Lusin's condition (N)*. Contemporary Mathematics. Amer. Math. Soc., 42 (1985), 55-61.

[F11] Foran,J.: *Fundamentals of Real Analysis*. Marcel Dekker Inc, New York, Basel, Hong Kong (1991).

[FO] Foran,J. and O'Malley,R.J.: *Integrability conditions for approximate derivatives*. Real Analysis Exchange, 10 (1984- 1985), 294-306.

[FM] Foran,J. and Meinershagen,S.: *Some answers to a question of P.Bullen*. Real Analysis Exchange, 13 (1987-1988), 265-278.

[FrH] Freilig,C. and Humke,P.D.: *The exact Borel class where a density completeness axiom holds*. Real Analysis Exchange, 17 (1991-1992), 272-281.

[Fu] Fu,S.: *S-Henstock integration and the approximately strong Lusin condition.* Real Analysis Exchange, 19 (1993-1994), 312-316.

[Ga] Garg,K.M.: *A new notion of derivative.* Real Analysis Exchange, 7 (1981-1982), 65-84.

[Ge] Genquian,L.: *The measurability of δ in Henstock integration.* Real Analysis Exchange, 13 (1987-1988), 446-451.

[Go1] Goodman,G.S.: *N-functions and integration by substitution.* Rend.Sem. Mat. Fis. Milano, 47 (1978), 123-134.

[Go2] Goodman,G.S.: *Integration by substitution.* Proc. Amer. Math. Soc. 70 (1978), 89-91.

[G1] Gordon,R.: *Equivalence of the generalized Riemann and restricted Denjoy integral.* Real Analysis Exchange, 12 (1986-1987), 551-574.

[G2] Gordon,R.: *A descriptive characterization of the generalization of the generalized Riemann integral.* Real Analysis Exchange, 15 (1989-1990), 397-400.

[G3] Gordon,R.: *Another proof of the measurability of δ for the generalized Riemann integral.* Real Analysis Exchange, 15 (1989-1990), 389-390.

[G4] Gordon,R.: *Another approach to the controlled convergence theorem.* Real Analysis Exchange, 16 (1990-1991), 306-311.

[G5] Gordon,R.: *The inversion of approximate and dyadic derivatives using an extension of the Henstock integral.* Real Analysis Exchange, 16 (1990-1991), 154-168.

[G6] Gordon,R.: *Riemann tails and the Lebesgue and Henstock integrals.* Real Analysis Exchange, 17(1991-1992), 789-796.

[G7] Gordon,R.: *On the equivalence of two convergence theorems for the Henstock integral.* Real Analysis Exchange, 18 (1992-1993), 261-267.

[G8] Gordon,R.: *Baire one functions and perfect sets.* Real Analysis Exchange, 18 (1993), 612-614.

[Gr] Gross,W.: *Über das Flächenmass von Punctmengen.* Monatsh. Math. Phys., 29 (1918), 145-176.

[H1] Henstock,R.: *On Ward's Perron Stieltjes integral.* Canad. J. Math. 9 (1957), 96-109.

[H2] Henstock,R.: *A new descriptive definition of the Ward integral.* J.London Math. Soc., 35 (1960), 43-48.

[H3] Henstock,R.: *Definitions of Riemann type of the variational integrals.* Proc. London Math. Soc., (3) 11 (1961), 402-418.

[H4] Henstock,R.: *A Riemann type integral of Lebesgue power.* Can. J. Math., 20 (1968), 79-87.

[H5] Henstock,R.: *Integration by parts.* Aequationes Math.,9 (1973), 1-18.

[H6] Henstock,R.: *Theory of integration.* Butterworth London (1963).

[H7] Henstock,R.: *Linear Analysis.* Butterworth London (1968).

[H8] Henstock,R.: *Lectures on the theory of integration.* World Scientific Singapore (1988).

[H9] Henstock,R.: *The general theory of integration.* Clarendon-Press-Oxford (1991)

[I1] Iseki,K.: *On quasi-Denjoy integration.* Proc. Japan. Acad., 38 (1962), 252-257.

[I2] Iseki,K.: *An attempt to generalize the Denjoy integration.* Nat. Sci. Rep. Ochanomizu Univ., 34 (1983), 19-33.

[I3] Iseki,K.: *On the normal integration.* Nat. Sci. Rep. Ochanomizu Univ. 37 (1986) 1-34

[I4] Iseki,K.: *On the sparse integration.* Nat. Sci. Rep. Ochanomizu Univ., 37 (1986), 91-99.

[I5] Iseki,K.: *On two theorems of Nina Bary type.* Nat. Sci. Rep. Ochanomizu Univ., 38 (1987), 33-98.

[IM] Iseki,K. and Maeda,M.: *On a generalization of the Denjoy integration.* Nat. Sci. Rep. Ochanomizu Univ., 22 (1971), 101-110.

[JR] Jayne,J.E. and Rogers,C.A.: *First level Borel functions and isomorphisms.* Jour. Math. Pures et.Appl., 61 (1982), 177-205.

[Ki] Kirchheim,B.: *Baire one star functions.* Real Analysis Exchange, 18 (1993), 385-400.

[Kr1] Krzyzewski,K.: *On change of variable in the Denjoy-Perron integral (I),(II).* Coll. Math., 9 (1962), 99-104; 317-323.

[Kr2] Krzyzewski,K.: *A note on the Denjoy integral.* Colloq. Math., 19 (1968), 121-130.

[Kub1] Kubota,Y.: *A characterization of the approximately continuous Denjoy integral.* Can. J. Math., 22 (1970), 219-226.

[Kub2] Kubota,Y.: *An approximately continuous Perron integral.* Can. Math. Bull., 14 (1971), 261-263.

[Kub3] Kubota,Y.: *An elementary theory of the special Denjoy integral.* Math. Japan., 24 (1979-1980), 507-520.

[Kub4] Kubota,Y.: *A direct proof that the RC-integral is equivalent to the D^*-integral.* Proc. Amer. Math. Soc., 80 (1980), 293-296.

[Kub5] Kubota,Y.: *A characterization of the Denjoy integral.* Math. Japan., 26 (1981), 389-392.

[Kub6] Kubota,Y.: *Extensions of the Denjoy integral.* Real Analysis Exchange, 14 (1988-1989), 72-73.

[Ku] Kuratowski,K.: *Topology.* New York - London - Warszawa (1966).

[K] Kurzweil,J.: *Generalized ordinary differential equations and continuous dependence on a parameter.* Czech. Math.J., 7 (1957), 418-449.

[KJ1] Kurzweil,J. and Jarnik,J.: *The PU-integral and its properties.* Real Analysis Exchange, 14 (1988-1989), 34-43.

[KJ2] Kurzweil,J. and Jarnik,J.: *Equiintegrability and controlled convergence of Perron-type integrable functions.* Real Analysis Exchange, 17 (1991-1992), 110-140.

[Le1] Lee,C.M.: *An analogue of the theorem Hake-Alexandroff-Looman.* Fund. Math. C (1978), 69-74.

[Le2] Lee,C.M.: *On Baire one Darboux functions with Lusin's condition (N).* Real Analysis Exchange, 7 (1981), 61-64.

[Le3] Lee,C.M.: *Some Hausdorff variants of absolute continuity, Banach's condition (S) and Lusin's condition (N).* Real Analysis Exchange, 13 (1987-1988), 404-420.

[L1] Lee,P.Y.: *Generalized convergence theorems for Denjoy-Perron integrals.* Real Analysis Exchange, 14 (1988-1989), 48-49.

[L2] Lee,P.Y.: *On ACG^* functions.* Real Analysis Exchange, 15 (1989-1990), 754-760.

[L3] Lee,P.Y.: *Kurzweil-Henstock integration and the strong Lusin condition.* Real Analysis Exchange, 17 (1991-1992), 25-26.

[LC] Lee,P.Y. and Chew,T.S.: *A Riesz-type definition of the Denjoy integral.* Real Analysis Exchange, 11 (1985-1986), 221-227.

[LW] Lee,P.Y. and Wittoya,N.I.: *A direct proof that the Henstock and Denjoy integrals are equivalent.* Bull. Malaysian Math. Soc., 5 (2) (1988), 43-47.

[LSM] Lee,P.Y., Seng,C. and Ma Zheng-Min: *Absolute integration using Vitali covers.* Real Analysis Exchange, 18 (1993), 409-419.

[Lco] Leonard,J.: *Some conditions implying the monotonicity of a real function.* Rev.Roumaine Pures Appl., 12 (1972), 757-780.

[Li] Liao,K.: *On the descriptive definition of the Burkill approximately continuous integral.* Real Analysis Exchange, 18 (1992-1993), 253-261.

[LuMZ] Lukes,J., Maly,J. and Zajiček,L.: *Fine Topology Methods in Real Analysis and Potential Theory.* Lect. Notes in Math., 1189, Springer (1986).

[Mal] Maliszewski,A.: *Algebra generated by nondegenerate derivatives.* Real Analysis Exchange, 18 (1993), 599-611.

[MPZ] Maly,J., Preiss,D. and Zajiček,L.: *An unusual monotonicity theorem with applications.* Proc. Amer. Math. Soc., 102 (4), (1988), 925-932.

[Ma1] Marcus,S.: *Sur la limite approximative qualitative. Sur la continuité approximative qualitative. Sur la dérivée approximative qualitative.* Com. Acad. R.P. Române, 3 (1953), 9-12, 117-120 and 361-364.

[Ma2] Marcus,S: *Sur une theorie du type Lebesgue pour l'intégrale de Riemann.* Bull. Math. de la Société des Sciences Mathématiques et Physiques de Roumanie, 2 (50) no. 2 (1958), 187-197.

[Ma3] Marcus,S.: *La mesure de Jordan at l'intégrale de Riemann dans un espace mesuré topologique.* Acta Scientiarum Mathematicarum Szeged, 20, no.2-3, (1959), 156-163.

[Ma4] Marcus,S.: *On a theorem of Denjoy and on approximate derivatives.* Monatsh. Mat., 66 (1962), 435-440.

[Maz] Mazurkiewicz,S: *Sur les fonctions qui satisfont a la condition (N).* Fund. Math., 16 (1930), 348-352.

[McL] Mc Leod,R.M.: *The generalized Riemann integral.* Carus Math. Monographs, 20 (1980), Association of America.

[McS1] Mc Shane,E.J.: *On Perron integration.* Bull. Am. Math. Soc., 481 (1942), 718-726.

[McS2] Mc Shane,E.J.: *Integration.* Princeton University Press, Princeton (1944), 6-43.

[McS3] Mc Shane,E.J.: *A unified theory of integration.* Am. Math. Monthly, 80 (1973), 349-359.

[Mu] Mukhopadhyay,S.N.: *On a certain property of the derivative.* Fund.Math., 67 (1970), 279-284.

[N] Natanson,I.P.: *Theory of functions of a real variable.* 2nd. rev. ed. Ungar, New York (1961).

[O1] O'Malley,R.J.: *A density property and applications.* Trans. Amer. Math. Soc., 199 (1974), 75-87.

[O2] O'Malley,R.J.: *Baire* 1 functions.* Proc. Amer. Math. Soc., 60 (1976), 187-192.

[O3] O'Malley,R.J.: *Selective derivatives.* Acta Math. Acad. Sci. Hungar. 29 (1977), 77-97.

[O4] O'Malley,R.J.: *Selective derivatives and the M_2 or Denjoy-Clarkson properties.* Acta Math. Soc. Sci. Hungar., 36 (1980), 195-199.

[O5] O'Malley,R.J.: *Some consequences of the Freiling-Humke result on the density property.* Real Analysis Exchange, 19 (1993-1994), 242-247.

[Oxt] Oxtoby,J.C.: *Measure and category.* Springer-Verlag, New York Heidelberg Berlin (1971).

[P1] Pfeffer,W.F.: *A Riemann-type integration and the fundamental theorem of calculus.* Rend. Circ. Math. Palermo, 36 (1987), 482-506.

[P2] Pfeffer,W.F.: *Upper semicontinuous gages and the Riemann complete integral.* Real Analysis Exchange, 13 (1987-1988), 70.

[P3] Pfeffer,W.F.: *A note on the generalized Riemann integral.* Proc. Amer. Math. Soc., 103 (1988), 1161-1166.

[P4] Pfeffer,W.F.: *On the generalized Riemann integral defined by means of special partitions.* Real Analysis Exchange, 14 (1988-1989), 506-511.

[P5] Pfeffer,W.F.: *The Riemann Approach to Integration.* Cambrige Univ. Press, New York, 1993.

[Pr1] Preiss,D.: *Approximate derivatives and Baire classes.* Czech. Math. J., 21 (96), (1971), 373-382.

[Pr2] Preiss,D.: *Algebra generated by derivatives.* Real Analysis Exchange, 8 (1982- 1983), 208-216.

[Pu] Pu,H.W.: *On the derivative of indefinite RC-integral.* Colloq. Math., 28 (1973), 105-110.

[R1] Ridder,J.: *Über den Perronschen Integralbegriff und seine Beziehung zu den R-, L-, und D- Integralen.* Math. Zeit., 34 (1931), 234-269.

[R2] Ridder,J.: *Über approximativ stetige Denjoy-Integrale.* Fund. Math., 21 (1933), 1-10.

[R3] Ridder,J.: *Über die gegenseitigen Beziehungen verschiedener stetigen Denjoy-Perron Integrale.* Fund. Math., 22 (1934), 136-162.

[R4] Ridder,J.: *Über die T- und N- Bedingungen und die approximativ stetigen Denjoy-Perron Integrale.* Fund. Math., 22 (1934), 163-179.

[Rog] Rogers,C.A.: *Hausdorff measures.* Camb. Univ. Press, Cambridge (1970).

[Rom] Romanovski,P.: *Essai d'une exposition de l'integrale de Denjoy sans nombres transfini.* Fund. Math., 19 (1932), 38-44.

[Ros] Rosen,H.: *Darboux Baire 5 functions.* Proc. Amer. Math. Soc., 110 (1990), 285-286.

[S1] Saks,S.: *Sur l'integrale de M.Denjoy.* Fund. Math., 15 (1930), 242-262.

[S2] Saks,S.: *Sur certaines classes de fonctions continues.* Fund. Math., 17 (1931), 124-151.

[S3] Saks,S.: *Theory of the integral.* 2nd. rev. ed. Monografie Matematyczne, PWN Warsaw (1937).

[Sa] Sarkhel,D.N.: *A criterion for Perron integrability.* Proc. Amer. Math. Soc., 71 (1978), 109-112.

[Se] Seng,C.: *On the equivalence of Henstock-Kurzweil and restricted Denjoy integrals in* $\Re^n$. Real Analysis Exchange, 15 (1989-1990), 259-269.

[SerV] Serrin,J. and Varberg,D.E.: *A general chain rule for derivatives and the change of variable formula for the Lebesgue integral.* Amer. Math. Monthly, 76(1969),514-520.

[Sc1] Schurle,A.W.: *Perron integrability versus Lebesgue integrability.* Can. Bull., 28 (1985), 463-468.

[Sc2] Schurle,A.W.: *A function is Perron integrable if it has locally small Riemann sums.* J. Australian Math. Soc., Ser.A, 41 (1986), 224-232.

[Sc3] Schurle,A.W.: *A new property equivalent to the Lebesgue integrability.* Proc. Am. Math. Soc., 96 (1986), 103-106.

[Sch] Schuss,Z.: *A new proof of a theorem concerning the relationship between the Lebesgue and RC-integral.* J. London Math. Soc., 44 (1969), 365-368.

[Skl] Skljarenko,V.A.: *Integration by parts in the SCP Burkill integral.* Math. USSR Sbornik, 40 (1981), 567-583.

[Skv] Skvortsov,V.: *On some questions of R.Gordon related to approximate and dyadic Henstock integrals.* Real Analysis Exchange, 18 (1992-1993), 267-270.

[SU] Sunouchi,G. and Utagawa,M.: *The generalized Perron integrals.* Tôhoku Math. J., 1 (2) (1949), 95-99.

[Sw] Swiatkowski,T.: *On the conditions of monotonicity of functions.* Fund. Math., 59 (1966), 189-201.

[T1] Thomson,B.S.: *Monotonicity theorems.* Proc. Amer. Math. Soc., 83 (1981), 547-552.

[T2] Thomson,B.S.: *Monotonicity theorems.* Real Analysis Exchange, 6 (1981), 209-234.

[T3] Thomson,B.S.: *Some properties of generalized derivatives.* Real Analysis Exchange, 8 (1982-1983), 58-59.

[T4] Thomson,B.S.: *Derivation bases on the real line, I and II.* Real Analysis Exchange, 8 (1982-1983), 67-208, 280-442

[T5] Thomson,B.S.: *Real functions.* Lecture Notes in Mathematics, 1170, Springer- Verlag (1985).

[To1] Tolstov,G.P.(Tolstoff): *Sur quelques propriétés des fonctions approximativement continues.* Mat. Sb. 5 (47), (1939), 637-645.

[To2] Tolstov,G.P.(Tolstoff): *Sur l'integrale de Perron.* Mat. Sb. 5 (47), (1939), 647-660.

[V] Varberg,D.E.: *On absolutely continues functions.* Amer. Math. Monthly, 72 (1965), 831-841.

[WD] Wang,C-S. and Ding,C-S.: *An integral involving Thomson's local systems.* Real Analysis Exchange, 19 (1993-1994), 248-253.

[Wa] Ward,A.J.: *The Perron Stieltjes integral.* Math. Zeit., 41 (1936), 578-604.

[We1] Weil,C.E.: *On properties of derivatives.* Trans. Amer. Math. Soc., 114 (1965),363-376.

[We2] Weil,C.E.: *Monotonicity, convexity and symmetric derivatives.* Trans. Amer. Math. Soc., 221 (1976), 225-237.

[Z1] Zahorski,Z.: *Sur la premiere derivée.* Trans. Amer. Math. Soc., 69 (1950), 1-54.

[Z2] Zahorski,Z.: *Über die Menge der Puncte in welchen die Ableitung unendlich ist.* Tôhoku Math. J., 48 (1941), 321-330.

Index

$\overline{A}$ 1
AC 51
$\underline{AC}$ 51
AC_D 88
$\overline{AC}_D$ 88
$\underline{AC}_D$ 88
AC_{D_o} 88
$\overline{AC}_{D^o}$ 88
$\underline{AC}_{D^o}$ 88
$AC_{D^\#}$ 88
$\overline{AC}_{D^\#}$ 88
$\underline{AC}_{D^\#}$ 88
$AC(P;Q)$ 51
$\underline{AC}(P;Q)$ 51
$AC(P \wedge Q)$ 51
$\overline{AC}(P \wedge Q)$ 51
$\underline{AC}(P \wedge Q)$ 51
AC^* 54
$\overline{AC}^*$ 54
$\underline{AC}^*$ 54
$(AC^* \cup DW_1)G$ 116
AC_n 93
$\overline{AC}_n$ 93
$\underline{AC}_n$ 93
AC_ω 93
$\overline{AC}_\omega$ 93
$\underline{AC}_\omega$ 93
AC_∞ 93
$\overline{AC}_\infty$ 93
$\underline{AC}_\infty$ 93
ACG 51
$[ACG]$ 51
$\underline{AC}G$ 51
AC^*G 54
$\underline{AC}^*G$ 54
$[AC^*G]$ 54
AC_nG 93
$[AC_nG]$ 93
$AC_\omega G$ 93
$AC_\infty G$ 93

$a.e.$ 1
$A(n)$ 93
analytic set 9
(AP)-integral 200
approximately continuous 14
approximately quasi-derivable 127

bAC^* 54
$b\overline{AC}^*$ 54
$b\underline{AC}^*$ 54
B_d 141
$B(F;X)$ 1
B_i 141
bilaterally c-dense in themselves 38
bilaterally dense in themselves 38
Borel function 9
Borel set 8
$(BRD^\#S)$ (bounded $RD^\#$ sum) 173
BRS (bounded Riemann sum) 192
$\mathcal{B}$ 97
$\overline{\mathcal{B}}$ 97
$\underline{\mathcal{B}}$ 97
$[\overline{\mathcal{B}}]$ 97
$[\underline{\mathcal{B}}]$ 97
$[\mathcal{B}]$ 97
$B(n)$ 97
$\mathcal{B}_1$ 27
$\underline{\mathcal{B}}_1$ 27
$\overline{\mathcal{B}}_1$ 27
$\mathcal{B}_1^*$ 33
$\beta_\delta[P]$ 87
$\beta_\delta^o[P]$ 87, 188
$\beta_\delta^\#[P]$ 87, 169

CM 35
compact components 1, 38
components of P 1
components of a sparse figure 111
connected graph 38
contiguous interval 1
C 1

$\mathcal{C}_d$ 31
$\mathcal{C}_d^*$ 31
$\mathcal{C}_i$ 31
$\mathcal{C}_i^*$ 31
$\mathcal{C}_iG$ 31
$\mathcal{C}_i^*G$ 31
$[\mathcal{C}G]$ 31

*decreasing** 47
Denjoy Clarkson Property 40
density property of O'Malley 21
$D[P]$ 87
$D^o[P]$ 87
$D^\#[P]$ 87
$D(P;Q)$ 47
$D(P \wedge Q)$ 47
DW_n 116
$\overline{DW}_n$ 116
$\underline{DW}_n$ 116
DW_ω 116
$\overline{DW}_\omega$ 116
$\underline{DW}_\omega$ 116
DW_∞ 116
$\overline{DW}_\infty$ 116
$\underline{DW}_\infty$ 116
DW_nG 116
$[DW_nG]$ 116
DW^* 116
DW^*G 116
$\underline{d}^i_+(A;x_o)$ 9
$\overline{d}^i_+(A;x_o)$ 9
$\underline{D}F$ 13
$D^+F(x_o)$ 13
D_+F, D^-F, D_-F 13
$D^\#F$ 15
$\underline{D}^\#F$ 15
$\overline{D}^\#F$ 15
$\overline{D}F$ 13
$d(A;x_o)$ 9
$\mathcal{D}$ 25
$\mathcal{D}_-$ 25
$\mathcal{D}_+$ 25
$\mathcal{D}^*$-integral (Denjoy* - integral) 175
δ-decomposition 2
Δ-fine partition of $[a,b]$ 121
Δ-fine partial partition of X 121

E_n 118
$\overline{E}_n$ 118
$\underline{E}_n$ 118
E_ω 118
$\overline{E}_\omega$ 118
$\underline{E}_\omega$ 118
E_∞ 118
$\overline{E}_\infty$ 118
$\underline{E}_\infty$ 118
E_nG 118
$[E_nG]$ 118
everywhere dense 10
exact $\mathcal{S}$-derivative 16
$\mathcal{E}$ 118
$\overline{\mathcal{E}}$ 118
$\underline{\mathcal{E}}$ 118
$[\mathcal{E}]$ 118
$[\overline{\mathcal{E}}]$ 118
$[\underline{\mathcal{E}}]$ 118

$\underline{F}'$ 13
$\overline{F}'$ 13
$F^*(x)$ 175
F'_{ap} 15
$F^*_{ap}(x)$ 203
F'_+ 13
F'_- 13
filtering 17
first category 10
$f.l.$ 4
F_P 1
$F_{/P}$ 1
$\mathcal{F}$ 94
$\underline{\mathcal{F}}$ 94
$\overline{\mathcal{F}}$ 94
$[\mathcal{F}]$ 94
$[\underline{\mathcal{F}}]$ 94
$[\overline{\mathcal{F}}]$ 94
$L\mathcal{F}$- integrable 207
$\mathcal{F}$-integral 207
F_σ-type 8

G_δ-type 8

hereditary property 11

$\overline{I}(b)$ 198, 209
$\underline{I}(b)$ 198, 209
$\overline{I}_j(b)$ 175
$\underline{I}_j(b)$ 175

$\overline{I}_k(b)$ 162
$\underline{I}_k(b)$ 162
*increasing** 47
*increasing*G* 47
internal 33
internal* 35
intersection conditions
 (I.C.) 16
 (E.I.C.) 16
 (E.I.C.[m]) 17
$I(P;Q)$ 47
$I(P \wedge Q)$ 47
Iseki's sparse integral 210
$i = \overline{1,n}$ 1
$i = \overline{1, infty}$ 1

$\overline{J}_j(b)$ 166, 186
$\underline{J}_j(b)$ 166, 186

K_P the characteristic function 1

L 58
$\underline{L}$ 58
$< L >$ 197, 207, 210
L^* 58
L_n 104
$\overline{L}_n$ 104
$\underline{L}_n$ 104
L_ω 104
$\overline{L}_\omega$ 104
$\underline{L}\omega$ 104
L_∞ 104
$\overline{L}_\infty$ 104
$\underline{L}\infty$ 104
L_nG 104
$\overline{L}_nG$ 104
$[L_nG]$ 104
local system 16
lower internal 33
lower *internal** 35
LG 58
$\underline{L}G$ 58
$[LG]$ 58
$< LPG >$ integral 198
$< LPG >$ major function 197
$< LPG >$ minor function 197
$L(P;Q)$ 58
$\underline{L}(P;Q)$ 58 $L(P \wedge Q)$ 58
$\overline{L}(P \wedge Q)$ 58
$\underline{L}(P \wedge Q)$ 58
$LS\mathcal{F}$- integrable 210
$(LSRS)$ 192
$(LSRD^{\#}S)$ 173
ℓCM 35
$\mathcal{L}$ 105
$[\mathcal{L}]$ 105
$\underline{\mathcal{L}}$ 105
$\overline{\mathcal{L}}$ 105
$[\underline{\mathcal{L}}]$ 105
$[\overline{\mathcal{L}}]$ 105
λ_n^β 2
λ_∞^β 2
λ_ω^β 2
$\lambda_n(P)$ 2
$\lambda_\omega(P)$ 2
$\lambda_\infty(P)$ 2
ΛZ 108
$\Lambda \overline{Z}$ 108
$\Lambda \underline{Z}$ 108

(M) 84
$\overline{M}$ 84
$\underline{M}$ 84
M^* 82
$\overline{M}^*$ 82
$\underline{M}^*$ 82
$(-)$ 71
$(-)'$ 145
*monotone** 47
$M(P;Q)$ 47
$M(P \wedge Q)$ 47
$m(X)$ the Lebesgue measure 1
$\overline{\mathcal{M}}_1(f)$ 174
$\overline{\mathcal{M}}_2(f)$ 174
$\overline{\mathcal{M}}_3(f)$ 174
$\overline{\mathcal{M}}_4(f)$ 174
$\overline{\mathcal{M}}_5(f)$ 174
$\overline{\mathcal{M}}_6(f)$ 175
$\overline{\mathcal{M}}_7(f)$ 175
$\overline{\mathcal{M}}_8(f)$ 175
$\overline{\mathcal{M}}_9(f)$ 175
$\overline{\mathcal{M}}_{10}(f)$ 175
$\overline{\mathcal{M}}_{11}(f)$ 175
$\overline{\mathcal{M}}_{12}(f)$ 175
$\overline{\mathcal{M}}_{13}(f)$ 175
$\overline{\mathcal{M}}_{14}(f)$ 175

$\overline{\mathcal{M}}_{15}(f)$ 175
$\overline{\mathcal{M}}_{16}(f)$ 175
$\underline{\mathcal{M}}_j(f)$, $j=\overline{1,16}$ 175
$\overline{\mathcal{M}}_1^{\#}(f)$ 161
$\overline{\mathcal{M}}_2^{\#}(f)$ 161
$\overline{\mathcal{M}}_3^{\#}(f)$ 161
$\overline{\mathcal{M}}_4^{\#}(f)$ 161
$\overline{\mathcal{M}}_5^{\#}(f)$ 161
$\overline{\mathcal{M}}_6^{\#}(f)$ 161
$\overline{\mathcal{M}}_7^{\#}(f)$ 161
$\overline{\mathcal{M}}_8^{\#}(f)$ 161
$\overline{\mathcal{M}}_9^{\#}(f)$ 161
$\overline{\mathcal{M}}_{10}^{\#}(f)$ 161
$\overline{\mathcal{M}}_{11}^{\#}(f)$ 161
$\overline{\mathcal{M}}_{12}^{\#}(f)$ 161
$\overline{\mathcal{M}}_{13}^{\#}(f)$ 162
$\underline{\mathcal{M}}_k^{\#}(f)$, $k=\overline{1,13}$ 162

(N) 71
$(\overline{N})$ 78
$(\underline{N})$ 78
$N_g^{-\infty}$ 84
$N_g^{+\infty}$ 84
N_g^{∞} 84
N^{∞} 79
$N^{+\infty}$ 79
$N^{-\infty}$ 79
n.e. 1
nowhere dense 10

$\mathcal{O}(F;P)$ 4
$\mathcal{O}(F;x_o)$ 4
$\mathcal{O}_-(F;P)$ 4
$\mathcal{O}_+(F;P)$ 4
$\mathcal{O}^n(F;P)$ 6
$\mathcal{O}_+^n(F;P)$ 6
$\mathcal{O}_-^n(F;P)$ 6
$\mathcal{O}^{\infty}(F;P)$ 6
$\mathcal{O}_+^{\infty}(F;P)$ 6
$\mathcal{O}_-^{\infty}(F;P)$ 6
$\mathcal{O}^{\omega}(F;P)$ 6
$\mathcal{O}_+^{\omega}(F;P)$ 6
$\mathcal{O}_-^{\omega}(F;P)$ 6
$\Omega(F;Y \wedge X)$ 4
$\Omega_-(F;Y \wedge X)$ 4
$\Omega_+(F;Y \wedge X)$ 4
$\Omega(F;x_o)$ 4

path leading to x 124
$(+)$ 71
$(+)'$ 145
PG 11
$[PG]$ 11
$P_1;P_2G$ 11
$\mathcal{P}(E;\delta)$ 188
$\mathcal{P}_1([a,b];\delta)$ 188
$\mathcal{PF}$-integrable 209
$\mathcal{PF}$-major function 209
$\mathcal{PF}$-minor function 209
$(\mathcal{P}_{j,k})$-integral 175
$(\mathcal{P}_{j,k}^{\#})$-integral 162
$\mathcal{P}^{\#}(E;\delta)$ 169
$\mathcal{P}_1^{\#}([a,b];\delta)$ 169
portion 1
$\Pi([a,b];\Delta)$ 121
$\Pi_p(X;\Delta)$ 121

quasi-derivable 127

R_{ap}-integral 200
(RD^o)-integrable 188
$(RD^{\#})$-integrable 169
Ridder's β-integral 200
residual set 10

(S) 71
SAC 112
SAC_n 112
$S\underline{AC}_n$ 112
$S\overline{AC}_n$ 112
SAC_ω 112
SAC_∞ 112
SAC_nG 112
$[SAC_nG]$ 112
$S\underline{AC}_nG$ 112
$S\underline{AC}_\omega$ 112
sCM 35
$S(f;P)$ 169, 188
$sD(P;Q)$ 47
$sD(P \wedge Q)$ 47
second category 10
semicontinuity 21
$S\mathcal{F}$ 112, 210
$[S\mathcal{F}]$ 112
$S\underline{\mathcal{F}}$ 112

$[S\mathcal{E}]$ 112
singular 60
SG 71
$sI(P;Q)$ 47
$sI(P \wedge Q)$ 47
$sM(P;Q)$ 47
S_o 103
sparse figure 111
sparse figure for P 111
sparsely continuous 112
strictly increasing$^*G$ 47
strictly $\mathcal{S}$-increasing 16
strictly $\mathcal{S}$-decreasing 16
strong property 11
summable 60
SVB 114
SVB_n 114
$S\underline{VB}_n$ 114
$S\overline{VB}_n$ 114
SVB_ω 114
$S\underline{VB}_\omega$ 114
SVB_∞ 114
SVB_nG 114
$[SVB_nG]$ 114
$S\underline{VB}_nG$ 114
$s(y)$ - Banach indicatrix function 101
$\mathcal{S}AC$ 122
$\mathcal{S}\underline{AC}$ 122
$\mathcal{S}\overline{AC}$ 122
$\mathcal{S}ACG$ 122
$\mathcal{S}\underline{AC}G$ 122
$[\mathcal{S}ACG]$ 122
$\mathcal{S}$-closed 19
$\mathcal{S}$-decreasing 16
$\mathcal{S}_E(x)$ 124
$\mathcal{S}_E$ continuous (E-continuous) 124
$\mathcal{S}$-increasing 16
$\mathcal{S}$-integrable 193
$\mathcal{S} - \limsup$ 16
$\mathcal{S} - \liminf$ 16
$\mathcal{S}$-open 19
$\mathcal{S}$-semicontinuity 21
$\mathcal{S} - \underline{D}F$ 16
$\mathcal{S} - \overline{D}F$ 16
$\mathcal{S}$-DF 16
$\mathcal{S}^*$ 18
$\mathcal{S}_o$ 18
$\mathcal{S}_o^+$ 18
$\mathcal{S}_o^-$ 18
$\mathcal{S}_\infty$ 18
$\mathcal{S}_\infty^+$ 18
$\mathcal{S}_\infty^-$ 18
$\mathcal{S}_q$ 18
$\mathcal{S}_q^+$ 18
$\mathcal{S}_q^-$ 18
$\mathcal{S}_c$ 18
$\mathcal{S}_c^+$ 18
$\mathcal{S}_c^-$ 18
$\mathcal{S}_2$ 18
$\mathcal{S}_2^+$ 18
$\mathcal{S}_2^-$ 18
$\mathcal{S}_{ap}$ 18
$\mathcal{S}_{ap}^+$ 18
$\mathcal{S}_{ap}^-$ 18
$\mathcal{S}_3$ 18
$\mathcal{S}_3^+$ 18
$\mathcal{S}VB$ 124
$\mathcal{S}VBG$ 124
$\mathcal{S}Y$ 122
$\mathcal{S}\underline{Y}$ 122
$\mathcal{S}\overline{Y}$ 122
$\sigma f.l.$ 4
$\sigma(f;P)$ 169, 188
$\sigma(P)$ 169, 188

T_1 71
T_2 71
T_1G 71
typical continuous function 134

uCM 35
$< uL >$ 197
$U[P]$ 87
$U^o[P]$ 87
$U^\#[P]$ 87
upper internal 33
upper *internal** 35

$V^*(F;P)$ 44
VB 41
$\underline{VB}$ 41
VB^* 44
$\overline{VB}^*$ 44
$\underline{VB}^*$ 44
VBG 41

$[VBG]$ 41
VB^*G 44
$[VB^*G]$ 44
$VB(P \wedge Q)$ 41
$\overline{VB}(P \wedge Q)$ 41
$\underline{VB}(P \wedge Q)$ 41
VB_n 96
$\overline{VB}_n$ 96
$\underline{VB}_n$ 96
VB_ω 96
$\overline{VB}_\omega$ 96
$\underline{VB}_\omega$ 96
VB_∞ 96
$\overline{VB}_\infty$ 96
$\underline{VB}_\infty$ 96
VB_nG 96
$VB_\omega G$ 96
$VB_\infty G$ 96
$[VB_nG]$ 96
Vitali cover 11
$V_n(F;P)$ 100
$V_\omega(F;P)$ 100
$V_\infty(F;P)$ 100
$(\mathcal{V}_1)$- integrable 187
$(\mathcal{V}_2)$- integrable 187
$(\mathcal{V}_3)$-integrable 187
indefinite $(\mathcal{V}_i)$-integral 187
$(\mathcal{V}_1^\#)$-integrable 167
$(\mathcal{V}_2^\#)$-integrable 167
$(\mathcal{V}_3^\#)$-integrable 167
$(\mathcal{V}_i^\#)$-indefinite integral 167

(W) (wrinkled) 131
W^* 134
$w\mathcal{B}_1$ (wide $\mathcal{B}_1$) 39
wN 78
wS 78
wS_o 103
$\overline{\mathcal{W}}_1(f)$ 186
$\overline{\mathcal{W}}_2(f)$ 186
$\overline{\mathcal{W}}_3(f)$ 186
$\underline{\mathcal{W}}_j(f)$ 186
$(\mathcal{W}_{j,k})$-integral 186
$\overline{\mathcal{W}_1^\#}(f)$ 166
$\overline{\mathcal{W}_2^\#}(f)$ 166
$\overline{\mathcal{W}_3^\#}(f)$ 166
$\underline{\mathcal{W}_j^\#}(f)$ 166
$(\mathcal{W}_{j,k}^\#)$-integral 166

$|X|$ the outer Lebesgue measure 1
$|X|^i$ 9
$|X|^e$ 9

Y_D 89
$\overline{Y}_D$ 89
$\underline{Y}_D$ 89
Y_{D° 89
$\overline{Y}_{D^\circ}$ 89
$\underline{Y}_{D^\circ}$ 89
$Y_{D\#}$ 89
$\overline{Y}_{D\#}$ 89
$\underline{Y}_{D\#}$ 89

Zahorski's classes M_o, M_1, M_2, M_3, M_5 19
$\overline{\mathcal{Z}}_i$ 34
$\underline{\mathcal{Z}}_i$ 34
$\overline{\mathcal{Z}}_d$ 34
$\underline{\mathcal{Z}}_d$ 34

Lecture Notes in Mathematics

For information about Vols. 1–1425
please contact your bookseller or Springer-Verlag

Vol. 1426: J. Azéma, P.A. Meyer, M. Yor (Eds.), Séminaire de Probabilités XXIV, 1988/89. V, 490 pages. 1990.

Vol. 1427: A. Ancona, D. Geman, N. Ikeda, École d'Eté de Probabilités de Saint Flour XVIII, 1988. Ed.: P.L. Hennequin. VII, 330 pages. 1990.

Vol. 1428: K. Erdmann, Blocks of Tame Representation Type and Related Algebras. XV. 312 pages. 1990.

Vol. 1429: S. Homer, A. Nerode, R.A. Platek, G.E. Sacks, A. Scedrov, Logic and Computer Science. Seminar, 1988. Editor: P. Odifreddi. V, 162 pages. 1990.

Vol. 1430: W. Bruns, A. Simis (Eds.), Commutative Algebra. Proceedings. 1988. V, 160 pages. 1990.

Vol. 1431: J.G. Heywood, K. Masuda, R. Rautmann, V.A. Solonnikov (Eds.), The Navier-Stokes Equations – Theory and Numerical Methods. Proceedings, 1988. VII, 238 pages. 1990.

Vol. 1432: K. Ambos-Spies, G.H. Müller, G.E. Sacks (Eds.), Recursion Theory Week. Proceedings, 1989. VI, 393 pages. 1990.

Vol. 1433: S. Lang, W. Cherry, Topics in Nevanlinna Theory. II, 174 pages. 1990.

Vol. 1434: K. Nagasaka, E. Fouvry (Eds.), Analytic Number Theory. Proceedings, 1988. VI. 218 pages. 1990.

Vol. 1435: St. Ruscheweyh, E.B. Saff, L.C. Salinas, R.S. Varga (Eds.), Computational Methods and Function Theory. Proceedings, 1989. VI, 211 pages. 1990.

Vol. 1436: S. Xambó-Descamps (Ed.). Enumerative Geometry. Proceedings, 1987. V, 303 pages. 1990.

Vol. 1437: H. Inassaridze (Ed.), K-theory and Homological Algebra. Seminar, 1987–88. V, 313 pages. 1990.

Vol. 1438: P.G. Lemarié (Ed.) Les Ondelettes en 1989. Seminar. IV, 212 pages. 1990.

Vol. 1439: E. Bujalance, J.J. Etayo, J.M. Gamboa, G. Gromadzki. Automorphism Groups of Compact Bordered Klein Surfaces: A Combinatorial Approach. XIII, 201 pages. 1990.

Vol. 1440: P. Latiolais (Ed.), Topology and Combinatorial Groups Theory. Seminar, 1985–1988. VI, 207 pages. 1990.

Vol. 1441: M. Coornaert, T. Delzant, A. Papadopoulos. Géométrie et théorie des groupes. X, 165 pages. 1990.

Vol. 1442: L. Accardi, M. von Waldenfels (Eds.), Quantum Probability and Applications V. Proceedings, 1988. VI, 413 pages. 1990.

Vol. 1443: K.H. Dovermann, R. Schultz, Equivariant Surgery Theories and Their Periodicity Properties. VI, 227 pages. 1990.

Vol. 1444: H. Korezlioglu, A.S. Ustunel (Eds.), Stochastic Analysis and Related Topics VI. Proceedings, 1988. V, 268 pages. 1990.

Vol. 1445: F. Schulz, Regularity Theory for Quasilinear Elliptic Systems and – Monge Ampère Equations in Two Dimensions. XV, 123 pages. 1990.

Vol. 1446: Methods of Nonconvex Analysis. Seminar, 1989. Editor: A. Cellina. V, 206 pages. 1990.

Vol. 1447: J.-G. Labesse, J. Schwermer (Eds), Cohomology of Arithmetic Groups and Automorphic Forms. Proceedings, 1989. V, 358 pages. 1990.

Vol. 1448: S.K. Jain, S.R. López-Permouth (Eds.), Non-Commutative Ring Theory. Proceedings, 1989. V, 166 pages. 1990.

Vol. 1449: W. Odyniec, G. Lewicki, Minimal Projections in Banach Spaces. VIII, 168 pages. 1990.

Vol. 1450: H. Fujita, T. Ikebe, S.T. Kuroda (Eds.), Functional-Analytic Methods for Partial Differential Equations. Proceedings, 1989. VII, 252 pages. 1990.

Vol. 1451: L. Alvarez-Gaumé, E. Arbarello, C. De Concini, N.J. Hitchin, Global Geometry and Mathematical Physics. Montecatini Terme 1988. Seminar. Editors: M. Francaviglia, F. Gherardelli. IX, 197 pages. 1990.

Vol. 1452: E. Hlawka, R.F. Tichy (Eds.), Number-Theoretic Analysis. Seminar, 1988–89. V, 220 pages. 1990.

Vol. 1453: Yu.G. Borisovich, Yu.E. Gliklikh (Eds.), Global Analysis – Studies and Applications IV. V, 320 pages. 1990.

Vol. 1454: F. Baldassari, S. Bosch, B. Dwork (Eds.), p-adic Analysis. Proceedings, 1989. V, 382 pages. 1990.

Vol. 1455: J.-P. Françoise, R. Roussarie (Eds.), Bifurcations of Planar Vector Fields. Proceedings, 1989. VI, 396 pages. 1990.

Vol. 1456: L.G. Kovács (Ed.), Groups – Canberra 1989. Proceedings. XII, 198 pages. 1990.

Vol. 1457: O. Axelsson, L.Yu. Kolotilina (Eds.), Preconditioned Conjugate Gradient Methods. Proceedings, 1989. V, 196 pages. 1990.

Vol. 1458: R. Schaaf, Global Solution Branches of Two Point Boundary Value Problems. XIX, 141 pages. 1990.

Vol. 1459: D. Tiba, Optimal Control of Nonsmooth Distributed Parameter Systems. VII, 159 pages. 1990.

Vol. 1460: G. Toscani, V. Boffi, S. Rionero (Eds.), Mathematical Aspects of Fluid Plasma Dynamics. Proceedings, 1988. V, 221 pages. 1991.

Vol. 1461: R. Gorenflo, S. Vessella, Abel Integral Equations. VII, 215 pages. 1991.

Vol. 1462: D. Mond, J. Montaldi (Eds.), Singularity Theory and its Applications. Warwick 1989, Part I. VIII, 405 pages. 1991.

Vol. 1463: R. Roberts, I. Stewart (Eds.), Singularity Theory and its Applications. Warwick 1989, Part II. VIII, 322 pages. 1991.

Vol. 1464: D. L. Burkholder, E. Pardoux, A. Sznitman, Ecole d'Eté de Probabilités de Saint- Flour XIX-1989. Editor: P. L. Hennequin. VI, 256 pages. 1991.

Vol. 1465: G. David, Wavelets and Singular Integrals on Curves and Surfaces. X, 107 pages. 1991.

Vol. 1466: W. Banaszczyk, Additive Subgroups of Topological Vector Spaces. VII, 178 pages. 1991.

Vol. 1467: W. M. Schmidt, Diophantine Approximations and Diophantine Equations. VIII, 217 pages. 1991.

Vol. 1468: J. Noguchi, T. Ohsawa (Eds.), Prospects in Complex Geometry. Proceedings, 1989. VII, 421 pages. 1991.

Vol. 1469: J. Lindenstrauss, V. D. Milman (Eds.), Geometric Aspects of Functional Analysis. Seminar 1989-90. XI, 191 pages. 1991.

Vol. 1470: E. Odell, H. Rosenthal (Eds.), Functional Analysis. Proceedings, 1987-89. VII, 199 pages. 1991.

Vol. 1471: A. A. Panchishkin, Non-Archimedean L-Functions of Siegel and Hilbert Modular Forms. VII, 157 pages. 1991.

Vol. 1472: T. T. Nielsen, Bose Algebras: The Complex and Real Wave Representations. V, 132 pages. 1991.

Vol. 1473: Y. Hino, S. Murakami, T. Naito, Functional Differential Equations with Infinite Delay. X, 317 pages. 1991.

Vol. 1474: S. Jackowski, B. Oliver, K. Pawałowski (Eds.), Algebraic Topology, Poznań 1989. Proceedings. VIII, 397 pages. 1991.

Vol. 1475: S. Busenberg, M. Martelli (Eds.), Delay Differential Equations and Dynamical Systems. Proceedings, 1990. VIII, 249 pages. 1991.

Vol. 1476: M. Bekkali, Topics in Set Theory. VII, 120 pages. 1991.

Vol. 1477: R. Jajte, Strong Limit Theorems in Noncommutative L_2-Spaces. X, 113 pages. 1991.

Vol. 1478: M.-P. Malliavin (Ed.), Topics in Invariant Theory. Seminar 1989-1990. VI, 272 pages. 1991.

Vol. 1479: S. Bloch, I. Dolgachev, W. Fulton (Eds.), Algebraic Geometry. Proceedings, 1989. VII, 300 pages. 1991.

Vol. 1480: F. Dumortier, R. Roussarie, J. Sotomayor, H. Żoładek, Bifurcations of Planar Vector Fields: Nilpotent Singularities and Abelian Integrals. VIII, 226 pages. 1991.

Vol. 1481: D. Ferus, U. Pinkall, U. Simon, B. Wegner (Eds.), Global Differential Geometry and Global Analysis. Proceedings, 1991. VIII, 283 pages. 1991.

Vol. 1482: J. Chabrowski, The Dirichlet Problem with L^2-Boundary Data for Elliptic Linear Equations. VI, 173 pages. 1991.

Vol. 1483: E. Reithmeier, Periodic Solutions of Nonlinear Dynamical Systems. VI, 171 pages. 1991.

Vol. 1484: H. Delfs, Homology of Locally Semialgebraic Spaces. IX, 136 pages. 1991.

Vol. 1485: J. Azéma, P. A. Meyer, M. Yor (Eds.), Séminaire de Probabilités XXV. VIII, 440 pages. 1991.

Vol. 1486: L. Arnold, H. Crauel, J.-P. Eckmann (Eds.), Lyapunov Exponents. Proceedings, 1990. VIII, 365 pages. 1991.

Vol. 1487: E. Freitag, Singular Modular Forms and Theta Relations. VI, 172 pages. 1991.

Vol. 1488: A. Carboni, M. C. Pedicchio, G. Rosolini (Eds.), Category Theory. Proceedings, 1990. VII, 494 pages. 1991.

Vol. 1489: A. Mielke, Hamiltonian and Lagrangian Flows on Center Manifolds. X, 140 pages. 1991.

Vol. 1490: K. Metsch, Linear Spaces with Few Lines. XIII, 196 pages. 1991.

Vol. 1491: E. Lluis-Puebla, J.-L. Loday, H. Gillet, C. Soulé, V. Snaith, Higher Algebraic K-Theory: an overview. IX, 164 pages. 1992.

Vol. 1492: K. R. Wicks, Fractals and Hyperspaces. VIII, 168 pages. 1991.

Vol. 1493: E. Benoît (Ed.), Dynamic Bifurcations. Proceedings, Luminy 1990. VII, 219 pages. 1991.

Vol. 1494: M.-T. Cheng, X.-W. Zhou, D.-G. Deng (Eds.), Harmonic Analysis. Proceedings, 1988. IX, 226 pages. 1991.

Vol. 1495: J. M. Bony, G. Grubb, L. Hörmander, H. Komatsu, J. Sjöstrand, Microlocal Analysis and Applications. Montecatini Terme, 1989. Editors: L. Cattabriga, L. Rodino. VII, 349 pages. 1991.

Vol. 1496: C. Foias, B. Francis, J. W. Helton, H. Kwakernaak, J. B. Pearson, H_∞-Control Theory. Como, 1990. Editors: E. Mosca, L. Pandolfi. VII, 336 pages. 1991.

Vol. 1497: G. T. Herman, A. K. Louis, F. Natterer (Eds.), Mathematical Methods in Tomography. Proceedings 1990. X, 268 pages. 1991.

Vol. 1498: R. Lang, Spectral Theory of Random Schrödinger Operators. X, 125 pages. 1991.

Vol. 1499: K. Taira, Boundary Value Problems and Markov Processes. IX, 132 pages. 1991.

Vol. 1500: J.-P. Serre, Lie Algebras and Lie Groups. VII, 168 pages. 1992.

Vol. 1501: A. De Masi, E. Presutti, Mathematical Methods for Hydrodynamic Limits. IX, 196 pages. 1991.

Vol. 1502: C. Simpson, Asymptotic Behavior of Monodromy. V, 139 pages. 1991.

Vol. 1503: S. Shokranian, The Selberg-Arthur Trace Formula (Lectures by J. Arthur). VII, 97 pages. 1991.

Vol. 1504: J. Cheeger, M. Gromov, C. Okonek, P. Pansu, Geometric Topology: Recent Developments. Editors: P. de Bartolomeis, F. Tricerri. VII, 197 pages. 1991.

Vol. 1505: K. Kajitani, T. Nishitani, The Hyperbolic Cauchy Problem. VII, 168 pages. 1991.

Vol. 1506: A. Buium, Differential Algebraic Groups of Finite Dimension. XV, 145 pages. 1992.

Vol. 1507: K. Hulek, T. Peternell, M. Schneider, F.-O. Schreyer (Eds.), Complex Algebraic Varieties. Proceedings, 1990. VII, 179 pages. 1992.

Vol. 1508: M. Vuorinen (Ed.), Quasiconformal Space Mappings. A Collection of Surveys 1960-1990. IX, 148 pages. 1992.

Vol. 1509: J. Aguadé, M. Castellet, F. R. Cohen (Eds.), Algebraic Topology - Homotopy and Group Cohomology. Proceedings, 1990. X, 330 pages. 1992.

Vol. 1510: P. P. Kulish (Ed.), Quantum Groups. Proceedings, 1990. XII, 398 pages. 1992.

Vol. 1511: B. S. Yadav, D. Singh (Eds.), Functional Analysis and Operator Theory. Proceedings, 1990. VIII, 223 pages. 1992.

Vol. 1512: L. M. Adleman, M.-D. A. Huang, Primality Testing and Abelian Varieties Over Finite Fields. VII, 142 pages. 1992.

Vol. 1513: L. S. Block, W. A. Coppel, Dynamics in One Dimension. VIII, 249 pages. 1992.

Vol. 1514: U. Krengel, K. Richter, V. Warstat (Eds.), Ergodic Theory and Related Topics III. Proceedings, 1990. VIII, 236 pages. 1992.

Vol. 1515: E. Ballico, F. Catanese, C. Ciliberto (Eds.), Classification of Irregular Varieties. Proceedings, 1990. VII, 149 pages. 1992.

Vol. 1516: R. A. Lorentz, Multivariate Birkhoff Interpolation. IX, 192 pages. 1992.

Vol. 1517: K. Keimel, W. Roth, Ordered Cones and Approximation. VI, 134 pages. 1992.

Vol. 1518: H. Stichtenoth, M. A. Tsfasman (Eds.), Coding Theory and Algebraic Geometry. Proceedings, 1991. VIII, 223 pages. 1992.

Vol. 1519: M. W. Short, The Primitive Soluble Permutation Groups of Degree less than 256. IX, 145 pages. 1992.

Vol. 1520: Yu. G. Borisovich, Yu. E. Gliklikh (Eds.), Global Analysis – Studies and Applications V. VII, 284 pages. 1992.

Vol. 1521: S. Busenberg, B. Forte, H. K. Kuiken, Mathematical Modelling of Industrial Process. Bari, 1990. Editors: V. Capasso, A. Fasano. VII, 162 pages. 1992.

Vol. 1522: J.-M. Delort, F. B. I. Transformation. VII, 101 pages. 1992.

Vol. 1523: W. Xue, Rings with Morita Duality. X, 168 pages. 1992.

Vol. 1524: M. Coste, L. Mahé, M.-F. Roy (Eds.), Real Algebraic Geometry. Proceedings, 1991. VIII, 418 pages. 1992.

Vol. 1525: C. Casacuberta, M. Castellet (Eds.), Mathematical Research Today and Tomorrow. VII, 112 pages. 1992.

Vol. 1526: J. Azéma, P. A. Meyer, M. Yor (Eds.), Séminaire de Probabilités XXVI. X, 633 pages. 1992.

Vol. 1527: M. I. Freidlin, J.-F. Le Gall, Ecole d'Eté de Probabilités de Saint-Flour XX – 1990. Editor: P. L. Hennequin. VIII, 244 pages. 1992.

Vol. 1528: G. Isac, Complementarity Problems. VI, 297 pages. 1992.

Vol. 1529: J. van Neerven, The Adjoint of a Semigroup of Linear Operators. X, 195 pages. 1992.

Vol. 1530: J. G. Heywood, K. Masuda, R. Rautmann, S. A. Solonnikov (Eds.), The Navier-Stokes Equations II – Theory and Numerical Methods. IX, 322 pages. 1992.

Vol. 1531: M. Stoer, Design of Survivable Networks. IV, 206 pages. 1992.

Vol. 1532: J. F. Colombeau, Multiplication of Distributions. X, 184 pages. 1992.

Vol. 1533: P. Jipsen, H. Rose, Varieties of Lattices. X, 162 pages. 1992.

Vol. 1534: C. Greither, Cyclic Galois Extensions of Commutative Rings. X, 145 pages. 1992.

Vol. 1535: A. B. Evans, Orthomorphism Graphs of Groups. VIII, 114 pages. 1992.

Vol. 1536: M. K. Kwong, A. Zettl, Norm Inequalities for Derivatives and Differences. VII, 150 pages. 1992.

Vol. 1537: P. Fitzpatrick, M. Martelli, J. Mawhin, R. Nussbaum, Topological Methods for Ordinary Differential Equations. Montecatini Terme, 1991. Editors: M. Furi, P. Zecca. VII, 218 pages. 1993.

Vol. 1538: P.-A. Meyer, Quantum Probability for Probabilists. X, 287 pages. 1993.

Vol. 1539: M. Coornaert, A. Papadopoulos, Symbolic Dynamics and Hyperbolic Groups. VIII, 138 pages. 1993.

Vol. 1540: H. Komatsu (Ed.), Functional Analysis and Related Topics, 1991. Proceedings. XXI, 413 pages. 1993.

Vol. 1541: D. A. Dawson, B. Maisonneuve, J. Spencer, Ecole d' Eté de Probabilités de Saint-Flour XXI - 1991. Editor: P. L. Hennequin. VIII, 356 pages. 1993.

Vol. 1542: J.Fröhlich, Th.Kerler, Quantum Groups, Quantum Categories and Quantum Field Theory. VII, 431 pages. 1993.

Vol. 1543: A. L. Dontchev, T. Zolezzi, Well-Posed Optimization Problems. XII, 421 pages. 1993.

Vol. 1544: M.Schürmann, White Noise on Bialgebras. VII, 146 pages. 1993.

Vol. 1545: J. Morgan, K. O'Grady, Differential Topology of Complex Surfaces. VIII, 224 pages. 1993.

Vol. 1546: V. V. Kalashnikov, V. M. Zolotarev (Eds.), Stability Problems for Stochastic Models. Proceedings, 1991. VIII, 229 pages. 1993.

Vol. 1547: P. Harmand, D. Werner, W. Werner, M-ideals in Banach Spaces and Banach Algebras. VIII, 387 pages. 1993.

Vol. 1548: T. Urabe, Dynkin Graphs and Quadrilateral Singularities. VI, 233 pages. 1993.

Vol. 1549: G. Vainikko, Multidimensional Weakly Singular Integral Equations. XI, 159 pages. 1993.

Vol. 1550: A. A. Gonchar, E. B. Saff (Eds.), Methods of Approximation Theory in Complex Analysis and Mathematical Physics IV, 222 pages, 1993.

Vol. 1551: L. Arkeryd, P. L. Lions, P.A. Markowich, S.R. S. Varadhan. Nonequilibrium Problems in Many-Particle Systems. Montecatini, 1992. Editors: C. Cercignani, M. Pulvirenti. VII, 158 pages 1993.

Vol. 1552: J. Hilgert, K.-H. Neeb, Lie Semigroups and their Applications. XII, 315 pages. 1993.

Vol. 1553: J.-L- Colliot-Thélène, J. Kato, P. Vojta. Arithmetic Algebraic Geometry. Trento, 1991. Editor: E. Ballico. VII, 223 pages. 1993.

Vol. 1554: A. K. Lenstra, H. W. Lenstra, Jr. (Eds.), The Development of the Number Field Sieve. VIII, 131 pages. 1993.

Vol. 1555: O. Liess, Conical Refraction and Higher Microlocalization. X, 389 pages. 1993.

Vol. 1556: S. B. Kuksin, Nearly Integrable Infinite-Dimensional Hamiltonian Systems. XXVII, 101 pages. 1993.

Vol. 1557: J. Azéma, P. A. Meyer, M. Yor (Eds.), Séminaire de Probabilités XXVII. VI, 327 pages. 1993.

Vol. 1558: T. J. Bridges, J. E. Furter, Singularity Theory and Equivariant Symplectic Maps. VI, 226 pages. 1993.

Vol. 1559: V. G. Sprindžuk, Classical Diophantine Equations. XII, 228 pages. 1993.

Vol. 1560: T. Bartsch, Topological Methods for Variational Problems with Symmetries. X, 152 pages. 1993.

Vol. 1561: I. S. Molchanov, Limit Theorems for Unions of Random Closed Sets. X, 157 pages. 1993.

Vol. 1562: G. Harder, Eisensteinkohomologie und die Konstruktion gemischter Motive. XX, 184 pages. 1993.

Vol. 1563: E. Fabes, M. Fukushima, L. Gross, C. Kenig, M. Röckner, D. W. Stroock, Dirichlet Forms. Varenna, 1992. Editors: G. Dell'Antonio, U. Mosco. VII, 245 pages. 1993.

Vol. 1564: J. Jorgenson, S. Lang, Basic Analysis of Regularized Series and Products. IX, 122 pages. 1993.

Vol. 1565: L. Boutet de Monvel, C. De Concini, C. Procesi, P. Schapira, M. Vergne. D-modules, Representation Theory, and Quantum Groups. Venezia, 1992. Editors: G. Zampieri, A. D'Agnolo. VII, 217 pages. 1993.

Vol. 1566: B. Edixhoven, J.-H. Evertse (Eds.), Diophantine Approximation and Abelian Varieties. XIII, 127 pages. 1993.

Vol. 1567: R. L. Dobrushin, S. Kusuoka, Statistical Mechanics and Fractals. VII, 98 pages. 1993.

Vol. 1568: F. Weisz, Martingale Hardy Spaces and their Application in Fourier Analysis. VIII, 217 pages. 1994.

Vol. 1569: V. Totik, Weighted Approximation with Varying Weight. VI, 117 pages. 1994.

Vol. 1570: R. deLaubenfels, Existence Families, Functional Calculi and Evolution Equations. XV, 234 pages. 1994.

Vol. 1571: S. Yu. Pilyugin, The Space of Dynamical Systems with the C^0-Topology. X, 188 pages. 1994.

Vol. 1572: L. Göttsche, Hilbert Schemes of Zero-Dimensional Subschemes of Smooth Varieties. IX, 196 pages. 1994.

Vol. 1573: V. P. Havin, N. K. Nikolski (Eds.), Linear and Complex Analysis – Problem Book 3 – Part I. XXII, 489 pages. 1994.

Vol. 1574: V. P. Havin, N. K. Nikolski (Eds.), Linear and Complex Analysis – Problem Book 3 – Part II. XXII, 507 pages. 1994.

Vol. 1575: M. Mitrea, Clifford Wavelets, Singular Integrals, and Hardy Spaces. XI, 116 pages. 1994.

Vol. 1576: K. Kitahara, Spaces of Approximating Functions with Haar-Like Conditions. X, 110 pages. 1994.

Vol. 1577: N. Obata, White Noise Calculus and Fock Space. X, 183 pages. 1994.

Vol. 1578: J. Bernstein, V. Lunts, Equivariant Sheaves and Functors. V, 139 pages. 1994.

Vol. 1579: N. Kazamaki, Continuous Exponential Martingales and *BMO*. VII, 91 pages. 1994.

Vol. 1580: M. Milman, Extrapolation and Optimal Decompositions with Applications to Analysis. XI, 161 pages. 1994.

Vol. 1581: D. Bakry, R. D. Gill, S. A. Molchanov, Lectures on Probability Theory. Editor: P. Bernard. VIII, 420 pages. 1994.

Vol. 1582: W. Balser, From Divergent Power Series to Analytic Functions. X, 108 pages. 1994.

Vol. 1583: J. Azéma, P. A. Meyer, M. Yor (Eds.), Séminaire de Probabilités XXVIII. VI, 334 pages. 1994.

Vol. 1584: M. Brokate, N. Kenmochi, I. Müller, J. F. Rodriguez, C. Verdi, Phase Transitions and Hysteresis. Montecatini Terme, 1993. Editor: A. Visintin. VII. 291 pages. 1994.

Vol. 1585: G. Frey (Ed.), On Artin's Conjecture for Odd 2-dimensional Representations. VIII, 148 pages. 1994.

Vol. 1586: R. Nillsen, Difference Spaces and Invariant Linear Forms. XII, 186 pages. 1994.

Vol. 1587: N. Xi, Representations of Affine Hecke Algebras. VIII, 137 pages. 1994.

Vol. 1588: C. Scheiderer, Real and Étale Cohomology. XXIV, 273 pages. 1994.

Vol. 1589: J. Bellissard, M. Degli Esposti, G. Forni, S. Graffi, S. Isola, J. N. Mather, Transition to Chaos in Classical and Quantum Mechanics. Montecatini Terme, 1991. Editor: S. Graffi. VII, 192 pages. 1994.

Vol. 1590: P. M. Soardi, Potential Theory on Infinite Networks. VIII, 187 pages. 1994.

Vol. 1591: M. Abate, G. Patrizio, Finsler Metrics – A Global Approach. IX, 180 pages. 1994.

Vol. 1592: K. W. Breitung, Asymptotic Approximations for Probability Integrals. IX, 146 pages. 1994.

Vol. 1593: J. Jorgenson & S. Lang, D. Goldfeld, Explicit Formulas for Regularized Products and Series. VIII, 154 pages. 1994.

Vol. 1594: M. Green, J. Murre, C. Voisin, Algebraic Cycles and Hodge Theory. Torino, 1993. Editors: A. Albano, F. Bardelli. VII, 275 pages. 1994.

Vol. 1595: R.D.M. Accola, Topics in the Theory of Riemann Surfaces. IX, 105 pages. 1994.

Vol. 1596: L. Heindorf, L. B. Shapiro, Nearly Projective Boolean Algebras. X, 202 pages. 1994.

Vol. 1597: B. Herzog, Kodaira-Spencer Maps in Local Algebra. XVII, 176 pages. 1994.

Vol. 1598: J. Berndt, F. Tricerri, L. Vanhecke, Generalized Heisenberg Groups and Damek-Ricci Harmonic Spaces. VIII, 125 pages. 1995.

Vol. 1599: K. Johannson, Topology and Combinatorics of 3-Manifolds. XVIII, 446 pages. 1995.

Vol. 1600: W. Narkiewicz, Polynomial Mappings. VII, 130 pages. 1995.

Vol. 1601: A. Pott, Finite Geometry and Character Theory. VII, 181 pages. 1995.

Vol. 1602: J. Winkelmann, The Classification of Three-dimensional Homogeneous Complex Manifolds. XI, 230 pages. 1995.

Vol. 1603: V. Ene, Real Functions – Current Topics. XIII, 310 pages. 1995.